Mathematical Engineering

Today, the development of high-tech systems is unthinkable without mathematical modeling and analysis of system behavior. As such, many fields in the modern engineering sciences (e.g. control engineering, communications engineering, mechanical engineering, and robotics) call for sophisticated mathematical methods in order to solve the tasks at hand.

The series Mathematical Engineering presents new or heretofore little-known methods to support engineers in finding suitable answers to their questions, presenting those methods in such manner as to make them ideally comprehensible and applicable in practice.

Therefore, the primary focus is—without neglecting mathematical accuracy—on comprehensibility and real-world applicability.

To submit a proposal or request further information, please use the PDF Proposal Form or contact directly: Dr. Thomas Ditzinger (thomas.ditzinger@springer.com)

Indexed by SCOPUS, zbMATH, SCImago.

Steven L. Crouch · Sofia G. Mogilevskaya

A First Course in Boundary Element Methods

Springer

Steven L. Crouch
Minneapolis, MN, USA

Sofia G. Mogilevskaya
Minneapolis, MN, USA

ISSN 2192-4732 ISSN 2192-4740 (electronic)
Mathematical Engineering
ISBN 978-3-031-63340-9 ISBN 978-3-031-63341-6 (eBook)
https://doi.org/10.1007/978-3-031-63341-6

This Springer imprint is published by the registered company Springer Nature Switzerland AG
The registered company address is: Gewerbestrasse 11, 6330 Cham, Switzerland

Preface

This book is a compilation of material taught to beginning graduate students at the University of Minnesota over a period of several decades. The material has been expanded and revised over time and is intended as a textbook for an introductory course on the boundary element approach, both its mathematical basis and its numerical implementation. One objective of the work is to give students the ability to write their own boundary element computer programs; another is to give them a background sufficient to follow the extensive—and growing—body of literature on the subject.

We have tried to make the book as self-contained as possible and in so doing have included details that some readers might initially want to pass over in favor of studying the numerical implementations of the underlying mathematics that are at the heart of the boundary element approach. By the same token, other readers will want to focus on the theoretical aspects in order to gain an in-depth understanding of the subject needed to follow the literature in the field. We have assumed that readers will have had some exposure to potential theory and also to the mathematical theory of elasticity, and for completeness we have included relevant details of these subjects as reference material. Most of the discussion deals with two dimensions, even though the general theoretical considerations are also applicable to three-dimensional problems. This allows us to delve into numerical details that would be impractical to present for three dimensions, given the additional geometrical complexity involved in such problems.

One of us (SLC) was a co-author of an early book on boundary element methods for elasticity problems (Crouch, S. L., and A. M. Starfield. 1983. *Boundary element methods in solid mechanics*. George Allen and Unwin.). The present book can be regarded as an updated and expanded version of that work, including mathematical details that were omitted previously. We have chosen to begin our book—Chap. 1—with a discussion of two elementary boundary element methods for Laplace's equation, one based on the solution for a single layer potential and the other for a double layer potential (this terminology is explained in the text). Chapter 2 deals with mathematical considerations of potential theory, including Green's representation formula, which is the basis of the direct boundary element method for Laplace's equation discussed in Chap. 3. Aspects of the theory of elasticity are covered in Chap. 4, and numerical implementations of the theory are given in the next three chapters—the stress discontinuity method in Chap. 5, the displacement discontinuity method in Chap. 6, and the direct boundary integral method for elasticity in Chap. 7. Several extensions of these numerical methods for elasticity are discussed in Chap. 8, including development of special singular solutions, handling of piecewise homogeneous regions, and applications to problems in fracture mechanics.

As indicated by the title of this book, we have tried to limit ourselves to topics that might reasonably be covered in an introductory course on the subject. It is important to recognize, however, that boundary element methods can also be applied in other fields, including acoustics, transient heat conduction and thermoelasticity, viscoelasticity, and elastodynamics. We hope that some readers will be encouraged to pursue the literature on boundary element methods for these subjects after having set aside this book.

Many people have contributed to the contents of this book, both directly and indirectly, and we want to express our thanks to them all. In particular, we are pleased to acknowledge the important early contributions to the displacement discontinuity method (Chap. 6) made by A. M. Starfield. In the mid-1960s Prof. Starfield devised a specialized version of the method and introduced it to a group of University of Minnesota graduate students who were learning about numerical analysis and computer programming. One of us (SLC) was fortunate to have been one of those students. Parts of the present book are patterned after material in the book by Crouch and Starfield referenced previously, and we wish to acknowledge Prof. Starfield's part in that, too. Dozens of University of Minnesota graduate students have made many helpful suggestions over the years, and we are grateful for their input. Any mistakes, inaccuracies, or errors of omission are of course the authors' responsibility, and while we have done our best to eliminate these we apologize for any that we might have missed.

Minneapolis, USA

Steven L. Crouch
Sofia G. Mogilevskaya

Contents

Chapter 1
An Illustration of the Boundary Element Approach

1.1 Introduction

Many important practical problems in science and engineering can be reduced to mathematical models known as *boundary value problems*. These are all characterized by a region of interest V enclosed within a boundary S. A partial differential equation models the physics of the problem in V, and this equation has to be solved subject to certain constraints or conditions that describe what is happening on the boundary S. If V is a three-dimensional region, or volume, then S is a bounding surface. In two dimensions, V is a planar region, or area, and S is a bounding curve.

Examples of boundary value problems can be found in fields such as heat conduction, mass transfer, fluid flow, electrostatics, and solid mechanics. For instance, in two-dimensional, steady-state heat conduction the governing partial differential equation is Laplace's equation, $\nabla^2\phi(x, y) = \dfrac{\partial^2\phi}{\partial x^2} + \dfrac{\partial^2\phi}{\partial y^2} = 0$, where $\phi(x, y)$ is the temperature, and in any well-posed problem the temperature, the component of the temperature gradient normal to the boundary, or a combination of these quantities is specified at every point on S. In elasticity (elastostatics) the partial differential equations are the displacement equations of equilibrium,[1] and the normal and shear stresses, or the normal and shear displacements, or combinations of these quantities are specified at every point on S.

It is worth noting that in any boundary value problem the constraints define only some of the relevant parameters on the boundary; the others emerge as part of the solution to the problem. If, for example, the displacements are specified on the boundary for a problem in linear elasticity, then the stresses on the boundary (along with the displacements and stresses anywhere in region V) can eventually be found as part of the solution to the problem. Similarly, if the temperature is specified on the boundary for a problem in linear steady-state heat conduction, the components of the

[1] Also known as the Navier-Cauchy equations, these expressions are given in Chap. 4.

S. L. Crouch and S. G. Mogilevskaya, *A First Course in Boundary Element Methods*,
Mathematical Engineering, https://doi.org/10.1007/978-3-031-63341-6_1

temperature gradient on the boundary (and the temperature and temperature gradient anywhere in V) can eventually be found as part of the solution to that problem. If the boundary S is sufficiently smooth and is subjected to adequate data constraints then the solution to the problem everywhere in V is uniquely assured by these constraints.

Closed form, analytical solutions to boundary value problems can generally only be derived when the region V is homogeneous (i.e., the material properties do not change from point to point in V), when the geometry is simple, when the boundary conditions on S are relatively straightforward, and when the governing partial differential equations are linear, which implies that certain solutions can be added together to construct new solutions in V. Analytical solutions have been found in many disciplines for the case corresponding to a disturbance at a point in an *infinite* homogeneous region. This disturbance could represent a heat source or sink in a heat conduction problem, for example, or it could represent a point force applied within a homogeneous elastic solid for a problem in linear elasticity. These solutions are usually called *singular solutions* because, mathematically speaking, they are well-behaved everywhere in the region of interest except at the point of the disturbance, where there is a mathematical anomaly, or singularity. The solutions can, if necessary, be summed (in a linear problem) to find solutions for cases in which more than one point disturbance is present.

More often than not, analytical solutions cannot be found for problems of practical interest. Even when the governing partial differential equations are linear, the region V may be inhomogeneous, the geometry irregular, and the boundary conditions difficult to describe by simple mathematical functions. Suitable approximate solutions can almost always be constructed, however, using numerical methods, and this is the principal focus of this book.

Numerical methods for solving boundary value problems can be divided into two distinct classes: those that require approximations to be made throughout the region V, and those that require approximations to be made only on the boundary S. The finite element method typifies the first class,[2] while boundary element methods represent the second class.

In the finite element approach the whole region V is divided into a network of elements of varying sizes. The objective is to evaluate the solution to the problem at the nodes or mesh points of the network; the solution within each element is expressed in an approximate form in terms of the values at these points. Relating this approximate solution to the original partial differential equation (or equations) eventually leads to a system of algebraic equations in which the unknown parameters—the nodal values in V—are expressed in terms of the prescribed nodal values at the mesh points on the boundary S. This system of equations is large but sparse, i.e., although there is a large number of unknown parameters, and hence a correspondingly large number of linear equations, each equation contains only a few of the unknown parameters explicitly.

[2] Related approaches include finite difference and meshfree methods.

In boundary element methods only the boundary S is divided into elements.[3] The numerical approach builds on the analytical solutions that have already been obtained for simple singular problems in such a way as to satisfy, approximately, the specified boundary conditions at selected points along each element on S. Because these singular solutions satisfy the governing partial differential equations in V, there is no need to divide V itself into a network of elements. The system of algebraic equations to be solved is typically much smaller than the system needed to solve the same boundary value problem by the finite element method but, because each singular solution affects all the others, the equations are no longer sparse.[4]

This book is concerned with an exposition of boundary element methods for problems involving Laplace's equation (e.g., steady-state heat conduction or fluid flow in porous media) and linear elasticity. Detailed analyses are presented for two-dimensional problems but, as we will see, certain general formulas are also valid in three dimensions.

The remainder of this chapter describes two numerical implementations of a simple boundary element method for Laplace's equation in two dimensions. These simplified methods contain features common to other boundary element methods described in this book and are a comparatively easy way to develop an understanding of the basic approach. We will expand on this methodology in subsequent chapters, providing the mathematical foundations for the work as we go along.

1.2 Singular Solution to Laplace's Equation

We consider the two-dimensional version of Laplace's equation in Cartesian coordinates:

$$\nabla^2\phi(x, y) = \frac{\partial^2\phi}{\partial x^2} + \frac{\partial^2\phi}{\partial y^2} = 0 \tag{1.1}$$

A solution of Laplace's equation is called a *harmonic function*, often referred to in the literature as a *potential*. The solution to Eq. (1.1) for the infinite plane with a unit source at the point $\boldsymbol{\xi} = (\xi_x, \xi_y)$ is given as

$$\phi(\boldsymbol{x}) = G(\boldsymbol{x}, \boldsymbol{\xi}) = -\frac{1}{2\pi}\ln r \tag{1.2}$$

in which $\boldsymbol{x} = (x, y)$ and

---

[3] Volume discretization is required for certain nonlinear problems (e.g., elastoplasticity) but we do not consider these in this book.

[4] For very large problems, the fully-populated matrices inherent to the boundary element approach make boundary element methods less efficient than other numerical methods. Special equation-solving techniques can be implemented, however, to overcome this shortcoming (Rokhlin 1985; Greengard and Rokhlin 1987).

$$r = [(x - \xi_x)^2 + (y - \xi_y)^2]^{1/2} \tag{1.3}$$

is the distance between points $\boldsymbol{x}$ and $\boldsymbol{\xi}$. Equation (1.2) reveals that the potential $\phi(x, y)$ tends to plus infinity at the point of application of the unit source (where $r = 0$) and to minus infinity at large distances from this point. The latter behavior occurs because a two-dimensional "point" source for the x, y plane is actually a unit cross section of an infinitely long line source in the z direction perpendicular to the plane (otherwise, the problem would not be two-dimensional).

Function $G(\boldsymbol{x}, \boldsymbol{\xi})$ represents the potential at an arbitrary point $\boldsymbol{x} = (x, y)$ in the plane due to the unit source at point $\boldsymbol{\xi} = (\xi_x, \xi_y)$, and the function $\phi(x, y) = G(\boldsymbol{x}, \boldsymbol{\xi})$ satisfies (1.1) everywhere except at point $\boldsymbol{\xi}$. (We will derive this solution and its three-dimensional counterpart in the next chapter.) In the sequel, we will refer to point $\boldsymbol{x}$ as the **sample point** and point $\boldsymbol{\xi}$ as the **source point**, and use the notation $(\boldsymbol{x}, \boldsymbol{\xi})$ to mean $(x - \xi_x, y - \xi_y)$. It can be seen from (1.2) and (1.3) that function $G(\boldsymbol{x}, \boldsymbol{\xi})$ is symmetric with respect to its arguments, i.e.,

$$G(\boldsymbol{x}, \boldsymbol{\xi}) = G(\boldsymbol{\xi}, \boldsymbol{x}) \tag{1.4}$$

If the source has magnitude ρ, where ρ is a constant, the function $\phi(x, y) = \rho G(\boldsymbol{x}, \boldsymbol{\xi})$ again satisfies (1.1) except at point $\boldsymbol{\xi}$. If two or more sources are present the individual singular solutions can be added together because each source acts independently of the others. The solution to the problem shown in Fig. 1.1a, for example, can be written as follows:

$$\phi(\boldsymbol{x}) = \overset{1}{\rho} G(\boldsymbol{x}, \overset{1}{\boldsymbol{\xi}}) + \overset{2}{\rho} G(\boldsymbol{x}, \overset{2}{\boldsymbol{\xi}}) \tag{1.5}$$

In this expression $\overset{1}{\rho}$ and $\overset{2}{\rho}$ are the strengths of the sources located at points $\boldsymbol{x} = \overset{1}{\boldsymbol{\xi}}$ and $\boldsymbol{x} = \overset{2}{\boldsymbol{\xi}}$, respectively.

A useful application of this process occurs when the two sources have opposite signs and are positioned symmetrically with respect to the x axis, as shown in Fig. 1.1b. If the positive source is located at the point $x = \xi_x$, $y = \xi_y < 0$, the distance r from this point to an arbitrary point $\boldsymbol{x} = (x, y)$ is given by (1.3). The distance r' from the location of the negative source $(x = \xi_x, y = -\xi_y)$ to point $\boldsymbol{x} = (x, y)$ is

$$r' = [(x - \xi_x)^2 + (y + \xi_y)^2]^{1/2} \tag{1.6}$$

and the potential due to the two sources is

$$\phi(\boldsymbol{x}) = -\frac{\rho}{2\pi} \ln r + \frac{\rho}{2\pi} \ln r' = -\frac{\rho}{2\pi} \ln \frac{r}{r'} \tag{1.7}$$

It follows that the potential is zero on the line $y = 0$ for $-\infty < x < \infty$ because $r = r'$ all along this line. If we confine our attention to the lower half-plane, $y \leq 0$, for example, the potential $\phi(\boldsymbol{x})$ in (1.7) automatically satisfies a zero potential boundary

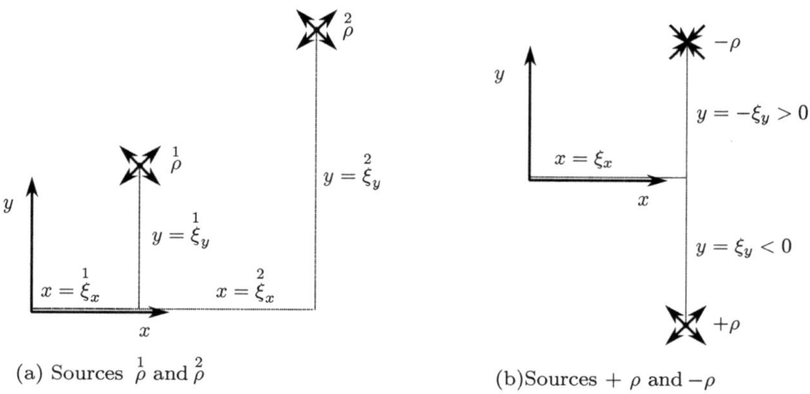

Fig. 1.1 Superposition of two sources

condition along the line $y = 0$ for $-\infty < x < \infty$. This superposition process is a simple example of the *method of images*, and we will return to this topic later in this chapter and on several other occasions in this book.

1.3 A Constant Strength Source Along a Line Segment

A generalization of the above results can be used to develop a solution to Laplace's equation for a distributed source along a line segment in the plane rather than acting at a point. A constant strength source P along the line segment $\mid x \mid \leq a$, $y = 0$, for example, can be thought of as a number of point sources of strength $P \mathrm{d}\xi_x$ acting on infinitesimal elements of length $\mathrm{d}\xi_x$ along the segment. The effect of a source located at the point $\boldsymbol{\xi} = (\xi_x, 0)$ can be written as $\phi(\boldsymbol{x}) = PG(\boldsymbol{x}, \boldsymbol{\xi})\mathrm{d}\xi_x$, and the combined effect of a continuous distribution of such sources is obtained by integration:

$$\phi(\boldsymbol{x}) = P \int_{-a}^{a} G(\boldsymbol{x}, \boldsymbol{\xi})\, \mathrm{d}\xi_x \tag{1.8}$$

The integral in this expression is easy to evaluate and can be written as

$$\begin{aligned} \int_{-a}^{a} G(\boldsymbol{x}, \boldsymbol{\xi})\, \mathrm{d}\xi_x &= -\frac{1}{2\pi} \int_{-a}^{a} \ln[(x - \xi_x)^2 + y^2]^{1/2}\, \mathrm{d}\xi_x \\ &= -\frac{1}{2\pi} \Bigg[y \bigg(\arctan \frac{y}{x - a} - \arctan \frac{y}{x + a} \bigg) \\ &\quad - (x - a) \ln[(x - a)^2 + y^2]^{1/2} \\ &\quad + (x + a) \ln[(x + a)^2 + y^2]^{1/2} - 2a \Bigg] \end{aligned} \tag{1.9}$$

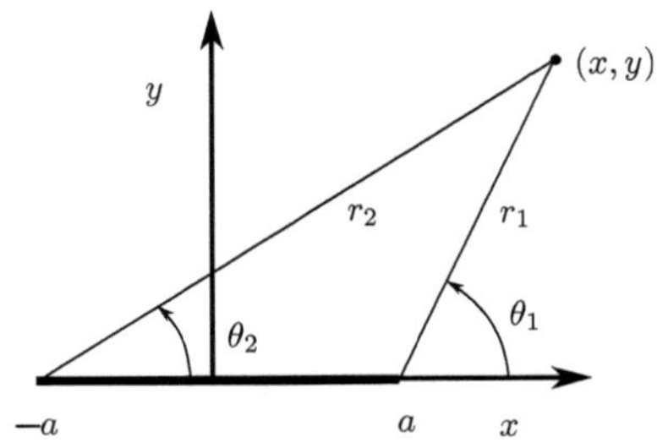

Fig. 1.2 Distances and angles from endpoints of line segment

so that the potential due to the line source is

$$\phi(x, y) = -\frac{P}{2\pi}[y(\theta_1 - \theta_2) - (x - a)\ln r_1 + (x + a)\ln r_2 - 2a] \tag{1.10}$$

The quantities r_1, r_2 and θ_1, θ_2 give the distances and angles from the endpoints of the line segment to the point $\boldsymbol{x} = (x, y)$, i.e.,

$$\begin{aligned} r_1 &= \sqrt{(x - a)^2 + y^2} \qquad & r_2 &= \sqrt{(x + a)^2 + y^2} \\ \theta_1 &= \arctan\frac{y}{x - a} \qquad & \theta_2 &= \arctan\frac{y}{x + a} \end{aligned} \tag{1.11}$$

as shown in Fig. 1.2. The angles θ_1 and θ_2 lie between $-\pi$ and $+\pi$ and are reckoned positive as shown in the figure.

It is interesting to note that the function defined by (1.10) is finite everywhere, including at the endpoints of the line segment, $x = \pm a$, $y = 0$. For instance, if we set $y = 0$ the following expression is found for $\phi(x, 0)$:

$$\phi(x, 0) = \frac{P}{2\pi}[(x - a)\ln \mid x - a \mid -(x + a)\ln \mid x + a \mid +2a] \tag{1.12}$$

When $x \to a$ the term $(x - a)\ln \mid x - a \mid$ is of the form $\varepsilon \ln \varepsilon$, which goes to zero as $\varepsilon \to 0$. The term $(x + a)\ln \mid x + a \mid$ similarly is zero for $x \to -a$. Thus, we have

$$\phi(\pm a, 0) = \frac{aP}{\pi}(1 - \ln 2a) \tag{1.13}$$

We also note that the value of ϕ at the midpoint of the line segment is

$$\phi(0, 0) = \frac{aP}{\pi}(1 - \ln a) \tag{1.14}$$

In applications, the gradient of function ϕ is frequently required at points within the region of interest. In steady-state heat conduction, for example, Fourier's law states that the heat flux density vector at a point is proportional to the negative gradient of the temperature. Similarly, in steady-state groundwater flow Darcy's law relates the specific discharge vector at a point to the negative gradient of the piezometric head. In both cases the direction of flow is from high to low (temperature or piezometric head), hence the use of negative gradients in the two relationships.

The component of the gradient in the direction of the outward normal to the boundary is of special interest because it is often used as a boundary condition instead of values of the function ϕ. For example, if a portion of the boundary S in a heat conduction problem is insulated, the derivative of ϕ in the direction normal to S is set equal to zero, i.e., $\partial\phi/\partial \boldsymbol{n} = 0$.[5] This ensures that no heat can cross the boundary at these locations.

The components of the gradient vector for the potential $\phi(x, y)$ are

$$
\begin{aligned}
q_x(x, y) &= \frac{\partial}{\partial x}\phi(x, y) \\
q_y(x, y) &= \frac{\partial}{\partial y}\phi(x, y)
\end{aligned}
\tag{1.15}
$$

For the constant strength line source specified by (1.10) these components are given as follows:

$$
\begin{aligned}
q_x(x, y) &= \frac{P}{2\pi}\left[\ln[(x-a)^2 + y^2]^{1/2} - \ln[(x+a)^2 + y^2]^{1/2}\right] = \frac{P}{2\pi}\ln\frac{r_1}{r_2} \\
q_y(x, y) &= -\frac{P}{2\pi}\left[\arctan\frac{y}{x-a} - \arctan\frac{y}{x+a}\right] = -\frac{P}{2\pi}(\theta_1 - \theta_2)
\end{aligned}
\tag{1.16}
$$

We see from these expressions that the gradient components are defined everywhere except at the endpoints of the line segment ($x = \pm a$, $y = 0$).

The x component of the gradient $q_x(x, y)$ is continuous along the segment where the source is prescribed, i.e., on $y = 0$ and $\mid x \mid < a$,

$$
q_x(x, 0) = \frac{P}{2\pi}\ln\mid\frac{x-a}{x+a}\mid \tag{1.17}
$$

The y component $q_y(x, y)$ represents the gradient in the direction perpendicular to the line segment and its boundary value depends on the way in which the segment is approached, i.e.,

$$
q_y(x, 0) = -\frac{P}{2\pi}\lim_{y\to 0_\pm}\left[\arctan\frac{y}{x-a} - \arctan\frac{y}{x+a}\right] \tag{1.18}
$$

The notation $y = 0_+$ means that we are considering the *positive* side of the line segment (i.e., $y \to 0$ through positive values), and similarly for the notation $y = 0_-$. The limits in expression (1.18) are necessary because the arctangent functions are multiple-valued. Referring to Fig. 1.2 we can easily obtain the following results

[5] As discussed further in Chap. 2, $\boldsymbol{n}$ is a vector, perpendicular to S, with components n_x and n_y.

$$\lim_{y \to 0} \left(\arctan \frac{y}{x-a} - \arctan \frac{y}{x+a} \right) = \begin{cases} 0 & \mid x \mid > a, \quad y = 0_{\pm} \\ +\pi & \mid x \mid < a, \quad y = 0_{+} \\ -\pi & \mid x \mid < a, \quad y = 0_{-} \end{cases} \tag{1.19}$$

Combining (1.18) and (1.19) we see that the value of q_y along the line $y = 0$ is zero at points away from the source (i.e., for $\mid x \mid > a$), while at the source itself $q_y(x, 0)$ is discontinuous:

$$q_y(x, 0) = -\frac{P}{2} \qquad \mid x \mid < a, \quad y = 0_{+}$$
$$q_y(x, 0) = +\frac{P}{2} \qquad \mid x \mid < a, \quad y = 0_{-} \tag{1.20}$$

That is,

$$q_y(x, 0_{+}) - q_y(x, 0_{-}) = -P \qquad \mid x \mid < a \tag{1.21}$$

but

$$q_y(x, 0_{+}) + q_y(x, 0_{-}) = 0 \qquad \mid x \mid < a \tag{1.22}$$

Equation (1.21) shows that the component of the gradient perpendicular to the line segment is discontinuous and undergoes a "jump" of constant value when crossing from one side of the line source to the other.

It is easy to show from (1.16) that for positive values of P the gradient vectors point toward the line segment from all directions. As previously mentioned, however, the negative of the gradient vector at a point governs the direction of flow in a heat transfer or groundwater flow problem, so that $P > 0$ physically represents a source, and not a sink.

1.4 Numerical Procedure

The preceding results provide the basis of a numerical method for solving boundary value problems for Laplace's equation for a two-dimensional region V with boundary S. If, for example, the potential ϕ is a constant, $\phi = \phi_0$, all along S, we wish to solve

$$\nabla^2 \phi(\boldsymbol{x}) = 0 \qquad \boldsymbol{x} \in V \tag{1.23}$$

subject to the boundary condition

$$\phi(\boldsymbol{x}) = \phi_0 \qquad \boldsymbol{x} \in S \tag{1.24}$$

A simple discretized version of the boundary value problem defined by (1.23) and (1.24) is depicted in Fig. 1.3a for a region V with boundary S. The boundary

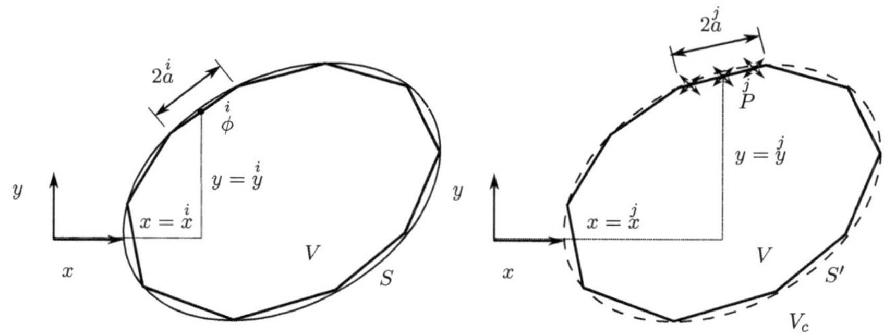

(a) Physical problem (b) Numerical model

Fig. 1.3 Illustration of a boundary element method

curve S is approximated by N straight line segments joined end to end. The length of a typical *boundary element* i is denoted as $2\overset{i}{a}$ in the figure.[6] The coordinates of the midpoint of the element are $x = \overset{i}{x}$ and $y = \overset{i}{y}$, and the value of the potential at this point is $\overset{i}{\phi}$. If the boundary elements are small they will approximate the curve S quite closely. We may then consider that the value of ϕ for each element is ϕ_0, so that boundary condition (1.24) can be represented as

$$\overset{i}{\phi} = \phi_0, \qquad i = 1 \text{ to } N \tag{1.25}$$

A numerical solution to the discretized problem described above can be found using the model depicted in Fig. 1.3b. The dashed curve S' in the figure has the same shape as the curve S that defines the boundary of the *finite* region V in Fig. 1.3a. Curve S', however, does not represent a boundary; it merely marks locations of line segments in an *infinite region* that are coincident with the line segments along the discretized boundary S of region V. Region V in Fig. 1.3b can therefore be regarded as a subregion of the infinite plane; the complementary region V_c in the figure is introduced to make $V + V_c$ infinite. Our solution for a constant line source is valid in region $V + V_c$. In the same way, V could be taken as the region outside a hole (or holes) of area (in two dimensions) V_c. In this case, V is an infinite region with a "cut-out," and $V + V_c$ is again the uncut infinite plane.

We next imagine that line sources of the form described in Sect. 1.3 are applied to each of the N line segments along curve S'. For simplicity, only the source $\overset{j}{P}$ applied to the jth segment is shown in Fig. 1.3b. The potential $\overset{i}{\phi}$ at the ith element due to a source $\overset{j}{P}$ at the jth element can be expressed as

[6] When a letter is used above a symbol in this book it will always represent a *number* and not a coordinate index as it normally will when it is used as a subscript.

$$\overset{i}{\phi} = \overset{ij}{\mathbb{G}}\overset{j}{P} \tag{1.26}$$

where $\overset{ij}{\mathbb{G}}$ is a *boundary influence coefficient* for the problem. Because each source acts independently and each one exerts an influence on all of the other sources we have

$$\overset{i}{\phi} = \sum_{j=1}^{N} \overset{ij}{\mathbb{G}}\overset{j}{P} \tag{1.27}$$

for $i = 1$ to N.

Now, if we are able to find the values of the N line sources such that $\overset{i}{\phi} = \phi_0$ for $i = 1$ to N, then we will have solved, approximately, the problem specified by (1.23) and (1.24). We therefore require that

$$\sum_{j=1}^{N} \overset{ij}{\mathbb{G}}\overset{j}{P} = \phi_0, \qquad i = 1 \text{ to } N \tag{1.28}$$

which is a system of N simultaneous linear algebraic equations in N unknowns. These equations can be solved by standard numerical methods.

It is important to recognize that, in general, the line sources $\overset{j}{P}$ in the above equations are *fictitious quantities*.[7] They were introduced to facilitate the numerical solution to a particular problem and have no physical significance with respect to this problem. The *linear combinations* of sources specified by (1.27) do, however, have physical meaning for the problem in question and we have taken advantage of this fact to establish the system of algebraic equations (1.28). Once these equations have been solved, we can express the potential $\phi(\boldsymbol{x}) = \phi(x, y)$ at any point $\boldsymbol{x} = (x, y)$ in the region of interest as other combinations of the sources $\overset{j}{P}$ at all N boundary elements, $j = 1$ to N.

The approach described above is an example of an *indirect* boundary element method, so called because the numerical solution is constructed in terms of quantities that do not themselves have physical meaning. Further details of the method and some examples of its use are given in the succeeding discussion. We will show in Chap. 3 how it is possible to use certain integral identities to develop a more elaborate, and more general, *direct* boundary element method for the same problem.

[7] An exception to this statement occurs if the boundary S is not closed, e.g., if it represents a thin membrane, or "screen." For such problems, the line sources might physically describe a discontinuity, or jump, in the component of the gradient perpendicular to the screen.

1.5 Coordinate Transformations

The foregoing numerical method is based on the solution to the problem of a constant line source over an arbitrarily oriented line segment in an infinite region. This solution can be found directly from the results of Sect. 1.3 by means of simple coordinate transformations. We will consider this problem below for the $\bar{x}$, $\bar{y}$ coordinate system shown in Fig. 1.4. The line segment in question is specified by the conditions $|\bar{x}| \leq a$, $\bar{y} = 0$, and a constant strength source P acts along this segment, as indicated in the figure.

The *local* $\bar{x}$, $\bar{y}$ coordinate system in Fig. 1.4 is obtained by a translation and rotation of the *global* x, y system. The components of translation are ξ_x in the x direction and ξ_y in the y direction and the angle of rotation is β, positive in the counterclockwise sense as shown. This coordinate transformation is defined by the formulas

$$\begin{aligned} \bar{x} &= (x - \xi_x)\cos\beta + (y - \xi_y)\sin\beta \\ \bar{y} &= -(x - \xi_x)\sin\beta + (y - \xi_y)\cos\beta \end{aligned} \tag{1.29}$$

Expressions for the potential and the components of the gradient vector in the $\bar{x}$, $\bar{y}$ coordinate system due to a constant strength source P over the line segment $|\bar{x}| \leq a$, $\bar{y} = 0$ can be obtained by simply replacing x and y in (1.10) and (1.16) with $\bar{x}$ and $\bar{y}$. (This is so because we have tacitly assumed that region V is homogeneous and isotropic.) We therefore have that

$$\phi(x, y) = -\frac{P}{2\pi}\left[\bar{y}(\bar{\theta}_1 - \bar{\theta}_2) - (\bar{x} - a)\ln\bar{r}_1 + (\bar{x} + a)\ln\bar{r}_2 - 2a\right] \tag{1.30}$$

and

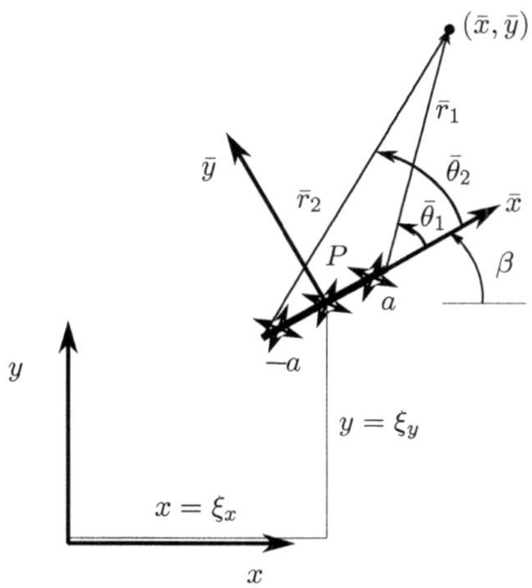

Fig. 1.4 Constant line source over an arbitrarily oriented line segment

Fig. 1.5 Transformation of gradient vector components

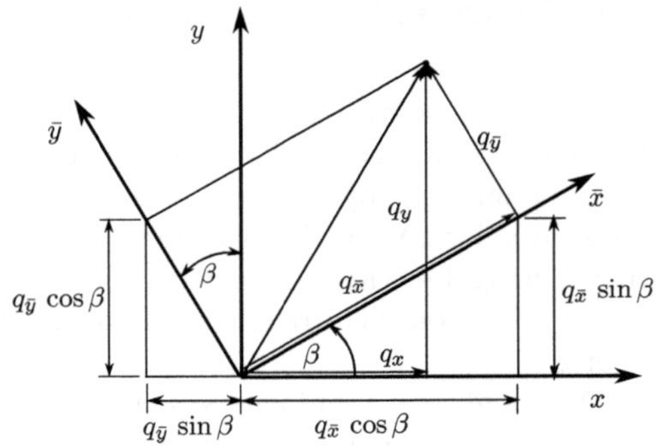

$$q_{\bar{x}}(x, y) = \frac{P}{2\pi}\left[\ln\left[(\bar{x}-a)^2+\bar{y}^2\right]^{1/2} - \ln\left[(\bar{x}+a)^2+\bar{y}^2\right]^{1/2}\right] = \frac{P}{2\pi}\ln\frac{\bar{r}_1}{\bar{r}_2} \tag{1.31}$$
$$q_{\bar{y}}(x, y) = -\frac{P}{2\pi}\left[\arctan\frac{\bar{y}}{\bar{x}-a} - \arctan\frac{\bar{y}}{\bar{x}+a}\right] = -\frac{P}{2\pi}(\bar{\theta}_1 - \bar{\theta}_2)$$

The quantities $\bar{r}_1$, $\bar{r}_2$ and $\bar{\theta}_1$, $\bar{\theta}_2$ in these equations give the distances and angles from the endpoints of the line segment to an arbitrary point $(\bar{x}, \bar{y})$ and are defined in a similar way as r_1, r_2 and θ_1, θ_2 in Fig. 1.2.

We have written $\phi(x, y)$ instead of $\phi(\bar{x}, \bar{y})$ in (1.30) and $q_{\bar{x}}(x, y)$ and $q_{\bar{y}}(x, y)$ instead of $q_{\bar{x}}(\bar{x}, \bar{y})$ and $q_{\bar{y}}(\bar{x}, \bar{y})$ in (1.31) because these equations are typically used to compute the potential and the components of the gradient vector at selected points in the global coordinate system. For specified values of x and y (and given values of ξ_x, ξ_y, and β) the associated values of $\bar{x}$ and $\bar{y}$ are defined by (1.29).

The components of the gradient vector in the global x, y coordinate system can be obtained by resolving the components $q_{\bar{x}}$ and $q_{\bar{y}}$ in the x and y directions, as shown in Fig. 1.5. Using the geometric relations indicated in the figure we find the transformation formulas

$$q_x = q_{\bar{x}}\cos\beta - q_{\bar{y}}\sin\beta \tag{1.32}$$
$$q_y = q_{\bar{x}}\sin\beta + q_{\bar{y}}\cos\beta$$

Substituting (1.31) in these equations we have

$$q_x(x, y) = \frac{P}{2\pi}\left[\ln\frac{\bar{r}_1}{\bar{r}_2}\cos\beta + (\bar{\theta}_1 - \bar{\theta}_2)\sin\beta\right] \tag{1.33}$$
$$q_y(x, y) = \frac{P}{2\pi}\left[\ln\frac{\bar{r}_1}{\bar{r}_2}\sin\beta - (\bar{\theta}_1 - \bar{\theta}_2)\cos\beta\right]$$

1.6 Computation of the Boundary Influence Coefficients

It follows from the discussion in Sect. 1.5 that the boundary influence coefficients $\overset{ij}{\mathbb{G}}$ relate the strength of a line source $\overset{j}{P}$ at the jth boundary element (the "sending" element) to the potential $\overset{i}{\phi}$ it produces at the ith element (the "receiving" element). These coefficients can be calculated from results given previously by using standard coordinate transformation formulas. The required expressions are given below.

Referring to Fig. 1.6, the local coordinates for element j are found from (1.29) by setting $\xi_x = \overset{j}{x}$, $\xi_y = \overset{j}{y}$, and $\beta = \overset{j}{\beta}$, i.e.,

$$\begin{aligned} \bar{x} &= (x - \overset{j}{x}) \cos \overset{j}{\beta} + (y - \overset{j}{y}) \sin \overset{j}{\beta} \\ \bar{y} &= -(x - \overset{j}{x}) \sin \overset{j}{\beta} + (y - \overset{j}{y}) \cos \overset{j}{\beta} \end{aligned} \tag{1.34}$$

Using these coordinates in the results of Sects. 1.4 and 1.5 allows us to calculate the effects of a source $\overset{j}{P}$ at element j on the potential $\phi(x, y)$ and the gradient components $q_x(x, y)$ and $q_y(x, y)$ anywhere in the region of interest. When point (x, y) is at the center of element i Eq. (1.34) become

$$\begin{aligned} \bar{x} &= (\overset{i}{x} - \overset{j}{x}) \cos \overset{j}{\beta} + (\overset{i}{y} - \overset{j}{y}) \sin \overset{j}{\beta} \\ \bar{y} &= -(\overset{i}{x} - \overset{j}{x}) \sin \overset{j}{\beta} + (\overset{i}{y} - \overset{j}{y}) \cos \overset{j}{\beta} \end{aligned} \tag{1.35}$$

which define the local coordinates of point i relative to point j in Fig. 1.6.

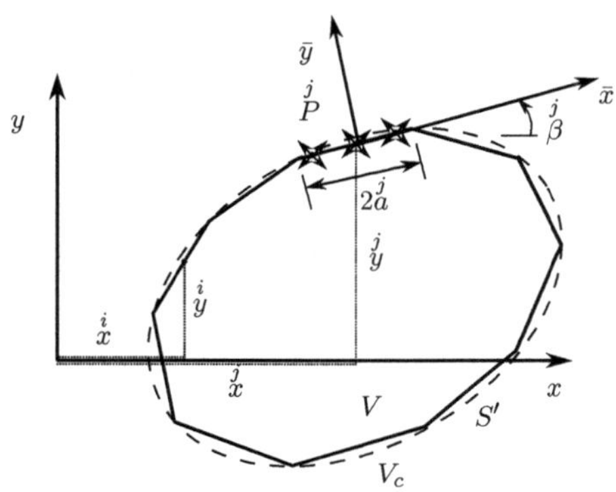

Fig. 1.6 Positions and orientations of boundary elements

Substitution of (1.35) into (1.30) defines the value of $\overset{ij}{\mathbb{G}}$ in the system (1.27) as

$$\overset{ij}{\mathbb{G}} = -\frac{1}{2\pi}\left[\bar{y}(\bar{\theta}_1 - \bar{\theta}_2) - (\bar{x} - \overset{j}{a})\ln \bar{r}_1 + (\bar{x} + \overset{j}{a})\ln \bar{r}_2 - 2\overset{j}{a}\right] \tag{1.36}$$

in which the local $\bar{x}$, $\bar{y}$ coordinates are associated with element j. We emphasize that the local coordinate systems are different for *all* of the boundary elements, whether or not their inclinations $\overset{j}{\beta}$ and $\overset{i}{\beta}$ are different, because, as shown in Fig. 1.6, they each have different midpoint coordinates $(\overset{j}{x}, \overset{j}{y})$ and $(\overset{i}{x}, \overset{i}{y})$. Also, while in principle the elements may have different lengths, it is usually found that the most accurate numerical results are obtained when the element sizes are approximately the same. This requirement can be relaxed for line sources that are allowed to vary linearly or quadratically over the elements, but we will not pursue this possibility in our introductory treatment of the numerical approach.

We are now in a position to explore numerical solutions to Laplace's equation (1.23) subject to boundary conditions of the form (1.24). One simple additional step will allow us to extend the numerical method to consider flux-type boundary conditions, too, and we consider this next.[8]

A flux-type boundary condition at the ith element requires us to compute the component of the gradient of ϕ perpendicular to the element, i.e., the component normal to the boundary. Denoting this quantity as $\overset{i}{q}$ and letting $\overset{i}{q}_x$ and $\overset{i}{q}_y$ be the gradient components at element i as calculated from (1.33) and (1.35), it can be shown that

$$\overset{i}{q} = -\overset{i}{q}_x \sin\overset{i}{\beta} + \overset{i}{q}_y \cos\overset{i}{\beta} \tag{1.37}$$

Substituting (1.33) into this expression and adding the contributions from all the elements, we find that

$$\overset{i}{q} = -\frac{1}{2\pi}\sum_{j=1}^{N}\left[\ln\frac{\bar{r}_1}{\bar{r}_2}\sin(\overset{i}{\beta} - \overset{j}{\beta}) + (\bar{\theta}_1 - \bar{\theta}_2)\cos(\overset{i}{\beta} - \overset{j}{\beta})\right]\overset{j}{P} \tag{1.38}$$

or

$$\overset{i}{q} = \sum_{j=1}^{N}\overset{ij}{\mathbb{F}}\overset{j}{P} \tag{1.39}$$

[8] A problem for which values of the potential ϕ are prescribed on the boundary S is called a Dirichlet boundary value problem. If the normal derivative $\partial\phi/\partial\boldsymbol{n}$ is prescribed instead the problem is called a Neumann boundary value problem. If ϕ is prescribed over part of S and $\partial\phi/\partial\boldsymbol{n}$ over the remainder the problem is called a mixed boundary value problem.

where $\overset{ij}{\mathbb{F}}$ is a boundary influence coefficient that relates the normal component of the gradient of ϕ at the ith element to a constant line source $\overset{j}{P}$ at the jth element. This equation should be compared with (1.27), which gives a similar expression for the potential $\overset{i}{\phi}$ at the ith element in terms of a boundary influence coefficient $\overset{ij}{\mathbb{G}}$.

Using Eqs. (1.27) and (1.39) we can incorporate mixed boundary conditions in our numerical model. If the normal derivative of the potential is prescribed at element i we use (1.39); if the potential is prescribed we use (1.27). Considering all elements in this way we obtain a system of simultaneous linear algebraic equations of the form

$$\sum_{j=1}^{N} \overset{ij}{\mathbb{A}} \overset{j}{P} = \overset{i}{b}, \qquad i = 1 \text{ to } N \tag{1.40}$$

where $\overset{ij}{\mathbb{A}}$ is either $\overset{ij}{\mathbb{F}}$ or $\overset{ij}{\mathbb{G}}$ depending on the type of boundary condition at the ith element; $\overset{i}{b}$ is the associated prescribed value of $\overset{i}{q}$ or $\overset{i}{\phi}$.

The influence of the line source $\overset{i}{P}$ on the ith element itself is called the *element self effect*. The self-effect terms represent the diagonal terms of the influence coefficients, $\overset{ii}{\mathbb{A}}$, and can be calculated from either (1.36) or (1.38), depending on the type of boundary condition being applied at the element. If the potential ϕ is prescribed it follows from (1.36) that $\overset{ii}{\mathbb{A}}$ is given as

$$\overset{ii}{\mathbb{A}} = \frac{\overset{i}{a}}{\pi}(1 - \ln \overset{i}{a}) \tag{1.41}$$

cf. (1.14). On the other hand, if the normal derivative q is prescribed we find from (1.38) that

$$\overset{ii}{\mathbb{A}} = +\frac{1}{2} \tag{1.42}$$

This value of $\overset{ii}{\mathbb{A}}$ is obtained from (1.38) by considering the limiting case in which $\bar{x} = 0$ and $\bar{y} \to 0_{-}$, i.e., from *inside* the region of interest. This result was obtained previously as (1.20).

1.7 Integral Representation I: The Single Layer Potential

A generalization of the results of Sect. 1.3 leads us to consider a continuous distribution of sources along a closed curve S in the infinite plane. The following integral expression then applies for a general harmonic function $\phi(\boldsymbol{x})$ for the case that the sample point $\boldsymbol{x}$ is not on S (usually written as $\boldsymbol{x} \notin S$):

$$\phi(\boldsymbol{x}) = \int_S \rho(\boldsymbol{\xi}) G(\boldsymbol{x}, \boldsymbol{\xi}) \, \mathrm{d}S(\boldsymbol{\xi}) \tag{1.43}$$

This expression can be thought of as the limiting case of the summation of a sufficiently smooth distribution of sources of strength $\rho(\boldsymbol{\xi})\mathrm{d}S(\boldsymbol{\xi})$ at points $\boldsymbol{\xi}$ along the curve S, where $\rho(\boldsymbol{\xi})$ is a *density function* that specifies the nature of the source distribution along S. Representation (1.43) defines the potential $\phi(\boldsymbol{x})$ for a distribution of sources throughout a volume whereby all of the sources lie along a single curve, i.e., a layer. For this reason, $\phi(\boldsymbol{x})$ is called a *single layer potential*, or sometimes a *simple layer potential*.

By construction, integral expression (1.43) automatically satisfies (1.23) provided we always consider a complementary region V_c together with the physical region of interest V such that the region $V + V_c$ is infinite. It will be tacitly assumed throughout this book that this is the case.

Equation (1.43) can be differentiated as in (1.15) to calculate the components of the gradient vector $q_x(\boldsymbol{x}) = \partial\phi(\boldsymbol{x})/\partial x$ and $q_y(\boldsymbol{x}) = \partial\phi(\boldsymbol{x})/\partial y$ at point $\boldsymbol{x} \notin S$. It is often required to find the *normal derivative* of the potential across a curve (not necessarily curve S) at point $\boldsymbol{x}$, and this is accomplished by combining the x and y components of the gradient vector as follows[9]:

$$\frac{\partial\phi(\boldsymbol{x})}{\partial\boldsymbol{n}(\boldsymbol{x})} = \frac{\partial\phi(\boldsymbol{x})}{\partial x} n_x(\boldsymbol{x}) + \frac{\partial\phi(\boldsymbol{x})}{\partial y} n_y(\boldsymbol{x}) \tag{1.44}$$

where $n_x(\boldsymbol{x})$ and $n_y(\boldsymbol{x})$ are the components of the normal to the curve. We therefore have the following equation for the normal derivative of $\phi(\boldsymbol{x})$ in (1.43):

$$\frac{\partial\phi(\boldsymbol{x})}{\partial\boldsymbol{n}(\boldsymbol{x})} = \int_S \rho(\boldsymbol{\xi}) \frac{\partial G(\boldsymbol{x}, \boldsymbol{\xi})}{\partial\boldsymbol{n}(\boldsymbol{x})} \, \mathrm{d}S(\boldsymbol{\xi}) \tag{1.45}$$

Or, using the abbreviation $Q(\boldsymbol{x}, \boldsymbol{\xi}) = \partial G(\boldsymbol{x}, \boldsymbol{\xi})/\partial\boldsymbol{n}(\boldsymbol{x})$,

$$\frac{\partial\phi(\boldsymbol{x})}{\partial\boldsymbol{n}(\boldsymbol{x})} = \int_S \rho(\boldsymbol{\xi}) Q(\boldsymbol{x}, \boldsymbol{\xi}) \, \mathrm{d}S(\boldsymbol{\xi}) \tag{1.46}$$

We emphasize again that the preceding integral expressions pertain to the case that $\boldsymbol{x} \notin S$. If $\boldsymbol{x} = \boldsymbol{x}_0$ is on the boundary S the integrals on the right-hand sides of (1.43) and (1.46) are called the *direct values* of these integrals. We will show in Chap. 2 that the direct value of integral (1.43) is the same as the limit value as $\boldsymbol{x} \to \boldsymbol{x}_0$ whereas the direct value of integral (1.46) differs from its corresponding limit value.

[9] The right-hand side of this expression is a scalar, which leads some researchers to write the normal derivative as $\partial\phi/\partial n$. Others (including the authors) use the notation $\partial\phi/\partial\boldsymbol{n}$ to emphasize that the derivative is taken *in the direction of the normal vector*.

1.7.1 Numerical Approximations

The preceding integrals can be discretized by dividing the boundary contour S into N straight-line elements and representing the density function $\rho(\boldsymbol{\xi})$ over these elements as constant strength sources $\overset{j}{P}$, $j = 1$ to N. Referring to Fig. 1.6 we write

$$\phi(x, y) = \sum_{j=1}^{N} \overset{j}{P} \int_{-\overset{j}{a}}^{\overset{j}{a}} G(\bar{x} - \xi_{\bar{x}}, \bar{y})\, d\xi_{\bar{x}} \tag{1.47}$$

where $\bar{x}$ and $\bar{y}$ are defined as in (1.34) so that (see (1.9))

$$\phi(x, y) = -\frac{1}{2\pi} \sum_{j=1}^{N} \overset{j}{P} \Big[\bar{y}(\bar{\theta}_1 - \bar{\theta}_2) - (\bar{x} - \overset{j}{a}) \ln \bar{r}_1 + (\bar{x} + \overset{j}{a}) \ln \bar{r}_2 - 2\overset{j}{a} \Big] \tag{1.48}$$

When $x = \overset{i}{x}$ and $y = \overset{i}{y}$ we use (1.35) and (1.36) to obtain

$$\overset{i}{\phi} = \sum_{j=1}^{N} \overset{ij}{\mathbb{G}} \overset{j}{P} \tag{1.49}$$

which is the same as the approximation for the potential $\phi(\boldsymbol{x})$ at the midpoint of element i represented by (1.27).

To approximate (1.45) we first write, using (1.44),

$$\frac{\partial \phi(\boldsymbol{x})}{\partial \boldsymbol{n}(\boldsymbol{x})} = \int_S \rho(\boldsymbol{\xi}) \Big[\frac{\partial}{\partial x} G(\boldsymbol{x}, \boldsymbol{\xi}) n_x(\boldsymbol{x}) + \frac{\partial}{\partial y} G(\boldsymbol{x}, \boldsymbol{\xi}) n_y(\boldsymbol{x}) \Big]\, dS(\boldsymbol{\xi}) \tag{1.50}$$

We next consider a local $\bar{x}$, $\bar{y}$ coordinate system originating at point $\boldsymbol{\xi}$ and with $\bar{y}$ pointing in the direction of the normal to S and passing through the point $\mathbf{x} \notin S$. Taking into account that $n_x(\boldsymbol{x}) = -\sin\beta$ and $n_y(\boldsymbol{x}) = \cos\beta$, where angle β defines the tangent to a curve passing through point $\boldsymbol{x}$, Eq. (1.50) can be written as

$$q(\boldsymbol{x}) = \int_S \rho(\boldsymbol{\xi}) \Big[-\frac{\partial}{\partial x} G(\boldsymbol{x}, \boldsymbol{\xi}) \sin\beta + \frac{\partial}{\partial y} G(\boldsymbol{x}, \boldsymbol{\xi}) \cos\beta \Big]\, dS(\boldsymbol{\xi}) \tag{1.51}$$

where we have written $q(\boldsymbol{x})$ for the normal derivative $\partial\phi(\boldsymbol{x})/\partial\boldsymbol{n}(\boldsymbol{x})$. Notice that if point $\boldsymbol{x}$ is on the line parallel to the ith element in Fig. 1.6 then $\beta = \overset{i}{\beta}$.

Proceeding as before, we approximate (1.51) for element i as

$$q(x,y) = \sum_{j=1}^{N} \overset{j}{P} \int_{-\overset{j}{a}}^{\overset{j}{a}} \left[-\frac{\partial}{\partial x} G(\bar{x} - \xi_{\bar{x}}, \bar{y}) \sin \overset{i}{\beta} + \frac{\partial}{\partial y} G(\bar{x} - \xi_{\bar{x}}, \bar{y}) \cos \overset{i}{\beta} \right] \mathrm{d}\xi_{\bar{x}} \tag{1.52}$$

Using the chain rule for partial differentiation and (1.34) we have that

$$\begin{aligned}
\frac{\partial}{\partial x} G(\bar{x} - \xi_{\bar{x}}, \bar{y}) &= \frac{\partial \bar{x}}{\partial x} \frac{\partial}{\partial \bar{x}} G(\bar{x} - \xi_{\bar{x}}, \bar{y}) + \frac{\partial \bar{y}}{\partial x} \frac{\partial}{\partial \bar{y}} G(\bar{x} - \xi_{\bar{x}}, \bar{y}) \\
&= \cos \overset{j}{\beta} \frac{\partial}{\partial \bar{x}} G(\bar{x} - \xi_{\bar{x}}, \bar{y}) - \sin \overset{j}{\beta} \frac{\partial}{\partial \bar{y}} G(\bar{x} - \xi_{\bar{x}}, \bar{y}) \\
\frac{\partial}{\partial y} G(\bar{x} - \xi_{\bar{x}}, \bar{y}) &= \frac{\partial \bar{x}}{\partial y} \frac{\partial}{\partial \bar{x}} G(\bar{x} - \xi_{\bar{x}}, \bar{y}) + \frac{\partial \bar{y}}{\partial y} \frac{\partial}{\partial \bar{y}} G(\bar{x} - \xi_{\bar{x}}, \bar{y}) \\
&= \sin \overset{j}{\beta} \frac{\partial}{\partial \bar{x}} G(\bar{x} - \xi_{\bar{x}}, \bar{y}) + \cos \overset{j}{\beta} \frac{\partial}{\partial \bar{y}} G(\bar{x} - \xi_{\bar{x}}, \bar{y})
\end{aligned} \tag{1.53}$$

and (1.52) can be written as

$$q(x,y) = \sum_{j=1}^{N} \overset{j}{P} \int_{-\overset{j}{a}}^{\overset{j}{a}} \left[-\frac{\partial}{\partial \bar{x}} G(\bar{x} - \xi_{\bar{x}}, \bar{y}) \sin(\overset{i}{\beta} - \overset{j}{\beta}) + \frac{\partial}{\partial \bar{y}} G(\bar{x} - \xi_{\bar{x}}, \bar{y}) \cos(\overset{i}{\beta} - \overset{j}{\beta}) \right] \mathrm{d}\xi_{\bar{x}} \tag{1.54}$$

With $G(\bar{x} - \xi_{\bar{x}}, \bar{y}) = -\ln\left[(\bar{x} - \xi_{\bar{x}})^2 + \bar{y}^2\right]^{1/2}/(2\pi)$ it is easy to show that, after taking the limit to the boundary,

$$\overset{i}{q} = -\frac{1}{2\pi} \sum_{j=1}^{N} \overset{j}{P} \left[\ln \frac{\bar{r}_1}{\bar{r}_2} \sin(\overset{i}{\beta} - \overset{j}{\beta}) + (\bar{\theta}_1 - \bar{\theta}_2) \cos(\overset{i}{\beta} - \overset{j}{\beta}) \right] \tag{1.55}$$

which is the same as (1.38).

According to the preceding discussion, we conclude that the heuristic approach adopted in Sects. 1.4–1.6 is equivalent to a numerical solution of the *boundary integral equation* for a single layer potential.

1.8 Another Approach: A Dipole Source

Another singular solution to Laplace's equation can be constructed by considering two point sources, one of strength ρ and the other of strength $-\rho$, separated by a distance $2c$, in the limit as $c \to 0$. By analogy with (1.5), the potential for the two-dimensional problem depicted in Fig. 1.7 is given as

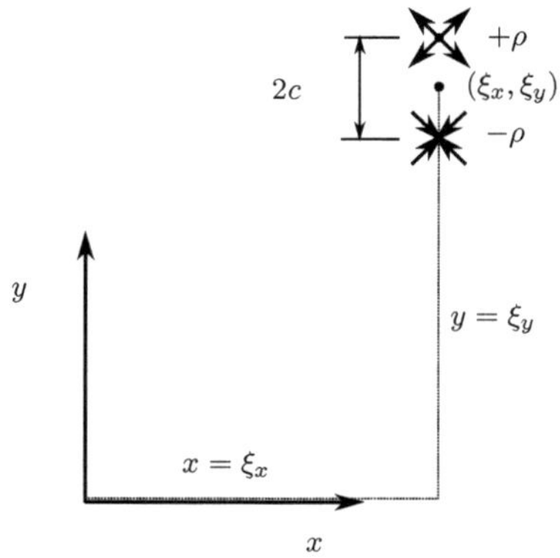

Fig. 1.7 Arrangement of sources used to define a dipole

$$\rho G(x - \xi_x, y - \xi_y - c) - \rho G(x - \xi_x, y - \xi_y + c) \tag{1.56}$$

If we require that the product $2c\rho = \psi$ be a constant as $c \to 0$ we write

$$\phi(x, y) = \psi \lim_{c \to 0} \frac{G(x - \xi_x, y - \xi_y - c) - G(x - \xi_x, y - \xi_y + c)}{2c} \tag{1.57}$$

Using the results $\partial G(x - \xi_x, y - \xi_y - c)/\partial c = \partial G(x - \xi_x, y - \xi_y - c)/\partial \xi_y$ and $\partial G(x - \xi_x, y - \xi_y + c)/\partial c = -\partial G(x - \xi_x, y - \xi_y + c)/\partial \xi_y$, a simple application of L'Hopital's rule gives

$$\phi(x, y) = \psi \frac{\partial}{\partial \xi_y} G(x - \xi_x, y - \xi_y) \tag{1.58}$$

The source represented by this equation is called a *dipole*. It can be integrated in the same manner as was done for a simple source in Sect. 1.3 to develop the potential $\phi(x, y)$ due to a constant strength dipole source D along the line segment $| x | \leq a,\ y = 0$. Noting that $\partial G(x - \xi_x, y - \xi_y)/\partial \xi_y = -\partial G(x - \xi_x, y - \xi_y)/\partial y$ we find that, after setting $\xi_y = 0$,

$$\begin{aligned} \phi(x, y) &= -D \int_{-a}^{a} \frac{\partial}{\partial y} G(x - \xi_x, y) \mathrm{d}\xi_x \\ &= -D \int_{-a}^{a} \frac{y}{(x - \xi_x)^2 + y^2} \mathrm{d}\xi_x \\ &= \frac{D}{2\pi}(\theta_1 - \theta_2) \end{aligned} \tag{1.59}$$

where θ_1 and θ_2 are defined in (1.11). The potential for this dipole source therefore exhibits a constant discontinuity along the line segment in question, i.e.,

$$\phi(x, 0_+) - \phi(x, 0_-) = D \qquad | x |< a \tag{1.60}$$

because

$$\begin{aligned}\phi(x, 0) &= +\frac{D}{2} \qquad | x |< a, \quad y = 0_+ \\ \phi(x, 0) &= -\frac{D}{2} \qquad | x |< a, \quad y = 0_-\end{aligned} \tag{1.61}$$

(Notice that the signs for these expressions are the opposite of those observed for the gradient vector component $q_y(x, 0)$ for a constant line source P over the same line segment; see (1.20).)

The components of the gradient vector for the potential $\phi(x, y)$ in (1.59) are

$$\begin{aligned}q_x(x, y) &= \frac{\partial}{\partial x}\phi(x, y) = -\frac{D}{2\pi}\left[\frac{y}{r_1^2} - \frac{y}{r_2^2}\right] \\ q_y(x, y) &= \frac{\partial}{\partial y}\phi(x, y) = \frac{D}{2\pi}\left[\frac{(x-a)}{r_1^2} - \frac{(x+a)}{r_2^2}\right]\end{aligned} \tag{1.62}$$

Not considering the points $x = \pm a,\ y = 0$ the gradient components are continuous everywhere in the plane whereas the potential $\phi(\boldsymbol{x})$ itself is discontinuous and undergoes a "jump" of constant value when crossing from one side of the line source to the other. This circumstance is the reverse of what we previously observed for the simple line source (see Sect. 1.3).

For a constant dipole source D over an arbitrarily oriented line segment $| \bar{x} |\leq a$, $\bar{y} = 0$ we follow the same procedure as in Sect. 1.5 to obtain the results

$$\phi(x, y) = \frac{D}{2\pi}(\bar{\theta}_1 - \bar{\theta}_2) \tag{1.63}$$

and

$$\begin{aligned}q_x(x, y) &= -\frac{D}{2\pi}\left[\left(\frac{\bar{y}}{\bar{r}_1^2} - \frac{\bar{y}}{\bar{r}_2^2}\right)\cos\beta + \left\{\frac{(\bar{x}-a)}{\bar{r}_1^2} - \frac{(\bar{x}+a)}{\bar{r}_2^2}\right\}\sin\beta\right] \\ q_y(x, y) &= -\frac{D}{2\pi}\left[\left(\frac{\bar{y}}{\bar{r}_1^2} - \frac{\bar{y}}{\bar{r}_2^2}\right)\sin\beta - \left\{\frac{(\bar{x}-a)}{\bar{r}_1^2} - \frac{(\bar{x}+a)}{\bar{r}_2^2}\right\}\cos\beta\right]\end{aligned} \tag{1.64}$$

It is now a straightforward matter to develop numerical procedure that is in all respects comparable to the one given in Sects. 1.4–1.6 for a single layer potential. For example, in place of (1.26) we could write

$$\overset{i}{\phi} = \sum_{j=1}^{N} \overset{ij}{\mathbb{Q}} \overset{j}{D} \tag{1.65}$$

in which, from (1.63), the boundary influence coefficient $\overset{ij}{\mathbb{Q}}$ is

$$\overset{ij}{\mathbb{Q}} = \frac{1}{2\pi}(\bar{\theta}_1 - \bar{\theta}_2) \tag{1.66}$$

Similarly, a formula comparable to (1.39) for the normal derivative $\overset{i}{q}$ at element i due to constant strength dipole sources at all N boundary elements can be written as

$$\overset{i}{q} = \sum_{j=1}^{N} \overset{ij}{\mathbb{E}}\overset{j}{D} \tag{1.67}$$

where the boundary influence coefficient $\overset{ij}{\mathbb{E}}$ is

$$\overset{ij}{\mathbb{E}} = \frac{1}{2\pi}\left[\left(\frac{\bar{y}}{\bar{r}_1^2} - \frac{\bar{y}}{\bar{r}_2^2}\right)\sin(\overset{i}{\beta} - \overset{j}{\beta}) + \left\{\frac{(\bar{x} - a)}{\bar{r}_1^2} - \frac{(\bar{x} + a)}{\bar{r}_2^2}\right\}\cos(\overset{i}{\beta} - \overset{j}{\beta})\right] \tag{1.68}$$

The system of simultaneous linear algebraic equations for this numerical procedure can be written as

$$\sum_{j=1}^{N} \overset{ij}{\mathbb{B}}\overset{j}{D} = \overset{i}{b}, \qquad i = 1 \text{ to } N \tag{1.69}$$

where $\overset{ij}{\mathbb{B}}$ is either $\overset{ij}{\mathbb{E}}$ or $\overset{ij}{\mathbb{Q}}$ depending on the type of boundary condition at the ith element; $\overset{i}{b}$ is again the associated value of $\overset{i}{q}$ or $\overset{i}{\phi}$. It is easy to show that the self-effect terms for the boundary influence coefficient matrix $\overset{ij}{\mathbb{B}}$ are given as follows:

$$\overset{ii}{\mathbb{B}} = -\frac{1}{2} \qquad \text{if } \overset{i}{\phi} \text{ is prescribed at element } i \tag{1.70}$$

$$\overset{ii}{\mathbb{B}} = -\frac{1}{\pi \overset{i}{a}} \qquad \text{if } \overset{i}{q} \text{ is prescribed at element } i$$

Just as was noted previously for line sources $\overset{j}{P}$, the line dipoles $\overset{j}{D}$ in the above expressions are fictitious quantities; they were introduced to obtain a numerical solution to a problem but have no physical significance for that problem. An exception occurs when the boundary S is not closed, in which case S can be treated as a crack with a jump in the potential across its two surfaces.

1.9 Integral Representation II: The Double Layer Potential

By analogy with (1.43) and with reference to (1.58), an integral expression can be written for a general harmonic function $\phi(\boldsymbol{x})$ associated with a sufficiently smooth distribution of dipole sources along a closed curve S in an infinite plane, again assuming that $\boldsymbol{x} \notin S$:

$$\phi(\boldsymbol{x}) = \int_S \psi(\boldsymbol{\xi}) \frac{\partial G(\boldsymbol{x}, \boldsymbol{\xi})}{\partial \boldsymbol{n}(\boldsymbol{\xi})} \, \mathrm{d}S(\boldsymbol{\xi}) \tag{1.71}$$

In this equation, $\psi(\boldsymbol{\xi})$ is the density function for the dipole source distribution and $\partial G(\boldsymbol{x}, \boldsymbol{\xi})/\partial \boldsymbol{n}(\boldsymbol{\xi})$ is the derivative of the function $G(\boldsymbol{x}, \boldsymbol{\xi})$ perpendicular (normal) to the boundary S at point $\boldsymbol{\xi}$. This representation for $\phi(\boldsymbol{x})$ is called a *double layer potential* because, by construction, it derives from the limiting case of two "layers" of sources along the boundary. A source associated with one of the layers can be thought of as being an infinitesimal distance away from a companion source of opposite sign on the other layer, with the line connecting them perpendicular to the boundary, hence the use of the normal derivative in (1.71).

Equation (1.71) can be rewritten using the definition introduced in Sect. 1.7 for function $Q(\boldsymbol{x}, \boldsymbol{\xi}) = \partial G(\boldsymbol{x}, \boldsymbol{\xi})/\partial \boldsymbol{n}(\boldsymbol{x})$. If $\boldsymbol{x}$ and $\boldsymbol{\xi}$ are interchanged this expression becomes $Q(\boldsymbol{\xi}, \boldsymbol{x}) = \partial G(\boldsymbol{\xi}, \boldsymbol{x})/\partial \boldsymbol{n}(\boldsymbol{\xi})$, or, because $G(\boldsymbol{x}, \boldsymbol{\xi}) = G(\boldsymbol{\xi}, \boldsymbol{x})$, $Q(\boldsymbol{\xi}, \boldsymbol{x}) = \partial G(\boldsymbol{x}, \boldsymbol{\xi})/\partial \boldsymbol{n}(\boldsymbol{\xi})$. Using this expression in (1.71) the double layer potential can also be written as

$$\phi(\boldsymbol{x}) = \int_S \psi(\boldsymbol{\xi}) Q(\boldsymbol{\xi}, \boldsymbol{x}) \, \mathrm{d}S(\boldsymbol{\xi}) \tag{1.72}$$

It can be seen from (1.46) that the kernel $Q(\boldsymbol{x}, \boldsymbol{\xi})$ on its right-hand side can be obtained from that of (1.72) by interchanging the arguments $\boldsymbol{\xi}$ and $\boldsymbol{x}$. For this reason, the potential specified by (1.46) is called the **adjoint double layer potential**.

The normal derivative of the double layer potential (1.71) at the sample point $\boldsymbol{x}$ is

$$\frac{\partial \phi(\boldsymbol{x})}{\partial \boldsymbol{n}(\boldsymbol{x})} = \int_S \psi(\boldsymbol{\xi}) \frac{\partial^2 G(\boldsymbol{x}, \boldsymbol{\xi})}{\partial \boldsymbol{n}(\boldsymbol{x}) \partial \boldsymbol{n}(\boldsymbol{\xi})} \, \mathrm{d}S(\boldsymbol{\xi}) \tag{1.73}$$

or

$$\frac{\partial \phi(\boldsymbol{x})}{\partial \boldsymbol{n}(\boldsymbol{x})} = \int_S \psi(\boldsymbol{\xi}) H(\boldsymbol{x}, \boldsymbol{\xi}) \, \mathrm{d}S(\boldsymbol{\xi}) \tag{1.74}$$

in which it can easily be seen that the kernel $H(\boldsymbol{x}, \boldsymbol{\xi})$ is symmetric with respect to its arguments, i.e., $H(\boldsymbol{x}, \boldsymbol{\xi}) = H(\boldsymbol{\xi}, \boldsymbol{x})$. This follows directly from the symmetry of $G(\boldsymbol{x}, \boldsymbol{\xi})$.

Just as for the single layer potential in Sect. 1.7, the preceding integral expressions pertain to the case that $\boldsymbol{x} \notin S$. The direct values of these integrals are again obtained by setting $\boldsymbol{x} = \boldsymbol{x}_0$. We will see in Chap. 2 that the direct value of integral (1.74) is the same as the limit value as $\boldsymbol{x} \to \boldsymbol{x}_0$ whereas the direct value of integral (1.72) differs from its corresponding limit value.

1.9.1 Numerical Approximations

Proceeding as in Sect. 1.7 we divide the boundary contour S into N straight-line elements and represent the density function $\psi(\boldsymbol{\xi})$ over these elements by constant strength sources $\overset{j}{D}$, $j = 1$ to N. Noting that for element j the normal derivative $\partial G(\boldsymbol{x}, \boldsymbol{\xi})/\partial \boldsymbol{n}(\boldsymbol{\xi})$ can be written as $\partial G(\bar{x} - \xi_{\bar{x}}, \bar{y} - \xi_{\bar{y}})/\partial \xi_{\bar{y}} = -\partial G(\bar{x} - \xi_{\bar{x}}, \bar{y} - \xi_{\bar{y}})/\partial \bar{y}$, where $\bar{x}$ and $\bar{y}$ are defined in (1.34), we have that

$$\phi(x, y) = -\sum_{j=1}^{N} \overset{j}{D} \int_{-\overset{j}{a}}^{\overset{j}{a}} \frac{\partial}{\partial \bar{y}} G(\bar{x} - \xi_{\bar{x}}, \bar{y})\, \mathrm{d}\xi_{\bar{x}} \tag{1.75}$$

or

$$\begin{aligned} \phi(x, y) &= \frac{1}{2\pi} \sum_{j=1}^{N} \overset{j}{D} \int_{-\overset{j}{a}}^{\overset{j}{a}} \frac{\bar{y}}{(\bar{x} - \xi_{\bar{x}})^2 + \bar{y}^2}\, \mathrm{d}\xi_{\bar{x}} \\ &= \frac{1}{2\pi} \sum_{j=1}^{N} \overset{j}{D} (\bar{\theta}_1 - \bar{\theta}_2) \end{aligned} \tag{1.76}$$

When $x = \overset{i}{x}$ and $y = \overset{i}{y}$ we write

$$\overset{i}{\phi} = \sum_{j=1}^{N} \overset{ij}{\mathbb{Q}} \overset{j}{D} \tag{1.77}$$

just as in (1.65).

A numerical approximation for the normal derivative $q(\boldsymbol{x}) = \partial \phi(\boldsymbol{x})/\partial \boldsymbol{n}(\boldsymbol{x})$ of the double layer potential at the point $(\overset{i}{x}, \overset{i}{y})$ can be written by recalling that, on a line parallel to the ith element, $\partial \phi(\boldsymbol{x})/\partial \boldsymbol{n}(\boldsymbol{x}) = -\sin \overset{i}{\beta}\, \partial \phi(\boldsymbol{x})/\partial x + \cos \overset{i}{\beta}\, \partial \phi(\boldsymbol{x})/\partial y$ and using (1.75) to write

$$q(\overset{i}{x}, \overset{i}{y}) = -\sum_{j=1}^{N} \overset{j}{D} \int_{-\overset{j}{a}}^{\overset{j}{a}} \frac{\partial}{\partial \bar{y}} \Big[-\frac{\partial}{\partial x} G(\bar{x} - \xi_{\bar{x}}, \bar{y}) \sin \overset{i}{\beta} + \frac{\partial}{\partial y} G(\bar{x} - \xi_{\bar{x}}, \bar{y}) \cos \overset{i}{\beta} \Big]\, \mathrm{d}\xi_{\bar{x}} \tag{1.78}$$

Using the chain rule formulas (1.53) this expression can be written as

$$q(\overset{i}{x},\overset{i}{y}) = \sum_{j=1}^{N} \overset{j}{D} \int_{-\overset{j}{a}}^{\overset{j}{a}} \Big[\frac{\partial^2}{\partial \bar{x} \partial \bar{y}} G(\bar{x}-\xi_{\bar{x}},\bar{y}) \sin(\overset{i}{\beta}-\overset{j}{\beta}) - \frac{\partial^2}{\partial \bar{y}^2} G(\bar{x}-\xi_{\bar{x}},\bar{y}) \cos(\overset{i}{\beta}-\overset{j}{\beta}) \Big] \mathrm{d}\xi_{\bar{x}} \tag{1.79}$$

from which we obtain the result

$$\overset{i}{q} = \frac{1}{2\pi} \sum_{j=1}^{N} \overset{j}{D} \Big[\Big(\frac{\bar{y}}{\bar{r}_1^2} - \frac{\bar{y}}{\bar{r}_2^2} \Big) \sin(\overset{i}{\beta}-\overset{j}{\beta}) + \Big\{ \frac{(\bar{x}-\overset{j}{a})}{\bar{r}_1^2} - \frac{(\bar{x}+\overset{j}{a})}{\bar{r}_2^2} \Big\} \cos(\overset{i}{\beta}-\overset{j}{\beta}) \Big] \tag{1.80}$$

which agrees with (1.67) and (1.68).

1.9.2 Special Case for a Multiply Connected Region

For reasons beyond the scope of this discussion the numerical approach outlined above for dipole sources must be modified to solve problems involving multiply connected regions. According to Mikhlin (1964) the following modification to (1.71) is needed for the case of a finite region V with boundary S_0 enclosing M holes with boundaries S_k, $k = 1$ to M:

$$\phi(\boldsymbol{x}) = \int_S \psi(\boldsymbol{\xi}) \frac{\partial G(\boldsymbol{x},\boldsymbol{\xi})}{\partial \boldsymbol{n}(\boldsymbol{\xi})} \mathrm{d}S(\boldsymbol{\xi}) - 2 \sum_{k=1}^{M} A_k \ln | \boldsymbol{x} - \boldsymbol{x}_k | \tag{1.81}$$

where S is the totality of all boundaries, $S_0 + S_1 +$, ..., S_M, A_k are constants, and $\boldsymbol{x}_k$ is point *inside* boundary S_k (i.e., outside the region of interest). The constants A_k satisfy the constraint equations

$$\int_{S_k} \psi(\boldsymbol{\xi}) \, \mathrm{d}S(\boldsymbol{\xi}) = 0, \qquad k = 1, ..., M \tag{1.82}$$

The normal derivative of $\phi(\boldsymbol{x})$ in (1.81) for a point $\boldsymbol{x}$ on a curve with normal $\boldsymbol{n}(\boldsymbol{x})$ is

$$\begin{aligned} \frac{\partial \phi(\boldsymbol{x})}{\partial \boldsymbol{n}(\boldsymbol{x})} &= \int_S \psi(\boldsymbol{\xi}) \frac{\partial^2 G(\boldsymbol{x},\boldsymbol{\xi})}{\partial \boldsymbol{n}(\boldsymbol{x}) \partial \boldsymbol{n}(\boldsymbol{\xi})} \mathrm{d}S(\boldsymbol{\xi}) - 2 \sum_{k=1}^{M} A_k \frac{\partial}{\partial \boldsymbol{n}(\boldsymbol{x})} \ln | \boldsymbol{x} - \boldsymbol{x}_k | \\ &= \int_S \psi(\boldsymbol{\xi}) H(\boldsymbol{x},\boldsymbol{\xi}) \, \mathrm{d}S(\boldsymbol{\xi}) + 2 \sum_{k=1}^{M} A_k \Big[\frac{\partial}{\partial x} \ln | \boldsymbol{x} - \boldsymbol{x}_k | \sin\beta - \frac{\partial}{\partial y} \ln | \boldsymbol{x} - \boldsymbol{x}_k | \cos\beta \Big] \end{aligned} \tag{1.83}$$

where angle β defines the tangent to the curve; $\beta = \overset{i}{\beta}$ for a line parallel to boundary element i.

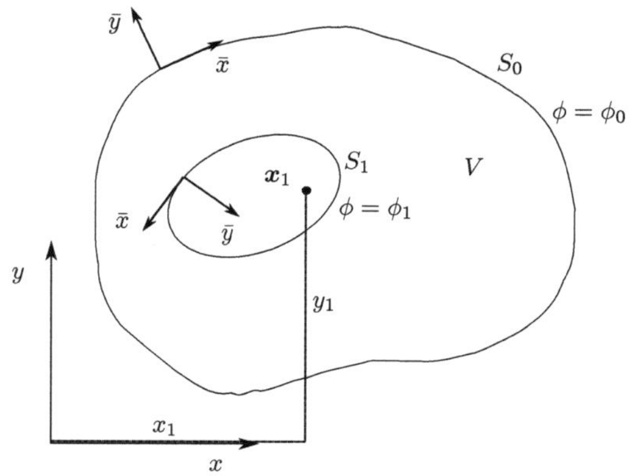

Fig. 1.8 Finite region V with boundary S_0 containing a hole with boundary S_1

The parts of (1.81) and (1.83) containing boundary integrals are the same as before so we will focus our attention on the incorporation of conditions (1.82) and the terms involving constants A_k in the numerical procedure.

For purposes of illustration we will consider Mikhlin's approach for the simple case of a finite region containing a single hole, as depicted in Fig. 1.8 for the case that boundaries S_0 and S_1 are subjected to constant potentials ϕ_0 and ϕ_1. For a more general and elaborate treatment of the problem the reader is referred to the paper by Greenbaum et al. (1993).

We will suppose that the two boundaries S_0 and S_1 are divided into N_0 and N_1 piecewise constant dipole sources. Then, letting $N = N_0 + N_1 + 1$, we seek values of the sources $\overset{j}{D}$ for $j = 1$ to $N - 1$ together with the constant A_1 from (1.81). A numerical solution to this problem involves adding one row and one column to the influence coefficient matrix $\overset{ij}{\mathbb{Q}}$ in (1.77) (which, in the present terminology, is of size $(N - 1) \times (N - 1)$), as well as including constant A_1 in the vector of sources, $\overset{j}{D}$. Letting $\boldsymbol{X} = \left[\overset{1}{D}, \overset{2}{D}, ..., \overset{N-1}{D}, A_1\right]^T$ we will write the revised $N \times N$ system of equations as

$$\sum_{j=1}^{N} \overset{ij}{\mathbb{C}} \overset{j}{X} = \overset{i}{b} \qquad i = 1 \text{ to } N \tag{1.84}$$

in which, for the present problem,

$$\overset{i}{b} = \begin{cases} \phi_0 & i = 1 \text{ to } N_0 \\ \phi_1 & i = N_0 + 1 \text{ to } N_0 + N_1 \\ 0 & i = N \end{cases} \tag{1.85}$$

By construction, entries in the coefficient matrix $\overset{ij}{\mathbb{C}}$ are the same as those for $\overset{ij}{\mathbb{Q}}$ when $i = 1$ to $N - 1$ and $j = 1$ to $N - 1$. Otherwise, we have

$$\overset{iN}{\mathbb{C}} = -2\ln\left[(\overset{i}{x} - x_1)^2 + (\overset{i}{y} - y_1)^2\right]^{1/2} \tag{1.86}$$

for the N rows of column N, and, using the discretized version of (1.82) for $k = 1$,

$$\overset{Nj}{\mathbb{C}} = \begin{cases} 0 & j = 1 \text{ to } N_0 \\ 2\overset{j}{a} & j = N_0 + 1 \text{ to } N_0 + N_1 \\ 0 & j = N \end{cases} \tag{1.87}$$

for the N columns of row N.

After (1.84) has been solved the values of $\phi(\boldsymbol{x})$ and $\partial\phi(\boldsymbol{x})/\partial\boldsymbol{n}(\boldsymbol{x})$ at points $\boldsymbol{x}$ within the region of interest V can be found by discretizing (1.81) and (1.83) in the same way as before. Values of the normal derivative $\overset{i}{q}$ for the $N_0 + N_1$ boundary elements are of special interest and these can be calculated using (1.80) to approximate the integrals in (1.83) and then adding Mikhlin's correction terms. For a typical element i the correction term is

$$2A_1\left[\frac{(\overset{i}{x} - x_1)\sin\overset{i}{\beta} - (\overset{i}{y} - y_1)\cos\overset{i}{\beta}}{(\overset{i}{x} - x_1)^2 + (\overset{i}{y} - y_1)^2}\right] \tag{1.88}$$

1.10 Half-Plane Solutions

We saw in Sect. 1.2 that the potential is zero along a line midway between two point sources with opposite signs. If both two sources have the same sign it can be shown (see exercise (3) in Sect. 1.13) that the normal derivative of the potential is zero along this line instead of the potential itself. These observations make it possible to extend our numerical approach to solve problems for the half-plane $y \leq 0$, say, with boundary conditions of the form $\phi = 0$ or $\partial\phi/\partial y = 0$ on $y = 0$ for $-\infty < x < \infty$ without the need to discretize this boundary. We will provide solutions in this section for a half-plane with a zero potential boundary that contains an arbitrarily oriented line segment with a constant strength source P as well as one with a dipole source D. Similar considerations apply for the case that the normal derivative of the potential is zero on the surface of the half-plane.

Considering first the case of a constant strength source $+P$ over the line segment $|\, \bar{x} \,| \leq a$, $\bar{y} = 0$ in a half-plane with a zero potential boundary, we use the method of images and represent the solution as the sum of an "actual" source in $y \leq 0$ and an "image" source in $y > 0$, as depicted in Fig. 1.9. The line segment for the actual source is centered at the point $x = \xi_x$, $y = \xi_y < 0$ and is inclined from the x

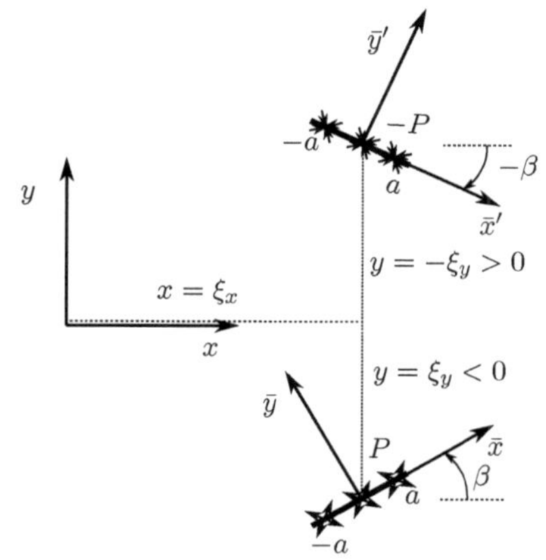

Fig. 1.9 Constant line sources arranged to make $\phi = 0$ on $y = 0$

direction by an angle β, just as defined in Sect. 1.5. The image source occurs over the line segment segment $| \bar{x}' | \leq a$, $\bar{y}' = 0$, where

$$\begin{aligned} \bar{x}' &= (x - \xi_x) \cos \beta' + (y + \xi_y) \sin \beta' \\ \bar{y}' &= -(x - \xi_x) \sin \beta' + (y + \xi_y) \cos \beta' \end{aligned} \tag{1.89}$$

or, since $\beta' = -\beta$,

$$\begin{aligned} \bar{x}' &= (x - \xi_x) \cos \beta - (y + \xi_y) \sin \beta \\ \bar{y}' &= (x - \xi_x) \sin \beta + (y + \xi_y) \cos \beta \end{aligned} \tag{1.90}$$

Referring to (1.30) and (1.31), the potential and the gradient components of the actual source can be written as

$$\overset{A}{\phi}(x, y) = -\frac{P}{2\pi}\left[\bar{y}(\bar{\theta}_1 - \bar{\theta}_2) - (\bar{x} - a) \ln \bar{r}_1 + (\bar{x} + a) \ln \bar{r}_2 - 2a\right] \tag{1.91}$$

and

$$\begin{aligned} \overset{A}{q}_{\bar{x}}(x, y) &= \frac{P}{2\pi} \ln \frac{\bar{r}_1}{\bar{r}_2} \\ \overset{A}{q}_{\bar{y}}(x, y) &= -\frac{P}{2\pi}(\bar{\theta}_1 - \bar{\theta}_2) \end{aligned} \tag{1.92}$$

where $\bar{r}_1$, $\bar{r}_2$ and $\bar{\theta}_1$, $\bar{\theta}_2$ are the distances and angles for the endpoints of the line segment to an arbitrary point (x, y). The x, y components of the gradient are found from (1.32), i.e.,

$$\overset{A}{q}_x = \overset{A}{q}_{\bar{x}} \cos\beta - \overset{A}{q}_{\bar{y}} \sin\beta \tag{1.93}$$
$$\overset{A}{q}_y = \overset{A}{q}_{\bar{x}} \sin\beta + \overset{A}{q}_{\bar{y}} \cos\beta$$

The local coordinates $\bar{x}', \bar{y}'$ of the image source are as in (1.90) and, noting that the strength of this source is $-P$, the potential and the gradient components can be written as

$$\overset{I}{\phi}(x, y) = +\frac{P}{2\pi}\left[\bar{y}'(\bar{\theta}_1' - \bar{\theta}_2') - (\bar{x}' - a)\ln \bar{r}_1' + (\bar{x}' + a)\ln \bar{r}_2' - a\right] \tag{1.94}$$

and

$$\overset{I}{q}_{\bar{x}'}(x, y) = -\frac{P}{2\pi} \ln \frac{\bar{r}_1'}{\bar{r}_2'} \tag{1.95}$$
$$\overset{I}{q}_{\bar{y}'}(x, y) = +\frac{P}{2\pi}(\bar{\theta}_1' - \bar{\theta}_2')$$

where $\bar{r}_1', \bar{r}_2'$ and $\bar{\theta}_1', \bar{\theta}_2'$ are defined similarly as in (1.92). The x, y components of the gradient are

$$\overset{I}{q}_x = \overset{I}{q}_{\bar{x}'} \cos\beta + \overset{I}{q}_{\bar{y}'} \sin\beta \tag{1.96}$$
$$\overset{I}{q}_y = -\overset{I}{q}_{\bar{x}'} \sin\beta + \overset{I}{q}_{\bar{y}'} \cos\beta$$

The complete solution for the half plane is found by adding (1.91) and (1.94), and (1.93) and (1.96) as follows:

$$\phi(x, y) = \overset{A}{\phi}(x, y) + \overset{I}{\phi}(x, y) \tag{1.97}$$

and

$$q_x = \overset{A}{q}_x + \overset{I}{q}_x \tag{1.98}$$
$$q_y = \overset{A}{q}_y + \overset{I}{q}_y$$

Using (1.29) in (1.91) and (1.90) in (1.94) it is easy to show that $\phi(x, 0) = 0$, as required.

A similar procedure can be followed for the case of a constant strength dipole source $+D$ over the line segment $|\ \bar{x}\ | \le a, \bar{y} = 0$ in the half-plane $y \le 0$. In this case, however, the image source must also have strength $+D$ to ensure that the potential $\phi(x, y)$ is zero on $y = 0$. By inspection of (1.59) and (1.62) equations comparable to (1.91) and (1.92) for the actual dipole source are as follows:

$$\overset{A}{\phi}(x, y) = \frac{D}{2\pi}(\bar{\theta}_1 - \bar{\theta}_2) \tag{1.99}$$

and

$$\begin{aligned} \overset{A}{q}_{\bar{x}}(x, y) &= -\frac{D}{2\pi}\Big[\frac{\bar{y}}{\bar{r}_1^2} - \frac{\bar{y}}{\bar{r}_2^2}\Big] \\ \overset{A}{q}_{\bar{y}}(x, y) &= \frac{D}{2\pi}\Big[\frac{(\bar{x} - a)}{\bar{r}_1^2} - \frac{(\bar{x} + a)}{\bar{r}_2^2}\Big] \end{aligned} \tag{1.100}$$

The x, y components of the gradient $\overset{A}{q}_x$ and $\overset{A}{q}_y$ are found by substituting (1.100) into (1.93). In the same way, the potential and gradient components for the image source D are

$$\overset{I}{\phi}(x, y) = \frac{D}{2\pi}(\bar{\theta}_1' - \bar{\theta}_2') \tag{1.101}$$

and

$$\begin{aligned} \overset{I}{q}_{\bar{x}'}(x, y) &= -\frac{D}{2\pi}\Big[\frac{\bar{y}'}{\bar{r}_1'^2} - \frac{\bar{y}'}{\bar{r}_2'^2}\Big] \\ \overset{I}{q}_{\bar{y}'}(x, y) &= \frac{D}{2\pi}\Big[\frac{(\bar{x}' - a)}{\bar{r}_1'^2} - \frac{(\bar{x}' + a)}{\bar{r}_2'^2}\Big] \end{aligned} \tag{1.102}$$

The x, y components $\overset{I}{q}_x$ and $\overset{I}{q}_y$ are found by substituting (1.102) into (1.96). The complete solution for the problem is obtained by adding the results for the actual and image dipole sources as in (1.97) and (1.98). Again, it is easy to show that the potential $\phi(x, y) = \overset{A}{\phi}(x, y) + \overset{I}{\phi}(x, y)$ is zero on $y = 0$ for $-\infty < x < \infty$.

1.11 Examples

Analytical solutions to Laplace's equation are available for boundary value problems involving relatively simple geometries. As illustrations of the numerical methods described in this chapter we will consider problems for (1) a rectangular region, (2) an annulus, and (3) a circular hole in a half-plane.

1.11.1 A Rectangular Plate

We consider first a boundary value problem for a rectangular plate of width W and height H. If the potential at the top of the plate ($y = H$) is $\phi = \phi_0$ and the potentials on the sides ($x = 0$ and $x = W$) and bottom ($y = 0$) are $\phi = 0$, the potential $\phi(x, y)$

in the finite region $0 \leq x \leq W, 0 \leq y \leq H$ is given by the following infinite series (Incropera et al. 2007):

$$\phi(x, y) = \frac{4}{\pi}\phi_0 \sum_{m=1}^{\infty} \frac{1}{2m-1} \frac{\sin((2m-1)\pi x/W)\sinh((2m-1)\pi y/W)}{\sinh((2m-1)\pi H/W)} \tag{1.103}$$

It can be seen from this expression that the potential is zero on the sides ($x = 0$ and $x = W$) and the bottom ($y = 0$) of the plate, as required. The prescribed potential at the top ($y = H$) of the plate is approximated by the series

$$\phi(x, H) = \frac{4}{\pi}\phi_0 \sum_{m=1}^{\infty} \frac{1}{2m-1} \sin((2m-1)\pi x/W) \tag{1.104}$$

The series (1.103) is convergent (Incropera et al. 2007) but the rate of convergence depends on the location of the point (x, y); many more terms are needed in the vicinity of the upper corners than elsewhere in the plate. For the case $W/H = 2$, for example, it was found by trial that convergence of the series (1.103) was achieved for 75 terms when $(x, y) = (0.0125W, 0.975H)$, but only 5 terms were required when $(x, y) = (0.0125W, 0.025H)$.

Differentiation of (1.103) gives expressions for the gradient components $q_x = \partial\phi(x, y)/\partial x$ and $q_y = \partial\phi(x, y)/\partial y$ in the plate, i.e.,

$$q_x(x, y) = \frac{4}{W}\phi_0 \sum_{m=1}^{\infty} \frac{\cos((2m-1)\pi x/W)\sinh((2m-1)\pi y/W)}{\sinh((2m-1)\pi H/W)} \tag{1.105}$$

$$q_y(x, y) = \frac{4}{W}\phi_0 \sum_{m=1}^{\infty} \frac{\sin((2m-1)\pi x/W)\cosh((2m-1)\pi y/W)}{\sinh((2m-1)\pi H/W)}$$

Two boundary element models of the rectangular plate problem were considered, one using the single layer potential formulation (Section 1.6) and the other using the double layer approach (Sect. 1.8). In both cases, 20 equal sized elements were placed along the sides of the plate and 40 elements along the top and bottom. Figure 1.10 shows a comparison of selected numerical results for the quantity $q_y = \partial\phi(x, y)/\partial y$ along the bottom of the plate ($0 \leq x/W \leq 1,\ y = 0$) with the analytical solution (1.105) for the case that $m = 5$. It can be seen that the numerical results are generally within plotting accuracy of the analytical solution, except that the single layer model exhibits a rather large relative error near the corners of the plate.

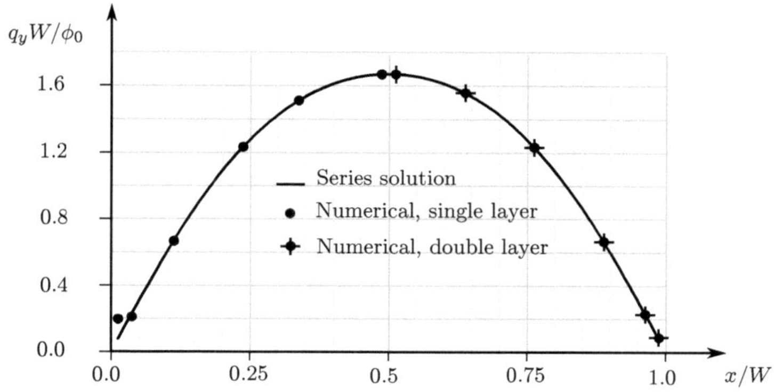

Fig. 1.10 Comparison of boundary element results with series solution for rectangular plate problem

1.11.2 An Annulus

We consider next the problem of an annulus with inner radius R_{in} and outer radius R_{out}. The potential is $\phi = \phi_{in}$ at the inner boundary and $\phi = \phi_{out}$ at the outer boundary. For this simple geometry and boundary conditions the problem is radially symmetric (i.e., one-dimesional) and the general solution of Laplace's equation is given as[10]

$$\phi(r) = C_1 \ln r + C_2 \tag{1.106}$$

so that, for the problem under consideration,

$$C_1 = \frac{\phi_{in} - \phi_{out}}{\ln(R_{in}/R_{out})} \tag{1.107}$$
$$C_2 = \frac{\phi_{out} \ln R_{in} - \phi_{in} \ln R_{out}}{\ln(R_{in}/R_{out})}$$

The normal derivative of potential (1.106) is

$$q(r) = C_1 \frac{1}{r} \tag{1.108}$$

The single layer potential boundary element method was used to solve this problem for the special case $R_{in} = 1$ and $R_{out} = 2$, with $\phi_{in} = 10$ and $\phi_{out} = 0$. The circular boundaries were each divided into 80 straight line elements, making $a = \pi/80$ for the inner boundary and $a = \pi/40$ for the outer one. For these parameters, (1.107) and (1.108) give $q(R_{in}) = -14.4270$ and $q(R_{out}) = -7.2135$; the corresponding

[10] This solution is derived in Sect. 2.2.

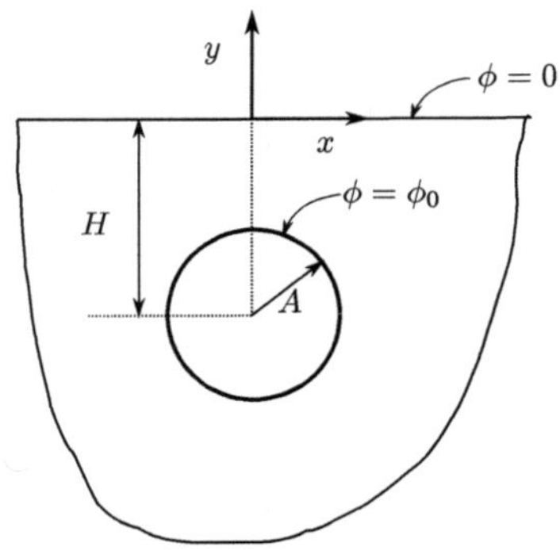

Fig. 1.11 Circular hole in a half-plane

values obtained by the boundary element model were −14.3094 and −7.2798, representing errors of less than 1% in each case.

A numerical solution to the problem was also obtained using the double layer potential method with Mikhlin's modification (Sect. 1.9). Taking $(x_1, y_1) = (0, 0)$ in (1.86) and (1.88) and keeping all the other parameters the same, we find $q(R_{in}) = -14.4380$ and $q(R_{out}) = -7.2191$, representing errors of less than 0.1%.

1.11.3 A Circular Hole in a Half-Plane

Barone and Caulk (1981) give the potential $\phi(x, y)$ for the problem of a half-plane $y \leq 0$ containing a circular hole of radius A centered a distance H below the surface, as depicted in Fig. 1.11. The potential is ϕ_0 on the boundary of the hole and zero on the surface of the half-plane. The solution to this problem, expressed in terms of bipolar coordinates (u, v), is[11]

$$\phi(u, v) = \phi_0 v \cosh^{-1}(H/A) \tag{1.109}$$

The Cartesian coordinates (x, y) are related to the bipolar coordinates (u, v) by the transformation

$$\begin{aligned} x &= \frac{(H^2 - A^2)^{1/2} \sinh v}{\cosh v - \cos u} \\ y &= \frac{(H^2 - A^2)^{1/2} \sin u}{\cosh v - \cos u} \end{aligned} \tag{1.110}$$

Barone and Caulk (1981) also give the following expression for the total flux across the boundary of the hole:

[11] The authors do not give a derivation or a reference to this solution, simply saying that it is "well known."

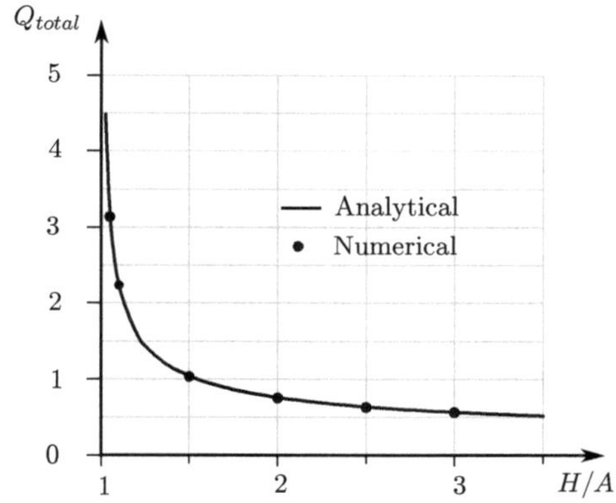

Fig. 1.12 Total flux across boundary of hole; analytical solution and numerical results

$$Q_{total} = \int_0^{2\pi} q(\theta) A \, \mathrm{d}\theta = 2\pi\phi_0 \frac{1}{\cosh^{-1}(H/A)} \tag{1.111}$$

A plot of this relationship is given in Fig. 1.12, together with numerical results for various values of H/A. The boundary element results were obtained using the single layer potential formulation with 100 equal size elements along the circumference of the circle. The results are within approximately 1% of the exact solution, even for relatively small values of H/A, and can be made more accurate by increasing the number of boundary elements. For example, when H/A = 1.05 the numerical error in Q_{total} is 1.1% for 100 elements but is only 0.2% for 360 elements.

1.12 Direct Versus Indirect Methods

Our numerical approach has features in common with other boundary element methods considered in this book but has the disadvantage that it requires us to determine quantities—the strengths of N line sources or dipoles for the N boundary elements—that are only indirectly related to the problem being solved. Determination of these quantities can therefore be viewed as an intermediate step in the solution process.

The direct boundary integral method eliminates this intermediate step by expressing the numerical solution directly in terms of the actual boundary parameters for the problem: at each point on the boundary where the potential ϕ is prescribed the normal derivative of the potential $q = \partial\phi/\partial\boldsymbol{n}$ is the unknown quantity, and *vice versa*. The basic result that makes this possible is a boundary integral representation formula originally developed by Green (1828)[12]:

$$\phi(\boldsymbol{\xi}) = \int_S q(\boldsymbol{x}) G(\boldsymbol{x}, \boldsymbol{\xi}) \mathrm{d}S(\boldsymbol{x}) - \int_S \phi(\boldsymbol{x}) \frac{\partial G(\boldsymbol{x}, \boldsymbol{\xi})}{\partial \boldsymbol{n}(\boldsymbol{x})} \mathrm{d}S(\boldsymbol{x}) \tag{1.112}$$

[12] This formula is derived in Chap. 2.

Green's representation formula should be compared with the equations for single and double layer potentials in (1.43) and (1.71). In the latter equations, the sample point $\boldsymbol{x}$ is inside the region V and the source point $\boldsymbol{\xi}$ is on the boundary S of this region whereas in (1.112) these roles are reversed. With this in mind, and recalling that $G(\boldsymbol{x}, \boldsymbol{\xi}) = G(\boldsymbol{\xi}, \boldsymbol{x})$ (see (1.4)), it can be seen that the first integral in Green's formula (1.112) represents a single layer potential for a density function $q(\boldsymbol{x})$ while the second integral represents a double layer potential for a density function $\phi(\boldsymbol{x})$. By dividing the boundary S into elements and approximating the values of ϕ and q along each element we can obtain a system of simultaneous linear algebraic equations in which the unknown quantity at each element is either the potential ϕ or its normal derivative q, whichever was not used to specify the boundary condition for that element. The solution at points inside the region of interest is expressed directly in terms of integrals over the boundary of the parameters ϕ and q.

If the boundary S is approximated by N straight line segments joined end to end all integrals over the individual boundary elements can be evaluated analytically. This allows us to develop closed-form expressions for the potential ϕ at an arbitrary point $\boldsymbol{\xi}$ inside the region of interest V. We then set up the required system of algebraic equations by taking limits of the resulting expressions as the interior point goes to the boundary, i.e., as $\boldsymbol{\xi} \to \boldsymbol{x}$. This approach is similar to the ones explained above for the two indirect methods involving single and double layer potentials where we first expressed the values of ϕ and its derivatives q_x and q_y at an arbitrary point within the region of interest in terms of N unknown line sources and then used the limiting values of the resulting expressions to obtain a system of algebraic equations from which to find the strengths of the line sources.

For the simplest version of the direct formulation the values of ϕ and q are assumed to be constant along the elements. These quantities can also be approximated by polynomials, however, and for straight line elements this still allows all the integrals to be evaluated analytically. More elaborate representations of the boundary S can be used, for example, by allowing the elements to possess curvature, but it is then generally not possible to evaluate the integrals in closed form and numerical integration must be used. In such cases closed-form expressions are not available and limits cannot be taken analytically. Instead, a special version of (1.112) is used that is obtained by allowing the interior point $\boldsymbol{\xi}$ to go to the boundary *before* the integrations are carried out. This approach, which is the one most often adopted in the literature, involves special handling of singular integrals and requires additional mathematical sophistication to explain. The necessary details for this are presented in the next chapter.

1.13 Exercises/Projects

1. Use L'Hopital's rule to show that $\lim_{\varepsilon \to 0} \varepsilon \ln \varepsilon = 0$.
2. Show that the potential $\phi(\boldsymbol{x})$ for the problem depicted in Fig. 1.1b is zero at infinity.

3. Derive expressions for the potential $\phi(\boldsymbol{x})$ and the components of the gradient vector $q_x(\boldsymbol{x})$ and $q_y(\boldsymbol{x})$ for two sources of magnitude $+\rho$ at the points $x = \xi_x$, $y = \pm\xi_y$ in an infinite region. What are the values of these quantities along the line $y = 0$?
   Answer: $\phi(x,0) = -\rho \ln\big[(x-\xi_x)^2 + \xi_y^2\big]^{1/2}/\pi$; $q_x(x,0) = -\rho(x-\xi_x)\big[(x-\xi_x)^2 + \xi_y^2\big]^{-1}/\pi$; $q_y(x,0) = 0$.
4. Consider the case of four point sources of unit strength ($\rho = 1$) at the corners of a square of side length w in an infinite region. What are the values of the potential and the components of the gradient vector at the center of the square?
   Answer: $\phi = -2\ln(w/\sqrt{2})/\pi$; $q_x = q_y = 0$.
5. Derive an expression analagous to (1.10) for a linearly varying source $P(x)$ over the line segment $|\, x \,| \leq a$, $y = 0$ in an infinite region.
   Answer: $\phi(x,y) = \dfrac{1}{4\pi a}\big[(x-a)P(-a) - (x+a)P(a)\big]\big[y(\theta_1-\theta_2) - (x-a)\ln r_1 + (x+a)\ln r_2 - 2a\big] + \dfrac{1}{8\pi a}\big[P(-a) - P(a)\big]\big[r_1^2\ln r_1 - r_2^2 \ln r_2 + 2ax\big]$.
6. Derive an expression analagous to (1.63) for a linearly varying dipole source $D(x)$ over the line segment $|\, x \,| \leq a$, $y = 0$ in an infinite region.
   Answer: $\phi(x,y) = -\dfrac{1}{4\pi a}\big[(x-a)D(-a) - (x+a)D(a)\big](\theta_1 - \theta_2) - \dfrac{1}{4\pi a}\big[D(-a) - D(a)\big]y \ln r_1/r_2$.
7. Consider a half-plane $y \leq 0$ that contains a point source ρ at $x = 0$, $y = -c$ (where $c > 0$) and is subject to the condition $\phi(x,0) = 0$. Derive expressions for the gradient vector components $q_x(x,0)$ and $q_y(x,0)$.
   Answer: $q_x(x,0) = 0$; $q_y(x,0) = -\rho c(x^2 + c^2)^{-1}/\pi$
8. Consider a constant strength dipole source $D = 1$ over the line segment $|\, x \,| \leq a$, $y = -c$ (where $c > 0$) in the half-plane $y \leq 0$ with a zero potential boundary. Confirm that $\phi(x,0) = 0$ for this situation and obtain expressions for the gradient vector components $q_x(x,0)$ and $q_y(x,0)$.
   Answer: $q_x(x,0) = 0$; $q_y(x,0) = \dfrac{1}{\pi}\left[\dfrac{x-a}{(x-a)^2 + c^2} - \dfrac{x+a}{(x+a)^2 + c^2}\right]$.
9. Write expressions for the inverse of coordinate transformation (1.29).
   Answer: $x - \xi_x = \bar{x}\cos\beta - \bar{y}\sin\beta$; $y - \xi_y = \bar{x}\sin\beta + \bar{y}\cos\beta$.
10. Write expressions for the inverse of Eq. (1.32).
   Answer: $q_{\bar{x}} = q_x \cos\beta + q_y \sin\beta$; $q_{\bar{y}} = -q_x \sin\beta + q_y \cos\beta$.

References

Barone, M.R., and D.A. Caulk. 1981. Special boundary integral equations for approximate solution of Laplace's equation in two-dimensional regions with circular holes. *The Quarterly Journal of Mechanics and Applied Mathematics* 34: 265–86.

Green, G. 1828. *An essay on the application of mathematical analysis to the theories of electricity and magnetism*. Nottingham: Printed for the author by T. Wheelhouse.

Greenbaum, A., L. Greengard, and G.B. McFadden. 1993. Laplace's equation and the Dirichlet-Neuman map in multiple connected domains. *Journal of Computational Physics* 105: 267–78.

Greengard, L., and V. Rokhlin. 1987. A fast algorithm for particle simulations. *Journal of Computational Physics* 73: 325–48.

Incropera, F.P., D.P. Dewitt, T.L. Begrman, and A.S. Lavine. 2007. *Fundamentals of heat and mass transfer*, 6th ed. Wiley.

Mikhlin, S.G. 1964. *Integral equations and their applications to certain problems in mechanics, mathematical physics, and technology*. Pergamon Press.

Rokhlin, V. 1985. Rapid solution of integral equations of classical potential theory. *Journal of Computational Physics* 60: 187–207.

Chapter 2
Potential Theory

2.1 Mathematical Preliminaries

We saw in the previous chapter that a simple boundary element method for potential problems can be developed in a straightforward manner starting from knowledge of the fundamental (singular) solution to Laplace's equation. This simplified approach has a certain physical appeal but it is limited in its generality and practical usefulness. Refinements and extensions of the basic methodology necessarily involve more mathematical details, and it is the purpose of this chapter to present this material and set the stage for the development of more general boundary element methods. Most of this material is specific to potential problems, but some of it is also relevant to our later treatment of problems in elasticity.

2.1.1 Index Notation

It is common practice in the scientific literature to replace subscripts or coordinate indices x, y, and z by the numbers 1, 2, and 3 as a form of shorthand. Two further conventions in the use of index notation will be used throughout this book to write equations in a compact form. The first is that a comma preceding an index denotes partial differentiation with respect to the variable represented by that index. Thus, for example, $\phi_{,j}$ denotes $\partial\phi/\partial x_j$ and $\phi,_{jk}$ denotes $\partial^2\phi/\partial x_j \partial x_k$, where j and k are 1 or 2 in two dimensions, and 1, 2, or 3 in three dimensions.

The second convention in the use of index notation is that a repeated literal index in any term of an expression implies summation. For example, by this so-called summation convention we understand the expression $\sigma_{ji}n_j$ to mean

S. L. Crouch and S. G. Mogilevskaya, *A First Course in Boundary Element Methods*, Mathematical Engineering, https://doi.org/10.1007/978-3-031-63341-6_2

$$\sigma_{ji}n_j = \sum_{j=1}^{3} \sigma_{ji}n_j = \sigma_{1i}n_1 + \sigma_{2i}n_2 + \sigma_{3i}n_3$$

because the index j is repeated. The particular symbol used to indicate summation in this way is unimportant. We could, for instance, write $\sigma_{ki}n_k$ and achieve exactly the same effect. For this reason, a repeated index is usually called a "dummy" index. It is not permissible for an index to appear more than twice in one term of an expression.

Combining the two conventions, we see that $\phi_{,jj}$ is a compact way of writing the Laplacian operator:

$$\phi_{,jj} = \frac{\partial^2 \phi}{\partial x_1^2} + \frac{\partial^2 \phi}{\partial x_2^2} + \frac{\partial^2 \phi}{\partial x_3^2} = \nabla^2 \phi$$

We will find it convenient in this book to use both index notation and unabridged notation where x, y, and z refer to the individual Cartesian coordinates. It should be clear from context whether or not index notation is being used, and also whether the representation pertains to two or three dimensions. The letters x, y, and z will never be used as general subscripts in index notation. Thus, for example, σ_{xx} will always represent just one component, the xx component, of the stress tensor (see Sect. 4.2), whereas σ_{kk} will represent (by the summation convention) $\sigma_{11} + \sigma_{22} + \sigma_{33}$ or, in two dimensions, $\sigma_{11} + \sigma_{22}$.

2.1.2 The Normal Derivative of a Function at a Boundary

The concept of the normal derivative of a function at a boundary appears frequently in this book. For example, we saw in Sect. 1.6 that the normal component q of the gradient of the potential can be obtained by a simple coordinate transformation in which the components of the gradient vector $\phi_{,i} = q_i$ are resolved in a local coordinate system with one axis perpendicular to the boundary at the point of interest. This procedure is equivalent to defining q as the dot product of the vector $\mathbf{q}$ with the unit outward normal to the boundary $\mathbf{n}$, as depicted in Fig. 2.1, i.e.,

$$q = q_i n_i \tag{2.1}$$

or, equivalently,

$$q = \phi_{,i} n_i \tag{2.2}$$

The quantity $\phi_{,i}n_i$ represents the normal derivative of function ϕ at the boundary, i.e.,

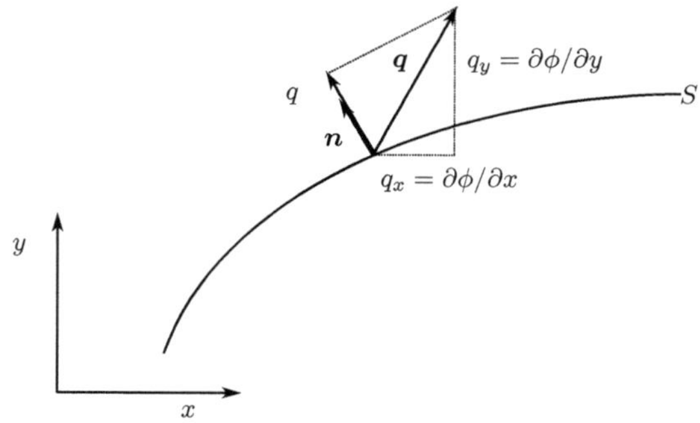

Fig. 2.1 Normal derivative of the potential ϕ at boundary S

$$\frac{\partial \phi}{\partial \boldsymbol{n}} = \frac{\partial \phi}{\partial x_i} n_i$$
$$= q \tag{2.3}$$

also called the *flux*. In some physical applications the flux represents the negative of the normal derivative of the potential multiplied by a material-specific constant. We will use such a representation in Sects. 3.11–3.14 but will otherwise use definition (2.3).

More generally, we write the normal derivative for any sufficiently smooth function f at a point along a boundary as

$$\frac{\partial f}{\partial \boldsymbol{n}} = \frac{\partial f}{\partial x_i} n_i \tag{2.4}$$

This result applies in both two and three dimensions.

As mentioned in Sect. 1.7, some authors write the normal derivative of a function as $\partial f/\partial n$, in recognition of the fact that the right-hand side of (2.4) is a scalar. We will use the notation $\partial f/\partial \boldsymbol{n}$ to emphasize that the derivative is taken in the direction of the normal vector.

2.1.3 The Divergence Theorem

The divergence of a vector field with components F_i is defined as $\partial F_i/\partial x_i = F_{i,i}$. It is a scalar quantity that gives a measure of the total amount of flux emanating from (or flowing into) an infinitesimal volume at a point in the vector field. The *divergence theorem*, also known as Gauss's theorem or Ostrogradsky's theorem, states that the integral of the divergence of a vector field over a closed volume V is equal to the flux of the field passing through the bounding surface S of that volume, i.e.,

$$\int_V F_{i,i}\,\mathrm{d}V = \int_S F_i n_i\,\mathrm{d}S \tag{2.5}$$

where n_i is the ith component of the unit outward normal vector to the boundary. This theorem is valid in both two and three dimensions and is of fundamental importance to the development of boundary element methods for potential problems as well as for elasticity.

As an illustration of the divergence theorem suppose that the vector field represents the gradient of a potential ϕ, or

$$F_i = \phi_{,i} \tag{2.6}$$

Then, using (2.4)–(2.6), we get

$$\begin{aligned}\int_V \phi_{,ii}\,\mathrm{d}V &= \int_S \phi_{,i} n_i\,\mathrm{d}S \\ &= \int_S \frac{\partial\phi}{\partial \boldsymbol{n}}\,\mathrm{d}S \\ &= \int_S q\,\mathrm{d}S \end{aligned} \tag{2.7}$$

If ϕ is harmonic within region V, (2.7) states that the integral of the normal derivative of the potential (the flux q) across the entire boundary S must vanish.

2.1.4 Green's First and Second Identities

Green's identities are integral expressions that link any two sufficiently smooth functions $u(\boldsymbol{x})$ and $v(\boldsymbol{x})$ in a region V enclosed by a sufficiently smooth boundary S. If we set $F_i = u\partial v/\partial x_i$ the divergence theorem gives

$$\int_V (uv_{,i})_{,i}\,\mathrm{d}V = \int_S uv_{,i} n_i\,\mathrm{d}S$$

or

$$\int_V (u_{,i}v_{,i} + uv_{,ii})\,\mathrm{d}V = \int_S u\frac{\partial v}{\partial \boldsymbol{n}}\,\mathrm{d}S$$

which can be written as

$$\int_S u(\boldsymbol{x})\frac{\partial v(\boldsymbol{x})}{\partial \boldsymbol{n}(\boldsymbol{x})}\,\mathrm{d}S(\boldsymbol{x}) = \int_V \Big[\frac{\partial u(\boldsymbol{x})}{\partial x_i}\frac{\partial v(\boldsymbol{x})}{\partial x_i} + u(\boldsymbol{x})\nabla^2 v(\boldsymbol{x})\Big]\,\mathrm{d}V(\boldsymbol{x}) \tag{2.8}$$

where $\partial v(\boldsymbol{x})/\partial \boldsymbol{n}(\boldsymbol{x})$ is the derivative of $v(\boldsymbol{x})$ normal to the boundary S at point $\boldsymbol{x}$. This expression is known as *Green's first identity*.

Proceeding similarly for a vector field $F_i = v\partial u/\partial x_i$ gives the result

$$\int_S v(\boldsymbol{x})\frac{\partial u(\boldsymbol{x})}{\partial \boldsymbol{n}(\boldsymbol{x})}\,\mathrm{d}S(\boldsymbol{x}) = \int_V \Big[\frac{\partial v(\boldsymbol{x})}{\partial x_i}\frac{\partial u(\boldsymbol{x})}{\partial x_i} + v(\boldsymbol{x})\nabla^2 u(\boldsymbol{x})\Big]\,\mathrm{d}V(\boldsymbol{x}) \tag{2.9}$$

Subtracting (2.9) from (2.8) gives *Green's second identity*:

$$\int_S \Big[u(\boldsymbol{x})\frac{\partial v(\boldsymbol{x})}{\partial \boldsymbol{n}(\boldsymbol{x})} - v(\boldsymbol{x})\frac{\partial u(\boldsymbol{x})}{\partial \boldsymbol{n}(\boldsymbol{x})}\Big]\,\mathrm{d}S(\boldsymbol{x}) = \int_V \Big[u(\boldsymbol{x})\nabla^2 v(\boldsymbol{x}) - v(\boldsymbol{x})\nabla^2 u(\boldsymbol{x})\Big]\,\mathrm{d}V(\boldsymbol{x}) \tag{2.10}$$

2.1.5 Regular, Singular, and Hypersingular Integrals

We next introduce several types of integrals associated with the use of fundamental solutions in boundary element methods. To fix ideas, we first consider the following integral over the segment $[-a, b]$ of the x axis:

$$\int_{-a}^{b} \frac{1}{x^p}\,\mathrm{d}x \tag{2.11}$$

in which $a > 0$ and $b > 0$.

The function $1/x^p$ in (2.11) is called the *kernel* of the integral. If $p \leq 0$ the kernel is continuous on the interval $[-a, b]$ and has only finite values, i.e., it does not have any singularities. In this case, (2.11) is referred to as a *regular integral*, the so-called Riemann integral. It can be geometrically represented by the signed area of the region bounded by the lines $x = -a$ and $x = b$, the segment $[-a, b]$, and the curve $y = 1/x^p$, as illustrated in Fig. 2.2a for $p = -1/3$. The integral can be directly calculated using the fundamental theorem of calculus (also called the Newton–Leibniz rule):

$$\int_{-a}^{b} \frac{1}{x^p}\,\mathrm{d}x = \int_{-a}^{b} x^{-p}\,\mathrm{d}x = \frac{x^{-p+1}}{-p+1}\Bigg|_{-a}^{b} = \frac{b^{-p+1}}{-p+1} - \frac{(-a)^{-p+1}}{-p+1} \tag{2.12}$$

For the case $p > 0$, the kernel $1/x^p$ is *singular* because it tends to infinity when $x \to 0$ (see Fig. 2.2b where $p = 1/3$). However, for $0 < p < 1$, it is still possible to interpret the integral geometrically as the sum of the signed areas of the regions bounded by the same lines as before but excluding the intervals $[-\varepsilon_1, 0)$ and $(0, \varepsilon_2]$ from the interval $[-a, b]$ and taking the limits $\varepsilon_1 \to 0$ and $\varepsilon_2 \to 0$.

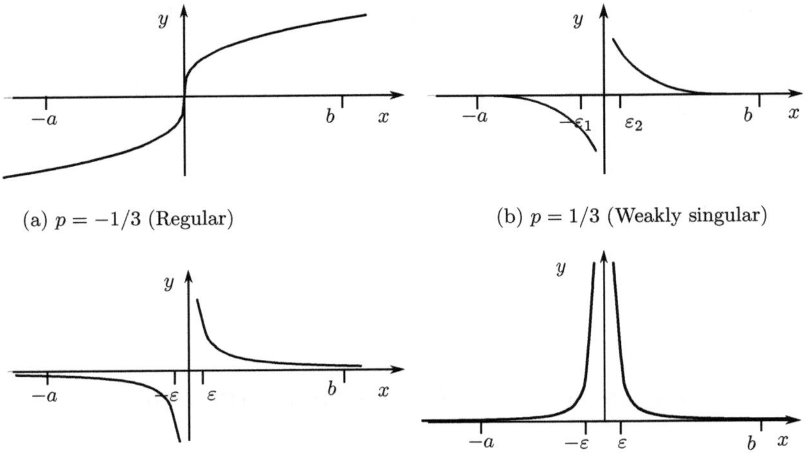

(a) $p = -1/3$ (Regular)

(b) $p = 1/3$ (Weakly singular)

(c) $p = 1$ (Singular; Cauchy principal value)

(d) $p = 2$ (Hypersingular)

Fig. 2.2 Function $y = 1/x^p$ for various values of p that define the order of singularity of the corresponding integral

To demonstrate this, consider the following expression obtained by applying the Newton–Leibniz rule to each of the two regular integrals

$$\int_{-a}^{-\varepsilon_1} x^{-p}\,\mathrm{d}x + \int_{\varepsilon_2}^{b} x^{-p}\,\mathrm{d}x = \frac{(-\varepsilon_1)^{-p+1}}{-p+1} - \frac{(-a)^{-p+1}}{-p+1} + \frac{b^{-p+1}}{-p+1} - \frac{(\varepsilon_2)^{-p+1}}{-p+1} \tag{2.13}$$

Because for $-p+1 > 0$ the quantities $(-\varepsilon_1)^{-p+1}$ and $(\varepsilon_2)^{-p+1}$ both go to zero as $\varepsilon_1 \to 0$ and $\varepsilon_2 \to 0$, the integral (2.11) could be defined as

$$\int_{-a}^{b} \frac{1}{x^p}\,\mathrm{d}x = \lim_{\varepsilon_1 \to 0} \int_{-a}^{-\varepsilon_1} x^{-p}\,\mathrm{d}x + \lim_{\varepsilon_2 \to 0} \int_{\varepsilon_2}^{b} x^{-p}\,\mathrm{d}x = \frac{b^{-p+1}}{-p+1} - \frac{(-a)^{-p+1}}{-p+1} \tag{2.14}$$

This result is the same as that obtained by formal application of the Newton–Leibniz rule to the integral on the left-hand side of (2.14). When $0 < p < 1$, (2.14) is said to be a *weakly singular* integral.

Another example of a weakly singular integral is

$$\int_{-a}^{b} \ln|x|\,\mathrm{d}x = \lim_{\varepsilon_1 \to 0} \int_{-a}^{-\varepsilon_1} \ln|x|\,\mathrm{d}x + \lim_{\varepsilon_2 \to 0} \int_{\varepsilon_2}^{b} \ln|x|\,\mathrm{d}x \tag{2.15}$$

Applying the Newton–Leibniz rule to each of the regular integrals in this expression gives

$$\int_{-a}^{b} \ln|x| \, \mathrm{d}x = (\ln a - 1)\, a - \lim_{\varepsilon_1 \to 0} \varepsilon_1 (\ln \varepsilon_1 - 1) + (\ln b - 1)\, b - \lim_{\varepsilon_2 \to 0} \varepsilon_2 (\ln \varepsilon_2 - 1) \tag{2.16}$$

The quantities $\varepsilon_1 (\ln \varepsilon_1 - 1)$ and $\varepsilon_2 (\ln \varepsilon_2 - 1)$ both go to zero as $\varepsilon_1 \to 0$ and $\varepsilon_2 \to 0$ so that (2.16) becomes

$$\int_{-a}^{b} \ln|x| \, \mathrm{d}x = (\ln a - 1)\, a + (\ln b - 1)\, b \tag{2.17}$$

It is easily verified that this result can be obtained by formal application of the Newton–Leibniz rule to the integral on the left-hand side of the equation, without going through the limiting process.

Now consider integral (2.11) with $p = 1$ (Fig. 2.2c) and apply the Newton–Leibniz rule to the following sum of two integrals:

$$\begin{aligned} \int_{-a}^{-\varepsilon_1} x^{-1} \, \mathrm{d}x + \int_{\varepsilon_2}^{b} x^{-1} \, \mathrm{d}x &= \ln|-\varepsilon_1| - \ln|-a| + \ln b - \ln \varepsilon_2 \\ &= \ln \varepsilon_1 - \ln a + \ln b - \ln \varepsilon_2 \end{aligned} \tag{2.18}$$

Unlike in the procedure described above for $p < 1$, the terms $\ln \varepsilon_1$ and $\ln \varepsilon_2$ in (2.18) diverge for $p = 1$ when $\varepsilon_1 \to 0$ and $\varepsilon_2 \to 0$, and the simple geometrical interpretation used for regular and weakly singular integrals is no longer applicable. If, however, $\varepsilon_1 = \varepsilon_2 = \varepsilon$, the divergent terms cancel out and it is possible to define integral (2.11) (with $p = 1$) as

$$⨍_{-a}^{b} \frac{1}{x} \, \mathrm{d}x = \lim_{\varepsilon \to 0} \left(\int_{-a}^{-\varepsilon} x^{-1} dx + \int_{\varepsilon}^{b} x^{-1} dx \right) = \ln b - \ln a = \ln (b/a) \tag{2.19}$$

An integral evaluated this way is called a *Cauchy principal value integral* and the symbol $⨍$ is used to identify it as such. It can be seen that the final result of the integration is again the same as that obtained by a formal application of the Newton–Leibniz rule to integral (2.11) with $p = 1$.

Finally, consider the case $p > 1$. The terms $(-\varepsilon_1)^{-p+1}/(-p+1)$ and $(\varepsilon_2)^{-p+1}/(-p+1)$ on the right-hand side of Eq. (2.13) diverge for this case and they in general do not cancel out even when $\varepsilon_1 = \varepsilon_2$. Such an integral is said to be *hypersingular* and can be evaluated as what is termed a *Hadamard finite part integral*, denoted by the symbol $⨎$. For the case $p = 2$ (Fig. 2.2d), the finite part integral is defined as

$$⨎_{-a}^{b} \frac{1}{x^2} \, \mathrm{d}x = \lim_{\varepsilon \to 0} \left[\int_{-a}^{-\varepsilon} x^{-2} \, \mathrm{d}x + \int_{\varepsilon}^{b} x^{-2} \, \mathrm{d}x - \frac{2}{\varepsilon} \right] \tag{2.20}$$

It is clear from (2.12) and (2.20) that, as in all the other cases considered above, the result of formal integration is

$$\int\!\!\!\!\!\!=_{-a}^{b} \frac{1}{x^2}\,\mathrm{d}x = -\frac{1}{b} - \frac{1}{a} \tag{2.21}$$

which is again the same as that obtained by a formal application of the Newton–Leibniz rule to integral (2.11) with $p = 2$.

Despite artificiality of such a construction, divergent integrals play an important role in some theoretical developments of boundary element methods. These integrals were "legitimized" in what is known as the *theory of distributions* or *generalized functions* developed by Sobolev and Schwartz (see Zemanian 1987). This theory made it possible to define differentiation and integration for functions whose derivatives or integrals do not exist in the classical sense.

We have so far considered integrals with a specific kernel $(1/x^p)$ or $\ln|x|$, a specific density (unit density), and a specific interval of integration $[-a, b]$ along the x axis. In the following sections we will encounter more general integrals of the kind

$$\int_S f(\boldsymbol{x}) K(\boldsymbol{x}, \boldsymbol{\xi})\,\mathrm{d}S(\boldsymbol{x}) \tag{2.22}$$

where S is a curve (for two-dimensional problems) or surface (for three-dimensional problems), $\boldsymbol{x}$ is a point on that curve or surface, $f(\boldsymbol{x})$ is density function, and $K(\boldsymbol{x}, \boldsymbol{\xi})$ is a kernel, which typically includes various functions of distance r between the points $\boldsymbol{x}$ and $\boldsymbol{\xi}$ and their derivatives. The kernel may also include some characteristics of the curve (or surface) of integration, e.g., the components of its normal vectors. The kind of integral (regular, singular, or hypersingular) that expression (2.22) represents depends on the smoothness conditions for the density function, the kernel, the curve (or surface) of integration, and on the location of the point $\boldsymbol{\xi}$.

2.1.6 Smoothness Conditions for Functions and Curves

On several occasions we have used the words "sufficiently smooth" when referring to functions and curves. In mathematics, smoothness conditions characterize how a function behaves when its argument changes. An example of a smoothness condition is the *continuity condition*, which states that a real function $f(\boldsymbol{x})$ is continuous at the point $\boldsymbol{x}_0$ if there exists an interval of $\boldsymbol{x}_0$ such that for all its inner points the values of $f(\boldsymbol{x})$ fall within a chosen interval of $f(\boldsymbol{x}_0)$. More precisely, for all $\varepsilon > 0$ there exists a $\delta > 0$ such that

$$|f(\boldsymbol{x}) - f(\boldsymbol{x}_0)| < \varepsilon \tag{2.23}$$

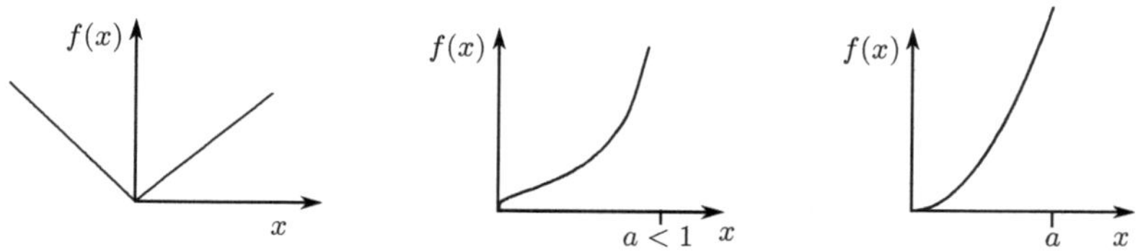

(a) $f(x) = |x|$ (b) $f(x) = -1/\ln x$ (c) $f(x) = x^2$

Fig. 2.3 Examples of functions that are **a** Lipschitz continuous, **b** continuous but not Hölder continuous, or **c** differentiable

when $|\boldsymbol{x} - \boldsymbol{x}_0| < \delta$. (The notation $|\boldsymbol{x} - \boldsymbol{y}|$ is used to identify the length of the interval $[\boldsymbol{x}, \boldsymbol{y}]$.) A function that is continuous at every point of a curve is said to be continuous on that curve. Continuous functions that have derivatives with respect to their arguments are *smoother* than those that do not, and the more derivatives a function has, the smoother it is.

Another smoothness condition is represented by the so-called *Hölder condition* (sometimes referred to as an H-condition). The function $f(\boldsymbol{x})$ satisfies a Hölder condition on a curve S if for any points $\boldsymbol{x}_1$ and $\boldsymbol{x}_2$ of the curve

$$|f(\boldsymbol{x}_2) - f(\boldsymbol{x}_1)| \leq C\, |\boldsymbol{x}_2 - \boldsymbol{x}_1|^{\alpha} \tag{2.24}$$

where C and $0 < \alpha \leq 1$ are real positive constants. The particular case for which $\alpha = 1$ is called a *Lipschitz condition.*

A function that satisfies a Hölder or Lipschitz condition is continuous on S (e.g., the Lipschitz-continuous real function $f(x) = |x|$ depicted in Fig. 2.3a. However, condition (2.24) is stricter than the continuity condition (2.23) because it constrains how the function $f(x)$ behaves inside a particular interval. For example, the function of real variable x defined as

$$f(x) = \begin{cases} -1/\ln(x) & x > 0 \\ 0 & x = 0 \end{cases} \tag{2.25}$$

is continuous on $[0, a]$ when $a < 1$ (Fig. 2.3b) but it does not satisfy a Hölder condition for any α. While it can be proved that differentiable functions with bounded derivatives (e.g., the real function $f(x) = x^2$ on $[0, a]$ in Fig. 2.3c) satisfy a Hölder condition, the opposite is not true.

A curve is said to be *smooth* if the coordinates of any point along the curve can be represented in parametric form

$$x = x(s), \quad y = y(s) \tag{2.26}$$

where $x(s)$ and $y(s)$ are continuous functions of the parameter $s \in [s_a, s_b]$ such that $\mathrm{d}x/\mathrm{d}s$ and $\mathrm{d}y/\mathrm{d}s$ are continuous on $[s_a, s_b]$. This means that the direction of

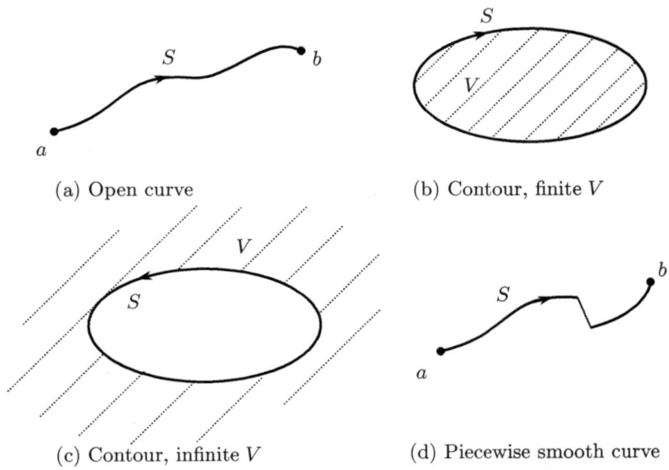

Fig. 2.4 Examples of curves: open, contour (finite and infinite regions), and piecewise smooth

the tangent is a continuous function of s, and the curve does not intersect itself anywhere on (s_a, s_b) (a so-called *simple curve*). The points **a** and **b** with coordinates $x(s_a)$, $y(s_a)$ and $x(s_b)$, $y(s_b)$ are said to be the endpoints of the curve. Each curve has associated with it a prescribed direction that corresponds to an increase of the parameter s.

A curve S is called *open* if its endpoints do not coincide, as in Fig. 2.4a. Otherwise, the curve is closed (Fig. 2.4b and c) and is referred to as a *contour*. The curve S is said to be *piecewise smooth* if it consists of a finite number of smooth curves S_i $(i = 1, 2, \ldots)$ that do not intersect each other and the endpoint of one curve serves as the beginning point of another (e.g., Fig. 2.4d). A piecewise smooth curve is also called a *regular curve*.

Lyapunov curves represent a special class of smooth curves. The required conditions for a Lyapunov curve are stricter than those for a smooth curve. Whereas a smooth curve is required just to have a unique tangent and normal at every point along it, an additional condition for a Lyapunov curve is that the angle between the normals at any of its two points must satisfy a Hölder condition similar to that given by (2.24) (in which the parameters have to be the same for the whole curve). Any smooth curve that possesses curvature (a second derivative of the parametrized equation of the curve) at every point along it is also a Lyapunov curve. However, the opposite is not true, as there are Lyapunov curves that have unique tangents and normals at every point but might not possess curvature at some points. Lyapunov curves play an important role in classical potential theory. However, the concept is too restrictive for applications because it does not allow for consideration of curves and surfaces that have corners and edges. The latter problems can be formulated with the use of *Lipschitz curves* whose graphs locally can be represented by Lipschitz functions (such as shown in Fig. 2.3a, for example).

2.1.7 The Dirac Delta Function

The Dirac delta function (which is an example of a generalized function) is a useful concept in the development of boundary element methods. This quantity is denoted as $\delta(\boldsymbol{x} - \boldsymbol{\xi})$ and is assigned the following formal properties:

$$\delta(\boldsymbol{x} - \boldsymbol{\xi}) = \begin{cases} \infty & \boldsymbol{x} = \xi \\ 0 & \boldsymbol{x} \neq \boldsymbol{\xi} \end{cases} \tag{2.27}$$

$$\int_V \delta(\boldsymbol{x} - \boldsymbol{\xi})\, \mathrm{d}V(\boldsymbol{x}) = \begin{cases} 1 & \boldsymbol{\xi} \in V \\ 0 & \boldsymbol{\xi} \notin V \end{cases} \tag{2.28}$$

$$\int_V \delta(\boldsymbol{x} - \boldsymbol{\xi}) f(\boldsymbol{x})\, \mathrm{d}V(\boldsymbol{x}) = \begin{cases} f(\boldsymbol{\xi}) & \boldsymbol{\xi} \in V \\ 0 & \boldsymbol{\xi} \notin V \end{cases} \tag{2.29}$$

where V is the region of integration and $\boldsymbol{x}$ is a point inside that region.

The last of these properties is known as the *selection property* of the Dirac delta function: when a continuous function $f(\boldsymbol{x})$ appears in the integrand as a multiplier for $\delta(\boldsymbol{x} - \boldsymbol{\xi})$, the value of the function at the point of the singularity ($\boldsymbol{x} = \boldsymbol{\xi}$) is "selected." We will make use of this property in derivations of certain boundary integral formulas.

2.2 The Fundamental Solution $G(x, \xi)$

According to the theory of distributions, the fundamental solution of Laplace's equation (1.1) is defined as the solution of the equation

$$\nabla^2 G(\boldsymbol{x}, \boldsymbol{\xi}) = -\delta(\boldsymbol{x} - \boldsymbol{\xi}) \tag{2.30}$$

subject to certain conditions (described below) when point $\boldsymbol{x}$ tends to infinity.

The differentiation in (2.30) is carried out with respect to the coordinates at point $\boldsymbol{x}$, i.e., $\nabla^2 = \partial^2/\partial x_i \partial x_i$, and the Dirac delta function $\delta(\boldsymbol{x} - \boldsymbol{\xi})$ represents a unit source at point $\boldsymbol{\xi}$. In terminology used previously (Sect. 1.2), $\boldsymbol{x}$ is the *sample point* (sometimes called the *field point*) and $\boldsymbol{\xi}$ is the *source point* (sometimes called the *load point*). Function $G(\boldsymbol{x}, \boldsymbol{\xi})$ gives the response at point $\boldsymbol{x}$ due to a source at point $\boldsymbol{\xi}$. The minus sign on the right-hand side of (2.30) is chosen for theoretical reasons that are beyond the scope of this discussion.

The fundamental solution $G(\boldsymbol{x}, \boldsymbol{\xi})$ can be obtained by solving the problem of a circular (or spherical) hole in a homogeneous, infinite region, subject to a uniform boundary condition $f = f_0$, and then letting the radius of the hole r_0 go to zero while requiring the total flux emanating from the source to have unit magnitude.

To derive this solution, we first reformulate Laplace's equation in plane polar coordinates ($x = r\cos\theta,\ y = r\sin\theta$) as

$$\nabla^2 f = \frac{1}{r}\frac{\partial}{\partial r}\left(r\frac{\partial f}{\partial r}\right) + \frac{1}{r^2}\frac{\partial^2 f}{\partial\theta^2} = 0 \tag{2.31}$$

and seek its solution for the condition

$$f = f_0 \text{ at } r = r_0 \tag{2.32}$$

This problem is independent of the polar angle θ and hence (2.31) reduces to the ordinary differential equation

$$\frac{1}{r}\frac{\mathrm{d}}{\mathrm{d}r}\left(r\frac{\mathrm{d}f}{\mathrm{d}r}\right) = 0 \tag{2.33}$$

from which it follows that

$$r\frac{\mathrm{d}f}{\mathrm{d}r} = C_1 \tag{2.34}$$

where C_1 is an arbitrary constant. Writing this expression as

$$\mathrm{d}f = C_1\frac{\mathrm{d}r}{r}$$

we find that

$$f = C_1 \ln r + C_2 \tag{2.35}$$

If we require that f behaves as $\ln r$ at large distances from the origin, i.e., $f = O(\ln r)$ when $r \to \infty$, we set $C_2 = 0$ and use boundary condition (2.32) to find the value of C_1 as

$$C_1 = \frac{f_0}{\ln r_0} \tag{2.36}$$

so that

$$f = \frac{f_0}{\ln r_0}\ln r \tag{2.37}$$

The flux q associated with potential f is given as $q = \partial f/\partial\mathbf{n}$, which in the present case can be written as $q = -\partial f/\partial r$ because the outward normal to the region outside the hole (our region of interest) points in the direction opposite to r, i.e., toward the origin. We therefore have that

$$q = -\frac{f_0}{\ln r_0}\frac{1}{r} \tag{2.38}$$

We next require that the total flux across any closed boundary surrounding the origin have unit magnitude. Taking this boundary as a circle centered at the origin gives, for any value of r,

$$\int_0^{2\pi} qr\,\mathrm{d}\theta = 1 \tag{2.39}$$

from which it follows that $f_0/\ln r_0 = -1/2\pi$ so that

$$f = -\frac{1}{2\pi}\ln r \tag{2.40}$$

By construction, function f satisfies Laplace's equation everywhere except at the origin, where it is singular.[1]

Considering the case for which the singularity occurs at an arbitrary point $\boldsymbol{\xi}$ and denoting this particular value of f as $G(\boldsymbol{x}, \boldsymbol{\xi})$ we obtain the desired result:

$$G(\boldsymbol{x}, \boldsymbol{\xi}) = -\frac{1}{2\pi}\ln r \tag{2.41}$$

where, as before, $r = [(x - \xi_x)^2 + (y - \xi_y)^2]^{1/2}$ is the distance between points $\boldsymbol{x} = (x, y)$ and $\boldsymbol{\xi} = (\xi_x, \xi_y)$.

A similar analysis for a spherical hole leads to the results

$$f = f_0\frac{r_0}{r} \tag{2.42}$$

and

$$q = -\frac{\partial f}{\partial r} = f_0\frac{r_0}{r^2} \tag{2.43}$$

The counterpart of (2.39) for a sphere centered at the origin is

$$\int_0^{2\pi}\mathrm{d}\theta\int_0^{\pi} qr^2\sin\varphi\,\mathrm{d}\varphi = 1 \tag{2.44}$$

which, together with (2.43), gives $f_0r_0 = 1/4\pi$. It then follows from (2.42) that

$$f = \frac{1}{4\pi}\frac{1}{r} \tag{2.45}$$

The required singular solution for three dimensions is therefore

[1] In some textbooks (2.40) is derived by considering the interior of a circle as the region of interest, taking the flux in the direction of the outward normal (i.e., the opposite of what we used in (2.38)), and then requiring that the total flux be equal to -1. The resulting expression for the fundamental solution is the same, Eq. (2.41).

$$G(\boldsymbol{x}, \boldsymbol{\xi}) = \frac{1}{4\pi} \frac{1}{r} \tag{2.46}$$

where now $r = [(x - \xi_x)^2 + (y - \xi_y)^2 + (z - \xi_z)^2]^{1/2}$ is the distance between points $\boldsymbol{x} = (x, y, z)$ and $\boldsymbol{\xi} = (\xi_x, \xi_y, \xi_z)$.

2.3 Potentials and Their Direct and Limit Values

In Sects. 1.7 and 1.9, we introduced four integral expressions (potentials) that play important roles in potential theory. These potentials are given by the right-hand sides of the following expressions:

- single layer potential

$$\phi(\boldsymbol{x}) = \int_S \rho(\boldsymbol{\xi}) G(\boldsymbol{x}, \boldsymbol{\xi}) \mathrm{d}S(\boldsymbol{\xi}) \tag{2.47}$$

- adjoint double layer potential

$$\frac{\partial \phi(\boldsymbol{x})}{\partial \boldsymbol{n}(\boldsymbol{x})} = \int_S \rho(\boldsymbol{\xi}) Q(\boldsymbol{x}, \boldsymbol{\xi}) \mathrm{d}S(\boldsymbol{\xi}) \tag{2.48}$$

- double layer potential

$$\phi(\boldsymbol{x}) = \int_S \psi(\boldsymbol{\xi}) Q(\boldsymbol{\xi}, \boldsymbol{x}) \mathrm{d}S(\boldsymbol{\xi}) \tag{2.49}$$

- hypersingular potential

$$\frac{\partial \phi(\boldsymbol{x})}{\partial \boldsymbol{n}(\boldsymbol{x})} = \int_S \psi(\boldsymbol{\xi}) H(\boldsymbol{x}, \boldsymbol{\xi}) \mathrm{d}S(\boldsymbol{\xi}) \tag{2.50}$$

in which $\rho(\boldsymbol{\xi})$ and $\psi(\boldsymbol{\xi})$ are density functions and S is a sufficiently smooth open or closed curve (in two-dimensions) or surface (in three dimensions).

In (2.47)–(2.50), it is assumed that the point $\boldsymbol{x}$ is located outside of S on a curve (surface) characterized by the normal vector $\boldsymbol{n}(\boldsymbol{x})$ and

$$Q(\boldsymbol{x}, \boldsymbol{\xi}) = \frac{\partial G(\boldsymbol{x}, \boldsymbol{\xi})}{\partial \boldsymbol{n}(\boldsymbol{x})} \tag{2.51}$$

$$H(\boldsymbol{x}, \boldsymbol{\xi}) = \frac{\partial^2 G(\boldsymbol{x}, \boldsymbol{\xi})}{\partial \boldsymbol{n}(\boldsymbol{x}) \partial \boldsymbol{n}(\boldsymbol{\xi})} \tag{2.52}$$

Expressions (2.47)–(2.52) are the same for two- and three-dimensional potentials, provided the corresponding fundamental solutions $G(\boldsymbol{x}, \boldsymbol{\xi})$ are used.

The rationale for the terminology "adjoint double layer potential" was given in Sect. 1.9; the terminology "hypersingular potential" requires some additional explanation, given below.

2.3.1 *Limit and Direct Values of the Potentials*

Equations (2.47)–(2.50) pertain specifically to the case that the sample point $\boldsymbol{x}$ does not lie along boundary S and therefore does not coincide with the source point $\boldsymbol{\xi}$. Because of this, the integrals involved in (2.47)–(2.50) are all regular. The limiting procedure in which $\boldsymbol{x}$ is allowed to approach a point $\boldsymbol{x}_0$ on S results in the *limit values* of the potentials. It was demonstrated by Eqs. (1.10), (1.12), and (1.21) of Sect. 1.3, for the case when S is a line segment and $\rho(\boldsymbol{\xi})$ is constant, that the single layer potential (2.47) is continuous across S while the adjoint double layer potential (2.48) undergoes a jump across it. It can be shown that (2.47) and (2.48) exhibit similar behavior in more general cases of sufficiently smooth S and $\rho(\boldsymbol{\xi})$. That is, if $\boldsymbol{x}_0$ is located on a smooth portion of S, the limit values of (2.47) and (2.48) are given as follows:

$$\lim_{\boldsymbol{x}\to\boldsymbol{x}_0} \phi(\boldsymbol{x}) = \int_S \rho(\boldsymbol{\xi}) G(\boldsymbol{x}_0, \boldsymbol{\xi})\, \mathrm{d}S(\boldsymbol{\xi}) \tag{2.53}$$

and

$$\lim_{\boldsymbol{x}\to\boldsymbol{x}_0} \frac{\partial\phi(\boldsymbol{x})}{\partial\boldsymbol{n}(\boldsymbol{x})} = \pm\frac{1}{2}\rho(\boldsymbol{x}_0) + \int_S \rho(\boldsymbol{\xi}) Q(\boldsymbol{x}_0, \boldsymbol{\xi})\, \mathrm{d}S(\boldsymbol{\xi}) \tag{2.54}$$

in which the $\pm$ sign depends on whether point $\boldsymbol{x}_0$ is approached from the direction of the normal vector to boundary S at that point or from the opposite direction.

It is instructive to compare (2.54) with (1.20) in Sect. 1.3. It can be seen that the limit values of the flux q_y represented by (1.20) are obtained without reference to the normal vector to the line segment. However, if the normal vector to that line segment is selected as $\boldsymbol{n} = (0, 1)$, Eq. (1.20) can be retrieved from (2.54) by assuming that S represents the line segment and $\rho(\boldsymbol{x}_0) = P$.

Equations (2.53) and (2.54) are valid both in two and three dimensions. The integrals on the right-hand sides of these equations are referred to as *direct values* of the single layer potential and its normal derivative. As seen from (2.53), the limit value of (2.47) coincides with its direct value, while as can be seen from Eq. (2.54) the limit value of (2.48) differs from its direct value.

Just as in the case of the single layer potential it is possible to rewrite (2.49) and (2.50) for the limiting case that $\boldsymbol{x}$ is allowed to approach a point $\boldsymbol{x}_0$ on a smooth portion of S. The results are

$$\lim_{\boldsymbol{x}\to\boldsymbol{x}_0} \phi(\boldsymbol{x}) = \mp\frac{1}{2}\psi(\boldsymbol{x}_0) + \int_S \psi(\boldsymbol{\xi}) Q(\boldsymbol{\xi}, \boldsymbol{x}_0)\,\mathrm{d}S(\boldsymbol{\xi}) \tag{2.55}$$

and

$$\lim_{\boldsymbol{x}\to\boldsymbol{x}_0} \frac{\partial\phi(\boldsymbol{x})}{\partial\boldsymbol{n}(\boldsymbol{x})} = \int_S \psi(\boldsymbol{\xi}) H(\boldsymbol{x}_0, \boldsymbol{\xi})\,\mathrm{d}S(\boldsymbol{\xi}) \tag{2.56}$$

The $\mp$ sign in (2.55) depends on whether point $\boldsymbol{x}_0$ is approached from the direction of the normal vector to boundary S at that point or from the opposite direction.

Comparing (2.55) with (1.61) in Sect. 1.8, it can be seen that the limit values of the potential ϕ represented by (1.61) are obtained without reference to the normal vector to the line segment. However, if the normal vector to that line segment is selected as $\boldsymbol{n} = (0, 1)$, Eq. (1.61) can be retrieved from (2.55) by assuming that S represents the line segment and $\psi(\boldsymbol{x}_0) = D$.

The integrals in the above expressions are direct values of the double layer potential (2.49) and its normal derivative (2.50). Equation (2.56) shows that the limit value of (2.50) coincides with its direct value, while the limit value of (2.49) is different from its direct value. For the case that S is a line segment and $\psi(\boldsymbol{\xi})$ is constant, this result was demonstrated in Sect. 1.8, see (1.60) and (1.62).

The kernels $G(\boldsymbol{x}_0, \boldsymbol{\xi})$, $Q(\boldsymbol{x}_0, \boldsymbol{\xi})$, $Q(\boldsymbol{\xi}, \boldsymbol{x}_0)$ and $H(\boldsymbol{x}_0, \boldsymbol{\xi})$ of integrals (2.53), (2.54), (2.55), and (2.56), are singular when $\boldsymbol{\xi} = \boldsymbol{x}_0$, both in two and three dimensions. To illustrate this, consider the two-dimensional case for which the kernels of (2.47)–(2.50) are given by (2.41) and the following expressions:

$$\begin{aligned}
Q(\boldsymbol{x}, \boldsymbol{\xi}) &= \frac{\partial G(\boldsymbol{x}, \boldsymbol{\xi})}{\partial \boldsymbol{n}(\boldsymbol{x})} = \frac{\partial}{\partial x_j} G(\boldsymbol{x}, \boldsymbol{\xi}) n_j(\boldsymbol{x}) = -\frac{1}{2\pi}\frac{r_{,j}}{r} n_j(\boldsymbol{x}) \\
Q(\boldsymbol{\xi}, \boldsymbol{x}) &= \frac{\partial G(\boldsymbol{\xi}, \boldsymbol{x})}{\partial \boldsymbol{n}(\boldsymbol{\xi})} = \frac{\partial}{\partial \xi_j} G(\boldsymbol{x}, \boldsymbol{\xi}) n_j(\boldsymbol{\xi}) = \frac{1}{2\pi}\frac{r_{,j}}{r} n_j(\boldsymbol{\xi}) \\
H(\boldsymbol{x}, \boldsymbol{\xi}) = H(\boldsymbol{\xi}, \boldsymbol{x}) &= \frac{\partial^2 G(\boldsymbol{x}, \boldsymbol{\xi})}{\partial \boldsymbol{n}(\boldsymbol{x})\partial \boldsymbol{n}(\boldsymbol{\xi})} = \frac{\partial^2}{\partial \xi_j \partial x_k} G(\boldsymbol{x}, \boldsymbol{\xi}) n_j(\boldsymbol{\xi}) n_k(\boldsymbol{x}) \\
&= \frac{1}{2\pi}\left(\frac{\delta_{jk}}{r^2} - 2\frac{r_{,j}r_{,k}}{r^2}\right) n_j(\boldsymbol{\xi}) n_k(\boldsymbol{x})
\end{aligned} \tag{2.57}$$

where $\delta_{jk} = r_{j},_{k}$ is known as the Kronecker delta, which is defined as

$$\delta_{jk} = \begin{cases} 1 & \text{if } j = k \\ 0 & \text{if } j \neq k \end{cases} \tag{2.58}$$

If $\boldsymbol{x} = \boldsymbol{x}_0 \in S$, fundamental solution (2.41) has a logarithmic singularity. Thus, if the density function $\rho(\boldsymbol{\xi})$ is continuous and bounded the direct value of potential (2.47) is given by a weakly singular integral; this can be seen from (2.15), if one takes r instead of $\boldsymbol{x}$ and $\mathrm{d}r$ instead of $\mathrm{d}x$.

The behavior of the direct values of potentials (2.48) and (2.49) depend on the smoothness conditions for the density functions $\rho(\boldsymbol{\xi})$ and $\psi(\boldsymbol{\xi})$ and the curve S. For the case when S is at least of Lyapunov's type or smoother and the density functions are continuous bounded functions, the integrals of (2.48) and (2.49) are weakly singular due to the additional requirement on the angles between the normals at the points of a Lyapunov curve. If S is just smooth (see the definition following (2.26)) and the density functions satisfy a Hölder condition (2.24), the integrals of (2.48) and (2.49) are of Cauchy principal value types.

For the direct value of potential (2.50) to exist, the first *derivative* of density function $\psi(\boldsymbol{\xi})$ has to satisfy a Hölder condition and S should be of Lyapunov's type or smoother. In such a case, the direct value of (2.50) is given by a hypersingular integral, as it is clear from the last expression of (2.57) that that kernel behaves as $1/r^2$, see (2.20). For this reason (2.50) is called the hypersingular potential.

2.4 Green's Representation Formula

Green's representation formula (1.112) plays an important role in potential theory. To obtain this formula, we use Green's second identity (2.10) and set $u(\boldsymbol{x}) = G(\boldsymbol{x}, \boldsymbol{\xi})$ and $v(\boldsymbol{x}) = \phi(\boldsymbol{x})$, where $G(\boldsymbol{x}, \boldsymbol{\xi})$ satisfies (2.30), $\phi(\boldsymbol{x})$ satisfies (1.23), and $\boldsymbol{x} \in S$. We then find that

$$\int_S \left[G(\boldsymbol{x}, \boldsymbol{\xi}) \frac{\partial \phi(\boldsymbol{x})}{\partial \boldsymbol{n}(\boldsymbol{x})} - \phi(\boldsymbol{x}) \frac{\partial G(\boldsymbol{x}, \boldsymbol{\xi})}{\partial \boldsymbol{n}(\boldsymbol{x})} \right] \mathrm{d}S(\boldsymbol{x}) = \int_V \phi(\boldsymbol{x}) \delta(\boldsymbol{x}, \boldsymbol{\xi}) \, \mathrm{d}V(\boldsymbol{x}) \qquad (2.59)$$

Using the selection property of the Dirac delta function, Eq. (2.29), we arrive at *Green's third identity* or *Green's representation formula*:

$$\phi(\boldsymbol{\xi}) = \int_S \left[G(\boldsymbol{x}, \boldsymbol{\xi}) \frac{\partial \phi(\boldsymbol{x})}{\partial \boldsymbol{n}(\boldsymbol{x})} - \phi(\boldsymbol{x}) \frac{\partial G(\boldsymbol{x}, \boldsymbol{\xi})}{\partial \boldsymbol{n}(\boldsymbol{x})} \right] \mathrm{d}S(\boldsymbol{x}) \qquad (2.60)$$

It is important to recognize that this equation is valid for a point $\boldsymbol{\xi}$ *inside* region V but not on its boundary S.

Equation (2.60) forms the basis for the direct boundary element method for Laplace's equation, which is the subject of Chap. 3. The equation is valid in both two and three dimensions, provided, of course, the appropriate version of the fundamental solution $G(\boldsymbol{x}, \boldsymbol{\xi})$ is used (see Sect. 2.2). Referring to (2.3), the normal derivative of the potential across the boundary at point $\boldsymbol{x}$, $\partial \phi(\boldsymbol{x}) / \partial \boldsymbol{n}(\boldsymbol{x})$, can be written as $q(\boldsymbol{x})$, and (2.60) becomes

$$\phi(\boldsymbol{\xi}) = \int_S q(\boldsymbol{x}) G(\boldsymbol{x}, \boldsymbol{\xi}) \, \mathrm{d}S(\boldsymbol{x}) - \int_S \phi(\boldsymbol{x}) \frac{\partial G(\boldsymbol{x}, \boldsymbol{\xi})}{\partial \boldsymbol{n}(\boldsymbol{x})} \, \mathrm{d}S(\boldsymbol{x}) \tag{2.61}$$

Using notation introduced in (1.46) and (2.51), Eq. (2.61) can also be written as

$$\phi(\boldsymbol{\xi}) = \int_S q(\boldsymbol{x}) G(\boldsymbol{x}, \boldsymbol{\xi}) \, \mathrm{d}S(\boldsymbol{x}) - \int_S \phi(\boldsymbol{x}) Q(\boldsymbol{x}, \boldsymbol{\xi}) \, \mathrm{d}S(\boldsymbol{x}) \tag{2.62}$$

Comparing the integrals on the right-hand side of (2.62) with potentials (2.47) and (2.49), noting that the roles of the points $\boldsymbol{\xi}$ and $\boldsymbol{x}$ are reversed in the two sets of expressions, and taking into account that $G(\boldsymbol{x}, \boldsymbol{\xi}) = G(\boldsymbol{\xi}, \boldsymbol{x})$, we conclude that the first integral in Green's formula (2.62) represents a single layer potential for a density function $q(\boldsymbol{x})$, while the second integral represents a double layer potential for a density function $\phi(\boldsymbol{x})$, as pointed out in Sect. 1.12.

An additional result stemming from the use of the selection property of the Dirac delta function in (2.59) is that the integral over V in this equation vanishes if point $\boldsymbol{\xi}$ is *outside* region V. We can therefore write

$$\int_S q(\boldsymbol{x}) G(\boldsymbol{x}, \boldsymbol{\xi}) \, \mathrm{d}S(\boldsymbol{x}) - \int_S \phi(\boldsymbol{x}) Q(\boldsymbol{x}, \boldsymbol{\xi}) \, \mathrm{d}S(\boldsymbol{x}) = \begin{cases} \phi(\boldsymbol{\xi}) & \boldsymbol{\xi} \in V, \ \ \boldsymbol{\xi} \notin S \\ 0 & \boldsymbol{\xi} \notin V \end{cases} \tag{2.63}$$

In Sect. 2.7 we will add another condition to this expression for the case that point $\boldsymbol{\xi}$ is on the boundary S.

2.5 Analysis for an Infinite Region

Green's representation formula (2.61) was derived for the case of a finite domain V, but this restriction can be removed by enforcing the appropriate *regularity conditions* at infinity.

For three-dimensional problems these conditions require that

$$\phi = O\left(\frac{1}{r}\right), \ q = O(1/r^2) \tag{2.64}$$

when $r \to \infty$.

Regularity conditions (2.64) are too strict for two-dimensional problems, because these problems may have solutions that do not vanish at infinity. Two types of regularity conditions exist for two-dimensional problems:

- ϕ is bounded at infinity, i.e.,

$$\phi = O(1), \ q = O(1/r^2) \tag{2.65}$$

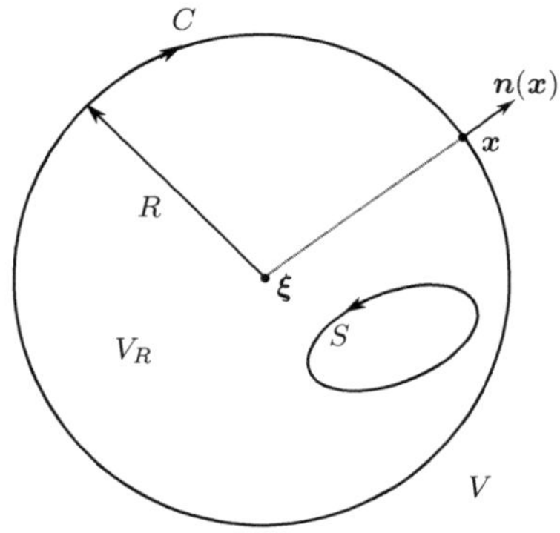

Fig. 2.5 A cavity with boundary S in an infinite region V

- ϕ is unbounded at infinity, i.e.,

$$\phi = O(\ln r), \; q = O(1/r) \tag{2.66}$$

To see how these conditions can be imposed, consider a cavity within an infinite two-dimensional domain V and assume that the boundary S of the cavity is surrounded by the circular boundary C with center at the point $\boldsymbol{\xi}$ and with radius R, as shown in Fig. 2.5. The point $\boldsymbol{\xi}$ may be located either inside inside or outside the cavity. The region V_R bounded by curves S and C is a finite region, and hence representation formula (2.63) applies in the following form:

$$\begin{aligned}&\int_S q(\boldsymbol{x})G(\boldsymbol{x},\boldsymbol{\xi})\,\mathrm{d}S(\boldsymbol{x}) + \int_C q(\boldsymbol{x})G(\boldsymbol{x},\boldsymbol{\xi})\,\mathrm{d}C(\boldsymbol{x})\\ &-\int_S \phi(\boldsymbol{x})Q(\boldsymbol{x},\boldsymbol{\xi})\,\mathrm{d}S(\boldsymbol{x}) - \int_C \phi(\boldsymbol{x})Q(\boldsymbol{x},\boldsymbol{\xi})\,\mathrm{d}C(\boldsymbol{x}) = \begin{cases}\phi(\boldsymbol{\xi}) & \boldsymbol{\xi}\in V_R, \;\; \boldsymbol{\xi}\notin S\\ 0 & \boldsymbol{\xi}\notin V_R\end{cases}\end{aligned} \tag{2.67}$$

Using (2.41) and the first expression in (2.57), the two integrals over the circular boundary C in this expression can be written in terms of polar coordinates $x = R\cos\theta$ and $y = R\sin\theta$ as

$$\int_C q\,(\boldsymbol{x})\,G(\boldsymbol{x},\boldsymbol{\xi})\,\mathrm{d}C(\boldsymbol{x}) = -\frac{1}{2\pi}\int_0^{2\pi} q\,(\boldsymbol{x})\,(\ln R)R\,\mathrm{d}\theta \tag{2.68}$$

$$\begin{aligned}\int_C \phi(\boldsymbol{x})Q(\boldsymbol{x},\boldsymbol{\xi})\,\mathrm{d}C(\boldsymbol{x}) &= -\frac{1}{2\pi}\int_0^{2\pi}\phi(\boldsymbol{x})\frac{R_{,i}n_i(\boldsymbol{x})}{R}R\,\mathrm{d}\theta\\ &= -\frac{1}{2\pi}\int_0^{2\pi}\phi(\boldsymbol{x})\,\mathrm{d}\theta\end{aligned} \tag{2.69}$$

which follows from the results $n_i(\boldsymbol{x}) = (x_i - \xi_i)/R$, $R^2 = (x_i - \xi_i)(x_i - \xi_i)$, and $R_{,i} = (x_i - \xi_i)/R$. We seek the limiting values of these integrals as $R \to \infty$.

Assume first that $\phi(\boldsymbol{x})$ is *bounded* at infinity, i.e., assume that $\lim_{\boldsymbol{x}\to\infty} \phi(\boldsymbol{x}) = \phi_0$. Then, regularity conditions (2.65) suggest that $q = O(1/R)^2$ and, using L'Hopital's rule, the limiting behavior of the integrand of the right-hand side of (2.68) is

$$\lim_{R\to\infty} \frac{\ln R}{R} = \lim_{R\to\infty} \frac{1/R}{1} = 0 \tag{2.70}$$

The limiting value of (2.69) is

$$\begin{aligned}
\lim_{R\to\infty} \int_C \phi(\boldsymbol{x}) Q(\boldsymbol{x}, \boldsymbol{\xi})\, \mathrm{d}C(\boldsymbol{x}) &= -\frac{1}{2\pi} \lim_{R\to\infty} \int_0^{2\pi} \phi(\boldsymbol{x}) \mathrm{d}\theta \\
&= -\frac{1}{2\pi} \lim_{R\to\infty} \int_0^{2\pi} [\phi(\boldsymbol{x}) - \phi_0]\, \mathrm{d}\theta - \frac{\phi_0}{2\pi} \int_0^{2\pi} \mathrm{d}\theta \\
&= -\phi_0
\end{aligned} \tag{2.71}$$

Using these results in (2.67) we obtain

$$\int_S q(\boldsymbol{x}) G(\boldsymbol{x}, \boldsymbol{\xi})\, \mathrm{d}S(\boldsymbol{x}) - \int_S \phi(\boldsymbol{x}) Q(\boldsymbol{x}, \boldsymbol{\xi})\, \mathrm{d}S(\boldsymbol{x}) + \phi_0 = \begin{cases} \phi(\boldsymbol{\xi}) & \boldsymbol{\xi} \in V,\ \ \boldsymbol{\xi} \notin S \\ \\ 0 & \boldsymbol{\xi} \notin V \end{cases} \tag{2.72}$$

If $\phi(\boldsymbol{x})$ is *unbounded* at infinity, regularity conditions (2.66) can be treated in a similar fashion, but in this case it can be shown that integrals (2.68) and (2.69) do not vanish. However, both integrals are equal to $-\ln R$, so their difference is zero. We therefore find that representation formula (2.72) still holds, but with $\phi_0 = 0$.

Derivations for the three-dimensional case follow the same procedure, using a spherical boundary of radius R, and regularity conditions (2.64). Representation formula (2.72) holds for this case, too, with $\phi_0 = 0$.

2.6 Integral Formula for the Gradient and Normal Derivative of $\phi(\boldsymbol{\xi})$

It is often necessary to determine values of the gradient of $\phi(\boldsymbol{\xi})$ at points within the region of interest V. This is accomplished by differentiation of (2.62) with respect to the parameters $\boldsymbol{\xi} = \xi_j$, where $j = 1$ to 3 (or 1 to 2 in two dimensions). This differentiation can be carried out under the integral sign as long as point $\boldsymbol{\xi}$ is not on the boundary S, i.e., provided $\boldsymbol{\xi} \neq \boldsymbol{x}$.

We then have

$$\frac{\partial\phi(\boldsymbol{\xi})}{\partial\xi_j} = \int_S q(\boldsymbol{x})\frac{\partial}{\partial\xi_j}G(\boldsymbol{x},\boldsymbol{\xi})\,\mathrm{d}S(\boldsymbol{x}) - \int_S \phi(\boldsymbol{x})\frac{\partial}{\partial\xi_j}Q(\boldsymbol{x},\boldsymbol{\xi})\,\mathrm{d}S(\boldsymbol{x}) \tag{2.73}$$

where by definition $\partial\phi(\boldsymbol{\xi})/\partial\xi_j = q_j(\boldsymbol{\xi})$. Noting that $\partial G(\boldsymbol{x},\boldsymbol{\xi})/\partial\xi_j = -\partial G(\boldsymbol{x},\boldsymbol{\xi})/\partial x_j$ and referring to the first expression of (2.57) we see that (2.73) involves the first and second derivatives of function $G(\boldsymbol{x},\boldsymbol{\xi}) = G(\boldsymbol{\xi},\boldsymbol{x})$. For the two-dimensional case with $G(\boldsymbol{x},\boldsymbol{\xi})$ given by (2.41) these derivatives are

$$\begin{aligned}\frac{\partial}{\partial x_j}G(\boldsymbol{x},\boldsymbol{\xi}) &= -\frac{1}{2\pi r}\frac{\partial r}{\partial x_j} = -\frac{1}{2\pi}\frac{r_{,j}}{r}\\ \frac{\partial^2}{\partial x_j\partial x_k}G(\boldsymbol{x},\boldsymbol{\xi}) &= -\frac{1}{2\pi r^2}(\delta_{jk} - 2r_{,j}r_{,k})\end{aligned} \tag{2.74}$$

Similar expressions in three dimensions are

$$\begin{aligned}\frac{\partial}{\partial x_j}G(\boldsymbol{x},\boldsymbol{\xi}) &= -\frac{1}{4\pi}\frac{r_{,j}}{r^2}\\ \frac{\partial^2}{\partial x_j\partial x_k}G(\boldsymbol{x},\boldsymbol{\xi}) &= -\frac{1}{4\pi r^3}(\delta_{jk} - 3r_{,j}r_{,k})\end{aligned} \tag{2.75}$$

The normal derivative of the potential across a curve (surface) with local normal vector $\boldsymbol{n}(\boldsymbol{\xi})$ is defined as $\partial\phi(\boldsymbol{\xi})/\partial\boldsymbol{n}(\boldsymbol{\xi}) = q_j(\boldsymbol{\xi})n_j(\boldsymbol{\xi})$ so that

$$n_j(\boldsymbol{\xi})q_j(\boldsymbol{\xi}) = \int_S q(\boldsymbol{x})n_j(\boldsymbol{\xi})\frac{\partial}{\partial\xi_j}G(\boldsymbol{x},\boldsymbol{\xi})\,\mathrm{d}S(\boldsymbol{x}) - \int_S \phi(\boldsymbol{x})n_j(\boldsymbol{\xi})\frac{\partial}{\partial\xi_j}Q(\boldsymbol{x},\boldsymbol{\xi})\,\mathrm{d}S(\boldsymbol{x}) \tag{2.76}$$

which, using (2.57), can be written as

$$\frac{\partial\phi(\boldsymbol{\xi})}{\partial\boldsymbol{n}(\boldsymbol{\xi})} = \int_S q(\boldsymbol{x})Q(\boldsymbol{\xi},\boldsymbol{x})\,\mathrm{d}S(\boldsymbol{x}) - \int_S \phi(\boldsymbol{x})H(\boldsymbol{x},\boldsymbol{\xi})\,\mathrm{d}S(\boldsymbol{x}) \tag{2.77}$$

where $H(\boldsymbol{x},\boldsymbol{\xi}) = H(\boldsymbol{\xi},\boldsymbol{x})$ is given by (2.52). It can be seen from (2.48) and (2.50) that the first integral in (2.77) represents the adjoint double layer potential[2] for density function $q(\boldsymbol{x})$, while the second integral represents the hypersingular potential for density function $\phi(\boldsymbol{x})$.

---

[2] The definition of the adjoint double layer potential $Q(\boldsymbol{x},\boldsymbol{\xi})$ in (1.46) assumed that point $\boldsymbol{\xi}$ was on boundary S and point $\boldsymbol{x}$ was in region V. These roles are reversed in (2.77).

2.7 Limit Values of Green's Representation Formula and Its Normal Derivative

Limit values of potentials (2.47)–(2.50) were given without proof in Sect. 2.3. Similar limit values can be written for Green's representation formula (2.62) and its normal derivative (2.77) and we next demonstrate how this can be accomplished.

To begin, we consider the situation depicted in Fig. 2.6a, which shows a portion of a finite, two-dimensional region V with a sufficiently smooth boundary S divided into two parts: one part, S_ε, is a circular arc of radius ε centered at point $\boldsymbol{\xi}$, and the other, $S - S_\varepsilon$, represents the remainder of the boundary. In this arrangement, we have effectively "bumped out" the boundary S so as to keep point $\boldsymbol{\xi}$ inside region V, with the idea that in the limit, as $\varepsilon \to 0$, point $\boldsymbol{\xi}$ will lie on the boundary, as desired. For this boundary configuration (2.62) becomes

$$\begin{aligned}\phi(\boldsymbol{\xi}) = &\lim_{\varepsilon\to 0}\int_{S-S_\varepsilon} q(\boldsymbol{x})G(\boldsymbol{x},\boldsymbol{\xi})\,\mathrm{d}S(\boldsymbol{x}) + \lim_{\varepsilon\to 0}\int_{S_\varepsilon} q(\boldsymbol{x})G(\boldsymbol{x},\boldsymbol{\xi})\,\mathrm{d}S(\boldsymbol{x}) \\ &- \lim_{\varepsilon\to 0}\int_{S-S_\varepsilon} \phi(\boldsymbol{x})Q(\boldsymbol{x},\boldsymbol{\xi})\,\mathrm{d}S(\boldsymbol{x}) - \lim_{\varepsilon\to 0}\int_{S_\varepsilon} \phi(\boldsymbol{x})Q(\boldsymbol{x},\boldsymbol{\xi})\,\mathrm{d}S(\boldsymbol{x}) \qquad (2.78)\end{aligned}$$

The integrals labeled $S - S_\varepsilon$ cover the entire boundary S except for an infinitesimal portion of length 2ε centered at point $\boldsymbol{\xi}$.

The second term on the right-hand side of (2.78) can be written as

$$\lim_{\varepsilon\to 0}\int_{S_\varepsilon} q(\boldsymbol{x})G(\boldsymbol{x},\boldsymbol{\xi})\,\mathrm{d}S(\boldsymbol{x}) = \lim_{\varepsilon\to 0}\left\{\int_{S_\varepsilon}[q(\boldsymbol{x}) - q(\boldsymbol{\xi})]\,G(\boldsymbol{x},\boldsymbol{\xi})\,\mathrm{d}S(\boldsymbol{x}) + q(\boldsymbol{\xi})\int_{S_\varepsilon} G(\boldsymbol{x},\boldsymbol{\xi})\,\mathrm{d}S(\boldsymbol{x})\right\} \qquad (2.79)$$

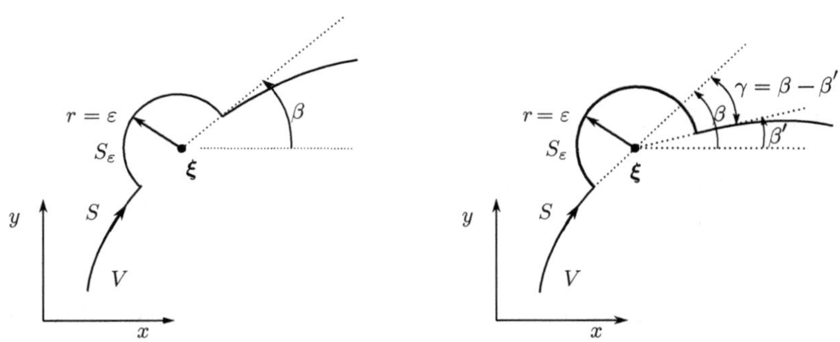

(a) $\boldsymbol{\xi}$ at smooth part of S. (b) $\boldsymbol{\xi}$ at sharp corner of S.

Fig. 2.6 Limit cases for load point $\boldsymbol{\xi}$ on boundary

If $q(\boldsymbol{x})$ is a continuous function (cf. (2.23)) the limit of the integral involving $q(\boldsymbol{x}) - q(\boldsymbol{\xi})$ is zero as $\varepsilon \to 0$. The other integral in (2.79) can be evaluated using polar coordinates $x = \varepsilon \cos\theta$ $y = \varepsilon \sin\theta$, as follows[3]:

$$\int_{S_\varepsilon} G(\boldsymbol{x}, \boldsymbol{\xi})\, \mathrm{d}S(\boldsymbol{x}) = -\frac{1}{2\pi} \int_{\beta+\pi}^{\beta} \ln\varepsilon(-\varepsilon \mathrm{d}\theta) = -\frac{1}{2}\varepsilon \ln\varepsilon \tag{2.80}$$

which is also zero as $\varepsilon \to 0$. For a smooth boundary S we therefore have that

$$\lim_{\varepsilon\to 0} \int_{S-S\varepsilon} q(\boldsymbol{x}) G(\boldsymbol{x}, \boldsymbol{\xi})\, \mathrm{d}S(\boldsymbol{x}) + \lim_{\varepsilon\to 0} \int_{S_\varepsilon} q(\boldsymbol{x}) G(\boldsymbol{x}, \boldsymbol{\xi})\, \mathrm{d}S(\boldsymbol{x}) = \int_S q(\boldsymbol{x}) G(\boldsymbol{x}, \boldsymbol{\xi})\, \mathrm{d}S(\boldsymbol{x}) \tag{2.81}$$

This result bears a clear resemblance to relation (2.53) for a single layer potential.

In light of these results (2.78) can be simplified to read

$$\begin{aligned} \phi(\boldsymbol{\xi}) = & \int_S q(\boldsymbol{x}) G(\boldsymbol{x}, \boldsymbol{\xi})\, \mathrm{d}S(\boldsymbol{x}) - \lim_{\varepsilon\to 0} \int_{S-S\varepsilon} \phi(\boldsymbol{x}) Q(\boldsymbol{x}, \boldsymbol{\xi})\, \mathrm{d}S(\boldsymbol{x}) \\ & - \lim_{\varepsilon\to 0} \int_{S_\varepsilon} \phi(\boldsymbol{x}) Q(\boldsymbol{x}, \boldsymbol{\xi})\, \mathrm{d}S(\boldsymbol{x}) \end{aligned} \tag{2.82}$$

The last integral in this expression can be written as

$$\lim_{\varepsilon\to 0} \int_{S_\varepsilon} \phi(\boldsymbol{x}) Q(\boldsymbol{x}, \boldsymbol{\xi})\, \mathrm{d}S(\boldsymbol{x}) = \lim_{\varepsilon\to 0} \left\{ \int_{S_\varepsilon} [\phi(\boldsymbol{x}) - \phi(\boldsymbol{\xi})]\, Q(\boldsymbol{x}, \boldsymbol{\xi})\, \mathrm{d}S(\boldsymbol{x}) + \phi(\boldsymbol{\xi}) \int_{S_\varepsilon} Q(\boldsymbol{x}, \boldsymbol{\xi})\, \mathrm{d}S(\boldsymbol{x}) \right\} \tag{2.83}$$

For a continuous function $\phi(\boldsymbol{x})$, the two terms on the right-hand side of this expression can be evaluated using the definition of $Q(\boldsymbol{x}, \boldsymbol{\xi}) = \partial G(\boldsymbol{x}, \boldsymbol{\xi})/\partial \boldsymbol{n}(\boldsymbol{x})$, as follows. The first term is

$$\lim_{\varepsilon\to 0} \int_{S_\varepsilon} [\phi(\boldsymbol{x}) - \phi(\boldsymbol{\xi})] Q(\boldsymbol{x}, \boldsymbol{\xi})\, \mathrm{d}S(\boldsymbol{x}) = -\frac{1}{2\pi} \lim_{\varepsilon\to 0} \int_{\beta+\pi}^{\beta} [\phi(\boldsymbol{x}) - \phi(\boldsymbol{\xi})] \left(\frac{1}{\epsilon}\right) (-\varepsilon\, \mathrm{d}\theta) = 0 \tag{2.84}$$

because the potential is continuous, i.e., $\phi(\boldsymbol{x}) - \phi(\boldsymbol{\xi}) \to 0$ as $\boldsymbol{\xi} \to \boldsymbol{x}$. The second term in (2.83) is

$$\phi(\boldsymbol{\xi}) \lim_{\varepsilon\to 0} \int_{S_\varepsilon} Q(\boldsymbol{x}, \boldsymbol{\xi})\, \mathrm{d}S(\boldsymbol{x}) = -\frac{1}{2\pi} \phi(\boldsymbol{\xi}) \lim_{\varepsilon\to 0} \int_{\beta+\pi}^{\beta} \frac{1}{\varepsilon} (-\varepsilon\, \mathrm{d}\theta) = -\frac{1}{2} \phi(\boldsymbol{\xi}) \tag{2.85}$$

[3] The minus sign for the element of arc length ($\varepsilon\, \mathrm{d}\theta$) appears because the direction of integration is opposite to the counterclockwise sense for angle θ.

We therefore have that

$$\lim_{\varepsilon\to 0}\int_{S-S\varepsilon}\phi(\boldsymbol{x})Q(\boldsymbol{x},\boldsymbol{\xi})\,\mathrm{d}S(\boldsymbol{x})+\lim_{\varepsilon\to 0}\int_{S_\varepsilon}\phi(\boldsymbol{x})Q(\boldsymbol{x},\boldsymbol{\xi})\,\mathrm{d}S(\boldsymbol{x}) = -\frac{1}{2}\phi(\boldsymbol{\xi})+\int_S\phi(\boldsymbol{x})Q(\boldsymbol{x},\boldsymbol{\xi})\,\mathrm{d}S(\boldsymbol{x}) \tag{2.86}$$

which is comparable to relation (2.55) for a double layer potential.

Combining (2.82) and (2.86), we write the alternate form of Green's representation formula as

$$\frac{1}{2}\phi(\boldsymbol{\xi})=\int_S q(\boldsymbol{x})G(\boldsymbol{x},\boldsymbol{\xi})\,\mathrm{d}S(\boldsymbol{x})-\int_S\phi(\boldsymbol{x})Q(\boldsymbol{x},\boldsymbol{\xi})\,\mathrm{d}S(\boldsymbol{x}) \tag{2.87}$$

The first integral on the right-hand side of this expression (the direct value of a single layer potential) is weakly singular, while the type of singularity for the second integral (the direct value of a double layer potential) depends on the smoothness conditions for the density function $\phi(\boldsymbol{x})$ and the curve S. As explained in Sect. 2.3, depending on those conditions, the second integral in (2.87) can be identified as either weakly singular or a Cauchy principal value type integral. In the latter case, the notation $⨍$ is used, so (2.87) can be written as

$$\frac{1}{2}\phi(\boldsymbol{\xi})=\int_S q(\boldsymbol{x})G(\boldsymbol{x},\boldsymbol{\xi})\,\mathrm{d}S(\boldsymbol{x})-⨍_S\phi(\boldsymbol{x})Q(\boldsymbol{x},\boldsymbol{\xi})\,\mathrm{d}S(\boldsymbol{x}) \tag{2.88}$$

Formulas (2.87) and (2.88) apply when points $\boldsymbol{\xi}$ and $\boldsymbol{x}$ lie on a smooth portion of the boundary S, as shown in Fig. 2.6a.

If point $\boldsymbol{\xi}$ is at a sharp corner, as illustrated in Fig. 2.6b, the limits of integration over the circular arc range from $\theta=\beta+\pi$ to $\theta=\beta'$, where β' and β are the angles of the tangents to the undeflected boundary S on either side of point $\boldsymbol{\xi}$. Then, assuming appropriate smoothness conditions, the integrals corresponding to (2.84) and (2.85) become

$$\lim_{\varepsilon\to 0}\int_{S_\varepsilon}[\phi(\boldsymbol{x})-\phi(\boldsymbol{\xi})]Q(\boldsymbol{x},\boldsymbol{\xi})\,\mathrm{d}S(\boldsymbol{x})=-\frac{1}{2\pi}\lim_{\varepsilon\to 0}\int_{\beta+\pi}^{\beta'}[\phi(\boldsymbol{x})-\phi(\boldsymbol{\xi})]\left(\frac{1}{\epsilon}\right)(-\varepsilon\,\mathrm{d}\theta)=0 \tag{2.89}$$

and

$$\phi(\boldsymbol{\xi})\lim_{\varepsilon\to 0}\int_{S_\varepsilon}Q(\boldsymbol{x},\boldsymbol{\xi})\,\mathrm{d}S(\boldsymbol{x})=-\frac{1}{2\pi}\phi(\boldsymbol{\xi})\lim_{\varepsilon\to 0}\int_{\beta+\pi}^{\beta'}\frac{1}{\varepsilon}(-\varepsilon\,\mathrm{d}\theta)=-\frac{\pi-\gamma}{2\pi}\phi(\boldsymbol{\xi}) \tag{2.90}$$

where $\gamma=\beta-\beta'$, leading to the following more general result:

$$\frac{\pi - \gamma}{2\pi}\phi(\boldsymbol{\xi}) = \int_S q(\boldsymbol{x})G(\boldsymbol{x},\boldsymbol{\xi})\,\mathrm{d}S(\boldsymbol{x}) - ⨍_S \phi(\boldsymbol{x})Q(\boldsymbol{x},\boldsymbol{\xi})\,\mathrm{d}S(\boldsymbol{x}) \tag{2.91}$$

If boundary S is smooth at point $\boldsymbol{\xi}$ then $\gamma = 0$ and (2.91) reduces to (2.88).

Similarly, using a limiting procedure of the same type it can be shown (e.g., Wrobel 2002) that, under appropriate smoothness assumptions for boundary S and density functions $\phi(\boldsymbol{x})$ and $q(\boldsymbol{x})$,

$$\frac{\pi - \gamma}{2\pi}q(\boldsymbol{\xi}) = ⨍_S q(\boldsymbol{x})Q(\boldsymbol{\xi},\boldsymbol{x})\,\mathrm{d}S(\boldsymbol{x}) - ⨎_S \phi(\boldsymbol{x})H(\boldsymbol{x},\boldsymbol{\xi})\,\mathrm{d}S(\boldsymbol{x}) \tag{2.92}$$

where the notation $⨎$ identifies the Hadamard finite part integral.

2.8 Summary of Integral Representations

In this section, we summarize the boundary integral representations used in boundary element methods for potential problems. They are

- Indirect formulation with a single layer potential

$$\phi(\boldsymbol{x}) = \int_S \rho(\boldsymbol{\xi})G(\boldsymbol{x},\boldsymbol{\xi})\,\mathrm{d}S(\boldsymbol{\xi}) \tag{2.93}$$

$$q_k(\boldsymbol{x}) = \int_S \rho(\boldsymbol{\xi})\frac{\partial}{\partial x_k}G(\boldsymbol{x},\boldsymbol{\xi})\,\mathrm{d}S(\boldsymbol{\xi}) \tag{2.94}$$

$$\frac{\partial \phi(\boldsymbol{x})}{\partial \boldsymbol{n}(\boldsymbol{x})} = \int_S \rho(\boldsymbol{\xi})Q(\boldsymbol{x},\boldsymbol{\xi})\,\mathrm{d}S(\boldsymbol{\xi}) \tag{2.95}$$

- Indirect formulation with a double layer potential

$$\phi(\boldsymbol{x}) = \int_S \psi(\boldsymbol{\xi})Q(\boldsymbol{\xi},\boldsymbol{x})\,\mathrm{d}S(\boldsymbol{\xi}) \tag{2.96}$$

$$q_k(\boldsymbol{x}) = \int_S \psi(\boldsymbol{\xi})\frac{\partial}{\partial x_k}Q(\boldsymbol{\xi},\boldsymbol{x})\,\mathrm{d}S(\boldsymbol{\xi}) \tag{2.97}$$

$$\frac{\partial \phi(\boldsymbol{x})}{\partial \boldsymbol{n}(\boldsymbol{x})} = \int_S \psi(\boldsymbol{\xi})H(\boldsymbol{x},\boldsymbol{\xi})\,\mathrm{d}S(\boldsymbol{\xi}) \tag{2.98}$$

The limit values of the single and double layer potentials are given by expressions (2.53)–(2.56) when point $\boldsymbol{x}$ is located on a smooth portion of the boundary. If these points are at a corner, the multiplier $1/2$ in the free terms of the latter equations should be replaced by $(\pi - \gamma)/2\pi$, where the value of γ represents the angle shown in Fig. 2.6b.

- Direct formulation with Green's representation formula

$$\int_S q(\boldsymbol{x})G(\boldsymbol{x},\boldsymbol{\xi})\,\mathrm{d}S(\boldsymbol{x}) - \int_S \phi(\boldsymbol{x})Q(\boldsymbol{x},\boldsymbol{\xi})\,\mathrm{d}S(\boldsymbol{x}) = \begin{cases} \phi(\boldsymbol{\xi}) & \boldsymbol{\xi} \in V, \quad \boldsymbol{\xi} \notin S \\ \dfrac{\pi - \gamma}{2\pi}\phi(\boldsymbol{\xi}) & \boldsymbol{\xi} \in S \\ 0 & \boldsymbol{\xi} \notin V \end{cases} \tag{2.99}$$

$$\int_S q(\boldsymbol{x})Q(\boldsymbol{\xi},\boldsymbol{x})\,\mathrm{d}S(\boldsymbol{x}) - \int_S \phi(\boldsymbol{x})H(\boldsymbol{x},\boldsymbol{\xi})\,\mathrm{d}S(\boldsymbol{x}) = \begin{cases} q(\boldsymbol{\xi}) & \boldsymbol{\xi} \in V, \quad \boldsymbol{\xi} \notin S \\ \dfrac{\pi - \gamma}{2\pi}q(\boldsymbol{\xi}) & \boldsymbol{\xi} \in S \\ 0 & \boldsymbol{\xi} \notin V \end{cases} \tag{2.100}$$

Equations (2.99) and (2.100) contain regular integrals when $\boldsymbol{\xi} \notin S$. If $\boldsymbol{\xi} \in S$, the integrals become singular at points where $\boldsymbol{x} = \boldsymbol{\xi}$. The type of singularity for each integral involved depends on the smoothness conditions for the functions $\phi(\boldsymbol{x})$ and $q(\boldsymbol{x})$ as well as those for the curve S; these conditions are similar to the conditions discussed in Sects. 2.3 and 2.7.

2.8.1 Discussion

The preceding integral representations can be used for numerical solutions of various boundary value problems using the boundary element technique. The indirect integral representations are convenient to use in problems involving thin layers or cracks located inside infinite domains. For such problems, it can be shown, by subtracting the limit values of the corresponding fields from the opposite sides of S in (2.54) and (2.55), that the density functions $\rho(\boldsymbol{\xi})$ and $\psi(\boldsymbol{\xi})$ have physical meanings of the jumps in either the normal derivative of the potential or the potential itself. The direct integral representation is convenient to use in problems involving finite domains, because the density functions explicitly represent the potential and its normal derivative at the boundaries. For the case of an infinite domain, all integral representations should be augmented by adding terms due to prescribed conditions at infinity, as was done in Sect. 2.5.

Other considerations might need to be taken into account when choosing an appropriate integral representation for a specific boundary value problem. In some cases, the representation must be supplemented by additional conditions for the solution to exist. For example, in the case of a Neumann problem for a finite region the integral over the contour of the prescribed normal derivative of the potential must vanish. In other cases it is necessary to include additional terms to ensure the correct behavior of the solution at infinity. An illustration of this for a multiply connected region was described in Sect. 1.9.

The boundary element methods discussed in Chap. 1 were based on integral expressions (2.93) and (2.96) for single and double layer potentials. In both cases we approximated the boundary S by N straight line boundary elements joined end to end and assumed that the density functions $\rho(\boldsymbol{\xi})$ and $\psi(\boldsymbol{\xi})$ were piecewise constant over the elements. This allowed us to analytically integrate the kernels $G(\boldsymbol{x}, \boldsymbol{\xi})$ and $Q(\boldsymbol{x}, \boldsymbol{\xi})$ over each boundary element for the general case that point $\boldsymbol{x}$ does not lie on the boundary. We then considered the limiting case that point $\boldsymbol{x}$ was at the midpoint of a boundary element, leading to a system of $N \times N$ simultaneous linear equations in N unknowns.

Instead of this "limit after discretization" approach it is possible to follow a "limit before discretization" approach using (2.53) and (2.55), which represent versions of (2.93) and (2.96) for which the limits are taken before any integration is carried out. Discretization of these boundary integral equations proceeds in essentially the same manner as before, and, assuming that the density functions $\rho(\boldsymbol{\xi})$ and $\psi(\boldsymbol{\xi})$ are piecewise constant over the elements, leads to the same results. We will demonstrate this equivalence for the direct boundary integral method in the next chapter.

Most of the boundary integral equations for potential problems contain only weakly singular integrals (for sufficiently smooth density functions and boundary S, as discussed at the end of Sect. 2.3) and are of the so-called Fredholm type. The theory of such equations is described by the Fredholm alternatives which are used to prove the existence of solutions of various boundary value problems. Description of this theory is outside of the scope of this book and the interested reader is referred to the literature for details (see, for example, Bonnet 1999).

2.9 Generalizations

The integral representations summarized in the previous section involve the fundamental solution $G(\boldsymbol{x}, \boldsymbol{\xi})$ of Laplace's equation for an infinite domain. For some boundary value problems, it is possible to construct solutions that, in addition to satisfying Laplace's equation in the problem domain, also satisfy prescribed conditions on some parts of its boundary. Such solutions are called *Green's functions* for the corresponding boundary value problems. For example, in Sect. 1.10 we obtained, using the method of images, Green's function for the boundary value problem of a half-plane with prescribed zero potential at its boundary. This function, given by (1.97), represents a sum of the two-dimensional fundamental solution $G(\boldsymbol{x}, \boldsymbol{\xi})$ of (2.41) and an additional harmonic function that has no singularities inside the half-plane.

In general, Green's function for any boundary value problem governed by Laplace's equation can be represented by the following sum:

$$G^*(\boldsymbol{x}, \boldsymbol{\xi}) = G(\boldsymbol{x}, \boldsymbol{\xi}) + v\,(\boldsymbol{x}, \boldsymbol{\xi}) \tag{2.101}$$

in which $G(\boldsymbol{x}, \boldsymbol{\xi})$ is given by either (2.41) or (2.46) and $v\,(\boldsymbol{x}, \boldsymbol{\xi})$ is a harmonic function that is regular inside the domain under study.

We list below several Green's functions for potential problems. The interested reader can find more examples of Green's functions in Duffy (2001).

- A half-plane with zero potential at its boundary $y = x_2 = 0$:

$$G^*(\boldsymbol{x}, \boldsymbol{\xi}) = G(\boldsymbol{x}, \boldsymbol{\xi}) - G(\boldsymbol{x}, \boldsymbol{\xi}') = \frac{1}{2\pi} \ln \frac{r'}{r} \tag{2.102}$$

in which $G(\boldsymbol{x}, \boldsymbol{\xi})$ is given by (2.41), $\boldsymbol{\xi} = (\xi_1, \xi_2)$ and $\boldsymbol{x} = (x_1, x_2)$ are located in the interior of the half-plane; point $\boldsymbol{\xi}' = (\xi_1, -\xi_2)$ is located outside the half-plane, and $r = |\boldsymbol{x} - \boldsymbol{\xi}|$ and $r' = |\boldsymbol{x} - \boldsymbol{\xi}'|$.

- A half-plane with zero normal derivative of the potential at its boundary $y = x_2 = 0$:

$$G^*(\boldsymbol{x}, \boldsymbol{\xi}) = G(\boldsymbol{x}, \boldsymbol{\xi}) + G(\boldsymbol{x}, \boldsymbol{\xi}') = -\frac{1}{2\pi} \ln(r'r) \tag{2.103}$$

The meanings of the symbols $\boldsymbol{\xi}'$ and r' are the same as in (2.102).

- A circular disk of radius ρ_0 centered at the origin of the coordinate system with zero potential at its boundary:

$$G^*(\boldsymbol{x}, \boldsymbol{\xi}) = G(\boldsymbol{x}, \boldsymbol{\xi}) - G(\boldsymbol{x}, \boldsymbol{\xi}') - \frac{1}{2\pi} \ln \frac{\rho_0}{\rho} \tag{2.104}$$

in which $G(\boldsymbol{x}, \boldsymbol{\xi})$ is given by (2.41), point $\boldsymbol{\xi} = (\rho, \theta)$ is located inside the disk, and point $\boldsymbol{\xi}' = (\rho_0^2/\rho, \theta)$ is located outside it, as shown in Fig. 2.7.

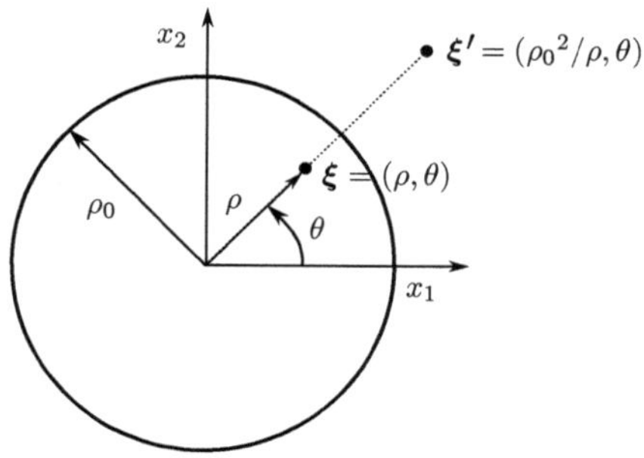

Fig. 2.7 Geometry for Green's function for a circular disk

Finally, we list a few Green's functions for three-dimensional problems. They are:

- A half-space with zero potential at its boundary $z = x_3 = 0$:

$$G^*(\boldsymbol{x}, \boldsymbol{\xi}) = G(\boldsymbol{x}, \boldsymbol{\xi}) - G(\boldsymbol{x}, \boldsymbol{\xi}') = \frac{1}{4\pi}\left(\frac{1}{r} - \frac{1}{r'}\right) \tag{2.105}$$

in which $G(\boldsymbol{x}, \boldsymbol{\xi})$ is given by (2.46), $\boldsymbol{\xi} = (\xi_1, \xi_2, \xi_3)$ and $\boldsymbol{x} = (x_1, x_2, x_3)$ are located in the interior of the half-space, point $\boldsymbol{\xi}' = (\xi_1, \xi_2, -\xi_3)$ is located outside the half-space, and $r = |\boldsymbol{x} - \boldsymbol{\xi}|$, $r' = |\boldsymbol{x} - \boldsymbol{\xi}'|$.

- A half-space with zero normal derivative of the potential at its boundary $z = x_3 = 0$:

$$G^*(\boldsymbol{x}, \boldsymbol{\xi}) = G(\boldsymbol{x}, \boldsymbol{\xi}) + G(\boldsymbol{x}, \boldsymbol{\xi}') = \frac{1}{4\pi}\left(\frac{1}{r} + \frac{1}{r'}\right) \tag{2.106}$$

The meanings of the symbols $\boldsymbol{\xi}'$ and r' are the same as in (2.105).

The use of the above expressions for Green's functions $G^*(\boldsymbol{x}, \boldsymbol{\xi})$ instead of $G(\boldsymbol{x}, \boldsymbol{\xi})$ in (2.51) and (2.52) yields expressions for the corresponding kernels $Q^*(\boldsymbol{x}, \boldsymbol{\xi})$ and $H^*(\boldsymbol{x}, \boldsymbol{\xi})$. The use of $G^*(\boldsymbol{x}, \boldsymbol{\xi})$, $Q^*(\boldsymbol{x}, \boldsymbol{\xi})$, and $H^*(\boldsymbol{x}, \boldsymbol{\xi})$ in place of $G(\boldsymbol{x}, \boldsymbol{\xi})$, $Q(\boldsymbol{x}, \boldsymbol{\xi})$, and $H(\boldsymbol{x}, \boldsymbol{\xi})$ in (2.93)–(2.100) leads to integral representations that automatically satisfy prescribed conditions for the part of a boundary in a problem associated with this specific Green's function. This means that there is no need to discretize this part of the boundary or to approximate unknown boundary data on that part. Moreover, because the additional functions involved in $G^*(\boldsymbol{x}, \boldsymbol{\xi})$ have no singularities in the domain under study, the limiting and direct values of the obtained integral representations are connected by the same relations (2.53)–(2.56) provided the kernels in those relations are replaced by their counterparts with "*" superscripts.

2.10 Exercises

1. Use the chain rule for partial differentiation to write Laplace's equation $\nabla^2 f = \partial^2 f/\partial x^2 + \partial^2 f/\partial y^2 = 0$ in plane polar coordinates, $x = r\cos\theta$ and $y = r\sin\theta$ (see (2.31)).
   Answer: Note that, for example, because $r = (x^2 + y^2)^{1/2}$ and $\theta = \arctan(y/x)$, $\partial/\partial x = (\partial r/\partial x)(\partial/\partial r) + (\partial\theta/\partial x)(\partial/\partial\theta) = \cos\theta\partial/\partial r - \sin\theta r^{-1}\partial/\partial\theta$, with a similar expression for $\partial/\partial y$. A second differentiation of each expression by x and y establishes the required result.
2. Show that (2.3) can be derived using the chain rule for partial differentiation and a local $\bar{x}, \bar{y}$ coordinate system, with $\bar{y}$ parallel to n_2.
   Answer: Note that the normal vector $\boldsymbol{n}$ has the components $(-\sin\beta, \cos\beta)$ in the x, y global coordinate system and the angle between x and $\bar{x}$ is equal to β.
3. Consider the case when S is the line segment $| x_1 | \leq a, x_2 = 0, | \xi_1 | < a, \xi_2 = 0$, assume that $\rho(\boldsymbol{\xi}) = P,\ \psi(\boldsymbol{\xi}) = D$, and take the components of the normal vectors to S to be $n_1(\boldsymbol{x}) = n_1(\boldsymbol{\xi}) = 0$ and $n_2(\boldsymbol{x}) = n_2(\boldsymbol{\xi}) = -1$. Evaluate the direct and limit values of potentials (2.47)–(2.50) and prove the relations (1.20) and (1.61).
   Answer: The limit and direct values of (2.48) coincide and are equal to

   $$P\left[(x_1 - a)\ln|x_1 - a| - (x_1 + a)\ln|x_1 + a| + 2a\right]/2\pi.$$

   The limit values of (2.49) are $\pm P/2$, when $x_2 \to \pm 0$; its direct value is zero. The limit values of (2.50) are $\mp Q/2$, when $x_2 \to \pm 0$; its direct value is zero. The limit and direct values of (2.51) coincide and are equal to $Q\left[1/(x_1 - a) - 1/(x_1 + a)\right]/2\pi$.
4. Show that $G^*(\boldsymbol{x}, \boldsymbol{\xi})$ of (2.102) and (2.105) satisfy the condition of zero potential at the boundaries $y = x_2 = 0$ and $z = x_3 = 0$, respectively.
5. Show that $G^*(\boldsymbol{x}, \boldsymbol{\xi})$ of (2.103) and (2.106) satisfy the condition of zero normal derivatives at the boundaries $y = x_2 = 0$ and $z = x_3 = 0$, respectively.
6. Show that $G^*(\boldsymbol{x}, \boldsymbol{\xi})$ of (2.104) satisfies the condition of zero potential at the boundary of a circle of radius ρ_0 centered at the origin of the coordinate system.
   Answer: Assume that at the boundary $\boldsymbol{x} = (\rho_0, \theta_0)$ and use the fact that $\cos(\theta - \theta_0) = \cos(\theta)\cos(\theta_0) + \sin(\theta)\sin(\theta_0)$.

References

Bonnet, M. 1999. *Boundary integral methods for solids and fluids*. Wiley.

Duffy, D.G. 2001. *Green's functions with applications*. Chapman and Hall.

Wrobel, L.C. 2002. *The boundary element method, Volume 1: Applications in thermo-fluids and acoustics*. Wiley.

Zemanian, A.H. 1987. *Distribution theory and transform analysis: An introduction to generalized functions, with applications*. Courier Corporation.

Chapter 3
The Direct Boundary Integral Method for Laplace's Equation

3.1 Introduction

Green's representation formula (2.61) is the starting point for our version of the direct boundary integral method for Laplace's equation. This formula expresses the potential $\phi(\boldsymbol{\xi})$ at a point $\boldsymbol{\xi}$ within a homogeneous region V in terms of integrals of the potential $\phi(\boldsymbol{x})$ and its normal derivative $q(\boldsymbol{x}) = \partial\phi(\boldsymbol{x})/\partial\boldsymbol{n}(\boldsymbol{x})$ over the boundary S of this region, as follows:

$$\phi(\boldsymbol{\xi}) = \int_S q(\boldsymbol{x})G(\boldsymbol{x},\boldsymbol{\xi})\,\mathrm{d}S(\boldsymbol{x}) - \int_S \phi(\boldsymbol{x})\frac{\partial G(\boldsymbol{x},\boldsymbol{\xi})}{\partial\boldsymbol{n}(\boldsymbol{x})}\,\mathrm{d}S(\boldsymbol{x}) \tag{3.1}$$

where $\boldsymbol{\xi} \notin S$. Recall that $G(\boldsymbol{x},\boldsymbol{\xi})$ is the fundamental solution to Laplace's equation that gives the potential at the sample point $\boldsymbol{x}$ due to a point source at $\boldsymbol{\xi}$. The function $\partial G(\boldsymbol{x},\boldsymbol{\xi})/\partial\boldsymbol{n}(\boldsymbol{x})$ is the flux across the boundary S at point $\boldsymbol{x}$ associated with the source at point $\boldsymbol{\xi}$. The two integrals represent, respectively, a single layer potential with a density function $q(\boldsymbol{x})$—the actual normal derivative of the potential across the boundary—and a double layer potential with a density function $\phi(\boldsymbol{x})$—the actual potential on the boundary.

Equation (3.1) can also be written as (see (2.62))

$$\phi(\boldsymbol{\xi}) = \int_S q(\boldsymbol{x})G(\boldsymbol{x},\boldsymbol{\xi})\,\mathrm{d}S(\boldsymbol{x}) - \int_S \phi(\boldsymbol{x})Q(\boldsymbol{x},\boldsymbol{\xi})\,\mathrm{d}S(\boldsymbol{x}) \tag{3.2}$$

in which

$$Q(\boldsymbol{x},\boldsymbol{\xi}) = \frac{\partial G(\boldsymbol{x},\boldsymbol{\xi})}{\partial\boldsymbol{n}(\boldsymbol{x})} \tag{3.3}$$

(see (2.51)). In practice, the integrals in (3.2) are evaluated by dividing the boundary contour S into separate pieces—boundary elements—and approximating $q(\boldsymbol{x})$ and $\phi(\boldsymbol{x})$ over each element by polynomials. The polynomials are fitted to a specified

S. L. Crouch and S. G. Mogilevskaya, *A First Course in Boundary Element Methods*, Mathematical Engineering, https://doi.org/10.1007/978-3-031-63341-6_3

number of points along the elements so that the potential and its normal derivative are approximated in piecewise fashion along the entire boundary.

In order to focus attention on the basic structure of the numerical approach, we will initially assume that q and ϕ are constant over each element; we will later consider a more elaborate (and more accurate) "higher order" approximation of the boundary parameters. We will also assume that the boundary S can be represented with sufficient accuracy by N straight line segments joined end to end. This approximation allows us to evaluate all of the integrals over the individual boundary elements analytically.

We present in this chapter two formulations of the direct boundary integral method for Laplace's equation in two dimensions. Both formulations use the same boundary discretization, and both are based on Green's representation formula (3.2); they differ only in the manner in which limits are taken as the source point $\boldsymbol{\xi}$ in the region V approaches the sample point $\boldsymbol{x}$ on the boundary S of that region. In the first approach, the limits are taken *after* the integrations have been performed; it is thus assumed at the outset that points $\boldsymbol{x}$ and $\boldsymbol{\xi}$ do not coincide. The second approach uses a version of Green's formula (2.99) that has been modified to account for the special terms that arise as $\boldsymbol{\xi} \to \boldsymbol{x}$ *before* any integrations have been performed. The two approaches are only subtly different and give exactly the same results for cases in which the integrations can be done analytically, which generally requires the boundary to be approximated in piecewise fashion by straight line segments. If the integrations are done numerically then the second approach must be used.[1]

3.2 Numerical Procedure

As shown in Fig. 3.1, we take the sample point $\boldsymbol{x}$ at the center of a typical element j on the discretized boundary contour. The local $\overline{x}$, $\overline{y}$ coordinate system originating at this point is chosen such that $\overline{y}$ points in the direction of the local outward normal to the boundary. We are considering a finite region V and the boundary S of this region is traversed in a clockwise direction so that $\bar{x}$ always points in the direction of traversal. The local coordinates are then defined by transformation formulas (1.34), i.e.,

$$\begin{aligned} \overline{x} &= (x - \overset{j}{x}) \cos \overset{j}{\beta} + (y - \overset{j}{y}) \sin \overset{j}{\beta} \\ \overline{y} &= -(x - \overset{j}{x}) \sin \overset{j}{\beta} + (y - \overset{j}{y}) \cos \overset{j}{\beta} \end{aligned} \tag{3.4}$$

where $\overset{j}{\beta}$ is the inclination of the element with respect to the global x direction, and where $x = \overset{j}{x}$ and $y = \overset{j}{y}$ are the global coordinates of the element midpoint. In terms

[1] This topic is discussed in Sect. 7.17 for elasticity. Similar considerations apply for potential problems.

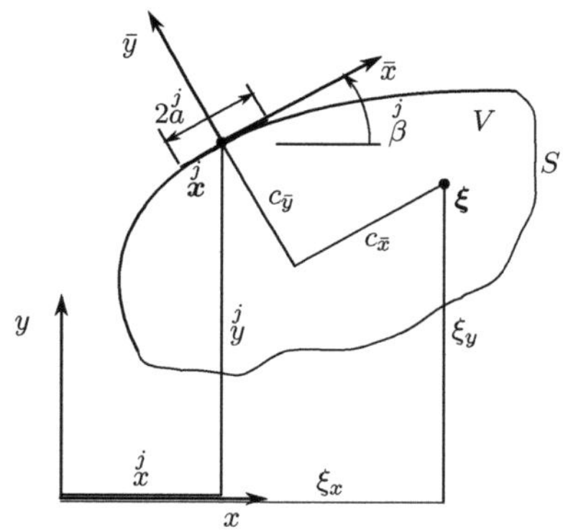

Fig. 3.1 Locations of sample point $\overset{j}{\boldsymbol{x}}$ and source point $\boldsymbol{\xi}$

of these variables the source point $\boldsymbol{\xi}$ has coordinates $\overline{x} = c_{\overline{x}}$ and $\overline{y} = c_{\overline{y}}$, and element j is the straight line segment $-\overset{j}{a} \leqslant \overline{x} \leqslant \overset{j}{a}$, $\overline{y} = 0$. We note that the parameters $c_{\overline{x}}$ and $c_{\overline{y}}$ are related to the global coordinates $x = \xi_x$ and $y = \xi_y$ of point $\boldsymbol{\xi}$ by the expressions

$$
\begin{aligned}
c_{\overline{x}} &= (\xi_x - \overset{j}{x}) \cos \overset{j}{\beta} + (\xi_y - \overset{j}{y}) \sin \overset{j}{\beta} = c_x \cos \overset{j}{\beta} + c_y \sin \overset{j}{\beta} \\
c_{\overline{y}} &= -(\xi_x - \overset{j}{x}) \sin \overset{j}{\beta} + (\xi_y - \overset{j}{y}) \cos \overset{j}{\beta} = -c_x \sin \overset{j}{\beta} + c_y \cos \overset{j}{\beta}
\end{aligned} \tag{3.5}
$$

as depicted in Fig. 3.1. Combining (3.4) and (3.5), we obtain

$$
\begin{aligned}
\overline{x} - c_{\overline{x}} &= (x - \xi_x) \cos \overset{j}{\beta} + (y - \xi_y) \sin \overset{j}{\beta} \\
\overline{y} - c_{\overline{y}} &= -(x - \xi_x) \sin \overset{j}{\beta} + (y - \xi_y) \cos \overset{j}{\beta}
\end{aligned} \tag{3.6}
$$

so that

$$
\begin{aligned}
x - \xi_x &= (\overline{x} - c_{\overline{x}}) \cos \overset{j}{\beta} - (\overline{y} - c_{\overline{y}}) \sin \overset{j}{\beta} \\
y - \xi_y &= (\overline{x} - c_{\overline{x}}) \sin \overset{j}{\beta} + (\overline{y} - c_{\overline{y}}) \cos \overset{j}{\beta}
\end{aligned} \tag{3.7}
$$

from which it easily follows that

$$
\begin{aligned}
r &= \left[(x - \xi_x)^2 + (y - \xi_y)^2\right]^{1/2} \\
&= \left[(\bar{x} - c_{\bar{x}})^2 + (\bar{y} - c_{\bar{y}})^2\right]^{1/2}
\end{aligned} \tag{3.8}
$$

It is important to emphasize that each of the N boundary elements has a unique local coordinate system associated with it, and hence has different values of $c_{\bar{x}}$ and

$c_{\bar{y}}$. Using the above definitions, we write (3.2) for the discretized boundary contour as

$$\phi(\boldsymbol{\xi}) = \sum_{j=1}^{N} \int_{-\overset{j}{a}}^{\overset{j}{a}} q(\bar{x}) G(\bar{x} - c_{\bar{x}}, \bar{y} - c_{\bar{y}})\, \mathrm{d}\bar{x} - \sum_{j=1}^{N} \int_{-\overset{j}{a}}^{\overset{j}{a}} \phi(\bar{x}) Q(\bar{x} - c_{\bar{x}}, \bar{y} - c_{\bar{y}})\, \mathrm{d}\bar{x} \tag{3.9}$$

The expression for function $G(\bar{x} - c_{\bar{x}}, \bar{y} - c_{\bar{y}})$ follows from (1.2) and (3.8):

$$G(\bar{x} - c_{\bar{x}}, \bar{y} - c_{\bar{y}}) = -\frac{1}{2\pi} \ln\left[(\bar{x} - c_{\bar{x}})^2 + (\bar{y} - c_{\bar{y}})^2\right]^{1/2} \tag{3.10}$$

The function Q is defined as the normal derivative of G at the boundary (see (3.3)). For the local $\bar{x}$, $\bar{y}$ coordinate system we are considering $\bar{y}$ is normal to the element and hence

$$\begin{aligned} Q(\bar{x} - c_{\bar{x}}, \bar{y} - c_{\bar{y}}) &= \frac{\partial}{\partial \bar{y}} G(\bar{x} - c_{\bar{x}}, \bar{y} - c_{\bar{y}}) \\ &= -\frac{1}{2\pi} \frac{\bar{y} - c_{\bar{y}}}{(\bar{x} - c_{\bar{x}})^2 + (\bar{y} - c_{\bar{y}})^2} \end{aligned} \tag{3.11}$$

We next make the assumption that $q(\bar{x})$ and $\phi(\bar{x})$ are constant along the elements, with values $\overset{j}{q}$ and $\overset{j}{\phi}$ for $j = 1$ to N. Element j is defined by the conditions $-\overset{j}{a} \leq \bar{x} \leq \overset{j}{a}$, $\bar{y} = 0$ so we set $\bar{y} = 0$ in (3.9) to obtain

$$\phi(\boldsymbol{\xi}) = \sum_{j=1}^{N} \overset{j}{q} \int_{-\overset{j}{a}}^{\overset{j}{a}} G(\bar{x} - c_{\bar{x}}, c_{\bar{y}})\, \mathrm{d}\bar{x} - \sum_{j=1}^{N} \overset{j}{\phi} \int_{-\overset{j}{a}}^{\overset{j}{a}} Q(\bar{x} - c_{\bar{x}}, c_{\bar{y}})\, \mathrm{d}\bar{x} \tag{3.12}$$

The integrations in this equation can be performed analytically and the result put in the form

$$\phi(\boldsymbol{\xi}) = \sum_{j=1}^{N} \mathbb{G}(\overset{j}{\boldsymbol{x}}, \boldsymbol{\xi}) \overset{j}{q} - \sum_{j=1}^{N} \mathbb{Q}(\overset{j}{\boldsymbol{x}}, \boldsymbol{\xi}) \overset{j}{\phi} \tag{3.13}$$

in which we have used the notation

$$\begin{aligned} \mathbb{G}(\overset{j}{\boldsymbol{x}}, \boldsymbol{\xi}) &= \int_{-\overset{j}{a}}^{\overset{j}{a}} G(\bar{x} - c_{\bar{x}}, c_{\bar{y}})\, \mathrm{d}\bar{x} \\ &= -\frac{1}{2\pi} \left[c_{\bar{y}}(\bar{\theta}_1 - \bar{\theta}_2) - (c_{\bar{x}} - \overset{j}{a}) \ln \bar{r}_1 + (c_{\bar{x}} + \overset{j}{a}) \ln \bar{r}_2 - 2\overset{j}{a}\right] \end{aligned} \tag{3.14}$$

$$
\begin{aligned}
\mathbb{Q}(\overset{j}{\boldsymbol{x}}, \boldsymbol{\xi}) &= \int_{-\overset{j}{a}}^{\overset{j}{a}} Q(\bar{x} - c_{\bar{x}}, c_{\bar{y}})\, \mathrm{d}\bar{x} \\
&= \frac{1}{2\pi}(\bar{\theta}_1 - \bar{\theta}_2)
\end{aligned}
\tag{3.15}
$$

The quantities $\bar{r}_1, \bar{r}_2$ and $\bar{\theta}_1, \bar{\theta}_2$ in these expressions give the distances and angles from the endpoints of element j to the source point $\boldsymbol{\xi}$, i.e.,

$$
\begin{aligned}
\bar{r}_1 &= \left[(c_{\bar{x}} - \overset{j}{a})^2 + c_{\bar{y}}^2\right]^{1/2} \\
\bar{r}_2 &= \left[(c_{\bar{x}} + \overset{j}{a})^2 + c_{\bar{y}}^2\right]^{1/2}
\end{aligned}
\tag{3.16}
$$

$$
\begin{aligned}
\bar{\theta}_1 &= \arctan \frac{c_{\bar{y}}}{c_{\bar{x}} - \overset{j}{a}} \\
\bar{\theta}_2 &= \arctan \frac{c_{\bar{y}}}{c_{\bar{x}} + \overset{j}{a}}
\end{aligned}
\tag{3.17}
$$

Admissible values of $\bar{\theta}_1$ and $\bar{\theta}_2$ lie between $-\pi$ and $+\pi$.

When point $\boldsymbol{\xi}$ is taken at the midpoint of a boundary element, say point $\overset{i}{\boldsymbol{x}}$, Eq. (3.13) becomes

$$
\phi(\overset{i}{\boldsymbol{x}}) = \sum_{j=1}^{N} \mathbb{G}(\overset{j}{\boldsymbol{x}}, \overset{i}{\boldsymbol{x}})\overset{j}{q} - \sum_{j=1}^{N} \mathbb{Q}(\overset{j}{\boldsymbol{x}}, \overset{i}{\boldsymbol{x}})\overset{j}{\phi} \tag{3.18}
$$

or, setting $\phi(\overset{i}{\boldsymbol{x}}) = \overset{i}{\phi}$, $\mathbb{G}(\overset{j}{\boldsymbol{x}}, \overset{i}{\boldsymbol{x}}) = \overset{ij}{\mathbb{G}}$, and $\mathbb{Q}(\overset{j}{\boldsymbol{x}}, \overset{i}{\boldsymbol{x}}) = \overset{ij}{\mathbb{Q}}$,

$$
\overset{i}{\phi} = \sum_{j=1}^{N} \overset{ij}{\mathbb{G}}\overset{j}{q} - \sum_{j=1}^{N} \overset{ij}{\mathbb{Q}}\overset{j}{\phi} \tag{3.19}
$$

(The reason why $\overset{ij}{\mathbb{G}}$ and $\overset{ij}{\mathbb{Q}}$ are defined this way is explained in the next section). Equation (3.19) can also be written as

$$
\sum_{j=1}^{N} \overset{ij}{\mathbb{G}}\overset{j}{q} = \sum_{j=1}^{N} (\overset{ij}{\mathbb{Q}} + \overset{ij}{\mathbb{I}})\overset{j}{\phi} \tag{3.20}
$$

where $\overset{ij}{\mathbb{I}}$ is the $N \times N$ identity matrix, in which the entry $\overset{ij}{\mathbb{I}} = 1$ if $i = j$ and $\overset{ij}{\mathbb{I}} = 0$ if $i \neq j$.

If we consider N possible positions for point $\overset{i}{\boldsymbol{x}}$ on the discretized boundary contour, (3.20) represents a system of N simultaneous linear algebraic equations in N

unknowns—the unspecified values of the potential or its normal derivative at the mid-points of the N boundary elements. The form of this system of equations depends on the nature of the boundary conditions for a particular problem. If, for example, the quantities $\overset{i}{\phi} = (\overset{i}{\phi})_0$ are known for all N elements, the right-hand side of the equation

$$\sum_{j=1}^{N} \overset{ij}{\mathbb{G}} \overset{j}{q} = \sum_{j=1}^{N} (\overset{ij}{\mathbb{Q}} + \overset{ij}{\mathbb{I}})(\overset{j}{\phi})_0 \tag{3.21}$$

can be evaluated directly, leading to a set of algebraic equations of the form $\mathbf{AX} = \mathbf{b}$, which can be solved by standard numerical techniques.

A similar process is used to set up the system of algebraic equations for the case that the normal derivative of the potential $\overset{i}{q} = (\overset{i}{q})_0$ is prescribed for all N boundary elements; the potentials $\overset{i}{\phi}$ are then the unknowns in the system, i.e.,

$$\sum_{j=1}^{N} (\overset{ij}{\mathbb{Q}} + \overset{ij}{\mathbb{I}}) \overset{j}{\phi} = \sum_{j=1}^{N} \overset{ij}{\mathbb{G}} (\overset{j}{q})_0 \tag{3.22}$$

For a mixed boundary value problem, the system of equations is formed by rearranging (3.20) such that the N known boundary parameters are on the right-hand side of the equals sign and the N unknown parameters are on the left.

3.3 Boundary Influence Coefficients

We have now encountered two sets of boundary influence coefficients for potential problems. One set of coefficients is for the single and double layer indirect methods in Chap. 1. Referring to (1.49) and (1.77), these coefficients, which we will temporarily designate as $\overset{ij}{\mathbb{G}}_{indirect}$ and $\overset{ij}{\mathbb{Q}}_{indirect}$, give the potential at element i due to unit constant single layer and double layer sources at element j.

The other set of boundary influence coefficients is for the direct boundary integral method. Referring to (3.19), these coefficients, which we will temporarily designate as $\overset{ij}{\mathbb{G}}_{direct}$ and $\overset{ij}{\mathbb{Q}}_{direct}$, give the potential at element i due to (a) a unit constant flux and (b) a unit constant potential over element j.

The way that these two sets of influence coefficients are constructed is depicted in Fig. 3.2. The coefficients for the direct method are developed from Eqs. (3.14) and (3.15). Coefficients for the two indirect methods are obtained using similar expressions for single and double layer potential line sources in an infinite region. These results, Eqs. (1.36) and (1.66), can be put in the form

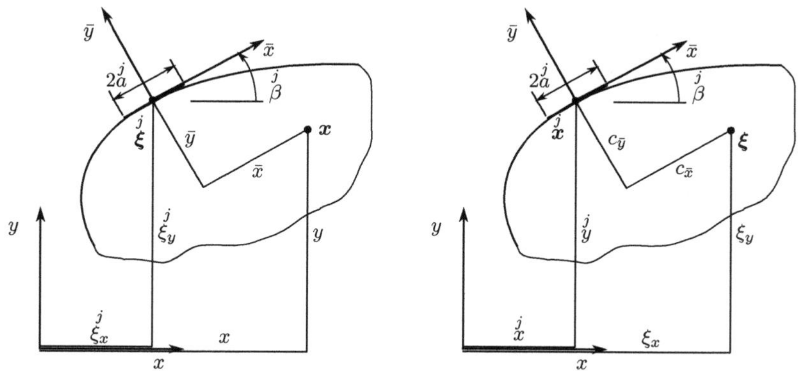

(a) Indirect formulation (b) Direct formulation

Fig. 3.2 Depiction of calculation of influence function coefficients for indirect and direct boundary element formulations

$$\mathbb{G}(\boldsymbol{x}, \overset{j}{\boldsymbol{\xi}}) = -\frac{1}{2\pi}\left[\bar{y}(\bar{\theta}_1 - \bar{\theta}_2) - (\bar{x} - \overset{j}{a})\ln\bar{r}_1 + (\bar{x} + \overset{j}{a})\ln\bar{r}_2 - 2\overset{j}{a}\right] \tag{3.23}$$

and

$$\mathbb{Q}(\boldsymbol{x}, \overset{j}{\boldsymbol{\xi}}) = \frac{1}{2\pi}(\bar{\theta}_1 - \bar{\theta}_2) \tag{3.24}$$

It can be seen from Fig. 3.2a and b that the coordinates $c_{\bar{x}}$ and $c_{\bar{y}}$ in (3.23) are geometrically the same as the coordinates $\bar{x}$ and $\bar{y}$ in (3.14), and therefore influence functions $\mathbb{G}(\overset{j}{\boldsymbol{x}}, \boldsymbol{\xi})$ and $\mathbb{G}(\boldsymbol{x}, \overset{j}{\boldsymbol{\xi}})$ are equivalent. Comparison of (3.15) and (3.24) shows that influence functions $\mathbb{Q}(\overset{j}{\boldsymbol{x}}, \boldsymbol{\xi})$ and $\mathbb{Q}(\boldsymbol{x}, \overset{j}{\boldsymbol{\xi}})$ are also equivalent.

Calculation of influence coefficients for the two indirect boundary element methods is depicted in Fig. 3.2a. The functions $\mathbb{G}(\boldsymbol{x}, \overset{j}{\boldsymbol{\xi}})$ and $\mathbb{Q}(\boldsymbol{x}, \overset{j}{\boldsymbol{\xi}})$ in (3.23) and (3.24) give the potential $\phi(\boldsymbol{x})$ at a *sample point* $\boldsymbol{x}$ inside region V due to a unit (a) single layer or (b) double layer source applied to boundary element j, which is centered at point $\overset{j}{\boldsymbol{\xi}}$. If point $\boldsymbol{x}$ is placed at the center of boundary element i, i.e., $\boldsymbol{x} = \overset{i}{\boldsymbol{\xi}}$, we obtain the boundary influence coefficients

$$\begin{aligned}\mathbb{G}(\overset{i}{\boldsymbol{\xi}}, \overset{j}{\boldsymbol{\xi}}) &= \overset{ij}{\mathbb{G}}_{indirect}\\ \mathbb{Q}(\overset{i}{\boldsymbol{\xi}}, \overset{j}{\boldsymbol{\xi}}) &= \overset{ij}{\mathbb{Q}}_{indirect}\end{aligned} \tag{3.25}$$

In the direct formulation, Fig. 3.2b, the source and sample points are interchanged. The functions $\mathbb{G}(\overset{j}{\boldsymbol{x}}, \boldsymbol{\xi})$ and $\mathbb{Q}(\overset{j}{\boldsymbol{x}}, \boldsymbol{\xi})$ in (3.14) and (3.15) represent the *effects* of a unit source at point $\boldsymbol{\xi}$ integrated over element j, which is centered at point $\overset{j}{\boldsymbol{x}}$. If point $\boldsymbol{\xi}$ is moved to point $\overset{i}{\boldsymbol{x}}$ (the midpoint of element i) we obtain the boundary influence coefficients

$$\mathbb{G}(\overset{j}{\boldsymbol{x}}, \overset{i}{\boldsymbol{x}}) = \overset{ij}{\mathbb{G}}_{direct}$$
$$\mathbb{Q}(\overset{j}{\boldsymbol{x}}, \overset{i}{\boldsymbol{x}}) = \overset{ij}{\mathbb{Q}}_{direct} \tag{3.26}$$

Now, because $\mathbb{G}(\overset{j}{\boldsymbol{x}}, \overset{i}{\boldsymbol{x}})$, $\mathbb{G}(\overset{i}{\boldsymbol{\xi}}, \overset{j}{\boldsymbol{\xi}})$ and $\mathbb{Q}(\overset{j}{\boldsymbol{x}}, \overset{i}{\boldsymbol{x}})$, $\mathbb{Q}(\overset{i}{\boldsymbol{\xi}}, \overset{j}{\boldsymbol{\xi}})$ are numerically equivalent, we conclude that

$$\overset{ij}{\mathbb{G}}_{direct} = \overset{ij}{\mathbb{G}}_{indirect} = \overset{ij}{\mathbb{G}}$$

$$\overset{ij}{\mathbb{Q}}_{direct} = \overset{ij}{\mathbb{Q}}_{indirect} = \overset{ij}{\mathbb{Q}}.$$

3.4 Computation of the Gradient of the Potential

The components of the gradient vector $q_k(\boldsymbol{\xi}) = \partial\phi(\boldsymbol{\xi})/\partial\xi_k$ at point $\boldsymbol{\xi} \neq \boldsymbol{x}$ can be calculated from the results of Sect. 3.2 in two ways. In the first, Eq. (3.12) can be used to write

$$q_x(\boldsymbol{\xi}) = \sum_{j=1}^{N} \overset{j}{q} \int_{-\overset{j}{a}}^{\overset{j}{a}} \frac{\partial}{\partial\xi_x} G(\bar{x} - c_{\bar{x}}, c_{\bar{y}})\, \mathrm{d}\bar{x} - \sum_{j=1}^{N} \overset{j}{\phi} \int_{-\overset{j}{a}}^{\overset{j}{a}} \frac{\partial}{\partial\xi_x} Q(\bar{x} - c_{\bar{x}}, c_{\bar{y}})\, \mathrm{d}\bar{x}$$
$$q_y(\boldsymbol{\xi}) = \sum_{j=1}^{N} \overset{j}{q} \int_{-\overset{j}{a}}^{\overset{j}{a}} \frac{\partial}{\partial\xi_y} G(\bar{x} - c_{\bar{x}}, c_{\bar{y}})\, \mathrm{d}\bar{x} - \sum_{j=1}^{N} \overset{j}{\phi} \int_{-\overset{j}{a}}^{\overset{j}{a}} \frac{\partial}{\partial\xi_y} Q(\bar{x} - c_{\bar{x}}, c_{\bar{y}})\, \mathrm{d}\bar{x} \tag{3.27}$$

in which we have assumed as before that the potential and its normal derivative are constant over the individual boundary elements. The quantities $G(\bar{x} - c_{\bar{x}}, c_{\bar{y}})$ and $Q(\bar{x} - c_{\bar{x}}, c_{\bar{y}})$ can be expressed in terms of ξ_x and ξ_y by means of coordinate transformation formulas (3.6) and the differentiations carried out directly. Then, the integrals can be evaluated to obtain expressions for the gradient components.

The second approach for calculating the gradient components—and the one we will pursue here—takes advantage of the fact that the integrals in (3.12) have already been evaluated. Expressions for the gradient vector components can be written from (3.13) as follows:

$$q_x(\boldsymbol{\xi}) = \sum_{j=1}^{N} \frac{\partial}{\partial \xi_x} \mathbb{G}(\overset{j}{\boldsymbol{x}}, \boldsymbol{\xi}) \overset{j}{q} - \sum_{j=1}^{N} \frac{\partial}{\partial \xi_x} \mathbb{Q}(\overset{j}{\boldsymbol{x}}, \boldsymbol{\xi}) \overset{j}{\phi}$$

$$q_y(\boldsymbol{\xi}) = \sum_{j=1}^{N} \frac{\partial}{\partial \xi_y} \mathbb{G}(\overset{j}{\boldsymbol{x}}, \boldsymbol{\xi}) \overset{j}{q} - \sum_{j=1}^{N} \frac{\partial}{\partial \xi_y} \mathbb{Q}(\overset{j}{\boldsymbol{x}}, \boldsymbol{\xi}) \overset{j}{\phi} \tag{3.28}$$

The derivatives $\partial/\partial\xi_x$ and $\partial/\partial\xi_y$ can be expressed in terms of $c_{\bar{x}}$ and $c_{\bar{y}}$ using (3.5) and the chain rule for partial differentiation, i.e.,

$$\begin{aligned} \frac{\partial}{\partial \xi_x} &= \frac{\partial c_{\bar{x}}}{\partial \xi_x} \frac{\partial}{\partial c_{\bar{x}}} + \frac{\partial c_{\bar{y}}}{\partial \xi_x} \frac{\partial}{\partial c_{\bar{y}}} \\ &= \cos \overset{j}{\beta} \frac{\partial}{\partial c_{\bar{x}}} - \sin \overset{j}{\beta} \frac{\partial}{\partial c_{\bar{y}}} \end{aligned}$$

$$\begin{aligned} \frac{\partial}{\partial \xi_y} &= \frac{\partial c_{\bar{x}}}{\partial \xi_y} \frac{\partial}{\partial c_{\bar{x}}} + \frac{\partial c_{\bar{y}}}{\partial \xi_y} \frac{\partial}{\partial c_{\bar{y}}} \\ &= \sin \overset{j}{\beta} \frac{\partial}{\partial c_{\bar{x}}} + \cos \overset{j}{\beta} \frac{\partial}{\partial c_{\bar{y}}} \end{aligned} \tag{3.29}$$

We then write

$$\frac{\partial}{\partial \xi_x} \mathbb{G}(\overset{j}{\boldsymbol{x}}, \boldsymbol{\xi}) = \frac{\partial}{\partial c_{\bar{x}}} \mathbb{G}(\overset{j}{\boldsymbol{x}}, \boldsymbol{\xi}) \cos \overset{j}{\beta} - \frac{\partial}{\partial c_{\bar{y}}} \mathbb{G}(\overset{j}{\boldsymbol{x}}, \boldsymbol{\xi}) \sin \overset{j}{\beta}$$

$$\frac{\partial}{\partial \xi_y} \mathbb{G}(\overset{j}{\boldsymbol{x}}, \boldsymbol{\xi}) = \frac{\partial}{\partial c_{\bar{x}}} \mathbb{G}(\overset{j}{\boldsymbol{x}}, \boldsymbol{\xi}) \sin \overset{j}{\beta} + \frac{\partial}{\partial c_{\bar{y}}} \mathbb{G}(\overset{j}{\boldsymbol{x}}, \boldsymbol{\xi}) \cos \overset{j}{\beta} \tag{3.30}$$

with similar expressions for $\partial \mathbb{Q}(\overset{j}{\boldsymbol{x}}, \boldsymbol{\xi})/\partial \xi_x$ and $\partial \mathbb{Q}(\overset{j}{\boldsymbol{x}}, \boldsymbol{\xi})/\partial \xi_y$.

It follows from (3.14), (3.15), and (3.30) that the gradient vector components can be written as

$$\frac{\partial}{\partial \xi_x} \mathbb{G}(\overset{j}{\boldsymbol{x}}, \boldsymbol{\xi}) = \frac{1}{2\pi} \left[\ln \frac{\bar{r}_1}{\bar{r}_2} \cos \overset{j}{\beta} + (\bar{\theta}_1 - \bar{\theta}_2) \sin \overset{j}{\beta} \right]$$

$$\frac{\partial}{\partial \xi_y} \mathbb{G}(\overset{j}{\boldsymbol{x}}, \boldsymbol{\xi}) = \frac{1}{2\pi} \left[\ln \frac{\bar{r}_1}{\bar{r}_2} \sin \overset{j}{\beta} - (\bar{\theta}_1 - \bar{\theta}_2) \cos \overset{j}{\beta} \right]$$

$$\frac{\partial}{\partial \xi_x} \mathbb{Q}(\overset{j}{\boldsymbol{x}}, \boldsymbol{\xi}) = -\frac{1}{2\pi} \left[\left(\frac{c_{\bar{y}}}{\bar{r}_1^2} - \frac{c_{\bar{y}}}{\bar{r}_2^2} \right) \cos \overset{j}{\beta} + \left(\frac{(c_{\bar{x}} - \overset{j}{a})}{\bar{r}_1^2} - \frac{(c_{\bar{x}} + \overset{j}{a})}{\bar{r}_2^2} \right) \sin \overset{j}{\beta} \right]$$

$$\frac{\partial}{\partial \xi_y} \mathbb{Q}(\overset{j}{\boldsymbol{x}}, \boldsymbol{\xi}) = -\frac{1}{2\pi} \left[\left(\frac{c_{\bar{y}}}{\bar{r}_1^2} - \frac{c_{\bar{y}}}{\bar{r}_2^2} \right) \sin \overset{j}{\beta} - \left(\frac{(c_{\bar{x}} - \overset{j}{a})}{\bar{r}_1^2} - \frac{(c_{\bar{x}} + \overset{j}{a})}{\bar{r}_2^2} \right) \cos \overset{j}{\beta} \right]. \tag{3.31}$$

3.5 Element Self Effects

The integrals (3.14) and (3.15) are well-defined everywhere in the region of interest V and its boundary S, except at the endpoints of the boundary elements, where some of the terms needed to calculate the gradient vector components exhibit singularities (see (3.31)). We will exclude these points from our analysis and consider results for the boundary *only* at the element midpoints, say points $\boldsymbol{\xi} = \overset{i}{\boldsymbol{x}}$ for $i = 1$ to N.

The case for which $\overset{i}{\boldsymbol{x}} = \overset{j}{\boldsymbol{x}}$ is of special interest because this gives the diagonal (self-effect) terms of the coefficient matrices $\overset{ij}{\mathbb{G}}$ and $\overset{ij}{\mathbb{Q}}$ in (3.20), i.e., $\overset{jj}{\mathbb{G}}$ and $\overset{jj}{\mathbb{Q}}$. These quantities are obtained by considering the limiting case in which point $\boldsymbol{\xi} \to \overset{j}{\boldsymbol{x}}$, i.e., the midpoint of the jth element. At this location we have that $c_{\bar{x}} = 0$, $c_{\bar{y}} \to 0_{-}$, $\bar{r}_1 = \bar{r}_2 = \overset{j}{a}$, $\bar{\theta}_1 = -\pi$, and $\bar{\theta}_2 = 0$, so that, from (3.14),

$$\mathbb{G}(\overset{j}{\boldsymbol{x}}, \overset{j}{\boldsymbol{x}}) = \frac{\overset{j}{a}}{\pi}(1 - \ln \overset{j}{a}) \tag{3.32}$$

and, from (3.15),

$$\mathbb{Q}(\overset{j}{\boldsymbol{x}}, \overset{j}{\boldsymbol{x}}) = -\frac{1}{2}. \tag{3.33}$$

3.6 Alternative Formulation of the Method

We saw in Sect. 2.7 that Eq. (3.1) can be put in a slightly different form by taking the limit as $\boldsymbol{\xi} \to \boldsymbol{x}$ before any integration is performed. This process extracts the "jump term" from the integral involving the quantity $\partial G(\boldsymbol{x}, \boldsymbol{\xi})/\partial \boldsymbol{n}(\boldsymbol{x})$ (or, equivalently, the quantity $Q(\boldsymbol{x}, \boldsymbol{\xi})$ in (3.3). The integral should be accurately evaluated to account for the singularity when $\boldsymbol{\xi} = \boldsymbol{x}$. The relevant equation is (2.87), repeated here for convenience:

$$\frac{1}{2}\phi(\boldsymbol{\xi}) = \int_S q(\boldsymbol{x}) G(\boldsymbol{x}, \boldsymbol{\xi}) \, \mathrm{d}S(\boldsymbol{x}) - \int_S \phi(\boldsymbol{x}) Q(\boldsymbol{x}, \boldsymbol{\xi}) \, \mathrm{d}S(\boldsymbol{x}) \tag{3.34}$$

The integrals in this expression differ from those in (3.1) only when points $\boldsymbol{\xi}$ and $\boldsymbol{x}$ coincide. Following the same approach as in Sect. 3.2 we can therefore write the following discretized version of (3.34) as

$$\frac{1}{2}\phi(\boldsymbol{\xi}) = \sum_{j=1}^{N} \mathbb{G}^*(\overset{j}{\boldsymbol{x}}, \boldsymbol{\xi}) \overset{j}{q} - \sum_{j=1}^{N} \mathbb{Q}^*(\overset{j}{\boldsymbol{x}}, \boldsymbol{\xi}) \overset{j}{\phi} \tag{3.35}$$

(cf. (3.13)), where

$$\mathbb{G}^*(\overset{j}{\boldsymbol{x}}, \boldsymbol{\xi}) = \int_{-\overset{j}{a}}^{\overset{j}{a}} G(\bar{x} - c_{\bar{x}}, c_{\bar{y}})\,\mathrm{d}\bar{x}$$

$$\mathbb{Q}^*(\overset{j}{\boldsymbol{x}}, \boldsymbol{\xi}) = \int_{-\overset{j}{a}}^{\overset{j}{a}} Q(\bar{x} - c_{\bar{x}}, c_{\bar{y}})\,\mathrm{d}\bar{x} \tag{3.36}$$

and the notation $\mathbb{G}^*$ and $\mathbb{Q}^*$ is used to distinguish these quantities from their counterparts $\mathbb{G}$ and $\mathbb{Q}$ used previously.

Going a step further, we can replace (3.19) with the result

$$\frac{1}{2}\overset{i}{\phi} = \sum_{j=1}^{N} \overset{ij}{\mathbb{G}^*}\overset{j}{q} - \sum_{j=1}^{N} \overset{ij}{\mathbb{Q}^*}\overset{j}{\phi} \tag{3.37}$$

or

$$\sum_{j=1}^{N} \overset{ij}{\mathbb{G}^*}\overset{j}{q} = \sum_{j=1}^{N} (\overset{ij}{Q^*} + \frac{1}{2}\overset{ij}{\mathbb{I}})\overset{j}{\phi} \tag{3.38}$$

The difference between this version of the direct boundary integral method for Laplace's equation and the one introduced previously lies in the way that the element self-effect terms are calculated. Integrals (3.36) are the same as (3.14) and (3.15) except when $i = j$ (i.e., $\boldsymbol{\xi} = \overset{j}{\boldsymbol{x}}$), in which case we set $c_{\bar{x}} = 0$ and $c_{\bar{y}} = 0$ and evaluate the integrals directly. The quantity $\mathbb{G}^*(\overset{j}{\boldsymbol{x}}, \overset{j}{\boldsymbol{x}}) = \overset{jj}{\mathbb{G}^*}$ is given as

$$\overset{jj}{\mathbb{G}^*} = \frac{\overset{j}{a}}{\pi}(1 - \ln \overset{j}{a}) \tag{3.39}$$

which is the same as the result obtained previously for $\overset{jj}{\mathbb{G}}$ (see (3.32)). Substituting (3.11) into the second one of (3.36) and setting $c_{\bar{x}} = 0$, $c_{\bar{y}} = 0$, and $\bar{y} = 0$ shows that

$$\mathbb{Q}^*(\overset{j}{\boldsymbol{x}}, \overset{j}{\boldsymbol{x}}) = \overset{jj}{\mathbb{Q}^*} = 0 \tag{3.40}$$

We conclude from these calculations that (3.38) is equivalent to (3.20): $\overset{ij}{\mathbb{G}^*} = \overset{ij}{\mathbb{G}}$ and $\overset{ij}{\mathbb{Q}^*} = \overset{ij}{\mathbb{Q}}$ if $i \neq j$, and $\overset{jj}{\mathbb{G}^*} = \overset{jj}{\mathbb{G}}$ and $\overset{jj}{\mathbb{Q}^*} + \frac{1}{2}\overset{jj}{\mathbb{I}} = \overset{jj}{\mathbb{Q}} + \overset{jj}{\mathbb{I}} = \frac{1}{2}$. Thus, both versions of Green's representation formula—(3.2) or (3.34)—lead to the same result.

3.7 Incorporation of Symmetry Conditions

If the geometry and boundary conditions for a problem are symmetric with respect to a line, the size of the system of algebraic equations for a boundary element model can be reduced by half using a simple imaging technique. In this technique an element on one side of the line of symmetry—the "actual" element—is paired with an image element on the other side such that the system of equations involves only those unknowns associated with the actual elements; the contributions from the image elements are built into the influence coefficient matrix. We will explain this procedure below with respect to the calculation of the boundary influence coefficients. Calculation of coefficients for arbitrary points in the region of interest proceeds in an analogous fashion.

The case for which $x = x^*$ is a line of symmetry is illustrated in Fig. 3.3. The actual element is centered at the point $x = \overset{j}{x} > x^*$, $y = \overset{j}{y}$ and is oriented at an angle $-\overset{j}{\beta}$ with respect to the positive x direction. (We assume that the region of interest is finite and that the boundary is traversed in a clockwise sense). The potential at this element is $\overset{j}{\phi}$ and its normal derivative is $\overset{j}{q}$.

The midpoint of the image element is located at the point $x = \overset{j}{x}{}'$, $y = \overset{j}{y}{}'$. This element is inclined at an angle $\overset{j}{\beta}{}' = \overset{j}{\beta}$ and has associated with it a potential $\overset{j}{\phi}{}' = \overset{j}{\phi}$ and normal derivative $\overset{j}{q}{}' = \overset{j}{q}$. The y coordinates of the two elements are equal, i.e.,

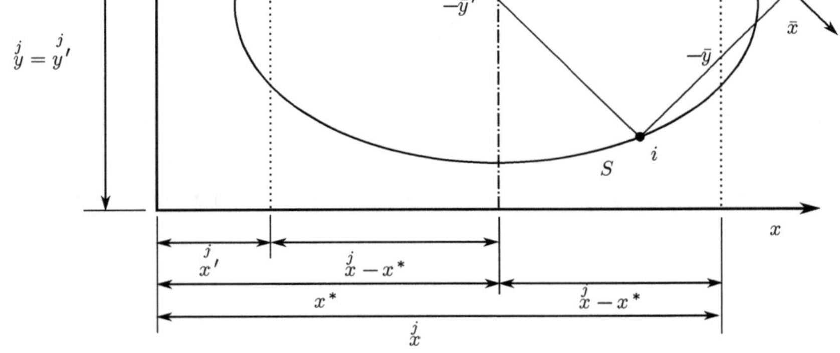

Fig. 3.3 Symmetry condition for the line $x = x^*$

$\overset{j}{y}' = \overset{j}{y}$. The x coordinate of the image element is found by noting that the actual and image elements are equidistant from the line of symmetry, $x = x^*$. Referring to Fig. 3.3 we write

$$\begin{aligned}\overset{j}{x}' &= x^* - (\overset{j}{x} - x^*)\\ &= 2x^* - \overset{j}{x}\\ \overset{j}{y}' &= \overset{j}{y}\end{aligned} \tag{3.41}$$

The effects of the image element on the potential and its normal derivative at the point $x = \overset{i}{x}$, $y = \overset{i}{y}$ can be computed in terms of the local coordinates

$$\begin{aligned}\bar{x}' &= (\overset{i}{x} - \overset{j}{x}') \cos \overset{j}{\beta}' + (\overset{i}{y} - \overset{j}{y}') \sin \overset{j}{\beta}'\\ &= (\overset{i}{x} - 2x^* + \overset{j}{x}) \cos \overset{j}{\beta} + (\overset{i}{y} - \overset{j}{y}) \sin \overset{j}{\beta}\end{aligned}$$

$$\begin{aligned}\bar{y}' &= -(\overset{i}{x} - \overset{j}{x}') \sin \overset{j}{\beta}' + (\overset{i}{y} - \overset{j}{y}') \cos \overset{j}{\beta}'\\ &= -(\overset{i}{x} - 2x^* + \overset{j}{x}) \sin \overset{j}{\beta} + (\overset{i}{y} - \overset{j}{y}) \cos \overset{j}{\beta}\end{aligned} \tag{3.42}$$

cf. (3.4). The coordinates $\bar{x}'$ and $\bar{y}'$ define the position of point i relative to the midpoint of the image element j', as shown in Fig. 3.3. Boundary influence coefficients $\overset{ij}{\mathbb{G}}$ and $\overset{ij}{\mathbb{Q}}$ that automatically incorporate the effects of the image element can be computed using the procedure outlined in Sect. 3.3; it is merely necessary to include *both* the actual and image elements in the computations.

A similar procedure can be followed for the case that $y = y^*$ is a line of symmetry, as illustrated in Fig. 3.4. The x and y coordinates of the midpoint of the image element in this case are

$$\begin{aligned}\overset{j}{x}' &= \overset{j}{x}\\ \overset{j}{y}' &= y^* - (\overset{j}{y} - y^*)\\ &= 2y^* - \overset{j}{y}\end{aligned} \tag{3.43}$$

The image element is inclined at an angle $\overset{j}{\beta}' = \overset{j}{\beta} - \pi$ from the positive x direction and again has associated with it a potential $\overset{j}{\phi}' = \overset{j}{\phi}$ and normal derivative $\overset{j}{q}' = \overset{j}{q}$. It

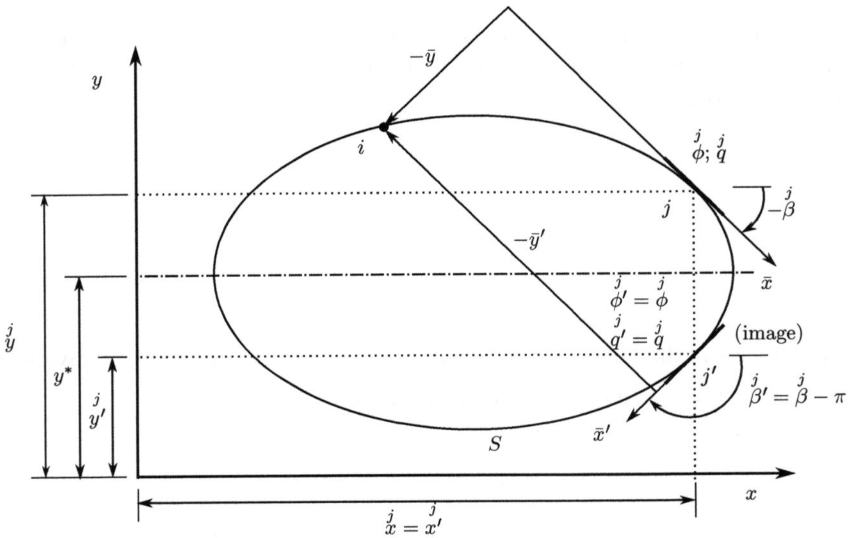

Fig. 3.4 Symmetry condition for the line $y = y^*$

is easy to show that the local coordinates of the point $x = \overset{i}{x}$, $y = \overset{i}{y}$ with respect to this element are

$$\begin{aligned}\bar{x}' &= (\overset{i}{x} - \overset{j}{x})\cos(\overset{j}{\beta} - \pi) + (\overset{i}{y} - 2y^* + \overset{j}{y})\sin(\overset{j}{\beta} - \pi)\\ &= -(\overset{i}{x} - \overset{j}{x})\cos\overset{j}{\beta} - (\overset{i}{y} - 2y^* + \overset{j}{y})\sin\overset{j}{\beta}\end{aligned}$$

$$\begin{aligned}\bar{y}' &= -(\overset{i}{x} - \overset{j}{x})\sin(\overset{j}{\beta} - \pi) + (\overset{i}{y} - 2y^* + \overset{j}{y})\cos(\overset{j}{\beta} - \pi)\\ &= (\overset{i}{x} - \overset{j}{x})\sin\overset{j}{\beta} - (\overset{i}{y} - 2y^* + \overset{j}{y})\cos\overset{j}{\beta}\end{aligned} \tag{3.44}$$

The influence coefficients $\overset{ij}{\mathbb{G}}$ and $\overset{ij}{\mathbb{Q}}$ can now be computed using the same procedure as before.

If both $x = x^*$ and $y = y^*$ are lines of symmetry the influence coefficients are computed by supplying *three* image elements. The local coordinates associated with each of the image elements can be written in the same way as discussed above.

3.8 Examples

As illustrations of the direct boundary integral method for Laplace's equation we consider again two of the examples from Chap. 1: a rectangular plate with the potential $\phi(\boldsymbol{x})$ prescribed on its four sides and an annulus with the potential $\phi(\boldsymbol{x})$ prescribed on one boundary and its normal derivative $q(\boldsymbol{x}) = \partial\phi(\boldsymbol{x})/\partial\boldsymbol{n}(\boldsymbol{x})$ on the other. The latter problem is a variation of the one considered in Chap. 1 where the potential was prescribed on both boundaries.

3.8.1 A Rectangular Plate

This problem was treated in Sect. 1.11. It was found that the indirect boundary element methods based on the single and double layer potentials gave results that were in good agreement with the analytical solution, except that the single layer model exhibited a rather large relative error near the corners of the plate (see Fig. 1.10). A similar calculation was done with the direct boundary integral method using the same parameters as before. Figure 3.5 shows a comparison of numerical results for the quantity $q_y = \partial\phi(x, y)/\partial y$ at selected points along the bottom of the plate with the analytical solution, (1.105). The accuracy of the solution for the direct boundary integral method is comparable to that for the double layer potential approach, also shown in the figure. In particular, the two models exhibit similar results in the vicinity of the corners of the plate.

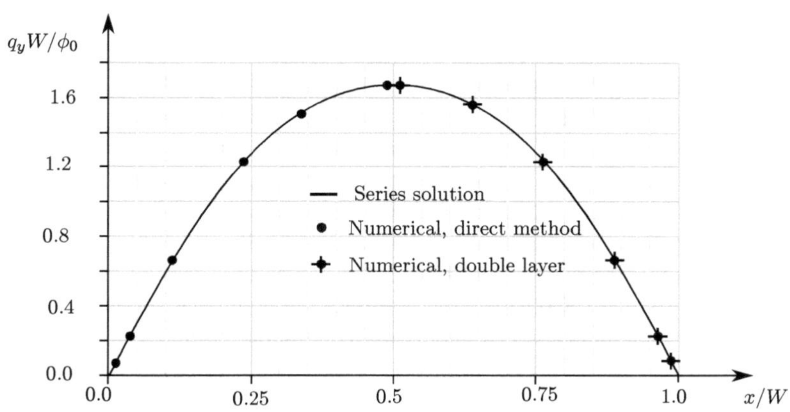

Fig. 3.5 Comparison of boundary element results with series solution for rectangular plate problem

3.8.2 An Annulus

It is easy to show using (1.106) and (1.108) that the analytical solution to the problem of an annulus with inner and outer boundaries R_{in} and R_{out} subjected to the conditions $\phi = \phi_{out}$ at $r = R_{out}$ and $q = q_{in}$ at $r = R_{in}$ is given by

$$\phi(r) = \phi_{out} + R_{in} q_{in} \ln \frac{r}{R_{out}}$$

$$q(r) = \frac{R_{in} q_{in}}{r}. \tag{3.45}$$

As was done for a similar example in Sect. 1.11, we consider the special case that $R_{in} = 1$ and $R_{out} = 2$, but now with boundary conditions $\phi_{out} = 0$ and $q_{in} = 10$. For these parameters (3.45) give $\phi(R_{in}) = -6.931$ and $q(R_{out}) = 5.000$. The corresponding values found for a boundary element model with each of the two boundaries divided into 80 straight line elements were $\phi(R_{in}) = -6.930$ and $q(R_{out}) = 4.998$.

3.9 Piecewise Linear Approximation of Boundary Parameters

A more elaborate treatment of the direct boundary integral method allows the potential and its normal derivative to vary along the individual boundary elements. We will assume in this section that $\phi(\boldsymbol{x})$ and $q(\boldsymbol{x})$ vary linearly along the elements. We could assume other "higher order" approximations of the boundary parameters—quadratic, for example—but the case of linear variations is sufficient to explain the general approach. As in the case of constant strength elements, the advantage of using straight line segments to represent the boundary is that all integrations can be carried out analytically. If the boundary elements are allowed to possess curvature the integrations must in general be done numerically. (See Sect. 7.17 in this connection.)

We note at the outset that the use of straight line segments to represent a boundary introduces a complication at the junction between two adjacent boundary elements with different inclinations. Such a connection point—called a *nodal point*—does not have a unique outward normal and hence two values of q can be associated with it, one for each element. Consider, for example, the geometry depicted in Fig. 3.6, which shows two elements, $J - 1$ and J, joined at nodal point j.

The normal derivative $\overset{J}{q}(\bar{x})$ for element J varies linearly from a value $(\overset{j}{q})_J$ at nodal point j to a value $(\overset{j+1}{q})_J$ at nodal point $j + 1$. Similarly, the normal derivative $\overset{J-1}{q}(\bar{x})$ for element $J - 1$ varies linearly from a value $(\overset{j-1}{q})_{J-1}$ at nodal point $j - 1$ to a value $(\overset{j}{q})_{J-1}$ at nodal point j. Assuming that the potential is continuous from element to element, i.e., assuming that $\overset{j}{\phi} = (\overset{j}{\phi})_{J-1} = (\overset{j}{\phi})_J$, three boundary parameters can be

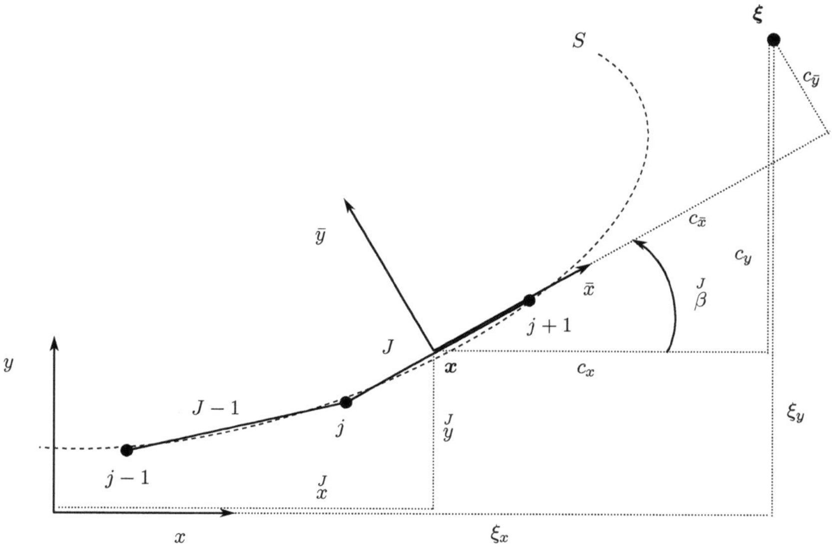

Fig. 3.6 Discretized boundary contour

associated with each node: two normal derivatives, $(\overset{j}{q})_{J-1}$ and $(\overset{j}{q})_J$, and a potential, $\overset{j}{\phi}$. Two of these three parameters must be prescribed because the system of algebraic equations developed from (3.2) allows for only one equation per nodal point.

The simplest situation occurs when the normal derivatives $(\overset{j}{q})_J$ and $(\overset{j}{q})_{J-1}$ are prescribed for $j = 1$ to N. Equations comparable to (3.22) can then be set up and solved for the unknown values of the potential $\overset{j}{\phi}$. Mixed boundary conditions for any particular nodal point can be handled in a similar fashion. If, for example, the potential $\overset{j}{\phi}$ and the normal derivative $(\overset{j}{q})_J$ are prescribed at nodal point j then $(\overset{j}{q})_{J-1}$ is unknown at that point.

An obvious difficulty arises when the potential $\overset{j}{\phi}$ is prescribed and both $(\overset{j}{q})_J$ and $(\overset{j}{q})_{J-1}$ are unknown, which is perhaps in practice the most common occurrence. A solution could be found by assuming that the normal derivatives are continuous, i.e., that $(\overset{j}{q})_J$ and $(\overset{j}{q})_{J-1}$ are equal, but this assumption is open to question when the two elements have different inclinations. This situation can be handled by introducing double nodes at the connection points between those elements where the boundary parameters are expected to be discontinuous. A convenient way of accomplishing this is to express linear variations of the potentials and their normal derivatives along elements in terms of their values at points away from the element ends, the so-called *collocation points*. This is shown in Fig. 3.7, again for elements $J - 1$ and J. Each

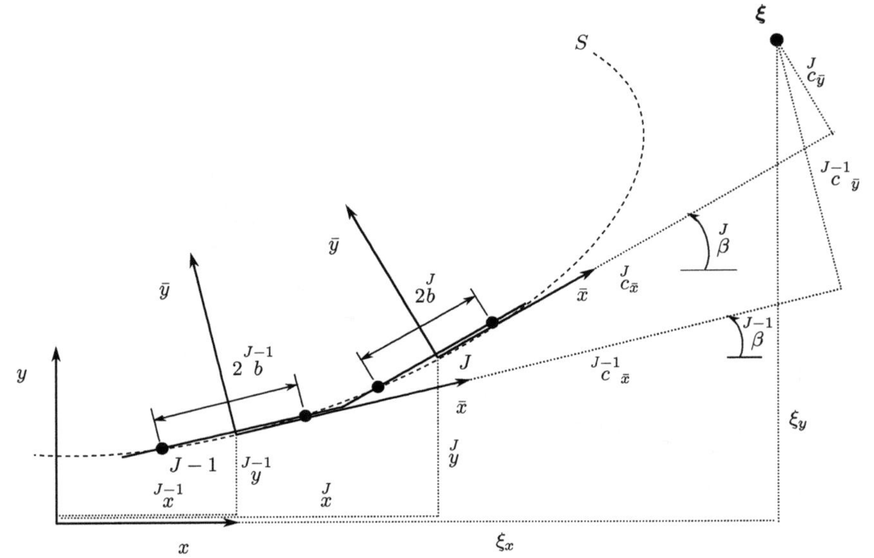

Fig. 3.7 Discretized boundary contour with discontinuous elements

element has two collocation points, $\pm \overset{J-1}{b}$ and $\pm\overset{J}{b}$, and a potential and a normal derivative are associated with each point. Because both the potentials and the normal derivatives are discontinuous at the element junctions, boundary elements of this type are called *non-conforming elements*. (The constant strength elements introduced previously are the simplest example of a non-conforming boundary element.)

A problem defined with N linear strength, non-conforming boundary elements has $2N$ collocation points and will therefore involve a system of $2N$ algebraic equations in $2N$ unknowns. We will demonstrate later that the locations of the collocation points can be taken very close to the element ends, effectively coinciding with these points and creating double nodes.

The local $\overline{x}$, $\overline{y}$ coordinate system originating at the midpoint of element J is defined by the coordinate transformation formulas

$$
\begin{aligned}
\bar{x} &= (x - \overset{J}{x}) \cos \overset{J}{\beta} + (y - \overset{J}{y}) \sin \overset{J}{\beta} \\
\bar{y} &= -(x - \overset{J}{x}) \sin \overset{J}{\beta} + (y - \overset{J}{y}) \cos \overset{J}{\beta}
\end{aligned}
\tag{3.46}
$$

where $\overset{J}{\beta}$ is the inclination of the element with respect to the global x-direction, and where $x = \overset{J}{x}$ and $y = \overset{J}{y}$ are the global coordinates of the element midpoint. As before, the source point $\boldsymbol{\xi}$ has coordinates $\bar{x} = c_{\bar{x}}$ and $\bar{y} = c_{\bar{y}}$, and element J is the straight line segment $-\overset{J}{a} \leqslant \bar{x} \leqslant \overset{J}{a}$, $\bar{y} = 0$. In the present notation, the parameters $c_{\bar{x}}$ and $c_{\bar{y}}$ are related to the global coordinates $x = \xi_x$ and $y = \xi_y$ of point $\boldsymbol{\xi}$ by the expressions

$$\begin{aligned} c_{\bar{x}} &= (\xi_x - \overset{J}{x}) \cos \overset{J}{\beta} + (\xi_y - \overset{J}{y}) \sin \overset{J}{\beta} = c_x \cos \overset{J}{\beta} + c_y \sin \overset{J}{\beta} \\ c_{\bar{y}} &= -(\xi_x - \overset{J}{x}) \sin \overset{J}{\beta} + (\xi_y - \overset{J}{y}) \cos \overset{J}{\beta} = -c_x \sin \overset{J}{\beta} + c_y \cos \overset{J}{\beta} \end{aligned} \tag{3.47}$$

The parameters $c_{\bar{x}}$ and $c_{\bar{y}}$ are particular to element J, and could as well be designated as $\overset{J}{c}_{\bar{x}}$ and $\overset{J}{c}_{\bar{y}}$ to emphasize this fact. It will usually be clear from context which element is being considered, however, so we will adopt the simpler notation for convenience. In cases where a possibility for confusion exists we will use the unabridged notation to specifically identify the element number.

Combining (3.46) and (3.47), we obtain

$$\begin{aligned} \bar{x} - c_{\bar{x}} &= (x - \xi_x) \cos \overset{J}{\beta} + (y - \xi_y) \sin \overset{J}{\beta} \\ \bar{y} - c_{\bar{y}} &= -(x - \xi_x) \sin \overset{J}{\beta} + (y - \xi_y) \cos \overset{J}{\beta} \end{aligned} \tag{3.48}$$

Using the above definitions, (3.9) becomes

$$\phi(\boldsymbol{\xi}) = \sum_{j=1}^{N} \int_{-\overset{J}{a}}^{\overset{J}{a}} \overset{J}{q}(\bar{x}) G(\bar{x} - c_{\bar{x}}, c_{\bar{y}})\, \mathrm{d}\bar{x} - \sum_{j=1}^{N} \int_{-\overset{J}{a}}^{\overset{J}{a}} \overset{J}{\phi}(\bar{x}) Q(\bar{x} - c_{\bar{x}}, c_{\bar{y}})\, \mathrm{d}\bar{x} \tag{3.49}$$

Expressions for $G(\bar{x} - c_{\bar{x}}, c_{\bar{y}})$ and $Q(\bar{x} - c_{\bar{x}}, c_{\bar{y}})$ are given in Sect. 3.2. In the present case, however, the angle $\overset{j}{\beta}$ appearing in those expressions is designated as $\overset{J}{\beta}$.

We next assume that the potential $\overset{J}{\phi}(\bar{x})$ and its normal derivative $\overset{J}{q}(\bar{x})$ along element J can be expressed as linear functions of $\bar{x}$, i.e.,

$$\begin{aligned} \overset{J}{\phi}(\bar{x}) &= \frac{1}{2}(1 - \frac{\bar{x}}{\overset{J}{b}})\phi(-\overset{J}{b}) + \frac{1}{2}(1 + \frac{\bar{x}}{\overset{J}{b}})\phi(\overset{J}{b}) \\ \overset{J}{q}(\bar{x}) &= \frac{1}{2}(1 - \frac{\bar{x}}{\overset{J}{b}})q(-\overset{J}{b}) + \frac{1}{2}(1 + \frac{\bar{x}}{\overset{J}{b}})q(\overset{J}{b}) \end{aligned} \tag{3.50}$$

If we let

$$\begin{aligned} \alpha_1(\bar{x}) &= \frac{1}{2}(1 - \frac{\bar{x}}{\overset{J}{b}}) \\ \alpha_2(\bar{x}) &= \frac{1}{2}(1 + \frac{\bar{x}}{\overset{J}{b}}) \end{aligned} \tag{3.51}$$

and set $\phi(-\overset{J}{b}) = (\overset{J}{\phi})_1$ and $\phi(\overset{J}{b}) = (\overset{J}{\phi})_2$, and $q(-\overset{J}{b}) = (\overset{J}{q})_1$ and $q(\overset{J}{b}) = (\overset{J}{q})_2$, Eq. (3.50) can be written as

$$
\begin{aligned}
\overset{J}{\phi}(\bar{x}) &= \alpha_1(\bar{x})(\overset{J}{\phi})_1 + \alpha_2(\bar{x})(\overset{J}{\phi})_2 \\
\overset{J}{q}(\bar{x}) &= \alpha_1(\bar{x})(\overset{J}{q})_1 + \alpha_2(\bar{x})(\overset{J}{q})_2
\end{aligned}
\tag{3.52}
$$

The quantities $\alpha_1(\bar{x})$ and $\alpha_2(\bar{x})$ in the above expressions are the so-called element *shape functions* for non-conforming elements. Note that, by construction, $\alpha_1 = 1$ and $\alpha_2 = 0$ for $\bar{x} = -\overset{J}{b}$, and $\alpha_1 = 0$ and $\alpha_2 = 1$ for $\bar{x} = +\overset{J}{b}$.

If definitions (3.52) are substituted into (3.49) one obtains a discretized version of Green's representation formula involving four integrals of kernel-shape function products, as follows:

$$
\begin{aligned}
\phi(\boldsymbol{\xi}) = &\sum_{J=1}^{N}\Big[(\overset{J}{q})_1 \int_{-\overset{J}{a}}^{\overset{J}{a}} G(\bar{x} - c_{\bar{x}}, c_{\bar{y}})\alpha_1(\bar{x})\, \mathrm{d}\bar{x} + (\overset{J}{q})_2 \int_{-\overset{J}{a}}^{\overset{J}{a}} G(\bar{x} - c_{\bar{x}}, c_{\bar{y}})\alpha_2(\bar{x})\, \mathrm{d}\bar{x}\Big] \\
&- \sum_{J=1}^{N}\Big[(\overset{J}{\phi})_1 \int_{-\overset{J}{a}}^{\overset{J}{a}} Q(\bar{x} - c_{\bar{x}}, c_{\bar{y}})\alpha_1(\bar{x})\, \mathrm{d}\bar{x} + (\overset{J}{\phi})_2 \int_{-\overset{J}{a}}^{\overset{J}{a}} Q(\bar{x} - c_{\bar{x}}, c_{\bar{y}})\alpha_2(\bar{x})\, \mathrm{d}\bar{x}\Big]
\end{aligned}
\tag{3.53}
$$

We therefore seek to evaluate the expressions

$$
\int_{-\overset{J}{a}}^{\overset{J}{a}} G(\bar{x} - c_{\bar{x}}, c_{\bar{y}})\alpha_m(\bar{x})\, \mathrm{d}\bar{x} \tag{3.54}
$$

and

$$
\int_{-\overset{J}{a}}^{\overset{J}{a}} Q(\bar{x} - c_{\bar{x}}, c_{\bar{y}})\alpha_m(\bar{x})\, \mathrm{d}\bar{x} \tag{3.55}
$$

where m is 1 or 2. These integrals can be evaluated analytically, as described below.

3.9.1 Integration Formulas

Evaluation of expressions (3.54) and (3.55) can be facilitated by writing the shape functions (3.51) in the form

$$
\begin{aligned}
\alpha_1(\bar{x}) &= -\frac{1}{2\overset{J}{b}}\big[(c_{\bar{x}} - \overset{J}{b}) + (\bar{x} - c_{\bar{x}})\big] \\
\alpha_2(\bar{x}) &= \frac{1}{2\overset{J}{b}}\big[(c_{\bar{x}} + \overset{J}{b}) + (\bar{x} - c_{\bar{x}})\big]
\end{aligned}
\tag{3.56}
$$

Then, letting $(\bullet)$ stand for $G(\bar{x} - c_{\bar{x}}, c_{\bar{y}})$ or $Q(\bar{x} - c_{\bar{x}}, c_{\bar{y}})$, we have

$$
\begin{aligned}
\int_{-\overset{J}{a}}^{\overset{J}{a}} (\bullet)\, \alpha_1(\bar{x})\, \mathrm{d}\bar{x} &= -\frac{1}{2\overset{J}{b}}(c_{\bar{x}} - \overset{J}{b}) \int_{-\overset{J}{a}}^{\overset{J}{a}} (\bullet)\, \mathrm{d}\bar{x} - \frac{1}{2\overset{J}{b}} \int_{-\overset{J}{a}}^{\overset{J}{a}} (\bar{x} - c_{\bar{x}})\, (\bullet)\, \mathrm{d}\bar{x} \\
\int_{-\overset{J}{a}}^{\overset{J}{a}} (\bullet)\, \alpha_2(\bar{x})\, \mathrm{d}\bar{x} &= \frac{1}{2\overset{J}{b}}(c_{\bar{x}} + \overset{J}{b}) \int_{-\overset{J}{a}}^{\overset{J}{a}} (\bullet)\, \mathrm{d}\bar{x} + \frac{1}{2\overset{J}{b}} \int_{-\overset{J}{a}}^{\overset{J}{a}} (\bar{x} - c_{\bar{x}})\, (\bullet)\, \mathrm{d}\bar{x}
\end{aligned} \tag{3.57}
$$

The first integrals on the right-hand sides of these expressions have already been evaluated (see (3.14) and (3.15)) and, in the present notation, are given as

$$
\begin{aligned}
\int_{-\overset{J}{a}}^{\overset{J}{a}} G(\bar{x} - c_{\bar{x}}, c_{\bar{y}})\, \mathrm{d}\bar{x} &= -\frac{1}{2\pi}\left[c_{\bar{y}}(\bar{\theta}_1 - \bar{\theta}_2) - (c_{\bar{x}} - \overset{J}{a}) \ln \bar{r}_1 + (c_{\bar{x}} + \overset{J}{a}) \ln \bar{r}_2 - 2\overset{J}{a}\right] \\
\int_{-\overset{J}{a}}^{\overset{J}{a}} Q(\bar{x} - c_{\bar{x}}, c_{\bar{y}})\, \mathrm{d}\bar{x} &= \frac{1}{2\pi}(\bar{\theta}_1 - \bar{\theta}_2)
\end{aligned} \tag{3.58}
$$

It should be recognized that the values of $c_{\bar{x}}$, $c_{\bar{y}}$, etc. in these results are specific to element J. The other integrals in (3.57) are found to be

$$
\begin{aligned}
\int_{-\overset{J}{a}}^{\overset{J}{a}} (\bar{x} - c_{\bar{x}}) G(\bar{x} - c_{\bar{x}}, c_{\bar{y}})\, \mathrm{d}\bar{x} &= -\frac{1}{2\pi}\left[\frac{1}{2}(\bar{r}_1^2 \ln \bar{r}_1 - \bar{r}_2^2 \ln \bar{r}_2) + \overset{J}{a} c_{\bar{x}}\right] \\
\int_{-\overset{J}{a}}^{\overset{J}{a}} (\bar{x} - c_{\bar{x}}) Q(\bar{x} - c_{\bar{x}}, c_{\bar{y}})\, \mathrm{d}\bar{x} &= \frac{1}{2\pi} c_{\bar{y}} \ln \frac{\bar{r}_1}{\bar{r}_2}
\end{aligned} \tag{3.59}
$$

Using the preceding results we write

$$
\begin{aligned}
\int_{-\overset{J}{a}}^{\overset{J}{a}} G(\bar{x} - c_{\bar{x}}, c_{\bar{y}}) \alpha_1(\bar{x})\, \mathrm{d}\bar{x} = {} & \frac{1}{4\pi \overset{J}{b}}(c_{\bar{x}} - \overset{J}{b})\left[c_{\bar{y}}(\bar{\theta}_1 - \bar{\theta}_2) - (c_{\bar{x}} - \overset{J}{a}) \ln \bar{r}_1 + (c_{\bar{x}} + \overset{J}{a}) \ln \bar{r}_2 - 2\overset{J}{a}\right] \\
& + \frac{1}{4\pi \overset{J}{b}}\left[\frac{1}{2}(\bar{r}_1^2 \ln \bar{r}_1 - \bar{r}_2^2 \ln \bar{r}_2) + \overset{J}{a} c_{\bar{x}}\right] \\
\int_{-\overset{J}{a}}^{\overset{J}{a}} G(\bar{x} - c_{\bar{x}}, c_{\bar{y}}) \alpha_2(\bar{x})\, \mathrm{d}\bar{x} = {} & -\frac{1}{4\pi \overset{J}{b}}(c_{\bar{x}} + \overset{J}{b})\left[c_{\bar{y}}(\bar{\theta}_1 - \bar{\theta}_2) - (c_{\bar{x}} - \overset{J}{a}) \ln \bar{r}_1 + (c_{\bar{x}} + \overset{J}{a}) \ln \bar{r}_2 - 2\overset{J}{a}\right] \\
& - \frac{1}{4\pi \overset{J}{b}}\left[\frac{1}{2}(\bar{r}_1^2 \ln \bar{r}_1 - \bar{r}_2^2 \ln \bar{r}_2) + \overset{J}{a} c_{\bar{x}}\right]
\end{aligned} \tag{3.60}
$$

and

$$\int_{-\overset{J}{a}}^{\overset{J}{a}} Q(\bar{x} - c_{\bar{x}}, c_{\bar{y}})\alpha_1(\bar{x})\,\mathrm{d}\bar{x} = -\frac{1}{4\pi \overset{J}{b}}\left[(c_{\bar{x}} - \overset{J}{b})(\bar{\theta}_1 - \bar{\theta}_2) + c_{\bar{y}} \ln \frac{\bar{r}_1}{\bar{r}_2}\right]$$

$$\int_{-\overset{J}{a}}^{\overset{J}{a}} Q(\bar{x} - c_{\bar{x}}, c_{\bar{y}})\alpha_2(\bar{x})\,\mathrm{d}\bar{x} = \frac{1}{4\pi \overset{J}{b}}\left[(c_{\bar{x}} + \overset{J}{b})(\bar{\theta}_1 - \bar{\theta}_2) + c_{\bar{y}} \ln \frac{\bar{r}_1}{\bar{r}_2}\right]. \tag{3.61}$$

3.9.2 Numerical Procedure

After substituting (3.60) and (3.61) into (3.53) the potential $\phi(\boldsymbol{\xi})$ at a point $\boldsymbol{\xi}$ can be written in terms of summations over the $2N$ collocation points of the N elements. If the two collocation points for element J are designated as $j = 2J - 1$ and $j + 1 = 2J$, Eq. (3.13) can be adapted for present purposes by extending the summations from $j = 1$ to $2N$ and, of course, suitably revising the expressions for the quantities $\mathbb{G}(\overset{j}{\boldsymbol{x}}, \boldsymbol{\xi})$ and $\mathbb{Q}(\overset{j}{\boldsymbol{x}}, \boldsymbol{\xi})$. Thus, we write

$$\phi(\boldsymbol{\xi}) = \sum_{j=1}^{2N} \mathbb{G}(\overset{j}{\boldsymbol{x}}, \boldsymbol{\xi})\overset{j}{q} - \sum_{j=1}^{2N} \mathbb{Q}(\overset{j}{\boldsymbol{x}}, \boldsymbol{\xi})\overset{j}{\phi} \tag{3.62}$$

Just as was done previously for constant strength elements, a special version of (3.62) can be written for the case that point $\boldsymbol{\xi}$ is taken at one of the collocation points of a boundary element. For instance, letting $\overset{i}{\boldsymbol{x}}$ be the location of one of the two collocation points for element I, we write

$$\phi(\overset{i}{\boldsymbol{x}}) = \sum_{j=1}^{2N} \mathbb{G}(\overset{j}{\boldsymbol{x}}, \overset{i}{\boldsymbol{x}})\overset{j}{q} - \sum_{j=1}^{2N} \mathbb{Q}(\overset{j}{\boldsymbol{x}}, \overset{i}{\boldsymbol{x}})\overset{j}{\phi} \tag{3.63}$$

or, in the notation of (3.20),

$$\sum_{j=1}^{2N} \overset{ij}{\mathbb{G}}\overset{j}{q} = \sum_{j=1}^{2N} (\overset{ij}{\mathbb{Q}} + \overset{ij}{\mathbb{I}})\overset{j}{\phi} \tag{3.64}$$

For a well-posed problem either the potential $\overset{i}{\phi}$ or its normal derivative $\overset{i}{q}$ will be prescribed at collocation point i so that (3.64) can be rearranged so as to represent a system of $2N$ algebraic equations in $2N$ unknowns of the form $\boldsymbol{AX} = \boldsymbol{b}$.

3.9.3 Example

As an example of the foregoing numerical procedure we consider the case of a square block of side width $2W$ containing a square hole of side width W, both centered at the origin. The boundary conditions for the problem were taken as follows: $q = -10$ on the walls of the hole and $\phi = 0$ on the sides of the block. The distribution of the normal derivative q on a side of the block ($-W \leq x \leq W,\ y = W$) is shown by the smooth curve in Fig. 3.8. This numerical solution was obtained using 50 constant strength elements along each side of both the block and the hole, i.e., 400 elements total.

Two separate sets of results for the linear variation elements are also shown in the figure, both obtained using 6 elements per side of the block and the hole. The values of q for $-W < x < 0$ were obtained by taking $b/a = \sqrt{2}/2$ while those for $0 < x < W$ were found using $b/a = 0.999$. The latter results are somewhat less accurate than those with collocation points away from the element endpoints, but both sets of results are in good agreement with the reference calculation using more than 8 times as many constant strength elements.

The results obtained using linear variations of boundary parameters suggest that the value of b is relatively unimportant and the collocation points can just as well be taken very near the element ends, thereby effectively creating double nodes at these points. The reason for this is that the function $G(\boldsymbol{x}, \boldsymbol{\xi})$ is weakly singular (cf. (2.17)). We will see later that the situation is different for elasticity, where the singularities are much stronger, and where the collocation points cannot be taken close to the element endpoints.

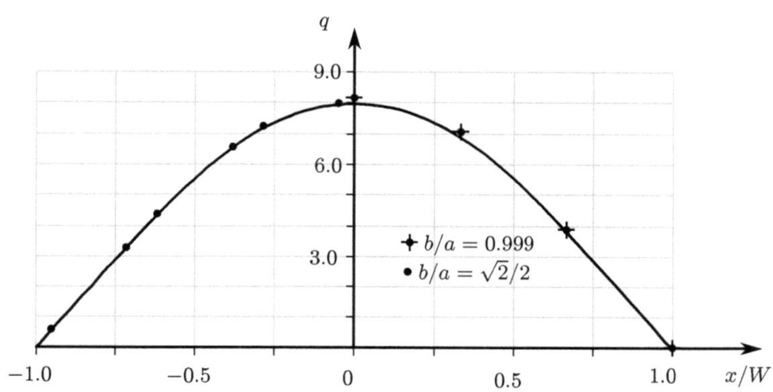

Fig. 3.8 Boundary element results for the value of q along an edge of a square block with a square hole

3.10 Formulas for the Gradient of the Potential

Equations comparable to (3.27) for the components of the gradient vector can be written from (3.53) as

$$
\begin{aligned}
q_x(\boldsymbol{\xi}) = & \sum_{J=1}^{N}\Big[(\overset{J}{q})_1 \int_{-\overset{J}{a}}^{\overset{J}{a}} \frac{\partial}{\partial \xi_x} G(\bar{x}-c_{\bar{x}}, c_{\bar{y}})\alpha_1(\bar{x})\,\mathrm{d}\bar{x} + (\overset{J}{q})_2 \int_{-\overset{J}{a}}^{\overset{J}{a}} \frac{\partial}{\partial \xi_x} G(\bar{x}-c_{\bar{x}}, c_{\bar{y}})\alpha_2(\bar{x})\,\mathrm{d}\bar{x}\Big] \\
& - \sum_{J=1}^{N}\Big[(\overset{J}{\phi})_1 \int_{-\overset{J}{a}}^{\overset{J}{a}} \frac{\partial}{\partial \xi_x} Q(\bar{x}-c_{\bar{x}}, c_{\bar{y}})\alpha_1(\bar{x})\,\mathrm{d}\bar{x} + (\overset{J}{\phi})_2 \int_{-\overset{J}{a}}^{\overset{J}{a}} \frac{\partial}{\partial \xi_x} Q(\bar{x}-c_{\bar{x}}, c_{\bar{y}})\alpha_2(\bar{x})\,\mathrm{d}\bar{x}\Big]
\end{aligned}
$$

$$
\begin{aligned}
q_y(\boldsymbol{\xi}) = & \sum_{J=1}^{N}\Big[(\overset{J}{q})_1 \int_{-\overset{J}{a}}^{\overset{J}{a}} \frac{\partial}{\partial \xi_y} G(\bar{x}-c_{\bar{x}}, c_{\bar{y}})\alpha_1(\bar{x})\,\mathrm{d}\bar{x} + (\overset{J}{q})_2 \int_{-\overset{J}{a}}^{\overset{J}{a}} \frac{\partial}{\partial \xi_y} G(\bar{x}-c_{\bar{x}}, c_{\bar{y}})\alpha_2(\bar{x})\,\mathrm{d}\bar{x}\Big] \\
& - \sum_{J=1}^{N}\Big[(\overset{J}{\phi})_1 \int_{-\overset{J}{a}}^{\overset{J}{a}} \frac{\partial}{\partial \xi_y} Q(\bar{x}-c_{\bar{x}}, c_{\bar{y}})\alpha_1(\bar{x})\,\mathrm{d}\bar{x} + (\overset{J}{\phi})_2 \int_{-\overset{J}{a}}^{\overset{J}{a}} \frac{\partial}{\partial \xi_y} Q(\bar{x}-c_{\bar{x}}, c_{\bar{y}})\alpha_2(\bar{x})\,\mathrm{d}\bar{x}\Big]
\end{aligned}
\tag{3.65}
$$

The derivatives with respect to ξ_x and ξ_y in these expressions can be taken outside the integrals and written in terms of $c_{\bar{x}}$ and $c_{\bar{y}}$ using (3.29), i.e.,

$$
\begin{aligned}
\frac{\partial}{\partial \xi_x} &= \cos\overset{J}{\beta}\frac{\partial}{\partial c_{\bar{x}}} - \sin\overset{J}{\beta}\frac{\partial}{\partial c_{\bar{y}}} \\
\frac{\partial}{\partial \xi_y} &= \sin\overset{J}{\beta}\frac{\partial}{\partial c_{\bar{x}}} + \cos\overset{J}{\beta}\frac{\partial}{\partial c_{\bar{y}}}
\end{aligned}
\tag{3.66}
$$

The derivatives needed to calculate the gradient components are listed below. First, for the terms involving $G(\bar{x}-c_{\bar{x}}, c_{\bar{y}})$ we use (3.60) to write

$$
\begin{aligned}
\frac{\partial}{\partial \xi_x}\int_{-\overset{J}{a}}^{\overset{J}{a}} G(\bar{x}-c_{\bar{x}}, c_{\bar{y}})\alpha_1(\bar{x})\,\mathrm{d}\bar{x} = & -\frac{1}{4\pi\overset{J}{b}}\Big[(\bar{\theta}_1-\bar{\theta}_2)\big\{(c_{\bar{x}}-\overset{J}{b})\sin\overset{J}{\beta} - c_{\bar{y}}\cos\overset{J}{\beta}\big\} \\
& + \ln\frac{\bar{r}_1}{\bar{r}_2}\big\{(c_{\bar{x}}-\overset{J}{b})\cos\overset{J}{\beta} + c_{\bar{y}}\sin\overset{J}{\beta}\big\} + 2\overset{J}{a}\cos\overset{J}{\beta}\Big]
\end{aligned}
$$

$$
\begin{aligned}
\frac{\partial}{\partial \xi_x}\int_{-\overset{J}{a}}^{\overset{J}{a}} G(\bar{x}-c_{\bar{x}}, c_{\bar{y}})\alpha_2(\bar{x})\,\mathrm{d}\bar{x} = & \frac{1}{4\pi\overset{J}{b}}\Big[(\bar{\theta}_1-\bar{\theta}_2)\big\{(c_{\bar{x}}+\overset{J}{b})\sin\overset{J}{\beta} - c_{\bar{y}}\cos\overset{J}{\beta}\big\} \\
& + \ln\frac{\bar{r}_1}{\bar{r}_2}\big\{(c_{\bar{x}}+\overset{J}{b})\cos\overset{J}{\beta} + c_{\bar{y}}\sin\overset{J}{\beta}\big\} + 2\overset{J}{a}\cos\overset{J}{\beta}\Big]
\end{aligned}
$$

$$
\begin{aligned}
\frac{\partial}{\partial \xi_y} \int_{-\overset{J}{a}}^{\overset{J}{a}} G(\bar{x}-c_{\bar{x}}, c_{\bar{y}})\alpha_1(\bar{x})\,\mathrm{d}\bar{x} = \frac{1}{4\pi \overset{J}{b}}\Big[&(\bar{\theta}_1-\bar{\theta}_2)\{(c_{\bar{x}}-\overset{J}{b})\cos\overset{J}{\beta}+c_{\bar{y}}\sin\overset{J}{\beta}\} \\
&-\ln\frac{\bar{r}_1}{\bar{r}_2}\{(c_{\bar{x}}-\overset{J}{b})\sin\overset{J}{\beta}-c_{\bar{y}}\cos\overset{J}{\beta}\}-2\overset{J}{a}\sin\overset{J}{\beta}\Big] \\
\frac{\partial}{\partial \xi_y} \int_{-\overset{J}{a}}^{\overset{J}{a}} G(\bar{x}-c_{\bar{x}}, c_{\bar{y}})\alpha_2(\bar{x})\,\mathrm{d}\bar{x} = -\frac{1}{4\pi \overset{J}{b}}\Big[&(\bar{\theta}_1-\bar{\theta}_2)\{(c_{\bar{x}}+\overset{J}{b})\cos\overset{J}{\beta}+c_{\bar{y}}\sin\overset{J}{\beta}\} \\
&-\ln\frac{\bar{r}_1}{\bar{r}_2}\{(c_{\bar{x}}+\overset{J}{b})\sin\overset{J}{\beta}-c_{\bar{y}}\cos\overset{J}{\beta}\}-2\overset{J}{a}\sin\overset{J}{\beta}\Big]
\end{aligned}
\tag{3.67}
$$

Similarly, for the terms involving $Q(\bar{x}-c_{\bar{x}}, c_{\bar{y}})$ we use (3.61) to find

$$
\begin{aligned}
\frac{\partial}{\partial \xi_x} \int_{-\overset{J}{a}}^{\overset{J}{a}} Q(\bar{x}-c_{\bar{x}}, c_{\bar{y}})\alpha_1(\bar{x})\,\mathrm{d}\bar{x} = -\frac{1}{4\pi \overset{J}{b}}\Big[&\Big\{(\bar{\theta}_1-\bar{\theta}_2)\cos\overset{J}{\beta}-\ln\frac{\bar{r}_1}{\bar{r}_2}\sin\overset{J}{\beta}-\frac{2\overset{J}{a}}{\bar{r}_2}\sin(\bar{\theta}_2+\overset{J}{\beta})\Big\} \\
&-(\overset{J}{a}-\overset{J}{b})\Big\{\frac{1}{\bar{r}_1}\sin(\bar{\theta}_1+\overset{J}{\beta})-\frac{1}{\bar{r}_2}\sin(\bar{\theta}_2+\overset{J}{\beta})\Big\}\Big] \\
\frac{\partial}{\partial \xi_x} \int_{-\overset{J}{a}}^{\overset{J}{a}} Q(\bar{x}-c_{\bar{x}}, c_{\bar{y}})\alpha_2(\bar{x})\,\mathrm{d}\bar{x} = \frac{1}{4\pi \overset{J}{b}}\Big[&\Big\{(\bar{\theta}_1-\bar{\theta}_2)\cos\overset{J}{\beta}-\ln\frac{\bar{r}_1}{\bar{r}_2}\sin\overset{J}{\beta}-\frac{2\overset{J}{a}}{\bar{r}_1}\sin(\bar{\theta}_1+\overset{J}{\beta})\Big\} \\
&+(\overset{J}{a}-\overset{J}{b})\Big\{\frac{1}{\bar{r}_1}\sin(\bar{\theta}_1+\overset{J}{\beta})-\frac{1}{\bar{r}_2}\sin(\bar{\theta}_2+\overset{J}{\beta})\Big\}\Big] \\
\frac{\partial}{\partial \xi_y} \int_{-\overset{J}{a}}^{\overset{J}{a}} Q(\bar{x}-c_{\bar{x}}, c_{\bar{y}})\alpha_1(\bar{x})\,\mathrm{d}\bar{x} = -\frac{1}{4\pi \overset{J}{b}}\Big[&\Big\{(\bar{\theta}_1-\bar{\theta}_2)\sin\overset{J}{\beta}+\ln\frac{\bar{r}_1}{\bar{r}_2}\cos\overset{J}{\beta}+\frac{2\overset{J}{a}}{\bar{r}_2}\cos(\bar{\theta}_2+\overset{J}{\beta})\Big\} \\
&+(\overset{J}{a}-\overset{J}{b})\Big\{\frac{1}{\bar{r}_1}\cos(\bar{\theta}_1+\overset{J}{\beta})-\frac{1}{\bar{r}_2}\cos(\bar{\theta}_2+\overset{J}{\beta})\Big\}\Big] \\
\frac{\partial}{\partial \xi_y} \int_{-\overset{J}{a}}^{\overset{J}{a}} Q(\bar{x}-c_{\bar{x}}, c_{\bar{y}})\alpha_2(\bar{x})\,\mathrm{d}\bar{x} = +\frac{1}{4\pi \overset{J}{b}}\Big[&\Big\{(\bar{\theta}_1-\bar{\theta}_2)\sin\overset{J}{\beta}+\ln\frac{\bar{r}_1}{\bar{r}_2}\cos\overset{J}{\beta}+\frac{2\overset{J}{a}}{\bar{r}_1}\cos(\bar{\theta}_1+\overset{J}{\beta})\Big\} \\
&-(\overset{J}{a}-\overset{J}{b})\Big\{\frac{1}{\bar{r}_1}\cos(\bar{\theta}_1+\overset{J}{\beta})-\frac{1}{\bar{r}_2}\cos(\bar{\theta}_2+\overset{J}{\beta})\Big\}\Big]
\end{aligned}
\tag{3.68}
$$

After a numerical solution to a given boundary value problem has been found, the values of the potential and its normal derivatives at both collocation points of element J will be known. Equations (3.65), (3.67), and (3.68) can then be evaluated to obtain the components of the gradient vector $q_x(\boldsymbol{\xi})$ and $q_y(\boldsymbol{\xi})$ at points $\boldsymbol{\xi}$ within the region of interest. The potential $\phi(\boldsymbol{\xi})$ at these same points can be found using (3.53), (3.60), and (3.61).

3.11 Incorporation of Material Properties

We previously mentioned that boundary value problems of potential theory describe important physical phenomena such as steady-state heat conduction, groundwater flow, electrostatics, and more. In all such problems the region of interest is characterized by some material-specific parameter pertinent to the phenomenon under study. If the material is homogeneous (its properties do not vary with position) and isotropic (its properties are independent of direction) this parameter is a constant. For example, in steady-state heat conduction for a homogeneous, isotropic domain the relationship between the component of the heat flux density vector q_i and the temperature gradient $\partial\phi/\partial x_i$ is governed by Fourier's law:

$$q_i = -k\frac{\partial \phi}{\partial x_i} \tag{3.69}$$

where k is the thermal conductivity. Notice that the definition for q_i in (3.69) differs from that used previously (see (1.15)) by the presence of a multiplicative constant, $-k$. It is worth mentioning that (3.69) is also used for steady-state groundwater flow and is known as Darcy's law for that application. For groundwater flow, q_i is a component of the specific discharge vector, k is the hydraulic conductivity, and ϕ is the hydraulic head.

The component of the heat flux density vector q_i normal to a boundary is called the (heat) *flux*. In the present context the flux is defined for a point $\boldsymbol{x}$ on a boundary as

$$q^*(\boldsymbol{x}) = -k\frac{\partial \phi(\boldsymbol{x})}{\partial \boldsymbol{n}(\boldsymbol{x})} \tag{3.70}$$

The symbol $q^*(\boldsymbol{x})$ is used for the flux to distinguish it from the symbol used previously to denote the normal derivative of the potential, $q(\boldsymbol{x}) = \partial\phi(\boldsymbol{x})/\partial\boldsymbol{n}(\boldsymbol{x})$. We therefore have that

$$q^*(\boldsymbol{x}) = -kq(\boldsymbol{x}) \tag{3.71}$$

The governing equation for potential ϕ is found by using the continuity equation (see, for example, Strack 1989)

$$\frac{\partial q_x}{\partial x} + \frac{\partial q_y}{\partial y} = 0 \tag{3.72}$$

and (3.69), which gives

$$\frac{\partial^2 \phi}{\partial x^2} + \frac{\partial^2 \phi}{\partial y^2} = 0 \tag{3.73}$$

Because the potential for an isotropic material satisfies Laplace's equation, Green's representation formula can be used directly:

$$\phi(\boldsymbol{\xi}) = \int\limits_S q(\boldsymbol{x})G(\boldsymbol{x},\boldsymbol{\xi})\,\mathrm{d}S(\boldsymbol{x}) - \int_S \phi(\boldsymbol{x})Q(\boldsymbol{x},\boldsymbol{\xi})\,\mathrm{d}S(\boldsymbol{x}) \tag{3.74}$$

Therefore, an analysis for a homogeneous, isotropic material can be carried out without reference to the material constant k. After a problem has been solved and the potential $\phi(\boldsymbol{\xi})$ is known throughout region V, Fourier's law (for example) can be used to find the heat flux density vector with component $q_i = -k\partial\phi(\boldsymbol{\xi})/\xi_i$ at selected points $\boldsymbol{\xi}$.

3.12 Piecewise Homogeneous Regions I

In some applications, the region of interest V might contain several finite subregions, each represented by its own material parameter k. Such a region is called *piecewise homogeneous* and we consider next the particular case for which V is composed of two subdomains, V_1 and V_2, as depicted in Fig. 3.9. The material constants for the two regions are k_1 and k_2 and the boundary contours, S_1 for subregion V_1 and S_2 for subregion V_2, are traversed in a clockwise sense. The local $\bar{x}$, $\bar{y}$ coordinates associated with the two contours then appear as shown in the figure. The common portions of the two contours define the *interface* between the two subregions. The local coordinates $(\bar{x})_1$, $(\bar{y})_1$ and $(\bar{x})_2$, $(\bar{y})_2$ are oppositely directed all along the interface, i.e., $(\bar{x})_1 = -(\bar{x})_2$ and $(\bar{y})_1 = -(\bar{y})_2$.

A boundary value problem for the piecewise homogeneous body of Fig. 3.9 is defined by prescribing either the potential (the temperature) $\phi^{[p]}(\boldsymbol{x})$ or the flux $q^{*[p]}(\boldsymbol{x}) = -k_p\partial\phi^{[p]}(\boldsymbol{x})/\partial\boldsymbol{n}^{[p]}(\boldsymbol{x})$ on the "free" portions of the boundary contour S_p, where $p = 1$ or 2, as well as by *continuity conditions* for the interface between

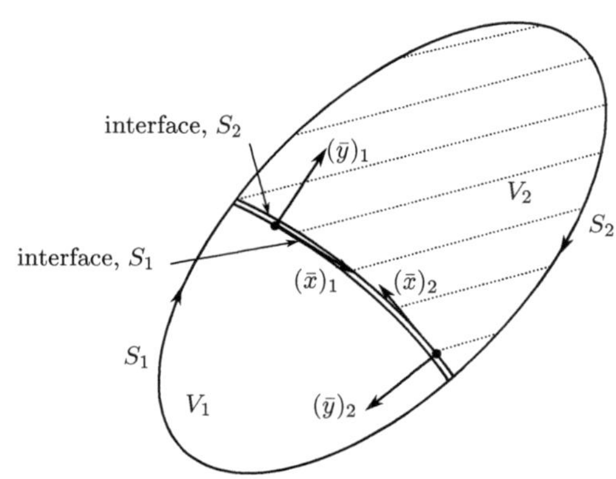

Fig. 3.9 Piecewise homogeneous body consisting of two homogeneous, isotropic subregions

the two subregions. The latter conditions can be written as

$$\phi^{[1]}(\boldsymbol{x}) = \phi^{[2]}(\boldsymbol{x}) \tag{3.75}$$

for the potentials and

$$k_1 \frac{\partial \phi^{[1]}(\boldsymbol{x})}{\partial \boldsymbol{n}^{[1]}(\boldsymbol{x})} = -k_2 \frac{\partial \phi^{[2]}(\boldsymbol{x})}{\partial \boldsymbol{n}^{[2]}(\boldsymbol{x})} \tag{3.76}$$

or

$$q^{*[1]}(\boldsymbol{x}) = -q^{*[2]}(\boldsymbol{x}) \tag{3.77}$$

for the fluxes. The minus signs in (3.76) and (3.77) are a consequence of the opposite directions of the local coordinates along the interface.

With this terminology, Green's representation formula (3.74) for the two subregions can be written as

$$\begin{aligned}
\phi^{[1]}(\boldsymbol{\xi}) &= -\frac{1}{k_1}\int_{S_1} q^{*[1]}(\boldsymbol{x}) G(\boldsymbol{x},\boldsymbol{\xi})\,\mathrm{d}S_1(\boldsymbol{x}) - \int_{S_1} \phi^{[1]}(\boldsymbol{x}) Q(\boldsymbol{x},\boldsymbol{\xi})\,\mathrm{d}S_1(\boldsymbol{x}) \\
\phi^{[2]}(\boldsymbol{\xi}) &= -\frac{1}{k_2}\int_{S_2} q^{*[2]}(\boldsymbol{x}) G(\boldsymbol{x},\boldsymbol{\xi})\,\mathrm{d}S_2(\boldsymbol{x}) - \int_{S_2} \phi^{[2]}(\boldsymbol{x}) Q(\boldsymbol{x},\boldsymbol{\xi})\,\mathrm{d}S_2(\boldsymbol{x})
\end{aligned} \tag{3.78}$$

If we assume that the boundary parameters are piecewise constant along $N_1 + N_2 = N$ straight line elements on the two boundary contours, (3.78) can be discretized in the same way as was done in Sect. 3.2 for (3.2), leading directly to the expressions

$$-\frac{1}{k_1}\sum_{j=1}^{N_1} \overset{ij}{\mathbb{G}} (\overset{j}{q})^{*[1]} = \sum_{j=1}^{N_1} (\overset{ij}{\mathbb{Q}} + \overset{ij}{\mathbb{I}})(\overset{j}{\phi})^{[1]} \tag{3.79}$$

where i ranges from 1 to N_1, and

$$-\frac{1}{k_2}\sum_{j=N_1+1}^{N} \overset{ij}{\mathbb{G}} (\overset{j}{q})^{*[2]} = \sum_{j=N_1+1}^{N} (\overset{ij}{\mathbb{Q}} + \overset{ij}{\mathbb{I}})(\overset{j}{\phi})^{[2]} \tag{3.80}$$

where i ranges from $N_1 + 1$ to N. Elements along the two sides of the interface must match exactly.[2] The influence coefficients $\overset{ij}{\mathbb{G}}$ and $\overset{ij}{\mathbb{Q}}$ in (3.79) and (3.80) do not depend on the material constants k_1 and k_2; they are given by (3.14) and (3.15) with $\boldsymbol{\xi} = \overset{i}{\boldsymbol{x}}$.

Equations (3.79) and (3.80) are N relations involving $2N$ boundary parameters: the potential ϕ and the flux q^* at each of the N boundary elements. For an element

[2] Matching elements will have the same x, y coordinates but their orientations will differ by π.

on a free portion of one of the boundary contours one of these parameters will be specified as a boundary condition, leaving the other as an unknown. Both boundary parameters are unknown for an element on the interface, but by using continuity conditions (3.75) and (3.76) we can associate one parameter with an element on one side of the interface and the other with a matching element on the other side. Equations (3.79) and (3.80) can therefore be used to set up the following system of N algebraic equations in N unknowns:

$$\sum_{j=1}^{N} \overset{ij}{A}\overset{j}{X} = \overset{i}{b} \qquad i = 1 \text{ to } N \tag{3.81}$$

The quantities $\overset{i}{b}$ in these equations are certain linear combinations of the known boundary parameters, and $\overset{ij}{A}$ are the appropriate boundary influence coefficients associated with the unknown boundary parameters $\overset{j}{X}$.

As an illustration of the procedure used to construct the system (3.81) from Eqs. (3.79) and (3.80), we will suppose that the flux $(\overset{j}{q})^{*[1]} = \overset{j}{q}_0$ is specified on the free portion of contour S_1 and the potential $(\overset{j}{\phi})^{[2]} = \overset{j}{\phi}_0$ is specified on the free portion of contour S_2. Then, if we treat the potential $\phi^{[1]}$ as the unknown boundary parameter on the S_1 side of the interface and the flux $q^{*[2]}$ as the unknown parameter on the S_2 side, the quantity $\overset{j}{X}$ will be defined as follows:

$$\overset{j}{X} = \begin{cases} (\overset{j}{\phi})^{[1]} & j = 1 \text{ to } N_1 \\ \\ (\overset{j}{q})^{*[2]} & j = N_1 + 1 \text{ to } N \end{cases} \tag{3.82}$$

If there are M boundary elements on each side of the interface, then, by virtue of our assumptions about the boundary conditions for the other elements, the flux $(\overset{j}{q})^{*[1]}$ is known for $j = 1$ to $N_1 - M$ and the potential $(\overset{j}{\phi})^{[2]}$ is known for $j = N_1 + 1 + M$ to N. Using (3.79) for all elements of contour S_1 ($i = 1$ to N_1) and (3.80) for all elements of contour S_2 ($i = N_1 + 1$ to N), the quantities $\overset{i}{b}$ in (3.81) can be written as

$$\overset{i}{b} = \frac{1}{k_1} \sum_{j=1}^{N_1 - M} \overset{ij}{\mathbb{G}} \overset{j}{q}_0 \tag{3.83}$$

for $i = 1$ to $N_1 - M$ and

$$\overset{i}{b} = \sum_{j=N_1+1+M}^{N} (\overset{ij}{\mathbb{Q}} + \overset{ij}{\mathbb{I}})\overset{j}{\phi}_0 \tag{3.84}$$

for $i = N_1 + M + 1$ to N. For elements along the interface, $i = N_1 - M + 1$ to $N_1 + M$, the value of $\overset{i}{b}$ is zero because the interface potentials and fluxes are unknown (they are included in (3.82)).

The coefficients $\overset{ij}{A}$ for the ith equation of the system (3.81) are found by considering the boundary conditions and interface continuity conditions along the contours S_1 and S_2. According to definitions (3.82), the potentials $(\overset{j}{\phi})^{[1]}$ are the unknown boundary parameters along contour S_1 (where $i \leq N_1$) and the fluxes $(\overset{j}{q})^{*[2]}$ are the unknown boundary parameters along contour S_2 (where $N_1 + 1 \leq i \leq N$). In setting up the system (3.81) from (3.79) and (3.80), therefore, we must eliminate reference to the fluxes $(\overset{j}{q})^{*[1]}$ along the S_1 side of the interface (where $j = N_1 - M + 1$ to N_1), and also to the potential $(\overset{j}{\phi})^{[2]}$ along the S_2 side (where $j = N_1 + 1$ to $N_1 + M$). This is achieved by using the interface continuity conditions.

An element j on the S_1 side of the interface is necessarily connected to another element j' on the S_2 side. Continuity conditions (3.77) for the flux can then be written as

$$(\overset{j}{q})^{*[1]} = -(\overset{j'}{q})^{*[2]} \tag{3.85}$$

Similarly, if element j is on the S_2 side of the interface and the matching element j' is on the S_1 side, continuity conditions (3.75) for the potential are

$$(\overset{j}{\phi})^{[2]} = (\overset{j'}{\phi})^{[1]} \tag{3.86}$$

It is now a simple matter to specify the coefficients $\overset{ij}{A}$ in (3.81). Combining (3.79), (3.82), and (3.85) we have, for $i \leq N_1$,

$$\overset{ij}{A} = \begin{cases} -\left(\overset{ij}{\mathbb{Q}} + \overset{ij}{\mathbb{I}}\right) & j \leq N_1 \\ -\dfrac{1}{k_1}\overset{ij'}{\mathbb{G}} & N_1 + 1 \leq j \leq N_1 + M \\ 0 & N_1 + M + 1 \leq j \leq N \end{cases} \tag{3.87}$$

The values of the coefficients for $N_1 + 1 \leq i \leq N$ are found by combining (3.80), (3.82), and (3.86):

$$\overset{ij}{A} = \begin{cases} 0 & j \leq N_1 - M \\ (\overset{ij'}{\mathbb{Q}} + \overset{ij'}{\mathbb{I}}) & N_1 - M + 1 \leq j \leq N_1 \\ -\dfrac{1}{k_2}\overset{ij}{\mathbb{G}} & N_1 + 1 \leq j \leq N \end{cases} \tag{3.88}$$

Finally, we note that the potential $\phi(\boldsymbol{\xi})$ and the gradient components $q_x(\boldsymbol{\xi})$ and $q_y(\boldsymbol{\xi})$ for points $\boldsymbol{\xi}$ inside either one of the two subregions can be obtained by using (3.13) and (3.28), noting that the values of $\overset{j}{q}$ in these expressions are given by $\overset{j}{q} = -(\overset{j}{q})^{*[1]}/k_1$ for elements in subregion V_1 and $\overset{j}{q} = -(\overset{j}{q})^{*[2]}/k_2$ for those in subregion V_2.

3.12.1 Example

A simple example of a boundary value problem for a piecewise homogeneous body is shown in Fig. 3.10. The region of interest consists of an annulus $a \leq r \leq b$ bonded to another annulus $b \leq r \leq c$; the inner annulus has conductivity k_1 and the outer one has conductivity k_2. The potential (temperature) is ϕ_{in} at the inner boundary ($r = a$) and ϕ_{out} at the outer boundary ($r = c$). The solution to this problem, satisfying continuity of both the potential and the flux at the interface $r = b$, can be constructed using (1.106) for each annulus. Setting

$$C = \frac{(\phi_{out} - \phi_{in})}{k_2 \ln b/a + k_1 \ln c/b}$$

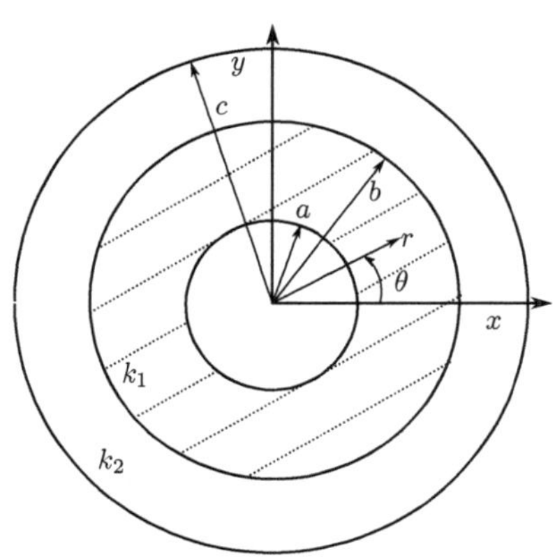

Fig. 3.10 A bimaterial annulus

expressions for the potential and flux are as follows:

$$\phi(r) = \begin{cases} \phi_{in} + k_2 C \ln \dfrac{r}{a} & a \le r \le b \\ \\ \phi_{in} + k_2 C \ln \dfrac{b}{a} + k_1 C \ln \dfrac{r}{b} & b \le r \le c \end{cases}$$

$$q(r) = -k_1 k_2 C \frac{1}{r} \quad a \le r \le c \tag{3.89}$$

In deriving the expression for the flux we used the relations $q(r) = -k_p \partial\phi(r)/\partial r$ for each of the two subregions.

A numerical solution to this problem was obtained for the case that $a = 10$, $b = 20$, and $c = 30$ (arbitrary length units), with $k_1 = 1$, $k_2 = 5$, $\phi_{in} = 10$, and $\phi_{out} = 0$. The inner and outer boundaries and the two sides of the interface between $\theta = 0$ and $\theta = \pi/2$ were each divided into 25 elements, i.e., 100 elements total, and symmetry conditions were imposed for both the x and y directions. The numerical results were in close agreement with the analytical solution; the computed values of the unspecified boundary parameters were $q(a) = 1.2918$, $\phi(b) = 1.0474$, $q(b) = 0.6459$, and $q(c) = 0.4306$ compared with $q(a) = 1.2916$, $\phi(b) = 1.0480$, $q(b) = 0.6458$, and $q(c) = 0.4305$ for the analytical solution.

3.13 Piecewise Homogeneous Regions II

Another approach for piecewise homogeneous regions can be developed from (3.78) by assuming that $\boldsymbol{\xi}$ is inside region V_1 so that $\phi^{[2]}(\boldsymbol{\xi}) = 0$, a result that follows from (2.63). If we let L_1 refer to the portion of boundary contour S_1 that represents the interface, with L_2 defined similarly for boundary contour S_2, (3.78) can be written as

$$\begin{aligned} \phi^{[1]}(\boldsymbol{\xi}) = &-\frac{1}{k_1}\int_{S_1-L_1} q^{*[1]}(\boldsymbol{x})G(\boldsymbol{x},\boldsymbol{\xi})\,\mathrm{d}S_1(\boldsymbol{x}) - \int_{S_1-L_1} \phi^{[1]}(\boldsymbol{x})Q(\boldsymbol{x},\boldsymbol{\xi})\,\mathrm{d}S_1(\boldsymbol{x}) \\ &-\frac{1}{k_1}\int_{L_1} q^{*[1]}(\boldsymbol{x})G(\boldsymbol{x},\boldsymbol{\xi})\,\mathrm{d}S_1(\boldsymbol{x}) - \int_{L_1} \phi^{[1]}(\boldsymbol{x})Q(\boldsymbol{x},\boldsymbol{\xi})\,\mathrm{d}S_1(\boldsymbol{x}) \\ 0 = &-\frac{1}{k_2}\int_{S_2-L_2} q^{*[2]}(\boldsymbol{x})G(\boldsymbol{x},\boldsymbol{\xi})\,\mathrm{d}S_2(\boldsymbol{x}) - \int_{S_2-L_2} \phi^{[2]}(\boldsymbol{x})Q(\boldsymbol{x},\boldsymbol{\xi})\,\mathrm{d}S_2(\boldsymbol{x}) \\ &-\frac{1}{k_2}\int_{L_2} q^{*[2]}(\boldsymbol{x})G(\boldsymbol{x},\boldsymbol{\xi})\,\mathrm{d}S_2(\boldsymbol{x}) - \int_{L_2} \phi^{[2]}(\boldsymbol{x})Q(\boldsymbol{x},\boldsymbol{\xi})\,\mathrm{d}S_2(\boldsymbol{x}) \end{aligned} \tag{3.90}$$

If we multiply the first equation by k_1 and the second by k_2 and add the results we obtain

$$\begin{aligned}
k_1\phi^{[1]}(\boldsymbol{\xi}) = &-\int_{S_1-L_1} q^{*[1]}(\boldsymbol{x})G(\boldsymbol{x},\boldsymbol{\xi})\,\mathrm{d}S_1(\boldsymbol{x}) - k_1\int_{S_1-L_1}\phi^{[1]}(\boldsymbol{x})Q(\boldsymbol{x},\boldsymbol{\xi})\,\mathrm{d}S_1(\boldsymbol{x})\\
&-\int_{S_2-L_2} q^{*[2]}(\boldsymbol{x})G(\boldsymbol{x},\boldsymbol{\xi})\,\mathrm{d}S_2(\boldsymbol{x}) - k_2\int_{S_2-L_2}\phi^{[2]}(\boldsymbol{x})Q(\boldsymbol{x},\boldsymbol{\xi})\,\mathrm{d}S_2(\boldsymbol{x})\\
&-\left\{\int_{L_1} q^{*[1]}(\boldsymbol{x})G(\boldsymbol{x},\boldsymbol{\xi})\,\mathrm{d}S_1(\boldsymbol{x}) + \int_{L_2} q^{*[2]}(\boldsymbol{x})G(\boldsymbol{x},\boldsymbol{\xi})\,\mathrm{d}S_2(\boldsymbol{x})\right\}\\
&-\left\{k_1\int_{L_1}\phi^{[1]}(\boldsymbol{x})Q(\boldsymbol{x},\boldsymbol{\xi})\,\mathrm{d}S_1(\boldsymbol{x}) + k_2\int_{L_2}\phi^{[2]}(\boldsymbol{x})Q(\boldsymbol{x},\boldsymbol{\xi})\,\mathrm{d}S_2(\boldsymbol{x})\right\}
\end{aligned} \tag{3.91}$$

The sum of the two integrals in the first set of curly brackets is zero by virtue of (3.77). The two integrals in the second set of curly brackets can be combined by using (3.75) and noting that the values of $Q(\boldsymbol{x},\boldsymbol{\xi})$ on L_2 are the negatives of those of $Q(\boldsymbol{x},\boldsymbol{\xi})$ on L_1 because $\boldsymbol{n}^{[2]}(\boldsymbol{x}) = -\boldsymbol{n}^{[1]}(\boldsymbol{x})$. We therefore have that

$$\begin{aligned}
k_1\phi^{[1]}(\boldsymbol{\xi}) = &-\int_{S_1-L_1} q^{*[1]}(\boldsymbol{x})G(\boldsymbol{x},\boldsymbol{\xi})\,\mathrm{d}S_1(\boldsymbol{x}) - k_1\int_{S_1-L_1}\phi^{[1]}(\boldsymbol{x})Q(\boldsymbol{x},\boldsymbol{\xi})\,\mathrm{d}S_1(\boldsymbol{x})\\
&-\int_{S_2-L_2} q^{*[2]}(\boldsymbol{x})G(\boldsymbol{x},\boldsymbol{\xi})\,\mathrm{d}S_2(\boldsymbol{x}) - k_2\int_{S_2-L_2}\phi^{[2]}(\boldsymbol{x})Q(\boldsymbol{x},\boldsymbol{\xi})\,\mathrm{d}S_2(\boldsymbol{x})\\
&-(k_1-k_2)\int_{L_1}\phi^{[1]}(\boldsymbol{x})Q(\boldsymbol{x},\boldsymbol{\xi})\,\mathrm{d}S_1(\boldsymbol{x})
\end{aligned} \tag{3.92}$$

for $\boldsymbol{\xi} \in V_1$.

If we again consider the case that the flux $q^{*[1]}(\boldsymbol{x})$ is prescribed on $S_1 - L_1$ and the potential (temperature) $\phi^{[2]}(\boldsymbol{x})$ is prescribed on $S_2 - L_2$, (3.92) can be rearranged as follows:

$$\begin{aligned}
k_1\phi^{[1]}(\boldsymbol{\xi}) = &-\left\{k_1\int_{S_1-L_1}\phi^{[1]}(\boldsymbol{x})Q(\boldsymbol{x},\boldsymbol{\xi})\,\mathrm{d}S_1(\boldsymbol{x}) + (k_1-k_2)\int_{L_1}\phi^{[1]}(\boldsymbol{x})Q(\boldsymbol{x},\boldsymbol{\xi})\,\mathrm{d}S_1(\boldsymbol{x})\right.\\
&\left.\quad+\int_{S_2-L_2} q^{*[2]}(\boldsymbol{x})G(\boldsymbol{x},\boldsymbol{\xi})\,\mathrm{d}S_2(\boldsymbol{x})\right\}\\
&-\left\{\int_{S_1-L_1} q^{*[1]}(\boldsymbol{x})G(\boldsymbol{x},\boldsymbol{\xi})\,\mathrm{d}S_1(\boldsymbol{x}) + k_2\int_{S_2-L_2}\phi^{[2]}(\boldsymbol{x})Q(\boldsymbol{x},\boldsymbol{\xi})\,\mathrm{d}S_2(\boldsymbol{x})\right\}
\end{aligned} \tag{3.93}$$

The first set of terms on the right of the equals sign represent the unknown quantities—the potential $\phi^{[1]}(\boldsymbol{x})$ on S_1 and the flux $q^{*[2]}(\boldsymbol{x})$ on $S_2 - L_2$—while the second set are the prescribed values of the flux $q^{*[1]}(\boldsymbol{x})$ on $S_1 - L_1$ and potential $\phi^{[2]}(\boldsymbol{x})$ on $S_2 - L_2$. This equation can be discretized in the usual way and used to set up the system (3.81). After the unknown boundary parameters have been determined, values of $\phi^{[1]}(\boldsymbol{\xi})$ anywhere in V_1, including the boundary $S_1 - L_1$ and the interface L_1, can be calculated using the discretized version of (3.93).

Values of the potential $\phi^{[2]}(\boldsymbol{\xi})$ can be found by repeating the preceding analysis with the assumption that point $\boldsymbol{\xi}$ is inside region V_2 instead of region V_1. It is easy to show that a similar version of (3.93) also holds for the quantity $k_2\phi^{[2]}(\boldsymbol{\xi})$, i.e.,

$$
\begin{aligned}
k_2\phi^{[2]}(\boldsymbol{\xi}) = &-\left\{k_1 \int_{S_1-L_1} \phi^{[1]}(\boldsymbol{x})Q(\boldsymbol{x},\boldsymbol{\xi})\,\mathrm{d}S_1(\boldsymbol{x}) + (k_1-k_2)\int_{L_1} \phi^{[1]}(\boldsymbol{x})Q(\boldsymbol{x},\boldsymbol{\xi})\,\mathrm{d}S_1(\boldsymbol{x}) \right. \\
&\left. + \int_{S_2-L_2} q^{*[2]}(\boldsymbol{x})G(\boldsymbol{x},\boldsymbol{\xi})\,\mathrm{d}S_2(\boldsymbol{x})\right\} \\
&-\left\{\int_{S_1-L_1} q^{*[1]}(\boldsymbol{x})G(\boldsymbol{x},\boldsymbol{\xi})\,\mathrm{d}S_1(\boldsymbol{x}) + k_2\int_{S_2-L_2} \phi^{[2]}(\boldsymbol{x})Q(\boldsymbol{x},\boldsymbol{\xi})\,\mathrm{d}S_2(\boldsymbol{x})\right\}
\end{aligned}
\tag{3.94}
$$

for $\boldsymbol{\xi} \in V_2$. Discretization of this equation as before directly provides the desired result because all of the boundary parameters are known.

The foregoing solution procedure does not directly provide values of the flux $q^{*[1]}$ at points along L_1, but these values can be found by calculating the gradient components q_x and q_y at these points using the procedure described in Sect. 3.4 and then finding the component normal to an element on the interface using the formula $q_{\bar{y}}^{[1]} = -q_x^{[1]} \sin\beta + q_y^{[1]} \cos\beta$ (cf. (1.37)) . The flux associated with gradient component $q_{\bar{y}}^{[1]}$ is $q^{*[1]} = -k_1 q_{\bar{y}}^{[1]}$; continuity condition (3.77) provides the flux $q^{*[2]}(\boldsymbol{x})$ on L_2.

3.14 Anisotropy

In some applications the material parameter k is direction-dependent. For example, a soil consisting of alternating thin layers of sand and clay has different hydraulic conductivities in directions parallel and perpendicular to the layers. We will show in this section that such an anisotropic material can be analyzed using a modification of the method already given for isotropic materials.

The general case of an anisotropic material is represented by a hydraulic (or thermal) conductivity tensor with components $k_{ij} = k_{ji}$ and (3.69) is replaced by

$$
q_i = -k_{ij}\frac{\partial\phi}{\partial x_j}
\tag{3.95}
$$

in which the summation convention applies to repeated index j. We will consider the two-dimensional case where indices i and j are x or y, so that, with $k_{xy} = k_{yx}$, (3.95) becomes

$$q_x = -k_{xx}\frac{\partial \phi}{\partial x} - k_{xy}\frac{\partial \phi}{\partial y}$$
$$q_y = -k_{xy}\frac{\partial \phi}{\partial x} - k_{yy}\frac{\partial \phi}{\partial y} \tag{3.96}$$

The governing equation for the potential ϕ is found by using the continuity Eqs. (3.72) and (3.96), which gives

$$k_{xx}\frac{\partial^2 \phi}{\partial x^2} + 2k_{xy}\frac{\partial^2 \phi}{\partial x \partial y} + k_{yy}\frac{\partial^2 \phi}{\partial y^2} = 0 \tag{3.97}$$

or, in index notation,

$$k_{ij}\phi_{,ij} = 0 \tag{3.98}$$

The conductivities in directions parallel (the component $k_{\bar{x}}$) and perpendicular (the component $k_{\bar{y}}$) to a line inclined at an angle β from the x axis are given as

$$k_{\bar{x}} = -\frac{1}{2}(k_{xx} - k_{yy}) \sin 2\beta + k_{xy} \cos 2\beta$$
$$k_{\bar{y}} = k_{xx} \sin^2 \beta - k_{xy} \sin 2\beta + k_{yy} \cos^2 \beta \tag{3.99}$$

These expressions are similar to ones derived in Sect. 4.2 for the stresses acting on a line segment of arbitrary orientation with respect to the x, y coordinate axes.

Some simplification in the analysis is achieved by choosing the x and y coordinate axes to coincide with the principal axes of conductivity, which can always be done by means of a simple coordinate transformation. Then, setting $k_{xy} = k_{yx} = 0$ and letting $k_{xx} = k_1$ and $k_{yy} = k_2$, where k_1 and k_2 are respectively the major and minor conductivities,[3] (3.96), (3.97), and (3.99) become

$$q_x = -k_1\frac{\partial \phi}{\partial x}$$
$$q_y = -k_2\frac{\partial \phi}{\partial y} \tag{3.100}$$

$$k_{\bar{x}} = -(k_1 - k_2) \sin \beta \cos \beta$$
$$k_{\bar{y}} = k_1 \sin^2 \beta + k_2 \cos^2 \beta \tag{3.101}$$

[3] For notational convenience we use the same symbols k_1 and k_2 for the principal conductivities as we used for the conductivities of subregions V_1 and V_2 in Sects. 3.12 and 3.13.

$$k_1 \frac{\partial^2 \phi}{\partial x^2} + k_2 \frac{\partial^2 \phi}{\partial y^2} = 0 \tag{3.102}$$

Now, we define scaled x and y coordinates

$$\begin{aligned} \hat{x} &= \frac{1}{\sqrt{k_1}} x \\ \hat{y} &= \frac{1}{\sqrt{k_2}} y \end{aligned} \tag{3.103}$$

such that

$$\begin{aligned} \frac{\partial}{\partial x} &= \frac{\partial \hat{x}}{\partial x} \frac{\partial}{\partial \hat{x}} = \frac{1}{\sqrt{k_1}} \frac{\partial}{\partial \hat{x}} \\ \frac{\partial}{\partial y} &= \frac{\partial \hat{y}}{\partial y} \frac{\partial}{\partial \hat{y}} = \frac{1}{\sqrt{k_2}} \frac{\partial}{\partial \hat{y}} \end{aligned} \tag{3.104}$$

Combining (3.102) and (3.104) we find

$$\frac{\partial^2 \phi}{\partial \hat{x}^2} + \frac{\partial^2 \phi}{\partial \hat{y}^2} = 0 \tag{3.105}$$

so that the potential ϕ for an anisotropic material is harmonic in *scaled coordinates* $\hat{x}$ and $\hat{y}$.

3.14.1 Fundamental Solution

The fundamental solution $G^*(\boldsymbol{x}, \boldsymbol{\xi})$ for an anisotropic material can be found in the same way as was done in Chap. 2, but now in terms of scaled coordinates. Proceeding as in Sect. 2.2 for a harmonic function f representing a point source at the origin, we write Laplace's Eq. (2.33) in terms of a scaled radial coordinate $\hat{r}$ as

$$\frac{1}{\hat{r}} \frac{\mathrm{d}}{\mathrm{d}\hat{r}} \left(\hat{r} \frac{\mathrm{d}f}{\mathrm{d}\hat{r}} \right) = 0 \tag{3.106}$$

where, from (3.103),

$$\begin{aligned} \hat{r} &= \sqrt{\hat{x}^2 + \hat{y}^2} \\ &= \frac{1}{\sqrt{k_1 k_2}} \sqrt{k_2 x^2 + k_1 y^2} \end{aligned} \tag{3.107}$$

By analogy with (2.37), the solution to (3.106) is

$$f = \frac{f_0}{\ln \hat{r}_0} \ln \hat{r} \tag{3.108}$$

The flux components $q_x^* = -k_1 \partial f / \partial x$ and $q_y^* = -k_2 \partial f / \partial y$ associated with this potential are

$$\begin{aligned}
q_x^* &= -k_1 \frac{f_0}{\ln \hat{r}_0} \frac{\partial}{\partial x} \ln \hat{r} \\
&= -\frac{f_0}{\ln \hat{r}_0} \frac{x}{x^2/k_1 + y^2/k_2} \\
q_y^* &= -k_2 \frac{f_0}{\ln \hat{r}_0} \frac{\partial}{\partial y} \ln \hat{r} \\
&= -\frac{f_0}{\ln \hat{r}_0} \frac{y}{x^2/k_1 + y^2/k_2}
\end{aligned} \tag{3.109}$$

and the flux q^* normal to a boundary is

$$q^* = q_x^* n_x + q_y^* n_y \tag{3.110}$$

For a circular boundary surrounding the origin we have that $n_x = -\cos\theta = -x/r$ and $n_y = -\sin\theta = -y/r$ (the outward normal points toward the origin), and, combining (3.109) and (3.110) gives

$$q^* = \frac{f_0}{\ln \hat{r}_0} \frac{r}{x^2/k_1 + y^2/k_2} \tag{3.111}$$

Because the direction of the flux defined by (3.111) is opposite to that of the outward normal to the boundary, we require the total flux over the circular boundary to be of negative unit magnitude, i.e.,

$$\int_{-\pi}^{\pi} q^* r \, \mathrm{d}\theta = -1 \tag{3.112}$$

so that

$$\begin{aligned}
\frac{\ln \hat{r}_0}{f_0 k_1 k_2} &= -\int_{-\pi}^{\pi} \frac{1}{k_2 \cos^2\theta + k_1 \sin^2\theta} \, \mathrm{d}\theta \\
&= -\frac{1}{\sqrt{k_1 k_2}} \arctan\left(\sqrt{\frac{k_1}{k_2}} \tan\theta \right) \Bigg|_{-\pi}^{\pi}
\end{aligned} \tag{3.113}$$

(see, for example, Petit Bois 1961). The limits of integration are chosen as $\pm\pi$ to account for the branch cut for the arctangent function along the negative x axis. The function $\arctan(\sqrt{k_1/k_2}\tan\theta)$ is equivalent to $\arctan(\sqrt{k_1/k_2}\,y/x)$ and the conditions $\theta = \pi$ and $\theta = -\pi$ are equivalent to $x < 0,\ y = 0_+$ and $x < 0,\ y = 0_-$, so (3.113) gives the result

$$\frac{f_0}{\ln \hat{r}_0} = -\frac{1}{2\pi\sqrt{k_1 k_2}} \tag{3.114}$$

Therefore, referring to (3.108), for a source at point $\boldsymbol{\xi}$ we write the fundamental solution as

$$G^*(\boldsymbol{x}, \boldsymbol{\xi}) = -\frac{1}{2\pi\sqrt{k_1 k_2}} \ln \hat{r} \tag{3.115}$$

where

$$\hat{r} = \left[(\hat{x} - \hat{\xi}_x)^2 + (\hat{y} - \hat{\xi}_y)^2\right]^{1/2} \tag{3.116}$$

in which $\hat{\xi}_x$ and $\hat{\xi}_y$ are the scaled coordinates of point $\boldsymbol{\xi}$, i.e.,

$$\begin{aligned} \hat{\xi}_x &= \frac{1}{\sqrt{k_1}}\xi_x \\ \hat{\xi}_y &= \frac{1}{\sqrt{k_2}}\xi_y \end{aligned} \tag{3.117}$$

In the more general case the fundamental solution is given by (Chang et al. 1973)

$$G^*(\boldsymbol{x}, \boldsymbol{\xi}) = -\frac{1}{2\pi\sqrt{\mid k_{ij} \mid}} \ln \hat{r} \tag{3.118}$$

in which $\mid k_{ij} \mid$ is the determinant of the matrix with components k_{ij} and

$$\hat{r} = \left[k^{ij}(x_i - \xi_i)(x_j - \xi_j)\right]^{1/2} \tag{3.119}$$

where k^{ij} is the inverse of conductivity matrix k_{ij}. The fundamental solution satisfies the equation

$$k_{ij} G^*_{,ij}(\boldsymbol{x}, \boldsymbol{\xi}) = -\delta(\boldsymbol{x} - \boldsymbol{\xi}). \tag{3.120}$$

3.14.2 Representation Formula for an Anisotropic Material

The governing equation of steady state heat conduction for an anisotropic material is given by Eq. (3.98):

$$k_{ij}\phi_{,ij}(\boldsymbol{x}) = 0 \tag{3.121}$$

Because the function $\phi(\boldsymbol{x})$ is not harmonic, Green's representation formula used previously for an isotropic material is not valid, and a new relation must be found. As shown below, this can be accomplished using the *method of weighted residuals*, which consists in multiplying both sides of (3.121) by a weight function and integrating the resulting equation over the domain of interest. Taking $G^*(\boldsymbol{x}, \boldsymbol{\xi})$ as the weight function we have

$$\int_V k_{ij}\phi_{,ij}(\boldsymbol{x})G^*(\boldsymbol{x}, \boldsymbol{\xi})\,\mathrm{d}V(\boldsymbol{x}) = 0 \tag{3.122}$$

Equation (3.122) can be written as

$$\int_V k_{ij}\phi_{,ij}(\boldsymbol{x})G^*(\boldsymbol{x}, \boldsymbol{\xi})\,\mathrm{d}V(\boldsymbol{x}) = \int_V k_{ij}\left[\phi_{,i}(\boldsymbol{x})G^*(\boldsymbol{x}, \boldsymbol{\xi})\right]_{,j}\,\mathrm{d}V(\boldsymbol{x}) - \int_V k_{ij}\phi_{,i}(\boldsymbol{x})G^*_{,j}(\boldsymbol{x}, \boldsymbol{\xi})\,\mathrm{d}V(\boldsymbol{x}) = 0 \tag{3.123}$$

or, using the divergence theorem (see Sect. 2.1.3),

$$\int_V k_{ij}\phi_{,ij}(\boldsymbol{x})G^*(\boldsymbol{x}, \boldsymbol{\xi})\,\mathrm{d}V(\boldsymbol{x}) = \int_S k_{ij}\phi_{,i}(\boldsymbol{x})G^*(\boldsymbol{x}, \boldsymbol{\xi})\,n_j(\boldsymbol{x})\,\mathrm{d}S(\boldsymbol{x}) - \int_V k_{ij}\phi_{,i}(\boldsymbol{x})G^*_{,j}(\boldsymbol{x}, \boldsymbol{\xi})\,\mathrm{d}V(\boldsymbol{x}) = 0 \tag{3.124}$$

The second integral on the right-hand side of this expression can be written as

$$\int_V k_{ij}\phi_{,i}(\boldsymbol{x})G^*_{,j}(\boldsymbol{x}, \boldsymbol{\xi})\,\mathrm{d}V(\boldsymbol{x}) = \int_V k_{ij}\left[\phi(\boldsymbol{x})G^*_{,j}(\boldsymbol{x}, \boldsymbol{\xi})\right]_{,i}\,\mathrm{d}V(\boldsymbol{x}) - \int_V k_{ij}\phi(\boldsymbol{x})G^*_{,ij}(\boldsymbol{x}, \boldsymbol{\xi})\,\mathrm{d}V(\boldsymbol{x}) \tag{3.125}$$

Taking into account (3.120) and (2.29) and using the divergence theorem again, we find that

$$\int_V k_{ij}\phi_{,i}(\boldsymbol{x})G^*_{,j}(\boldsymbol{x},\boldsymbol{\xi})\,\mathrm{d}V(\boldsymbol{x}) = \phi(\boldsymbol{\xi}) + \int_S k_{ij}\phi(\boldsymbol{x})G^*_{,j}(\boldsymbol{x},\boldsymbol{\xi})\,n_i(\boldsymbol{x})\,\mathrm{d}S(\boldsymbol{x}) \quad (3.126)$$

We therefore have that

$$\int_V k_{ij}\phi_{,ij}(\boldsymbol{x})G^*(\boldsymbol{x},\boldsymbol{\xi})\,\mathrm{d}V(\boldsymbol{x}) = \int_S k_{ij}\phi_{,i}(\boldsymbol{x})G^*(\boldsymbol{x},\boldsymbol{\xi})\,n_j(\boldsymbol{x})\,\mathrm{d}S(\boldsymbol{x}) - \phi(\boldsymbol{\xi}) - \int_S k_{ij}\phi(\boldsymbol{x})G^*_{,j}(\boldsymbol{x},\boldsymbol{\xi})\,n_i(\boldsymbol{x})\,\mathrm{d}S(\boldsymbol{x}) = 0 \quad (3.127)$$

or

$$\phi(\boldsymbol{\xi}) = \int_S k_{ij}\phi_{,i}(\boldsymbol{x})G^*(\boldsymbol{x},\boldsymbol{\xi})\,n_j(\boldsymbol{x})\,\mathrm{d}S(\boldsymbol{x}) - \int_S k_{ij}\phi(\boldsymbol{x})G^*_{,j}(\boldsymbol{x},\boldsymbol{\xi})\,n_i(\boldsymbol{x})\,\mathrm{d}S(\boldsymbol{x}) \quad (3.128)$$

We next recall that the normal flux is defined as

$$q(\boldsymbol{x}) = q_1(\boldsymbol{x})n_1(\boldsymbol{x}) + q_2(\boldsymbol{x})n_2(\boldsymbol{x}) = q_j(\boldsymbol{x})n_j(\boldsymbol{x}) \quad (3.129)$$

where

$$q_j(\boldsymbol{x}) = -k_{ij}\phi_{,i}(\boldsymbol{x}) \quad (3.130)$$

and introduce the following expressions

$$\begin{aligned} q(\boldsymbol{x}) &= -k_{ij}\phi_{,i}(\boldsymbol{x})n_j(\boldsymbol{x}) \\ Q^*(\boldsymbol{x},\boldsymbol{\xi}) &= -k_{ij}G^*_{,j}(\boldsymbol{x},\boldsymbol{\xi})n_i(\boldsymbol{x}) = -k_{ij}G^*_{,i}(\boldsymbol{x},\boldsymbol{\xi})n_j(\boldsymbol{x}) \end{aligned} \quad (3.131)$$

We then rewrite (3.128) as

$$\phi(\boldsymbol{\xi}) = -\int_S q(\boldsymbol{x})G^*(\boldsymbol{x},\boldsymbol{\xi})\,\mathrm{d}S(\boldsymbol{x}) + \int_S \phi(\boldsymbol{x})Q^*(\boldsymbol{x},\boldsymbol{\xi})\,\mathrm{d}S(\boldsymbol{x}) \quad (3.132)$$

which is called the *representation formula* for differential Eq. (3.121).

For the case that $k_{12} = k_{21} = 0$, $k_{11} = k_1$, and $k_{22} = k_2$ (the case we will consider in detail below) the fundamental solution (3.118) is

$$G^*(\boldsymbol{x},\boldsymbol{\xi}) = -\frac{1}{2\pi\sqrt{k_1k_2}}\ln\hat{r} \quad (3.133)$$

Further,

$$q(\boldsymbol{x}) = -k_1\frac{\partial\phi(\boldsymbol{x})}{\partial x}n_x(\boldsymbol{x}) - k_2\frac{\partial\phi(\boldsymbol{x})}{\partial y}n_y(\boldsymbol{x}) \quad (3.134)$$

and

$$Q^*(\boldsymbol{x}, \boldsymbol{\xi}) = -k_1 \frac{\partial G^*(\boldsymbol{x}, \boldsymbol{\xi})}{\partial x} n_x(\boldsymbol{x}) - k_2 \frac{\partial G^*(\boldsymbol{x}, \boldsymbol{\xi})}{\partial y} n_y(\boldsymbol{x}). \tag{3.135}$$

3.14.3 Numerical Procedure

Discretization of (3.132) proceeds in the same way as described in Sect. 3.2 for a closed boundary contour divided into N straight line segments. Now, however, the relationship (3.7) between the local coordinates for an element, $\bar{x} - c_{\bar{x}}$ and $\bar{y} - c_{\bar{y}}$, and the global coordinates, $x - \xi_x$ and $y - \xi_y$, must be expressed in terms of scaled coordinates

$$\begin{aligned} \hat{x} - \hat{\xi}_x &= \frac{1}{\sqrt{k_1}}(\bar{x} - c_{\bar{x}}) \cos \overset{j}{\beta} - \frac{1}{\sqrt{k_1}}(\bar{y} - c_{\bar{y}}) \sin \overset{j}{\beta} \\ \hat{y} - \hat{\xi}_y &= \frac{1}{\sqrt{k_2}}(\bar{x} - c_{\bar{x}}) \sin \overset{j}{\beta} + \frac{1}{\sqrt{k_2}}(\bar{y} - c_{\bar{y}}) \cos \overset{j}{\beta} \end{aligned} \tag{3.136}$$

Substitution of these expressions into (3.116) gives

$$\begin{aligned} \hat{r} = \frac{1}{\sqrt{k_1 k_2}} \Big[&(k_2 \cos^2 \overset{j}{\beta} + k_1 \sin^2 \overset{j}{\beta})(\bar{x} - c_{\bar{x}})^2 + 2(k_1 - k_2) \sin \overset{j}{\beta} \cos \overset{j}{\beta} (\bar{x} - c_{\bar{x}})(\bar{y} - c_{\bar{y}}) \\ &+ (k_2 \sin^2 \overset{j}{\beta} + k_1 \cos^2 \overset{j}{\beta})(\bar{y} - c_{\bar{y}})^2 \Big]^{1/2} \end{aligned} \tag{3.137}$$

This expression can be written more compactly by setting

$$\begin{aligned} C_1 &= \frac{1}{k_1 k_2}(k_2 \cos^2 \overset{j}{\beta} + k_1 \sin^2 \overset{j}{\beta}) \\ C_2 &= \frac{1}{k_1 k_2}(k_1 - k_2) \sin \overset{j}{\beta} \cos \overset{j}{\beta} \\ C_3 &= \frac{1}{k_1 k_2}(k_2 \sin^2 \overset{j}{\beta} + k_1 \cos^2 \overset{j}{\beta}) \end{aligned} \tag{3.138}$$

so that

$$\hat{r} = \Big[C_1(\bar{x} - c_{\bar{x}})^2 + 2C_2(\bar{x} - c_{\bar{x}})(\bar{y} - c_{\bar{y}}) + C_3(\bar{y} - c_{\bar{y}})^2 \Big]^{1/2} \tag{3.139}$$

It then follows from (3.133) that

$$G^*(\bar{x}-c_{\bar{x}},\bar{y}-c_{\bar{y}}) = -\frac{1}{2\pi\sqrt{k_1k_2}}\ln\Big[C_1(\bar{x}-c_{\bar{x}})^2+2C_2(\bar{x}-c_{\bar{x}})(\bar{y}-c_{\bar{y}})+C_3(\bar{y}-c_{\bar{y}})^2\Big]^{1/2} \tag{3.140}$$

The function $Q^*(\bar{x}-c_{\bar{x}},\bar{y}-c_{\bar{y}})$ is defined from (3.135) as

$$Q^*(\bar{x}-c_{\bar{x}},\bar{y}-c_{\bar{y}}) = -k_1\frac{\partial G^*}{\partial x}(\bar{x}-c_{\bar{x}},\bar{y}-c_{\bar{y}})n_x(\boldsymbol{x}) - k_2\frac{\partial G^*}{\partial y}(\bar{x}-c_{\bar{x}},\bar{y}-c_{\bar{y}})n_y(\boldsymbol{x}) \tag{3.141}$$

The derivatives with respect to x and y in this expression can be expressed in terms of $\bar{x}$ and $\bar{y}$ by using the chain rule for partial differentiation in combination with (3.4), viz.,

$$\begin{aligned}\frac{\partial}{\partial x} &= \frac{\partial\bar{x}}{\partial x}\frac{\partial}{\partial\bar{x}}+\frac{\partial\bar{y}}{\partial x}\frac{\partial}{\partial\bar{y}} = \cos\overset{j}{\beta}\frac{\partial}{\partial\bar{x}}-\sin\overset{j}{\beta}\frac{\partial}{\partial\bar{y}}\\ \frac{\partial}{\partial y} &= \frac{\partial\bar{x}}{\partial y}\frac{\partial}{\partial\bar{x}}+\frac{\partial\bar{y}}{\partial y}\frac{\partial}{\partial\bar{y}} = \sin\overset{j}{\beta}\frac{\partial}{\partial\bar{x}}+\cos\overset{j}{\beta}\frac{\partial}{\partial\bar{y}}\end{aligned} \tag{3.142}$$

and, noting that $n_x(\boldsymbol{x}) = -\sin\overset{j}{\beta}$ and $n_y(\boldsymbol{x}) = \cos\overset{j}{\beta}$, (3.141) can be written as

$$\begin{aligned}Q^*(\bar{x}-c_{\bar{x}},\bar{y}-c_{\bar{y}}) &= (k_1-k_2)\sin\overset{j}{\beta}\cos\overset{j}{\beta}\frac{\partial G^*}{\partial\bar{x}}(\bar{x}-c_{\bar{x}},\bar{y}-c_{\bar{y}})\\ &\quad-(k_1\sin^2\overset{j}{\beta}+k_2\cos^2\overset{j}{\beta})\frac{\partial G^*}{\partial\bar{y}}(\bar{x}-c_{\bar{x}},\bar{y}-c_{\bar{y}})\end{aligned} \tag{3.143}$$

Finally, using (3.101) and carrying out the differentiations in (3.143) we have

$$\begin{aligned}Q^*(\bar{x}-c_{\bar{x}},\bar{y}-c_{\bar{y}}) &= \frac{k_{\bar{x}}}{2\pi\sqrt{k_1k_2}}\left[\frac{C_1(\bar{x}-c_{\bar{x}})+C_2(\bar{y}-c_y)}{C_1(\bar{x}-c_{\bar{x}})^2+2C_2(\bar{x}-c_{\bar{x}})(\bar{y}-c_{\bar{y}})+C_3(\bar{y}-c_{\bar{y}})^2}\right]\\ &\quad+\frac{k_{\bar{y}}}{2\pi\sqrt{k_1k_2}}\left[\frac{C_2(\bar{x}-c_{\bar{x}})+C_3(\bar{y}-c_y)}{C_1(\bar{x}-c_{\bar{x}})^2+2C_2(\bar{x}-c_{\bar{x}})(\bar{y}-c_{\bar{y}})+C_3(\bar{y}-c_{\bar{y}})^2}\right].\end{aligned} \tag{3.144}$$

Following the same procedure as in Sect. 3.2, Eq. (3.132) can be written for a discretized boundary contour in a form similar to (3.9):

$$\phi(\boldsymbol{\xi}) = -\sum_{j=1}^{N}\int_{-\overset{j}{a}}^{\overset{j}{a}} q(\bar{x})G^*(\bar{x}-c_{\bar{x}},\bar{y}-c_{\bar{y}})\,\mathrm{d}\bar{x} + \sum_{j=1}^{N}\int_{-\overset{j}{a}}^{\overset{j}{a}}\phi(\bar{x})Q^*(\bar{x}-c_{\bar{x}},\bar{y}-c_{\bar{y}})\,\mathrm{d}\bar{x} \tag{3.145}$$

We assume that the potential ϕ and the normal flux q are piecewise constant over the elements with values $\overset{j}{\phi}$ and $\overset{j}{q}$, and set $\bar{y} = 0$ (its value for element j) to obtain

$$\phi(\boldsymbol{\xi}) = -\sum_{j=1}^{N} \overset{j}{q} \int_{-\overset{j}{a}}^{\overset{j}{a}} G^*(\bar{x} - c_{\bar{x}}, c_{\bar{y}})\,\mathrm{d}\bar{x} + \sum_{j=1}^{N} \overset{j}{\phi} \int_{-\overset{j}{a}}^{\overset{j}{a}} Q^*(\bar{x} - c_{\bar{x}}, c_{\bar{y}})\,\mathrm{d}\bar{x} \tag{3.146}$$

The indicated integrations can be performed analytically (see, for example, Petit Bois 1961) and the result put in the same form as (3.13):

$$\phi(\boldsymbol{\xi}) = -\sum_{j=1}^{N} \mathbb{G}^*(\overset{j}{\boldsymbol{x}}, \boldsymbol{\xi})\overset{j}{q} + \sum_{j=1}^{N} \mathbb{Q}^*(\overset{j}{\boldsymbol{x}}, \boldsymbol{\xi})\overset{j}{\phi} \tag{3.147}$$

where now

$$\begin{aligned}
\mathbb{G}^*(\overset{j}{\boldsymbol{x}}, \boldsymbol{\xi}) &= \int_{-\overset{j}{a}}^{\overset{j}{a}} G^*(\bar{x} - c_{\bar{x}}, c_{\bar{y}})\,\mathrm{d}\bar{x} \\
&= -\frac{1}{2\pi\sqrt{k_1k_2}}\Bigg[\frac{c_{\bar{y}}}{\sqrt{k_1k_2}C_1}\left\{\arctan\frac{c_{\bar{y}}/\sqrt{k_1k_2}}{C_1(c_{\bar{x}} - \overset{j}{a}) + C_2c_{\bar{y}}} - \arctan\frac{c_{\bar{y}}/\sqrt{k_1k_2}}{C_1(c_{\bar{x}} + \overset{j}{a}) + C_2c_{\bar{y}}}\right\} \\
&\quad - \left[(c_{\bar{x}} - \overset{j}{a}) + \frac{C_2}{C_1}c_{\bar{y}}\right]\ln\left[C_1(c_{\bar{x}} - \overset{j}{a})^2 + 2C_2(c_{\bar{x}} - \overset{j}{a})c_{\bar{y}} + C_3c_{\bar{y}}^2\right]^{1/2} \\
&\quad + \left[(c_{\bar{x}} + \overset{j}{a}) + \frac{C_2}{C_1}c_{\bar{y}}\right]\ln\left[C_1(c_{\bar{x}} + \overset{j}{a})^2 + 2C_2(c_{\bar{x}} + \overset{j}{a})c_{\bar{y}} + C_3c_{\bar{y}}^2\right]^{1/2} - 2\overset{j}{a}\Bigg]
\end{aligned} \tag{3.148}$$

and

$$\begin{aligned}
\mathbb{Q}^*(\overset{j}{\boldsymbol{x}}, \boldsymbol{\xi}) &= \int_{-\overset{j}{a}}^{\overset{j}{a}} Q^*(\bar{x} - c_{\bar{x}}, c_{\bar{y}})\,\mathrm{d}\bar{x} \\
&= \frac{k_{\bar{x}}}{2\pi\sqrt{k_1k_2}}\Bigg[\ln\left[C_1(c_{\bar{x}} - \overset{j}{a})^2 + 2C_2(c_{\bar{x}} - \overset{j}{a})c_{\bar{y}} + C_3c_{\bar{y}}^2\right]^{1/2} \\
&\quad - \ln\left[C_1(c_{\bar{x}} + \overset{j}{a})^2 + 2C_2(c_{\bar{x}} + \overset{j}{a})c_{\bar{y}} + C_3c_{\bar{y}}^2\right]^{1/2}\Bigg] \\
&+ \frac{k_{\bar{y}}}{2\pi\sqrt{k_1k_2}}\Bigg[\frac{1}{\sqrt{k_1k_2}C_1}\left\{\arctan\frac{c_{\bar{y}}/\sqrt{k_1k_2}}{C_1(c_{\bar{x}} - \overset{j}{a}) + C_2c_{\bar{y}}} - \arctan\frac{c_{\bar{y}}/\sqrt{k_1k_2}}{C_1(c_{\bar{x}} + \overset{j}{a}) + C_2c_{\bar{y}}}\right\} \\
&\quad + \Bigg\{\frac{C_2}{C_1}\ln\left[C_1(c_{\bar{x}} - \overset{j}{a})^2 + 2C_2(c_{\bar{x}} - \overset{j}{a})c_{\bar{y}} + C_3c_{\bar{y}}^2\right]^{1/2} \\
&\quad - \frac{C_2}{C_1}\ln\left[C_1(c_{\bar{x}} + \overset{j}{a})^2 + 2C_2(c_{\bar{x}} + \overset{j}{a})c_{\bar{y}} + C_3c_{\bar{y}}^2\right]^{1/2}\Bigg\}\Bigg]
\end{aligned} \tag{3.149}$$

An $N \times N$ system of algebraic equations can be set up from (3.147) in the usual way by setting $\boldsymbol{\xi} = \overset{i}{\boldsymbol{x}}$ and transferring all of the unknown quantities to one side of the equals sign and all of the known quantities to the other.

3.14.4 Example

As an example, we will consider the problem of a square block of width $2W$ containing a square hole of width W, both centered at the origin. The conductivities in the x and y directions for the square block are k_1 and k_2, respectively, and the boundary conditions for the problem are $\phi = 10$ on the walls of the hole and $\phi = 0$ on the sides of the block. The objective is to find the flux going from the hole and away from the block for three different combinations of conductivity: $k_1 = 1.0$ and $k_2 = 1.0$, 0.5, and 0.1. Numerical solutions were obtained for the case that the problem domain was divided into 120 equal size boundary elements: 40 along the sides of the hole and 80 along the sides of the block.

The results in Fig. 3.11 show the normal flux q along the top of the block ($-W \leq x \leq W,\ y = W$) for various values of k_2, while those in Fig. 3.12 show the normal flux q along a side ($x = -W,\ -W \leq y \leq W$) for the same values of conductivity. Figure 3.11 shows that the flux decreases more or less in proportion to the value of conductivity in the y direction (k_2), while Fig. 3.12 indicates that the flux in the x direction is comparatively less influenced by this change of conductivity.

In each example problem the total flux emitted from the hole (the sum of the values for the elements around the boundary) was the same as that passing out of the block, which provides a consistency check on the numerical results. The total fluxes for the three separate cases were found to be as follows: 101.9, 76.0, and 53.2.

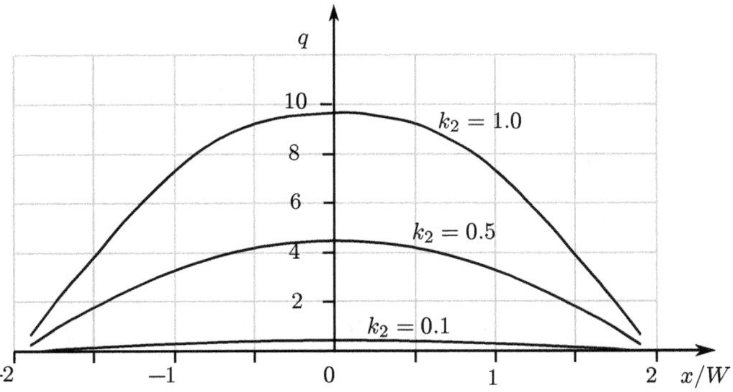

Fig. 3.11 Flux q along the top ($y = W$) of the block for $k_1 = 1$ and several values of k_2

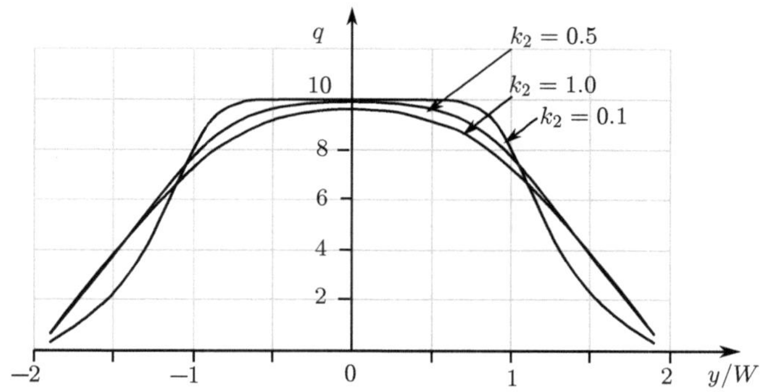

Fig. 3.12 Flux q along a side ($x = -W$) of the block for $k_1 = 1$ and several values of k_2

3.15 Exercises/Projects

The reader's mastery of the boundary element approach will be facilitated by writing computer programs to implement the algorithms described in this book. The particular programming language used is not important—the goal is to gain first-hand programming experience. Fully functional computer programs can be written in languages such as Python, Matlab, or, for more experienced programmers, C++ or FORTRAN. The exercises, or projects, given below are suggested with this thought in mind.

1. Prepare a computer program for the direct boundary integral method with constant strength elements, as outlined in Sect. 3.2, et seq. Verify your program by solving the example problems described in Sect. 3.8.
2. Modify your program from Exercise 1 to include an option for one or two lines of symmetry, as explained in Sect. 3.7.
3. Modify your program from Exercise 1 to include piecewise linear variation of the boundary parameters.
4. Include an option in your program from Exercise 3 to compute the components of the gradient of the potential, $q_x(\boldsymbol{\xi})$ and $q_y(\boldsymbol{\xi})$, at points within the region of interest, allowing for the possibility that point $\boldsymbol{\xi}$ might be at the midpoint of a boundary element.
5. Modify your program from Exercise 1 to handle piecewise homogeneous regions as explained in either Sects. 3.12 and 3.13.
6. Modify your program from Exercise 5 to allow the subregions to be anisotropic.
7. Show that (3.118) satisfies (3.121) everywhere except at the point $\boldsymbol{x} = \boldsymbol{\xi}$.

References

Chang, Y.P., C.S. Kang, and D.J. Chen. 1973. The use of fundamental Green's functions for the solution of problems of heat conduction in anisotropic media. *International Journal of Heat and Mass Transfer* 16: 1905–1918.

Petit Bois, G. 1961. *Tables of indefinite integrals*. Dover.

Strack, O.D.L. 1989. *Groundwater mechanics*. Prentice-Hall.

Chapter 4
Elasticity

4.1 Introduction

The concepts of stress, strain, and linear elasticity are fundamental to the boundary element methods discussed in the remainder of this book. It is assumed that the reader is familiar with these concepts as developed, for example, in a text on strength of materials or deformable body mechanics. In order to set our notation and sign conventions we will review these topics in this chapter. We will also present several basic elasticity solutions and integral formulas that are needed to help the reader develop a thorough understanding of the boundary element approach for elasticity. Although our principal focus will be on two-dimensional problems, the general results in this chapter are also applicable in three dimensions. We will make note of this fact whenever it is not clear from context that this is the case.

4.2 Stress

The concept of stress is used to specify the way in which forces are transmitted through a solid, continuous body. If we imagine an isolated, prismatic element of the body, we can consider that certain forces are exerted on it by the surrounding material. (Conversely, we can consider that the element exerts equal and opposite forces on the surrounding material.) If the element is small enough, these forces are distributed more or less uniformly over each of its faces. The *stress* on any particular face is then defined as the resultant force divided by the area. Stress therefore has units of Newtons per square meter (N/m^2).

S. L. Crouch and S. G. Mogilevskaya, *A First Course in Boundary Element Methods*, Mathematical Engineering, https://doi.org/10.1007/978-3-031-63341-6_4

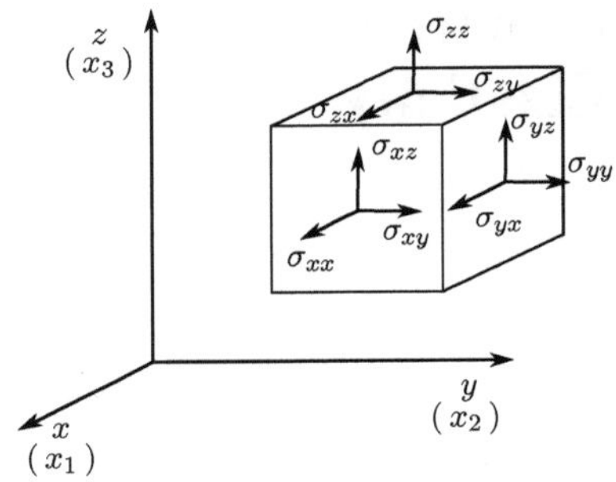

Fig. 4.1 Components of the stress tensor for Cartesian coordinates

4.2.1 Stresses on Coordinate Planes; Components of the Stress Tensor

The stresses acting on planes whose normals are parallel to the coordinate directions are known as the components of the stress tensor. For a Cartesian coordinate system, these stresses are identified as depicted in Fig. 4.1. Collectively, the stress components shown in the figure are denoted by the symbol σ_{ij}, where i and j may be either x, y, or z, sometimes represented as x_1, x_2, and x_3, or even simply 1, 2, and 3. The first index (i) refers to the direction of the *normal* to the plane on which the stress acts; the second index (j) refers to the direction of the stress component. Thus, for example, the component σ_{yz} is parallel to the z direction and acts on a plane whose normal is parallel to the y direction (note that there are two such planes for the prismatic element in Fig. 4.1).

The sign convention for stress used throughout this book is as follows: the stress component σ_{ij} is positive if it acts in the *positive* (or *negative*) j direction on a plane whose outward normal points in the *positive* (or *negative*) i direction. All of the stress components illustrated in Fig. 4.1 are positive under this definition. This sign convention means that the stresses normal to the faces of a prismatic element (i.e., $\sigma_{11} = \sigma_{xx}$, $\sigma_{22} = \sigma_{yy}$, and $\sigma_{33} = \sigma_{zz}$) are positive if they are tensile.

4.2.2 Equilibrium Equations

The components of the stress tensor at various points in a body cannot be specified arbitrarily. The requirement that the body be in equilibrium imposes certain conditions on these components. First, one finds from moment equilibrium conditions that the stress tensor is symmetric, i.e.,

$$\sigma_{ij} = \sigma_{ji} \tag{4.1}$$

for i and $j = 1, 2$, or 3. This equation says, for example, that if $i = 1$ and $j = 3$ then $\sigma_{13} = \sigma_{31}$ (or $\sigma_{xz} = \sigma_{zx}$). Second, it can be shown from force equilibrium considerations that the components of the stress tensor must satisfy the equations

$$\begin{aligned}
\frac{\partial \sigma_{xx}}{\partial x} + \frac{\partial \sigma_{yx}}{\partial y} + \frac{\partial \sigma_{zx}}{\partial z} + B_x = 0 \\
\frac{\partial \sigma_{xy}}{\partial x} + \frac{\partial \sigma_{yy}}{\partial y} + \frac{\partial \sigma_{zy}}{\partial z} + B_y = 0 \\
\frac{\partial \sigma_{xz}}{\partial x} + \frac{\partial \sigma_{yz}}{\partial y} + \frac{\partial \sigma_{zz}}{\partial z} + B_z = 0
\end{aligned} \tag{4.2}$$

where B_x, B_y, and B_z are body forces, which must be specified for a particular problem. (In most situations considered in this book, we will take the body forces to be zero.) The body force components have dimensions of force/volume (e.g., N/m^3). They form the components of a vector and are positive if they act in the positive coordinate directions. Equation (4.2) are known as the differential equations of equilibrium, or simply the equilibrium equations.

The stress tensor has nine components (see Fig. 4.1). According to (4.1), however, only six of these components are independent: σ_{xx}, σ_{yy}, σ_{zz}, $\sigma_{xy} = \sigma_{yx}$, $\sigma_{xz} = \sigma_{zx}$, and $\sigma_{yz} = \sigma_{zy}$. The equilibrium equation (4.2) give three equations for these six unknowns. Hence, we see that the equilibrium equations alone are not sufficient to calculate the stresses for a particular problem; three additional equations are needed. These additional equations will be found when we consider the deformations associated with the stresses (see Sect. 4.9).

4.2.3 *Plane Stress*

A two-dimensional stress state known as *plane stress* occurs when all of the stress components associated with a particular coordinate direction are zero. The components of the stress tensor for the case of plane stress for the x, y plane are illustrated in Fig. 4.2. The stresses σ_{zz}, $\sigma_{xz} = \sigma_{zx}$, and $\sigma_{yz} = \sigma_{zy}$ (see Fig. 4.1) are zero, as is

Fig. 4.2 Plane stress

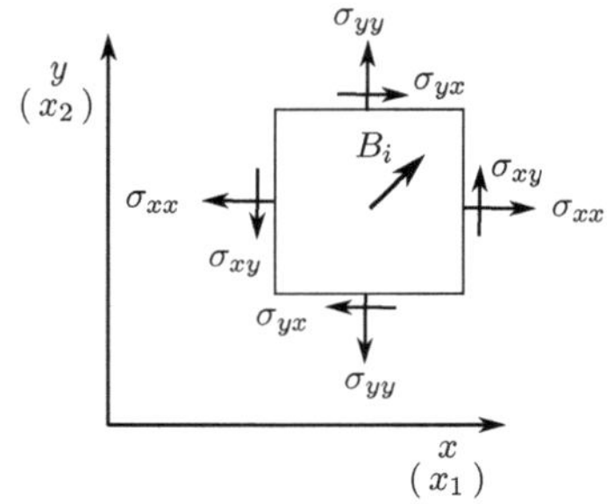

the body force component B_z. Equations (4.1) and (4.2) are applicable to the case of plane stress. Considering only the non-zero quantities, we write

$$\sigma_{xy} = \sigma_{yx} \tag{4.3}$$

and

$$\begin{aligned} \frac{\partial \sigma_{xx}}{\partial x} + \frac{\partial \sigma_{yx}}{\partial y} + B_x &= 0 \\ \frac{\partial \sigma_{xy}}{\partial x} + \frac{\partial \sigma_{yy}}{\partial y} + B_y &= 0 \end{aligned} \tag{4.4}$$

The stress tensor therefore has three independent components for plane stress: σ_{xx}, σ_{yy}, and $\sigma_{xy} = \sigma_{yx}$. Equilibrium equation (4.4) give two equations for these three unknowns.

4.2.4 Traction Vector

In applications, it is often necessary to calculate the stresses acting on a plane of arbitrary orientation with respect to the x_1, x_2, x_3 coordinate axes. These stresses can be represented by a vector, called the *traction vector*, which we will denote as $\boldsymbol{t} = (t_x, t_y, t_z)$, where t_x, t_y, and t_z are the components in the x_i coordinate directions. Notice that these stress components are denoted by the letter t and only *one* index. This is done to emphasize that t_i acts on a plane of arbitrary orientation and not, like the stresses σ_{ij}, on the particular planes whose normals are parallel to the x_i coordinate axes. The components of the traction vector are positive if they act in the positive coordinate directions.

The components t_i of traction can be expressed in terms of the components of the stress tensor σ_{ij} and the unit outward normal $\boldsymbol{n} = (n_x, n_y, n_z)$ to the plane in question. This is depicted in Fig. 4.3 for the case of plane stress, where $n_z = 0$. If

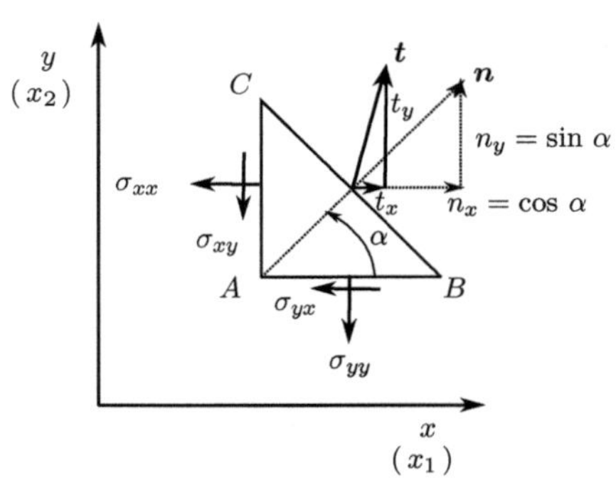

Fig. 4.3 Traction vector $\boldsymbol{t}$; plane stress

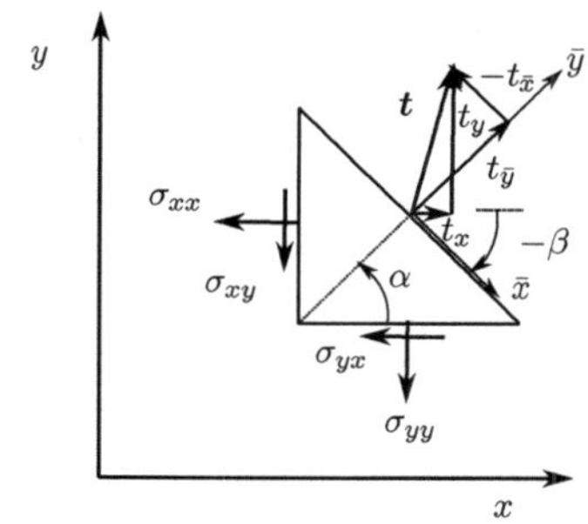

Fig. 4.4 Resolution of traction vector into $\bar{x}$ and $\bar{y}$ components; plane stress

the area of plane BC in the figure is $|BC|$, then the areas of coordinate planes AB and CA are $|AB| = |BC| \sin\alpha$ and $|CA| = |BC| \cos\alpha$, respectively, where angle α specifies the inclination of the outward normal $\boldsymbol{n} = (n_x, n_y)$ with respect to the x axis. It can be shown from force equilibrium considerations that t_x and t_y are given by the relations

$$\begin{aligned} t_x &= \sigma_{xx} \cos\alpha + \sigma_{yx} \sin\alpha \\ t_y &= \sigma_{xy} \cos\alpha + \sigma_{yy} \sin\alpha \end{aligned} \tag{4.5}$$

or, noting that $n_x = \cos\alpha$ and $n_y = \sin\alpha$,

$$\begin{aligned} t_x &= \sigma_{xx} n_x + \sigma_{yx} n_y \\ t_y &= \sigma_{xy} n_x + \sigma_{yy} n_y \end{aligned} \tag{4.6}$$

A similar set of equations can be written for an arbitrarily oriented plane in three dimensions:

$$\begin{aligned} t_x &= \sigma_{xx} n_x + \sigma_{yx} n_y + \sigma_{zx} n_z \\ t_y &= \sigma_{xy} n_x + \sigma_{yy} n_y + \sigma_{zy} n_z \\ t_z &= \sigma_{xz} n_x + \sigma_{yz} n_y + \sigma_{zz} n_z \end{aligned} \tag{4.7}$$

In two dimensions, t_x and t_y are the resultant stresses (in the x and y coordinate directions) acting on a plane whose outward normal $\boldsymbol{n} = (n_x, n_y)$ makes an angle α with the x axis. It is often more convenient to consider the components of traction in directions perpendicular and parallel to the plane in question. These components—called the *normal* and *shear tractions*—are shown in Fig. 4.4.[1] They are specified

[1] A similar resolution of the traction vector can be made in three dimensions. In this case, however, the traction vector has two shear components, so one would have to select local coordinate directions for both of them. Although the resultant of the shear traction is unambiguously defined,

with respect to the local $\overline{x}, \overline{y}$ coordinates shown in the figure. These coordinates are chosen such that $\overline{y}$ points in the direction of the outward normal to the plane; the direction of $\overline{x}$ is then fixed by the requirement that $\overline{x}$ and $\overline{y}$ form a right-handed system. The normal traction acts in the $\overline{y}$ direction and is denoted as $t_{\overline{y}}$; the shear traction acts in the $\overline{x}$ direction and is denoted as $t_{\overline{x}}$. These stresses are positive if they act in their respective positive coordinate directions.

The traction vector can be resolved into components in the $\overline{x}$ and $\overline{y}$ directions by simple statics. Omitting the details of the computation, we find from Fig. 4.4 that

$$\begin{aligned} t_{\bar{x}} &= t_x \sin\alpha - t_y \cos\alpha \\ t_{\bar{y}} &= t_x \cos\alpha + t_y \sin\alpha \end{aligned} \tag{4.8}$$

Combining these results with (4.5), we obtain the following expressions for the shear and normal tractions $t_{\overline{x}}$ and $t_{\overline{y}}$ in terms of the components of the stress tensor σ_{xx}, σ_{yy}, and $\sigma_{xy} = \sigma_{yx}$:

$$\begin{aligned} t_{\overline{x}} &= (\sigma_{xx} - \sigma_{yy}) \sin\alpha\cos\alpha - \sigma_{xy}(\cos^2\alpha - \sin^2\alpha) \\ t_{\overline{y}} &= \sigma_{xx}\cos^2\alpha + 2\sigma_{xy}\sin\alpha\cos\alpha + \sigma_{yy}\sin^2\alpha. \end{aligned} \tag{4.9}$$

Equation (4.9) can also be written in terms of the angle β shown in Fig. 4.4. (As was done in Sect. 1.5, we will customarily use this angle to define the transformation from x, y to $\overline{x}, \overline{y}$ coordinates, and *vice versa*.) Noting from the figure that $\alpha = \beta + \pi/2$ and using the relations $\sin(\beta + \pi/2) = \cos\beta$ and $\cos(\beta + \pi/2) = -\sin\beta$, we find that

$$\begin{aligned} t_{\overline{x}} &= -(\sigma_{xx} - \sigma_{yy}) \sin\beta\cos\beta + \sigma_{xy}(\cos^2\beta - \sin^2\beta) \\ t_{\overline{y}} &= \sigma_{xx}\sin^2\beta - 2\sigma_{xy}\sin\beta\cos\beta + \sigma_{yy}\cos^2\beta \end{aligned} \tag{4.10}$$

In terms of the double angle 2β these equations can be written as

$$\begin{aligned} t_{\bar{x}} &= -\frac{1}{2}(\sigma_{xx} - \sigma_{yy}) \sin 2\beta + \sigma_{xy}\cos 2\beta \\ t_{\bar{y}} &= \frac{1}{2}(\sigma_{xx} + \sigma_{yy}) - \frac{1}{2}(\sigma_{xx} - \sigma_{yy}) \cos 2\beta - \sigma_{xy}\sin 2\beta \end{aligned} \tag{4.11}$$

---

the components themselves depend upon the choice of the local coordinates. For this reason, it is not particularly useful in three dimensions to resolve the traction vector into normal and shear components.

We will occasionally refer to the shear and normal tractions as σ_s and σ_n, where the s, n coordinates are comparable to the $\bar{x}$, $\bar{y}$ coordinates. That is, we will use σ_s, σ_n and $t_{\bar{x}}$, $t_{\bar{y}}$ interchangeably in some applications (see, for example, Sects. 5.2 and 6.2).

4.3 Strain

The concept of strain is used to specify the way in which a solid, continuous body deforms when it is subjected to stress. Deformation represents a change of geometry and occurs when various points in the body displace relative to one another. The deformation of a body is characterized by two types of strain: *normal strain* and *shear strain*. Normal strain represents the change of length of a small line element divided by its original length; i.e., change of length per unit length. Shear strain specifies a change of angle (in radians) between two small line elements that were originally perpendicular to one another. Normal and shear strains are therefore dimensionless.

In most engineering problems, strains are very small quantities, negligible compared to unity. One can then neglect the product of two or more strains in comparison with the strains themselves. We will restrict our attention in this book to strains that meet this condition, so-called infinitesimal strains. The advantage of being able to work with infinitesimal strains is that one need not distinguish between the deformed and undeformed shapes of the body in defining the strain components. As a consequence, the strains produced at any point of a body by one system of stresses can be added to those produced by another system, with the final result independent of the order in which the stresses are applied. An operation of this kind is an illustration of the *principle of superposition*, which is of fundamental importance to the theory of linear elasticity.

4.3.1 Components of the Strain Tensor

The state of strain at a point in a body is specified by the components of the *strain tensor*. For a Cartesian coordinate system, these components give the normal and shear strains for infinitesimal line elements originally parallel to the x_i coordinate axes. They are defined in terms of the displacement components u_x, u_y, u_z at the point in question as follows:

$$\varepsilon_{xx} = \frac{\partial u_x}{\partial x}; \qquad \varepsilon_{yy} = \frac{\partial u_y}{\partial y}; \qquad \varepsilon_{zz} = \frac{\partial u_z}{\partial z}$$

$$\varepsilon_{xy} = \frac{1}{2}\left(\frac{\partial u_x}{\partial y} + \frac{\partial u_y}{\partial x}\right); \qquad \varepsilon_{xz} = \frac{1}{2}\left(\frac{\partial u_x}{\partial z} + \frac{\partial u_z}{\partial x}\right); \qquad \varepsilon_{yz} = \frac{1}{2}\left(\frac{\partial u_y}{\partial z} + \frac{\partial u_z}{\partial y}\right) \tag{4.12}$$

$$\varepsilon_{yx} = \varepsilon_{xy}; \qquad \varepsilon_{zx} = \varepsilon_{xz}; \qquad \varepsilon_{zy} = \varepsilon_{yz}$$

The components ε_{xx}, ε_{yy}, and ε_{zz} are the normal strains. They represent the changes in length per unit length of infinitesimal line elements in the x, y, and z directions, respectively. Normal strains are reckoned positive for lengthening (i.e., extension). The components $\varepsilon_{xy} = \varepsilon_{yx}$, $\varepsilon_{xz} = \varepsilon_{zx}$, and $\varepsilon_{yz} = \varepsilon_{zy}$ are the shear strains. The component ε_{ij} $(i \neq j)$ represents *one-half* the change in right angle between two infinitesimal line elements originally parallel to the x_i and x_j axes. Shear strains are reckoned positive for a decrease in right angle between any two positive (or any two negative) coordinate axes.

4.3.2 Plane Strain

A body is said to be in a state of plane strain parallel to the x, y plane if the component u_z of the displacement vector is zero and the components u_x and u_y are independent of the z coordinate. It then follows from (4.12) that the strain components ε_{zz}, $\varepsilon_{xz} = \varepsilon_{zx}$, and $\varepsilon_{yz} = \varepsilon_{zy}$ are zero and that the remaining components are independent of z. The non-zero strains are then

$$\varepsilon_{xx} = \frac{\partial u_x}{\partial x}; \qquad \varepsilon_{yy} = \frac{\partial u_y}{\partial y}; \qquad \varepsilon_{xy} = \varepsilon_{yx} = \frac{1}{2}\left(\frac{\partial u_x}{\partial y} + \frac{\partial u_y}{\partial x}\right) \tag{4.13}$$

These equations define three strain components in terms of two displacements, u_x and u_y. The strain components therefore cannot be specified independently of one another, and a relationship must exist among them. We can easily verify from (4.13) that this relationship is

$$\frac{\partial^2 \varepsilon_{yy}}{\partial x^2} + \frac{\partial^2 \varepsilon_{xx}}{\partial y^2} = 2\frac{\partial^2 \varepsilon_{xy}}{\partial x \partial y} \tag{4.14}$$

Equation (4.14) is known as the *compatibility condition* for plane strain. This name comes from the notion that (4.14) is a necessary and sufficient condition to ensure that the strains ε_{xx}, ε_{yy}, and ε_{xy} can be integrated to obtain displacements u_x and u_y that are continuous, single-valued functions of x and y (Sokolnikoff 1956). A continuous, single-valued displacement field (a "compatible" displacement field) precludes the possibility of gaps or overlaps developing within a deformed body. If we know that u_x and u_y are continuously differentiable functions of x and y, then (4.14) will be satisfied automatically.

Compatibility conditions can also be written for three-dimensional problems. It can be shown (Sokolnikoff 1956) that six relations connect the various strain components in this case, five in addition to (4.14). We do not require these relations in this book, however, so we will not quote them.[2]

4.4 Index Notation

The important equations introduced so far can be succinctly summarized using index notation (see Sect. 2.1). The reader can confirm that the equilibrium equation (4.2) can be written as

$$\sigma_{ji,j} + B_i = 0 \tag{4.15}$$

the components of the traction vector (4.7) are

$$t_i = \sigma_{ji} n_j \tag{4.16}$$

and the strain tensor (4.12) is defined by the relation

$$\varepsilon_{ij} = \tfrac{1}{2}(u_{i,j} + u_{j,i}) \tag{4.17}$$

where, in each case, indices i and j are 1, 2, or 3 in three dimensions and 1 or 2 in two dimensions and a repeated index signifies summation with respect to that index.

4.5 Generalized Hooke's Law

A stressed body is said to behave elastically if its strains are recovered when the stresses are removed. In other words, the behavior is elastic if there is a one-to-one correspondence between the stresses and strains in the body. A certain state of stress then corresponds to a certain state of strain, and vice versa, independent of the history of loading. A body is said to be linearly elastic if each stress component can be expressed as a linear combination of all of the strain components.

The classical theory of linear elasticity is based on a generalization of Hooke's law, which was originally formulated for a spring, or a "springy body" (Love 1944). The so-called generalized Hooke's law states that at each point of a three-dimensional, linearly elastic body the six components of the stress tensor ($\sigma_{ij} = \sigma_{ji}$) are linearly related to the six components of the strain tensor ($\varepsilon_{ij} = \varepsilon_{ji}$). Hooke's law therefore expresses a linear relationship between two second order tensors, as follows:

$$\sigma_{ij} = C_{ijkl}\varepsilon_{kl} \tag{4.18}$$

[2] See Sect. 4.9 in this connection.

where i, j, k, and l all range from 1 to 3 and summation is taken with respect to repeated indices k and l. The components of the fourth order tensor C_{ijkl} are known as the *elastic moduli*. We will assume that these properties are independent of position, i.e., we will assume that the body is*homogeneous*.

Formal expansion of (4.18) would give nine equations with a total of 81 coefficients (i.e., all combinations of four indices, each with three possible values: $3^4 = 81$). Three of these nine equations are redundant, however, because the stress and strain tensors are symmetric. The tensor C_{ijkl} is therefore symmetric with respect to the first two and the last two indices and, because σ_{ij} and ε_{kl} each have six independent components, (4.18) represents a set of six equations with 36 coefficients. Further simplification results from the observation that C_{ijkl} is completely symmetric, i.e., $C_{ijkl} = C_{klij}$. This result, which can be established from energy considerations (Sokolnikoff 1956), reduces the maximum number of independent elastic constants in (4.18) from 36 to 21.

It is sometimes convenient to write (4.18) in inverse form, expressing the strains in terms of the stresses as

$$\varepsilon_{ij} = S_{ijkl}\sigma_{kl} \tag{4.19}$$

Coefficients S_{ijkl}, of which 21 are independent for a general anisotropic material, are known as *compliances*; coefficients C_{ijkl} are called *moduli*, or *stiffnesses*.[3] Equations (4.18) and (4.19) are of course interchangeable. Equation (4.18) would be used, for example, if we knew the strains and wanted to compute the stresses, whereas (4.19) would be used for the opposite purpose.

A material with 21 elastic constants is said to possess the most general form of elastic anisotropy. Most engineering materials have a certain amount of structural symmetry, however, and their stress-strain relationships are consequently easier to specify. The two simplest forms of anisotropy are known as *orthotropy* and *transverse isotropy*. These conditions arise for materials having preferred directions of elastic symmetry. Many kinds of wood, composite materials, and sedimentary and metamorphic rocks can be considered as homogeneous, orthotropic or transversely isotropic elastic materials. Finally, we note that a material whose elastic properties are independent of direction is called *isotropic*. Aluminum, steel, glass, concrete, and igneous rocks are examples of materials that are commonly represented as isotropic elastic materials.

As we will see, an orthotropic material has nine independent elastic constants and a transversely isotropic material has five. Moreover, these two types of anisotropy are indistinguishable for plane problems, and the stress-strain relations for the x, y plane involve only four constants. An isotropic material has two independent elastic constants, usually taken as Young's modulus E and Poisson's ratio ν.

[3] It would appear preferable to use the symbol "S" for stiffness and "C" for compliance. The usage adopted here, however, has evolved in practice, and we will adhere to it to avoid confusion with the literature.

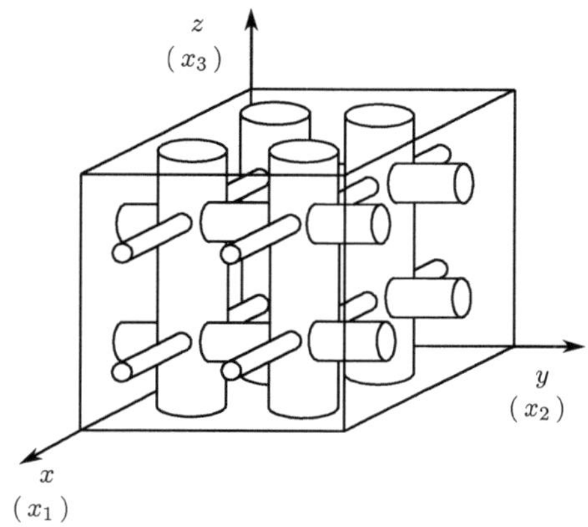

Fig. 4.5 Schematic diagram of an orthotropic body

4.5.1 *Orthotropic, Transversely Isotropic, and Isotropic Materials*

An orthotropic body can be imagined as a prism containing a lattice of "rods," all of different sizes and all aligned along the x, y, and z coordinate axes, as shown in Fig. 4.5. This representation of an orthotropic continuum is, of course, fictitious; it is merely a convenient way of visualizing a material that has different elastic properties in different directions. It appears from the figure that an orthotropic body would respond differently to normal and shear stresses applied to the faces of the prism with normals in the three coordinate directions. For example, the strain ε_{zz} associated with a stress σ_{zz} would evidently be less than the strain ε_{xx} associated with a stress σ_{xx} (assuming the the rods in these directions are made from the same material). Similarly, the shear strain ε_{yz} associated with a stress σ_{yz} would be less than the shear strain ε_{xy} associated with a stress σ_{xy}.

An orthotropic material has nine independent elastic constants. Three of these constants relate the normal strains ε_{xx}, ε_{yy}, and ε_{zz} to normal stresses σ_{xx}, σ_{yy}, and σ_{zz}, respectively. These constants are called *Young's moduli*. Three other independent elastic constants are called *Poisson's ratios*. These relate the normal strain in one direction, say ε_{xx}, to the normal strain in another direction, say ε_{zz}. The remaining three elastic constants for an orthotropic material relate the shear strains ε_{yz}, ε_{xz}, and ε_{xy} to the shear stresses σ_{yz}, σ_{xz}, and σ_{xy} that produce them. These constants are the *shear moduli*.

A transversely isotropic solid can be visualized in the same way as an orthotropic material, except that the perpendicular sets of rods in two of the coordinate directions have the same size. This is illustrated in Fig. 4.6, where it is seen that the x and y directions are equivalent. This equivalence allows one to describe a transversely isotropic material by five elastic constants instead of the nine needed for an orthotropic material. The five constants consist of two Young's moduli, two Poisson's ratios, and one shear modulus. Actually, two shear moduli can be defined, but

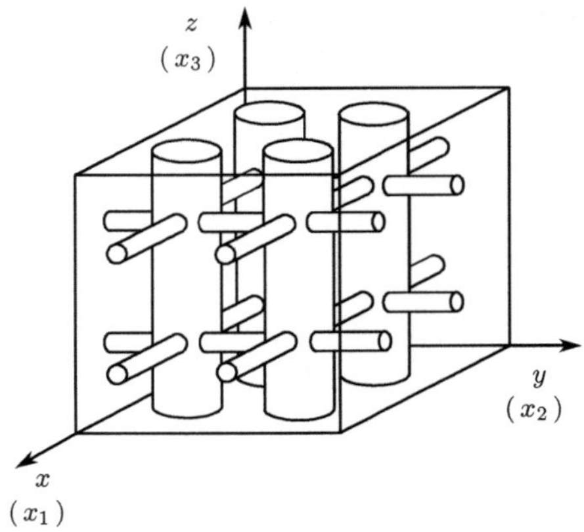

Fig. 4.6 Schematic diagram of a transversely isotropic body

one of them can be expressed in terms of one Young's modulus and one Poisson's ratio and is therefore not an independent quantity. A transversely isotropic material is said to be "isotropic in a plane"—in this case the x, y plane—because its elastic properties are independent of direction in one plane.

An isotropic material can be considered as the limiting case of the orthotropic and transversely isotropic materials illustrated above, i.e., the case for which the three sets of rods in the lattice are all of the same size. Only two independent elastic constants are needed to describe the behavior of an isotropic elastic material: Young's modulus E and Poisson's ratio ν. The shear modulus μ for an isotropic elastic material can be expressed in terms of E and ν as follows:

$$\mu = \frac{1}{2}\frac{E}{(1+\nu)} \tag{4.20}$$

(Timoshenko and Goodier 1970). We note that the symbol G is also commonly used in the literature to represent the shear modulus. To prevent confusion with the notation introduced previously for the fundamental solution to Laplace's equation, $G(\boldsymbol{x}, \boldsymbol{\xi})$, however, we will avoid this usage in this book.

4.5.2 Stress-Strain Relations—Orthotropic Material

Stress-strain relations for the orthotropic material depicted in Fig. 4.5 can be written as

$$\begin{aligned}
\varepsilon_{xx} &= \frac{1}{E_x}\sigma_{xx} - \frac{\nu_{yx}}{E_y}\sigma_{yy} - \frac{\nu_{zx}}{E_z}\sigma_{zz} \\
\varepsilon_{yy} &= -\frac{\nu_{xy}}{E_x}\sigma_{xx} + \frac{1}{E_y}\sigma_{yy} - \frac{\nu_{zy}}{E_z}\sigma_{zz}
\end{aligned} \tag{4.21}$$

$$\varepsilon_{zz} = -\frac{\nu_{xz}}{E_x}\sigma_{xx} - \frac{\nu_{yz}}{E_y}\sigma_{yy} + \frac{1}{E_z}\sigma_{zz}$$

$$\varepsilon_{yz} = \frac{1}{2\mu_{yz}}\sigma_{yz}; \qquad \varepsilon_{xz} = \frac{1}{2\mu_{xz}}\sigma_{xz}; \qquad \varepsilon_{xy} = \frac{1}{2\mu_{xy}}\sigma_{xy}$$

(Lekhnitskii 1963). The constants E_x, E_y, and E_z in these expressions are the Young's moduli for the x, y, and z coordinate directions (the directions of elastic symmetry for the orthotropic material under consideration). The constants ν_{yx}, ν_{zx}, etc., are Poisson's ratios, and μ_{yz}, μ_{xz}, and μ_{xy} are the three independent shear moduli. It is important to realize that the subscripts on all these quantities are used for purposes of identification only; E_x, E_y, and E_z, for example, do *not* form the components of a vector, and ν_{yx}, etc., and μ_{yz}, etc., are *not* components of a tensor.

The three Young's moduli in (4.21) govern the amount of normal strain in the direction of a *uniaxial applied normal stress*. For example, if $\sigma_{yy} = \sigma_{zz} = 0$, the first equation reads $\varepsilon_{xx} = \sigma_{xx}/E_x$, which is Hooke's law for simple tension or compression, and which defines the associated Young's modulus as $E_x = \sigma_{xx}/\varepsilon_{xx}$. The values of E_y and E_z are defined similarly for uniaxial stresses σ_{yy} and σ_{zz}.

The physical significance of the Poisson's ratios in (4.21) can be explained by considering the way that these quantities are defined. Poisson's ratios ν_{xy} and ν_{xz}, for example, relate the normal strains in the y and z directions (ε_{yy} and ε_{zz}) to the normal strain in the x direction (ε_{xx}) *for a uniaxial applied stress* σ_{xx}. Equations (4.21) in this case give

$$\varepsilon_{xx} = \sigma_{xx}/E_x; \qquad \varepsilon_{yy} = -\nu_{xy}\sigma_{xx}/E_x; \qquad \varepsilon_{zz} = -\nu_{xz}\sigma_{xx}/E_x$$

so that $\nu_{xy} = -\varepsilon_{yy}/\varepsilon_{xx}$ and $\nu_{xz} = -\varepsilon_{zz}/\varepsilon_{xx}$. In general, the constants ν_{ij} and ν_{ji} are *not equal*. Stress-strain relations (4.21), however, are symmetric, so that $\nu_{ij}/E_i = \nu_{ji}/E_j$ ($i \neq j$; no summation on repeated indices), and hence only three of the six different Poisson's ratios appearing in these equations are independent quantities.

The last three of (4.21) state that normal stresses produce no shear strains, just as the first three of these equations state that shear stresses produce no normal strains. Furthermore, the shear stresses (and strains) are all independent, and each component of shear strain is related to its corresponding component of shear stress by a different shear modulus.

Equations (4.21) can also be written in the form

$$\begin{aligned}
\varepsilon_{xx} &= s_{11}\sigma_{xx} + s_{12}\sigma_{yy} + s_{13}\sigma_{zz} \\
\varepsilon_{yy} &= s_{21}\sigma_{xx} + s_{22}\sigma_{yy} + s_{23}\sigma_{zz} \\
\varepsilon_{zz} &= s_{31}\sigma_{xx} + s_{32}\sigma_{yy} + s_{33}\sigma_{zz} \\
\varepsilon_{yz} &= \tfrac{1}{2}s_{44}\sigma_{yz} \\
\varepsilon_{xz} &= \tfrac{1}{2}s_{55}\sigma_{xz} \\
\varepsilon_{xy} &= \tfrac{1}{2}s_{66}\sigma_{xy}
\end{aligned} \tag{4.22}$$

The coefficients s_{ij} in these equations express a linear relationship between the six components of strain (ε_{xx}, ε_{yy}, ε_{zz}, ε_{yz}, ε_{xz}, ε_{xy}) and the six components of stress (σ_{xx}, σ_{yy}, σ_{zz}, σ_{yz}, σ_{xz}, σ_{xy}) at a point. Because the directions of elastic symmetry lie along the coordinate axes, however, many of the terms of the general 6×6 coefficient array are zero. In fact, taking into account that $s_{ij} = s_{ji}$, it appears that the orthotropic material under consideration has nine independent elastic constants. The relationships between the coefficients s_{ij} in (4.22) and the so-called engineering elastic constants E_x, E_z, ν_{xy}, etc., in (4.21) are easily established by inspection.

The inverse form of (4.22) can be written in terms of a set of coefficients $c_{ij} = c_{ji}$. These coefficients can also be expressed in terms of the engineering elastic constants but the results are unwieldy for the general three-dimensional case and will not be given here.

4.5.3 *Stress-Strain Relations—Transversely Isotropic Material*

Stress-strain relations for a transversely isotropic elastic material can be obtained from the foregoing results for an orthotropic material by noting from Fig. 4.6 that two coordinate directions (x and y) are equivalent. Because of this equivalence, the Young's moduli in the x and y directions are equal, i.e., $E_x = E_y$. Also, the following equalities hold for the Poisson's ratios and the shear moduli: $\nu_{xy} = \nu_{yx}$, $\nu_{xz} = \nu_{yz}$, $\nu_{zx} = \nu_{zy}$, $\mu_{xz} = \mu_{yz}$, $\mu_{xy} = \frac{1}{2}E_x/(1 + \nu_{xy}) = \frac{1}{2}E_y/(1 + \nu_{yx})$. The last of these results shows that the shear modulus in the x, y plane (the plane of isotropy) can be expressed in terms of the Young's modulus and Poisson's ratio associated with this plane, as in (4.20).

The non-zero constants s_{ij} in (4.22) for a transversely isotropic material can be written in terms of the engineering elastic constants E_x, E_z, ν_{xy}, etc., by using the equivalencies noted above. We thus have

$$
\begin{aligned}
s_{11} &= s_{22} = \frac{1}{E_x} \\
s_{12} &= s_{21} = -\frac{\nu_{xy}}{E_x} \\
s_{13} &= s_{31} = s_{23} = s_{32} = -\frac{\nu_{xz}}{E_x} = -\frac{\nu_{zx}}{E_z} \\
s_{33} &= \frac{1}{E_z} \\
s_{44} &= s_{55} = \frac{1}{\mu_{xz}} \\
s_{66} &= 2(s_{11} - s_{12}) = \frac{2(1 + \nu_{xy})}{E_x}
\end{aligned}
\tag{4.23}
$$

so that only five of the coefficients s_{ij} are independent. The full set of stress-strain relations for a transversely isotropic material can therefore be written as follows:

$$\begin{aligned}
\varepsilon_{xx} &= \frac{1}{E_x}\sigma_{xx} - \frac{\nu_{xy}}{E_x}\sigma_{yy} - \frac{\nu_{zx}}{E_z}\sigma_{zz} \\
\varepsilon_{yy} &= -\frac{\nu_{xy}}{E_x}\sigma_{xx} + \frac{1}{E_x}\sigma_{yy} - \frac{\nu_{zx}}{E_z}\sigma_{zz} \\
\varepsilon_{zz} &= -\frac{\nu_{xz}}{E_x}\sigma_{xx} - \frac{\nu_{xz}}{E_x}\sigma_{yy} + \frac{1}{E_z}\sigma_{zz} \\
\varepsilon_{yz} &= \varepsilon_{xz} = \frac{1}{2\mu_{xz}}\sigma_{xz}; \qquad \varepsilon_{xy} = \frac{1+\nu_{xy}}{E_x}\sigma_{xy}
\end{aligned} \tag{4.24}$$

As before, $\nu_{zx}/E_z = \nu_{xz}/E_x$.

4.5.4 Stress-Strain Relations—Isotropic Material

The stress-strain relations for an isotropic elastic material are obtained by noting that the x, y, and z coordinate directions are equivalent. The values of the non-zero coefficients s_{ij} in the preceding equations can then be written in terms of two elastic constants ν and E:

$$\begin{aligned}
s_{11} &= s_{22} = s_{33} = \frac{1}{E} \\
s_{12} &= s_{21} = s_{13} = s_{31} = s_{23} = s_{32} = -\frac{\nu}{E} \\
s_{44} &= s_{55} = s_{66} = 2(s_{11} - s_{12}) = \frac{2(1+\nu)}{E}
\end{aligned} \tag{4.25}$$

Substituting these results into (4.22) and using (4.20), we obtain

$$\begin{aligned}
\varepsilon_{xx} &= \frac{1}{E}\left[\sigma_{xx} - \nu(\sigma_{yy} + \sigma_{zz})\right] \\
\varepsilon_{yy} &= \frac{1}{E}\left[\sigma_{yy} - \nu(\sigma_{xx} + \sigma_{zz})\right] \\
\varepsilon_{zz} &= \frac{1}{E}\left[\sigma_{zz} - \nu(\sigma_{xx} + \sigma_{yy})\right] \\
\varepsilon_{yz} &= \frac{1}{2\mu}\sigma_{yz}; \qquad \varepsilon_{xz} = \frac{1}{2\mu}\sigma_{xz}; \qquad \varepsilon_{xy} = \frac{1}{2\mu}\sigma_{xy}
\end{aligned} \tag{4.26}$$

Notice that if we add and subtract the term $\nu\sigma_{xx}/E$ from the first of these equations, and the terms $\nu\sigma_{yy}/E$ and $\nu\sigma_{zz}/E$ from the second and third equations, respectively, we can write Hooke's law for an isotropic material as follows:

$$\begin{aligned}
\varepsilon_{xx} &= \frac{1}{2\mu}\left[\sigma_{xx} - \frac{\nu}{1+\nu}(\sigma_{xx}+\sigma_{yy}+\sigma_{zz})\right] \\
\varepsilon_{yy} &= \frac{1}{2\mu}\left[\sigma_{yy} - \frac{\nu}{1+\nu}(\sigma_{xx}+\sigma_{yy}+\sigma_{zz})\right] \\
\varepsilon_{zz} &= \frac{1}{2\mu}\left[\sigma_{zz} - \frac{\nu}{1+\nu}(\sigma_{xx}+\sigma_{yy}+\sigma_{zz})\right] \\
\varepsilon_{yz} &= \frac{1}{2\mu}\sigma_{yz}; \qquad \varepsilon_{xz} = \frac{1}{2\mu}\sigma_{xz}; \qquad \varepsilon_{xy} = \frac{1}{2\mu}\sigma_{xy}
\end{aligned} \tag{4.27}$$

Equation (4.27) are in a convenient form for computing the strains when the stresses are known. If, however, the strains are known and the stresses are to be computed, then these equations must be inverted, i.e., they must be solved for the stresses. The results of this simple calculation are

$$\begin{aligned}
\sigma_{xx} &= 2\mu\left[\varepsilon_{xx} + \frac{\nu}{1-2\nu}(\varepsilon_{xx}+\varepsilon_{yy}+\varepsilon_{zz})\right] \\
\sigma_{yy} &= 2\mu\left[\varepsilon_{yy} + \frac{\nu}{1-2\nu}(\varepsilon_{xx}+\varepsilon_{yy}+\varepsilon_{zz})\right] \\
\sigma_{zz} &= 2\mu\left[\varepsilon_{zz} + \frac{\nu}{1-2\nu}(\varepsilon_{xx}+\varepsilon_{yy}+\varepsilon_{zz})\right] \\
\sigma_{yz} &= 2\mu\varepsilon_{yz}; \qquad \sigma_{xz} = 2\mu\varepsilon_{xz}; \qquad \sigma_{xy} = 2\mu\varepsilon_{xy}
\end{aligned} \tag{4.28}$$

In index notation, Eqs. (4.27) and (4.28) can be represented as

$$\varepsilon_{ij} = \frac{1}{2\mu}\left(\sigma_{ij} - \frac{\nu}{1+\nu}\sigma_{kk}\delta_{ij}\right) \tag{4.29}$$

and

$$\sigma_{ij} = 2\mu\left(\varepsilon_{ij} + \frac{\nu}{1-2\nu}\varepsilon_{kk}\delta_{ij}\right) \tag{4.30}$$

where δ_{ij} is the Kronecker delta, defined as in (2.58).

4.6 Stress-Strain Relations for Plane Stress and Plane Strain

Many engineering problems in elasticity are essentially two-dimensional in nature and can be classified as problems in either plane stress or plane strain. Plane stress means that the stresses are restricted to a single plane, say the x, y plane. In this case, we define plane stress by the conditions $\sigma_{zz} = \sigma_{xz} = \sigma_{yz} = 0$. Plane strain means that the strains are restricted to a single plane. We specify a state of plane strain for the x, y plane by the conditions $\varepsilon_{zz} = \varepsilon_{xz} = \varepsilon_{yz} = 0$.

Stress-strain relations for plane stress and plane strain can be written by using the above definitions in conjunction with the general three-dimensional equations given in Sects. 4.5.2–4.5.4. Considering first an orthotropic material in a state of plane stress with respect to the x, y plane, we set $\sigma_{zz} = \sigma_{xz} = \sigma_{yz} = 0$ in (4.22) and write

$$\begin{aligned}
\varepsilon_{xx} &= s_{11}\sigma_{xx} + s_{12}\sigma_{yy} \\
\varepsilon_{yy} &= s_{12}\sigma_{xx} + s_{22}\sigma_{yy} \\
\varepsilon_{xy} &= \tfrac{1}{2}s_{66}\sigma_{xy}
\end{aligned} \tag{4.31}$$

with the subsidiary results $\varepsilon_{xz} = \varepsilon_{yz} = 0$ and $\varepsilon_{zz} = s_{31}\sigma_{xx} + s_{32}\sigma_{yy} = s_{13}\sigma_{xx} + s_{23}\sigma_{yy}$. Equations (4.31) can be inverted to give

$$\begin{aligned}
\sigma_{xx} &= (s_{22}\varepsilon_{xx} - s_{12}\varepsilon_{yy})/(s_{11}s_{22} - s_{12}^2) \\
\sigma_{yy} &= (-s_{12}\varepsilon_{xx} + s_{11}\varepsilon_{yy})/(s_{11}s_{22} - s_{12}^2) \\
\sigma_{xy} &= 2\varepsilon_{xy}/s_{66}
\end{aligned} \tag{4.32}$$

The stress-strain relations for the x, y plane therefore only involve four of the nine independent elastic constants for the material: s_{11}, s_{22}, s_{12}, and s_{66}.

A state of plane strain for the x, y plane is defined by the conditions $\varepsilon_{zz} = \varepsilon_{xz} = \varepsilon_{yz} = 0$. We find from (4.22) in this case that

$$\begin{aligned}
\sigma_{xz} &= \sigma_{yz} = 0 \\
\sigma_{zz} &= -(s_{13}\sigma_{xx} + s_{23}\sigma_{yy})/s_{33}
\end{aligned}$$

so that

$$\begin{aligned}
\varepsilon_{xx} &= (s_{11} - s_{13}^2/s_{33})\sigma_{xx} + (s_{12} - s_{13}s_{23}/s_{33})\sigma_{yy} \\
\varepsilon_{yy} &= (s_{12} - s_{13}s_{23}/s_{33})\sigma_{xx} + (s_{22} - s_{23}^2/s_{33})\sigma_{yy} \\
\varepsilon_{xy} &= \frac{1}{2}s_{66}\sigma_{xy}
\end{aligned} \tag{4.33}$$

The inverse forms of (4.33) can be written as

$$\begin{aligned}
\sigma_{xx} &= c_{11}\varepsilon_{xx} + c_{12}\varepsilon_{yy} \\
\sigma_{yy} &= c_{21}\varepsilon_{xx} + c_{22}\varepsilon_{yy} \\
\sigma_{xy} &= 2c_{66}\varepsilon_{xy}
\end{aligned} \tag{4.34}$$

where

$$\begin{aligned} c_{11} &= (s_{22} - s_{23}^2/s_{33})/s_0^2 \\ c_{12} &= c_{21} = -(s_{12} - s_{13}s_{23}/s_{33})/s_0^2 \\ c_{22} &= (s_{11} - s_{13}^2/s_{33})/s_0^2 \\ c_{66} &= 1/s_{66} \end{aligned} \tag{4.35}$$

in which

$$s_0^2 = s_{11}s_{22} - s_{12}^2 + (2s_{12}s_{13}s_{23} - s_{11}s_{23}^2 - s_{22}s_{13}^2)/s_{33} \tag{4.36}$$

Just as in the case of plane stress, the stress-strain relations for the x, y plane can be expressed in terms of four constants, here denoted as c_{11}, c_{22}, c_{12}, and c_{66}. In order to compute these constants for a problem in plane strain, however, seven of the nine independent elastic constants for the orthotropic material must be known.

Despite the physical differences between the cases of plane stress and plane strain, there is a formal mathematical equivalence between these two conditions with respect to the stress-strain relations for the x, y plane. If, for example, we set s_{13} and s_{23} equal to zero in (4.33), then these equations are formally equivalent to (4.31). Consequently, it is not necessary to consider elasticity problems separately for plane stress and plane strain; we can go from one case to the other merely by modifying the values of the elastic constants. It appears, in fact, that we can represent the results for *both* cases by (4.34). The constants c_{11}, c_{22}, c_{12}, and c_{66} for a problem in plane stress can then be obtained from definitions (4.35) and (4.36) by setting $s_{13} = s_{23} = 0$ (cf. (4.32)).

In developing two-dimensional stress-strain relations for a transversely isotropic material, we are free to choose any one of the three coordinate directions as an axis of elastic symmetry. The z direction was selected for this purpose in Sect. 4.5.3. (This choice is the usual one for three-dimensional problems.) Plane stress and plane strain conditions for the x, y plane are not very interesting in this case, however, because the material is isotropic in this plane. If we select, say, the y direction as the axis of elastic symmetry, then the material will be anisotropic in the x, y plane. Furthermore, the x and z axes will be equivalent, which means that $E_x = E_z$ and $\nu_{xy} = \nu_{zy}$. We then find from (4.21) and (4.22) that $s_{11} = s_{33}$ and $s_{13} = s_{23}$ for the transversely isotropic material in question. As a consequence, (4.35) and (4.36) could be simplified by eliminating reference to the constants s_{33} and s_{23}. It is preferable, however, to leave these equations in general form, noting that they include transverse isotropy as a special case.

The two-dimensional stress-strain relations for an isotropic elastic material are also a special case of the preceding results. The equations for plane stress are found by substituting (4.25) into (4.31) and (4.32), which gives

$$
\begin{aligned}
\varepsilon_{xx} &= \frac{1}{E}(\sigma_{xx} - \nu\sigma_{yy}) = \frac{1}{2\mu}\left[\sigma_{xx} - \frac{\nu}{1+\nu}(\sigma_{xx} + \sigma_{yy})\right] \\
\varepsilon_{yy} &= \frac{1}{E}(-\nu\sigma_{xx} + \sigma_{yy}) = \frac{1}{2\mu}\left[\sigma_{yy} - \frac{\nu}{1+\nu}(\sigma_{xx} + \sigma_{yy})\right] \\
\varepsilon_{xy} &= \frac{1}{2\mu}\sigma_{xy}
\end{aligned}
\tag{4.37}
$$

and

$$
\begin{aligned}
\sigma_{xx} &= \frac{E}{1-\nu^2}(\varepsilon_{xx} + \nu\varepsilon_{yy}) = 2\mu\left[\varepsilon_{xx} + \frac{\nu}{1-\nu}(\varepsilon_{xx} + \varepsilon_{yy})\right] \\
\sigma_{yy} &= \frac{E}{1-\nu^2}(\nu\varepsilon_{xx} + \varepsilon_{yy}) = 2\mu\left[\varepsilon_{yy} + \frac{\nu}{1-\nu}(\varepsilon_{xx} + \varepsilon_{yy})\right] \\
\sigma_{xy} &= 2\mu\varepsilon_{xy}
\end{aligned}
\tag{4.38}
$$

In writing these expressions, use has been made of (4.20). Given that $\sigma_{zz} = \sigma_{xz} = \sigma_{yz} = 0$ for the state of plane stress in question, it is easy to show from (4.27) that $\varepsilon_{zz} = -\nu(\sigma_{xx} + \sigma_{yy})/E = -\nu(\varepsilon_{xx} + \varepsilon_{yy})/(1-\nu)$ and $\varepsilon_{xz} = \varepsilon_{yz} = 0$.

For an isotropic material in plane strain, we similarly find from (4.25) and (4.33)–(4.35) that

$$
\begin{aligned}
\varepsilon_{xx} &= \frac{1}{2\mu}\left[\sigma_{xx} - \nu(\sigma_{xx} + \sigma_{yy})\right] \\
\varepsilon_{yy} &= \frac{1}{2\mu}\left[\sigma_{yy} - \nu(\sigma_{xx} + \sigma_{yy})\right] \\
\varepsilon_{xy} &= \frac{1}{2\mu}\sigma_{xy}
\end{aligned}
\tag{4.39}
$$

and

$$
\begin{aligned}
\sigma_{xx} &= 2\mu\left[\varepsilon_{xx} + \frac{\nu}{1-2\nu}(\varepsilon_{xx} + \varepsilon_{yy})\right] \\
\sigma_{yy} &= 2\mu\left[\varepsilon_{yy} + \frac{\nu}{1-2\nu}(\varepsilon_{xx} + \varepsilon_{yy})\right] \\
\sigma_{xy} &= 2\mu\varepsilon_{xy}
\end{aligned}
\tag{4.40}
$$

In addition, the plane strain conditions $\varepsilon_{zz} = \varepsilon_{xz} = \varepsilon_{yz} = 0$ can be used in (4.28) to show that $\sigma_{zz} = 2\mu\nu(\varepsilon_{xx} + \varepsilon_{yy})/(1-2\nu) = \nu(\sigma_{xx} + \sigma_{yy})/(1-\nu)$ and $\sigma_{xz} = \sigma_{yz} = 0$.

The preceding equations can be written more compactly using index notation. The stress-strain relations for the x, y plane for plane stress conditions are

$$\varepsilon_{ij} = \frac{1}{2\mu}\left(\sigma_{ij} - \frac{\nu}{1+\nu}\sigma_{kk}\delta_{ij}\right) \tag{4.41}$$

and

$$\sigma_{ij} = 2\mu\left(\varepsilon_{ij} + \frac{\nu}{1-\nu}\varepsilon_{kk}\delta_{ij}\right) \tag{4.42}$$

while those for plane strain are

$$\varepsilon_{ij} = \frac{1}{2\mu}\left(\sigma_{ij} - \nu\sigma_{kk}\delta_{ij}\right) \tag{4.43}$$

and

$$\sigma_{ij} = 2\mu\left(\varepsilon_{ij} + \frac{\nu}{1-2\nu}\varepsilon_{kk}\delta_{ij}\right) \tag{4.44}$$

We see from these results that the stress-strain relations for plane stress can be formally obtained from those for plane strain by replacing ν by $\nu/(1+\nu)$. It is also important to note that stress-strain relation (4.44) for plane strain has the same form as the corresponding relation for an isotropic material in three dimensions, Eq. (4.30).

4.7 Transformation Formulas for Vectors and Tensors

Boundary element methods make frequent use of coordinate transformation formulas to change back and forth from local to global Cartesian coordinates. In addition, boundary element methods for elasticity make repeated use of transformation formulas for both vectors (the displacements and tractions) and tensors (the stresses and strains). The important transformation formulas used in the remainder of this book are summarized next.

In two dimensions, the basic formulas for translation between the global x, y Cartesian coordinate system and a local $\bar{x}$, $\bar{y}$ system are given by (1.29)

$$\begin{aligned} \bar{x} &= (x - \xi_x)\cos\beta + (y - \xi_y)\sin\beta \\ \bar{y} &= -(x - \xi_x)\sin\beta + (y - \xi_y)\cos\beta \end{aligned} \tag{4.45}$$

with the inverse transformation

$$\begin{aligned} x - \xi_x &= \bar{x}\cos\beta - \bar{y}\sin\beta \\ y - \xi_y &= \bar{x}\sin\beta + \bar{y}\cos\beta \end{aligned} \tag{4.46}$$

In these equations ξ_x and ξ_y are x and y distances from the origin of the $x,\ y$ system to that of the $\bar{x},\ \bar{y}$ system, and β is the rotation, positive in the counterclockwise sense as shown, for example, in Fig. 1.4.

4.7.1 Direction Cosines

The foregoing results can be put in a slightly different form by using the trigonometric identities $\cos(\pi/2 - \beta) = \sin\beta$ and $\cos(\pi/2 + \beta) = -\sin\beta$. Equation (4.45), for example, become

$$
\begin{aligned}
\bar{x} &= (x - \xi_x)\cos\beta + (y - \xi_y)\cos(\pi/2 - \beta) \\
\bar{y} &= (x - \xi_x)\cos(\pi/2 + \beta) + (y - \xi_y)\cos\beta
\end{aligned}
\tag{4.47}
$$

or, equivalently,

$$
\begin{aligned}
\bar{x}_1 &= (x_1 - \xi_1)\cos\beta + (x_2 - \xi_2)\cos(\pi/2 - \beta) \\
\bar{x}_2 &= (x_1 - \xi_1)\cos(\pi/2 + \beta) + (x_2 - \xi_2)\cos\beta
\end{aligned}
\tag{4.48}
$$

In order to generalize this result we will denote the angle between the $\bar{x}_i$ and x_j coordinate axes as $(\bar{x}_i, x_j)$. Furthermore, we will define the cosine of this angle as λ_{ij}, i.e.,

$$\lambda_{ij} = \cos(\bar{x}_i, x_j) \tag{4.49}$$

The quantity λ_{ij} is called the *direction cosine* of the positive $\bar{x}_i$ axis with respect to the positive x_j axis. In unabridged notation, we have

$$
\begin{aligned}
\lambda_{11} &= \cos(\bar{x}_1, x_1) = \cos\beta \\
\lambda_{12} &= \cos(\bar{x}_1, x_2) = \cos(\pi/2 - \beta) = \sin\beta \\
\lambda_{21} &= \cos(\bar{x}_2, x_1) = \cos(\pi/2 + \beta) = -\sin\beta \\
\lambda_{22} &= \cos(\bar{x}_2, x_2) = \cos\beta
\end{aligned}
\tag{4.50}
$$

so that (4.47) can be written as

$$\begin{aligned}\bar{x}_1 &= (x_1 - \xi_1)\cos(\bar{x}_1, x_1) + (x_2 - \xi_2)\cos(\bar{x}_1, x_2)\\ &= \lambda_{11}(x_1 - \xi_1) + \lambda_{12}(x_2 - \xi_2)\\ \bar{x}_2 &= (x_1 - \xi_1)\cos(\bar{x}_2, x_1) + (x_2 - \xi_2)\cos(\bar{x}_2, x_2)\\ &= \lambda_{21}(x_1 - \xi_1) + \lambda_{22}(x_2 - \xi_2)\end{aligned} \tag{4.51}$$

or, in index notation,

$$\bar{x}_i = \lambda_{ij}(x_j - \xi_j) \tag{4.52}$$

where i and j are 1 or 2. The direction cosines λ_{ij} provide a linear relationship between two vectors, $\bar{x}_i$ and $x_j - \xi_j$. They therefore constitute the components of a second order tensor.

The reader should verify that the inverse of transformation (4.52) is

$$x_i - \xi_i = \lambda_{ji}\bar{x}_j \tag{4.53}$$

4.7.2 Transformation of Vectors and Tensors

The direction cosine tensor λ_{ij} can be used to transform *any* vector in the x_i coordinate system to a vector in the $\bar{x}_i$ system. For example, the displacement vector in the $\bar{x}_i$ system can be written in terms of the corresponding vector in the x_j system as

$$\bar{u}_i = \lambda_{ij}u_j \tag{4.54}$$

which has the inverse

$$u_i = \lambda_{ji}\bar{u}_i \tag{4.55}$$

The same general formulas apply to components of the traction vector, i.e.,

$$\bar{t}_i = \lambda_{ij}t_j \tag{4.56}$$

and

$$t_i = \lambda_{ji}\bar{t}_j \tag{4.57}$$

and these results can be used to develop transformation formulas for the stress tensor.

We begin with the result $t_i = \sigma_{ji}n_j$ (see (4.16)) and the corresponding expression in $\bar{x}_i$ coordinates, $\bar{t}_i = \bar{\sigma}_{ji}\bar{n}_j$. It follows from (4.56) that

$$\begin{aligned}\overline{t}_i &= \lambda_{ik} t_k \\ &= \lambda_{ik}(\sigma_{pk} n_p) \\ &= \lambda_{ik}\sigma_{pk}(\lambda_{jp}\overline{n}_j) \\ &= \lambda_{jp}\lambda_{ik}\sigma_{pk}\overline{n}_j \\ &= \overline{\sigma}_{ji}\overline{n}_j\end{aligned}$$

so that

$$\overline{\sigma}_{ji} = \lambda_{jp}\lambda_{ik}\sigma_{pk} \tag{4.58}$$

A similar development starting with the result $t_i = \lambda_{ki}\overline{t}_k$ can be used to show that the inverse transformation is

$$\sigma_{ji} = \lambda_{pj}\lambda_{ki}\overline{\sigma}_{pk} \tag{4.59}$$

Equations (4.58) and (4.59) are the general transformation formulas for second order Cartesian tensors (Jeffreys 1963). It can be shown, for example, that the strain tensor ε_{ij} transforms according to the same formulas, i.e.,

$$\overline{\varepsilon}_{ji} = \lambda_{jp}\lambda_{ik}\varepsilon_{pk} \tag{4.60}$$

and

$$\varepsilon_{ji} = \lambda_{pj}\lambda_{ki}\overline{\varepsilon}_{pk} \tag{4.61}$$

4.7.3 Generalization to Three Dimensions

We state without proof that the preceding transformation formulas in terms of direction cosines are valid for three dimensions, where all indices range from 1 to 3.

4.7.4 Summary of Results for Two Dimensions

We conclude this section by writing in unabridged notation the two-dimensional transformation formulas for the displacements, tractions, and stresses that are required elsewhere in this book. These results are obtained by expanding the preceding expressions for indices ranging from 1 to 2, using definitions (4.50) for the direction cosines in terms of angle β, and replacing x_1 and x_2 by x and y. In writing these transformation formulas, we will associate the "bar" used to denote the $\overline{x}, \overline{y}$ coordinates with the subscripts for the displacement, traction, and stress components rather than with the symbols u, t, and σ used to represent these quantities. Thus, for example, we will write $u_{\overline{x}}$ and $u_{\overline{y}}$ instead of $\overline{u}_x$ and $\overline{u}_y$, and so forth for the traction and stress components.

The results are as follows:

- Displacement vector components

$$
\begin{aligned}
u_x &= u_{\overline{x}} \cos\beta - u_{\overline{y}} \sin\beta \\
u_y &= u_{\overline{x}} \sin\beta + u_{\overline{y}} \cos\beta
\end{aligned}
\tag{4.62}
$$

$$
\begin{aligned}
u_{\overline{x}} &= u_x \cos\beta + u_y \sin\beta \\
u_{\overline{y}} &= -u_x \sin\beta + u_y \cos\beta
\end{aligned}
\tag{4.63}
$$

- Traction vector components:

$$
\begin{aligned}
t_x &= t_{\overline{x}} \cos\beta - t_{\overline{y}} \sin\beta \\
t_y &= t_{\overline{x}} \sin\beta + t_{\overline{y}} \cos\beta
\end{aligned}
\tag{4.64}
$$

$$
\begin{aligned}
t_{\overline{x}} &= t_x \cos\beta + t_y \sin\beta \\
t_{\overline{y}} &= -t_x \sin\beta + t_y \cos\beta
\end{aligned}
\tag{4.65}
$$

- Stress tensor components:

$$
\begin{aligned}
\sigma_{xx} &= \sigma_{\overline{x}\overline{x}} \cos^2\beta - \sigma_{\overline{x}\overline{y}} \sin 2\beta + \sigma_{\overline{y}\overline{y}} \sin^2\beta \\
\sigma_{yy} &= \sigma_{\overline{x}\overline{x}} \sin^2\beta + \sigma_{\overline{x}\overline{y}} \sin 2\beta + \sigma_{\overline{y}\overline{y}} \cos^2\beta \\
\sigma_{xy} &= \sigma_{yx} = \tfrac{1}{2}(\sigma_{\overline{x}\overline{x}} - \sigma_{\overline{y}\overline{y}}) \sin 2\beta + \sigma_{\overline{x}\overline{y}} \cos 2\beta
\end{aligned}
\tag{4.66}
$$

$$
\begin{aligned}
\sigma_{\overline{x}\overline{x}} &= \sigma_{xx} \cos^2\beta + \sigma_{xy} \sin 2\beta + \sigma_{yy} \sin^2\beta \\
\sigma_{\overline{y}\overline{y}} &= \sigma_{xx} \sin^2\beta - \sigma_{xy} \sin 2\beta + \sigma_{yy} \cos^2\beta \\
\sigma_{\overline{x}\overline{y}} &= \sigma_{\overline{y}\overline{x}} = -\tfrac{1}{2}(\sigma_{xx} - \sigma_{yy}) \sin 2\beta + \sigma_{xy} \cos 2\beta
\end{aligned}
\tag{4.67}
$$

4.8 Formulation of Elasticity Problems

The object of solving an elasticity problem is to find the distributions of the stresses and displacements in an elastic body V, subject to a known set of body forces and to prescribed tractions or displacements on the bounding surface S of region V. In the general three-dimensional case, this means finding six stress components, $\sigma_{ij} = \sigma_{ji}$, and three displacements, u_i, each as a function of position in the body. The equilibrium equation (4.15) and the stress-strain relations (4.18) give nine equations for these nine unknowns:

$$\sigma_{ji,j} + B_i = 0 \tag{4.68}$$

where the body forces B_i are presumed known, and

$$\begin{aligned} \sigma_{ij} = \sigma_{ji} &= C_{ijkl}\varepsilon_{kl} \\ &= \tfrac{1}{2}C_{ijkl}(u_{k,l} + u_{l,k}) \end{aligned} \tag{4.69}$$

in which use has been made of (4.17). If we stipulate that the displacements u_i be continuous functions of the coordinates x_i, except possibly at specified locations within the region of interest, these equations are sufficient to determine the nine unknown quantities (Sokolnikoff 1956).

We will be mainly concerned in this book with two-dimensional problems. For these, we must find three stresses, σ_{xx}, σ_{yy}, and $\sigma_{xy} = \sigma_{yx}$, and two displacements, u_x and u_y, each as a function of x and y in the region of interest.

The solution of every elasticity problem must satisfy (4.68) and (4.69) at all points of region of interest V. Boundary conditions on S then distinguish one problem from another. Several different kinds of *boundary value problems* can be defined, depending upon the type of boundary conditions imposed on the boundaries of the body in question. A *traction boundary value problem* is one for which the components t_i of traction are specified at all points of the boundary (or boundaries) of the body; a *displacement boundary value problem* is one for which the components u_i of displacement are specified at these points. A problem resulting from a combination of these two basic boundary value problems is called a *mixed boundary value problem*.

4.8.1 Handling of Initial Stresses

Many problems in applied mechanics involve loading (or unloading) of a body that has previously been deformed and is subject to an initial state of stress. The case of an underground excavation in rock provides a good example of such a problem. When the excavation is created, the initial stresses are disturbed, deformations are induced in the surrounding rock, and a new state of stress is developed. In analyzing such a problem, it is convenient to represent the *total stresses* σ_{ij} as the sum of the *initial stresses* $\sigma_{ij}^{(0)}$ and the *stress changes* $\sigma_{ij}^{(1)}$ due to mining:

$$\sigma_{ij} = \sigma_{ij}^{(0)} + \sigma_{ij}^{(1)} \tag{4.70}$$

The stress changes $\sigma_{ij}^{(1)}$ are usually called the *induced stresses* in the geomechanics literature. Similar relationships can be written for the displacement components, namely,

$$u_i = u_i^{(0)} + u_i^{(1)} \tag{4.71}$$

The initial displacements $u_i^{(0)}$ are ordinarily taken as zero, because deformations in the field are measured relative to a "pre-mining" reference state. The total displacements u_i and the induced displacements $u_i^{(1)}$ are identical in this case.

The initial state of stress is presumed to be known—it can be estimated from a series of measurements, postulated from local geological conditions, or calculated by solving another boundary value problem. We can therefore determine σ_{ij} (and u_i) by finding $\sigma_{ij}^{(1)}$ (and $u_i^{(1)}$).

Definition of a boundary value problem in terms of induced stresses is facilitated by introducing the concepts of total tractions $\boldsymbol{t}$, initial tractions $\boldsymbol{t}^0$, and induced tractions $\boldsymbol{t}^{(1)}$. For a surface with outward normal components n_j, the relationships between the components of the tractions and stresses are

$$\begin{aligned} t_i &= \sigma_{ji} n_j \\ t_i^{(0)} &= \sigma_{ji}^{(0)} n_j \\ t_i^{(1)} &= \sigma_{ji}^{(1)} n_j \end{aligned} \tag{4.72}$$

It follows from (4.70) and (4.72) that

$$t_i = t_i^{(0)} + t_i^{(1)} \tag{4.73}$$

and so

$$t_i^{(1)} = t_i - t_i^{(0)} \tag{4.74}$$

Equation (4.74) is used to specify traction boundary conditions for the induced stress problem. For example, if the excavation does not contain artificial support, the total (or resultant) tractions t_i at the walls of the opening are zero. The induced tractions are then $t_i^{(1)} = -t_i^{(0)} = -\sigma_{ji}^{(0)} n_j$, and hence are known in terms of the initial stresses. Or, stating this finding in words: the induced tractions at all points along the excavation boundary are the negatives of the tractions that existed at these points prior to mining.

The foregoing discussion has tacitly assumed that the region of interest can be represented as an infinite or semi-infinite elastic solid, but the methodology is not restricted to mining: a similar superposition process can be followed to find the displacements and stresses in any initially stressed structure. The procedure is most advantageous, however, when the region of interest can be represented as an infinite or semi-infinite solid, because the initial stress conditions at infinity are automatically included in the analysis. When solving a boundary value problem for an infinite or semi-infinite region, therefore, one does not need to explicitly model a boundary at infinity. This fact has obvious significance for a boundary element method.

4.8.2 *Removal of Body Force Terms*

Boundary value problems in linear elasticity are easier to solve when the body forces B_i do not appear in the governing differential equations, i.e., when (4.68) reduces to homogeneous form. This can be accomplished by including the effects of body forces with the components of initial stresses $\sigma_{ij}^{(0)}$ and identifying them as a particular solution of (4.68):

$$\sigma_{ji,j}^{(0)} + B_i = 0 \tag{4.75}$$

By (4.68), (4.70), and (4.75), it is seen that the components of induced stresses $\sigma_{ij}^{(1)}$ satisfy the homogeneous system

$$\sigma_{ji,j}^{(1)} = 0 \tag{4.76}$$

as desired. If the particular solution $\sigma_{ij}^{(0)}$ is known, the solution of (4.76) subject to the appropriate boundary conditions gives the solution to the original problem, by virtue of (4.70).

If the values of the resultant traction components t_i are prescribed on S, the traction boundary conditions for the induced stress problem must be defined as in (4.74). In order to specify traction boundary conditions for the homogeneous system (4.76), therefore, the prescribed tractions t_i for the original problem must be modified by subtracting the tractions associated with the particular solution, $t_i^{(0)} = \sigma_{ji}^{(0)} n_j$.

4.9 Displacement Potentials—Isotropic Elasticity

In this section, we will discuss methods for solving two-dimensional problems in isotropic elasticity using functions called *displacement potentials*. This approach is normally used only for three-dimensional problems, because general techniques based on complex variable representations of Airy's stress function are highly developed for plane problems (see, for example, Sokolnikoff 1956; Muskhelishvili 1963). An advantage of using displacement potentials for two-dimensional problems is that one can gain insight into the procedures required to solve comparable problems in three dimensions.

4.9.1 Equations of Equilibrium in Terms of Displacements

We begin by writing the equations of equilibrium in terms of displacements. These calculations are conveniently carried out using index notation and, as we will presently observe, the results are valid for both plane strain, where indices are either 1 or 2, and three dimensions, where indices range from 1 to 3.

The required results are obtained by combining the equilibrium equation (4.15), the stress-strain relations (4.30), and the definitions of strain (4.17). First, differentiation of (4.30) with respect to x_j gives

$$\sigma_{ij,j} = 2\mu\left(\varepsilon_{ij,j} + \frac{\nu}{1-2\nu}\varepsilon_{kk,j}\delta_{ij}\right) \tag{4.77}$$

which must be equal to $-B_i$, by (4.15). By definition, δ_{ij} is zero if $i \neq j$ and unity if $i = j$. The term $\varepsilon_{kk,j}\delta_{ij}$ in (4.77) is therefore zero unless j equals i, i.e., $\varepsilon_{kk,j}\delta_{ij} = \varepsilon_{kk,i}$. Substitution of this result into (4.77) gives

$$\sigma_{ij,j} = 2\mu\left(\varepsilon_{ij,j} + \frac{\nu}{1-2\nu}\varepsilon_{kk,i}\right)$$

or, because $\sigma_{ji,j} = \sigma_{ij,j} = -B_i$,

$$2\mu\left(\varepsilon_{ij,j} + \frac{\nu}{1-2\nu}\varepsilon_{kk,i}\right) = -B_i \tag{4.78}$$

Equation (4.78) can now be expressed in terms of the displacements by using (4.17):

$$2\mu\left[\frac{1}{2}(u_{i,jj} + u_{j,ij}) + \frac{\nu}{1-2\nu}u_{k,ki}\right] = -B_i \tag{4.79}$$

The repeated indices in the separate terms of this expression denote summation, and therefore act as dummy indices, which may be given any symbol (provided, of course, that no symbol is used more than twice in any one term). In particular, we may write $u_{k,ki}$ as $u_{j,ji}$ without changing the meaning of (4.79). Hence, the equilibrium equations in terms of displacements are

$$2\mu\left[\frac{1}{2}(u_{i,jj} + u_{j,ij}) + \frac{\nu}{1-2\nu}u_{j,ji}\right] = -B_i$$

or, because $u_{j,ji} = u_{j,ij}$,

$$\mu\left(u_{i,jj} + \frac{1}{1-2\nu}u_{j,ij}\right) = -B_i \tag{4.80}$$

The term $u_{i,jj}$ in (4.80) can also be written as $\nabla^2 u_i$, where $\nabla^2 = \partial^2/\partial x_1^2 + \partial^2/\partial x_2^2 + \partial^2/\partial x_3^2$ is the Laplacian operator. An alternate way of writing (4.80) is therefore

$$\mu\left(\nabla^2 u_i + \frac{1}{1-2\nu} u_{j,ij}\right) = -B_i \tag{4.81}$$

These equations are known as the *Navier-Cauchy equations*. They form a set of three partial differential equations that are coupled because they each contain u_1, u_2, and u_3. This can be seen by writing the equations out in full, i.e.,

$$\begin{aligned}
\mu\left[\nabla^2 u_1 + \frac{1}{1-2\nu}\frac{\partial}{\partial x_1}\left(\frac{\partial u_1}{\partial x_1} + \frac{\partial u_2}{\partial x_2} + \frac{\partial u_3}{\partial x_3}\right)\right] &= -B_1 \\
\mu\left[\nabla^2 u_2 + \frac{1}{1-2\nu}\frac{\partial}{\partial x_2}\left(\frac{\partial u_1}{\partial x_1} + \frac{\partial u_2}{\partial x_2} + \frac{\partial u_3}{\partial x_3}\right)\right] &= -B_2 \\
\mu\left[\nabla^2 u_3 + \frac{1}{1-2\nu}\frac{\partial}{\partial x_3}\left(\frac{\partial u_1}{\partial x_1} + \frac{\partial u_2}{\partial x_2} + \frac{\partial u_3}{\partial x_3}\right)\right] &= -B_3
\end{aligned} \tag{4.82}$$

By construction, the Navier-Cauchy equations simultaneously satisfy the equilibrium equations and the stress-strain relations for an isotropic, linearly elastic material. We therefore conclude that a problem in three-dimensional, isotropic elasticity is reduced to the problem of solving the Navier-Cauchy equations subject to the appropriate boundary conditions. As explained in Sect. 4.8, the effects of body forces can be taken into account by using the principle of superposition. One is therefore normally concerned with solving the homogeneous form of (4.81), i.e., the form in which $B_i = 0$.

Finally, we note that the preceding development would be unchanged if we had assumed plane strain conditions at the outset, because the required form of the Hooke's law is the same in plane strain and three dimensions (see (4.30) and (4.44)).

4.9.2 Papkovitch-Neuber Functions

It was observed independently by Papkovitch in 1932 and Neuber in 1934 (see Sokolnikoff 1956) that the general solution of the Navier-Cauchy equations with zero body forces can be written as[4]

$$u_i = \varphi_i - \frac{1}{4(1-\nu)}\frac{\partial}{\partial x_i}(x_k\varphi_k + \varphi_0) \tag{4.83}$$

where the four functions φ_i ($i = 1$ to 3) and φ_0 satisfy Laplace's equation:

[4] The functions φ_i and φ_0 in (4.83) are multiples of 2μ of similarly named functions given by Sokolnikoff.

$$\nabla^2 \varphi_i = 0$$
$$\nabla^2 \varphi_0 = 0 \tag{4.84}$$

These functions are known as the *Papkovitch-Neuber functions*, or *displacement potentials*. It is left as an exercise for the reader to verify that the representation (4.83) satisfies (4.81) with $B_i = 0$.

The stresses σ_{ij} corresponding to displacements u_i in (4.83) are found by computing the strains ε_{ij} according to definitions (4.17) and substituting them into the stress-strain relations (4.30) (or (4.44)). First, combining (4.17) and (4.83) gives

$$\varepsilon_{ij} = \tfrac{1}{2}\left[\varphi_{i,j} - \frac{1}{4(1-\nu)}(x_k\varphi_k + \varphi_0)_{,ij} + \varphi_{j,i} - \frac{1}{4(1-\nu)}(x_k\varphi_k + \varphi_0)_{,ji}\right]$$

or

$$\varepsilon_{ij} = \frac{1}{2}(\varphi_{i,j} + \varphi_{j,i}) - \frac{1}{4(1-\nu)}(x_k\varphi_k + \varphi_0)_{,ij} \tag{4.85}$$

It is instructive to evaluate the term $(x_k\varphi_k + \varphi_0)_{,ij}$ in this expression by carrying out the indicated differentiations in stages, i.e.,

$$(x_k\varphi_k + \varphi_0)_{,ij} = (x_{k,i}\varphi_k + x_k\varphi_{k,i} + \varphi_{0,i})_{,j}$$

or, because $x_{k,i}\varphi_k = \delta_{ik}\varphi_k = \varphi_i$,

$$(x_k\varphi_k + \varphi_0)_{,ij} = (\varphi_i + x_k\varphi_{k,i} + \varphi_{0,i})_{,j}$$

Now, differentiation with respect to x_j gives

$$\begin{aligned}(x_k\varphi_k + \varphi_0)_{,ij} &= \varphi_{i,j} + x_{k,j}\varphi_{k,i} + x_k\varphi_{k,ij} + \varphi_{0,ij}\\ &= \varphi_{i,j} + \varphi_{j,i} + x_k\varphi_{k,ij} + \varphi_{0,ij}\end{aligned}$$

and hence

$$\varepsilon_{ij} = \frac{1}{2}(\varphi_{i,j} + \varphi_{j,i}) - \frac{1}{4(1-\nu)}(\varphi_{i,j} + \varphi_{j,i} + x_k\varphi_{k,ij} + \varphi_{0,ij})$$

or

$$\varepsilon_{ij} = \frac{1}{4(1-\nu)}\left[(1-2\nu)(\varphi_{i,j} + \varphi_{j,i}) - x_k\varphi_{k,ij} - \varphi_{0,ij}\right] \tag{4.86}$$

The sum of the three normal strains $\varepsilon_{11} + \varepsilon_{22} + \varepsilon_{33}$ is found by setting j equal to i in (4.86), i.e.,

$$\varepsilon_{ii} = \frac{1}{4(1-\nu)}\left[2(1-2\nu)\varphi_{i,i} - x_k\varphi_{k,ii} - \varphi_{0,ii}\right]$$

The terms $x_k\varphi_{k,ii}$ and $\varphi_{0,ii}$ in this expression stand for $x_k\nabla^2\varphi_k$ and $\nabla^2\varphi_0$, which both are zero by virtue of (4.84). We therefore have that

$$\varepsilon_{ii} = \frac{1-2\nu}{2(1-\nu)}\varphi_{i,i} \tag{4.87}$$

Writing k in place of i in (4.87) and using the resulting expression together with (4.86) in the stress-strain relations (4.30), we obtain

$$\sigma_{ij} = \frac{\mu}{2(1-\nu)}\left[(1-2\nu)(\varphi_{i,j}+\varphi_{j,i}) - x_k\varphi_{k,ij} - \varphi_{0,ij} + 2\nu\varphi_{k,k}\delta_{ij}\right] \tag{4.88}$$

This equation defines the stress components σ_{ij} in terms of the Papkovitch-Neuber functions φ_i and φ_0, as desired. The reader should verify that these stresses satisfy the equilibrium equations with zero body forces, $\sigma_{ij,j} = \sigma_{ji,j} = 0$.

4.9.3 Particular Representations for Two Dimensions

In order to solve an elasticity problem using displacement potentials, it is necessary to find the Papkovitch-Neuber functions for the problem in question. This task is difficult in general, but can be simplified for certain kinds of problems. An example of such a problem is the case for which the displacements or tractions are applied along a single plane, e.g., the plane boundary of an elastic half-space.

In this section, we will determine two particular representations for the Papkovitch-Neuber functions in two dimensions (plane strain). The first representation concerns the case for which the surface of the half-plane $x_2 \leq 0$ is free from shear traction, i.e., $t_1 = \sigma_{12} = 0$ on $x_2 = 0$ for $-\infty < x_1 < \infty$. The second representation concerns the case for which the surface of the same half-plane is free from normal traction, i.e., $t_2 = \sigma_{22} = 0$ on $x_2 = 0$ for $-\infty < x_1 < \infty$. (Note that the components of the unit outward normal to the surface of the half-plane $x_2 \leq 0$ are $n_1 = 0$ and $n_2 = +1$. The tractions on this surface are thus $t_1 = \sigma_{11}n_1 + \sigma_{21}n_2 = \sigma_{21} = \sigma_{12}$ and $t_2 = \sigma_{12}n_1 + \sigma_{22}n_2 = \sigma_{22}$.)

The displacements and stresses in terms of the Papkovitch-Neuber functions for two dimensions are obtained from (4.83) and (4.88) by letting the indices be 1 or 2. Writing out the equations in full, we have

$$\begin{aligned} u_1 &= \varphi_1 - \frac{1}{4(1-\nu)}(\varphi_1 + x_1\varphi_{1,1} + x_2\varphi_{2,1} + \varphi_{0,1}) \\ u_2 &= \varphi_2 - \frac{1}{4(1-\nu)}(x_1\varphi_{1,2} + \varphi_2 + x_2\varphi_{2,2} + \varphi_{0,2}) \end{aligned} \tag{4.89}$$

and

$$\begin{aligned}\sigma_{11} &= \frac{\mu}{2(1-\nu)}\left[2(1-\nu)\varphi_{1,1} + 2\nu\varphi_{2,2} - x_1\varphi_{1,11} - x_2\varphi_{2,11} - \varphi_{0,11}\right]\\ \sigma_{22} &= \frac{\mu}{2(1-\nu)}\left[2(1-\nu)\varphi_{2,2} + 2\nu\varphi_{1,1} - x_1\varphi_{1,22} - x_2\varphi_{2,22} - \varphi_{0,22}\right]\\ \sigma_{12} &= \frac{\mu}{2(1-\nu)}\left[(1-2\nu)(\varphi_{1,2} + \varphi_{2,1}) - x_1\varphi_{1,12} - x_2\varphi_{2,12} - \varphi_{0,12}\right]\end{aligned} \tag{4.90}$$

We seek particular representations for the cases that, respectively, $\sigma_{12} = 0$ and $\sigma_{22} = 0$ on $x_2 = 0$ *for all values of* x_1. In order to eliminate the multiplications by x_1 in (4.90), we set φ_1 equal to zero, i.e.,

$$\varphi_1 = 0 \tag{4.91}$$

With this choice for φ_1, (4.89) and (4.90) become

$$\begin{aligned}u_1 &= -\frac{1}{4(1-\nu)}(x_2\varphi_{2,1} + \varphi_{0,1})\\ u_2 &= \varphi_2 - \frac{1}{4(1-\nu)}(\varphi_2 + x_2\varphi_{2,2} + \varphi_{0,2})\end{aligned} \tag{4.92}$$

and

$$\begin{aligned}\sigma_{11} &= \frac{\mu}{2(1-\nu)}\left[2\nu\varphi_{2,2} - x_2\varphi_{2,11} - \varphi_{0,11}\right]\\ \sigma_{22} &= \frac{\mu}{2(1-\nu)}\left[2(1-\nu)\varphi_{2,2} - x_2\varphi_{2,22} - \varphi_{0,22}\right]\\ \sigma_{12} &= \frac{\mu}{2(1-\nu)}\left[(1-2\nu)\varphi_{2,1} - x_2\varphi_{2,12} - \varphi_{0,12}\right]\end{aligned} \tag{4.93}$$

We now consider the two cases (i) $\sigma_{12} = 0$ on $x_2 = 0$ and (ii) $\sigma_{22} = 0$ on $x_2 = 0$ for $-\infty < x_1 < \infty$. For the first case, the third one of (4.93) shows that σ_{12} will be zero when $x_2 = 0$ provided $\varphi_{2,12}$ is finite for $x_2 = 0$, and provided $\varphi_{0,12} = (1-2\nu)\varphi_{2,1}$, i.e.,

$$\frac{\partial^2\varphi_0}{\partial x_1 \partial x_2} = (1-2\nu)\frac{\partial\varphi_2}{\partial x_1} \tag{4.94}$$

or

$$\frac{\partial\varphi_0}{\partial x_2} = (1-2\nu)\varphi_2 \tag{4.95}$$

in which a linear function of x_2 has been omitted. (See exercise (5) in Sect. 4.28 in this connection.) Using (4.95), we can now express the displacements and stresses in (4.92) and (4.93) in terms of φ_0 alone. For notational convenience, we introduce a new function $\Phi(x_1, x_2)$, defined in terms of φ_0 as

$$\varphi_0 = 4(1-\nu)(1-2\nu)\Phi \tag{4.96}$$

in which, by virtue of (4.84), function Φ must satisfy Laplace's equation in two dimensions:

$$\nabla^2\Phi = \frac{\partial^2\Phi}{\partial x_1^2} + \frac{\partial^2\Phi}{\partial x_2^2} = 0 \tag{4.97}$$

Combination of (4.95) and (4.96) then yields the result

$$\varphi_2 = 4(1-\nu)\frac{\partial\Phi}{\partial x_2} \tag{4.98}$$

Finally, substitution of (4.96) and (4.98) into (4.92) and (4.93) gives the following expressions for the displacements and stresses in terms of a single harmonic function Φ:

$$\begin{aligned} u_1 &= -(1-2\nu)\Phi_{,1} - x_2\Phi_{,12} \\ u_2 &= 2(1-\nu)\Phi_{,2} - x_2\Phi_{,22} \end{aligned} \tag{4.99}$$

and

$$\begin{aligned} \sigma_{11} &= 2\mu(\Phi_{,22} + x_2\Phi_{,222}) \\ \sigma_{22} &= 2\mu(\Phi_{,22} - x_2\Phi_{,222}) \\ \sigma_{12} &= -2\mu x_2\Phi_{,122} \end{aligned} \tag{4.100}$$

By construction, this representation is appropriate only for cases in which the shear traction $t_1 = \sigma_{12}$ is zero everywhere on $x_2 = 0$. Different choices for the harmonic function Φ will then correspond to certain values of normal displacement u_2 or normal traction $t_2 = \sigma_{22}$ along $x_2 = 0$.

In a similar way, inspection of the second one of (4.93) shows that σ_{22} is zero on $x_2 = 0$ provided $\varphi_{2,22}$ is finite on $x_2 = 0$, and provided $\varphi_{0,22} = 2(1-\nu)\varphi_{2,2}$, i.e.,

$$\frac{\partial^2\varphi_0}{\partial x_2^2} = 2(1-\nu)\frac{\partial\varphi_2}{\partial x_2} \tag{4.101}$$

or

$$\frac{\partial\varphi_0}{\partial x_2} = 2(1-\nu)\varphi_2 \tag{4.102}$$

in which a linear function of x_1 has been omitted. In order to obtain expressions for the displacements and streses in forms comparable to (4.99) and (4.100) we now express φ_0 in terms of a function $\Psi(x_1, x_2)$ as

$$\varphi_0 = 8(1-\nu)^2\int\frac{\partial\Psi}{\partial x_1}\,\mathrm{d}x_2 \tag{4.103}$$

in which

$$\nabla^2\Psi = \frac{\partial^2\Psi}{\partial x_1^2} + \frac{\partial^2\Psi}{\partial x_2^2} = 0 \tag{4.104}$$

We note from (4.103) and (4.104) that

$$\begin{aligned} \frac{\partial\varphi_0}{\partial x_2} &= 8(1-\nu)^2\frac{\partial\Psi}{\partial x_1} \\ \frac{\partial\varphi_0}{\partial x_1} &= -8(1-\nu)^2\frac{\partial\Psi}{\partial x_2} \end{aligned} \tag{4.105}$$

and hence, using (4.102),

$$\varphi_2 = 4(1-\nu)\frac{\partial\Psi}{\partial x_1} \tag{4.106}$$

Substitution of (4.105) and (4.106) into (4.92) and (4.93) gives the following expressions for the displacements and stresses in terms of a single harmonic function Ψ:

$$\begin{aligned} u_1 &= 2(1-\nu)\Psi_{,2} + x_2\Psi_{,22} \\ u_2 &= (1-2\nu)\Psi_{,1} - x_2\Psi_{,12} \end{aligned} \tag{4.107}$$

and

$$\begin{aligned} \sigma_{11} &= 2\mu(2\Psi_{,12} + x_2\Psi_{,122}) \\ \sigma_{22} &= -2\mu x_2\Psi_{,122} \\ \sigma_{12} &= 2\mu(\Psi_{,22} + x_2\Psi_{,222}) \end{aligned} \tag{4.108}$$

This representation is appropriate for cases in which the normal traction $t_2 = \sigma_{22}$ is zero everywhere on $x_2 = 0$. Different choices for the harmonic function Ψ will then correspond to certain values of shear displacement u_1 or shear traction $t_1 = \sigma_{12}$ along $x_2 = 0$.

The preceding two particular representations for the Papkovitch-Neuber functions relate to different boundary conditions for the surface of the half-plane $x_2 \leq 0$. We may therefore superimpose the results to obtain the following expressions for the displacements and stresses in terms of two harmonic functions, $\Phi(x_1, x_2)$ and $\Psi(x_1, x_2)$:

$$\begin{aligned} u_1 &= -(1-2\nu)\Phi_{,1} + 2(1-\nu)\Psi_{,2} - x_2(\Psi_{,1} + \Phi_{,2})_{,1} \\ u_2 &= 2(1-\nu)\Phi_{,2} + (1-2\nu)\Psi_{,1} - x_2(\Psi_{,1} + \Phi_{,2})_{,2} \end{aligned} \tag{4.109}$$

and

$$
\begin{aligned}
\sigma_{11} &= 2\mu\left[\Phi_{,22} + 2\Psi_{,12} + x_2(\Psi_{,1} + \Phi_{,2})_{,22}\right] \\
\sigma_{22} &= 2\mu\left[\Phi_{,22} - x_2(\Psi_{,1} + \Phi_{,2})_{,22}\right] \\
\sigma_{12} &= 2\mu\left[\Psi_{,22} - x_2(\Psi_{,1} + \Phi_{,2})_{,12}\right]
\end{aligned} \tag{4.110}
$$

In writing these expressions, use has been made of the relations $\Phi_{,11} = -\Phi_{,22}$ and $\Psi_{,11} = -\Psi_{,22}$, which follow directly from (4.97) and (4.104).

Equations (4.109) and (4.110) should be compared with (4.89) and (4.90), which give the displacements and stresses in terms of *three* harmonic functions. The difference between these two sets of equations is that (4.109) and (4.110) are restricted to problems in which boundary conditions can be specified with respect to the line $x_2 = 0$, whereas (4.89) and (4.90) can accommodate more general boundary conditions. Consequently, when using the solution to the Navier-Cauchy equations in the form (4.109) we achieve some simplification, but we do so at the expense of generality.

4.10 Displacement Potentials—Anisotropic Elasticity

Three-dimensional elasticity problems for transversely isotropic materials can be solved using a displacement potential approach analogous to the one described above for an isotropic material. This general approach was developed by Elliott (1948); Shield (1951), and Lodge (1955), among others, and is summarized in the book by Green and Zerna (1968). We will consider this approach below for the case of two dimensions. Other techniques are available for solving two-dimensional elasticity problems for anisotropic materials (see Green and Zerna 1968), but the displacement potential formulation has the advantage of being directly extensible to three dimensions. Further, as we will see, the solutions to certain problems in anisotropic elasticity can be deduced from the displacement potentials for the corresponding problems in isotropic elasticity.

We observed in Sect. 4.6 that the two-dimensional stress-strain relations for orthotropic and transversely isotropic elastic materials have the same general form. This permits us to develop the two-dimensional displacement potential approach for the case of an orthotropic material, even though the three-dimensional procedure is restricted to transversely isotropic materials.

4.10.1 Equations of Equilibrium in Terms of Displacements

As in Sect. 4.9, we begin by writing the equilibrium equations in terms of displacements. Index notation is not particularly helpful in this case, because the stress-strain relations (4.34) are expressed in terms of coefficients c_{ij} whose indices do not correspond to the coordinate directions. We will therefore carry out the computations in unabridged form.

Substitution of stress-strain relations (4.34) into the equilibrium equation (4.4) gives

$$
\begin{aligned}
c_{11}\frac{\partial \varepsilon_{xx}}{\partial x} + c_{12}\frac{\partial \varepsilon_{yy}}{\partial x} + 2c_{66}\frac{\partial \varepsilon_{xy}}{\partial y} &= -B_x \\
2c_{66}\frac{\partial \varepsilon_{xy}}{\partial x} + c_{12}\frac{\partial \varepsilon_{xx}}{\partial y} + c_{22}\frac{\partial \varepsilon_{yy}}{\partial y} &= -B_y
\end{aligned}
\tag{4.111}
$$

or, using the definitions of strain (4.13),

$$
\begin{aligned}
c_{11}\frac{\partial^2 u_x}{\partial x^2} + c_{66}\frac{\partial^2 u_x}{\partial y^2} + (c_{12} + c_{66})\frac{\partial^2 u_y}{\partial x \partial y} &= -B_x \\
c_{66}\frac{\partial^2 u_y}{\partial x^2} + c_{22}\frac{\partial^2 u_y}{\partial y^2} + (c_{12} + c_{66})\frac{\partial^2 u_x}{\partial x \partial y} &= -B_y
\end{aligned}
\tag{4.112}
$$

Equation (4.112) are the required results. These equations must be satisfied for every two-dimensional problem in orthotropic (transversely isotropic) elasticity. As mentioned previously, we will ordinarily be concerned with the homogeneous form of these equations, i.e., the form in which $B_x = B_y = 0$.

4.10.2 Solution in Terms of Potentials

Following Elliott (1948), we take $B_x = B_y = 0$ and provisionally assume that the solution of (4.112) can be found in form

$$
\begin{aligned}
u_x &= \frac{\partial}{\partial x}\phi(x, y) \\
u_y &= q\frac{\partial}{\partial y}\phi(x, y)
\end{aligned}
\tag{4.113}
$$

where q is a constant and ϕ is an as yet unrestricted function of x and y.[5] Substituting these expressions into (4.112), we find that

[5] The symbols q and ϕ were previously used for problems in potential theory. The different meanings for these quantities for problems in elasticity should be clear from context.

$$c_{11}\frac{\partial^3\phi}{\partial x^3}+[c_{66}+q(c_{12}+c_{66})]\frac{\partial^3\phi}{\partial x\partial y^2}=0$$
$$[(c_{12}+c_{66})+qc_{66}]\frac{\partial^3\phi}{\partial x^2\partial y}+qc_{22}\frac{\partial^3\phi}{\partial y^3}=0$$

or

$$\frac{\partial}{\partial x}\left\{\frac{\partial^2\phi}{\partial x^2}+\frac{[c_{66}+q(c_{12}+c_{66})]}{c_{11}}\frac{\partial^2\phi}{\partial y^2}\right\}=0$$
$$\frac{\partial}{\partial y}\left\{\frac{\partial^2\phi}{\partial x^2}+\frac{qc_{22}}{[(c_{12}+c_{66})+qc_{66}]}\frac{\partial^2\phi}{\partial y^2}\right\}=0$$

and hence

$$\begin{aligned}\frac{\partial^2\phi}{\partial x^2}+\frac{[c_{66}+q(c_{12}+c_{66})]}{c_{11}}\frac{\partial^2\phi}{\partial y^2}&=A_1(y)\\ \frac{\partial^2\phi}{\partial x^2}+\frac{qc_{22}}{[(c_{12}+c_{66})+qc_{66}]}\frac{\partial^2\phi}{\partial y^2}&=A_2(x)\end{aligned}\tag{4.114}$$

It can be shown that A_1 and A_2 are in fact constants, and furthermore that they represent a constant initial stress state in the region of interest.[6] Without loss in generality we may take the initial stresses as zero, so that

$$\begin{aligned}\frac{\partial^2\phi}{\partial x^2}+\frac{[c_{66}+q(c_{12}+c_{66})]}{c_{11}}\frac{\partial^2\phi}{\partial y^2}&=0\\ \frac{\partial^2\phi}{\partial x^2}+\frac{qc_{22}}{[(c_{12}+c_{66})+qc_{66}]}\frac{\partial^2\phi}{\partial y^2}&=0\end{aligned}\tag{4.115}$$

A nontrivial solution of these two equations can be found only if they are identical. This occurs if the expressions preceding the terms $\partial^2\phi/\partial y^2$ are equal to a constant, say γ^2. (We take the constant as γ^2 rather than γ to avoid having to use square roots later in the analysis.) Thus, we require that

---

[6] Equations (4.13), (4.34), and (4.113) show that the stresses associated with function $A_1(y)$, for example, in (4.114) are $\sigma_{xx}=c_{11}A_1(y)$, $\sigma_{yy}=c_{12}A_1(y)$, and $\sigma_{xy}=c_{66}(1+q)(x\,\mathrm{d}A_1(y)/\mathrm{d}y+\mathrm{d}F(y)/\mathrm{d}y)$, in which $F(y)$ is a function obtained by integrating $\partial^2\phi/\partial x^2$ in the first one of (4.114). Equilibrium equations (4.4) then require that $A_1(y)=B_1$ and $\mathrm{d}F(y)/\mathrm{d}y=C_1$, where B_1 and C_1 are constants. A similar analysis can be done to show that $A_2(x)$ is also a constant.

$$\frac{c_{66} + q(c_{12} + c_{66})}{c_{11}} = \gamma^2$$
$$\frac{qc_{22}}{(c_{12} + c_{66}) + qc_{66}} = \gamma^2 \tag{4.116}$$

so that (4.115) can be written as one equation:

$$\frac{\partial^2 \phi}{\partial x^2} + \gamma^2 \frac{\partial^2 \phi}{\partial y^2} = 0 \tag{4.117}$$

This is the governing differential equation for the function ϕ introduced in (4.113). We next seek a relation for the constant γ^2.

The first one of (4.116) can be solved for q to give

$$q = \frac{c_{11}\gamma^2 - c_{66}}{c_{12} + c_{66}} \tag{4.118}$$

Substitution of this expression into the second one of (4.116) then gives the following quadratic equation for γ^2:

$$c_{11}c_{66}\gamma^4 + [c_{12}(2c_{66} + c_{12}) - c_{11}c_{22}]\gamma^2 + c_{22}c_{66} = 0 \tag{4.119}$$

The two roots of this equation are designated as γ_1^2 and γ_2^2. It can be shown that these quantities are either both real and positive, or are complex conjugates with positive real parts (Elliott 1948; Green and Zerna 1968); we will henceforth consider only the case for which they are real. The values of q in (4.118) associated with γ_1^2 and γ_2^2 are denoted as q_1 and q_2, i.e.,

$$q_1 = \frac{c_{11}\gamma_1^2 - c_{66}}{c_{12} + c_{66}}$$
$$q_2 = \frac{c_{11}\gamma_2^2 - c_{66}}{c_{12} + c_{66}}$$

or

$$q_i = \frac{c_{11}\gamma_i^2 - c_{66}}{c_{12} + c_{66}}, \qquad i = 1 \text{ or } 2 \tag{4.120}$$

Note that index i in this expression refers to a number, and not a coordinate direction.

We have thus established that (4.117) is satisfied for two values of γ^2. Denoting the values of ϕ associated with γ_1^2 and γ_2^2 by ϕ_1 and ϕ_2, respectively, the possible functions $\phi(x, y)$ for representation (4.113) must satisfy the equations

$$\frac{\partial^2 \phi_1}{\partial x^2} + \gamma_1^2 \frac{\partial^2 \phi_1}{\partial y^2} = 0$$
$$\frac{\partial^2 \phi_2}{\partial x^2} + \gamma_2^2 \frac{\partial^2 \phi_2}{\partial y^2} = 0$$

or briefly,

$$\frac{\partial^2 \phi_i}{\partial x^2} + \gamma_i^2 \frac{\partial^2 \phi_i}{\partial y^2} = 0, \quad i = 1 \text{ or } 2 \text{ (no sum on } i) \tag{4.121}$$

We then take the solution of (4.112) as

$$\begin{aligned} u_x &= \frac{\partial \phi_1}{\partial x} + \frac{\partial \phi_2}{\partial x} \\ u_y &= q_1 \frac{\partial \phi_1}{\partial y} + q_2 \frac{\partial \phi_2}{\partial y} \end{aligned} \tag{4.122}$$

where q_1 and q_2 are given by (4.120), functions ϕ_1 and ϕ_2 satisfy (4.121), and $B_x = B_y = 0$.

Functions ϕ_1 and ϕ_2 are the displacement potentials for an orthotropic elastic body in a state of plane strain. These functions may be called "quasi-harmonic" because of the similarity between (4.121) and Laplace's equation.

4.10.3 Evaluation of the Stress Components

The stress components can be expressed in terms of functions ϕ_1 and ϕ_2 by computing the strains associated with displacements (4.122) and using them in stress-strain relations (4.34). We find in this way that

$$\begin{aligned} \sigma_{xx} &= (c_{12}q_1 - c_{11}\gamma_1^2)\frac{\partial^2 \phi_1}{\partial y^2} + (c_{12}q_2 - c_{11}\gamma_2^2)\frac{\partial^2 \phi_2}{\partial y^2} \\ \sigma_{yy} &= (c_{22}q_1 - c_{12}\gamma_1^2)\frac{\partial^2 \phi_1}{\partial y^2} + (c_{22}q_2 - c_{12}\gamma_2^2)\frac{\partial^2 \phi_2}{\partial y^2} \\ \sigma_{xy} &= c_{66}\left[(1+q_1)\frac{\partial^2 \phi_1}{\partial x \partial y} + (1+q_2)\frac{\partial^2 \phi_2}{\partial x \partial y}\right] \end{aligned} \tag{4.123}$$

where use has been made of the result $\partial^2 \phi_i / \partial x^2 = -\gamma_i^2 \partial^2 \phi_i / \partial y^2$ ($i = 1$ or 2; no sum on i). The terms $c_{12}q_i - c_{11}\gamma_i^2$ and $c_{22}q_i - c_{12}\gamma_i^2$ occurring in the expressions for σ_{xx} and σ_{yy} can be written in slightly different forms by using (4.116). The term $c_{12}q_i - c_{11}\gamma_i^2$, for example, can be written with the aid of the first one of (4.116) as

$$
\begin{aligned}
c_{12}q_i - c_{11}\gamma_i^2 &= c_{12}q_i - [(c_{12}+c_{66})q_i + c_{66}] \\
&= -c_{66}(1+q_i)
\end{aligned}
\tag{4.124}
$$

Similarly, the second one of (4.116) can be used to find an equivalent way of writing $c_{22}q_i - c_{12}\gamma_i^2$:

$$
\begin{aligned}
c_{22}q_i - c_{12}\gamma_i^2 &= \gamma_i^2\,[(c_{12}+c_{66}) + q_i c_{66}] - c_{12}\gamma_i^2 \\
&= c_{66}(1+q_i)\gamma_i^2
\end{aligned}
\tag{4.125}
$$

(no sum on i). Combining (4.123) through (4.125), we obtain the following general expressions for the stresses in terms of displacement potentials ϕ_1 and ϕ_2:

$$
\begin{aligned}
\sigma_{xx} &= -c_{66}\left[(1+q_1)\frac{\partial^2\phi_1}{\partial y^2} + (1+q_2)\frac{\partial^2\phi_2}{\partial y^2}\right] \\
\sigma_{yy} &= c_{66}\left[(1+q_1)\gamma_1^2\frac{\partial^2\phi_1}{\partial y^2} + (1+q_2)\gamma_2^2\frac{\partial^2\phi_2}{\partial y^2}\right] \\
\sigma_{xy} &= c_{66}\left[(1+q_1)\frac{\partial^2\phi_1}{\partial x\partial y} + (1+q_2)\frac{\partial^2\phi_2}{\partial x\partial y}\right]
\end{aligned}
\tag{4.126}
$$

4.10.4 Particular Representations

Equations (4.122) and (4.126) give the general solution to a plane problem in orthotropic (transversely isotropic) elasticity in terms of two quasi-harmonic functions ϕ_1 and ϕ_2. In order to solve a particular problem, we must find functions ϕ_1 and ϕ_2 such that the appropriate traction or displacement boundary conditions are met. As in the corresponding situation in isotropic elasticity, this task is difficult in general, but can be simplified for cases in which the tractions or displacements are specified along a straight line, e.g., the surface of an elastic half-plane.

In the remainder of this section we will develop two particular representations for the potentials ϕ_1 and ϕ_2, comparable to the results given in Sect. 4.9.3 for isotropic elasticity. The first representation concerns the case for which the surface of the half-plane $y \le 0$ is free from shear traction, i.e., $t_x = \sigma_{xy} = 0$ on $y = 0$ for $-\infty < x < \infty$, and the second concerns the case for which the surface of the same half-plane is free from normal traction, i.e., $t_y = \sigma_{yy} = 0$ on $y = 0$ for $-\infty < x < \infty$. Again, these two particular solutions can be combined to specify general boundary conditions along $y = 0$.

Before constructing the required particular solutions, we note that (4.122) and (4.126) can be written in yet another way by introducing two scaled y coordinates y_1 and y_2, defined by the relations $y_1 = y/\gamma_1$ and $y_2 = y/\gamma_2$, or briefly

$$y_i = \frac{y}{\gamma_i} \tag{4.127}$$

cf. Sect. 3.14. First and second derivatives with respect to y can then be expressed in terms of the scaled y coordinates by using the chain rule for partial differentiation, i.e.,

$$\begin{aligned} \frac{\partial}{\partial y} &= \frac{\partial y_i}{\partial y}\frac{\partial}{\partial y_i} = \frac{1}{\gamma_i}\frac{\partial}{\partial y_i} \\ \frac{\partial^2}{\partial y^2} &= \frac{\partial}{\partial y}\left(\frac{1}{\gamma_i}\frac{\partial}{\partial y_i}\right) = \frac{1}{\gamma_i^2}\frac{\partial^2}{\partial y_i^2} \end{aligned} \tag{4.128}$$

(no sum on i). Introducing these results into (4.122) and (4.126), we find

$$\begin{aligned} u_x &= \frac{\partial \phi_1}{\partial x} + \frac{\partial \phi_2}{\partial x} \\ u_y &= \frac{q_1}{\gamma_1}\frac{\partial \phi_1}{\partial y_1} + \frac{q_2}{\gamma_2}\frac{\partial \phi_2}{\partial y_2} \end{aligned} \tag{4.129}$$

and

$$\begin{aligned} \sigma_{xx} &= -c_{66}\left[\frac{(1+q_1)}{\gamma_1^2}\frac{\partial^2 \phi_1}{\partial y_1^2} + \frac{(1+q_2)}{\gamma_2^2}\frac{\partial^2 \phi_2}{\partial y_2^2}\right] \\ \sigma_{yy} &= c_{66}\left[(1+q_1)\frac{\partial^2 \phi_1}{\partial y_1^2} + (1+q_2)\frac{\partial^2 \phi_2}{\partial y_2^2}\right] \\ \sigma_{xy} &= c_{66}\left[\frac{(1+q_1)}{\gamma_1}\frac{\partial^2 \phi_1}{\partial x \partial y_1} + \frac{(1+q_2)}{\gamma_2}\frac{\partial^2 \phi_2}{\partial x \partial y_2}\right] \end{aligned} \tag{4.130}$$

The advantage of introducing scaled coordinates y_i becomes evident if we write the governing equation for ϕ_i, Eq. (4.121), in terms of these coordinates:

$$\frac{\partial^2 \phi_i}{\partial x^2} + \frac{\partial^2 \phi_i}{\partial y_i^2} = 0, \qquad i = 1 \text{ or } 2 \text{ (no sum on } i) \tag{4.131}$$

Thus, functions ϕ_i are harmonic in x, y_i coordinates.[7]

We now consider the problem of finding particular representations for ϕ_1 and ϕ_2 that respectively make $\sigma_{xy} = 0$ and $\sigma_{yy} = 0$ on $y = 0$ for $-\infty < x < \infty$. For the first case ($\sigma_{xy} = 0$ on $y = 0$ for $-\infty < x < \infty$), we adopt the procedure used by Shield (1951) and take ϕ_1 and ϕ_2 as follows:

[7] A similar result was found in Sect. 3.14 for the potential ϕ for a material with anisotropic hydraulic or thermal conductivity.

$$
\begin{aligned}
\phi_1 &= \frac{\gamma_1}{1+q_1}\Phi(x, y_1) \\
\phi_2 &= -\frac{\gamma_2}{1+q_2}\Phi(x, y_2)
\end{aligned}
\tag{4.132}
$$

where $\Phi(x, y)$ is harmonic in x, y coordinates, i.e.,

$$
\frac{\partial^2}{\partial x^2}\Phi(x, y) + \frac{\partial^2}{\partial y^2}\Phi(x, y) = 0 \tag{4.133}
$$

In view of (4.132), functions $\Phi(x, y_1)$ and $\Phi(x, y_2)$ are necessarily harmonic in x, y_1 and x, y_2 coordinates, respectively. Functions ϕ_1 and ϕ_2 therefore satisfy their respective governing equation, (4.131).

Substituting (4.132) into the third one of (4.130), we obtain

$$
\sigma_{xy} = c_{66}\left[\frac{\partial^2}{\partial x \partial y_1}\Phi(x, y_1) - \frac{\partial^2}{\partial x \partial y_2}\Phi(x, y_2)\right] \tag{4.134}
$$

which is equal to zero when y equals zero, as desired, because functions $\Phi(x, y_1)$ and $\Phi(x, y_2)$ and all their derivatives are equal when $y = 0$. This conclusion follows from the observation that scaled coordinates y_1 and y_2 are both equal to zero when $y = 0$. These coordinates are not equal to one another for any other value of y, provided $\gamma_1 \neq \gamma_2$. We will for now exclude the case that γ_1 equals γ_2, but will show later that this circumstance can be handled by adopting a limiting procedure.

The complete solution to the problem at hand is obtained by substituting (4.132) into (4.129) and (4.130). The results are as follows:

$$
\begin{aligned}
u_x &= \frac{\gamma_1}{1+q_1}\frac{\partial}{\partial x}\Phi(x, y_1) - \frac{\gamma_2}{1+q_2}\frac{\partial}{\partial x}\Phi(x, y_2) \\
u_y &= \frac{q_1}{1+q_1}\frac{\partial}{\partial y_1}\Phi(x, y_1) - \frac{q_2}{1+q_2}\frac{\partial}{\partial y_2}\Phi(x, y_2)
\end{aligned}
\tag{4.135}
$$

and

$$
\begin{aligned}
\sigma_{xx} &= -c_{66}\left[\frac{1}{\gamma_1}\frac{\partial^2}{\partial y_1^2}\Phi(x, y_1) - \frac{1}{\gamma_2}\frac{\partial^2}{\partial y_2^2}\Phi(x, y_2)\right] \\
\sigma_{yy} &= c_{66}\left[\gamma_1\frac{\partial^2}{\partial y_1^2}\Phi(x, y_1) - \gamma_2\frac{\partial^2}{\partial y_2^2}\Phi(x, y_2)\right] \\
\sigma_{xy} &= c_{66}\left[\frac{\partial^2}{\partial x \partial y_1}\Phi(x, y_1) - \frac{\partial^2}{\partial x \partial y_2}\Phi(x, y_2)\right]
\end{aligned}
\tag{4.136}
$$

The particular case for which $\sigma_{yy} = 0$ on $y = 0$ for $-\infty < x < \infty$ can be treated in a similar manner. In order to achieve notational similarity between this case and the corresponding one considered in Sect. 4.9.3 for an isotropic material, we will express the results in terms of a harmonic function $\Psi(x, y)$, i.e.,

$$\frac{\partial^2}{\partial x^2}\Psi(x, y) + \frac{\partial^2}{\partial y^2}\Psi(x, y) = 0 \tag{4.137}$$

Inspection of the second one of (4.130) shows that we could make $\sigma_{yy} = 0$ on $y = 0$ by choosing ϕ_1 and ϕ_2 as $(1 + q_1)^{-1}\Psi(x, y_1)$ and $-(1 + q_2)^{-1}\Psi(x, y_2)$, respectively. Although this choice is acceptable, some simplification will result later if we instead choose them as follows:

$$\begin{aligned} \phi_1 &= \frac{1}{1+q_1}\int \frac{\partial}{\partial y_1}\Psi(x, y_1)\,\mathrm{d}x \\ \phi_2 &= -\frac{1}{1+q_2}\int \frac{\partial}{\partial y_2}\Psi(x, y_2)\,\mathrm{d}x \end{aligned} \tag{4.138}$$

These functions satisfy Eq. (4.131), because $\Psi(x, y_1)$ and $\Psi(x, y_2)$ are harmonic in x, y_1 and x, y_2 coordinates by virtue of (4.137). Substitution of (4.138) into (4.129) and (4.130) then yields the following expressions for the displacements and stresses in terms of a single harmonic function $\Psi(x, y)$:

$$\begin{aligned} u_x &= \frac{1}{1+q_1}\frac{\partial}{\partial y_1}\Psi(x, y_1) - \frac{1}{1+q_2}\frac{\partial}{\partial y_2}\Psi(x, y_2) \\ u_y &= -\frac{q_1}{\gamma_1(1+q_1)}\frac{\partial}{\partial x}\Psi(x, y_1) + \frac{q_2}{\gamma_2(1+q_2)}\frac{\partial}{\partial x}\Psi(x, y_2) \end{aligned} \tag{4.139}$$

and

$$\begin{aligned} \sigma_{xx} &= c_{66}\left[\frac{1}{\gamma_1^2}\frac{\partial^2}{\partial x\partial y_1}\Psi(x, y_1) - \frac{1}{\gamma_2^2}\frac{\partial^2}{\partial x\partial y_2}\Psi(x, y_2)\right] \\ \sigma_{yy} &= -c_{66}\left[\frac{\partial^2}{\partial x\partial y_1}\Psi(x, y_1) - \frac{\partial^2}{\partial x\partial y_2}\Psi(x, y_2)\right] \\ \sigma_{xy} &= c_{66}\left[\frac{1}{\gamma_1}\frac{\partial^2}{\partial y_1^2}\Psi(x, y_1) - \frac{1}{\gamma_2}\frac{\partial^2}{\partial y_2^2}\Psi(x, y_2)\right] \end{aligned} \tag{4.140}$$

where use has been made of the relations $\partial^2\Psi(x, y_i)/\partial x^2 = -\partial^2\Psi(x, y_i)/\partial y_i^2$. The second one of (4.140) shows that $\sigma_{yy} = 0$ on $y = 0$ for $-\infty < x < \infty$, as desired.

4.11 Kelvin's Problem for Plane Strain

Boundary element methods in elasticity are based on singular solutions to the displacement equations of equilibrium that correspond to isolated disturbances at points within the region of interest. The solution to the problem of a concentrated force at a point in an infinite elastic solid is fundamental in this connection and is presented below for both isotropic and anisotropic elasticity. Additional singular solutions are recorded in Sects. 4.15, 8.3, and 8.5.

4.11.1 *Concentrated Force—Isotropic Elasticity*

The problem of a concentrated force in an infinite elastic solid is known as Kelvin's problem (Sokolnikoff 1956). We consider here the plane strain version of this problem in which a concentrated force $F_i = (F_x, F_y)$ is applied at the point $\boldsymbol{\xi} = (\xi_x, \xi_y)$. For this two-dimensional problem the force actually represents a *line* of force applied along the z axis, perpendicular to the x, y plane, and its components $F_x > 0$ and $F_y > 0$ have dimensions of force/length (e.g., Newtons/meter).

The solution to this problem can be expressed in terms of the fundamental harmonic function for two dimensions (see 2.41), $G(\boldsymbol{x}, \boldsymbol{\xi}) = -(1/2\pi)\ln r$, where $r = [(x-\xi_x)^2 + (y-\xi_y)^2]^{1/2}$ is the distance between an arbitrary point $\boldsymbol{x} = (x, y)$ in the plane and the point of application of the force, $\boldsymbol{\xi} = (\xi_x, \xi_y)$. Here, and in the sequel, we will represent this function multiplied by the constant $[2(1-\nu)]^{-1}$ as $g(\boldsymbol{x}, \boldsymbol{\xi})$, i.e.,

$$g(\boldsymbol{x}, \boldsymbol{\xi}) = \frac{1}{2(1-\nu)} G(\boldsymbol{x}, \boldsymbol{\xi}) \tag{4.141}$$

so that

$$g(x-\xi_x, y-\xi_y) = -\frac{1}{4\pi(1-\nu)} \ln\left[(x-\xi_x)^2 + (y-\xi_y)^2\right]^{1/2} \tag{4.142}$$

For the case that the concentrated force is applied at the origin, i.e., $\xi_x = 0, \xi_y = 0$, the displacements due to that force can be written in terms of the function $g(x, y)$ as

$$\begin{aligned} u_x &= \frac{F_x}{2\mu}\left[(3-4\nu)g - xg_{,x}\right] + \frac{F_y}{2\mu}\left[-yg_{,x}\right] \\ u_y &= \frac{F_x}{2\mu}\left[-xg_{,y}\right] + \frac{F_y}{2\mu}\left[(3-4\nu)g - yg_{,y}\right] \end{aligned} \tag{4.143}$$

and the stresses as

$$
\begin{aligned}
\sigma_{xx} &= F_x\left[2(1-\nu)g_{,x} - xg_{,xx}\right] + F_y\left[2\nu g_{,y} - yg_{,xx}\right] \\
\sigma_{yy} &= F_x\left[2\nu g_{,x} - xg_{,yy}\right] + F_y\left[2(1-\nu)g_{,y} - yg_{,yy}\right] \\
\sigma_{xy} &= F_x\left[(1-2\nu)g_{,y} - xg_{,xy}\right] + F_y\left[(1-2\nu)g_{,x} - yg_{,xy}\right]
\end{aligned}
\tag{4.144}
$$

It can be verified from (4.89) and (4.90) that these results are obtained from the following Papkovitch-Neuber functions:

$$
\begin{aligned}
\varphi_x &= \varphi_1 = \frac{2(1-\nu)F_x}{\mu} g(x,y) \\
\varphi_y &= \varphi_2 = \frac{2(1-\nu)F_y}{\mu} g(x,y) \\
\varphi_0 &= 0
\end{aligned}
\tag{4.145}
$$

It is instructive to confirm that stresses (4.144) indeed correspond to a (line of) concentrated force at the origin. To this end, we first record the required derivatives of function $g(x, y)$ in terms of polar coordinates $x = r\cos\theta$ and $y = r\sin\theta$:

$$
\begin{aligned}
g_{,x} &= -\frac{1}{4\pi(1-\nu)}\frac{x}{x^2+y^2} = -\frac{1}{4\pi(1-\nu)}\frac{\cos\theta}{r} \\
g_{,y} &= -\frac{1}{4\pi(1-\nu)}\frac{y}{x^2+y^2} = -\frac{1}{4\pi(1-\nu)}\frac{\sin\theta}{r} \\
g_{,xy} &= \frac{1}{4\pi(1-\nu)}\frac{2xy}{(x^2+y^2)^2} = \frac{1}{4\pi(1-\nu)}\frac{\sin 2\theta}{r^2} \\
g_{,xx} = -g_{,yy} &= \frac{1}{4\pi(1-\nu)}\frac{x^2-y^2}{(x^2+y^2)^2} = \frac{1}{4\pi(1-\nu)}\frac{\cos 2\theta}{r^2}
\end{aligned}
\tag{4.146}
$$

Substitution of these results into (4.144) then yields

$$
\begin{aligned}
\sigma_{xx} &= \frac{F_x}{4\pi(1-\nu)}\frac{\cos\theta}{r}\left[-2(1-\nu) - \cos 2\theta\right] + \frac{F_y}{4\pi(1-\nu)}\frac{\sin\theta}{r}\left[-2\nu - \cos 2\theta\right] \\
\sigma_{yy} &= \frac{F_x}{4\pi(1-\nu)}\frac{\cos\theta}{r}\left[-2\nu + \cos 2\theta\right] + \frac{F_y}{4\pi(1-\nu)}\frac{\sin\theta}{r}\left[-2(1-\nu) + \cos 2\theta\right] \\
\sigma_{xy} &= \frac{F_x}{4\pi(1-\nu)}\frac{\sin\theta}{r}\left[-(1-2\nu) - 2\cos^2\theta\right] + \frac{F_y}{4\pi(1-\nu)}\frac{\cos\theta}{r}\left[-(1-2\nu) - 2\sin^2\theta\right]
\end{aligned}
\tag{4.147}
$$

We see from these expressions that the stresses are defined everywhere except at the origin, $r = 0$.

In order to avoid dealing with infinite stresses at $r = 0$, we imagine that the material within a small circular hole of radius $r = \varepsilon$ is excluded from the analysis, as depicted in Fig. 4.7a. The resultant traction components on the boundary of this hole are $\int_0^{2\pi} t_x \varepsilon \, d\theta$ and $\int_0^{2\pi} t_y \varepsilon \, d\theta$. As shown below, these components are statically equivalent to the force $F_i = (F_x, F_y)$.

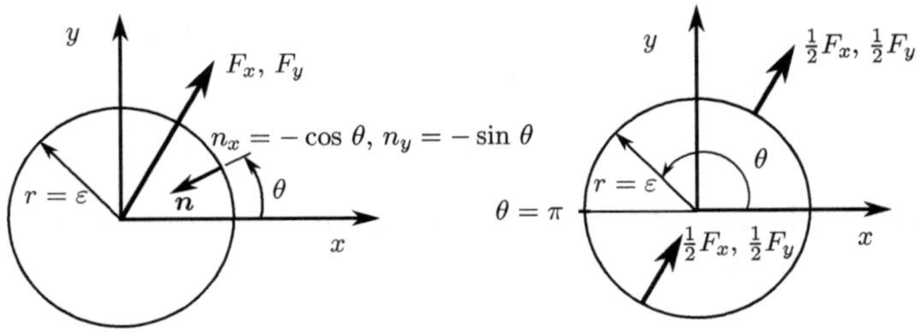

(a) Concentrated force at origin. (b) Resultant forces on two semicircles.

Fig. 4.7 Kelvin's problem, plane strain

The components of the unit outward normal to the boundary of the region $r \geq \varepsilon$ are $n_x = -\cos\theta$ and $n_y = -\sin\theta$ (see Fig. 4.7a). The tractions on $r = \varepsilon$ are obtained by using these results in (4.6) and (4.147). The computation is straightforward and, after some simple manipulation, we obtain the following expressions:

$$
\begin{aligned}
t_x &= \frac{F_x}{4\pi(1-\nu)}\frac{1}{\varepsilon}\left[2(1-\nu)+\cos 2\theta\right] + \frac{F_y}{4\pi(1-\nu)}\frac{1}{\varepsilon}\sin 2\theta \\
t_y &= \frac{F_x}{4\pi(1-\nu)}\frac{1}{\varepsilon}\sin 2\theta + \frac{F_y}{4\pi(1-\nu)}\frac{1}{\varepsilon}\left[2(1-\nu)-\cos 2\theta\right]
\end{aligned}
\tag{4.148}
$$

The resultant tractions are then

$$
\begin{aligned}
\int_0^{2\pi} t_x\varepsilon\,\mathrm{d}\theta &= \frac{F_x}{2\pi}\int_0^{2\pi}\mathrm{d}\theta + \frac{1}{4\pi(1-\nu)}\int_0^{2\pi}(F_x\cos 2\theta + F_y\sin 2\theta)\,\mathrm{d}\theta \\
\int_0^{2\pi} t_y\varepsilon\,\mathrm{d}\theta &= \frac{F_y}{2\pi}\int_0^{2\pi}\mathrm{d}\theta + \frac{1}{4\pi(1-\nu)}\int_0^{2\pi}(F_x\sin 2\theta - F_y\cos 2\theta)\,\mathrm{d}\theta
\end{aligned}
\tag{4.149}
$$

which are equal to F_x and F_y, respectively, independent of ε. We can therefore consider the limiting case in which ε is zero and imagine that the force $F_i = (F_x, F_y)$ is concentrated at the origin.

A physical interpretation of the manner in which the force is applied is shown in Fig. 4.7b. This interpretation is developed by performing the integrations in (4.149) over two semi-circles, say, 0 to π and π to 2π. We may then consider that one-half of the force is applied to the upper semi-circle, and one-half to the lower one, as depicted in the figure. As a result, one half of the circular boundary (the upper half) is being "pushed," while the other half is being "pulled." An important consequence of this is that the stress tensor components exhibit discontinuities, or "jumps," in the neighborhood of the applied force. As a simple example, suppose $F_x = 0$ and $F_y > 0$. Then, from (4.147), the values of σ_{yy} along the upper half of the small circular boundary $r = \varepsilon$ are negative, i.e., compressive, while those on the bottom half are positive. We will return to this topic later because discontinuities associated with an applied load are crucial to the development of boundary element methods.

Equations (4.143) and (4.144) give the displacements and stresses at an arbitrary point (x, y) due to a concentrated force at the origin. If the force is applied at the point (ξ_x, ξ_y) we replace x and y by $x - \xi_x$ and $y - \xi_y$, respectively, and write (4.143) and (4.144) as

$$
\begin{aligned}
u_x &= \frac{F_x}{2\mu}\left[(3-4\nu)g - (x-\xi_x)g_{,x}\right] + \frac{F_y}{2\mu}\left[-(y-\xi_y)g_{,x}\right] \\
u_y &= \frac{F_x}{2\mu}\left[-(x-\xi_x)g_{,y}\right] + \frac{F_y}{2\mu}\left[(3-4\nu)g - (y-\xi_y)g_{,y}\right]
\end{aligned}
\tag{4.150}
$$

and

$$
\begin{aligned}
\sigma_{xx} &= F_x\left[2(1-\nu)g_{,x} - (x-\xi_x)g_{,xx}\right] + F_y\left[2\nu g_{,y} - (y-\xi_y)g_{,xx}\right] \\
\sigma_{yy} &= F_x\left[2\nu g_{,x} - (x-\xi_x)g_{,yy}\right] + F_y\left[2(1-\nu)g_{,y} - (y-\xi_y)g_{,yy}\right] \\
\sigma_{xy} &= F_x\left[(1-2\nu)g_{,y} - (x-\xi_x)g_{,xy}\right] + F_y\left[(1-2\nu)g_{,x} - (y-\xi_y)g_{,xy}\right]
\end{aligned}
\tag{4.151}
$$

where $g(x - \xi_x, y - \xi_y)$ is defined as in (4.142).

The derivatives of function $g(x - \xi_x, y - \xi_y)$ all go to zero as $r = [(x - \xi_x)^2 + (y - \xi_y)^2]^{1/2} \to \infty$ and hence the stresses (4.151) vanish at infinity. The displacements (4.150), however, have terms with $g(x - \xi_x, y - \xi_y)$ alone, and these terms are unbounded at infinity. These infinite displacements are due to the nature of the applied loading. Although we have adopted a plane strain condition for the x, y plane, it is implicit that the line force is infinitely long in the z direction. (Otherwise, plane strain conditions would not apply for an arbitrary plane whose normal is parallel to the z axis.) The total applied force (i.e., force/length times length) is therefore infinite, and the unbounded displacements at infinity are needed to accommodate it.

4.11.2 Concentrated Force—Anisotropic Elasticity

In solving Kelvin's problem for an orthotropic (or transversely isotropic) material in two dimensions it is convenient to consider separately the two components of the concentrated line force, F_x and F_y. Guided by the results for the isotropic case, we will examine the two sets of displacement potentials

$$\phi_1 = C_x \frac{\gamma_1}{q_1} \int \ln \sqrt{x^2 + y_1^2}\, \mathrm{d}x$$
$$\phi_2 = -C_x \frac{\gamma_2}{q_2} \int \ln \sqrt{x^2 + y_2^2}\, \mathrm{d}x \tag{4.152}$$

and

$$\phi_1 = C_y \int \ln \sqrt{x^2 + y_1^2}\, \mathrm{d}y_1$$
$$\phi_2 = -C_y \int \ln \sqrt{x^2 + y_2^2}\, \mathrm{d}y_2 \tag{4.153}$$

where C_x and C_y are constants and where $\gamma_1,\ \gamma_2$ and $q_1,\ q_2$ are defined in (4.119) and (4.120). We will show below that C_x and C_y are proportional to the components F_x and F_y of a concentrated line of force at the origin.

One reason for selecting the potentials (4.152) and (4.153) in the indicated form is that, by symmetry, the displacement perpendicular to the direction of the applied force must be zero along the line of application of the force. (This requirement applies only to force components acting parallel to the directions of elastic symmetry, which here are the x and y coordinate axes.) Referring to (4.129), we see that potential (4.152) meets this requirement with respect to the displacement component u_y due to a force $F_x \neq 0,\ F_y = 0$ and that (4.153) meets it with respect to the component u_x due to a force $F_x = 0,\ F_y \neq 0$. Further, by analogy with the solution for the isotropic case, we know that the shear stress σ_{xy} is zero along the lines of action of the individual force components F_x and F_y, except at the point of application of the force, where all the stress components are undefined. The last one of (4.130) shows that potentials (4.152) and (4.153) also meet this condition.

We now determine the values of the constants C_x and C_y in terms of forces F_x and F_y by noting, as for the isotropic case, that the integrals of the traction components t_x and t_y along a closed contour surrounding the origin are statically equivalent to the force components F_x and F_y. To this end, we note that the tractions on the boundary of a small circular hole $r = \varepsilon$ can be written using (4.130) as

$$\begin{aligned} t_x = -\frac{c_{66}}{\varepsilon}\Bigg[(1+q_1)\left\{\frac{C_x\gamma_1/q_1 - C_y \sin\theta\cos\theta(1-\gamma_1^2)/\gamma_1}{\gamma_1^2\cos^2\theta + \sin^2\theta}\right\} \\ -(1+q_2)\left\{\frac{C_x\gamma_2/q_2 - C_y \sin\theta\cos\theta(1-\gamma_2^2)/\gamma_2}{\gamma_2^2\cos^2\theta + \sin^2\theta}\right\}\Bigg] \\ t_y = -\frac{c_{66}}{\varepsilon}\Bigg[\gamma_1(1+q_1)\left\{\frac{C_x \sin\theta\cos\theta(1-\gamma_1^2)/q_1 + C_y}{\gamma_1^2\cos^2\theta + \sin^2\theta}\right\} \\ -\gamma_2(1+q_2)\left\{\frac{C_x \sin\theta\cos\theta(1-\gamma_2^2)/q_2 + C_y}{\gamma_2^2\cos^2\theta + \sin^2\theta}\right\}\Bigg] \end{aligned} \tag{4.154}$$

The relations[8]

$$F_x = \int_{-\pi}^{\pi} t_x \varepsilon \, \mathrm{d}\theta$$
$$F_y = \int_{-\pi}^{\pi} t_y \varepsilon \, \mathrm{d}\theta \tag{4.155}$$

then give

$$F_x = 2\pi c_{66}(q_1 - q_2)C_x$$
$$F_y = -2\pi c_{66}(q_1 - q_2)C_y \tag{4.156}$$

because

$$\gamma_i \int_{-\pi}^{\pi} \frac{1}{\gamma_i^2 \cos^2\theta + \sin^2\theta} \mathrm{d}\theta = \left[\arctan(\frac{1}{\gamma_i} \tan\theta)\right]_{-\pi}^{\pi} = 2\pi$$
$$(1-\gamma_i^2) \int_{-\pi}^{\pi} \frac{\sin\theta\cos\theta}{\gamma_i^2 \cos^2\theta + \sin^2\theta} \mathrm{d}\theta = \left[\ln\sqrt{\gamma_i^2\cos^2\theta + \sin^2\theta}\right]_{-\pi}^{\pi} = 0 \tag{4.157}$$

(see, for example, Petit Bois 1961). The first of conditions (4.157) follows from the fact that $\arctan(\tan\theta/\gamma_i)$ can also be written as $\arctan((y/\gamma_i)/x)$. The limits of integration $\theta = \pi$ and $\theta = -\pi$ are then equivalent to the conditions $x < 0,\ y = 0_+$ and $x < 0,\ y = 0_-$, respectively, which directly lead to the indicated result. In writing the first one of (4.156) use was made of the relation $q_1 q_2 = 1$, a result that follows from (4.120) after noting from (4.119) that $\gamma_1^2\gamma_2^2 = c_{22}/c_{11}$.

Equations (4.156) determine the constants C_x and C_y as

$$C_x = \frac{1}{2\pi c_{66}(q_1 - q_2)} F_x$$
$$C_y = -\frac{1}{2\pi c_{66}(q_1 - q_2)} F_y \tag{4.158}$$

Using these results in (4.152) and (4.153) we find that the displacement potentials ϕ_1 and ϕ_2 for the subject problem are

$$\phi_1 = \frac{1}{2\pi c_{66}(q_1 - q_2)}\left[F_x \frac{\gamma_1}{q_1} \int \ln\sqrt{x^2 + y_1^2}\, \mathrm{d}x - F_y \int \ln\sqrt{x^2 + y_1^2}\, \mathrm{d}y_1\right]$$
$$\phi_2 = -\frac{1}{2\pi c_{66}(q_1 - q_2)}\left[F_x \frac{\gamma_2}{q_2} \int \ln\sqrt{x^2 + y_2^2}\, \mathrm{d}x - F_y \int \ln\sqrt{x^2 + y_2^2}\, \mathrm{d}y_2\right] \tag{4.159}$$

[8] The indicated limits of integration are needed to accommodate the discontinuity in the arctangent function along the negative x axis.

The displacements and stresses due to the concentrated force are found by using these potentials in (4.129) and (4.130). The results are

$$
\begin{aligned}
u_x &= \frac{1}{2\pi c_{66}(q_1-q_2)}F_x\left[\frac{\gamma_1}{q_1}\ln\sqrt{x^2+y_1^2}-\frac{\gamma_2}{q_2}\ln\sqrt{x^2+y_2^2}\right]\\
&\quad-\frac{1}{2\pi c_{66}(q_1-q_2)}F_y\left[\arctan\frac{y_1}{x}-\arctan\frac{y_2}{x}\right]\\
u_y &= -\frac{1}{2\pi c_{66}(q_1-q_2)}F_x\left[\arctan\frac{y_1}{x}-\arctan\frac{y_2}{x}\right]\\
&\quad-\frac{1}{2\pi c_{66}(q_1-q_2)}F_y\left[\frac{q_1}{\gamma_1}\ln\sqrt{x^2+y_1^2}-\frac{q_2}{\gamma_2}\ln\sqrt{x^2+y_2^2}\right]
\end{aligned}
\tag{4.160}
$$

and

$$
\begin{aligned}
\sigma_{xx} &= \frac{1}{2\pi(q_1-q_2)}F_x\left[\frac{(1+q_1)}{\gamma_1 q_1}\frac{x}{x^2+y_1^2}-\frac{(1+q_2)}{\gamma_2 q_2}\frac{x}{x^2+y_2^2}\right]\\
&\quad+\frac{1}{2\pi(q_1-q_2)}F_y\left[\frac{(1+q_1)}{\gamma_1^2}\frac{y_1}{x^2+y_1^2}-\frac{(1+q_2)}{\gamma_2^2}\frac{y_2}{x^2+y_2^2}\right]\\
\sigma_{yy} &= -\frac{1}{2\pi(q_1-q_2)}F_x\left[\frac{\gamma_1(1+q_1)}{q_1}\frac{x}{x^2+y_1^2}-\frac{\gamma_2(1+q_2)}{q_2}\frac{x}{x^2+y_2^2}\right]\\
&\quad-\frac{1}{2\pi(q_1-q_2)}F_y\left[(1+q_1)\frac{y_1}{x^2+y_1^2}-(1+q_2)\frac{y_2}{x^2+y_2^2}\right]\\
\sigma_{xy} &= \frac{1}{2\pi(q_1-q_2)}F_x\left[\frac{(1+q_1)}{q_1}\frac{y_1}{x^2+y_1^2}-\frac{(1+q_2)}{q_2}\frac{y_2}{x^2+y_2^2}\right]\\
&\quad-\frac{1}{2\pi(q_1-q_2)}F_y\left[\frac{(1+q_1)}{\gamma_1}\frac{x}{x^2+y_1^2}-\frac{(1+q_2)}{\gamma_2}\frac{x}{x^2+y_2^2}\right]
\end{aligned}
\tag{4.161}
$$

In these equations, γ_1^2 and γ_2^2 are the two roots of (4.119), q_1 and q_2 are defined as in (4.120), and $y_1 = y/\gamma_1$ and $y_2 = y/\gamma_2$. Equations (4.160) and (4.161) are written under the assumption that γ_1^2 and γ_2^2 are both real and positive. Similar equations could be written for the case that these two constants are complex conjugates (Green 1943), but this circumstance appears to be of limited practical significance, so we will not consider it here. In the event that γ_1 and γ_2 are equal, say $\gamma_1 = \gamma_2 = \gamma$, a limiting procedure using L'Hopital's rule can be used to obtain the solution from (4.160) and (4.161).[9] The special case of isotropy occurs when $\gamma_1 = \gamma_2 = 1$; (4.160) and (4.161) then reduce to (4.143) and (4.144).

For the case in which the concentrated force is applied at a point $(x, y) = (\xi_x, \xi_y)$, the solution is obtained by replacing x and y in (4.160) and (4.161) by $x - \xi_x$ and $y - \xi_y$, respectively. Note that a scaled value for ξ_y must now be used in the calculations,

[9] In each expression, set $\gamma_2 = \gamma$ and take limits as $\gamma_1 \to \gamma$, noting that, from (4.120), $q_1 - q_2 = c_{11}(\gamma_1^2 - \gamma_2^2)/(c_{12} - c_{66})$.

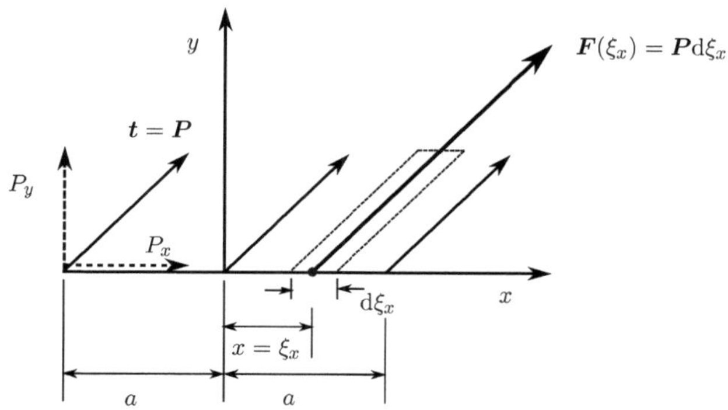

Fig. 4.8 Integration of Kelvin's solution for an isotropic material; line segment $| x | \leq a, \ y = 0$

i.e., ξ_y/γ_i for $i = 1$ and 2. Apart from differences in notation the preceding results agree with those given by Green (1943).

4.12 Loading Along a Line Segment I—Constant Tractions

The solution to the problem of constant tractions $t_x = P_x$ and $t_y = P_y$ applied to the line segment $| x | \leq a, \ y = 0$ in an infinite elastic solid is the basis for an indirect boundary element method called the *stress discontinuity method*, discussed in Chap. 5. This solution can be found using a procedure comparable to that described in Sect. 1.3 for the problem of a constant source P along a line segment. In the present case we divide the line segment into infinitesimal elements of length $d\xi_x$ as shown in Fig. 4.8. The components of force (per unit length perpendicular to the x, y plane) on an element centered at the point $x = \xi_x$, $y = 0$ are then

$$F_i = P_i \, d\xi_x \tag{4.162}$$

where i represents either x or y. The solution to the problem in question is found by substituting forces F_x and F_y into the appropriate expressions for the displacements and stresses for Kelvin's problem, replacing x by $x - \xi_x$, and integrating with respect to ξ_x between the limits $-a$ and $+a$.

This approach is suitable for the case of an infinite isotropic material, which has no preferred orientation, but is not sufficiently general for an anisotropic material. For such a material it is necessary to consider a line segment that is arbitrarily oriented with respect to the directions of elastic symmetry. This is easily accomplished using standard coordinate transformation formulas, as will be shown below (Sect. 4.12.2).

4.12.1 Isotropic Material

The calculations for an isotropic elastic material involve integration of equations (4.143) and (4.144). The task is made simpler by using the results

$$
\begin{aligned}
xg_{,y} &= yg_{,x} \\
xg_{,x} &= -yg_{,y} - \frac{1}{4\pi(1-\nu)}
\end{aligned}
\tag{4.163}
$$

to eliminate terms involving derivatives of function $g(x, y)$ multiplied by coordinate x. The constant term in the second one of these expressions can be disregarded because it corresponds to a rigid body translation of u_x (see (4.143)). The required solution can be expressed in terms of a function $f(x, y)$, defined as

$$
\begin{aligned}
f(x, y) &= \int_{-a}^{a} g(x - \xi_x, y)\, \mathrm{d}\xi_x \\
&= \frac{-1}{4\pi(1-\nu)} \left[y(\theta_1 - \theta_2) - (x - a)\ln r_1 + (x + a)\ln r_2 - 2a\right]
\end{aligned}
\tag{4.164}
$$

where r_1, r_2 and θ_1, θ_2 are the distances and angles from the endpoints of the line segment to an arbitrary point (x, y), as shown in Fig. 1.2.

The displacements and stresses due to the applied line loads P_x and P_y can be written as

$$
\begin{aligned}
u_x &= \frac{P_x}{2\mu}\left[(3-4\nu)f + yf_{,y}\right] + \frac{P_y}{2\mu}\left[-yf_{,x}\right] \\
u_y &= \frac{P_x}{2\mu}\left[-yf_{,x}\right] + \frac{P_y}{2\mu}\left[(3-4\nu)f - yf_{,y}\right]
\end{aligned}
\tag{4.165}
$$

and

$$
\begin{aligned}
\sigma_{xx} &= P_x\left[(3-2\nu)f_{,x} + yf_{,xy}\right] + P_y\left[2\nu f_{,y} + yf_{,yy}\right] \\
\sigma_{yy} &= P_x\left[-(1-2\nu)f_{,x} - yf_{,xy}\right] + P_y\left[2(1-\nu)f_{,y} - yf_{,yy}\right] \\
\sigma_{xy} &= P_x\left[2(1-\nu)f_{,y} + yf_{,yy}\right] + P_y\left[(1-2\nu)f_{,x} - yf_{,xy}\right]
\end{aligned}
\tag{4.166}
$$

The derivatives of function $f(x, y)$ in (4.165) and (4.166) are

$$
\begin{aligned}
f_{,x} &= \frac{1}{4\pi(1-\nu)} \ln \frac{r_1}{r_2} \\
f_{,y} &= \frac{-1}{4\pi(1-\nu)} (\theta_1 - \theta_2) \\
f_{,xy} &= \frac{1}{4\pi(1-\nu)} \left(\frac{y}{r_1^2} - \frac{y}{r_2^2} \right) \\
f_{,yy} &= \frac{-1}{4\pi(1-\nu)} \left(\frac{x-a}{r_1^2} - \frac{x+a}{r_2^2} \right)
\end{aligned}
\tag{4.167}
$$

As expected, the displacements and stresses contain terms that are undefined at the endpoints of the line segment, and these points must be avoided in all calculations. The values of the displacements and stresses elsewhere along the segment—and especially at its midpoint—are of special interest for the boundary element method based on the preceding equations that is the subject of Chap. 5, and we record these values next. Setting $y = 0$ in (4.165) to (4.167) gives

$$
\begin{aligned}
u_x(x,0) &= \frac{(3-4\nu)}{8\pi(1-\nu)\mu} P_x \left[(x-a)\ln \mid x-a \mid - (x+a)\ln \mid x+a \mid \right] + \frac{(3-4\nu)a}{4\pi(1-\nu)\mu} P_x \\
u_y(x,0) &= \frac{(3-4\nu)}{8\pi(1-\nu)\mu} P_y \left[(x-a)\ln \mid x-a \mid - (x+a)\ln \mid x+a \mid \right] + \frac{(3-4\nu)a}{4\pi(1-\nu)\mu} P_y
\end{aligned}
\tag{4.168}
$$

and

$$
\begin{aligned}
\sigma_{xx}(x,0_+) &= \frac{(3-2\nu)}{8\pi(1-\nu)} P_x \ln \left(\frac{x-a}{x+a} \right)^2 - \frac{\nu}{2(1-\nu)} P_y \\
\sigma_{yy}(x,0_+) &= \frac{-(1-2\nu)}{8\pi(1-\nu)} P_x \ln \left(\frac{x-a}{x+a} \right)^2 - \frac{1}{2} P_y \\
\sigma_{xy}(x,0_+) &= -\frac{1}{2} P_x + \frac{(1-2\nu)}{8\pi(1-\nu)} P_y \ln \left(\frac{x-a}{x+a} \right)^2
\end{aligned}
\tag{4.169}
$$

$$
\begin{aligned}
\sigma_{xx}(x,0_-) &= \frac{(3-2\nu)}{8\pi(1-\nu)} P_x \ln \left(\frac{x-a}{x+a} \right)^2 + \frac{\nu}{2(1-\nu)} P_y \\
\sigma_{yy}(x,0_-) &= \frac{-(1-2\nu)}{8\pi(1-\nu)} P_x \ln \left(\frac{x-a}{x+a} \right)^2 + \frac{1}{2} P_y \\
\sigma_{xy}(x,0_-) &= \frac{1}{2} P_x + \frac{(1-2\nu)}{8\pi(1-\nu)} P_y \ln \left(\frac{x-a}{x+a} \right)^2
\end{aligned}
\tag{4.170}
$$

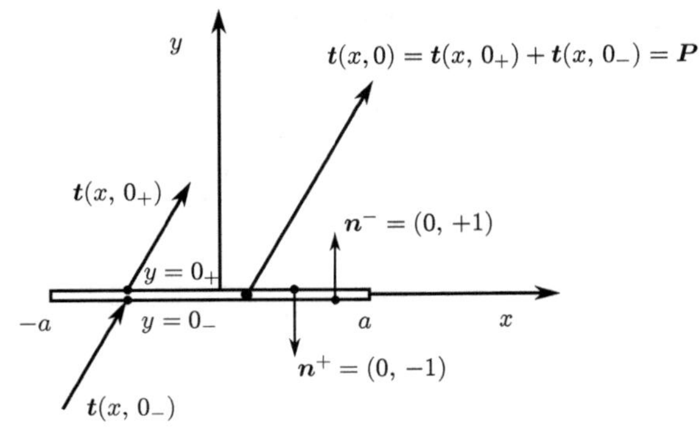

Fig. 4.9 Resultant tractions on line segment

The two sets of expressions for the stresses reflect the fact that the quantity $\theta_1 - \theta_2$ has different values on the positive and negative sides of the line segment (cf. (1.19)).

It follows from (4.169) and (4.170) that the stresses are discontinuous when crossing from one side of the line segment to the other. The values of the stress discontinuities are given as follows:

$$\begin{aligned} \Delta\sigma_{xx} &= \sigma_{xx}(x, 0_+) - \sigma_{xx}(x, 0_-) = -\frac{\nu}{1-\nu} P_y \\ \Delta\sigma_{yy} &= \sigma_{yy}(x, 0_+) - \sigma_{yy}(x, 0_-) = -P_y \\ \Delta\sigma_{xy} &= \sigma_{xy}(x, 0_+) - \sigma_{xy}(x, 0_-) = -P_x \end{aligned} \tag{4.171}$$

The last two of these equations show that P_x and P_y represent constant discontinuities in the shear and normal stresses σ_{xy} and σ_{yy} along the line segment.[10] We note, however, that this interpretation holds for the *stress tensor* components σ_{xy} and σ_{yy}; we will show below that P_x and P_y represent constant *resultant traction vector* components t_x and t_y over the line segment.

The physical significance of the stress discontinuities P_x and P_y can be understood by imagining that the two surfaces of the line segment $|\ x\ | \leq a,\ y = 0$ are separated, as depicted in Fig. 4.9. In reality these two surfaces, $y = 0_-$ and $y = 0_+$, coincide (i.e., they are joined together); they are shown separated in the figure for clarity.

The *outward* normal to the negative surface of the line segment $(y = 0_-)$ has components $n_i = (0, +1)$ and the outward normal to the positive surface $(y = 0_+)$ has components $n_i = (0, -1)$. The components of traction $t_i = \sigma_{ji} n_j$ on the two surfaces are then

$$\begin{aligned} t_x(x, 0_-) &= +\sigma_{xy}(x, 0_-) \\ t_y(x, 0_-) &= +\sigma_{yy}(x, 0_-) \end{aligned} \tag{4.172}$$

[10] The component σ_{xx} is also discontinuous but its value is dependent on the discontinuity in σ_{yy}.

and

$$t_x(x, 0_+) = -\sigma_{xy}(x, 0_+)$$
$$t_y(x, 0_+) = -\sigma_{yy}(x, 0_+) \tag{4.173}$$

The *resultant* stresses applied to the line segment are found by adding the traction components on the two surfaces, i.e.,

$$t_i(x, 0) = t_i(x, 0_-) + t_i(x, 0_+) \tag{4.174}$$

for $i = x$ or y. Combining (4.171) to (4.174), we find that

$$t_x(x, 0) = P_x$$
$$t_y(x, 0) = P_y \tag{4.175}$$

for $| x | \leq a$. The quantities P_x and P_y therefore represent constant resultant traction vector components over the line segment, as asserted.

According to (4.172) and (4.173), we could imagine that one surface of the line segment ($y = 0_+$) is being "pushed" while the other ($y = 0_-$) is being "pulled," because the normal and shear stresses on these surfaces are equal in magnitude but opposite in sign. As a consequence, the displacements u_x and u_y are continuous across the line segment, i.e., $u_i(x, 0_+) - u_i(x, 0_-) = 0$. This result follows easily from (4.168) and means that the two surfaces of the segment do not displace relative to one another.

4.12.2 Anisotropic Material

In developing an indirect boundary element method for an anisotropic elastic material in Chap. 5 we require the solution to the problem of constant tractions applied to a line segment that is arbitrarily oriented with respect to the directions of elastic symmetry, which we will assume coincide with the global x, y coordinate axes. Without loss in generality we consider that the line segment is centered at the origin and is inclined at an angle β from the x axis, as shown in Fig. 4.10. The two coordinate systems in the figure are related by the transformation equations

$$x = \bar{x} \cos \beta - \bar{y} \sin \beta$$
$$y = \bar{x} \sin \beta + \bar{y} \cos \beta \tag{4.176}$$

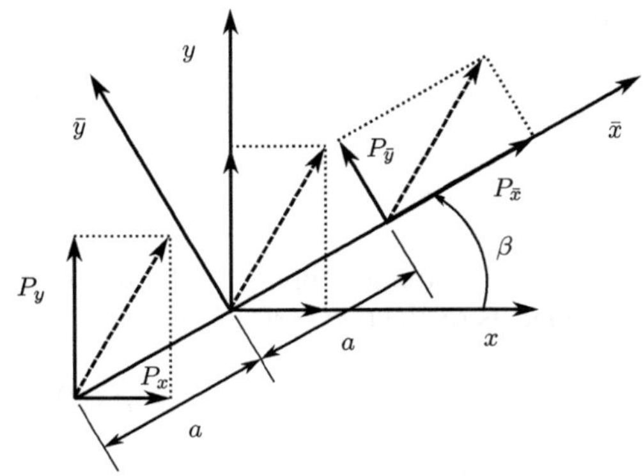

Fig. 4.10 Integration of Kelvin's solution for an anisotropic material; line segment $|\bar{x}| \le a,\ \bar{y} = 0$

and the line segment in question is defined by the conditions $|\bar{x}| \le a,\ \bar{y} = 0$. The applied tractions can be specified with reference to either of the two coordinate systems shown in the figure. We will elect to specify them initially with reference to the $x,\ y$ system, noting that they can later be referred to the $\bar{x},\ \bar{y}$ system by using the relations

$$\begin{aligned} P_x &= P_{\bar{x}} \cos\beta - P_{\bar{y}} \sin\beta \\ P_y &= P_{\bar{x}} \sin\beta + P_{\bar{y}} \cos\beta \end{aligned} \tag{4.177}$$

Inspection of (4.160) and (4.161) shows that the displacements and stresses for the plane strain version of Kelvin's problem depend on the functions $\ln(x^2 + y_i^2)^{1/2}$, $\arctan y_i/x,\ x/(x^2 + y_i^2)$, and $y_i/(x^2 + y_i^2)$, where $i = 1$ or 2. Using (4.176) and recalling that $y_i = y/\gamma_i$, we express these functions in terms of the $\bar{x},\ \bar{y}$ coordinates as follows:

$$\begin{aligned} \ln(x^2 + y_i^2)^{1/2} &= \ln(A_i\bar{x}^2 + B_i\bar{x}\,\bar{y} + C_i\bar{y}^2)^{1/2} \\ \arctan\frac{y_i}{x} &= \arctan\frac{1}{\gamma_i}\frac{\bar{x}\sin\beta + \bar{y}\cos\beta}{\bar{x}\cos\beta - \bar{y}\sin\beta} \\ \frac{x}{x^2 + y_i^2} &= \frac{\bar{x}\cos\beta - \bar{y}\sin\beta}{A_i\bar{x}^2 + B_i\bar{x}\,\bar{y} + C_i\bar{y}^2} \\ \frac{y_i}{x^2 + y_i^2} &= \frac{1}{\gamma_i}\frac{\bar{x}\sin\beta + \bar{y}\cos\beta}{A_i\bar{x}^2 + B_i\bar{x}\,\bar{y} + C_i\bar{y}^2} \end{aligned} \tag{4.178}$$

in which

$$
\begin{aligned}
A_i &= \frac{1}{\gamma_i^2}\left(\gamma_i^2 \cos^2\beta + \sin^2\beta\right) \\
B_i &= \frac{1}{\gamma_i^2}\left(1-\gamma_i^2\right)\sin 2\beta \\
C_i &= \frac{1}{\gamma_i^2}\left(\gamma_i^2 \sin^2\beta + \cos^2\beta\right)
\end{aligned}
\tag{4.179}
$$

for $i = 1$ or 2.

The displacements and stresses due to constant tractions P_x and P_y applied to the line segment can be found in the same way as was done for an isotropic material. That is, we sum the effects of forces $F_x = P_x \mathrm{d}\xi_{\bar{x}}$ and $F_y = P_y \mathrm{d}\xi_{\bar{x}}$ at a series of infinitesimal elements of the segment. The results involve the following integrals of expressions (4.178) (see Appendix A):

$$
\begin{aligned}
I_1(\bar{x},\bar{y},\gamma_i) &= \int_{-a}^{a} \ln\left[A_i(\bar{x}-\xi_{\bar{x}})^2 + B_i(\bar{x}-\xi_{\bar{x}})\,\bar{y} + C_i\bar{y}^2\right]^{1/2} \mathrm{d}\xi_{\bar{x}} \\
&= -\left(\bar{x}-a+\frac{B_i\bar{y}}{2A_i}\right)\ln\bar{r}_1(\gamma_i) + \left(\bar{x}+a+\frac{B_i\bar{y}}{2A_i}\right)\ln\bar{r}_2(\gamma_i) \\
&\quad + \frac{\bar{y}}{\gamma_i A_i}\left[\bar{\theta}_1(\gamma_i)-\bar{\theta}_2(\gamma_i)\right] - 2a \\
I_2(\bar{x},\bar{y},\gamma_i) &= \int_{-a}^{a} \arctan\frac{1}{\gamma_i}\frac{(\bar{x}-\xi_{\bar{x}})\sin\beta + \bar{y}\cos\beta}{(\bar{x}-\xi_{\bar{x}})\cos\beta - \bar{y}\sin\beta}\,\mathrm{d}\xi_{\bar{x}} \\
&= -\left(\bar{x}-a+\frac{B_i\bar{y}}{2A_i}\right)\bar{\theta}_1(\gamma_i) + \left(\bar{x}+a+\frac{B_i\bar{y}}{2A_i}\right)\bar{\theta}_2(\gamma_i) \\
&\quad - \frac{\bar{y}}{\gamma_i A_i}\ln\left[\bar{r}_1(\gamma_i)/\bar{r}_2(\gamma_i)\right] \\
I_3(\bar{x},\bar{y},\gamma_i) &= \int_{-a}^{a} \frac{(\bar{x}-\xi_{\bar{x}})\cos\beta - \bar{y}\sin\beta}{A_i(\bar{x}-\xi_{\bar{x}})^2 + B_i(\bar{x}-\xi_{\bar{x}})\,\bar{y} + C_i\bar{y}^2}\,\mathrm{d}\xi_{\bar{x}} \\
&= -\frac{\cos\beta}{A_i}\ln\left[\bar{r}_1(\gamma_i)/\bar{r}_2(\gamma_i)\right] - \frac{\sin\beta}{\gamma_i A_i}\left[\bar{\theta}_1(\gamma_i)-\bar{\theta}_2(\gamma_i)\right] \\
I_4(\bar{x},\bar{y},\gamma_i) &= \int_{-a}^{a} \frac{1}{\gamma_i}\frac{(\bar{x}-\xi_{\bar{x}})\sin\beta + \bar{y}\cos\beta}{A_i(\bar{x}-\xi_{\bar{x}})^2 + B_i(\bar{x}-\xi_{\bar{x}})\,\bar{y} + C_i\bar{y}^2}\,\mathrm{d}\xi_{\bar{x}} \\
&= -\frac{\sin\beta}{\gamma_i A_i}\ln\left[\bar{r}_1(\gamma_i)/\bar{r}_2(\gamma_i)\right] + \frac{\cos\beta}{A_i}\left[\bar{\theta}_1(\gamma_i)-\bar{\theta}_2(\gamma_i)\right]
\end{aligned}
\tag{4.180}
$$

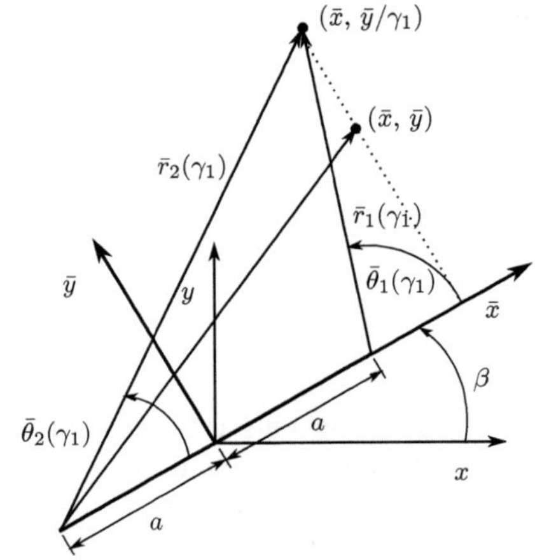

Fig. 4.11 Scaled distances and angles from endpoints of line segment $|\bar{x}| \leq a,\ \bar{y} = 0$

in which we have used the notation

$$\begin{aligned}
\bar{r}_1(\gamma_i) &= \left[A_i(\bar{x}-a)^2 + B_i(\bar{x}-a)\bar{y} + C_i\bar{y}^2\right]^{1/2} \\
\bar{r}_2(\gamma_i) &= \left[A_i(\bar{x}+a)^2 + B_i(\bar{x}+a)\bar{y} + C_i\bar{y}^2\right]^{1/2} \\
\bar{\theta}_1(\gamma_i) &= \arctan\frac{\bar{y}/\gamma_i}{A_i(\bar{x}-a) + B_i\bar{y}/2} \\
\bar{\theta}_2(\gamma_i) &= \arctan\frac{\bar{y}/\gamma_i}{A_i(\bar{x}+a) + B_i\bar{y}/2}
\end{aligned} \tag{4.181}$$

where A_i, B_i, and C_i are given by (4.179), and where $\bar{r}_1(\gamma_i),\ \bar{r}_2(\gamma_i)$ and $\bar{\theta}_1(\gamma_i),\ \bar{\theta}_2(\gamma_i)$ are the *scaled* distances and angles from the end points of the line segment to an arbitrary point $\bar{x},\ \bar{y}$, as indicated in Fig. 4.11 for the parameter γ_1 (for purposes of illustration we take $\gamma_1 < 1$).

In terms of the quantities (4.180) the displacements are, using (4.160),

$$\begin{aligned}
u_x &= \frac{1}{2\pi c_{66}(q_1-q_2)} P_x \left[\frac{\gamma_1}{q_1} I_1(\bar{x}, \bar{y}, \gamma_1) - \frac{\gamma_2}{q_2} I_1(\bar{x}, \bar{y}, \gamma_2)\right] \\
&\quad - \frac{1}{2\pi c_{66}(q_1-q_2)} P_y \left[I_2(\bar{x}, \bar{y}, \gamma_1) - I_2(\bar{x}, \bar{y}, \gamma_2)\right] \\
u_y &= \frac{-1}{2\pi c_{66}(q_1-q_2)} P_x \left[I_2(\bar{x}, \bar{y}, \gamma_1) - I_2(\bar{x}, \bar{y}, \gamma_2)\right] \\
&\quad - \frac{1}{2\pi c_{66}(q_1-q_2)} P_y \left[\frac{q_1}{\gamma_1} I_1(\bar{x}, \bar{y}, \gamma_1) - \frac{q_2}{\gamma_2} I_1(\bar{x}, \bar{y}, \gamma_2)\right]
\end{aligned} \tag{4.182}$$

and the stresses are, using (4.161),

$$
\begin{aligned}
\sigma_{xx} &= \frac{1}{2\pi(q_1-q_2)} P_x \left[\frac{(1+q_1)}{\gamma_1 q_1} I_3(\bar{x},\bar{y},\gamma_1) - \frac{(1+q_2)}{\gamma_2 q_2} I_3(\bar{x},\bar{y},\gamma_2) \right] \\
&+ \frac{1}{2\pi(q_1-q_2)} P_y \left[\frac{(1+q_1)}{\gamma_1^2} I_4(\bar{x},\bar{y},\gamma_1) - \frac{(1+q_2)}{\gamma_2^2} I_4(\bar{x},\bar{y},\gamma_2) \right] \\
\sigma_{yy} &= \frac{-1}{2\pi(q_1-q_2)} P_x \left[\frac{\gamma_1(1+q_1)}{q_1} I_3(\bar{x},\bar{y},\gamma_1) - \frac{\gamma_2(1+q_2)}{q_2} I_3(\bar{x},\bar{y},\gamma_2) \right] \\
&- \frac{1}{2\pi(q_1-q_2)} P_y \left[(1+q_1) I_4(\bar{x},\bar{y},\gamma_1) - (1+q_2) I_4(\bar{x},\bar{y},\gamma_2) \right] \qquad (4.183) \\
\sigma_{xy} &= \frac{1}{2\pi(q_1-q_2)} P_x \left[\frac{(1+q_1)}{q_1} I_4(\bar{x},\bar{y},\gamma_1) - \frac{(1+q_2)}{q_2} I_4(\bar{x},\bar{y},\gamma_2) \right] \\
&- \frac{1}{2\pi(q_1-q_2)} P_y \left[\frac{(1+q_1)}{\gamma_1} I_3(\bar{x},\bar{y},\gamma_1) - \frac{(1+q_2)}{\gamma_2} I_3(\bar{x},\bar{y},\gamma_2) \right]
\end{aligned}
$$

To obtain the values of the displacements and stresses along the line segment $|\bar{x}| \leq a,\ \bar{y}=0$ we require the limit values of $I_1(\bar{x},\bar{y},\gamma_i)$ through $I_4(\bar{x},\bar{y},\gamma_i)$ in (4.182) as $\bar{y} \to 0$. These are given as follows:

$$
\begin{aligned}
I_1(\bar{x},0_\pm,\gamma_i) &= -(\bar{x}-a)\ln\left[A_i(\bar{x}-a)^2\right]^{1/2} + (\bar{x}+a)\ln\left[A_i(\bar{x}+a)^2\right]^{1/2} - 2a \\
I_2(\bar{x},0_\pm,\gamma_i) &= \mp(\bar{x}-a)\pi \\
I_3(\bar{x},0_\pm,\gamma_i) &= -\frac{\cos\beta}{A_i}\ln\left(\frac{\bar{x}-a}{\bar{x}+a}\right)^2 \mp \frac{\pi\sin\beta}{\gamma_i A_i} \qquad (4.184) \\
I_4(\bar{x},0_\pm,\gamma_i) &= -\frac{\sin\beta}{\gamma_i A_i}\ln\left(\frac{\bar{x}-a}{\bar{x}+a}\right)^2 \pm \frac{\pi\cos\beta}{A_i}
\end{aligned}
$$

Introducing the first two of these equations into (4.182) it is seen that the displacements are continuous along the line segment, as expected, despite the discontinuity in function $I_2(\bar{x},\bar{y},\gamma_i)$. Expressions for the stress discontinuities in $\bar{x},\ \bar{y}$ coordinates can be calculated using Eqs. (4.177), (4.183), (4.184), and coordinate transformation formulas (4.67). After some straightforward but tedious algebra we find that

$$
\begin{aligned}
\sigma_{\bar{x}\bar{y}}(\bar{x},0_+) - \sigma_{\bar{x}\bar{y}}(\bar{x},0_-) &= -P_{\bar{x}} \\
\sigma_{\bar{y}\bar{y}}(\bar{x},0_+) - \sigma_{\bar{y}\bar{y}}(\bar{x},0_-) &= -P_{\bar{y}}
\end{aligned}
\qquad (4.185)
$$

which are equal to the shear and normal stress discontinuities along the line segment $|\bar{x}| \leq a,\ \bar{y}=0$. Following the same procedure as in isotropic elasticity, we find that $P_{\bar{x}}$ and $P_{\bar{y}}$ represent constant resultant traction vector components over the line segment.

The discontinuity in the stress component $\sigma_{\bar{x}\bar{x}}$ can be calculated in the same manner. The results are considerably more complicated than for the isotropic case and involve both components of the applied loads, $P_{\bar{x}}$ and $P_{\bar{y}}$. Letting $\Delta\sigma_{\bar{x}\bar{x}}$ be the value of the stress discontinuity $\sigma_{\bar{x}\bar{x}}(\bar{x}, 0_+) - \sigma_{\bar{x}\bar{x}}(\bar{x}, 0_-)$, it is found that

$$
\begin{aligned}
\Delta\sigma_{\bar{x}\bar{x}} = &-P_{\bar{x}}\frac{\sin\beta\cos\beta}{q_1 - q_2}\left[\frac{(1+q_1)(1-q_2)(1-\gamma_1^2)}{\gamma_1^2\cos^2\beta + \sin^2\beta} - \frac{(1+q_2)(1-q_1)(1-\gamma_2^2)}{\gamma_2^2\cos^2\beta + \sin^2\beta}\right] \\
&- P_{\bar{y}}\frac{1}{q_1 - q_2}\left[\frac{(1+q_1)}{\gamma_1^2\cos^2\beta + \sin^2\beta}\left\{(1+q_2)(1-\gamma_1^2)\sin^2\beta\cos^2\beta - q_2\gamma_1^2\sin^2\beta + \cos^2\beta\right\}\right. \\
&\left. - \frac{(1+q_2)}{\gamma_2^2\cos^2\beta + \sin^2\beta}\left\{(1+q_1)(1-\gamma_2^2)\sin^2\beta\cos^2\beta - q_1\gamma_2^2\sin^2\beta + \cos^2\beta\right\}\right]
\end{aligned}
\tag{4.186}
$$

The special case of isotropy occurs when $\gamma_1 \to \gamma_2 = 1$, and for this it can be shown using L'Hopital's rule that $\Delta\sigma_{\bar{x}\bar{x}} = -\nu P_{\bar{y}}/(1-\nu)$, in accord with the result found previously.

4.13 Loading Along a Line Segment II—Constant Displacement Discontinuity

A problem related to the one discussed in the previous section concerns the loading of a line segment that undergoes a constant discontinuity in displacement along its length. The basic equations for this problem are developed below for both isotropic and anisotropic elastic materials. These equations form the basis of a special boundary element method called the *displacement discontinuity method*, which is described in Chap. 6.

4.13.1 *Isotropic Material*

We begin by considering the case of a constant displacement discontinuity $D_y = u_y(x, 0_+) - u_y(x, 0_-) > 0$ over the line segment $-a \le x \le a$, $y = 0$ in an infinite, isotropic, linearly elastic solid, depicted in Fig. 4.12. The displacement u_y is continuous everywhere except across the line segment in question, and all other quantities—including the displacement u_x—are continuous everywhere in the infinite region. (The stresses are singular at the ends of the line segment, $x = \pm a$, $y = 0$, but we will not consider these points in our computations.) By symmetry, the shear stress σ_{xy} vanishes all along the line $y = 0$, and this allows us to obtain the solution to

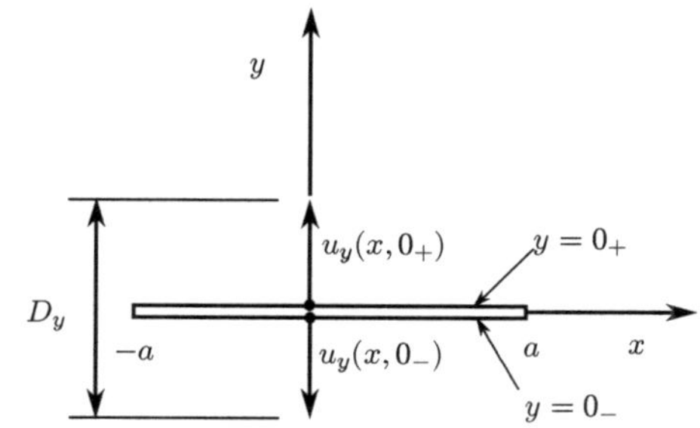

Fig. 4.12 Constant displacement discontinuity $D_y = u_y(x, 0_+) - u_y(x, 0_-)$ over the line segment $|x| \le a,\ y = 0$

the problem in terms of the special Papkovitch-Neuber function $\Phi(x, y)$ introduced in Sect. 4.9.3.

Guided by the results of the previous section, we note that the function $\partial f(x, y)/\partial y$ in (4.167) meets the conditions for the problem we seek to solve: (a) it satisfies Laplace's equation in two dimensions and (b) it undergoes a constant jump across the line segment in question. We are therefore led to consider the function

$$\frac{\partial}{\partial y}\Phi(x, y) = C\frac{\partial}{\partial y}f(x, y) = -\frac{C}{4\pi(1-\nu)}(\theta_1 - \theta_2) \tag{4.187}$$

According to the second one of (4.99), the displacement component $u_y(x, y)$ associated with this value of $\partial\Phi(x, y)/\partial y$ is

$$\begin{aligned} u_y(x, y) &= 2(1-\nu)\frac{\partial}{\partial y}\Phi(x, y) - y\frac{\partial^2}{\partial y^2}\Phi(x, y) \\ &= -\frac{C}{2\pi}(\theta_1 - \theta_2) + \frac{C}{4\pi(1-\nu)}y\left(\frac{x-a}{r_1^2} - \frac{x+a}{r_2^2}\right) \end{aligned} \tag{4.188}$$

Using this result in the definition $D_y = u_y(x, 0_+) - u_y(x, 0_-)$ and recalling that $(\theta_1 - \theta_2) = \pm\pi$ for $|x| \le a,\ y = 0_\pm$ gives directly that $C = -D_y$. We therefore find that the solution for a constant displacement discontinuity D_y over the line segment $-a \le x \le a$, $y = 0$ is obtained using the function

$$\frac{\partial}{\partial y}\Phi(x, y) = -\frac{D_y}{4\pi(1-\nu)}(\theta_1 - \theta_2) \tag{4.189}$$

The displacements and stresses associated with this function can easily be found using Eqs. (4.99) and (4.100).

A similar analysis can be done for the companion problem of a constant displacement discontinuity D_x over the line segment $-a \le x \le a$, $y = 0$, illustrated in Fig. 4.13. The solution can be written in terms of the special Papkovitch-Neuber function $\Psi(x, y)$ defined in Sect. 4.9.3 as follows:

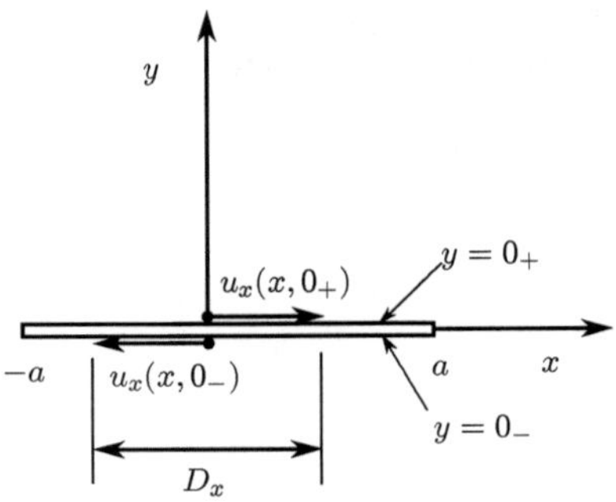

Fig. 4.13 Constant displacement discontinuity $D_x = u_x(x, 0_+) - u_x(x, 0_-)$ over the line segment $|x| \leq a,\ y = 0$

$$\frac{\partial}{\partial y}\Psi(x, y) = -\frac{D_x}{4\pi(1-\nu)}(\theta_1 - \theta_2) \tag{4.190}$$

The displacements and stresses associated with this function can be calculated using Eqs. (4.107) and (4.108).

By construction, the functions $\Phi(x, y)$ and $\Psi(x, y)$ are constant multiples of $-f(x, y)$ in (4.164), i.e.,

$$\begin{aligned} \Phi(x, y) &= -D_y f(x, y) \\ \Psi(x, y) &= -D_x f(x, y) \end{aligned} \tag{4.191}$$

It therefore follows that the complete solution for a constant displacement discontinuity (D_x, D_y) over the line segment $-a \leq x \leq a$, $y = 0$ can be put in the following form using (4.109) and (4.110):

$$\begin{aligned} u_x &= D_x\left[-2(1-\nu)f_{,y} - yf_{,yy}\right] + D_y\left[(1-2\nu)f_{,x} + yf_{,xy}\right] \\ u_y &= D_x\left[-(1-2\nu)f_{,x} + yf_{,xy}\right] + D_y\left[-2(1-\nu)f_{,y} + yf_{,yy}\right] \end{aligned} \tag{4.192}$$

$$\begin{aligned} \sigma_{xx} &= 2\mu D_x\left[-2f_{,xy} - yf_{,xyy}\right] + 2\mu D_y\left[-f_{,yy} - yf_{,yyy}\right] \\ \sigma_{yy} &= 2\mu D_x\left[yf_{,xyy}\right] + 2\mu D_y\left[-f_{,yy} + yf_{,yyy}\right] \\ \sigma_{xy} &= 2\mu D_x\left[-f_{,yy} - yf_{,yyy}\right] + 2\mu D_y\left[yf_{,xyy}\right] \end{aligned} \tag{4.193}$$

The derivatives of function $f(x, y)$ through second order are given in (4.167). The two third order derivatives needed to evaluate the stresses in (4.193) are

$$f_{,xyy} = \frac{1}{4\pi(1-\nu)}\left[\frac{(x-a)^2 - y^2}{r_1^4} - \frac{(x+a)^2 - y^2}{r_2^4}\right]$$
$$f_{,yyy} = \frac{2}{4\pi(1-\nu)}\left[\frac{(x-a)y}{r_1^4} - \frac{(x+a)y}{r_2^4}\right] \tag{4.194}$$

By inspection of (4.193) we see that the stresses along the line $y = 0$ depend on the values of $(\partial^2 f(x,y)/\partial x \partial y)_{y=0}$ and $(\partial^2 f(x,y)/\partial y^2)_{y=0}$. The stresses are

$$\sigma_{xx}(x,0) = \frac{\mu}{2\pi(1-\nu)} D_y \left(\frac{1}{x-a} - \frac{1}{x+a}\right) = \frac{a\mu}{\pi(1-\nu)} D_y \frac{1}{x^2 - a^2}$$
$$\sigma_{yy}(x,0) = \frac{a\mu}{\pi(1-\nu)} D_y \frac{1}{x^2 - a^2} \tag{4.195}$$
$$\sigma_{xy}(x,0) = \frac{a\mu}{\pi(1-\nu)} D_x \frac{1}{x^2 - a^2}$$

The stress components σ_{xx} and σ_{yy} on $y = 0$ therefore depend upon only the normal component of displacement discontinuity D_y whereas the component σ_{xy} depends upon only the shear component D_x. The stresses are singular at the end points of the line segment, $x = \pm a$, but are otherwise finite and continuous.

If we imagine that the line segment is a portion of a crack, the displacement discontinuity components D_x and D_y physically represent the relative displacements between the crack surfaces. Positive values of these quantities are shown in Figs. 4.12 and 4.13: the normal component, D_y, is positive for crack opening and the shear component, D_x, is positive when the positive surface of the crack ($y = 0_+$) moves to the right with respect to the negative surface ($y = 0_-$). The same sign conventions apply if the line segment is defined for a local $\bar{x}$, $\bar{y}$ coordinate system. In this case, $D_{\bar{y}}$ is the normal component of displacement discontinuity, positive for opening, and $D_{\bar{x}}$ is the shear component, positive when the positive crack surface ($\bar{y} = 0_+$) moves to the right relative to the negative surface ($\bar{y} = 0_-$).[11]

4.13.2 Anisotropic Material

The problem of a constant displacement discontinuity over an arbitrarily oriented, finite line segment in an infinite, orthotropic (or transversely isotropic), linearly elastic solid is more complicated because of the absence of symmetry. In order to

[11] In describing the movement of a strike-slip fault, geologists call this convention for $D_{\bar{x}}$ "right lateral slip".

gain insight into the form of the solution, it is useful to consider first the case in which the discontinuity occurs over a line segment parallel to the x axis, i.e., over $| x |\leq a,\ y = 0$. The line segment then lies along a direction of elastic symmetry, and the solution to the problem can be written in terms of the two harmonic functions $\Psi(x, y)$ and $\Phi(x, y)$, as explained in Sect. 4.10.4.

In analogy with the the isotropic case discussed above, consider the functions

$$\begin{aligned}\Psi(x, y) &= C_x w(x, y)\\ \Phi(x, y) &= C_y w(x, y)\end{aligned} \tag{4.196}$$

in which C_x and C_y are constants and where

$$w(x, y) = -y(\theta_1 - \theta_2) + (x - a)\ln r_1 - (x + a)\ln r_2 \tag{4.197}$$

cf. (4.164). The displacements u_x and u_y associated with functions (4.196) are then found using (4.135) and (4.139), i.e.,

$$\begin{aligned}u_x &= C_x\left[\frac{1}{1+q_1}\frac{\partial w(x, y_1)}{\partial y_1} - \frac{1}{1+q_2}\frac{\partial w(x, y_2)}{\partial y_2}\right]\\ &\quad + C_y\left[\frac{\gamma_1}{1+q_1}\frac{\partial w(x, y_1)}{\partial x} - \frac{\gamma_2}{1+q_2}\frac{\partial w(x, y_2)}{\partial x}\right]\\ u_y &= -C_x\left[\frac{q_1}{\gamma_1(1+q_1)}\frac{\partial w(x, y_1)}{\partial x} - \frac{q_2}{\gamma_2(1+q_2)}\frac{\partial w(x, y_2)}{\partial x}\right]\\ &\quad + C_y\left[\frac{q_1}{1+q_1}\frac{\partial w(x, y_1)}{\partial y_1} - \frac{q_2}{1+q_2}\frac{\partial w(x, y_2)}{\partial y_2}\right]\end{aligned} \tag{4.198}$$

Because

$$\begin{aligned}\frac{\partial}{\partial x} w(x, y_i) &= \ln\left[(x-a)^2 + y_i^2\right]^{1/2} - \ln\left[(x+a)^2 + y_i^2\right]^{1/2}\\ \frac{\partial}{\partial y_i} w(x, y_i) &= -\arctan\frac{y_i}{x-a} + \arctan\frac{y_i}{x+a}\end{aligned} \tag{4.199}$$

it is easy to show from these expressions that, for $| x |\leq a$,

$$
\begin{aligned}
D_x &= u_x(x, 0_+) - u_x(x, 0_-) \\
&= 2\pi C_x \frac{q_1 - q_2}{(1+q_1)(1+q_2)} \\
D_y &= u_y(x, 0_+) - u_y(x, 0_-) \\
&= -2\pi C_y \frac{q_1 - q_2}{(1+q_1)(1+q_2)}
\end{aligned}
\tag{4.200}
$$

The solution to the problem of a constant displacement discontinuity $(D_x,\ D_y)$ over the line segment $| x | \leq a,\ y = 0$ in an infinite, orthotropic elastic solid is therefore given by the harmonic functions

$$
\begin{aligned}
\Psi(x, y) &= \frac{D_x}{2\pi} \frac{(1+q_1)(1+q_2)}{q_1 - q_2} w(x, y) \\
\Phi(x, y) &= -\frac{D_y}{2\pi} \frac{(1+q_1)(1+q_2)}{q_1 - q_2} w(x, y)
\end{aligned}
\tag{4.201}
$$

The solution for the general case in which the displacement discontinuity occurs over a line segment that is inclined from a direction of elastic symmetry can be expressed in terms of the displacement potentials ϕ_1 and ϕ_2 introduced in Sect. 4.10.2. The particular representations of these potentials involving harmonic functions $\Psi(x, y)$ and $\Phi(x, y)$ are only applicable to problems in which boundary conditions are known for $y = 0$, and hence are not suitable for treating the general problem. The general solution, however, must reduce to the one obtained above when the line segment is parallel to the x axis, so it is useful to record here the potentials ϕ_1 and ϕ_2 associated with functions $\Psi(x, y)$ and $\Phi(x, y)$ in (4.201). These follow directly from (4.132) and (4.138) and are

$$
\begin{aligned}
\phi_1 &= \frac{1}{2\pi} \frac{1+q_2}{q_1 - q_2} \left[D_x \int \frac{\partial}{\partial y_1} w(x, y_1)\, \mathrm{d}x - D_y \gamma_1 w(x, y_1) \right] \\
\phi_2 &= -\frac{1}{2\pi} \frac{1+q_1}{q_1 - q_2} \left[D_x \int \frac{\partial}{\partial y_2} w(x, y_2)\, \mathrm{d}x - D_y \gamma_2 w(x, y_2) \right]
\end{aligned}
\tag{4.202}
$$

in which the function

$$
\begin{aligned}
w(x, y_i) = &-y_i \left(\arctan \frac{y_i}{x-a} - \arctan \frac{y_i}{x+a} \right) + (x-a) \ln \left[(x-a)^2 + y_i^2 \right]^{1/2} \\
&- (x+a) \ln \left[(x+a)^2 + y_i^2 \right]^{1/2}
\end{aligned}
\tag{4.203}
$$

is harmonic in $x,\ y_i$ coordinates ($i = 1$ or 2).

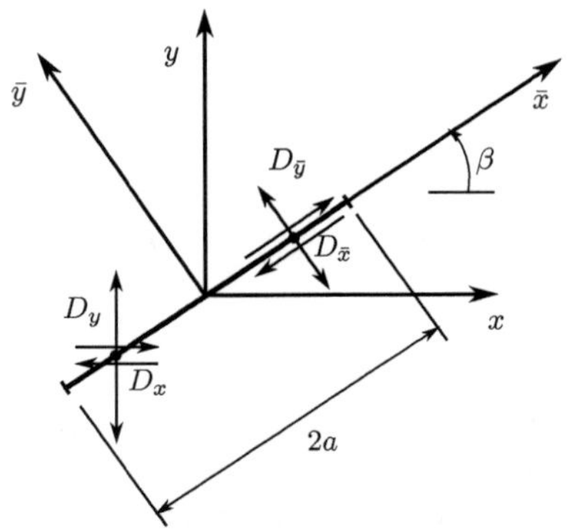

Fig. 4.14 Displacement discontinuity components over the line segment $|\bar{x}| \le a,\ \bar{y} = 0$

For the general problem we consider the line segment $|\bar{x}| \le a,\ \bar{y} = 0$ depicted in Fig. 4.14. The components of displacement discontinuity over the segment are initially specified with reference to the $x,\ y$ coordinate system but they can later be referred to the $\bar{x},\ \bar{y}$ system (4.176) by using the relations

$$\begin{aligned} D_x &= D_{\bar{x}} \cos\beta - D_{\bar{y}} \sin\beta \\ D_y &= D_{\bar{x}} \sin\beta + D_{\bar{y}} \cos\beta \end{aligned} \tag{4.204}$$

We seek a function that has a constant discontinuity (or jump) along $|\bar{x}| \le a,\ \bar{y} = 0$ but is otherwise continuous. Moreover, this function must be harmonic in $x,\ y_i$ coordinates, in accordance with (4.131). Guided by the results of Sect. 4.12.2, Eqs. (4.179) and (4.181), we see that the function

$$\psi(\bar{x}, \bar{y}, \gamma_i) = -\arctan\frac{\bar{y}/\gamma_i}{A_i(\bar{x} - a) + B_i\bar{y}/2} + \arctan\frac{\bar{y}/\gamma_i}{A_i(\bar{x} + a) + B_i\bar{y}/2} \tag{4.205}$$

exhibits the required discontinuity for $|\bar{x}| \le a,\ \bar{y} = 0$.

Using the identity

$$\arctan z_1 - \arctan z_2 = \arctan\frac{z_1 - z_2}{1 + z_1 z_2} \tag{4.206}$$

with

$$\begin{aligned} z_1 &= \frac{1}{\gamma_i}\frac{(\bar{x} \pm a)\sin\beta + \bar{y}\cos\beta}{(\bar{x} \pm a)\cos\beta - \bar{y}\sin\beta} \\ z_2 &= \frac{1}{\gamma_i}\frac{\sin\beta}{\cos\beta} \end{aligned} \tag{4.207}$$

and definitions (4.179), it is easy to verify that

$$\arctan \frac{\bar{y}/\gamma_i}{A_i(\bar{x} \pm a) + B_i \bar{y}/2} = \arctan \frac{1}{\gamma_i} \frac{(\bar{x} \pm a) \sin\beta + \bar{y} \cos\beta}{(\bar{x} \pm a)\cos\beta - \bar{y}\sin\beta} - \arctan \frac{1}{\gamma_i} \frac{\sin\beta}{\cos\beta} \tag{4.208}$$

Equation (4.205) can therefore be written as

$$\psi(\bar{x}, \bar{y}, \gamma_i) = -\arctan \frac{1}{\gamma_i} \frac{(\bar{x} - a)\sin\beta + \bar{y}\cos\beta}{(\bar{x} - a)\cos\beta - \bar{y}\sin\beta} + \arctan \frac{1}{\gamma_i} \frac{(\bar{x} + a)\sin\beta + \bar{y}\cos\beta}{(\bar{x} + a)\cos\beta - \bar{y}\sin\beta} \tag{4.209}$$

It follows from (4.176) that this function can be written in terms of x, y_i coordinates as

$$\psi(x, y_i) = -\arctan \frac{y_i - (a\sin\beta)/\gamma_i}{x - a\cos\beta} + \arctan \frac{y_i + (a\sin\beta)/\gamma_i}{x + a\cos\beta} \tag{4.210}$$

which satisfies (4.131), as required.

Function $\psi(x, y_i)$ reduces to the function $\partial w(x, y_i)/\partial y_i$ when $\beta = 0$ (see (4.199)). We therefore define

$$W(x, y_i) = \int \psi(x, y_i)\, \mathrm{d}y_i \tag{4.211}$$

and by analogy with (4.202) take

$$\begin{aligned}
\phi_1 &= \frac{1}{2\pi} \frac{1 + q_2}{q_1 - q_2} \left[D_x \int \frac{\partial}{\partial y_1} W(x, y_1)\, \mathrm{d}x - D_y \gamma_1 W(x, y_1) \right] \\
\phi_2 &= -\frac{1}{2\pi} \frac{1 + q_1}{q_1 - q_2} \left[D_x \int \frac{\partial}{\partial y_2} W(x, y_2)\, \mathrm{d}x - D_y \gamma_2 W(x, y_2) \right]
\end{aligned} \tag{4.212}$$

in which, from (4.210) and (4.211),

$$\begin{aligned}
W(x, y_i) = &-(y_i - (a\sin\beta)/\gamma_i) \arctan \frac{y_i - (a\sin\beta)/\gamma_i}{x - a\cos\beta} \\
&+ (y_i + (a\sin\beta)/\gamma_i) \arctan \frac{y_i + (a\sin\beta)/\gamma_i}{x + a\cos\beta} \\
&+ (x - a\cos\beta) \ln \left[(x - a\cos\beta)^2 + (y_i - (a\sin\beta)/\gamma_i)^2 \right]^{1/2} \\
&- (x + a\cos\beta) \ln \left[(x + a\cos\beta)^2 + (y_i + (a\sin\beta)/\gamma_i)^2 \right]^{1/2}
\end{aligned} \tag{4.213}$$

The displacements and stresses associated with potentials ϕ_1 and ϕ_2 are found using (4.129) and (4.130). The reader can verify that the first and second derivatives of function $W(x, y_i)$ involved in these expressions can be written in terms of $\bar{x}$, $\bar{y}$ coordinates as follows, using (4.176), (4.179), and (4.208):

$$\begin{aligned}\frac{\partial}{\partial x}W(x, y_i) = J_1(\bar{x}, \bar{y}, \gamma_i) &= \ln\left[A_i(\bar{x}-a)^2 + B_i(\bar{x}-a)\bar{y} + C_i\bar{y}^2\right]^{1/2} \\ &\quad - \ln\left[A_i(\bar{x}+a)^2 + B_i(\bar{x}+a)\bar{y} + C_i\bar{y}^2\right]^{1/2} \\ &= \ln\left[\bar{r}_1(\gamma_i)/\bar{r}_2(\gamma_i)\right]\end{aligned}$$

$$\begin{aligned}\frac{\partial}{\partial y_i}W(x, y_i) = J_2(\bar{x}, \bar{y}, \gamma_i) &= -\arctan\frac{\bar{y}/\gamma_i}{A_i(\bar{x}-a) + B_i\bar{y}/2} + \arctan\frac{\bar{y}/\gamma_i}{A_i(\bar{x}+a) + B_i\bar{y}/2} \\ &= -\left[\bar{\theta}_1(\gamma_i) - \bar{\theta}_2(\gamma_i)\right]\end{aligned} \quad (4.214)$$

$$\frac{\partial^2}{\partial x\partial y_i}W(x, y_i) = J_3(\bar{x}, \bar{y}, \gamma_i) = \frac{\left[(\bar{x}-a)\sin\beta + \bar{y}\cos\beta\right]/\gamma_i}{\bar{r}_1^2(\gamma_i)} - \frac{\left[(\bar{x}+a)\sin\beta + \bar{y}\cos\beta\right]/\gamma_i}{\bar{r}_2^2(\gamma_i)}$$

$$\frac{\partial^2}{\partial y_i^2}W(x, y_i) = J_4(\bar{x}, \bar{y}, \gamma_i) = -\frac{\left[(\bar{x}-a)\cos\beta - \bar{y}\sin\beta\right]}{\bar{r}_1^2(\gamma_i)} + \frac{\left[(\bar{x}+a)\cos\beta - \bar{y}\sin\beta\right]}{\bar{r}_2^2(\gamma_i)}$$

where $\bar{r}_1(\gamma_i)$, $\bar{\theta}_1(\gamma_i)$, etc., are defined in (4.181). In terms of functions $J_1(\bar{x}, \bar{y}, \gamma_i)$ and $J_2(\bar{x}, \bar{y}, \gamma_i)$ in (4.214) the displacements can then be written as

$$\begin{aligned}u_x &= \frac{q_3}{2\pi}D_x\left[\frac{1}{1+q_1}J_2(\bar{x}, \bar{y}, \gamma_1) - \frac{1}{1+q_2}J_2(\bar{x}, \bar{y}, \gamma_2)\right] \\ &\quad - \frac{q_3}{2\pi}D_y\left[\frac{\gamma_1}{1+q_1}J_1(\bar{x}, \bar{y}, \gamma_1) - \frac{\gamma_2}{1+q_2}J_1(\bar{x}, \bar{y}, \gamma_2)\right] \\ u_y &= \frac{q_3}{2\pi}D_x\left[-\frac{q_1}{\gamma_1(1+q_1)}J_1(\bar{x}, \bar{y}, \gamma_1) + \frac{q_2}{\gamma_2(1+q_2)}J_1(\bar{x}, \bar{y}, \gamma_2)\right] \\ &\quad - \frac{q_3}{2\pi}D_y\left[\frac{q_1}{1+q_1}J_2(\bar{x}, \bar{y}, \gamma_1) - \frac{q_2}{1+q_2}J_2(\bar{x}, \bar{y}, \gamma_2)\right]\end{aligned} \quad (4.215)$$

in which the constant q_3 is defined as

$$q_3 = \frac{(1+q_1)(1+q_2)}{(q_1-q_2)} \quad (4.216)$$

The first two of (4.214) show that, for $|\bar{x}| \leq a$,

$$
\begin{aligned}
J_1(\bar{x}, 0_+, \gamma_i) - J_1(\bar{x}, 0_-, \gamma_i) &= 0 \\
J_2(\bar{x}, 0_+, \gamma_i) - J_2(\bar{x}, 0_-, \gamma_i) &= -2\pi
\end{aligned}
\tag{4.217}
$$

so that the displacements exhibit the required discontinuity for $\bar{y} \to 0_{\pm}$, confirming that potentials (4.212) give the solution to the problem of a constant displacement discontinuity over an arbitrarily oriented, finite line segment in an infinite, orthotropic elastic body.

Finally, the stresses associated with displacements (4.215) are

$$
\begin{aligned}
\sigma_{xx} &= c_{66}\frac{q_3}{2\pi}D_x\left[\frac{1}{\gamma_1^2}J_3(\bar{x}, \bar{y}, \gamma_1) - \frac{1}{\gamma_2^2}J_3(\bar{x}, \bar{y}, \gamma_2)\right] \\
&\quad + c_{66}\frac{q_3}{2\pi}D_y\left[\frac{1}{\gamma_1}J_4(\bar{x}, \bar{y}, \gamma_1) - \frac{1}{\gamma_2}J_4(\bar{x}, \bar{y}, \gamma_2)\right] \\
\sigma_{yy} &= -c_{66}\frac{q_3}{2\pi}D_x\left[J_3(\bar{x}, \bar{y}, \gamma_1) - J_3(\bar{x}, \bar{y}, \gamma_2)\right] \\
&\quad - c_{66}\frac{q_3}{2\pi}D_y\left[\gamma_1 J_4(\bar{x}, \bar{y}, \gamma_1) - \gamma_2 J_4(\bar{x}, \bar{y}, \gamma_2)\right] \\
\sigma_{xy} &= c_{66}\frac{q_3}{2\pi}D_x\left[\frac{1}{\gamma_1}J_4(\bar{x}, \bar{y}, \gamma_1) - \frac{1}{\gamma_2}J_4(\bar{x}, \bar{y}, \gamma_2)\right] \\
&\quad - c_{66}\frac{q_3}{2\pi}D_y\left[J_3(\bar{x}, \bar{y}, \gamma_1) - J_3(\bar{x}, \bar{y}, \gamma_2)\right]
\end{aligned}
\tag{4.218}
$$

4.14 A Concentrated Displacement Discontinuity

A concentrated displacement discontinuity at a point is a limiting case of the problem of a constant displacement discontinuity over a finite line segment along the x axis in the x, y plane of an infinite elastic solid. If we let the "strength" of the concentrated discontinuity be defined by the quantity $\mathcal{D}_i = 2aD_i$, we require that $\mathcal{D}_i$ be a constant as $a \to 0$, thus representing a concentrated displacement discontinuity at the origin. (It is worth noting that $\mathcal{D}_i$ has units of area, or volume if we consider that the line segment $-a \le x \le a$, $y = 0$ has unit length in the third dimension.)

We present in this section the solution for a concentrated displacement discontinuity at the origin for both isotropic and anisotropic elasticity. These solutions are of fundamental importance in developing higher order versions of the displacement discontinuity method in which the displacement discontinuities vary along the elements, including the special crack tip elements discussed in Sect. 8.11.

4.14.1 Isotropic Elasticity

Functions Ψ and Φ for the subject problem are obtained from (4.191) as follows:

$$\Psi(x, y) = -\mathcal{D}_x \lim_{a\to 0} \frac{1}{2a} f(x, y)$$
$$\Phi(x, y) = -\mathcal{D}_y \lim_{a\to 0} \frac{1}{2a} f(x, y) \tag{4.219}$$

Noting the definition of function $f(x, y)$ in (4.164), a simple application of L'Hopital's rule gives the results

$$\Psi(x, y) = -\mathcal{D}_x g(x, y)$$
$$\Phi(x, y) = -\mathcal{D}_y g(x, y) \tag{4.220}$$

where function $g(x, y)$ is defined as in (4.142), with $\xi_x = 0$ and $\xi_y = 0$ for the problem in question. The displacements and stresses associated with these values of $\Psi(x, y)$ and $\Phi(x, y)$ are given by (4.109) and (4.110). Alternatively, we can obtain the required results by replacing $f(x, y)$ in (4.192) and (4.193) by $g(x, y)$ and writing $\mathcal{D}_x$ and $\mathcal{D}_y$ in place of D_x and D_y. We therefore have that

$$u_x = \mathcal{D}_x\left[-2(1-\nu)g_{,y} - yg_{,yy}\right] + \mathcal{D}_y\left[(1-2\nu)g_{,x} + yg_{,xy}\right]$$
$$u_y = \mathcal{D}_x\left[-(1-2\nu)g_{,x} + yg_{,xy}\right] + \mathcal{D}_y\left[-2(1-\nu)g_{,y} + yg_{,yy}\right] \tag{4.221}$$

and

$$\sigma_{xx} = 2\mu\mathcal{D}_x(-2g_{,xy} - yg_{,xyy}) + 2\mu\mathcal{D}_y(-g_{,yy} - yg_{,yyy})$$
$$\sigma_{yy} = 2\mu\mathcal{D}_x(yg_{,xyy}) + 2\mu\mathcal{D}_y(-g_{,yy} + yg_{,yyy}) \tag{4.222}$$
$$\sigma_{xy} = 2\mu\mathcal{D}_x(-g_{,yy} - yg_{,yyy}) + 2\mu\mathcal{D}_y(yg_{,xyy})$$

To obtain results for a concentrated displacement discontinuity $\mathcal{D}_{\bar{x}}$, $\mathcal{D}_{\bar{y}}$ on the line $\bar{y} = 0$ for a local $\bar{x}$, $\bar{y}$ coordinate system we simply replace all occurrences of x and y in (4.221) and (4.222) by $\bar{x}$ and $\bar{y}$. This is possible because there is no preferred orientation for an infinite, homogeneous, isotropic elastic solid. Transformation Eqs. (4.62) and (4.66) can then be used to express the local displacement and stress components in the global x, y system. We will follow this approach in Chap. 6

to solve the problem of a quadratically varying displacement discontinuity over an arbitrarily oriented straight line segment by integrating the above solution for a concentrated displacement discontinuity along the segment.

4.14.2 Anisotropic Elasticity

Referring to (4.212)–(4.214), the potentials ϕ_1 and ϕ_2 for the problem of a concentrated displacement discontinuity in an infinite, orthotropic (or transversely isotropic) elastic material can be written as

$$
\begin{aligned}
\phi_1 &= \frac{1}{2\pi}\frac{1+q_2}{q_1-q_2}\lim_{a\to 0}\frac{1}{2a}\left[\mathcal{D}_x\int\frac{\partial W(x,y_1)}{\partial y_1}dx - \mathcal{D}_y\gamma_1 W(x,y_1)\right] \\
\phi_2 &= -\frac{1}{2\pi}\frac{1+q_1}{q_1-q_2}\lim_{a\to 0}\frac{1}{2a}\left[\mathcal{D}_x\int\frac{\partial W(x,y_2)}{\partial y_2}dx - \mathcal{D}_y\gamma_2 W(x,y_2)\right]
\end{aligned}
\tag{4.223}
$$

It can be shown using L'Hopital's rule that

$$
\begin{aligned}
\lim_{a\to 0}\frac{1}{2a}W(x,y_i) &= \lim_{a\to 0}\frac{\partial W(x,y_i)/\partial a}{2} \\
&= -\cos\beta\ln(x^2+y_i^2)^{1/2} + \frac{\sin\beta}{\gamma_i}\arctan\frac{y_i}{x} \\
&= \Omega(x,y_i)
\end{aligned}
\tag{4.224}
$$

and therefore

$$
\begin{aligned}
\phi_1 &= \frac{1}{2\pi}\frac{1+q_2}{q_1-q_2}\left[\mathcal{D}_x\int\frac{\partial\Omega(x,y_1)}{\partial y_1}\,\mathrm{d}x - \mathcal{D}_y\gamma_1\Omega(x,y_1)\right] \\
\phi_2 &= -\frac{1}{2\pi}\frac{1+q_1}{q_1-q_2}\left[\mathcal{D}_x\int\frac{\partial\Omega(x,y_2)}{\partial y_2}\,\mathrm{d}x - \mathcal{D}_y\gamma_2\Omega(x,y_2)\right]
\end{aligned}
\tag{4.225}
$$

The displacements and stresses due to the concentrated displacement discontinuity $\mathcal{D}_x$, $\mathcal{D}_y$ at the origin are found by substituting (4.225) into (4.129) and (4.130), noting (4.216), to get

$$
\begin{aligned}
u_x = {} & \frac{q_3}{2\pi}\mathcal{D}_x\left[\frac{1}{1+q_1}\frac{\partial}{\partial y_1}\Omega(x, y_1) - \frac{1}{1+q_2}\frac{\partial}{\partial y_2}\Omega(x, y_2)\right] \\
& - \frac{q_3}{2\pi}\mathcal{D}_y\left[\frac{\gamma_1}{1+q_1}\frac{\partial}{\partial x}\Omega(x, y_1) - \frac{\gamma_2}{1+q_2}\frac{\partial}{\partial x}\Omega(x, y_2)\right]
\end{aligned}
\tag{4.226}
$$

$$
\begin{aligned}
u_y = {} & -\frac{q_3}{2\pi}\mathcal{D}_x\left[\frac{q_1}{\gamma_1(1+q_1)}\frac{\partial}{\partial x}\Omega(x, y_1) - \frac{q_2}{\gamma_2(1+q_2)}\frac{\partial}{\partial x}\Omega(x, y_2)\right] \\
& - \frac{q_3}{2\pi}\mathcal{D}_y\left[\frac{q_1}{1+q_1}\frac{\partial}{\partial y_1}\Omega(x, y_1) - \frac{q_2}{(1+q_2)}\frac{\partial}{\partial y_2}\Omega(x, y_2)\right]
\end{aligned}
$$

and

$$
\begin{aligned}
\sigma_{xx} = {} & c_{66}\frac{q_3}{2\pi}\mathcal{D}_x\left[\frac{1}{\gamma_1^2}\frac{\partial^2}{\partial x\partial y_1}\Omega(x, y_1) - \frac{1}{\gamma_2^2}\frac{\partial^2}{\partial x\partial y_2}\Omega(x, y_2)\right] \\
& + c_{66}\frac{q_3}{2\pi}\mathcal{D}_y\left[\frac{1}{\gamma_1}\frac{\partial^2}{\partial y_1^2}\Omega(x, y_1) - \frac{1}{\gamma_2}\frac{\partial^2}{\partial y_2^2}\Omega(x, y_2)\right]
\end{aligned}
$$

$$
\begin{aligned}
\sigma_{yy} = {} & -c_{66}\frac{q_3}{2\pi}\mathcal{D}_x\left[\frac{\partial^2}{\partial x\partial y_1}\Omega(x, y_1) - \frac{\partial^2}{\partial x\partial y_2}\Omega(x, y_2)\right] \\
& - c_{66}\frac{q_3}{2\pi}\mathcal{D}_y\left[\gamma_1\frac{\partial^2}{\partial y_1^2}\Omega(x, y_1) - \gamma_2\frac{\partial^2}{\partial y_2^2}\Omega(x, y_2)\right]
\end{aligned}
\tag{4.227}
$$

$$
\begin{aligned}
\sigma_{xy} = {} & c_{66}\frac{q_3}{2\pi}\mathcal{D}_x\left[\frac{1}{\gamma_1}\frac{\partial^2}{\partial y_1^2}\Omega(x, y_1) - \frac{1}{\gamma_2}\frac{\partial^2}{\partial y_2^2}\Omega(x, y_2)\right] \\
& - c_{66}\frac{q_3}{2\pi}\mathcal{D}_y\left[\frac{\partial^2}{\partial x\partial y_1}\Omega(x, y_1) - \frac{\partial^2}{\partial x\partial y_2}\Omega(x, y_2)\right]
\end{aligned}
$$

Similar to the case of isotropic elasticity, the above results can be integrated along an arbitrarily oriented line segment to obtain the displacements and stresses due to a variable displacement discontinuity along the segment. The procedure is more complicated than in the isotropic case, however, because an anisotropic material has preferred directions of elastic symmetry (we have assumed that these directions lie along the $x,\ y$ coordinate axes).

The first step in the analysis is to observe that the $\bar{x},\ \bar{y}$ components of a concentrated displacement discontinuity can be written in terms of its $x,\ y$ components by means of the transformation formulas (4.204). Integration of the displacements and stresses over the line segment is facilitated by writing the derivatives of function $\Omega(x, y_i)$ in terms of $\bar{x},\ \bar{y}$ coordinates. It is left as an exercise for the reader to verify that the results can be written as

$$\frac{\partial\Omega(x, y_i)}{\partial x} = \frac{-(A_i\bar{x} + B_i\bar{y}/2)}{A_i\bar{x}^2 + B_i\bar{x}\bar{y} + C_i\bar{y}^2}$$

$$\frac{\partial\Omega(x, y_i)}{\partial y_i} = \frac{-\bar{y}/\gamma_i}{A_i\bar{x}^2 + B_i\bar{x}\bar{y} + C_i\bar{y}^2}$$

$$\frac{\partial^2\Omega(x, y_i)}{\partial x\partial y_i} = \frac{1}{\gamma_i}\frac{[A_i\bar{x}^2 \sin\beta + 2A_i\bar{x}\bar{y}\cos\beta + (B_i\cos\beta - C_i\sin\beta)\bar{y}^2]}{(A_i\bar{x}^2 + B_i\bar{x}\bar{y} + C_i\bar{y}^2)^2} \tag{4.228}$$

$$\frac{\partial^2\Omega(x, y_i)}{\partial y_i^2} = \frac{[-A_i\bar{x}^2\cos\beta + 2A_i\bar{x}\bar{y}\sin\beta + (B_i\sin\beta + C_i\cos\beta)\bar{y}^2]}{(A_i\bar{x}^2 + B_i\bar{x}\bar{y} + C_i\bar{y}^2)^2}$$

where A_i, B_i, and C_i are defined in (4.179). Substitution of these expressions into (4.226) and (4.227) gives the required solution.

4.15 Nuclei of Strain

In the literature, singular solutions to the equations of elasticity are referred to as *nuclei of strain*. Examples of nuclei of strain are given by Love (1944) and Mindlin and Cheng (1950), including the so-called double force without moment, double force with moment, and center of dilatation. The fundamental solution in this connection is Kelvin's solution for a point force in an infinite elastic solid, because all of the other nuclei can be obtained from this.

Nuclei of strain for two dimensions are ordinarily derived by integrating the corresponding three-dimensional nuclei along the z axis from $-\infty$ to $+\infty$. In this way, one finds solutions for such nuclei as a *line* of double force without moment or a *line* of centers of dilatation. Equivalently, one can obtain the two-dimensional results directly from the plane strain version of Kelvin's problem, which is the approach adopted here. Also, in what follows, we assume that the material is isotropic; derivations of comparable results for orthotropic materials can be obtained in a similar manner.

4.15.1 *Line of Double Force Without Moment*

A line of double force without moment is defined with reference to Fig. 4.15, which shows two lines of force, equal in magnitude but opposite in direction, separated by a distance $2c$. The resultant force and moment of this system are zero. A concentrated line of double force without moment at the origin $x = 0$, $y = 0$ is represented by the

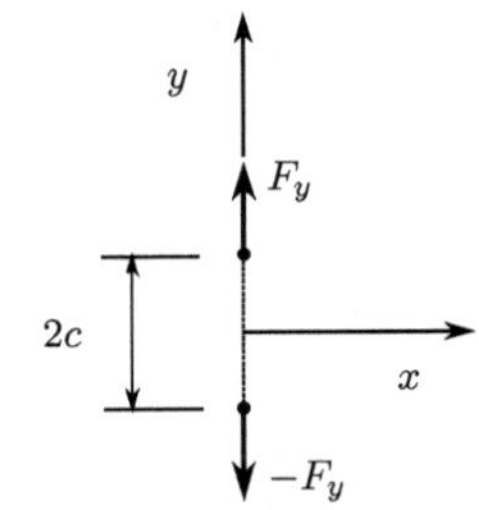

Fig. 4.15 Line of double force without moment, $\mathcal{F}_y$

situation in which the product $2cF_y$ is kept constant as c goes to zero. The quantity $\mathcal{F}_y = 2cF_y$ then represents the "strength" of the double force.[12]

As an illustration of the limiting procedure used to obtain the solution to the subject problem, we consider the x component of displacement for the force system depicted in Fig. 4.15. Using (4.143), we write

$$u_x = \frac{1}{2\mu}F_y\left[-(y-c)\frac{\partial}{\partial x}g(x, y-c)\right] - \frac{1}{2\mu}F_y\left[-(y+c)\frac{\partial}{\partial x}g(x, y+c)\right] \tag{4.229}$$

The value of u_x for a concentrated line of double force without moment in the y direction is obtained from this expression by replacing F_y with $\mathcal{F}_y/2c$ and passing to the limit as $c \to 0$, i.e.,

$$u_x = \frac{1}{2\mu}\mathcal{F}_y \lim_{c\to 0}\frac{1}{2c}\left[(y+c)\frac{\partial}{\partial x}g(x, y+c) - (y-c)\frac{\partial}{\partial x}g(x, y-c)\right] \tag{4.230}$$

Using the results $\partial g(x, y+c)/\partial c = \partial g(x, y+c)/\partial y$ and $\partial g(x, y-c)/\partial c = -\partial g(x, y-c)/\partial y$, a simple application of L'Hopital's rule gives

$$\begin{aligned} u_x = \frac{1}{2\mu}\mathcal{F}_y \lim_{c\to 0}\frac{1}{2}\Bigg[&\frac{\partial}{\partial x}g(x, y+c) + (y+c)\frac{\partial^2}{\partial x\partial y}g(x, y+c) \\ &+\frac{\partial}{\partial x}g(x, y-c) + (y-c)\frac{\partial^2}{\partial x\partial y}g(x, y-c)\Bigg] \end{aligned} \tag{4.231}$$

or

$$u_x = \frac{1}{2\mu}\mathcal{F}_y\left[\frac{\partial}{\partial x}g(x, y) + y\frac{\partial^2}{\partial x\partial y}g(x, y)\right] \tag{4.232}$$

Notice that this equation can be formally obtained from the expression for u_x in (4.143) by setting $F_x = 0$, replacing F_y by $-\mathcal{F}_y$, and differentiating with respect to y. Applying this process to u_y yields

[12] This approach is comparable to that used in Sect. 1.8 to derive the two-dimensional potential for a dipole source.

$$
\begin{aligned}
u_y &= \frac{1}{2\mu}(-\mathcal{F}_y)\frac{\partial}{\partial y}\left[(3-4\nu)g(x,y) - y\frac{\partial}{\partial y}g(x,y)\right] \\
&= \frac{1}{2\mu}\mathcal{F}_y\left[-2(1-2\nu)\frac{\partial}{\partial y}g(x,y) + y\frac{\partial^2}{\partial y^2}g(x,y)\right]
\end{aligned} \tag{4.233}
$$

Similarly, to obtain the displacements for a line of double force without moment in the x-direction, we set $F_y = 0$ in (4.143), replace F_x by $-\mathcal{F}_x$, and differentiate with respect to x, which gives

$$
\begin{aligned}
u_x &= \frac{1}{2\mu}\mathcal{F}_x\left[-2(1-2\nu)\frac{\partial}{\partial x}g(x,y) + x\frac{\partial^2}{\partial x^2}g(x,y)\right] \\
u_y &= \frac{1}{2\mu}\mathcal{F}_x\left[\frac{\partial}{\partial y}g(x,y) + x\frac{\partial^2}{\partial x \partial y}g(x,y)\right]
\end{aligned} \tag{4.234}
$$

In the notation of (4.143), the displacements due to both lines of double force without moment acting simultaneously are

$$
\begin{aligned}
u_x &= \frac{1}{2\mu}\mathcal{F}_x\left[-2(1-2\nu)g_{,x} + xg_{,xx}\right] + \frac{1}{2\mu}\mathcal{F}_y\left[g_{,x} + yg_{,xy}\right] \\
u_y &= \frac{1}{2\mu}\mathcal{F}_x\left[g_{,y} + xg_{,xy}\right] + \frac{1}{2\mu}\mathcal{F}_y\left[-2(1-2\nu)g_{,y} + yg_{,yy}\right]
\end{aligned} \tag{4.235}
$$

The stresses associated with these displacements can be found in the same manner, or else by computing the strains and using Hooke's law. The results can be summarized as follows:

$$
\begin{aligned}
\sigma_{xx} &= \mathcal{F}_x\left[-(1-2\nu)g_{,xx} - xg_{,xyy}\right] + \mathcal{F}_y\left[(1+2\nu)g_{,xx} + yg_{,xxy}\right] \\
\sigma_{yy} &= \mathcal{F}_x\left[-(1+2\nu)g_{,xx} + xg_{,xyy}\right] + \mathcal{F}_y\left[(1-2\nu)g_{,xx} - yg_{,xxy}\right] \\
\sigma_{xy} &= \mathcal{F}_x\left[2\nu g_{,xy} + xg_{,xxy}\right] + \mathcal{F}_y\left[2\nu g_{,xy} + yg_{,xyy}\right]
\end{aligned} \tag{4.236}
$$

4.15.2 Line of Centers of Dilatation

A special case of the above solution occurs when the strengths of the two components of double force are equal, i.e., $\mathcal{F}_x = \mathcal{F}_y = \mathcal{F}$. The singular solution for this case has the characteristic that the stress is purely normal on the boundary of a circular hole

$r = r_o > 0$ centered at the origin. The $r,\ \theta$ directions in plane polar coordinates are equivalent to transformed Cartesian coordinates $\bar{x},\ \bar{y}$ when angle β is equal to θ, so that $\sigma_{rr} = \sigma_{\bar{x}\,\bar{x}}$, $\sigma_{\theta\theta} = \sigma_{\bar{y}\,\bar{y}}$, and $\sigma_{r\theta} = \sigma_{\bar{x}\,\bar{y}}$.

To obtain the required results, it is convenient to use (4.163) to eliminate the terms in (4.236) involving derivatives of function $g(x, y)$ multiplied by coordinate x. Differentiating the first one of (4.163) twice wiith respect to y we find

$$xg_{,yyy} = 2g_{,xy} + yg_{,xyy} \tag{4.237}$$

or, because $g_{,xx} + g_{,yy} = 0$,

$$xg_{,xxy} + yg_{,xyy} = -2g_{,xy} \tag{4.238}$$

Similarly, starting with the the second one of (4.163) and differentiating twice with respect to x we obtain the formula

$$xg_{,xyy} - yg_{,xxy} = 2g_{,xx} \tag{4.239}$$

Substituting (4.237) and (4.239) into (4.236), with $\mathcal{F}_x = \mathcal{F}_y = \mathcal{F}$ and $x = r\cos\theta$, $y = r\sin\theta$, and using (4.146) gives

$$\begin{aligned}
\sigma_{xx} &= -2(1-2\nu)\mathcal{F}g_{,xx} \\
&= -\frac{C\mathcal{F}}{2\pi}\frac{x^2-y^2}{r^4} = -\frac{C\mathcal{F}}{2\pi}\frac{\cos 2\theta}{r^2} \\
\sigma_{yy} &= 2(1-2\nu)\mathcal{F}g_{,xx} \\
&= \frac{C\mathcal{F}}{2\pi}\frac{x^2-y^2}{r^4} = \frac{C\mathcal{F}}{2\pi}\frac{\cos 2\theta}{r^2} \\
\sigma_{xy} &= -2(1-2\nu)\mathcal{F}g_{,xy} \\
&= -\frac{C\mathcal{F}}{2\pi}\frac{2xy}{r^4} = -\frac{C\mathcal{F}}{2\pi}\frac{\sin 2\theta}{r^2}
\end{aligned} \tag{4.240}$$

where $C = (1-2\nu)/(1-\nu)$. Hence, from (4.67) with $\beta = \theta$,

$$\begin{aligned}
\sigma_{\bar{x}\,\bar{y}} &= \sigma_{r\theta} = 0 \\
\sigma_{\bar{x}\,\bar{x}} &= \sigma_{rr} = -\frac{C\mathcal{F}}{2\pi}\frac{1}{r^2}
\end{aligned} \tag{4.241}$$

The stress on $r = r_o$ is therefore purely normal, as asserted. When $\mathcal{F}$ is positive, the normal stress on the wall of the hole is compressive ($\sigma_{\bar{x}\,\bar{x}} = \sigma_{rr} < 0$) and the hole expands, or dilates. The solution for this case represents a line of centers of dilatation. When $\mathcal{F}$ is negative, the solution represents a line of centers of compression.

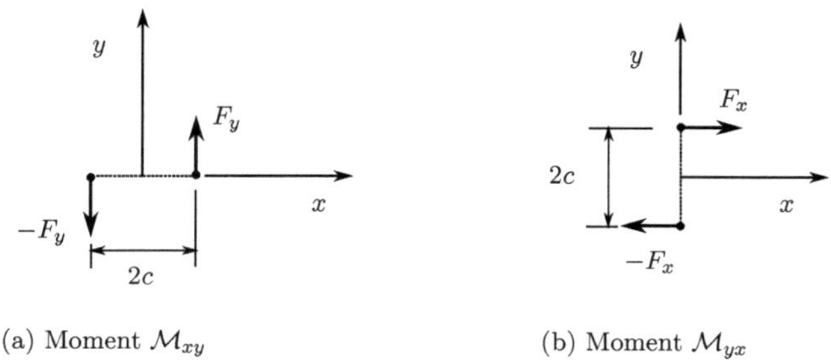

Fig. 4.16 Lines of double force with moment

A similar calculation for the displacements gives the results

$$\begin{aligned} u_{\bar{x}} &= u_r = \frac{C\mathcal{F}}{4\pi\mu}\frac{1}{r} \\ u_{\bar{y}} &= u_\theta = 0 \end{aligned} \tag{4.242}$$

confirming that the hole expands (i.e., $u_r > 0$) when $\mathcal{F} > 0$.

4.15.3 Line of Double Force with Moment

Another nucleus of strain can be derived for the case that a line of double force has a moment associated with it. This situation is depicted in Fig. 4.16a for two lines of force acting in the y direction. Again, the two forces are equal in magnitude but opposite in direction. Their lines of action are separated by a distance $2c$, giving a resultant moment (per unit length perpendicular to the x, y plane) of $2cF_y$. The singular solution for this problem is obtained by keeping the strength of the moment constant as the separation distance between the lines of action of the two forces goes to zero. The strength of the moment is designated as $\mathcal{M}_{xy}$, where the first subscript denotes the direction of the infinitesimal separation between the two forces used to create the concentrated moment and the second subscript denotes the direction in which the force pair acts.

In the same way, a moment of strength $\mathcal{M}_{yx} = 2cF_x$ is the limiting case of two lines of force in the x direction, separated by a distance $2c$ along the y axis, as shown in Fig. 4.16b. It is clear from the figure that $\mathcal{M}_{xy}$ represents a counterclockwise moment whereas $\mathcal{M}_{yx}$ represents a clockwise one. These quantities are both positive as shown, however, by virtue of the following convention (which also applies in three dimensions): the concentrated moment $\mathcal{M}_{ij}$ is positive if the infinitesimal distance from the point of application of the moment for the positive-directed force F_j in the force pair is in the positive x_i direction.

As in the previous cases for lines of double force without moment, the displacements and stresses due to a line of double force with moment can be obtained from (4.143) and (4.144) by a mathematical formalism: For the line of double force with moment $\mathcal{M}_{xy}$ (Fig. 4.16a), set $F_x = 0$, $F_y = -\mathcal{M}_{xy}$, and differentiate each equation with respect to x; for a line of double force with moment $\mathcal{M}_{yx}$ (Fig. 4.16b), set $F_x = -\mathcal{M}_{yx}$, $F_y = 0$, and differentiate each equation with respect to y. This process gives

$$
\begin{aligned}
u_x &= \frac{1}{2\mu}\mathcal{M}_{yx}\left[-(3-4\nu)g_{,y} + xg_{,xy}\right] + \frac{1}{2\mu}\mathcal{M}_{xy}\left[yg_{,xx}\right] \\
u_y &= \frac{1}{2\mu}\mathcal{M}_{yx}\left[-xg_{,xx}\right] + \frac{1}{2\mu}\mathcal{M}_{xy}\left[-(3-4\nu)g_{,x} + yg_{,xy}\right]
\end{aligned}
\tag{4.243}
$$

and

$$
\begin{aligned}
\sigma_{xx} &= \mathcal{M}_{yx}\left[-2(1-\nu)g_{,xy} + xg_{,xxy}\right] + \mathcal{M}_{xy}\left[-2\nu g_{,xy} - yg_{,xyy}\right] \\
\sigma_{yy} &= \mathcal{M}_{yx}\left[-2\nu g_{,xy} - xg_{,xxy}\right] + \mathcal{M}_{xy}\left[-2(1-\nu)g_{,xy} + yg_{,xyy}\right] \\
\sigma_{xy} &= \mathcal{M}_{yx}\left[(1-2\nu)g_{,xx} + xg_{,xyy}\right] + \mathcal{M}_{xy}\left[-(1-2\nu)g_{,xx} + yg_{,xxy}\right]
\end{aligned}
\tag{4.244}
$$

4.15.4 The Seismic Moment Tensor

In seismology, combinations of double forces with and without moment are used to represent displacement discontinuities across elemental areas of fault planes in three dimensions. The nuclei of strain involved in such analyses are components of the so-called seismic moment tensor, $\mathcal{M}_{ij}$ (see, for example, Aki and Richards 2002). The three diagonal components of this tensor ($\mathcal{M}_{xx}$, $\mathcal{M}_{yy}$, and $\mathcal{M}_{zz}$) are double forces without moment (also called vector dipoles) while the other six components are double forces with moment. In order to ensure equilibrium, the moments are arranged to occur as "double couples." For the two-dimensional situation considered here, this means that the two double forces with moment $\mathcal{M}_{xy}$ and $\mathcal{M}_{yx}$ are equal, and they must occur together. Therefore, the seismic moment tensor is symmetric: $\mathcal{M}_{ij} = \mathcal{M}_{ji}$.

The diagonal components of the seismic moment tensor are equivalent to what were previously called $\mathcal{F}_x$ and $\mathcal{F}_y$ (with an obvious extension to a third component, $\mathcal{F}_z$). The seismic moment terminology will be used in the ensuing discussion.

Taking $\mathcal{F}_x = \mathcal{M}_{xx}$ and $\mathcal{F}_y = \mathcal{M}_{yy}$ in (4.235), $\mathcal{M}_{yx} = \mathcal{M}_{xy}$ in (4.243), and noting from (4.163) that $xg_{,xx} = -g_{,x} - yg_{,xy}$ and $xg_{,xy} = -g_{,y} - yg_{,yy}$, the following expressions for the displacements are obtained for the two-dimensional seismic moment tensor:

$$
\begin{aligned}
u_x = & \frac{1}{2\mu}\mathcal{M}_{xx}\left[-(3-4\nu)g_{,x} - yg_{,xy}\right] + \frac{1}{2\mu}\mathcal{M}_{yy}\left[g_{,x} + yg_{,xy}\right] \\
& + \frac{1}{2\mu}2\mathcal{M}_{xy}\left[-2(1-\nu)g_{,y} - yg_{,yy}\right] \\
u_y = & \frac{1}{2\mu}\mathcal{M}_{xx}\left[-yg_{,yy}\right] + \frac{1}{2\mu}\mathcal{M}_{yy}\left[-2(1-2\nu)g_{,y} + yg_{,yy}\right] \\
& + \frac{1}{2\mu}2\mathcal{M}_{xy}\left[-(1-2\nu)g_{,x} + yg_{,xy}\right]
\end{aligned}
\tag{4.245}
$$

The stresses associated with displacements (4.245) are

$$
\begin{aligned}
\sigma_{xx} &= \mathcal{M}_{xx}\left[(3-2\nu)g_{,yy} + yg_{,yyy}\right] + \mathcal{M}_{yy}\left[-(1+2\nu)g_{,yy} - yg_{,yyy}\right] + 2\mathcal{M}_{xy}\left[-2g_{,xy} - yg_{,xyy}\right] \\
\sigma_{yy} &= \mathcal{M}_{xx}\left[-(1-2\nu)g_{,yy} - yg_{,yyy}\right] + \mathcal{M}_{yy}\left[-(1-2\nu)g_{,yy} + yg_{,yyy}\right] + 2\mathcal{M}_{xy}\left[yg_{,xyy}\right] \\
\sigma_{xy} &= \mathcal{M}_{xx}\left[-2(1-\nu)g_{,xy} - yg_{,xyy}\right] + \mathcal{M}_{yy}\left[2\nu g_{,xy} + yg_{,xyy}\right] + 2\mathcal{M}_{xy}\left[-g_{,yy} - yg_{,yyy}\right]
\end{aligned}
\tag{4.246}
$$

The forms of these equations were modified slightly by using (4.163) and (4.237) et seq. to eliminate terms involving derivatives of function $g(x, y)$ multiplied by coordinate x.

The relationship between the seismic moment tensor and a concentrated displacement discontinuity can be established by comparing Eqs. (4.245) and (4.246) with (4.221) and (4.222). It can easily be shown that the combination $\mathcal{M}_{xy} + \mathcal{M}_{yx} = 2\mathcal{M}_{xy}$ is equivalent to a concentrated displacement discontinuity $\mathcal{D}_x$ at the origin if

$$\mathcal{M}_{xy} = \mu\mathcal{D}_x \tag{4.247}$$

Also, the combination $\mathcal{M}_{xx} + \mathcal{M}_{yy}$ is equivalent to a concentrated displacement discontinuity $\mathcal{D}_y$ at the origin if

$$
\begin{aligned}
\mathcal{M}_{xx} &= 2\mu\frac{\nu}{1-2\nu}\mathcal{D}_y \\
\mathcal{M}_{yy} &= 2\mu\frac{1-\nu}{1-2\nu}\mathcal{D}_y
\end{aligned}
\tag{4.248}
$$

4.16 Concentrated Force—Index Notation

The singular solutions given in Sect. 4.11 for a concentrated force acting at a point in a plane can be succinctly represented using index notation. The essential idea is that these solutions give the displacements and stresses at a point $\boldsymbol{x} = (x_1, x_2)$ (the *sample point*, or *field point*) due to a concentrated force $F_i = (F_1, F_2)$ applied at another point $\boldsymbol{\xi} = (\xi_1, \xi_2)$ (the *load point*, or *source point*).

In general, the displacements at point $\boldsymbol{x}$ can be written as

$$u_i(\boldsymbol{x}) = U_{ji}(\boldsymbol{x}, \boldsymbol{\xi})F_j(\boldsymbol{\xi}) \tag{4.249}$$

where the quantity $U_{ji}(\boldsymbol{x}, \boldsymbol{\xi})$ gives the displacement in the i direction at point $\boldsymbol{x}$ due to a unit concentrated force in the j direction at point $\boldsymbol{\xi}$.

It is important to mention that other naming conventions can be found in the literature for what we have identified as $U_{ji}(\boldsymbol{x}, \boldsymbol{\xi})$; in other words, the ordering of the arguments and subscripts is not unique. We will introduce other comparable quantities later and will consistently adopt the following conventions for the generic second-order tensor quantity $X_{pq}(\mathbf{y}, \mathbf{z})$:

1. The first argument in parentheses ($\mathbf{y}$) gives the location of the sample point while the second one ($\mathbf{z}$) gives the location of the load point.
2. The first subscript (p) gives the direction of the *cause* (e.g., a force component) at the load point while the second one (q) gives the direction of the *effect* (a displacement or traction component) at the sample point.

By these conventions, if we were to write $u_j(\boldsymbol{\xi}) = U_{ij}(\boldsymbol{\xi}, \boldsymbol{x})F_i(\boldsymbol{x})$ the quantity $U_{ij}(\boldsymbol{\xi}, \boldsymbol{x})$ would give the displacements $u_j(\boldsymbol{\xi})$ at a sample point $\boldsymbol{\xi}$ due to a unit force $F_i(\boldsymbol{x})$ at a load point $\boldsymbol{x}$.

In a similar way, the stresses at point $\boldsymbol{x}$ can be represented as

$$\sigma_{ij}(\boldsymbol{x}) = B_{kij}(\boldsymbol{x}, \boldsymbol{\xi})F_k(\boldsymbol{\xi}) \tag{4.250}$$

Here, the the third-order tensor $B_{kij}(\boldsymbol{x}, \boldsymbol{\xi})$ gives the stress component σ_{ij} at point $\boldsymbol{x}$ due to a unit concentrated force in the k-direction at point $\boldsymbol{\xi}$.

For the case of a concentrated force in an isotropic plane (Kelvin's problem), inspection of (4.150) reveals that the individual terms of $U_{ji}(\boldsymbol{x}, \boldsymbol{\xi})$ in (4.249) are given as follows:

$$
\begin{aligned}
U_{11}(\boldsymbol{x}, \boldsymbol{\xi}) &= \frac{1}{2\mu}\left[(3-4\nu)g - (x_1 - \xi_1)\frac{\partial g}{\partial x_1}\right] \\
U_{21}(\boldsymbol{x}, \boldsymbol{\xi}) &= \frac{1}{2\mu}\left[-(x_2 - \xi_2)\frac{\partial g}{\partial x_1}\right] \\
U_{12}(\boldsymbol{x}, \boldsymbol{\xi}) &= \frac{1}{2\mu}\left[-(x_1 - \xi_1)\frac{\partial g}{\partial x_2}\right] \\
U_{22}(\boldsymbol{x}, \boldsymbol{\xi}) &= \frac{1}{2\mu}\left[(3-4\nu)g - (x_2 - \xi_2)\frac{\partial g}{\partial x_2}\right]
\end{aligned}
\tag{4.251}
$$

where function g is

$$
g(\boldsymbol{x}, \boldsymbol{\xi}) = -\frac{1}{4\pi(1-\nu)} \ln\left[(x_1 - \xi_1)^2 + (x_2 - \xi_2)^2\right]^{1/2} \tag{4.252}
$$

Equation (4.251) can be written in index notation as

$$
U_{ji}(\boldsymbol{x}, \boldsymbol{\xi}) = \frac{1}{2\mu}\left[(3-4\nu)g\delta_{ji} - (x_j - \xi_j)\frac{\partial g}{\partial x_i}\right] \tag{4.253}
$$

It is easy to verify that $U_{ji}(\boldsymbol{x}, \boldsymbol{\xi})$ is symmetric with respect to the coordinate indices i and j, i.e.,

$$
U_{ji}(\boldsymbol{x}, \boldsymbol{\xi}) = U_{ij}(\boldsymbol{x}, \boldsymbol{\xi}) \tag{4.254}
$$

and also that it is symmetric with respect to the arguments $\boldsymbol{x}$ and $\boldsymbol{\xi}$:

$$
U_{ji}(\boldsymbol{\xi}, \boldsymbol{x}) = U_{ij}(\boldsymbol{x}, \boldsymbol{\xi}) \tag{4.255}
$$

The individual terms of $B_{kij}(\boldsymbol{x}, \boldsymbol{\xi})$ associated with the values of $U_{ji}(\boldsymbol{x}, \boldsymbol{\xi})$ in (4.251) follow directly from (4.151). They are

$$
\begin{aligned}
B_{111}(\boldsymbol{x}, \boldsymbol{\xi}) &= 2(1-\nu)\frac{\partial g}{\partial x_1} - (x_1 - \xi_1)\frac{\partial^2 g}{\partial x_1^2} \\
B_{211}(\boldsymbol{x}, \boldsymbol{\xi}) &= 2\nu\frac{\partial g}{\partial x_2} - (x_2 - \xi_2)\frac{\partial^2 g}{\partial x_1^2} \\
B_{122}(\boldsymbol{x}, \boldsymbol{\xi}) &= 2\nu\frac{\partial g}{\partial x_1} - (x_1 - \xi_1)\frac{\partial^2 g}{\partial x_2^2} \\
B_{222}(\boldsymbol{x}, \boldsymbol{\xi}) &= 2(1-\nu)\frac{\partial g}{\partial x_2} - (x_2 - \xi_2)\frac{\partial^2 g}{\partial x_2^2} \\
B_{112}(\boldsymbol{x}, \boldsymbol{\xi}) &= (1-2\nu)\frac{\partial g}{\partial x_2} - (x_1 - \xi_1)\frac{\partial^2 g}{\partial x_1 \partial x_2} \\
B_{212}(\boldsymbol{x}, \boldsymbol{\xi}) &= (1-2\nu)\frac{\partial g}{\partial x_1} - (x_2 - \xi_2)\frac{\partial^2 g}{\partial x_1 \partial x_2}
\end{aligned}
\tag{4.256}
$$

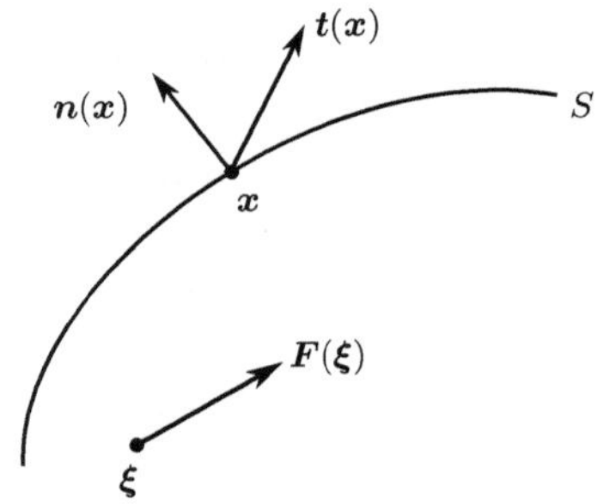

Fig. 4.17 Tractions at sample point $\boldsymbol{x}$ on curve S due to concentrated force at load point $\boldsymbol{\xi}$

These results can be written in index notation as

$$B_{kij}(\boldsymbol{x},\boldsymbol{\xi}) = (1-2\nu)\left(\frac{\partial g}{\partial x_i}\delta_{jk} + \frac{\partial g}{\partial x_j}\delta_{ik} - \frac{\partial g}{\partial x_k}\delta_{ij}\right) + \frac{\partial g}{\partial x_k}\delta_{ij} - (x_k - \xi_k)\frac{\partial^2 g}{\partial x_i \partial x_j} \tag{4.257}$$

from which it is easily seen that the quantity $B_{kij}(\boldsymbol{x},\boldsymbol{\xi})$ is symmetric with respect to its last two indices:

$$B_{kij}(\boldsymbol{x},\boldsymbol{\xi}) = B_{kji}(\boldsymbol{x},\boldsymbol{\xi}) \tag{4.258}$$

If we interchange $\boldsymbol{x}$ and $\boldsymbol{\xi}$ in (4.257)—that is, if we take $\boldsymbol{x}$ as the load point and $\boldsymbol{\xi}$ as the sample point—we write

$$B_{kij}(\boldsymbol{\xi},\boldsymbol{x}) = (1-2\nu)\left(\frac{\partial g}{\partial \xi_i}\delta_{jk} + \frac{\partial g}{\partial \xi_j}\delta_{ik} - \frac{\partial g}{\partial \xi_k}\delta_{ij}\right) + \frac{\partial g}{\partial \xi_k}\delta_{ij} - (\xi_k - x_k)\frac{\partial^2 g}{\partial \xi_i \partial \xi_j} \tag{4.259}$$

or, because $\partial g/\partial \xi_m = -\partial g/\partial x_m$,

$$B_{kij}(\boldsymbol{\xi},\boldsymbol{x}) = -B_{kij}(\boldsymbol{x},\boldsymbol{\xi}) \tag{4.260}$$

We will refer to this result and its companion, (4.255), later.

In addition to expressions (4.249) and (4.250) for the displacements and stresses at point $\boldsymbol{x}$, it is also useful to have a general formula for representing the tractions at this point, measured with reference to an arbitrary curve S, as depicted in Fig. 4.17. If the components of the unit normal to the curve at point $\boldsymbol{x}$ are $n_j(\boldsymbol{x})$, we use (4.16) and (4.250) to write

$$t_i(\boldsymbol{x}) = B_{kji}(\boldsymbol{x},\boldsymbol{\xi})n_j(\boldsymbol{x})F_k(\boldsymbol{\xi}) \tag{4.261}$$

or, after carrying out the summation with respect to repeated index j,

$$t_i(\boldsymbol{x}) = [B_{k1i}(\boldsymbol{x},\boldsymbol{\xi})n_1(\boldsymbol{x}) + B_{k2i}(\boldsymbol{x},\boldsymbol{\xi})n_2(\boldsymbol{x})]\, F_k(\boldsymbol{\xi}) \tag{4.262}$$

The repeated index k in this expression denotes summation and may be given any symbol without changing the meaning of the equation. In particular, (4.262) can be written as

$$t_i(\boldsymbol{x}) = T_{ji}(\boldsymbol{x}, \boldsymbol{\xi})F_j(\boldsymbol{\xi}) \tag{4.263}$$

in which

$$T_{ji}(\boldsymbol{x}, \boldsymbol{\xi}) = B_{j1i}(\boldsymbol{x}, \boldsymbol{\xi})n_1(\boldsymbol{x}) + B_{j2i}(\boldsymbol{x}, \boldsymbol{\xi})n_2(\boldsymbol{x}) \tag{4.264}$$

The quantity $T_{ji}(\boldsymbol{x}, \boldsymbol{\xi})$ gives the traction in the i direction at point $\boldsymbol{x}$ on curve S due to a unit concentrated force in the j direction at point $\boldsymbol{\xi}$. The values of the individual terms of $T_{ji}(\boldsymbol{x}, \boldsymbol{\xi})$ for the plane strain version of Kelvin's problem for an isotropic material are obtained by elaborating (4.264), taking into account (4.256):

$$\begin{aligned}
T_{11}(\boldsymbol{x}, \boldsymbol{\xi}) &= B_{111}(\boldsymbol{x}, \boldsymbol{\xi})n_1(\boldsymbol{x}) + B_{121}(\boldsymbol{x}, \boldsymbol{\xi})n_2(\boldsymbol{x}) \\
&= \left[2(1-\nu)\frac{\partial g}{\partial x_1} - (x_1 - \xi_1)\frac{\partial^2 g}{\partial x_1^2}\right] n_1(\boldsymbol{x}) + \left[(1-2\nu)\frac{\partial g}{\partial x_2} - (x_1 - \xi_1)\frac{\partial^2 g}{\partial x_1 \partial x_2}\right] n_2(\boldsymbol{x}) \\
T_{21}(\boldsymbol{x}, \boldsymbol{\xi}) &= B_{211}(\boldsymbol{x}, \boldsymbol{\xi})n_1(\boldsymbol{x}) + B_{221}(\boldsymbol{x}, \boldsymbol{\xi})n_2(\boldsymbol{x}) \\
&= \left[2\nu\frac{\partial g}{\partial x_2} - (x_2 - \xi_2)\frac{\partial^2 g}{\partial x_1^2}\right] n_1(\boldsymbol{x}) + \left[(1-2\nu)\frac{\partial g}{\partial x_1} - (x_2 - \xi_2)\frac{\partial^2 g}{\partial x_1 \partial x_2}\right] n_2(\boldsymbol{x}) \\
T_{12}(\boldsymbol{x}, \boldsymbol{\xi}) &= B_{112}(\boldsymbol{x}, \boldsymbol{\xi})n_1(\boldsymbol{x}) + B_{122}(\boldsymbol{x}, \boldsymbol{\xi})n_2(\boldsymbol{x}) \\
&= \left[(1-2\nu)\frac{\partial g}{\partial x_2} - (x_1 - \xi_1)\frac{\partial^2 g}{\partial x_1 \partial x_2}\right] n_1(\boldsymbol{x}) + \left[2\nu\frac{\partial g}{\partial x_1} - (x_1 - \xi_1)\frac{\partial^2 g}{\partial x_2^2}\right] n_2(\boldsymbol{x}) \\
T_{22}(\boldsymbol{x}, \boldsymbol{\xi}) &= B_{212}(\boldsymbol{x}, \boldsymbol{\xi})n_1(\boldsymbol{x}) + B_{222}(\boldsymbol{x}, \boldsymbol{\xi})n_2(\boldsymbol{x}) \\
&= \left[(1-2\nu)\frac{\partial g}{\partial x_1} - (x_2 - \xi_2)\frac{\partial^2 g}{\partial x_1 \partial x_2}\right] n_1(\boldsymbol{x}) + \left[2(1-\nu)\frac{\partial g}{\partial x_2} - (x_2 - \xi_2)\frac{\partial^2 g}{\partial x_2^2}\right] n_2(\boldsymbol{x})
\end{aligned} \tag{4.265}$$

It follows from these formulas that $T_{12}(\boldsymbol{x}, \boldsymbol{\xi}) \neq T_{21}(\boldsymbol{x}, \boldsymbol{\xi})$, and hence that $T_{ji}(\boldsymbol{x}, \boldsymbol{\xi})$ is *not* symmetric with respect to coordinate indices i and j, i.e.,

$$T_{ji}(\boldsymbol{x}, \boldsymbol{\xi}) \neq T_{ij}(\boldsymbol{x}, \boldsymbol{\xi}) \tag{4.266}$$

It can be seen from (4.252) that $T_{ji}(\boldsymbol{x}, \boldsymbol{\xi})$ is also not symmetric with respect to arguments $\boldsymbol{x}$ and $\boldsymbol{\xi}$:

$$T_{ji}(\boldsymbol{\xi}, \boldsymbol{x}) \neq T_{ij}(\boldsymbol{x}, \boldsymbol{\xi}) \tag{4.267}$$

When elaborating (4.265) for particular problems using local $\bar{x}$, $\bar{y}$ coordinates it should be kept in mind that the components of the unit normal to the curve S at point $\boldsymbol{x}$ are $n_1(\boldsymbol{x}) = -\sin\beta$ and $n_2(\boldsymbol{x}) = \cos\beta$. This result follows from the fact that the angle β defines the direction of the tangent to S at point $\boldsymbol{x}$, as illustrated in Fig. 4.18.

Comparable expressions for $U_{ji}(\boldsymbol{x}, \boldsymbol{\xi})$, etc., for an orthotropic elastic material are given in Appendix B.

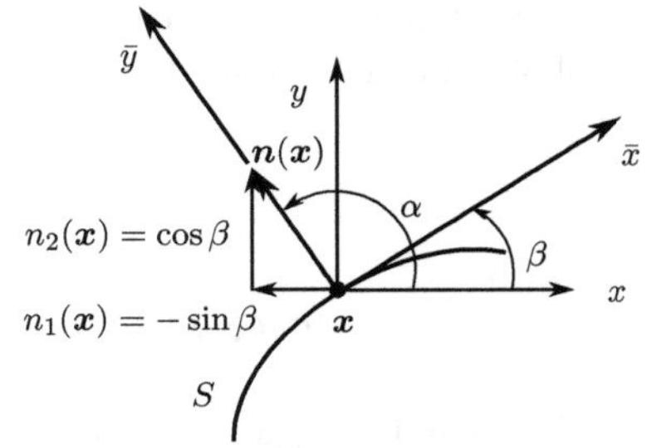

Fig. 4.18 Components of unit normal at sample point $\boldsymbol{x}$

4.16.1 Alternative Forms for $U_{ji}(x, \xi)$ and $T_{ji}(x, \xi)$

In the literature, the quantities $U_{ji}(\boldsymbol{x}, \boldsymbol{\xi})$ and $T_{ji}(\boldsymbol{x}, \boldsymbol{\xi})$ are most often written in terms of the distance $r = [(x - \xi_x)^2 + (y - \xi_y)^2]^{1/2}$ and its derivatives. We will conclude this section by showing how these expressions can be developed for the case of an isotropic elastic material.

Using the definition of $g(\boldsymbol{x}, \boldsymbol{\xi})$ in (4.252), the expression for $U_{ji}(\boldsymbol{x}, \boldsymbol{\xi})$ can be rewritten as

$$U_{ji}(\boldsymbol{x}, \boldsymbol{\xi}) = -\frac{1}{8\pi\mu(1-\nu)}\left[(3-4\nu)\delta_{ji}\ln r - (x_j - \xi_j)\frac{\partial}{\partial x_i}\ln r\right] \tag{4.268}$$

Noting that $\partial \ln r/\partial x_i = r^{-1}\partial r/\partial x_i$ and $\partial r/\partial x_j = (x_j - \xi_j)/r$, this becomes

$$U_{ji}(\boldsymbol{x}, \boldsymbol{\xi}) = -\frac{1}{8\pi\mu(1-\nu)}\left[(3-4\nu)\delta_{ji}\ln r - \frac{\partial r}{\partial x_j}\frac{\partial r}{\partial x_i}\right]$$

or

$$U_{ji}(\boldsymbol{x}, \boldsymbol{\xi}) = -\frac{1}{8\pi\mu(1-\nu)}\left[(3-4\nu)\delta_{ji}\ln r - r_{,j}r_{,i}\right] \tag{4.269}$$

It can be seen from (4.268) and (4.269) that the kernel $U_{ji}(\boldsymbol{x}, \boldsymbol{\xi})$ has a logarithmic singularity when $\boldsymbol{x} = \boldsymbol{\xi}$.

To obtain the corresponding expression for $T_{ji}(\boldsymbol{x}, \boldsymbol{\xi})$ we first write (4.257) as

$$B_{kij}(\boldsymbol{x}, \boldsymbol{\xi}) = -\frac{1}{4\pi(1-\nu)}\left[(1-2\nu)\left(\frac{\partial}{\partial x_i}\ln r\delta_{jk} + \frac{\partial}{\partial x_j}\ln r\delta_{ik} - \frac{\partial}{\partial x_k}\ln r\delta_{ij}\right) + \frac{\partial}{\partial x_k}\ln r\delta_{ij}\right.$$
$$\left. -(x_k - \xi_k)\frac{\partial^2}{\partial x_i \partial x_j}\ln r\right]$$

Using the result

$$\frac{\partial^2}{\partial x_i \partial x_j}\ln r = \frac{\partial}{\partial x_i}\left(\frac{1}{r}\frac{\partial r}{\partial x_j}\right) = \frac{1}{r}\left(\frac{\partial^2 r}{\partial x_i \partial x_j} - \frac{1}{r}\frac{\partial r}{\partial x_i}\frac{\partial r}{\partial x_j}\right)$$

and performing some simple algebra, we find that

$$B_{kij}(\boldsymbol{x}, \boldsymbol{\xi}) = -\frac{1}{4\pi(1-\nu)} \frac{1}{r} \left[(1-2\nu)\left(r_{,i}\delta_{jk} + r_{,j}\delta_{ik} - r_{,k}\delta_{ij}\right) + 2r_{,i}r_{,j}r_{,k}\right] \tag{4.270}$$

Then, using the definition $T_{kj}(\boldsymbol{x}, \boldsymbol{\xi}) = B_{kij}(\boldsymbol{x}, \boldsymbol{\xi})n_i(\boldsymbol{x})$ and summing the repeated index i,

$$T_{kj}(\boldsymbol{x}, \boldsymbol{\xi}) = -\frac{1}{4\pi(1-\nu)} \frac{1}{r} \left[(1-2\nu)\left(r_{,j}n_k(\boldsymbol{x}) - r_{,k}n_j(\boldsymbol{x}) + \frac{\partial r}{\partial \boldsymbol{n}(\boldsymbol{x})}\delta_{jk}\right) + 2r_{,j}r_{,k}\frac{\partial r}{\partial \boldsymbol{n}(\boldsymbol{x})}\right] \tag{4.271}$$

where we have used the additional result $r_{,i}n_i(\boldsymbol{x}) = \partial r/\partial \boldsymbol{n}(\boldsymbol{x})$ (see (2.4)). Finally, to conform with the notation of (4.263) we rearrange the indices in (4.271) and write

$$T_{ji}(\boldsymbol{x}, \boldsymbol{\xi}) = -\frac{1}{4\pi(1-\nu)} \frac{1}{r} \left[(1-2\nu)\left(r_{,i}n_j(\boldsymbol{x}) - r_{,j}n_i(\boldsymbol{x}) + \delta_{ij}\frac{\partial r}{\partial \boldsymbol{n}(\boldsymbol{x})}\right) + 2r_{,i}r_{,j}\frac{\partial r}{\partial \boldsymbol{n}(\boldsymbol{x})}\right] \tag{4.272}$$

It can be concluded from (4.271) and (4.272) that the kernels $B_{kij}(\boldsymbol{x}, \boldsymbol{\xi}) = B_{kji}(\boldsymbol{x}, \boldsymbol{\xi})$ and $T_{ji}(\boldsymbol{x}, \boldsymbol{\xi})$ have strong $1/r$ singularities when $\boldsymbol{x} = \boldsymbol{\xi}$.

Similar expressions for the quantities $U_{ji}(\boldsymbol{x}, \boldsymbol{\xi})$, $B_{kij}(\boldsymbol{x}, \boldsymbol{\xi})$, and $T_{ji}(\boldsymbol{x}, \boldsymbol{\xi})$ for three-dimensional isotropic elasticity can be found in the literature.

4.17 Concentrated Displacement Discontinuity—Index Notation

The displacements and stresses at a point $\boldsymbol{x} = (x_1, x_2)$ in an infinite isotropic plane due to a concentrated displacement discontinuity $\mathcal{D}_i = (\mathcal{D}_1, \mathcal{D}_2)$ at another point $\boldsymbol{\xi} = (\xi_1, \xi_2)$ can also be written compactly using index notation. In this case, however, the analysis is more involved because the displacement discontinuity components are defined with respect to the direction of the normal vector to a curve passing through point $\boldsymbol{\xi}$. We will outline the analysis procedure below, leaving the details of the calculations for the interested reader.

4.17.1 Computation of Displacements

To begin, we note that the displacements $u_{\bar{x}}$ and $u_{\bar{y}}$ due to a concentrated displacement discontinuity $(\mathcal{D}_{\bar{x}}, \mathcal{D}_{\bar{y}})$ at the origin of an x, y coordinate system are given by (4.221) with x and y replaced by $\bar{x}$ and $\bar{y}$. The $\bar{x}$, $\bar{y}$ coordinates are chosen such that they have the same orientation as the local tangent and normal to a curve S at point $\boldsymbol{\xi}$, which has coordinates $\bar{x} = \xi_{\bar{x}}$ and $\bar{y} = \xi_{\bar{y}}$.

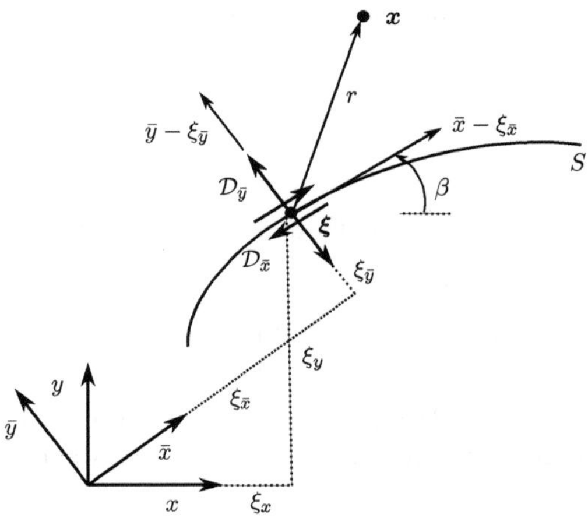

Fig. 4.19 Concentrated displacement discontinuity $(\mathcal{D}_{\bar{x}}, \mathcal{D}_{\bar{y}})$ at point $\boldsymbol{\xi}$

If the displacement discontinuity is located at point $\boldsymbol{\xi}$, as shown in Fig. 4.19, the displacements $u_{\bar{x}}$ and $u_{\bar{y}}$ are obtained by replacing x and y in (4.221) by $\bar{x}-\xi_{\bar{x}}$ and $\bar{y}-\xi_{\bar{y}}$, i.e.,

$$u_{\bar{x}} = \mathcal{D}_{\bar{x}}\left[-2(1-\nu)\frac{\partial g}{\partial \bar{y}} - (\bar{y}-\xi_{\bar{y}})\frac{\partial^2 g}{\partial \bar{y}^2}\right] + \mathcal{D}_{\bar{y}}\left[(1-2\nu)\frac{\partial g}{\partial \bar{x}} + (\bar{y}-\xi_{\bar{y}})\frac{\partial^2 g}{\partial \bar{x}\partial \bar{y}}\right]$$
$$u_{\bar{y}} = \mathcal{D}_{\bar{x}}\left[-(1-2\nu)\frac{\partial g}{\partial \bar{x}} + (\bar{y}-\xi_{\bar{y}})\frac{\partial^2 g}{\partial \bar{x}\partial \bar{y}}\right] + \mathcal{D}_{\bar{y}}\left[-2(1-\nu)\frac{\partial g}{\partial \bar{y}} + (\bar{y}-\xi_{\bar{y}})\frac{\partial^2 g}{\partial \bar{y}^2}\right] \tag{4.273}$$

in which we have used the relations $\partial g/\partial(\bar{x}-\xi_{\bar{x}}) = \partial g/\partial \bar{x}$ and $\partial g/\partial(\bar{y}-\xi_{\bar{y}}) = \partial g/\partial \bar{y}$. We want to transform (4.273) so that everything is expressed in terms of global $x,\ y$ coordinates. This is achieved in several steps. First, we use the coordinate transformation formulas (4.45) with $\bar{x}$ and $\bar{y}$ replaced by $\bar{x}-\xi_{\bar{x}}$ and $\bar{y}-\xi_{\bar{y}}$ to write

$$\bar{x}-\xi_{\bar{x}} = (x-\xi_x)\cos\beta + (y-\xi_y)\sin\beta$$
$$\bar{y}-\xi_{\bar{y}} = -(x-\xi_x)\sin\beta + (y-\xi_y)\cos\beta \tag{4.274}$$

and then employ the chain rule for partial differentiation to eliminate reference to $\bar{x}$ and $\bar{y}$ in the bracketed terms in (4.273). The expressions (see (4.62))

$$u_x = u_{\bar{x}}\cos\beta - u_{\bar{y}}\sin\beta$$
$$u_y = u_{\bar{x}}\sin\beta + u_{\bar{y}}\cos\beta \tag{4.275}$$

and (see (4.63))

$$\begin{aligned} \mathcal{D}_{\bar{x}} &= \mathcal{D}_x \cos\beta + \mathcal{D}_y \sin\beta \\ \mathcal{D}_{\bar{y}} &= -\mathcal{D}_x \sin\beta + \mathcal{D}_y \cos\beta \end{aligned} \tag{4.276}$$

can then be used to write

$$\begin{aligned} u_x &= K_{xx}\mathcal{D}_x + K_{yx}\mathcal{D}_y \\ u_y &= K_{xy}\mathcal{D}_x + K_{yy}\mathcal{D}_y \end{aligned} \tag{4.277}$$

In this way, the quantities K_{xx}, etc., are found to be

$$\begin{aligned} K_{xx} &= -2(1-\nu)\frac{\partial g}{\partial \bar{y}} + I_1 \\ K_{yx} &= (1-2\nu)\frac{\partial g}{\partial \bar{x}} - I_2 \\ K_{xy} &= -(1-2\nu)\frac{\partial g}{\partial \bar{x}} - I_2 \\ K_{yy} &= -2(1-\nu)\frac{\partial g}{\partial \bar{y}} - I_1 \end{aligned} \tag{4.278}$$

with

$$\begin{aligned} I_1 &= \left[(x-\xi_x)\sin\beta - (y-\xi_y)\cos\beta\right]\frac{\partial^2 g}{\partial y^2} \\ I_2 &= \left[(x-\xi_x)\sin\beta - (y-\xi_y)\cos\beta\right]\frac{\partial^2 g}{\partial x \partial y} \end{aligned} \tag{4.279}$$

We retained the first derivatives with respect to $\bar{x}$ and $\bar{y}$ in (4.278) for notational convenience.

Next, we write the preceding results in terms of the distance r and its derivatives, as in Sect. 4.16.1. The reader should verify that the quantities K_{xx}, etc., in (4.278) can be written as

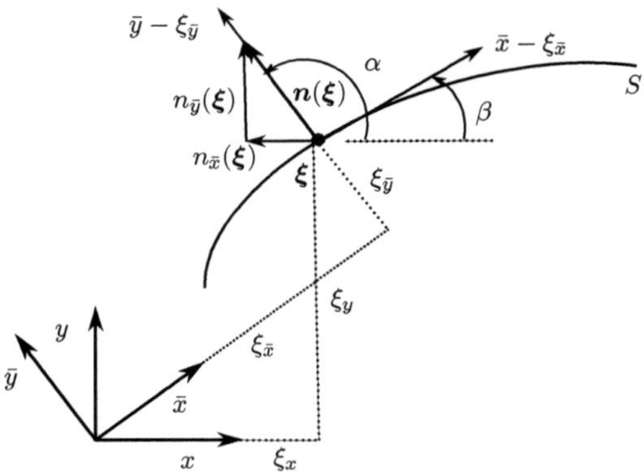

Fig. 4.20 Components of unit normal at point $\boldsymbol{\xi}$

$$
\begin{aligned}
K_{xx} &= \frac{1}{4\pi(1-\nu)}\frac{1}{r}\left[(1-2\nu)+2r_{,x}r_{,x}\right]\frac{\partial r}{\partial \bar{y}} \\
K_{yx} &= \frac{1}{4\pi(1-\nu)}\frac{1}{r}\left[-(1-2\nu)\frac{\partial r}{\partial \bar{x}}+2r_{,y}r_{,x}\frac{\partial r}{\partial \bar{y}}\right] \\
K_{xy} &= \frac{1}{4\pi(1-\nu)}\frac{1}{r}\left[(1-2\nu)\frac{\partial r}{\partial \bar{x}}+2r_{,x}r_{,y}\frac{\partial r}{\partial \bar{y}}\right] \\
K_{yy} &= \frac{1}{4\pi(1-\nu)}\frac{1}{r}\left[(1-2\nu)+2r_{,y}r_{,y}\right]\frac{\partial r}{\partial \bar{y}}
\end{aligned}
\tag{4.280}
$$

It can be seen from the geometry in Fig. 4.20 that the components of the normal to curve S at point $\boldsymbol{\xi}$ are $n_x(\boldsymbol{\xi}) = \cos\alpha = -\sin\beta$ and $n_y(\boldsymbol{\xi}) = \sin\alpha = \cos\beta$ so that, using the chain rule for partial differentiation,

$$
\begin{aligned}
\frac{\partial r}{\partial \bar{x}} &= \frac{\partial r}{\partial x}\cos\beta + \frac{\partial r}{\partial y}\sin\beta \\
&= \frac{\partial r}{\partial x}n_y(\boldsymbol{\xi}) - \frac{\partial r}{\partial y}n_x(\boldsymbol{\xi}) \\
\frac{\partial r}{\partial \bar{y}} &= -\frac{\partial r}{\partial x}\sin\beta + \frac{\partial r}{\partial y}\cos\beta \\
&= \frac{\partial r}{\partial x}n_x(\boldsymbol{\xi}) + \frac{\partial r}{\partial y}n_y(\boldsymbol{\xi})
\end{aligned}
\tag{4.281}
$$

The normal derivative of r at point $\boldsymbol{\xi}$ is defined as

$$\frac{\partial r}{\partial \boldsymbol{n}(\boldsymbol{\xi})} = \frac{\partial r}{\partial \xi_{\bar{y}}} = -\frac{\partial r}{\partial \bar{y}} \tag{4.282}$$

After introducing the preceding results into (4.280) we find

$$K_{11} = -\frac{1}{4\pi(1-\nu)}\frac{1}{r}\left[(1-2\nu) + 2r_{,1}r_{,1}\right]\frac{\partial r}{\partial \boldsymbol{n}(\boldsymbol{\xi})}$$

$$K_{21} = -\frac{1}{4\pi(1-\nu)}\frac{1}{r}\left[(1-2\nu)(r_{,1}n_2(\boldsymbol{\xi}) - r_{,2}n_1(\boldsymbol{\xi})) + 2r_{,2}r_{,1}\frac{\partial r}{\partial \boldsymbol{n}(\boldsymbol{\xi})}\right]$$

(4.283)

$$K_{12} = -\frac{1}{4\pi(1-\nu)}\frac{1}{r}\left[(1-2\nu)(r_{,2}n_1(\boldsymbol{\xi}) - r_{,1}n_2(\boldsymbol{\xi})) + 2r_{,1}r_{,2}\frac{\partial r}{\partial \boldsymbol{n}(\boldsymbol{\xi})}\right]$$

$$K_{22} = -\frac{1}{4\pi(1-\nu)}\frac{1}{r}\left[(1-2\nu) + 2r_{,2}r_{,2}\right]\frac{\partial r}{\partial \boldsymbol{n}(\boldsymbol{\xi})}$$

By inspection, these equations can be written in index notation as

$$K_{ji}(\boldsymbol{x},\boldsymbol{\xi}) = -\frac{1}{4\pi(1-\nu)}\frac{1}{r}\left[(1-2\nu)\left(r_{,i}n_j(\boldsymbol{\xi}) - r_{,j}n_i(\boldsymbol{\xi}) + \delta_{ij}\frac{\partial r}{\partial \boldsymbol{n}(\boldsymbol{\xi})}\right) + 2r_{,i}r_{,j}\frac{\partial r}{\partial \boldsymbol{n}(\boldsymbol{\xi})}\right] \tag{4.284}$$

so that, by (4.277), we have

$$u_i(\boldsymbol{x}) = K_{ji}(\boldsymbol{x},\boldsymbol{\xi})\mathcal{D}_j(\boldsymbol{\xi}) \tag{4.285}$$

Equations (4.284) and (4.285) give the displacements in the i direction at a point $\boldsymbol{x}$ due to a concentrated displacement discontinuity in the j direction at a point $\boldsymbol{\xi}$ on a curve S with unit normal $\boldsymbol{n}(\boldsymbol{\xi})$. Equation (4.285) should be compared to Eq. (4.249), which gives the displacements at point $\boldsymbol{x}$ due to a concentrated force applied at point $\boldsymbol{\xi}$. The tensor field $K_{ji}(\boldsymbol{x},\boldsymbol{\xi})$ in (4.284) is closely similar to $T_{ji}(\boldsymbol{x},\boldsymbol{\xi})$ in (4.272); these quantities only differ in regard to whether the unit outward normal to curve S is at the source point $\boldsymbol{\xi}$, as in (4.284), or at the field point $\boldsymbol{x}$, as in (4.272). Notice, however, that by interchanging $\boldsymbol{x}$ and $\boldsymbol{\xi}$ in (4.284) and (4.285)—that is, by taking the load point at $\boldsymbol{x}$ and the field point at $\boldsymbol{\xi}$—and also interchanging subscripts i and j then

$$K_{ij}(\boldsymbol{\xi},\boldsymbol{x}) = -\frac{1}{4\pi(1-\nu)}\frac{1}{r}\left[(1-2\nu)\left(\frac{\partial r}{\partial \xi_j}n_i(\boldsymbol{x}) - \frac{\partial r}{\partial \xi_i}n_j(\boldsymbol{x}) + \delta_{ji}\frac{\partial r}{\partial \boldsymbol{n}(\boldsymbol{x})}\right)\right.$$
$$\left. + 2\frac{\partial r}{\partial \xi_j}\frac{\partial r}{\partial \xi_i}\frac{\partial r}{\partial \boldsymbol{n}(\boldsymbol{x})}\right] \tag{4.286}$$

or, because $\partial r/\partial \xi_k = -\partial r/\partial x_k = -r_{,k}$,

$$K_{ij}(\boldsymbol{\xi}, \boldsymbol{x}) = -\frac{1}{4\pi(1-\nu)}\frac{1}{r}\left[(1-2\nu)\left(r_{,i}n_j(\boldsymbol{x}) - r_{,j}n_i(\boldsymbol{x}) + \delta_{ij}\frac{\partial r}{\partial \boldsymbol{n}(\boldsymbol{x})}\right) + 2r_{,i}r_{,j}\frac{\partial r}{\partial \boldsymbol{n}(\boldsymbol{x})}\right] \tag{4.287}$$

We therefore have

$$K_{ij}(\boldsymbol{\xi}, \boldsymbol{x}) = T_{ji}(\boldsymbol{x}, \boldsymbol{\xi}) \tag{4.288}$$

so that

$$\begin{aligned} u_i(\boldsymbol{\xi}) &= K_{ji}(\boldsymbol{\xi}, \boldsymbol{x})\mathcal{D}_j(\boldsymbol{x}) \\ &= T_{ij}(\boldsymbol{x}, \boldsymbol{\xi})\mathcal{D}_j(\boldsymbol{x}) \end{aligned} \tag{4.289}$$

We will return to this important result later. From (4.287), it follows that the kernel $K_{ij}(\boldsymbol{\xi}, \boldsymbol{x})$ has a strong $1/r$ singularity when $\boldsymbol{\xi} = \boldsymbol{x}$.

4.17.2 Computation of Stresses and Tractions

The stresses associated with a concentrated displacement discontinuity are found by computing the strains and using Hooke's law. By virtue of (4.285) and (4.289) we have two equivalent ways to proceed. First, using (4.285) we can differentiate displacements $u_i(\boldsymbol{x})$ with respect to coordinates x_k to find $\sigma_{ij}(\boldsymbol{x})$. Alternatively, using (4.289) we can differentiate $u_i(\boldsymbol{\xi})$ with respect to coordinates ξ_k to find $\sigma_{ij}(\boldsymbol{\xi})$. In the latter case, of course, we actually differentiate with respect to x_k, noting that $\partial r/\partial x_k = -\partial r/\partial \xi_k$. With the first approach, the reader should verify that the result can be written as

$$\sigma_{ij}(\boldsymbol{x}) = C_{kij}(\boldsymbol{x}, \boldsymbol{\xi})\mathcal{D}_k(\boldsymbol{\xi}) \tag{4.290}$$

in which

$$\begin{aligned} C_{kij}(\boldsymbol{x}, \boldsymbol{\xi}) = \frac{2\mu}{4\pi(1-\nu)}\frac{1}{r^2}\Big[&n_i(\boldsymbol{\xi})\delta_{jk} + n_j(\boldsymbol{\xi})\delta_{ik} - n_k(\boldsymbol{\xi})\delta_{ij} + 2r_{,i}r_{,j}n_k(\boldsymbol{\xi}) \\ &-2r_{,k}(\delta_{ij} - 4r_{,i}r_{,j})\frac{\partial r}{\partial \boldsymbol{n}(\boldsymbol{\xi})}\Big] \end{aligned} \tag{4.291}$$

which is symmetric with respect to its last two indices: $C_{kij}(\boldsymbol{x}, \boldsymbol{\xi}) = C_{kji}(\boldsymbol{x}, \boldsymbol{\xi})$.

For future reference, we note that an interchange of $\boldsymbol{x}$ and $\boldsymbol{\xi}$ in (4.291) gives

$$C_{kij}(\boldsymbol{\xi},\boldsymbol{x}) = \frac{2\mu}{4\pi(1-\nu)}\frac{1}{r^2}\left[n_i(\boldsymbol{x})\delta_{jk} + n_j(\boldsymbol{x})\delta_{ik} - n_k(\boldsymbol{x})\delta_{ij} + 2\frac{\partial r}{\partial \xi_i}\frac{\partial r}{\partial \xi_j}n_k(\boldsymbol{x}) \right.$$
$$\left. -2\frac{\partial r}{\partial \xi_k}(\delta_{ij} - 4\frac{\partial r}{\partial \xi_i}\frac{\partial r}{\partial \xi_j})\frac{\partial r}{\partial \boldsymbol{n}(\boldsymbol{x})}\right] \tag{4.292}$$

or, because $\partial r/\partial \xi_k = -\partial r/\partial x_k$,

$$C_{kij}(\boldsymbol{\xi},\boldsymbol{x}) = \frac{2\mu}{4\pi(1-\nu)}\frac{1}{r^2}\left[n_i(\boldsymbol{x})\delta_{jk} + n_j(\boldsymbol{x})\delta_{ik} - n_k(\boldsymbol{x})\delta_{ij} + 2r_{,i}r_{,j}n_k(\boldsymbol{x}) \right.$$
$$\left. +2r_{,k}(\delta_{ij} - 4r_{,i}r_{,j})\frac{\partial r}{\partial \boldsymbol{n}(\boldsymbol{x})}\right] \tag{4.293}$$

Finally, the tractions at point $\boldsymbol{x}$, measured with respect to the normal to a curve S at this point, are given as

$$t_i(\boldsymbol{x}) = C_{kji}(\boldsymbol{x},\boldsymbol{\xi})n_j(\boldsymbol{x})\mathcal{D}_k(\boldsymbol{\xi}) \tag{4.294}$$

cf. (4.261), or

$$t_i(\boldsymbol{x}) = H_{ki}(\boldsymbol{x},\boldsymbol{\xi})\mathcal{D}_k(\boldsymbol{\xi}) \tag{4.295}$$

in which

$$H_{ki}(\boldsymbol{x},\boldsymbol{\xi}) = C_{kij}(\boldsymbol{x},\boldsymbol{\xi})n_j(\boldsymbol{x}) \tag{4.296}$$

The tensor field $H_{ki}(\boldsymbol{x},\boldsymbol{\xi})$ plays the same role for a concentrated displacement discontinuity $\mathcal{D}_k(\boldsymbol{\xi})$ at point $\boldsymbol{\xi}$ as $T_{ki}(\boldsymbol{x},\boldsymbol{\xi})$ plays for a concentrated force $F_k(\boldsymbol{\xi})$ at that point. Introducing expression (4.291) for $C_{kij}(\boldsymbol{x},\boldsymbol{\xi}) = C_{kji}(\boldsymbol{x},\boldsymbol{\xi})$ into (4.296), and using index j in place of k for notational consistency, we obtain the following expression for $H_{ji}(\boldsymbol{x},\boldsymbol{\xi})$:

$$H_{ji}(\boldsymbol{x},\boldsymbol{\xi}) = \frac{2\mu}{4\pi(1-\nu)}\frac{1}{r^2}\left[n_m(\boldsymbol{\xi})n_m(\boldsymbol{x})\delta_{ij} + n_i(\boldsymbol{\xi})n_j(\boldsymbol{x}) - n_j(\boldsymbol{\xi})n_i(\boldsymbol{x}) \right.$$
$$\left. +2r_{,i}n_j(\boldsymbol{\xi})\frac{\partial r}{\partial \boldsymbol{n}(\boldsymbol{x})} - 2r_{,j}n_i(\boldsymbol{x})\frac{\partial r}{\partial \boldsymbol{n}(\boldsymbol{\xi})} + 8r_{,i}r_{,j}\frac{\partial r}{\partial \boldsymbol{n}(\boldsymbol{\xi})}\frac{\partial r}{\partial \boldsymbol{n}(\boldsymbol{x})}\right] \tag{4.297}$$

It is easy to confirm that $H_{ji}(\boldsymbol{x},\boldsymbol{\xi})$ has the following symmetry properties:

$$H_{ji}(\boldsymbol{x},\boldsymbol{\xi}) = H_{ij}(\boldsymbol{\xi},\boldsymbol{x}) \tag{4.298}$$

We conclude from inspection of (4.291) and (4.297) that the kernels $C_{kji}(\boldsymbol{x},\boldsymbol{\xi})$ and $H_{ji}(\boldsymbol{x},\boldsymbol{\xi})$ behave as $1/r^2$ when $\boldsymbol{x} = \boldsymbol{\xi}$ and hence are hypersingular.

4.18 Single and Double Layer Potential Analogues

In Chap. 1 we saw that the solution to Laplace's equation for a point source could be integrated along a boundary to develop a more general solution called a single layer potential. We also saw that a dipole source could be used in the same way to develop the solution for a double layer potential. We will show next that the singular solutions for a point force and a point displacement discontinuity play similar roles for elasticity.

The original work on this topic was done by Kupradze (1965), who used the term "elastic potentials" to describe the counterparts in elasticity for the single and double layer potentials in potential theory. This terminology is now firmly established in the literature—often without the modifier "elastic"— but it should be recognized that in elasticity the single and double layer entities do not themselves satisfy Laplace's equation, as do, for example, the Papkovitch-Neuber displacement *potentials*. In situations where confusion might arise we will refer to Kupradze's elastic potentials as single and double layer potential *analogues* to emphasize this distinction.

4.18.1 The Single Layer Analogue: A Distribution of Forces

By analogy with Eq. (1.43), we consider the limiting case of the summation of a sufficiently smooth distribution of forces of strength $\rho_j(\boldsymbol{\xi})\,\mathrm{d}S(\boldsymbol{\xi})$ along a closed curve S in the infinite plane and write, using (4.249), (4.250), and (4.263),

$$u_i(\boldsymbol{x}) = \int_S \rho_j(\boldsymbol{\xi}) U_{ji}(\boldsymbol{x}, \boldsymbol{\xi})\,\mathrm{d}S(\boldsymbol{\xi})$$

$$\sigma_{ij}(\boldsymbol{x}) = \int_S \rho_k(\boldsymbol{\xi}) B_{kij}(\boldsymbol{x}, \boldsymbol{\xi})\,\mathrm{d}S(\boldsymbol{\xi}) \tag{4.299}$$

$$t_i(\boldsymbol{x}) = \int_S \rho_j(\boldsymbol{\xi}) T_{ji}(\boldsymbol{x}, \boldsymbol{\xi})\,\mathrm{d}S(\boldsymbol{\xi})$$

where $\rho_j(\boldsymbol{\xi})$ is a density function that describes the boundary loading for the distributed forces. As written, these equations assume that points $\boldsymbol{x}$ and $\boldsymbol{\xi}$ do not coincide.

Equation (4.299) should be compared with Eqs. (2.93)–(2.95) to see that $u_i(\boldsymbol{x})$, $\sigma_{ij}(\boldsymbol{x})$, and $t_i(\boldsymbol{x})$ play the same roles for elasticity that $\phi(\boldsymbol{x})$, $q_k(\boldsymbol{x})$, and $\partial\phi(\boldsymbol{x})/\partial\boldsymbol{n}(\boldsymbol{x})$ play for potential theory. The comparison shows that the last integral in (4.299) represents an elastic analogue of the adjoint double layer potential (2.95).

4.18.2 The Double Layer Analogue: A Distribution of Displacement Discontinuities

Equations similar to (4.299) can be written for the case of a continuous distribution of displacement discontinuities along curve S. Assuming a sufficiently smooth distribution of concentrated displacement discontinuities of strength $\eta_j(\boldsymbol{\xi})\,\mathrm{d}S(\boldsymbol{\xi})$ along a curve S in the infinite plane, Eqs. (4.285), (4.290), and (4.295) can be used to write

$$u_i(\boldsymbol{x}) = \int_S \eta_j(\boldsymbol{\xi}) K_{ji}(\boldsymbol{x}, \boldsymbol{\xi})\,\mathrm{d}S(\boldsymbol{\xi})$$

$$\sigma_{ij}(\boldsymbol{x}) = \int_S \eta_k(\boldsymbol{\xi}) C_{kij}(\boldsymbol{x}, \boldsymbol{\xi})\,\mathrm{d}S(\boldsymbol{\xi}) \tag{4.300}$$

$$t_i(\boldsymbol{x}) = \int_S \eta_j(\boldsymbol{\xi}) H_{ji}(\boldsymbol{x}, \boldsymbol{\xi})\,\mathrm{d}S(\boldsymbol{\xi})$$

where $\eta_j(\boldsymbol{\xi})$ is a density function that describes the boundary loading for the distributed displacement discontinuities, and where again $\boldsymbol{x} \neq \boldsymbol{\xi}$.

Comparison of (4.300) with (2.96)–(2.98), taking into account that $K_{ji}(\boldsymbol{x}, \boldsymbol{\xi}) = T_{ij}(\boldsymbol{\xi}, \boldsymbol{x})$, shows the correspondence between $u_i(\boldsymbol{x})$, $\sigma_{ij}(\boldsymbol{x})$, and $t_i(\boldsymbol{x})$ for elasticity and $\phi(\boldsymbol{x})$, $q_k(\boldsymbol{x})$, and $\partial\phi(\boldsymbol{x})/\partial\boldsymbol{n}(\boldsymbol{x})$ for potential theory. The comparison also shows that the last integral in (4.300) represents an elastic analogue of the hypersingular potential (2.98).

4.18.3 Discussion

Similarly as in potential theory, the first and the third integrals in (4.299) and the corresponding integrals in (4.300) are referred to as direct values of the respective elastic potential analogues when $\boldsymbol{x} = \boldsymbol{\xi}$. It can be concluded from the discussions in Sects. 4.16 and 4.17 that, under the same smoothness conditions as those adopted in Sect. 2.8 in order to make the direct values of (2.95) and (2.96) be weakly singular, the singularities of the direct values of their elastic analogues are of the Cauchy principal value type. Just as in the case of potential theory, the direct values of the double layer and adjoint double layer potential analogues differ from their limit values when point $\boldsymbol{x}$ approaches curve S from different directions. We will discuss this in Sect. 4.27.

We will show in Chap. 5 that (4.299) form the basis of an indirect boundary element method called the stress discontinuity method. Also, (4.300) form the basis of another indirect method called the displacement discontinuity method. These methods are the elastic counterparts of the two indirect methods discussed in Chap. 1 for potential problems.

4.19 Treatment of Body Forces

A natural extension of the process used to write boundary integral formulas (4.299) leads us to consider a continuous distribution of forces acting throughout a body. If the force per unit volume at point $\boldsymbol{\xi} = (\xi_1, \xi_2)$ is denoted as $\mathbb{F}_i = (\mathbb{F}_1, \mathbb{F}_2)$, then the force F_i associated with an infinitesimal element of volume $\mathrm{d}V(\boldsymbol{\xi}) = \mathrm{d}\xi_1 \mathrm{d}\xi_2$ at this point is $F_i = \mathbb{F}_i(\boldsymbol{\xi})\,\mathrm{d}V(\boldsymbol{\xi})$. In the limit, the summation of a collection of such forces within volume V becomes an integral, and we write

$$u_i(\boldsymbol{x}) = \int_V U_{ji}(\boldsymbol{x}, \boldsymbol{\xi})\mathbb{F}_j(\boldsymbol{\xi})\,\mathrm{d}V(\boldsymbol{\xi}) \tag{4.301}$$

Using definitions (4.253) for the components of $U_{ji}(\boldsymbol{x}, \boldsymbol{\xi})$ for isotropic elasticity, it is easy to show that this equation can be written as

$$\begin{aligned} u_i(\boldsymbol{x}) = \frac{1}{2\mu}\bigg[&(3-4\nu)\int_V g(\boldsymbol{x}, \boldsymbol{\xi})\mathbb{F}_i(\boldsymbol{\xi})\,\mathrm{d}V(\boldsymbol{\xi}) \\ &- \int_V (x_j - \xi_j)\frac{\partial}{\partial x_i} g(\boldsymbol{x}, \boldsymbol{\xi})\mathbb{F}_j(\boldsymbol{\xi})\,\mathrm{d}V(\boldsymbol{\xi})\bigg] \end{aligned} \tag{4.302}$$

Following Sokolnikoff (1956), an interesting observation can be made by substituting the values of $u_i(\boldsymbol{x})$ from (4.302) into the Navier-Cauchy equations (4.81), namely, the distributed force per unit volume $\mathbb{F}_i(\boldsymbol{x})$ is actually a distributed *body force* (per unit volume), i.e.,

$$\mathbb{F}_i(\boldsymbol{x}) = B_i(\boldsymbol{x}) \tag{4.303}$$

(In obtaining this result care must be taken when differentiating under the integral signs in (4.302), because the integrals are improper when the field point $\boldsymbol{x}$ is in V.) Equation (4.302) can therefore be written as

$$\begin{aligned} u_i(\boldsymbol{x}) = \frac{1}{2\mu}\bigg[&(3-4\nu)\int_V g(\boldsymbol{x}, \boldsymbol{\xi})B_i(\boldsymbol{\xi})\,\mathrm{d}V(\boldsymbol{\xi}) \\ &- \int_V (x_j - \xi_j)\frac{\partial}{\partial x_i} g(\boldsymbol{x}, \boldsymbol{\xi})B_j(\boldsymbol{\xi})\,\mathrm{d}V(\boldsymbol{\xi})\bigg] \end{aligned} \tag{4.304}$$

or

$$\begin{aligned} u_i(\boldsymbol{x}) = \frac{1}{8\pi\mu(1-\nu)}\bigg[&(3-4\nu)\int_V \ln(1/r)B_i(\boldsymbol{\xi})\,\mathrm{d}V(\boldsymbol{\xi}) \\ &- \int_V (x_j - \xi_j)\frac{\partial}{\partial x_i} \ln(1/r)B_j(\boldsymbol{\xi})\,\mathrm{d}V(\boldsymbol{\xi})\bigg] \end{aligned} \tag{4.305}$$

In theory, this equation allows one to specify an arbitrarily complex distribution of body forces in volume V. Then, as explained in Sect. 4.8, a boundary value problem can always be defined which incorporates the effects of these body forces, but which relates to the homogeneous form of the Navier-Cauchy equations, i.e., the form in which $B_i = 0$.

As an aside, we note that (4.305) is the two-dimensional version of a general three-dimensional result developed by Lord Kelvin in 1848. (An exposition of Kelvin's approach is summarized in Love (1944).) The full three-dimensional solution can be written as

$$\begin{aligned} u_i(\boldsymbol{x}) = \frac{1}{16\pi\mu(1-\nu)} \Bigg[&(3-4\nu) \int_V \frac{B_i(\boldsymbol{\xi})}{r} \, \mathrm{d}V(\boldsymbol{\xi}) \\ &- \int_V (x_j - \xi_j) \frac{\partial}{\partial x_i} \left(\frac{1}{r}\right) B_j(\boldsymbol{\xi}) \, \mathrm{d}V(\boldsymbol{\xi}) \Bigg] \end{aligned} \tag{4.306}$$

where $r = r(\boldsymbol{x}, \boldsymbol{\xi})$ is the distance from the load point $\boldsymbol{\xi} = (\xi_1, \xi_2, \xi_3)$ to the sample point $\boldsymbol{x} = (x_1, x_2, x_3)$:

$$r(\boldsymbol{x}, \boldsymbol{\xi}) = \left[(x_1 - \xi_1)^2 + (x_2 - \xi_2)^2 + (x_3 - \xi_3)^2 \right]^{1/2} \tag{4.307}$$

The similarity between (4.305) and (4.306) is evident; the former uses the fundamental harmonic function in two dimensions, $\ln(1/r)$, while the latter uses the three-dimensional counterpart of this function, $1/r$.

Suppose now that the body forces are zero everywhere except within a subregion V_o of region V. The integrals in (4.305) or, for that matter, (4.306) are nonsingular for a point $\boldsymbol{x}$ in $V - V_o$, and it can easily be shown that the Navier-Cauchy equations with zero body forces are satisfied within this region, as well as within the rest of the infinite plane, V_c. If we pass to the limit as volume V_o shrinks to zero, we arrive at the notion of a concentrated body force at point $\boldsymbol{\xi}$.

It is customary in the literature to represent a concentrated body force B_i at point $\boldsymbol{\xi}$ by means of the Dirac delta function defined in Sect. 2.1.7, i.e.,

$$B_i(\boldsymbol{\xi}) = \delta(\boldsymbol{x} - \boldsymbol{\xi}) F_i(\boldsymbol{\xi}) \tag{4.308}$$

4.20 Strain Energy

A deformed elastic body stores energy, just as a compressed or extended spring stores the energy used to compress or extend it. An expression for the stored energy in an elastic solid can be deduced by pursuing the spring analogy in more detail.

Consider a linear spring with stiffness k. The force F required to stretch the spring by an amount u is given as $F = ku$, and the energy stored in the spring during the process of increasing the deformation u from zero to its final value is $\frac{1}{2}Fu$. If the spring represents an elastic bar of length L and cross-sectional area A, the stored energy is $\frac{1}{2}(F/A)(u/L)AL$. In terms of the nominal stress $\sigma = F/A$ and strain

$\varepsilon = u/L$, the stored energy in the bar is $\frac{1}{2}\sigma\varepsilon AL$, and the energy per unit volume is $\frac{1}{2}\sigma\varepsilon$. This quantity is referred to as the strain energy density, or strain energy per unit volume, denoted here by the symbol U.

A generalization of this simple one-dimensional situation leads us to consider an elemental volume $\mathrm{d}V$ subjected to principal stresses σ_1, σ_2, and σ_3 which are gradually increased from zero to their final values. If the associated principal strains are ε_1, ε_2, and ε_1, we invoke the principle of superposition and write the strain energy per unit volume as

$$U = \tfrac{1}{2}(\sigma_1\varepsilon_1 + \sigma_2\varepsilon_2 + \sigma_3\varepsilon_3)$$

In an arbitrary Cartesian axis system, this expression is

$$U = \tfrac{1}{2}\sigma_{ij}\varepsilon_{ij} \tag{4.309}$$

We next note that the total strain energy stored in an elastic solid is obtained by integrating the strain energy density over the entire volume V of the solid, i.e.,

$$W = \int_V U\,\mathrm{d}V = \tfrac{1}{2}\int_V \sigma_{ij}\varepsilon_{ij}\,\mathrm{d}V \tag{4.310}$$

This result can be put into several different forms. One form that is convenient for our purposes is obtained by noticing that $\varepsilon_{ij} = u_{i,j} - \omega_{ij}$, where $\omega_{ij} = \frac{1}{2}(u_{i,j} - u_{j,i})$, and recognizing that $\sigma_{ij}\omega_{ij} = 0$ because σ_{ij} is symmetric ($\sigma_{ij} = \sigma_{ij}$) while ω_{ij} is antisymmetric ($\omega_{ij} = -\omega_{ij}$). Equation (4.310) then becomes

$$W = \tfrac{1}{2}\int_V \sigma_{ij}u_{i,j}\,\mathrm{d}V \tag{4.311}$$

or

$$W = \tfrac{1}{2}\int_V (\sigma_{ij}u_i)_{,j}\,\mathrm{d}V - \tfrac{1}{2}\int_V \sigma_{ij,j}u_i\,\mathrm{d}V \tag{4.312}$$

The first of these volume integrals can be written as a surface integral by means of the divergence theorem (Sect. 2.1.3), and the second can be expressed in terms of the body force per unit volume B_i by using the equilibrium equations, $\sigma_{ij,j} + B_i = 0$. In this way, we obtain the following result:

$$W = \tfrac{1}{2}\int_S (\sigma_{ij}u_i)n_j\,\mathrm{d}S + \tfrac{1}{2}\int_V B_i u_i\,\mathrm{d}V \tag{4.313}$$

or, using the relation $t_i = \sigma_{ji}n_j$ $(= \sigma_{ij}n_j)$,

$$W = \tfrac{1}{2}\int_S t_i u_i\,\mathrm{d}S + \tfrac{1}{2}\int_V B_i u_i\,\mathrm{d}V \tag{4.314}$$

The stored strain energy in an elastic solid is therefore equal to one-half the work done by the surface tractions t_i and body forces B_i acting through the displacements u_i on the boundary S and in the interior of region V, respectively. If the body forces are zero, the strain energy stored in region V is determined by boundary data only.

Equation (4.314) is known as Clapeyron's theorem (Sokolnikoff 1956). It can be used to establish that boundary value problems in linear elasticity have unique solutions; the interested reader is referred to Sokolnikoff (1956) for details.

4.21 The Reciprocal Theorem

An integral identity known as the reciprocal theorem (Sokolnikoff 1956) is of fundamental importance to several of the boundary element methods for elasticity considered in this book. This theorem links the solutions to two different boundary value problems for a region V bounded by surface S (the same V and S for both problems). Suppose that the first problem is characterized by displacements $u_i^{(1)}$, stresses $\sigma_{ji}^{(1)}$, and body forces $B_i^{(1)}$, and that the second problem is characterized by displacements $u_i^{(2)}$, stresses $\sigma_{ji}^{(2)}$, and body forces $B_i^{(2)}$. The tractions on S for the two problems are $t_i^{(1)} = \sigma_{ji}^{(1)} n_j$ and $t_i^{(2)} = \sigma_{ji}^{(2)} n_j$, respectively. The reciprocal theorem then consists in the following integral identity:

$$\int_S t_i^{(1)} u_i^{(2)}\, \mathrm{d}S + \int_V B_i^{(1)} u_i^{(2)}\, \mathrm{d}V = \int_S t_i^{(2)} u_i^{(1)}\, \mathrm{d}S + \int_V B_i^{(2)} u_i^{(1)}\, \mathrm{d}V \tag{4.315}$$

The resemblance of this result to the right-hand side of (4.314) should be noted. Equation (4.315) states that the work done by the tractions $t_i^{(1)}$ and body forces $B_i^{(1)}$ of the first problem in acting through the displacements $u_i^{(2)}$ of the second problem is equal to the work done by the tractions $t_i^{(2)}$ and body forces $B_i^{(2)}$ of the second problem in acting through the displacements $u_i^{(1)}$ of the first problem.

The reciprocal theorem is a direct consequence of the equilibrium equations and generalized Hooke's law. We first note that the following relationship holds at each point of V

$$\sigma_{ij}^{(1)} \varepsilon_{ij}^{(2)} = \sigma_{ij}^{(2)} \varepsilon_{ij}^{(1)} \tag{4.316}$$

and hence

$$\int_V \sigma_{ij}^{(1)} \varepsilon_{ij}^{(2)}\, \mathrm{d}V = \int_V \sigma_{ij}^{(2)} \varepsilon_{ij}^{(1)}\, \mathrm{d}V \tag{4.317}$$

Expansion of both sides of (4.317) in the manner used to derive (4.314) proves the result. Thus, for example, we write

$$\begin{aligned}\int_V \sigma_{ij}^{(1)}\varepsilon_{ij}^{(2)}\,\mathrm{d}V &= \int_V \sigma_{ij}^{(1)}u_{i,j}^{(2)}\,\mathrm{d}V\\ &= \int_V (\sigma_{ij}^{(1)}u_i^{(2)})_{,j}\,\mathrm{d}V - \int_V \sigma_{ij,j}^{(1)}u_i^{(2)}\,\mathrm{d}V\\ &= \int_S (\sigma_{ij}^{(1)}u_i^{(2)})n_j\,\mathrm{d}S + \int_V B_i^{(1)}u_i^{(2)}\,\mathrm{d}V\\ &= \int_S t_i^{(1)}u_i^{(2)}\,\mathrm{d}S + \int_V B_i^{(1)}u_i^{(2)}\,\mathrm{d}V\end{aligned} \tag{4.318}$$

Similarly,

$$\begin{aligned}\int_V \sigma_{ij}^{(2)}\varepsilon_{ij}^{(1)}\,\mathrm{d}V &= \int_V \sigma_{ij}^{(2)}u_{i,j}^{(1)}\,\mathrm{d}V\\ &= \int_V (\sigma_{ij}^{(2)}u_i^{(1)})_{,j}\,\mathrm{d}V - \int_V \sigma_{ij,j}^{(2)}u_i^{(1)}\,\mathrm{d}V\\ &= \int_S (\sigma_{ij}^{(2)}u_i^{(1)})n_j\,\mathrm{d}S + \int_V B_i^{(2)}u_i^{(1)}\,\mathrm{d}V\\ &= \int_S t_i^{(2)}u_i^{(1)}\,\mathrm{d}S + \int_V B_i^{(2)}u_i^{(1)}\,\mathrm{d}V\end{aligned} \tag{4.319}$$

Substitution of (4.318) and (4.319) into (4.317) gives (4.315), as desired.

4.22 Somigliana's Formulas

As a specific application of the reciprocal theorem, we will consider that the first problem is Kelvin's problem, here represented as a concentrated body force $\delta(\boldsymbol{x} - \boldsymbol{\xi})F_k$ at a point $\boldsymbol{\xi}$ in V, while the second problem is the problem we wish to solve—i.e., a problem characterized by displacements u_i and tractions t_i on the boundary S of region V. We then have from (4.249), (4.263), and (4.308) that

$$\begin{aligned}u_j^{(1)} &= U_{kj}(\boldsymbol{x},\boldsymbol{\xi})F_k\\ t_j^{(1)} &= T_{kj}(\boldsymbol{x},\boldsymbol{\xi})F_k\\ B_j^{(1)} &= \delta(\boldsymbol{x}-\boldsymbol{\xi})F_j = \delta(\boldsymbol{x}-\boldsymbol{\xi})\delta_{kj}F_k\end{aligned} \tag{4.320}$$

and, by definition,

$$\begin{aligned}u_j^{(2)} &= u_j(\boldsymbol{x})\\ t_j^{(2)} &= t_j(\boldsymbol{x})\\ B_j^{(2)} &= B_j(\boldsymbol{x})\end{aligned} \tag{4.321}$$

Substitution of (4.320) and (4.321) into (4.315) then yields

$$\int_S F_k u_j(\boldsymbol{x}) T_{kj}(\boldsymbol{x},\boldsymbol{\xi})\,\mathrm{d}S(\boldsymbol{x}) + \int_V F_k u_j(\boldsymbol{x})\delta(\boldsymbol{x}-\boldsymbol{\xi})\delta_{kj}\,\mathrm{d}V(\boldsymbol{x}) = \int_S F_k t_j(\boldsymbol{x}) U_{kj}(\boldsymbol{x},\boldsymbol{\xi})\,\mathrm{d}S(\boldsymbol{x}) + \int_V F_k B_j(\boldsymbol{x}) U_{kj}(\boldsymbol{x},\boldsymbol{\xi})\,\mathrm{d}V(\boldsymbol{x}) \tag{4.322}$$

or, after some simple rearrangement,

$$\begin{aligned} F_k \int_V u_j(\boldsymbol{x})\delta(\boldsymbol{x}-\boldsymbol{\xi})\delta_{kj}\,\mathrm{d}V(\boldsymbol{x}) = {} & F_k \int_S t_j(\boldsymbol{x}) U_{kj}(\boldsymbol{x},\boldsymbol{\xi})\,\mathrm{d}S(\boldsymbol{x}) \\ & - F_k \int_S u_j(\boldsymbol{x}) T_{kj}(\boldsymbol{x},\boldsymbol{\xi})\,\mathrm{d}S(\boldsymbol{x}) \\ & + F_k \int_V B_j(\boldsymbol{x}) U_{kj}(\boldsymbol{x},\boldsymbol{\xi})\,\mathrm{d}V(\boldsymbol{x}) \end{aligned} \tag{4.323}$$

This result can be simplified by first noting that $\delta_{kj} u_j(\boldsymbol{x}) = u_k(\boldsymbol{x})$ and then using the selection property of the Dirac delta function, i.e.,

$$\int_V u_k(\boldsymbol{x})\delta(\boldsymbol{x}-\boldsymbol{\xi})\,\mathrm{d}V(\boldsymbol{x}) = \begin{cases} u_k(\boldsymbol{\xi}) & \boldsymbol{\xi} \in V \\ 0 & \boldsymbol{\xi} \notin V \end{cases} \tag{4.324}$$

to write

$$\begin{aligned} F_k u_k(\boldsymbol{\xi}) = {} & F_k \int_S t_j(\boldsymbol{x}) U_{kj}(\boldsymbol{x},\boldsymbol{\xi})\,\mathrm{d}S(\boldsymbol{x}) - F_k \int_S u_j(\boldsymbol{x}) T_{kj}(\boldsymbol{x},\boldsymbol{\xi})\,\mathrm{d}S(\boldsymbol{x}) \\ & + F_k \int_V B_j(\boldsymbol{x}) U_{kj}(\boldsymbol{x},\boldsymbol{\xi})\,\mathrm{d}V(\boldsymbol{x}) \end{aligned} \tag{4.325}$$

Note that if point $\boldsymbol{\xi}$ were outside region V then, by virtue of the second condition in (4.324), the left-hand side of this equation would be equal to zero.

Equation (4.325) represents the sum of two equations (three equations in three dimensions), one for $F_1 \neq 0,\ F_2 = 0$ and the other for $F_1 = 0,\ F_2 \neq 0$. Considering these two cases separately, and noting that the non-zero values of F_1 and F_2 are arbitrary, we obtain

$$
\begin{aligned}
u_1(\boldsymbol{\xi}) &= \int_S t_j(\boldsymbol{x}) U_{1j}(\boldsymbol{x}, \boldsymbol{\xi}) \,\mathrm{d}S(\boldsymbol{x}) - \int_S u_j(\boldsymbol{x}) T_{1j}(\boldsymbol{x}, \boldsymbol{\xi}) \,\mathrm{d}S(\boldsymbol{x}) \\
&\quad + \int_V B_j(\boldsymbol{x}) U_{1j}(\boldsymbol{x}, \boldsymbol{\xi}) \,\mathrm{d}V(\boldsymbol{x}) \\
u_2(\boldsymbol{\xi}) &= \int_S t_j(\boldsymbol{x}) U_{2j}(\boldsymbol{x}, \boldsymbol{\xi}) \,\mathrm{d}S(\boldsymbol{x}) - \int_S u_j(\boldsymbol{x}) T_{2j}(\boldsymbol{x}, \boldsymbol{\xi}) \,\mathrm{d}S(\boldsymbol{x}) \\
&\quad + \int_V B_j(\boldsymbol{x}) U_{2j}(\boldsymbol{x}, \boldsymbol{\xi}) \,\mathrm{d}V(\boldsymbol{x})
\end{aligned}
\tag{4.326}
$$

4.22.1 Formula for Displacements

Equations (4.326) are Somigliana's formulas for the displacements. Note that both of these equations can be recovered from the expression

$$
\begin{aligned}
u_k(\boldsymbol{\xi}) &= \int_S t_j(\boldsymbol{x}) U_{kj}(\boldsymbol{x}, \boldsymbol{\xi}) \,\mathrm{d}S(\boldsymbol{x}) - \int_S u_j(\boldsymbol{x}) T_{kj}(\boldsymbol{x}, \boldsymbol{\xi}) \,\mathrm{d}S(\boldsymbol{x}) \\
&\quad + \int_V B_j(\boldsymbol{x}) U_{kj}(\boldsymbol{x}, \boldsymbol{\xi}) \,\mathrm{d}V(\boldsymbol{x})
\end{aligned}
\tag{4.327}
$$

by taking $k = 1$ and $k = 2$. For the case that the body forces are zero—which we will typically assume in this book—the displacements at a point $\boldsymbol{\xi}$ inside region V are determined by integrals of the tractions and displacements over the boundary S of this region:

$$
u_i(\boldsymbol{\xi}) = \int_S t_j(\boldsymbol{x}) U_{ij}(\boldsymbol{x}, \boldsymbol{\xi}) \,\mathrm{d}S(\boldsymbol{x}) - \int_S u_j(\boldsymbol{x}) T_{ij}(\boldsymbol{x}, \boldsymbol{\xi}) \,\mathrm{d}S(\boldsymbol{x}) \tag{4.328}
$$

In writing this equation we have replaced the subscript k with i for notational convenience. The tensor fields $U_{ij}(\boldsymbol{x}, \boldsymbol{\xi})$ and $T_{ij}(\boldsymbol{x}, \boldsymbol{\xi})$ give, respectively, the displacements and tractions in the j direction at the point $\boldsymbol{x}$ due to a unit force in the i direction at the point $\boldsymbol{\xi}$.

In accordance with the comment following (4.325), if point $\boldsymbol{\xi}$ is *outside* region V then $u_i(\boldsymbol{\xi}) = 0$. Including this result in (4.328) we write

$$
\int_S t_j(\boldsymbol{x}) U_{ij}(\boldsymbol{x}, \boldsymbol{\xi}) \,\mathrm{d}S(\boldsymbol{x}) - \int_S u_j(\boldsymbol{x}) T_{ij}(\boldsymbol{x}, \boldsymbol{\xi}) \,\mathrm{d}S(\boldsymbol{x}) = \begin{cases} u_i(\boldsymbol{\xi}) & \boldsymbol{\xi} \in V, \quad \boldsymbol{\xi} \notin S \\ 0 & \boldsymbol{\xi} \notin V \end{cases} \tag{4.329}
$$

4.22.2 Relationship to Green's Representation Formula

Somigliana's formula for the displacements plays the same role for elasticity as Green's representation formula plays for potential theory. Recall that Green's formula (2.62) expresses the potential $\phi(\boldsymbol{\xi})$ at a source point $\boldsymbol{\xi}$ within the region of interest V as the sum of single and double layer potentials with density functions given, respectively, by the normal derivative of the potential $\partial\phi(\boldsymbol{x})/\partial\boldsymbol{n}(\boldsymbol{x}) = q(\boldsymbol{x})$ and the potential $\phi(\boldsymbol{x})$ at a sample point $\boldsymbol{x}$ on the boundary S of this region.

The single and double layer analogues for elasticity given in Sect. 4.18 are defined for a sample point $\boldsymbol{x}$ inside the region V and a source point $\boldsymbol{\xi}$ on the boundary S of this region. If we reverse the roles of these quantities the single and double layer analogues for the displacements given in (4.299) and (4.300) become

$$u_i(\boldsymbol{\xi}) = \int_S \rho_j(\boldsymbol{x}) U_{ji}(\boldsymbol{\xi}, \boldsymbol{x})\, \mathrm{d}S(\boldsymbol{x}) \tag{4.330}$$

$$u_i(\boldsymbol{\xi}) = \int_S \eta_j(\boldsymbol{x}) K_{ji}(\boldsymbol{\xi}, \boldsymbol{x})\, \mathrm{d}S(\boldsymbol{x}) \tag{4.331}$$

Because of the symmetry of $U_{ji}(\boldsymbol{\xi}, \boldsymbol{x})$ (see (4.255)) the first integral in Somigliana's formula for the displacements (4.328) is equivalent to a single layer with a density function $t_j(\boldsymbol{x})$ and, because $K_{ji}(\boldsymbol{\xi}, \boldsymbol{x}) = T_{ij}(\boldsymbol{x}, \boldsymbol{\xi})$ (see (4.288)), the second integral represents a double layer with a density function $u_j(\boldsymbol{x})$.

4.22.3 Formula for Stresses

Expressions for the stresses at point $\boldsymbol{\xi}$ are obtained by differentiating the displacements $u_i(\boldsymbol{\xi})$ and $u_m(\boldsymbol{\xi})$ with respect to ξ_m and ξ_i to compute the strains $\varepsilon_{im}(\boldsymbol{\xi}) = \frac{1}{2}\left[u_{i,m}(\boldsymbol{\xi}) + u_{m,i}(\boldsymbol{\xi})\right]$, and then using these in Hooke's law (4.44). The differentiations can be expressed in terms of coordinates x_m by using the result $\partial r/\partial x_m = -\partial r/\partial \xi_m$ and can be carried out under the integral signs provided $\boldsymbol{\xi} \neq \boldsymbol{x}$. The result can be written as

$$\sigma_{ij}(\boldsymbol{\xi}) = \int_S t_k(\boldsymbol{x}) D_{kij}(\boldsymbol{x}, \boldsymbol{\xi})\, \mathrm{d}S(\boldsymbol{x}) - \int_S u_k(\boldsymbol{x}) S_{kij}(\boldsymbol{x}, \boldsymbol{\xi})\, \mathrm{d}S(\boldsymbol{x}) \tag{4.332}$$

in which

$$D_{kij}(\boldsymbol{x}, \boldsymbol{\xi}) = \frac{1}{4\pi(1-\nu)} \frac{1}{r} \left[(1-2\nu)\left(r_{,i}\delta_{jk} + r_{,j}\delta_{ik} - r_{,k}\delta_{ij}\right) + 2r_{,i}r_{,j}r_{,k}\right] \tag{4.333}$$

and

$$S_{kij}(\boldsymbol{x}, \boldsymbol{\xi}) = \frac{2\mu}{4\pi(1-\nu)}\frac{1}{r^2}\Big[n_i(\boldsymbol{x})\delta_{jk} + n_j(\boldsymbol{x})\delta_{ik} - n_k(\boldsymbol{x})\delta_{ij} + 2r_{,i}r_{,j}n_k(\boldsymbol{x}) \\ +2r_{,k}(\delta_{ij} - 4r_{,i}r_{,j})\frac{\partial r}{\partial \boldsymbol{n}(\boldsymbol{x})}\Big] \tag{4.334}$$

It can be seen that D_{kij} and S_{kij} are symmetric with respect to the last two indices, i.e., $D_{kij} = D_{kji}$ and $S_{kij} = S_{kji}$. (These quantities can also be written in terms of function $g(\boldsymbol{x}, \boldsymbol{\xi})$; the results are summarized in Appendix B.)

We note that the kernel $D_{kij}(\boldsymbol{x}, \boldsymbol{\xi})$ in (4.333) is the *negative* of the quantity $B_{kij}(\boldsymbol{x}, \boldsymbol{\xi})$ in (4.270), i.e.,

$$D_{kij}(\boldsymbol{x}, \boldsymbol{\xi}) = -B_{kij}(\boldsymbol{x}, \boldsymbol{\xi}) \tag{4.335}$$

This result follows from the fact that $B_{kij}(\boldsymbol{x}, \boldsymbol{\xi})$ is obtained by differentiating $u_i(\boldsymbol{x})$ with respect to x_j while $D_{kij}(\boldsymbol{x}, \boldsymbol{\xi})$ is obtained by differentiating $u_i(\boldsymbol{\xi})$ with respect to $-x_j$. By interchanging the arguments $\boldsymbol{x}$ and $\boldsymbol{\xi}$ in (4.257), i.e., by taking the load point at $\boldsymbol{x}$ and the field point at $\boldsymbol{\xi}$, (4.260) and (4.335) give the result

$$D_{kij}(\boldsymbol{x}, \boldsymbol{\xi}) = B_{kij}(\boldsymbol{\xi}, \boldsymbol{x}) \tag{4.336}$$

Similarly, comparison of (4.334) and (4.293) shows that

$$S_{kij}(\boldsymbol{x}, \boldsymbol{\xi}) = C_{kij}(\boldsymbol{\xi}, \boldsymbol{x}) \tag{4.337}$$

4.22.4 Formula for Tractions

The tractions at point $\boldsymbol{\xi}$, measured with respect to the normal to curve S at that point, are given as $t_i(\boldsymbol{\xi}) = \sigma_{ij}(\boldsymbol{\xi})n_j(\boldsymbol{\xi})$, or

$$t_i(\boldsymbol{\xi}) = \int_S t_k(\boldsymbol{x})D_{kij}(\boldsymbol{x}, \boldsymbol{\xi})n_j(\boldsymbol{\xi})\,\mathrm{d}S(\boldsymbol{x}) - \int_S u_k(\boldsymbol{x})S_{kij}(\boldsymbol{x}, \boldsymbol{\xi})n_j(\boldsymbol{\xi})\,\mathrm{d}S(\boldsymbol{x}) \tag{4.338}$$

Using (4.333) and (4.334) and elaborating the quantities $D_{kij}(\boldsymbol{x}, \boldsymbol{\xi})n_j(\boldsymbol{\xi})$ and $S_{kij}(\boldsymbol{x}, \boldsymbol{\xi})n_j(\boldsymbol{\xi})$ gives

$$D_{kij}(\boldsymbol{x}, \boldsymbol{\xi})n_j(\boldsymbol{\xi}) = -\frac{1}{4\pi(1-\nu)}\frac{1}{r}\left[(1-2\nu)\left(r_{,k}n_i(\boldsymbol{\xi}) - r_{,i}n_k(\boldsymbol{\xi}) + \delta_{ik}\frac{\partial r}{\partial \boldsymbol{n}(\boldsymbol{\xi})}\right) + 2r_{,i}r_{,k}\frac{\partial r}{\partial \boldsymbol{n}(\boldsymbol{\xi})}\right] \tag{4.339}$$

and

$$S_{kij}(\boldsymbol{x},\boldsymbol{\xi})n_j(\boldsymbol{\xi}) = \frac{2\mu}{4\pi(1-\nu)}\frac{1}{r^2}\left[n_m(\boldsymbol{x})n_m(\boldsymbol{\xi})\delta_{ik} + n_i(\boldsymbol{x})n_k(\boldsymbol{\xi}) - n_k(\boldsymbol{x})n_i(\boldsymbol{\xi})\right.$$
$$\left.+2r_{,k}n_i(\boldsymbol{\xi})\frac{\partial r}{\partial \boldsymbol{n}(\boldsymbol{x})} - 2r_{,i}n_k(\boldsymbol{x})\frac{\partial r}{\partial \boldsymbol{n}(\boldsymbol{\xi})} + 8r_{,i}r_{,k}\frac{\partial r}{\partial \boldsymbol{n}(\boldsymbol{x})}\frac{\partial r}{\partial \boldsymbol{n}(\boldsymbol{\xi})}\right] \tag{4.340}$$

In writing these expressions, use has been made of the result (cf. (2.4))

$$r_{,j}n_j(\boldsymbol{\xi}) = \frac{\partial r}{\partial x_j}n_j(\boldsymbol{\xi}) = -\frac{\partial r}{\partial \xi_j}n_j(\boldsymbol{\xi}) = -\frac{\partial r}{\partial \boldsymbol{n}(\boldsymbol{\xi})} \tag{4.341}$$

By appropriately changing the subscripts in (4.284) it can be shown[13] that the right-hand side of (4.339) is equal to $K_{ik}(\boldsymbol{x},\boldsymbol{\xi})$. A similar change of subscripts in (4.297) shows that the right-hand side of (4.340) is equal to $H_{ik}(\boldsymbol{x},\boldsymbol{\xi})$. The traction Eq. (4.338) can therefore be written as

$$t_i(\boldsymbol{\xi}) = \int_S t_k(\boldsymbol{x})K_{ik}(\boldsymbol{x},\boldsymbol{\xi})\,\mathrm{d}S(\boldsymbol{x}) - \int_S u_k(\boldsymbol{x})H_{ik}(\boldsymbol{x},\boldsymbol{\xi})\,\mathrm{d}S(\boldsymbol{x}) \tag{4.342}$$

or, after replacing subscript k by j for notational consistency,

$$t_i(\boldsymbol{\xi}) = \int_S t_j(\boldsymbol{x})K_{ij}(\boldsymbol{x},\boldsymbol{\xi})\,\mathrm{d}S(\boldsymbol{x}) - \int_S u_j(\boldsymbol{x})H_{ij}(\boldsymbol{x},\boldsymbol{\xi})\,\mathrm{d}S(\boldsymbol{x}) \tag{4.343}$$

Using (4.299) and (4.300), it is easy to show that the two integrals in this expression represent, respectively, an adjoint double layer elastic potential with a density function $t_j(\boldsymbol{x})$ and a hypersingular elastic potential with a density function $u_j(\boldsymbol{x})$.

4.22.5 Concluding Remarks

It is worth emphasizing that the values of *both* $t_j(\boldsymbol{x})$ and $u_j(\boldsymbol{x})$ in Somigliana's formulas will be known for all points $\boldsymbol{x}$ of boundary S only *after* the boundary value problem in question has been solved. At the outset, we must specify one component of either traction or displacement at each point of the boundary. That is, for a well-posed problem, either t_1 or u_1 *and* either t_2 or u_2 must be prescribed at each point of S. If body forces are present, these must also be specified as part of the definition of the problem.

[13] First, interchange subscripts i and j in (4.284), then set $j = k$.

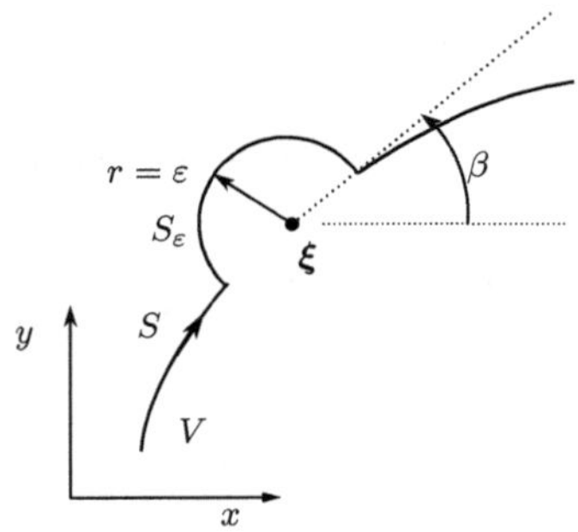

(a) $\boldsymbol{\xi}$ at smooth part of S.

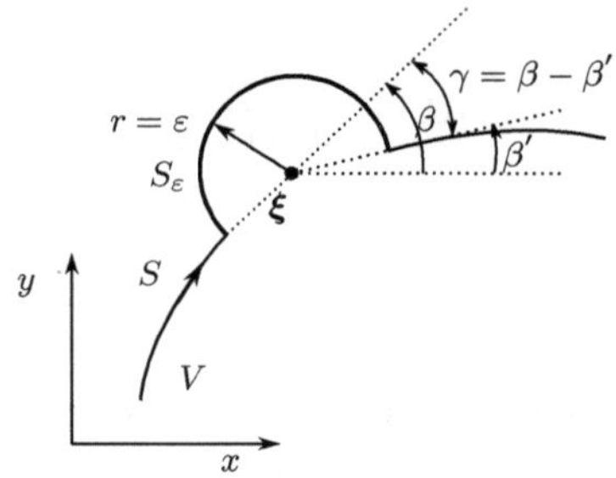

(b) $\boldsymbol{\xi}$ at sharp corner of S.

Fig. 4.21 Limit cases for load point $\boldsymbol{\xi}$ on boundary

The preceding expressions for $D_{kij}(\boldsymbol{x}, \boldsymbol{\xi})$, etc., were developed for the case of an isotropic elastic material. Comparable expressions for an orthotropic (transversely isotropic) elastic material are summarized in Appendix B.

4.23 Limit Values of Somigliana's Displacement Formula

The preceding integral formulas assume that the load point $\boldsymbol{\xi}$ is *not* located at the sample point $\boldsymbol{x}$ on the boundary S of region V. We will see in Chap. 7 that Somigliana's displacement Eq. (4.328) plays a special role in developing the direct boundary integral method for elasticity, and it is useful to have a version of this equation that is valid for the case that the load point $\boldsymbol{\xi}$ coincides with the sample point $\boldsymbol{x}$, i.e., the limiting case for which $\boldsymbol{\xi} \to \boldsymbol{x}$. Referring to Fig. 4.21a we accomplish this by considering a situation in which a smooth boundary S is locally deflected around point $\boldsymbol{\xi}$ by a circular arc S_ε of radius $r = \varepsilon$. The boundary integrals in (4.328) can be written as the sums of integrals over $S - S_\varepsilon$ and S_ε, with the limiting case $\varepsilon \to 0$ yielding the desired results. Detailed calculations for this are given below for the cases of both isotropic and anisotropic elastic materials.

4.23.1 Limit Terms for Isotropy

The integral in (4.328) involving the traction kernel $T_{ij}(\boldsymbol{x}, \boldsymbol{\xi})$, for example, can be written as

$$\int_S u_j(\boldsymbol{x}) T_{ij}(\boldsymbol{x}, \boldsymbol{\xi})\, \mathrm{d}S(\boldsymbol{x}) = \lim_{\varepsilon \to 0} \left[\int_{S-S_\varepsilon} u_j(\boldsymbol{x}) T_{ij}(\boldsymbol{x}, \boldsymbol{\xi})\, \mathrm{d}S(\boldsymbol{x}) + \int_{S_\varepsilon} u_j(\boldsymbol{x}) T_{ij}(\boldsymbol{x}, \boldsymbol{\xi})\, \mathrm{d}S(\boldsymbol{x}) \right] \tag{4.344}$$

The first integral in (4.344) covers the entire boundary except for a portion of length 2ε centered at point $\boldsymbol{\xi}$. The second integral must be interpreted in the sense of the Cauchy principal value and evaluated by excluding the point $\boldsymbol{\xi}$ (see Sect. 2.1.5). It can be written as

$$\lim_{\epsilon\to 0}\int_{S_\varepsilon} u_j(\boldsymbol{x})T_{ij}(\boldsymbol{x},\boldsymbol{\xi})\,\mathrm{d}S(\boldsymbol{x}) = \lim_{\varepsilon\to 0}\left[\int_{S_\varepsilon}\big(u_j(\boldsymbol{x})-u_j(\boldsymbol{\xi})\big)\,T_{ij}(\boldsymbol{x},\boldsymbol{\xi})\,\mathrm{d}S(\boldsymbol{x}) + u_j(\boldsymbol{\xi})\int_{S_\varepsilon} T_{ij}(\boldsymbol{x},\boldsymbol{\xi})\,\mathrm{d}S(\boldsymbol{x})\right] \tag{4.345}$$

The displacements are differentiable functions (and hence they satisfy a Hölder condition) and the limit of the integral involving $u_j(\boldsymbol{x}) - u_j(\boldsymbol{\xi})$ is zero as $\varepsilon \to 0$. The other integral can be evaluated directly by using (4.272) and noting that $r = \varepsilon$ so that $x = \varepsilon\cos\theta$ and $y = \varepsilon\sin\theta$. From (4.272), the individual terms of the tensor field $T_{ij}(\boldsymbol{x},\boldsymbol{\xi})$ for the boundary S_ε are

$$T_{xx}(\varepsilon,\theta) = -\frac{1}{4\pi(1-\nu)}\frac{1}{\varepsilon}\,[2(1-\nu)+\cos 2\theta]$$

$$T_{yx}(\varepsilon,\theta) = T_{xy}(\varepsilon,\theta) = -\frac{1}{4\pi(1-\nu)}\frac{1}{\varepsilon}\,[\sin 2\theta] \tag{4.346}$$

$$T_{yy}(\varepsilon,\theta) = -\frac{1}{4\pi(1-\nu)}\frac{1}{\varepsilon}\,[2(1-\nu)-\cos 2\theta]$$

For a smooth boundary as in Fig. 4.21a the limits of integration are from $\theta = \beta + \pi$ to $\theta = \beta$, where β is the angle that the tangent to the undeflected boundary S at point $\boldsymbol{\xi}$ makes with the x axis. In this way we find that[14]

$$\int_{\beta+\pi}^{\beta} T_{xx}(\varepsilon,\theta)(-\varepsilon\,\mathrm{d}\theta) = -\frac{1}{4\pi(1-\nu)}\frac{1}{\varepsilon}\int_{\beta+\pi}^{\beta}[2(1-\nu)+\cos 2\theta]\,(-\varepsilon\,\mathrm{d}\theta) = -\frac{1}{2}$$

$$\int_{\beta+\pi}^{\beta} T_{yx}(\varepsilon,\theta)(-\varepsilon\,\mathrm{d}\theta) = \int_{\beta+\pi}^{\beta} T_{xy}(\varepsilon,\theta)(-\varepsilon d\theta) = -\frac{1}{4\pi(1-\nu)}\frac{1}{\varepsilon}\int_{\beta+\pi}^{\beta}[\sin 2\theta]\,(-\varepsilon\,\mathrm{d}\theta) = 0 \tag{4.347}$$

$$\int_{\beta+\pi}^{\beta} T_{yy}(\varepsilon,\theta)(-\varepsilon\,\mathrm{d}\theta) = -\frac{1}{4\pi(1-\nu)}\frac{1}{\varepsilon}\int_{\beta+\pi}^{\beta}[2(1-\nu)-\cos 2\theta]\,(-\varepsilon\,\mathrm{d}\theta) = -\frac{1}{2}$$

---

[14] The minus sign for the element of arc length $(-\varepsilon\,\mathrm{d}\theta)$ appears because the direction of integration is opposite to the counterclockwise sense for angle θ.

Combining the preceding results, (4.344) can be written as

$$\int_S u_j(\boldsymbol{x}) T_{ij}(\boldsymbol{x}, \boldsymbol{\xi}) \, \mathrm{d}S(\boldsymbol{x}) = \fint_S u_j(\boldsymbol{x}) T_{ij}(\boldsymbol{x}, \boldsymbol{\xi}) \, \mathrm{d}S(\boldsymbol{x}) - \frac{1}{2} u_i(\boldsymbol{\xi}) \tag{4.348}$$

in which $\fint$ denotes a Cauchy principal value integral.

A similar process is used to find the limit value of the integral in Somigliana's formula (4.328) involving the displacement kernel $U_{ij}(\boldsymbol{x}, \boldsymbol{\xi})$, namely

$$\lim_{\epsilon \to 0} \int_{S_\varepsilon} t_j(\boldsymbol{x}) U_{ij}(\boldsymbol{x}, \boldsymbol{\xi}) \, \mathrm{d}S(\boldsymbol{x}) = \lim_{\varepsilon \to 0} \left[\int_{S_\varepsilon} \left(t_j(\boldsymbol{x}) - t_j(\boldsymbol{\xi}) \right) U_{ij}(\boldsymbol{x}, \boldsymbol{\xi}) \, \mathrm{d}S(\boldsymbol{x}) + t_j(\boldsymbol{\xi}) \int_{S_\varepsilon} U_{ij}(\boldsymbol{x}, \boldsymbol{\xi}) \, \mathrm{d}S(\boldsymbol{x}) \right] \tag{4.349}$$

The individual terms of the tensor field $U_{ij}(\boldsymbol{x}, \boldsymbol{\xi})$ for the boundary S_ε are, using (4.268),

$$\begin{aligned} U_{xx}(\varepsilon, \theta) &= -\frac{1}{8\pi\mu(1-\nu)} \left[(3 - 4\nu) \ln \varepsilon - \cos^2 \theta \right] \\ U_{yx}(\varepsilon, \theta) &= U_{xy}(\varepsilon, \theta) = \frac{1}{8\pi\mu(1-\nu)} \sin\theta \cos\theta \\ U_{yy}(\varepsilon, \theta) &= -\frac{1}{8\pi\mu(1-\nu)} \left[(3 - 4\nu) \ln \varepsilon - \sin^2 \theta \right] \end{aligned} \tag{4.350}$$

and the integrals of these expressions over the semi-circular boundary are zero (recall that $\varepsilon \ln \varepsilon \to 0$ as $\varepsilon \to 0$). By combining (4.328) and (4.348) we therefore find the following result for Somigliana's formula when the load point $\boldsymbol{\xi}$ is on a smooth portion of the boundary S:

$$\frac{1}{2} u_i(\boldsymbol{\xi}) = \int_S t_j(\boldsymbol{x}) U_{ij}(\boldsymbol{x}, \boldsymbol{\xi}) \, \mathrm{d}S(\boldsymbol{x}) - \fint_S u_j(\boldsymbol{x}) T_{ij}(\boldsymbol{x}, \boldsymbol{\xi}) \, \mathrm{d}S(\boldsymbol{x}) \tag{4.351}$$

If point $\boldsymbol{\xi}$ is at sharp corner, as illustrated in Fig. 4.21b, the limits of integration range from $\theta = \beta + \pi$ to $\theta = \beta'$, where β' and β are the angles of the tangents to the undeflected boundary S on either side of point $\boldsymbol{\xi}$. Then, assuming appropriate smoothness conditions, the integrals corresponding to (4.347) can be evaluated to give

$$\int_{\beta+\pi}^{\beta'} T_{xx}(\varepsilon,\theta)(-\varepsilon\, d\theta) = -\frac{1}{2} - \frac{\gamma}{2\pi} - \frac{1}{4\pi(1-\nu)}\sin\gamma\cos(\beta+\beta')$$

$$\int_{\beta+\pi}^{\beta'} T_{yx}(\varepsilon,\theta)(-\varepsilon\, d\theta) = \int_{\beta+\pi}^{\beta'} T_{xy}(\varepsilon,\theta)(-\varepsilon\, d\theta) = -\frac{1}{4\pi(1-\nu)}\sin\gamma\sin(\beta+\beta') \tag{4.352}$$

$$\int_{\beta+\pi}^{\beta'} T_{yy}(\varepsilon,\theta)(-\varepsilon\, d\theta) = -\frac{1}{2} - \frac{\gamma}{2\pi} + \frac{1}{4\pi(1-\nu)}\sin\gamma\cos(\beta+\beta')$$

where $\gamma = \beta - \beta'$. Instead of (4.351), we then have

$$c_{ij}u_j(\boldsymbol{\xi}) = \int_S t_j(\boldsymbol{x})U_{ij}(\boldsymbol{x},\boldsymbol{\xi})\, dS(\boldsymbol{x}) - \fint_S u_j(\boldsymbol{x})T_{ij}(\boldsymbol{x},\boldsymbol{\xi})\, dS(\boldsymbol{x}) \tag{4.353}$$

in which

$$\begin{aligned} c_{11} &= \frac{1}{2} - \frac{\gamma}{2\pi} - \frac{1}{4\pi(1-\nu)}\sin\gamma\cos(\beta+\beta') \\ c_{21} &= c_{12} = -\frac{1}{4\pi(1-\nu)}\sin\gamma\sin(\beta+\beta') \\ c_{22} &= \frac{1}{2} - \frac{\gamma}{2\pi} + \frac{1}{4\pi(1-\nu)}\sin\gamma\cos(\beta+\beta') \end{aligned} \tag{4.354}$$

If the boundary is smooth at point $\boldsymbol{\xi}$ then $\gamma = 0$ and (4.353) reduces to (4.351).

Including (4.354) in (4.329), we write

$$\int_S t_j(\boldsymbol{x})U_{ij}(\boldsymbol{x},\boldsymbol{\xi})\, dS(\boldsymbol{x}) - \int_S u_j(\boldsymbol{x})T_{ij}(\boldsymbol{x},\boldsymbol{\xi})\, dS(\boldsymbol{x}) = \begin{cases} u_i(\boldsymbol{\xi}) & \boldsymbol{\xi}\in V,\quad \boldsymbol{\xi}\notin S \\ c_{ij}u_j(\boldsymbol{\xi}) & \boldsymbol{\xi}\in S \\ 0 & \boldsymbol{\xi}\notin V \end{cases} \tag{4.355}$$

It can be shown that this result also holds in three dimensions, provided the three-dimensional versions of $U_{ij}(\boldsymbol{x},\boldsymbol{\xi})$, $T_{ij}(\boldsymbol{x},\boldsymbol{\xi})$, and c_{ij} are used. These quantities can be found in Bonnet (1999) and Gaul et al. (2003)

4.23.2 Limit Terms for Anisotropy

The preceding analysis can be extended to the case of an anisotropic elastic material by writing the tensor fields $U_{ij}(\boldsymbol{x}, \boldsymbol{\xi})$ and $T_{ij}(\boldsymbol{x}, \boldsymbol{\xi})$ given in Appendix B in forms comparable to (4.350) and (4.346). For the semicircular boundary S_ε in Fig. 4.21 we have that $x = \varepsilon \cos\theta$ and $y = \varepsilon \sin\theta$ and (B.7) become

$$
\begin{aligned}
U_{xx}(\varepsilon, \theta) &= \frac{1}{2\pi(q_1 - q_2)c_{66}} \left[\frac{\gamma_1}{q_1} \ln\left(\varepsilon\sqrt{\gamma_1^2 \cos^2\theta + \sin^2\theta}/\gamma_1\right) - \frac{\gamma_2}{q_2} \ln\left(\varepsilon\sqrt{\gamma_2^2 \cos^2\theta + \sin^2\theta}/\gamma_2\right)\right] \\
U_{yx}(\varepsilon, \theta) = U_{xy}(\varepsilon, \theta) &= -\frac{1}{2\pi(q_1 - q_2)c_{66}} \left[\arctan\left(\frac{1}{\gamma_1} \tan\theta\right) - \arctan\left(\frac{1}{\gamma_2} \tan\theta\right)\right] \\
U_{yy}(\varepsilon, \theta) &= -\frac{1}{2\pi(q_1 - q_2)c_{66}} \left[\frac{q_1}{\gamma_1} \ln\left(\varepsilon\sqrt{\gamma_1^2 \cos^2\theta + \sin^2\theta}/\gamma_1\right) - \frac{q_2}{\gamma_2} \ln\left(\varepsilon\sqrt{\gamma_2^2 \cos^2\theta + \sin^2\theta}/\gamma_2\right)\right]
\end{aligned}
\tag{4.356}
$$

Also, because $\sin\beta = -\cos\theta$ and $\cos\beta = \sin\theta$ on S_ε, (B.8) can be written as

$$
\begin{aligned}
T_{xx}(\varepsilon, \theta) &= \frac{1}{2\pi(q_1 - q_2)} \frac{1}{\varepsilon} \left[\frac{\gamma_1(1 + q_1)}{q_1} \frac{1}{\gamma_1^2 \cos^2\theta + \sin^2\theta} - \frac{\gamma_2(1 + q_2)}{q_2} \frac{1}{\gamma_2^2 \cos^2\theta + \sin^2\theta}\right] \\
T_{yx}(\varepsilon, \theta) &= \frac{1}{2\pi(q_1 - q_2)} \frac{1}{\varepsilon} \left[\frac{(1 + q_1)(1 - \gamma_1^2)}{\gamma_1} \frac{\sin\theta\cos\theta}{\gamma_1^2 \cos^2\theta + \sin^2\theta} \right. \\
&\qquad \left. - \frac{(1 + q_2)(1 - \gamma_2^2)}{\gamma_2} \frac{\sin\theta\cos\theta}{\gamma_2^2 \cos^2\theta + \sin^2\theta}\right] \\
T_{xy}(\varepsilon, \theta) &= \frac{1}{2\pi(q_1 - q_2)} \frac{1}{\varepsilon} \left[\frac{\gamma_1(1 + q_1)(1 - \gamma_1^2)}{q_1} \frac{\sin\theta\cos\theta}{\gamma_1^2 \cos^2\theta + \sin^2\theta} \right. \\
&\qquad \left. - \frac{\gamma_2(1 + q_2)(1 - \gamma_2^2)}{q_2} \frac{\sin\theta\cos\theta}{\gamma_2^2 \cos^2\theta + \sin^2\theta}\right] \\
T_{yy}(\varepsilon, \theta) &= -\frac{1}{2\pi(q_1 - q_2)} \frac{1}{\varepsilon} \left[\gamma_1(1 + q_1) \frac{1}{\gamma_1^2 \cos^2\theta + \sin^2\theta} - \gamma_2(1 + q_2) \frac{1}{\gamma_2^2 \cos^2\theta + \sin^2\theta}\right]
\end{aligned}
\tag{4.357}
$$

The integrals of expressions (4.356) over the semicircular boundary S_ε in Fig. 4.21b all involve multiples of ε and these terms are zero as $\varepsilon \to 0$ (again, recall that $\varepsilon \ln\varepsilon \to 0$ as $\varepsilon \to 0$). The integrals of (4.357) over S_ε can be evaluated using the formulas (Gradshteyn and Ryzhik 1980)

$$
\gamma_i \int \frac{1}{\gamma_i^2 \cos^2\theta + \sin^2\theta} \, d\theta = \arctan\left(\frac{1}{\gamma_i} \tan\theta\right) \tag{4.358}
$$

and

$$(1-\gamma_i^2)\int \frac{\sin\theta\cos\theta}{\gamma_i^2\cos^2\theta+\sin^2\theta}\,\mathrm{d}\theta = \ln\sqrt{\gamma_i^2\cos^2\theta+\sin^2\theta} - \ln\gamma_i \tag{4.359}$$

For these calculations it is noted that

$$\gamma_i\int_{\beta+\pi}^{\beta'} \frac{1}{\gamma_i^2\cos^2\theta+\sin^2\theta}\,\mathrm{d}\theta = \arctan\left(\frac{1}{\gamma_i}\tan\beta'\right) - \arctan\left(\frac{1}{\gamma_i}\tan\beta\right) - \pi \tag{4.360}$$

because $\arctan\left(\gamma_i^{-1}\tan(\beta+\pi)\right) = \arctan\left(\gamma_i^{-1}\tan\beta\right)+\pi$. Then, using (4.206) and noting that $q_1q_2 = 1$ (see exercise (12)), the required integrals are found to be as follows:

$$\begin{aligned}
\int_{\beta+\pi}^{\beta'} T_{xx}(\varepsilon,\theta)(-\varepsilon\,\mathrm{d}\theta) &= -\frac{1}{2} + \frac{1}{2\pi(q_1-q_2)}\left[\frac{(1+q_1)}{q_1}\arctan\frac{\gamma_1\sin\gamma}{\gamma_1^2\cos\beta\cos\beta'+\sin\beta\sin\beta'}\right.\\
&\qquad \left.-\frac{(1+q_2)}{q_2}\arctan\frac{\gamma_2\sin\gamma}{\gamma_2^2\cos\beta\cos\beta'+\sin\beta\sin\beta'}\right]\\
\int_{\beta+\pi}^{\beta'} T_{yx}(\varepsilon,\theta)(-\varepsilon\,\mathrm{d}\theta) &= \frac{1}{2\pi(q_1-q_2)}\left[\frac{1}{2}\frac{(1+q_1)}{\gamma_1}\ln\frac{\gamma_1^2\cos^2\beta+\sin^2\beta}{\gamma_1^2\cos^2\beta'+\sin^2\beta'}\right.\\
&\qquad \left.-\frac{1}{2}\frac{(1+q_2)}{\gamma_2}\ln\frac{\gamma_2^2\cos^2\beta+\sin^2\beta}{\gamma_2^2\cos^2\beta'+\sin^2\beta'}\right]\\
\int_{\beta+\pi}^{\beta'} T_{xy}(\varepsilon,\theta)(-\varepsilon\,\mathrm{d}\theta) &= \frac{1}{2\pi(q_1-q_2)}\left[\frac{1}{2}\frac{\gamma_1(1+q_1)}{q_1}\ln\frac{\gamma_1^2\cos^2\beta+\sin^2\beta}{\gamma_1^2\cos^2\beta'+\sin^2\beta'}\right.\\
&\qquad \left.-\frac{1}{2}\frac{\gamma_2(1+q_2)}{q_2}\ln\frac{\gamma_2^2\cos^2\beta+\sin^2\beta}{\gamma_2^2\cos^2\beta'+\sin^2\beta'}\right]\\
\int_{\beta+\pi}^{\beta'} T_{yy}(\varepsilon,\theta)(-\varepsilon\,\mathrm{d}\theta) &= -\frac{1}{2} - \frac{1}{2\pi(q_1-q_2)}\left[(1+q_1)\arctan\frac{\gamma_1\sin\gamma}{\gamma_1^2\cos\beta\cos\beta'+\sin\beta\sin\beta'}\right.\\
&\qquad \left.-(1+q_2)\arctan\frac{\gamma_2\sin\gamma}{\gamma_2^2\cos\beta\cos\beta'+\sin\beta\sin\beta'}\right]
\end{aligned} \tag{4.361}$$

Coefficients c_{ij} for anisotropic elasticity comparable to (4.354) can be found from these expressions by inspection.

The case of isotropic elasticity is found by taking the limits of (4.361) as $\gamma_1 \to \gamma_2$ and then setting $\gamma_2 = 1$. It can be shown that these operations lead to the same results as (4.352).

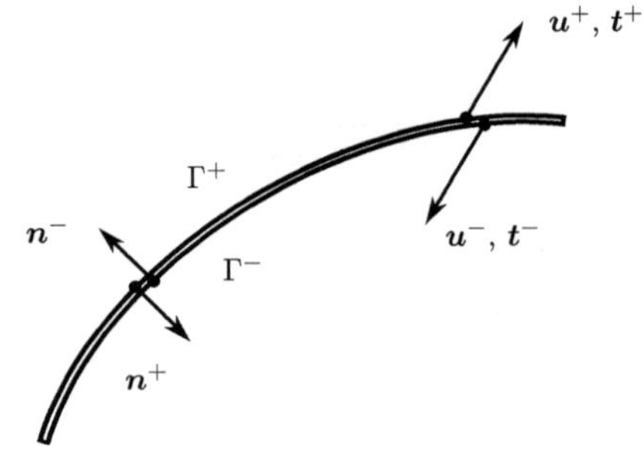

Fig. 4.22 Crack boundary S with surfaces Γ^+ and Γ^-

4.24 Volterra's Formulas

An important special case of Somigliana's formulas occurs when the boundary S represents a line crack Γ, as illustrated in Fig. 4.22. The crack has two surfaces, designated as Γ^+ and Γ^- in the figure, and the components of the displacements and tractions are $u_j^+(\boldsymbol{x})$, $t_j^+(\boldsymbol{x})$ and $u_j^-(\boldsymbol{x})$, $t_j^-(\boldsymbol{x})$, respectively. The superscripts + and − indicate the limit values as the two crack surfaces are approached in opposite directions from inside the region of interest. It can be seen from the figure that $n_j^+(\boldsymbol{x}) = -n_j^-(\boldsymbol{x})$. Equilibrium requires that $t_j^+(\boldsymbol{x}) = -t_j^-(\boldsymbol{x})$, and we define the components of *displacement discontinuity* at point $\boldsymbol{x}$ by the relation

$$\Delta u_j(\boldsymbol{x}) = u_j^+(\boldsymbol{x}) - u_j^-(\boldsymbol{x}) \tag{4.362}$$

Formulas comparable to (4.328) and (4.332) for this situation are attributed to Volterra (see, for example, Steketee 1958a, b) and are derived below.

We begin by considering a finite region V with boundary S, as shown in Fig. 4.23a. The complementary region V_c has boundary S_c, which by definition is coincident with S. (See Sect. 1.4 for additional discussion related to the representation of an infinite region by V and V_c.) We imagine that S (and S_c) is traversed in a clockwise sense, as shown in the figure, and that the normal $n_j(\boldsymbol{x})$ points away from region V. The normal to the complementary region V_c is denoted as $n_j'(\boldsymbol{x})$; this normal points in the direction opposite to $n_j(\boldsymbol{x})$, i.e., $n_j'(\boldsymbol{x}) = -n_j(\boldsymbol{x})$. The displacements and tractions on S are $u_i(\boldsymbol{x})$ and $t_i(\boldsymbol{x})$, while those on S_c are $u_i'(\boldsymbol{x})$ and $t_i'(\boldsymbol{x})$.

The displacements at a point $\boldsymbol{\xi}$ inside region V are given by Somigliana's formula (4.328):

$$u_i(\boldsymbol{\xi}) = \int_S t_j(\boldsymbol{x}) U_{ij}(\boldsymbol{x}, \boldsymbol{\xi})\, \mathrm{d}S(\boldsymbol{x}) - \int_S u_j(\boldsymbol{x}) T_{ij}(\boldsymbol{x}, \boldsymbol{\xi})\, \mathrm{d}S(\boldsymbol{x}) \tag{4.363}$$

Also, because point $\boldsymbol{\xi} \in V$ is *outside* the complementary region V_c (i.e., $\boldsymbol{\xi} \notin V_c$) we have

$$0 = \int_{S_c} t_j'(\boldsymbol{x}) U_{ij}'(\boldsymbol{x}, \boldsymbol{\xi})\, \mathrm{d}S(\boldsymbol{x}) - \int_{S_c} u_j'(\boldsymbol{x}) T_{ij}'(\boldsymbol{x}, \boldsymbol{\xi})\, \mathrm{d}S(\boldsymbol{x}) \tag{4.364}$$

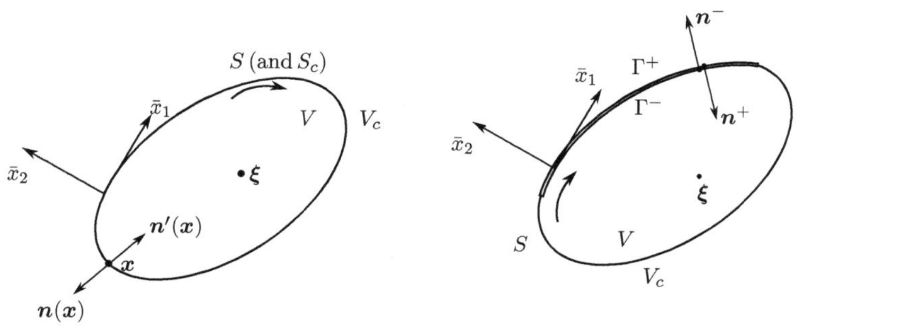

(a) Finite region V with complementary region V_c.

(b) Portion Γ of contour S represents a crack.

Fig. 4.23 Definition of boundaries for regions V and V_c

see (4.329). Noting that $S_c = S$ and adding Eqs. (4.363) and (4.364), we obtain

$$u_i(\boldsymbol{\xi}) = \int_S \left(t_j(\boldsymbol{x}) U_{ij}(\boldsymbol{x}, \boldsymbol{\xi}) + t_j^{'}(\boldsymbol{x}) U_{ij}^{'}(\boldsymbol{x}, \boldsymbol{\xi}) \right) \mathrm{d}S(\boldsymbol{x}) - \int_S \left(u_j(\boldsymbol{x}) T_{ij}(\boldsymbol{x}, \boldsymbol{\xi}) + u_j^{'}(\boldsymbol{x}) T_{ij}^{'}(\boldsymbol{x}, \boldsymbol{\xi}) \right) \mathrm{d}S(\boldsymbol{x}) \tag{4.365}$$

Recalling that tensor fields $U_{ij}(\boldsymbol{x}, \boldsymbol{\xi})$, $U_{ij}^{'}(\boldsymbol{x}, \boldsymbol{\xi})$ and $T_{ij}(\boldsymbol{x}, \boldsymbol{\xi})$, $T_{ij}^{'}(\boldsymbol{x}, \boldsymbol{\xi})$ represent the displacements and tractions in the j direction at point $\boldsymbol{x}$ of an *infinite region* $V + V_c$ due to a unit concentrated force in the i direction at point $\boldsymbol{\xi}$, we write

$$U_{ij}(\boldsymbol{x}, \boldsymbol{\xi}) = U_{ij}^{'}(\boldsymbol{x}, \boldsymbol{\xi}) \tag{4.366}$$

$$T_{ij}(\boldsymbol{x}, \boldsymbol{\xi}) = -T_{ij}^{'}(\boldsymbol{x}, \boldsymbol{\xi}) \tag{4.367}$$

These conditions stem from the observation that a concentrated load at point $\boldsymbol{\xi}$ in $V + V_c$ produces *continuous* displacements and stresses at point $\boldsymbol{x} \neq \boldsymbol{\xi}$. The minus sign in (4.367) then follows from (4.257) and (4.264) and the fact that the outward normals to boundaries S and S_c are in opposite directions, i.e.,

$$\begin{aligned} T_{ij}(\boldsymbol{x}, \boldsymbol{\xi}) &= B_{ijk}(\boldsymbol{x}, \boldsymbol{\xi}) n_k(\boldsymbol{x}) \\ &= -B_{ijk}(\boldsymbol{x}, \boldsymbol{\xi}) n_k^{'}(\boldsymbol{x}) \\ &= -B_{ijk}^{'}(\boldsymbol{x}, \boldsymbol{\xi}) n_k^{'}(\boldsymbol{x}) \\ &= -T_{ij}^{'}(\boldsymbol{x}, \boldsymbol{\xi}) \end{aligned} \tag{4.368}$$

Substitution of (4.366) and (4.367) into (4.365) then gives

$$u_i(\boldsymbol{\xi}) = \int_S \left(t_j(\boldsymbol{x}) + t_j^{'}(\boldsymbol{x})\right) U_{ij}(\boldsymbol{x}, \boldsymbol{\xi})\, \mathrm{d}S(\boldsymbol{x})$$
$$- \int_S \left(u_j(\boldsymbol{x}) - u_j^{'}(\boldsymbol{x})\right) T_{ij}(\boldsymbol{x}, \boldsymbol{\xi})\, \mathrm{d}S(\boldsymbol{x}) \tag{4.369}$$

Noting the equilibrium condition $t_j(\boldsymbol{x}) + t_j^{'}(\boldsymbol{x}) = 0$ and defining the components of displacement discontinuity along S as

$$\Delta u_j(\boldsymbol{x}) = u_j^{'}(\boldsymbol{x}) - u_j(\boldsymbol{x}) \tag{4.370}$$

Equation (4.369) becomes

$$u_i(\boldsymbol{\xi}) = \int_S \Delta u_j(\boldsymbol{x}) T_{ij}(\boldsymbol{x}, \boldsymbol{\xi})\, \mathrm{d}S(\boldsymbol{x}) \tag{4.371}$$

(The reader should realize that the definition of the displacement discontinuity used here is not unique; this quantity could alternatively be defined as $u_j(\boldsymbol{x}) - u_j^{'}(\boldsymbol{x})$.)

4.24.1 Formula for Displacements

We now suppose that the displacements are discontinuous along a portion Γ of contour S but are continuous elsewhere, i.e., $\Delta u_j(\boldsymbol{x}) \neq 0$ for $\boldsymbol{x} \in \Gamma$ but $\Delta u_j(\boldsymbol{x}) = 0$ for $\boldsymbol{x} \in S - \Gamma$. The contour Γ then represents a line crack in an infinite region, as depicted in Fig. 4.23b, and (4.371) becomes

$$u_i(\boldsymbol{\xi}) = \int_\Gamma \Delta u_j(\boldsymbol{x}) T_{ij}(\boldsymbol{x}, \boldsymbol{\xi})\, \mathrm{d}S(\boldsymbol{x}) \tag{4.372}$$

The positive and negative surfaces of Γ are identified with respect to a local $\overline{x}_j = (\overline{x}_1, \overline{x}_2)$ coordinate system along the crack, as shown in Fig. 4.23b. The $\overline{x}_1$ axis is positive in the direction of traversal of Γ (the choice for this direction is arbitrary); $\overline{x}_2$ points in the direction of the outward normal to the *negative* surface of the crack, i.e., Γ^-. Thus, Γ^- is defined as $\overline{x}_2 = 0_-$ while Γ^+ is $\overline{x}_2 = 0_+$.

Equation (4.372) is Volterra's formula for the displacements at point $\boldsymbol{\xi}$ of an infinite elastic region due to a displacement discontinuity at point $\boldsymbol{x}$ along a contour Γ. It is important to recognize that $T_{ij}(\boldsymbol{x}, \boldsymbol{\xi})$ relates to the negative side of the crack, Γ^-. One could also write (4.372) in terms of $T_{ij}^{'}(\boldsymbol{x}, \boldsymbol{\xi})$, in which case a minus sign would appear in the equation, by virtue of (4.367).

4.24.2 Formula for Stresses

The stresses at point $\boldsymbol{\xi}$ can be computed by differentiating (4.372) with respect to the coordinates of this point and using Hooke's law, as discussed previously for Somigliana's formula (4.328). Alternatively, the stresses could be obtained by repeating the procedure used to derive (4.372), but starting with (4.332). In either way, one finds

$$\sigma_{ij}(\boldsymbol{\xi}) = \int_{\Gamma} \Delta u_k(\boldsymbol{x}) S_{kij}(\boldsymbol{x}, \boldsymbol{\xi})\, \mathrm{d}S(\boldsymbol{x}) \tag{4.373}$$

where $S_{kij}(\boldsymbol{x}, \boldsymbol{\xi})$ is given by (4.334).

4.24.3 Formula for Tractions

The tractions at point $\boldsymbol{\xi}$, measured with respect to the normal to a curve at that point, are given as $t_i(\boldsymbol{\xi}) = \sigma_{ij}(\boldsymbol{\xi}) n_j(\boldsymbol{\xi})$. Following the same approach as that used to develop (4.343), we write Volterra's formula for the tractions as

$$t_i(\boldsymbol{\xi}) = \int_{\Gamma} \Delta u_j(\boldsymbol{x}) H_{ij}(\boldsymbol{x}, \boldsymbol{\xi})\, \mathrm{d}S(\boldsymbol{x}) \tag{4.374}$$

4.24.4 Alternative Forms

Using (4.289), Eq. (4.372) can be equivalently written as

$$u_i(\boldsymbol{\xi}) = \int_{\Gamma} \Delta u_j(\boldsymbol{x}) K_{ji}(\boldsymbol{\xi}, \boldsymbol{x})\, \mathrm{d}S(\boldsymbol{x}) \tag{4.375}$$

We have used the point displacement discontinuity $\Delta u_j(\boldsymbol{x})$ in this expression instead of $\mathcal{D}_j(\boldsymbol{x})$ as in (4.289), but of course these two quantities have the same meaning. According to the naming convention given in Sect. 4.17, $K_{ji}(\boldsymbol{\xi}, \boldsymbol{x})$ gives the displacement in the i direction at point $\boldsymbol{\xi}$ due to a unit concentrated displacement discontinuity in the j direction at point $\boldsymbol{x}$.

Similarly, using (4.290) and (4.337), Eq. (4.373) can be written as

$$\sigma_{ij}(\boldsymbol{\xi}) = \int_{\Gamma} \Delta u_k(\boldsymbol{x}) C_{kij}(\boldsymbol{\xi}, \boldsymbol{x})\, \mathrm{d}S(\boldsymbol{x}) \tag{4.376}$$

Finally, an expression for the tractions $t_i(\boldsymbol{\xi})$ can be developed from this equation by using the definition $t_i(\boldsymbol{\xi}) = \sigma_{ij}(\boldsymbol{\xi}) n_j(\boldsymbol{\xi})$ and elaborating the quantity $C_{kij}(\boldsymbol{\xi}, \boldsymbol{x}) n_j(\boldsymbol{\xi})$. The result is

$$t_i(\boldsymbol{\xi}) = \int_{\Gamma} \Delta u_j(\boldsymbol{x}) H_{ji}(\boldsymbol{\xi}, \boldsymbol{x})\, \mathrm{d}S(\boldsymbol{x}) \tag{4.377}$$

Equations (4.375) through (4.377) are therefore an alternative way of writing Volterra's formulas for the displacements, stresses, and tractions at a point $\boldsymbol{\xi}$ in region V due to a distribution of displacement discontinuity along a crack Γ. These equations represent the double layer formulas (4.300) with the load and sample points interchanged, and with density function $\eta_j(\boldsymbol{x})$ equal to $\Delta u_j(\boldsymbol{x})$.

4.25 Summary of Boundary Integral Formulas For Elasticity

A useful way to summarize the preceding boundary integral formulas for elasticity is to set $t_j(\boldsymbol{x}) + t_j'(\boldsymbol{x}) = \Sigma t_j(\boldsymbol{x})$ in (4.369) instead of taking this quantity to be zero. With this change we obtain the expression

$$u_i(\boldsymbol{\xi}) = \int_S \Sigma t_j(\boldsymbol{x}) U_{ij}(\boldsymbol{x}, \boldsymbol{\xi})\, \mathrm{d}S(\boldsymbol{x}) + \int_S \Delta u_j(\boldsymbol{x}) T_{ij}(\boldsymbol{x}, \boldsymbol{\xi})\, \mathrm{d}S(\boldsymbol{x}) \tag{4.378}$$

Companion expressions for the stresses $\sigma_{ij}(\boldsymbol{\xi})$ and tractions $t_i(\boldsymbol{\xi})$ can be derived in a similar way starting from (4.332) and (4.343). The results are

$$\sigma_{ij}(\boldsymbol{\xi}) = \int_S \Sigma t_k(\boldsymbol{x}) D_{kij}(\boldsymbol{x}, \boldsymbol{\xi})\, \mathrm{d}S(\boldsymbol{x}) + \int_S \Delta u_k(\boldsymbol{x}) S_{kij}(\boldsymbol{x}, \boldsymbol{\xi})\, \mathrm{d}S(\boldsymbol{x}) \tag{4.379}$$

and

$$t_i(\boldsymbol{\xi}) = \int_S \Sigma t_j(\boldsymbol{x}) K_{ij}(\boldsymbol{x}, \boldsymbol{\xi})\, \mathrm{d}S(\boldsymbol{x}) + \int_S \Delta u_j(\boldsymbol{x}) H_{ij}(\boldsymbol{x}, \boldsymbol{\xi})\, \mathrm{d}S(\boldsymbol{x}) \tag{4.380}$$

Somigliana's formulas (4.328), (4.332), and (4.343) can be recovered from these results by taking $u_j'(\boldsymbol{x}) = 0$ and $t_j'(\boldsymbol{x}) = 0$ so that $\Delta u_j(\boldsymbol{x}) = -u_j(\boldsymbol{x})$ and $\Sigma t_j(\boldsymbol{x}) = t_j(\boldsymbol{x})$. These equations are the basis for the direct boundary integral method discussed in Chap. 7.

Volterra's formulas (4.372), (4.373), and (4.374) are obtained by setting $\Sigma t_j(\boldsymbol{x}) = 0$ and letting $S = \Gamma$. As noted previously, these equations represent the double layer formulas (4.300) with the load and sample points interchanged (i.e., the load point is at $\boldsymbol{x}$ and the sample point is at $\boldsymbol{\xi}$), and with density function $\eta_j(\boldsymbol{x}) = \Delta u_j(\boldsymbol{x})$. The results are the basis of the displacement discontinuity method discussed in Chap. 6.

Finally, the case $\Delta u_j(\boldsymbol{x}) = 0$ represents the single layer formula (4.299) with the load and sample points interchanged, and with density function $\rho_j(\boldsymbol{x}) = \Sigma t_j(\boldsymbol{x})$. These results are the basis of an indirect boundary element method for elasticity called the stress discontinuity method, which is presented in Chap. 5.

It is useful to note that the quantity $\sum t_j(\boldsymbol{x}) = t_j(\boldsymbol{x}) + t'_j(\boldsymbol{x})$ can be written in terms of stress tensor discontinuities by using the relations $t_j(\boldsymbol{x}) = \sigma_{ij}(\boldsymbol{x})n_i(\boldsymbol{x})$, $t'_j(\boldsymbol{x}) = \sigma'_{ij}(\boldsymbol{x})n'_i(\boldsymbol{x})$, and $n'_i(\boldsymbol{x}) = -n_i(\boldsymbol{x})$:

$$\begin{aligned}\sum t_j(x) &= \sigma_{ij}(\boldsymbol{x})n_i(\boldsymbol{x}) + \sigma'_{ij}(\boldsymbol{x})n'_i(\boldsymbol{x}) \\ &= \left[\sigma_{ij}(\boldsymbol{x}) - \sigma'_{ij}(\boldsymbol{x})\right] n_i(\boldsymbol{x}) \\ &= \Delta\sigma_{ij}(\boldsymbol{x})n_i(\boldsymbol{x}) \end{aligned} \tag{4.381}$$

In the literature the quantity $\Delta\sigma_{ij}(\boldsymbol{x})n_i(\boldsymbol{x})$ is sometimes referred to as the jump in tractions, or the traction discontinuity, $\Delta t_j(\boldsymbol{x})$.

4.26 Correspondence with Direct Boundary Integral Formulas for Harmonic Functions

We observed in Sect. 4.18 that there is a close connection between indirect boundary integral formulas for elasticity and potential theory. In particular, we saw that displacements, stresses, and tractions play the same role in elasticity as the potential, its gradient, and its normal derivative play in potential theory. Similar parallels exist for Somigliana's formulas and the direct boundary integral formulation for potential theory. This can be seen by comparing integral formulas for (a) the displacements $u_i(\boldsymbol{\xi})$ and the potential $\phi(\boldsymbol{\xi})$ and (b) the tractions $t_i(\boldsymbol{\xi})$ and the normal derivative of the potential $q(\boldsymbol{\xi}) = \partial\phi(\boldsymbol{\xi})/\partial\boldsymbol{n}(\boldsymbol{\xi})$.

First, we compare formulas (4.328) for the displacements and (2.63) for the potential, i.e.,

$$u_i(\boldsymbol{\xi}) = \int_S t_j(\boldsymbol{x})U_{ij}(\boldsymbol{x},\boldsymbol{\xi})\,\mathrm{d}S(\boldsymbol{x}) - \int_S u_j(\boldsymbol{x})T_{ij}(\boldsymbol{x},\boldsymbol{\xi})\,\mathrm{d}S(\boldsymbol{x}) \tag{4.382}$$

and

$$\phi(\boldsymbol{\xi}) = \int_S q(\boldsymbol{x})G(\boldsymbol{x},\boldsymbol{\xi})\,\mathrm{d}S(\boldsymbol{x}) - \int_S \phi(\boldsymbol{x})Q(\boldsymbol{x},\boldsymbol{\xi})\,\mathrm{d}S(\boldsymbol{x}) \tag{4.383}$$

Notice that the kernels in these expressions represent those of the single and double layer potentials with kernels $G(\boldsymbol{x},\boldsymbol{\xi})$ and $Q(\boldsymbol{x},\boldsymbol{\xi})$ and their analogues with $U_{ij}(\boldsymbol{x},\boldsymbol{\xi})$ and $T_{ij}(\boldsymbol{x},\boldsymbol{\xi})$ for elasticity. Notice also the parallels between the parameters $\phi(\boldsymbol{x})$, $q(\boldsymbol{x})$ and $u_j(\boldsymbol{x})$, $t_j(\boldsymbol{x})$.

Similar comparisons can be drawn between formulas (4.343) for the tractions and (2.77) for the normal derivative of the potential, i.e.,

$$t_i(\boldsymbol{\xi}) = \int_S t_j(\boldsymbol{x})T_{ji}(\boldsymbol{\xi},\boldsymbol{x})\,\mathrm{d}S(\boldsymbol{x}) - \int_S u_j(\boldsymbol{x})H_{ij}(\boldsymbol{x},\boldsymbol{\xi})\,\mathrm{d}S(\boldsymbol{x}) \tag{4.384}$$

in which we have used (4.288), and

$$q(\boldsymbol{\xi}) = \int_S q(\boldsymbol{x}) Q(\boldsymbol{\xi}, \boldsymbol{x})\, \mathrm{d}S(\boldsymbol{x}) - \int_S \phi(\boldsymbol{x}) H(\boldsymbol{x}, \boldsymbol{\xi})\, \mathrm{d}S(\boldsymbol{x}) \tag{4.385}$$

Here, $Q(\boldsymbol{\xi}, \boldsymbol{x})$ is the kernel of the adjoint double layer potential and $H(\boldsymbol{x}, \boldsymbol{\xi})$ is the kernel of the hypersingular potential; $T_{ji}(\boldsymbol{\xi}, \boldsymbol{x})$ and $H_{ij}(\boldsymbol{x}, \boldsymbol{\xi})$ are their analogues for elasticity.

4.27 Direct and Limit Values of Elastic Potentials

The limit values of the four elastic potentials (given by the first and third equations in (4.299) and (4.300)) are obtained using the limiting procedure in which $\boldsymbol{x}$ is allowed to approach a point $\boldsymbol{x}_0$ on S. In Sect. 4.23, we in fact investigated those limit values because the Somigliana displacement identity combines single and double layer potentials and the Somigliana traction identity combines an adjoint double layer and a hypersingular elastic potential. Taking into account that the notations for the load and sample points are interchanged in the indirect formulations, we summarize the results for the case when a point $\boldsymbol{x}_0$ is located on a smooth portion of a curve S and the density functions $\rho_j(\boldsymbol{\xi})$ and $\eta_j(\boldsymbol{\xi})$ are sufficiently smooth. These values are given as follows:

- Single layer formulation

$$\lim_{\boldsymbol{x} \to \boldsymbol{x}_0} u_i(\boldsymbol{x}) = \int_S \rho_j(\boldsymbol{\xi}) U_{ji}(\boldsymbol{x}_0, \boldsymbol{\xi})\, \mathrm{d}S(\boldsymbol{\xi}) \tag{4.386}$$

$$\lim_{\boldsymbol{x} \to \boldsymbol{x}_0} t_i(\boldsymbol{x}) = \pm\frac{1}{2}\rho_i(\boldsymbol{x}_0) + \fint_S \rho_j(\boldsymbol{\xi}) T_{ji}(\boldsymbol{x}_0, \boldsymbol{\xi})\, \mathrm{d}S(\boldsymbol{\xi}) \tag{4.387}$$

- Double layer formulation

$$\begin{aligned}\lim_{\boldsymbol{x} \to \boldsymbol{x}_0} u_i(\boldsymbol{x}) &= \mp\frac{1}{2}\eta_i(\boldsymbol{x}_0) + \fint_S \eta_j(\boldsymbol{\xi}) K_{ji}(\boldsymbol{x}_0, \boldsymbol{\xi})\, \mathrm{d}S(\boldsymbol{\xi}) \\ &= \mp\frac{1}{2}\eta_i(\boldsymbol{x}_0) + \fint_S \eta_j(\boldsymbol{\xi}) T_{ij}(\boldsymbol{\xi}, \boldsymbol{x}_0)\, \mathrm{d}S(\boldsymbol{\xi})\end{aligned} \tag{4.388}$$

$$\lim_{x \to x_0} t_i(x) = \mp\!\!\!\!\!\!\int_S \eta_j(\xi) H_{ji}(x_0, \xi)\, dS(\xi) \tag{4.389}$$

in which the $\pm$ sign in (4.387) and the $\mp$ sign in (4.388) depend on whether point $\boldsymbol{x}_0$ is approached from the direction of the normal vector to boundary S at $\boldsymbol{x}_0$ (the upper sign) or from the opposite direction (the lower sign) and the integrals represent the direct values of the corresponding potentials. It can be seen that (4.386)–(4.389) are analogues of equations (2.53)–(2.56). However, unlike in potential theory, the integrals of (4.387) and (4.388) are of Cauchy principal values types, even under conditions for which the direct values of (2.54) and (2.55) are weakly singular. Just as in potential theory, the integral of (4.386) is weakly singular and that of (4.389) is hypersingular.

Similarly as was done in Sect. 2.3 for potential problems, one can establish the connection between (4.387) and its counterpart for a line segment given by the last two expressions of (4.171) by selecting the normal to the segment as $\boldsymbol{n} = (0, 1)$ and assuming that $\rho_i(\boldsymbol{x}_0) = P_j$. In the same way, if it is assumed that the boundary S in (4.388) is the line segment depicted in Fig. 4.12 and that $\eta_i(\boldsymbol{x}_0) = D_j$, we retrieve the limit values of the displacements at the line segment in terms of a constant displacement discontinuity.

If the domains from both sides of S have the same elastic properties then one can subtract the two limit values of (4.387) and (4.388) from one another, leading to the following values for $\rho_j(\boldsymbol{x}_0)$ and $\eta_j(\boldsymbol{x}_0)$:

$$\begin{aligned} \rho_j(\boldsymbol{x}_0) &= t_j^+(\boldsymbol{x}_0) - t_j^-(\boldsymbol{x}_0) \\ \eta_j(\boldsymbol{x}_0) &= -\left[u_j^+(\boldsymbol{x}_0) - u_j^-(\boldsymbol{x}_0)\right] \end{aligned} \tag{4.390}$$

where $t_j^-(\boldsymbol{x}_0)$ is defined using the same normal vector as for $t_j^+(\boldsymbol{x}_0)$. Therefore, if the normal is directed as $\boldsymbol{n}(\boldsymbol{x})$ in Fig. 4.23a, then $\rho_j(\boldsymbol{x}_0)$ represents a traction discontinuity, while $\eta_j(\boldsymbol{x}_0)$ represents a displacement discontinuity, as defined by (4.370).

4.28 Exercises

1. Express the constants c_{11}, c_{12}, c_{22}, and c_{66} in (4.34) in terms of Young's modulus E and Poisson's ratio ν for an isotropic elastic material and plane strain conditions.

   Answer: (a) $c_{11} = c_{22} = 2\mu(1-\nu)/(1-2\nu)$; $c_{12} = 2\mu\nu/(1-2\nu)$; $c_{66} = \mu$

2. Verify that the Papkovitch-Neuber displacement potential representation (4.83) satisfies the Navier-Cauchy Equation (4.81) with zero body forces.

3. Differentiate the Navier-Cauchy equations (4.81) with respect to to x_i to show that $\nabla^2 u_{j,j} = 0$ when the body forces are zero. (The quantity $u_{j,j} = \varepsilon_{jj}$ is called the *dilatation* in elasticity.) Use this result to show that, in the absence of body forces, the displacements are *biharmonic* functions, i.e. $\nabla^2\nabla^2 u_i = \nabla^4 u_i = 0$.
4. Consider the Papkovitch-Neuber functions $\varphi_0 = 0$ and $\varphi_i = a_{ij}x_j$, where a_{ij} are constants for i and $j = 1$ to 2. Show that these functions are associated with a homogeneous (i.e., constant) state of stress, and hence justify choosing $\varphi_1 = 0$ in (4.90).

   Answer: $\sigma_{ij} = \dfrac{\mu}{2(1-\nu)}\left[(1-2\nu)(a_{ij}+a_{ji}) + 2\nu a_{kk}\delta_{ij}\right]$

5. Use the condition $\nabla^2\varphi(x_1, x_2) = 0$ to show that the function of x_2 that was omitted from (4.95) is equal to $Ax_2 + B$, where A and B are constants. What do A and B represent physically and why can they be omitted from the analysis?
6. Show that the function $\dfrac{\partial^2\Phi}{\partial y^2} = -\dfrac{p_y}{2\pi\mu}\arctan\dfrac{y}{x}$ solves the problem of an isotropic, linearly elastic half-plane $y \leq 0$ whose surface is subjected to a normal stress $\sigma_{yy} = p_y$ along the entire negative x axis but is otherwise unloaded. Use this result and the principle of superposition to find the solution to the following problem for a half-plane $y \leq 0$: $\sigma_{xy} = 0$ for $-\infty < x < \infty,\ y = 0$; $\sigma_{yy} = p_y$ for $|\,x\,| \leq a,\ y = 0$; $\sigma_{yy} = 0$ for $|\,x\,| > a,\ y = 0$.

   Answer: $\dfrac{\partial^2\Phi}{\partial y^2} = -\dfrac{p_y}{2\pi\mu}\left(\arctan\dfrac{y}{x-a} - \arctan\dfrac{y}{x+a}\right) = -\dfrac{p_y}{2\pi\mu}(\theta_1 - \theta_2)$

7. Use the result from the previous exercise to find the function $\partial^2\Phi/\partial y^2$ that solves the problem of an isotropic, linearly elastic half-plane $y \leq 0$ subjected to a concentrated force $F_y > 0$ at the origin.

   Answer: $\dfrac{\partial^2\Phi}{\partial y^2} = -\dfrac{F_y}{2\pi\mu}\dfrac{y}{x^2+y^2}$

8. Derive expressions for the stresses σ_{ij} in an isotropic, linearly elastic half-plane $y \leq 0$ with the following boundary loading conditions: $\sigma_{yy} = 0$ for $-\infty < x < \infty,\ y = 0$; $\sigma_{xy} = p_x$ for $|\,x\,| \leq a,\ y = 0$; $\sigma_{xy} = 0$ for $|\,x\,| > a,\ y = 0$.

   Answer: $\dfrac{\partial^2\Psi}{\partial y^2} = -\dfrac{p_x}{2\pi\mu}(\theta_1 - \theta_2)$; $\sigma_{xx} = p_x\left[\ln r_1/r_2 + y^2(1/r_1^2 - 1/r_2^2)\right]/\pi$; $\sigma_{yy} = -p_x y^2(1/r_1^2 - 1/r_2^2)/\pi$; $\sigma_{xy} = -p_x\left[(\theta_1-\theta_2) + y((x-a)/r_1^2 - y(x+a)/r_2^2)\right]/\pi$

9. Given a harmonic function $\Omega(x_1, x_2, x_3)$, verify that $\nabla^2\left(x_i\dfrac{\partial\Omega}{\partial x_j} - x_j\dfrac{\partial\Omega}{\partial x_i}\right) = 0$ for i and $j = 1,\ 2,\ 3$. [This result was originally presented by Rongved (1957) and Rongved and Frasier (1958). Rongved attributes the result to some unpublished work of R. D. Mindlin.]

10. Use the result from the previous exercise to show that the function $x \arctan \dfrac{y}{x} + y \ln \sqrt{x^2 + y^2}$ satisfies Laplace's equation in two dimensions.
11. Verify that the function $\dfrac{\partial^2 \Phi}{\partial y^2} = \dfrac{p_y}{2\pi a \mu}\left(x \arctan \dfrac{y}{x} + y \ln \sqrt{x^2 + y^2}\right)$ solves the problem of an isotropic, linearly elastic half-plane $y \leq 0$ whose surface is subjected to a normal traction $\sigma_{yy} = t_y$ that is equal to p_y at the point $x = -a,\ y = 0$ and increases linearly with distance $-x$ from the origin.
12. Use (4.119) and (4.120) to show that $q_1 q_2 = 1$.
13. Derive expressions for the stresses in an orthotropic (transversely isotropic) linearly elastic half-plane $y \leq 0$ with the following boundary loading conditions: $\sigma_{xy} = 0$ for $-\infty < x < \infty,\ y = 0$; $\sigma_{yy} = p_y$ for $|\ x\ | \leq a,\ y = 0$; $\sigma_{yy} = 0$ for $|\ x\ | > a,\ y = 0$. [Hint: $\partial^2 \Phi(x, y)/\partial x \partial y = \int \partial^3 \Phi(x, y)/\partial x \partial y^2 \, \mathrm{d}y$]

Answer: $$\frac{\partial^2 \Phi}{\partial y^2} = -\frac{p_y}{c_{66}\pi(\gamma_1 - \gamma_2)}\left(\arctan \frac{y}{x-a} - \arctan \frac{y}{x+a}\right)$$

$$\sigma_{xx} = \frac{p_y}{\pi(\gamma_1 - \gamma_2)}\left[\frac{1}{\gamma_1}\left(\arctan \frac{y_1}{x-a} - \arctan \frac{y_1}{x+a}\right) - \frac{1}{\gamma_2}\left(\arctan \frac{y_2}{x-a} - \arctan \frac{y_2}{x+a}\right)\right]$$

$$\sigma_{yy} = \frac{-p_y}{\pi(\gamma_1 - \gamma_2)}\left[\gamma_1\left(\arctan \frac{y_1}{x-a} - \arctan \frac{y_1}{x+a}\right) - \gamma_2\left(\arctan \frac{y_2}{x-a} - \arctan \frac{y_2}{x+a}\right)\right]$$

$$\sigma_{xy} = \frac{p_y}{\pi(\gamma_1 - \gamma_2)}\left[\left(\ln \sqrt{(x-a)^2 + y_1^2} - \ln \sqrt{(x+a)^2 + y_1^2}\right)\right.$$
$$\left. - \left(\ln \sqrt{(x-a)^2 + y_2^2} - \ln \sqrt{(x+a)^2 + y_2^2}\right)\right]$$

14. A condition of anti-plane strain exists when $u_x = u_y = 0$ and $u_z = u_z(x, y)$. For the case of isotropic elasticity and zero body forces show that $\nabla^2 u_z(x, y) = 0$ for anti-plane strain. What is the corresponding equation for orthotropic elasticity?

Answer: $\dfrac{\partial^2}{\partial x^2} u_z(x, y) + \dfrac{c_{55}}{c_{44}} \dfrac{\partial^2}{\partial y^2} u_z(x, y) = 0$ where $c_{44} = \mu_{yz}$ and $c_{55} = \mu_{xz}$

15. Use Hooke's law (4.44) to verify (4.316) for a homogeneous, isotropic, linearly elastic material.
16. Consider a circular boundary $r = r_o$ subjected to tractions $t_i = (t_x, t_y)$. Show that the resultant couple of this stress system is $C = \int_0^{2\pi} r_o^2 (t_y \cos \theta - t_x \sin \theta) \mathrm{d}\theta$ and hence conclude that $C = 0$ for the plane strain version of Kelvin's problem.
17. Derive an expression for the displacement $u_y(x, y)$ due to a double force without moment $\mathcal{F}_y$ in an orthotropic (transversely isotropic) elastic body with directions of elastic symmetry along the x and y axes. Use L'Hopital's rule to show that your result reduces to (4.233) for the case of isotropic elasticity.

Answer: $$u_y(x, y) = \frac{1}{2\pi c_{66}(q_1 - q_2)} \mathcal{F}_y \left[\frac{q_1}{\gamma_1^2} \frac{y_1}{x^2 + y_1^2} - \frac{q_2}{\gamma_2^2} \frac{y_2}{x^2 + y_2^2}\right]$$

18. Derive expressions for the displacements $u_y(x, 0_\pm)$ for a constant displacement discontinuity $D_x,\ D_y$ over the line segment $|\ x\ | \leq a,\ y = 0$.

Answer: $|\,x\,| > a$: $u_y(x, 0_\pm) = -\dfrac{(1-2\nu)}{4\pi(1-\nu)} D_x \ln \left|\, \dfrac{x-a}{x+a} \,\right|$;

$$|\,x\,| < a : u_y(x, 0_\pm) = -\frac{(1-2\nu)}{4\pi(1-\nu)} D_x \ln \left|\, \frac{x-a}{x+a} \,\right| \pm \frac{1}{2} D_y$$

19. For a linearly elastic solid with volume V and surface S, and in the absence of body forces, use the divergence theorem (Sect. 2.1.3) show that $\int_V \sigma_{ij}\varepsilon_{ij}\,\mathrm{d}V = \int_S t_i u_i\,\mathrm{d}S$, where $\sigma_{ij} = \sigma_{ji}$ and $\varepsilon_{ij} = \varepsilon_{ji}$ are the stress and strain tensors, $t_i = \sigma_{ji} n_j$ are the components of traction at a point on the surface with normal n_j, and $\varepsilon_{ij} = \frac{1}{2}(\partial u_i/\partial x_j + \partial u_j/\partial x_i)$.

References

Aki, K., and P.G. Richards. 2002. *Quantitative seismology*, 2nd edn. University Science Books.

Bonnet, M. 1999. *Boundary integral methods for solids and fluids*. Wily.

Elliott, H.A. 1948. Three-dimensional stress distributions in hexagonal aeolotropic crystals. *Proceedings of the Cambridge Philosophical Society* 44: 522–33.

Gaul, L., M. Kögl, and M. Wagner. 2003. *Boundary element methods for engineers and scientists*. Springer.

Gradshteyn, I.S., and I.M. Ryzhik. 1980. *Table of integrals, series, and products*. Academic Press.

Green, A.E. 1943. A note on stress systems in aeolotropic materials. *Philosophical Magazine* 34: 416–18.

Green, A.E., and W. Zerna. 1968. *Theoretical elasticity*, 2nd edn. Dover.

Jeffreys, H. 1963. *Cartesian tensors*. Cambridge University Press.

Kupradze, V.D. 1965. *Potential methods in the theory of elasticity*. (Israeli translation of Russian). D. Davy Co.

Lekhnitskii, S. 1963. *Theory of elasticity of an anisotropic elastic body*. Holden Day.

Lodge, A.S. 1955. The transformation to isotropic form of the equilibrium equations for a class of anisotropic elastic solids. *The Quarterly Journal of Mechanics and Applied Mathematics* 8 (2): 211–25.

Love, A.E.H. 1944. *A treatise on the mathematical theory of elasticity*, 4th edn. Dover.

Mindlin, R.D., and D.H. Cheng. 1950. Nuclei of strain in the semi-infinite solid. *Journal of Applied Physics* 21: 926–30.

Muskhelishvili, N.I. 1963. *Some basic problems of the mathematical theory of elasticity*, 2nd (English) edn. P. Noordhoff.

Petit Bois, G. 1961. *Tables of indefinite integrals*. Dover.

Rongved, L. 1957. Dislocation over a bounded area in an infinite solid. *Journal of Applied Mechanics* 24: 253–4.

Rongved, L., and J.T. Frasier. 1958. Displacement discontinuity in the elastic half-space. *Journal of Applied Mechanics* 25: 125–8.

Shield, R.T. 1951. Notes on problems in hexagonal aeolotropic materials. *Proceedings of the Cambridge Philosophical Society* 47: 401–9.

Sokolnikoff, I.S. 1956. *Mathematical theory of elasticity*, 2nd edn. McGraw-Hill.

Steketee, J.A. 1958a. On Volterra's dislocations in a semi-infinite elastic medium. *Canadian Journal of Physics* 36: 192–205.

Steketee, J.A. 1958b. Some geophysical applications of the theory of dislocations. *Canadian Journal of Physics* 36: 1168–98.

Timoshenko, S.P., and J.N. Goodier. 1970. *Theory of elasticity*, 3rd edn. McGraw-Hill.

Chapter 5
The Stress Discontinuity Method

5.1 Introduction

We saw in Chaps. 1 and 3 that several different boundary element methods can be developed for potential problems, all starting with the fundamental singular solution to Laplace's equation. These methods can be categorized as either indirect or direct, depending on the form of the underlying boundary integral equation. Indirect methods make use of the solutions for either single or double layer potentials, while the direct method, based on Green's representation formula, makes use of both. The distinction between the different approaches is clear-cut: an indirect method determines a numerical solution to a boundary value problem in terms of quantities that in general do not have physical meaning, while a direct method involves only the physical parameters used to define the original problem, e.g., the potential and its normal derivative at the boundary.

A similar situation occurs for boundary element methods for elasticity, although the computations are necessarily more involved because the boundary parameters—the displacements and tractions— are vector quantities. In this chapter we develop an indirect method for two-dimensional elasticity problems that can be regarded as the counterpart to the boundary element method described in Sect. 1.4 for potential problems.

The simplest version of the method is based on the solution given in Sect. 4.12 for a point force (Kelvin's problem), integrated along a straight line segment so as to produce a line source with constant loading. Unlike the line source in Sect. 1.4, this source has two components, one parallel and the other perpendicular to the line segment: the components $P_{\bar{x}}$ and $P_{\bar{y}}$ represent constant resultant tractions $t_{\bar{x}}$ and $t_{\bar{y}}$ along the line segment $|\bar{x}| \leq a$, $\bar{y} = 0$. As noted in Sect. 4.12, however, $P_{\bar{x}}$ and $P_{\bar{y}}$ also represent discontinuities in the stress tensor components $\sigma_{\bar{x}\bar{y}}$ and $\sigma_{\bar{y}\bar{y}}$ along the segment for $\bar{y} = 0_+$ and $\bar{y} = 0_-$, that is, $\sigma_{\bar{x}\bar{y}}(x, 0_+) - \sigma_{\bar{x}\bar{y}}(x, 0_-) = -P_{\bar{x}}$ and $\sigma_{\bar{y}\bar{y}}(x, 0_+) - \sigma_{\bar{y}\bar{y}}(x, 0_-) = -P_{\bar{y}}$ (see (4.171), (4.185), and (4.186)).

S. L. Crouch and S. G. Mogilevskaya, *A First Course in Boundary Element Methods*, Mathematical Engineering, https://doi.org/10.1007/978-3-031-63341-6_5

In what follows we will refer to $P_{\bar{x}}$ and $P_{\bar{y}}$ as stress discontinuities and reserve the word "tractions" to define the loading conditions for traction boundary value problems. For this reason, the boundary element method discussed below will be referred to as the *stress discontinuity method*.[1]

In this chapter we will develop the numerical method for piecewise constant line sources (stress discontinuities) for both isotropic and anisotropic elastic materials. We will also show how the method can be extended in a straightforward manner to use piecewise quadratic variations of the line sources. In both cases, we will follow an intuitive, physical approach that uses superposition of the basic solutions to obtain a system of simultaneous linear algebraic equations from which to find the stress discontinuity components. We conclude the discussion by showing that the numerical approach is equivalent to a numerical solution of the equations given in Sect. 4.18 for the displacements, stresses, and tractions for the point force analogue of the single layer potential, Eq. (4.299).

5.2 Numerical Procedure

The results of Sect. 4.12 form the basis of an indirect boundary element method for elasticity. The physical considerations underlying this method are discussed below with reference to a particular plane strain problem for a cavity in an infinite region. (We will see later that the method also applies to boundary value problems for finite bodies.) For purposes of illustration we will suppose that the wall of the cavity is everywhere subjected to a uniform normal traction $\sigma_n = -p$ (i.e., compression) with the shear traction σ_s equal to zero. We wish to determine the displacements and stresses in the body due to this boundary loading.

To solve this problem numerically we proceed as follows. First, as was done in Sect. 1.4, we approximate the boundary S of the cavity by N straight line segments joined end to end. The length of a typical boundary element i is denoted as $2\overset{i}{a}$ in Fig. 5.1a. If these elements are small they will approximate the boundary quite closely and we may consider that each element is subjected to a normal stress $\sigma_n = -p$ along its entire length, but is free from shear stress. The boundary conditions for this *approximation* to the original problem then become

$$\overset{i}{\sigma}_n = -p, \qquad \overset{i}{\sigma}_s = 0 \tag{5.1}$$

for $i = 1$ to N.

[1] Crouch and Starfield (1983) called this method the "fictitious stress method" to emphasize that the loads applied to the boundary elements are not physically meaningful. The present terminology—the stress discontinuity method—is intended to emphasize the structural similarity of the approach to another boundary element method called the displacement discontinuity method, which is discussed in the next chapter.

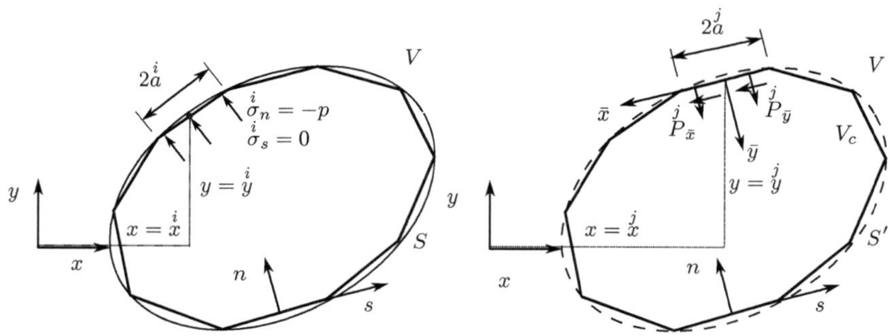

(a) Physical problem (b) Numerical model

Fig. 5.1 Illustration of the stress discontinuity method

The local coordinates s and n are respectively tangent and perpendicular to the boundary elements along curve S; they therefore vary from element to element along the boundary. These coordinates are selected such that n points in the direction of the local outward normal to the boundary and s points in the direction in which the boundary is traversed, in this case counterclockwise. It should be emphasized that each boundary element has its own local coordinate system so that, for example, the s, n coordinates for element i are different from those for element j. With this understanding, these coordinates can also be designated as $\bar{x}$ (for coordinate s) and $\bar{y}$ (for coordinate n), a choice we will make for notational convenience.

The problem specified by (5.1) can be solved numerically by means of the model depicted in Fig. 5.1b. The dashed curve S' in the figure has the same shape as the curve S used to define the boundary of the cavity in Fig. 5.1a and is also divided into N boundary elements. Curve S' encloses a complementary region V_c such that $V + V_c$ is infinite; our solution for a line source is valid in this region. We now imagine that constant stress discontinuities $P_{\bar{x}}$ and $P_{\bar{y}}$ are applied to each of the line segments approximating the dashed curve. For simplicity, only the discontinuities $\overset{j}{P}_{\bar{x}}$ and $\overset{j}{P}_{\bar{y}}$ applied to the jth segment are shown in Fig. 5.1b.

Using either of the solutions given in Sect. 4.12, together with suitable coordinate transformation equations to account for the locations and orientations of all the boundary elements, we can compute the tractions $\overset{i}{t}_{\bar{x}}$ and $\overset{i}{t}_{\bar{y}}$ at the midpoint of each element due to the N applied stress discontinuities. The results can be written in the form

$$
\begin{aligned}
\overset{i}{t}_{\bar{x}} &= \sum_{j=1}^{N} \overset{ij}{A}_{\bar{x}\,\bar{x}} \overset{j}{P}_{\bar{x}} + \sum_{j=1}^{N} \overset{ij}{A}_{\bar{x}\,\bar{y}} \overset{j}{P}_{\bar{y}} \\
\overset{i}{t}_{\bar{y}} &= \sum_{j=1}^{N} \overset{ij}{A}_{\bar{y}\,\bar{x}} \overset{j}{P}_{\bar{x}} + \sum_{j=1}^{N} \overset{ij}{A}_{\bar{y}\,\bar{y}} \overset{j}{P}_{\bar{y}}
\end{aligned}
\tag{5.2}
$$

for $i = 1$ to N. The quantities $\overset{ij}{A}_{\bar{x}\bar{x}}$, etc., in these equations are the boundary traction influence coefficients for the problem. The coefficient $\overset{ij}{A}_{\bar{x}\bar{y}}$, for example, gives the shear traction $\overset{i}{t}_{\bar{x}}$ at the midpoint of the ith element due to a constant unit normal stress discontinuity $\overset{j}{P}_{\bar{y}} = 1$ applied to the jth element.

Now, if we are able to find the stress discontinuities $\overset{j}{P}_{\bar{x}}$ and $\overset{j}{P}_{\bar{y}}$ for $j = 1$ to N such that the tractions $\overset{i}{t}_{\bar{x}} = \overset{i}{\sigma}_s$ and $\overset{i}{t}_{\bar{y}} = \overset{i}{\sigma}_n$ in (5.2) have the values specified by (5.1), we will have solved, approximately, the physical problem depicted in Fig. 5.1a. We therefore require that

$$
\begin{aligned}
\sum_{j=1}^{N} \overset{ij}{A}_{\bar{x}\bar{x}} \overset{j}{P}_{\bar{x}} + \sum_{j=1}^{N} \overset{ij}{A}_{\bar{x}\bar{y}} \overset{j}{P}_{\bar{y}} &= 0 \\
\sum_{j=1}^{N} \overset{ij}{A}_{\bar{y}\bar{x}} \overset{j}{P}_{\bar{x}} + \sum_{j=1}^{N} \overset{ij}{A}_{\bar{y}\bar{y}} \overset{j}{P}_{\bar{y}} &= -p
\end{aligned}
\tag{5.3}
$$

for $i = 1$ to N, which is a system of $2N$ simultaneous linear algebraic equations in $2N$ unknowns. These equations can be solved by standard numerical methods.

It is important to recognize that the stress discontinuities $\overset{j}{P}_{\bar{x}}$ and $\overset{j}{P}_{\bar{y}}$ in the above equations are not in general physically meaningful[2]; they were introduced to facilitate a numerical solution to a particular problem and have no physical significance with respect to this problem. The linear combinations of these quantities specified by (5.2) do, however, have physical meaning for the problem in question, and we have taken advantage of this fact to establish the system of algebraic equations (5.3). Once these equations have been solved, we can express the displacements and stresses at any point in the body in terms of the stress discontinuities $\overset{j}{P}_{\bar{x}}$ and $\overset{j}{P}_{\bar{y}}$.

Further details of this method and some examples of its use are discussed in the remainder of this chapter.

5.3 Coordinate Transformations

The stress discontinuity method is based on the analytical solution to the problem of constant loading of an arbitrarily oriented, finite line segment in an infinite elastic solid. We will consider this problem below for the $\bar{x}$, $\bar{y}$ coordinate system shown in Fig. 5.2. The line segment is specified by the conditions $|\ \bar{x}\ | \leq a$, $\bar{y} = 0$, and the loads applied to this segment are $P_{\bar{x}}$ and $P_{\bar{y}}$, as indicated in the figure.

---

[2] But see Sect. 5.11.1 in this connection.

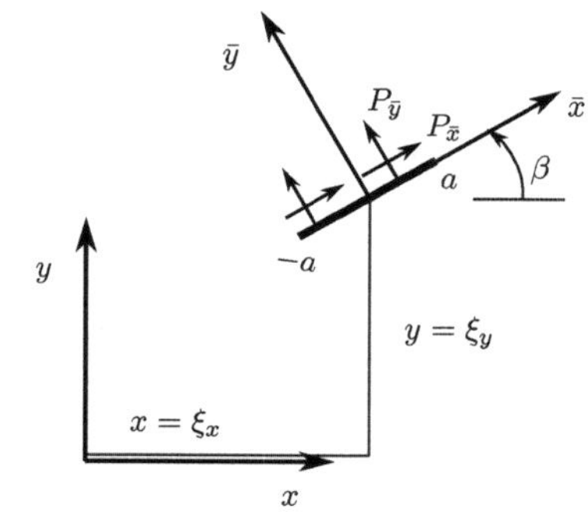

Fig. 5.2 Constant loading of an arbitrarily oriented line segment

The local coordinate system in the Fig. 5.2 is obtained by a translation and rotation of the global x, y system, just as in the case of the local coordinate system in Fig. 1.4. The components of translation are ξ_x in the x direction and ξ_y in the y direction and the rotation is defined by the angle β, positive in the counterclockwise sense as shown. This coordinate transformation is given by (1.29), repeated here for easy reference:

$$\begin{aligned}\bar{x} &= (x - \xi_x)\cos\beta + (y - \xi_y)\sin\beta \\ \bar{y} &= -(x - \xi_x)\sin\beta + (y - \xi_y)\cos\beta\end{aligned} \tag{5.4}$$

We next record the expressions for the displacements and stresses in the global coordinate system for both isotropic and orthotropic (transversely isotropic) materials.

5.3.1 Isotropic Material

The displacements and stresses in the $\bar{x}$, $\bar{y}$ coordinate system due to constant loads $P_{\bar{x}}$ and $P_{\bar{y}}$ on the line segment $|\,\bar{x}\,| \le a$, $\bar{y} = 0$ in an isotropic material can be written by inspection of (4.165) and (4.166) because there is no preferred direction for an infinite, homogeneous, isotropic elastic body. Thus, we replace all occurrences of x and y in these equations by $\bar{x}$ and $\bar{y}$ and obtain the following results:

$$\begin{aligned}u_{\bar{x}} &= \frac{P_{\bar{x}}}{2\mu}\Big[(3-4\nu)f + \bar{y}f_{,\bar{y}}\Big] + \frac{P_{\bar{y}}}{2\mu}\big[-\bar{y}f_{,\bar{x}}\big] \\ u_{\bar{y}} &= \frac{P_{\bar{x}}}{2\mu}\big[-\bar{y}f_{,\bar{x}}\big] + \frac{P_{\bar{y}}}{2\mu}\big[(3-4\nu)f - \bar{y}f_{,\bar{y}}\big]\end{aligned} \tag{5.5}$$

and

$$\begin{aligned}
\sigma_{\bar{x}\bar{x}} &= P_{\bar{x}}\big[(3-2\nu)f_{,\bar{x}} + \bar{y}f_{,\bar{x}\bar{y}}\big] + P_{\bar{y}}\big[2\nu f_{,\bar{y}} + \bar{y}f_{,\bar{y}\bar{y}}\big] \\
\sigma_{\bar{y}\bar{y}} &= P_{\bar{x}}\big[-(1-2\nu)f_{,\bar{x}} - \bar{y}f_{,\bar{x}\bar{y}}\big] + P_{\bar{y}}\big[2(1-\nu)f_{,\bar{y}} - \bar{y}f_{,\bar{y}\bar{y}}\big] \\
\sigma_{\bar{x}\bar{y}} &= P_{\bar{x}}\big[2(1-\nu)f_{,\bar{y}} + \bar{y}f_{,\bar{y}\bar{y}}\big] + P_{\bar{y}}\big[(1-2\nu)f_{,\bar{x}} - \bar{y}f_{,\bar{x}\bar{y}}\big]
\end{aligned} \tag{5.6}$$

The function $f(\bar{x},\bar{y})$ in these equations is

$$f(\bar{x},\bar{y}) = F_1(\bar{x},\bar{y}) = \frac{-1}{4\pi(1-\nu)}\big[\bar{y}(\bar{\theta}_1 - \bar{\theta}_2) - (\bar{x}-a)\ln\bar{r}_1 + (\bar{x}+a)\ln\bar{r}_2 - 2a\big] \tag{5.7}$$

and its derivatives through second order are (cf. (4.167))

$$\begin{aligned}
f_{,\bar{x}} &= F_2(\bar{x},\bar{y}) = \frac{1}{4\pi(1-\nu)}\ln\frac{\bar{r}_1}{\bar{r}_2} \\
f_{,\bar{y}} &= F_3(\bar{x},\bar{y}) = \frac{-1}{4\pi(1-\nu)}(\bar{\theta}_1 - \bar{\theta}_2) \\
f_{,\bar{x}\bar{y}} &= F_4(\bar{x},\bar{y}) = \frac{1}{4\pi(1-\nu)}\left(\frac{\bar{y}}{\bar{r}_1^2} - \frac{\bar{y}}{\bar{r}_2^2}\right) \\
f_{\bar{x}\bar{x}} &= -f_{,\bar{y}\bar{y}} = F_5(\bar{x},\bar{y}) = \frac{1}{4\pi(1-\nu)}\left(\frac{\bar{x}-a}{\bar{r}_1^2} - \frac{\bar{x}+a}{\bar{r}_2^2}\right)
\end{aligned} \tag{5.8}$$

in which $\bar{r}_1$, $\bar{r}_2$ and $\bar{\theta}_1$, $\bar{\theta}_2$ are the distances and angles from the endpoints of the line segment to an arbitrary point in the $\bar{x}$, $\bar{y}$ coordinate system, as depicted in Fig. 1.4.

Functions $F_1(\bar{x},\bar{y})$ through $F_5(\bar{x},\bar{y})$ are introduced in (5.7) and (5.8) for notational convenience. Letting $\bar{F}_k$ stand for $F_k(\bar{x},\bar{y})$, $k = 1$ to 5, we rewrite (5.5) and (5.6) as

$$\begin{aligned}
u_{\bar{x}} &= \frac{P_{\bar{x}}}{2\mu}\Big[(3-4\nu)\bar{F}_1 + \bar{y}\bar{F}_3\Big] + \frac{P_{\bar{y}}}{2\mu}\big[-\bar{y}\bar{F}_2\big] \\
u_{\bar{y}} &= \frac{P_{\bar{x}}}{2\mu}\big[-\bar{y}\bar{F}_2\big] + \frac{P_{\bar{y}}}{2\mu}\big[(3-4\nu)\bar{F}_1 - \bar{y}\bar{F}_3\big]
\end{aligned} \tag{5.9}$$

and

$$\begin{aligned}
\sigma_{\bar{x}\bar{x}} &= P_{\bar{x}}\big[(3-2\nu)\bar{F}_2 + \bar{y}\bar{F}_4\big] + P_{\bar{y}}\big[2\nu\bar{F}_3 - \bar{y}\bar{F}_5\big] \\
\sigma_{\bar{y}\bar{y}} &= P_{\bar{x}}\big[-(1-2\nu)\bar{F}_2 - \bar{y}\bar{F}_4\big] + P_{\bar{y}}\big[2(1-\nu)\bar{F}_3 + \bar{y}\bar{F}_5\big] \\
\sigma_{\bar{x}\bar{y}} &= P_{\bar{x}}\big[2(1-\nu)\bar{F}_3 - \bar{y}\bar{F}_5\big] + P_{\bar{y}}\big[(1-2\nu)\bar{F}_2 - \bar{y}\bar{F}_4\big]
\end{aligned} \tag{5.10}$$

The displacements and stresses at an arbitrary point (x, y) can be found by computing the $\bar{x}$, $\bar{y}$ coordinates of this point from (5.4) and then using (5.7)–(5.10). These displacements and stresses relate to a particular local coordinate system and are therefore not in a convenient form for general computations. We can refer the results to the global x, y system, however, by using the transformation formulas given in Sect. 4.7. Substitution of (5.9) into (4.62) gives

$$
\begin{aligned}
u_x &= \frac{P_{\bar{x}}}{2\mu}\Big[(3-4\nu)\cos\beta\bar{F}_1 + \bar{y}(\sin\beta\bar{F}_2 + \cos\beta\bar{F}_3)\Big] \\
&\quad \frac{P_{\bar{y}}}{2\mu}\big[-(3-4\nu)\sin\beta\bar{F}_1 - \bar{y}(\cos\beta\bar{F}_2 - \sin\beta\bar{F}_3)\big] \\
u_y &= \frac{P_{\bar{x}}}{2\mu}\Big[(3-4\nu)\sin\beta\bar{F}_1 - \bar{y}(\cos\beta\bar{F}_2 - \sin\beta\bar{F}_3)\Big] \\
&\quad + \frac{P_{\bar{y}}}{2\mu}\big[(3-4\nu)\cos\beta\bar{F}_1 - \bar{y}(\sin\beta\bar{F}_2 + \cos\beta\bar{F}_3)\big]
\end{aligned} \tag{5.11}
$$

and substitution of (5.10) into (4.66) gives

$$
\begin{aligned}
\sigma_{xx} &= P_{\bar{x}}\big[\bar{F}_2 + 2(1-\nu)(\cos 2\beta\bar{F}_2 - \sin 2\beta\bar{F}_3) + \bar{y}(\cos 2\beta\bar{F}_4 + \sin 2\beta\bar{F}_5)\big] \\
&\quad P_{\bar{y}}\big[\bar{F}_3 - (1-2\nu)(\sin 2\beta\bar{F}_2 + \cos 2\beta\bar{F}_3) + \bar{y}(\sin 2\beta\bar{F}_4 - \cos 2\beta\bar{F}_5)\big] \\
\sigma_{yy} &= P_{\bar{x}}\big[\bar{F}_2 - 2(1-\nu)(\cos 2\beta\bar{F}_2 - \sin 2\beta\bar{F}_3) - \bar{y}(\cos 2\beta\bar{F}_4 + \sin 2\beta\bar{F}_5)\big] \\
&\quad P_{\bar{y}}\big[\bar{F}_3 + (1-2\nu)(\sin 2\beta\bar{F}_2 + \cos 2\beta\bar{F}_3) - \bar{y}(\sin 2\beta\bar{F}_4 - \cos 2\beta\bar{F}_5)\big] \\
\sigma_{xy} &= P_{\bar{x}}\big[2(1-\nu)(\sin 2\beta\bar{F}_2 + \cos 2\beta\bar{F}_3) + \bar{y}(\sin 2\beta\bar{F}_4 - \cos 2\beta\bar{F}_5)\big] \\
&\quad P_{\bar{y}}\big[(1-2\nu)(\cos 2\beta\bar{F}_2 - \sin 2\beta\bar{F}_3) - \bar{y}(\cos 2\beta\bar{F}_4 + \sin 2\beta\bar{F}_5)\big]
\end{aligned} \tag{5.12}
$$

5.3.2 *Orthotropic Material*

The counterparts of Eqs. (5.11) and (5.12) for anisotropic elasticity are obtained by substituting (4.177) into (4.182) and (4.183) to obtain

$$
\begin{aligned}
u_x &= \frac{1}{2\pi c_{66}(q_1-q_2)}P_{\bar{x}}\Big[\Big\{\frac{\gamma_1}{q_1}I_1(\bar{x},\bar{y},\gamma_1)\cos\beta - I_2(\bar{x},\bar{y},\gamma_1)\sin\beta\Big\} \\
&\qquad - \Big\{\frac{\gamma_2}{q_2}I_1(\bar{x},\bar{y},\gamma_2)\cos\beta - I_2(\bar{x},\bar{y},\gamma_2)\sin\beta\Big\}\Big] \\
&\quad + \frac{1}{2\pi c_{66}(q_1-q_2)}P_{\bar{y}}\Big[-\Big\{\frac{\gamma_1}{q_1}I_1(\bar{x},\bar{y},\gamma_1)\sin\beta + I_2(\bar{x},\bar{y},\gamma_1)\cos\beta\Big\} \\
&\qquad + \Big\{\frac{\gamma_2}{q_2}I_1(\bar{x},\bar{y},\gamma_2)\sin\beta + I_2(\bar{x},\bar{y},\gamma_2)\cos\beta\Big\}\Big] \\
u_y &= \frac{1}{2\pi c_{66}(q_1-q_2)}P_{\bar{x}}\Big[-\Big\{I_2(\bar{x},\bar{y},\gamma_1)\cos\beta + \frac{q_1}{\gamma_1}I_1(\bar{x},\bar{y},\gamma_1)\sin\beta\Big\} \\
&\qquad + \Big\{I_2(\bar{x},\bar{y},\gamma_2)\cos\beta + \frac{q_2}{\gamma_2}I_1(\bar{x},\bar{y},\gamma_2)\sin\beta\Big\}\Big] \\
&\quad + \frac{1}{2\pi c_{66}(q_1-q_2)}P_{\bar{y}}\Big[\Big\{I_2(\bar{x},\bar{y},\gamma_1)\sin\beta - \frac{q_1}{\gamma_1}I_1(\bar{x},\bar{y},\gamma_1)\cos\beta\Big\} \\
&\qquad - \Big\{I_2(\bar{x},\bar{y},\gamma_2)\sin\beta - \frac{q_2}{\gamma_2}I_1(\bar{x},\bar{y},\gamma_2)\cos\beta\Big\}\Big]
\end{aligned} \tag{5.13}
$$

and

$$
\begin{aligned}
\sigma_{xx} &= \frac{1}{2\pi(q_1-q_2)} P_{\bar{x}} \Big[\frac{1+q_1}{\gamma_1} \Big\{ \frac{1}{q_1} I_3(\bar{x}, \bar{y}, \gamma_1) \cos\beta + \frac{1}{\gamma_1} I_4(\bar{x}, \bar{y}, \gamma_1) \sin\beta \Big\} \\
&\quad - \frac{1+q_2}{\gamma_2} \Big\{ \frac{1}{q_2} I_3(\bar{x}, \bar{y}, \gamma_2) \cos\beta + \frac{1}{\gamma_2} I_4(\bar{x}, \bar{y}, \gamma_2) \sin\beta \Big\} \Big] \\
&\quad + \frac{1}{2\pi(q_1-q_2)} P_{\bar{y}} \Big[-\frac{1+q_1}{\gamma_1} \Big\{ \frac{1}{q_1} I_3(\bar{x}, \bar{y}, \gamma_1) \sin\beta - \frac{1}{\gamma_1} I_4(\bar{x}, \bar{y}, \gamma_1) \cos\beta \Big\} \\
&\quad + \frac{1+q_2}{\gamma_2} \Big\{ \frac{1}{q_2} I_3(\bar{x}, \bar{y}, \gamma_2) \sin\beta - \frac{1}{\gamma_2} I_4(\bar{x}, \bar{y}, \gamma_2) \cos\beta \Big\} \Big] \\
\sigma_{yy} &= \frac{1}{2\pi(q_1-q_2)} P_{\bar{x}} \Big[-(1+q_1) \Big\{ \frac{\gamma_1}{q_1} I_3(\bar{x}, \bar{y}, \gamma_1) \cos\beta + I_4(\bar{x}, \bar{y}, \gamma_1) \sin\beta \Big\} \\
&\quad + (1+q_2) \Big\{ \frac{\gamma_2}{q_2} I_3(\bar{x}, \bar{y}, \gamma_2) \cos\beta + I_4(\bar{x}, \bar{y}, \gamma_2) \sin\beta \Big\} \Big] \\
&\quad + \frac{1}{2\pi(q_1-q_2)} P_{\bar{y}} \Big[(1+q_1) \Big\{ \frac{\gamma_1}{q_1} I_3(\bar{x}, \bar{y}, \gamma_1) \sin\beta - I_4(\bar{x}, \bar{y}, \gamma_1) \cos\beta \Big\} \\
&\quad - (1+q_2) \Big\{ \frac{\gamma_2}{q_2} I_3(\bar{x}, \bar{y}, \gamma_2) \sin\beta - I_4(\bar{x}, \bar{y}, \gamma_2) \cos\beta \Big\} \Big] \\
\sigma_{xy} &= \frac{1}{2\pi(q_1-q_2)} P_{\bar{x}} \Big[(1+q_1) \Big\{ \frac{1}{q_1} I_4(\bar{x}, \bar{y}, \gamma_1) \cos\beta - \frac{1}{\gamma_1} I_3(\bar{x}, \bar{y}, \gamma_1) \sin\beta \Big\} \\
&\quad - (1+q_2) \Big\{ \frac{1}{q_2} I_4(\bar{x}, \bar{y}, \gamma_2) \cos\beta - \frac{1}{\gamma_2} I_3(\bar{x}, \bar{y}, \gamma_2) \sin\beta \Big\} \Big] \\
&\quad + \frac{1}{2\pi(q_1-q_2)} P_{\bar{y}} \Big[-(1+q_1) \Big\{ \frac{1}{q_1} I_4(\bar{x}, \bar{y}, \gamma_1) \sin\beta + \frac{1}{\gamma_1} I_3(\bar{x}, \bar{y}, \gamma_1) \cos\beta \Big\} \\
&\quad + (1+q_2) \Big\{ \frac{1}{q_2} I_4(\bar{x}, \bar{y}, \gamma_2) \sin\beta + \frac{1}{\gamma_2} I_3(\bar{x}, \bar{y}, \gamma_2) \cos\beta \Big\} \Big]
\end{aligned}
\tag{5.14}
$$

The preceding equations can now be used to compute general influence coefficients for the stress discontinuity method for both isotropic and orthotropic elastic materials. For a point $(x,\ y)$ that lies along the boundary (that is, at the midpoint of a boundary element), we obtain expressions for the boundary influence coefficients introduced in Sect. 5.2; for any other point we obtain the effects of the applied loads (stress discontinuities) $P_{\bar{x}}$ and $P_{\bar{y}}$ on the displacements and stresses at that point. The resultant values of these quantities are obtained by summing the effects of all the boundary elements.

5.4 Computation of the Boundary Influence Coefficients

It follows from the discussion in Sect. 5.2 that the boundary influence coefficients for the tractions relate the stress discontinuities $\overset{j}{P}_{\bar{x}}$ and $\overset{j}{P}_{\bar{y}}$ at the jth boundary element (the "sending" element) to the tractions $\overset{i}{t}_{\bar{x}}$ and $\overset{i}{t}_{\bar{y}}$ they produce at the ith element (the "receiving" element). Equations similar to (5.2) can be written for the boundary displacements, i.e.,

$$\begin{aligned}
\overset{i}{u}_{\bar{x}} &= \sum_{j=1}^{N} \overset{ij}{B}_{\bar{x}\bar{x}} \overset{j}{P}_{\bar{x}} + \sum_{j=1}^{N} \overset{ij}{B}_{\bar{x}\bar{y}} \overset{j}{P}_{\bar{y}} \\
\overset{i}{u}_{\bar{y}} &= \sum_{j=1}^{N} \overset{ij}{B}_{\bar{y}\bar{x}} \overset{j}{P}_{\bar{x}} + \sum_{j=1}^{N} \overset{ij}{B}_{\bar{y}\bar{y}} \overset{j}{P}_{\bar{y}}
\end{aligned} \tag{5.15}$$

The boundary influence coefficients for the tractions ($\overset{ij}{A}_{\bar{x}\bar{x}}$, etc.) and the displacements ($\overset{ij}{B}_{\bar{x}\bar{x}}$, etc.) can be calculated from the results of the previous section by using standard coordinate transformation formulas. The required expressions are given below.

Referring to Fig. 5.3, the local coordinates for element j are found from (5.4) by setting $\xi_x = \overset{j}{x}$, $\xi_y = \overset{j}{y}$, and $\beta = \overset{j}{\beta}$, i.e.,

$$\begin{aligned}
\bar{x} &= (x - \overset{j}{x}) \cos \overset{j}{\beta} + (y - \overset{j}{y}) \sin \overset{j}{\beta} \\
\bar{y} &= -(x - \overset{j}{x}) \sin \overset{j}{\beta} + (y - \overset{j}{y}) \cos \overset{j}{\beta}
\end{aligned} \tag{5.16}$$

Using these coordinates in the results of the previous section allows us to calculate the effects of element j on the displacements u_x and u_y and stresses σ_{xx}, σ_{yy}, and σ_{xy} anywhere in the infinite region. When point (x, y) is at the midpoint of element i Eqs. (5.16) become

$$\begin{aligned}
\bar{x} &= (\overset{i}{x} - \overset{j}{x}) \cos \overset{j}{\beta} + (\overset{i}{y} - \overset{j}{y}) \sin \overset{j}{\beta} \\
\bar{y} &= -(\overset{i}{x} - \overset{j}{x}) \sin \overset{j}{\beta} + (\overset{i}{y} - \overset{j}{y}) \cos \overset{j}{\beta}
\end{aligned} \tag{5.17}$$

which define the local coordinates of point i relative to point j in Fig. 5.3.

We again emphasize that the local $\bar{x}$, $\bar{y}$ coordinate systems are different for *all* of the boundary elements, whether or not their inclinations $\overset{j}{\beta}$ and $\overset{i}{\beta}$ are different because, as shown in the figure, the elements have different midpoint coordinates $(\overset{j}{x}, \overset{j}{y})$ and $(\overset{i}{x}, \overset{i}{y})$. Also, while in principle the elements may all have different lengths, it is found

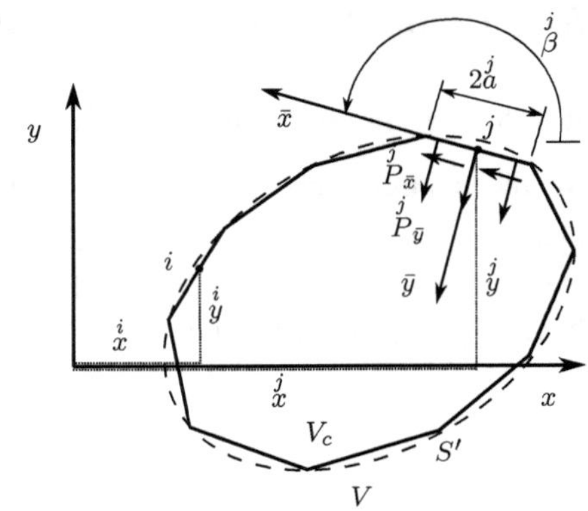

Fig. 5.3 Positions and orientations of boundary elements

in practice that the most accurate numerical results are obtained when the element sizes are nearly all the same. This requirement can be relaxed for stress discontinuities that are allowed to vary linearly or quadratically, say, over the elements. We will consider this possibility in Sect. 5.10.

5.4.1 Traction Coefficients

The tractions $\overset{i}{t}_x$ and $\overset{i}{t}_y$ at point i are found using (4.5):

$$
\begin{aligned}
\overset{i}{t}_x &= \sigma_{xx}(\overset{j}{P}_{\bar{x}}, \overset{j}{P}_{\bar{y}}) \cos \overset{i}{\alpha} + \sigma_{xy}(\overset{j}{P}_{\bar{x}}, \overset{j}{P}_{\bar{y}}) \sin \overset{i}{\alpha} \\
\overset{i}{t}_y &= \sigma_{xy}(\overset{j}{P}_{\bar{x}}, \overset{j}{P}_{\bar{y}}) \cos \overset{i}{\alpha} + \sigma_{yy}(\overset{j}{P}_{\bar{x}}, \overset{j}{P}_{\bar{y}}) \sin \overset{i}{\alpha}
\end{aligned} \tag{5.18}
$$

where $\overset{i}{\alpha}$ is the inclination of the normal to element i and where we have written $\sigma_{xx}(\overset{j}{P}_{\bar{x}}, \overset{j}{P}_{\bar{y}})$, etc., to emphasize that these stress tensor components depend on the stress discontinuities for element j. The components of traction at element i in terms of the local coordinates for the element are found by using transformation formulas (4.8) to write

$$
\begin{aligned}
\overset{i}{t}_{\bar{x}} &= \overset{i}{t}_x \sin \overset{i}{\alpha} - \overset{i}{t}_y \cos \overset{i}{\alpha} \\
\overset{i}{t}_{\bar{y}} &= \overset{i}{t}_x \cos \overset{i}{\alpha} + \overset{i}{t}_y \sin \overset{i}{\alpha}
\end{aligned} \tag{5.19}
$$

Finally, noting that $\overset{i}{\alpha} = \overset{i}{\beta} + \pi/2$ and combining (5.18) and (5.19), we obtain the results

$$\overset{i}{t}_{\bar{x}} = \left[\sigma_{yy}(\overset{j}{P}_{\bar{x}}, \overset{j}{P}_{\bar{y}}) - \sigma_{xx}(\overset{j}{P}_{\bar{x}}, \overset{j}{P}_{\bar{y}})\right] \sin \overset{i}{\beta} \cos \overset{i}{\beta} + \sigma_{xy}(\overset{j}{P}_{\bar{x}}, \overset{j}{P}_{\bar{y}}) \cos 2\overset{i}{\beta}$$

$$\overset{i}{t}_{\bar{y}} = \sigma_{xx}(\overset{j}{P}_{\bar{x}}, \overset{j}{P}_{\bar{y}}) \sin^2 \overset{i}{\beta} - \sigma_{xy}(\overset{j}{P}_{\bar{x}}, \overset{j}{P}_{\bar{y}}) \sin 2\,\overset{i}{\beta} + \sigma_{yy}(\overset{j}{P}_{\bar{x}}, \overset{j}{P}_{\bar{y}}) \cos^2 \overset{i}{\beta} \tag{5.20}$$

The boundary influence coefficients for the tractions—$\overset{ij}{A}_{\bar{x}\bar{x}}$, etc., in (5.2)—can be calculated directly from these equations by using the appropriate expressions for the stress tensor components from either (5.12) or (5.14). The coefficients for isotropic elasticity, for example, are

$$\begin{aligned}
\overset{ij}{A}_{\bar{x}\bar{x}} &= -2(1-\nu)\big(\bar{F}_2 \sin 2\gamma - \bar{F}_3 \cos 2\gamma\big) - \bar{y}\big(\bar{F}_4 \sin 2\gamma + \bar{F}_5 \cos 2\gamma\big) \\
\overset{ij}{A}_{\bar{x}\bar{y}} &= (1-2\nu)\big(\bar{F}_2 \cos 2\gamma + \bar{F}_3 \sin 2\gamma\big) - \bar{y}\big(\bar{F}_4 \cos 2\gamma - \bar{F}_5 \sin 2\gamma\big) \\
\overset{ij}{A}_{\bar{y}\bar{x}} &= \bar{F}_2 - 2(1-\nu)\big(\bar{F}_2 \cos 2\gamma + \bar{F}_3 \sin 2\gamma\big) - \bar{y}\big(\bar{F}_4 \cos 2\gamma - \bar{F}_5 \sin 2\gamma\big) \\
\overset{ij}{A}_{\bar{y}\bar{y}} &= \bar{F}_3 - (1-2\nu)\big(\bar{F}_2 \sin 2\gamma - \bar{F}_3 \cos 2\gamma\big) + \bar{y}\big(\bar{F}_4 \sin 2\gamma + \bar{F}_5 \cos 2\gamma\big)
\end{aligned} \tag{5.21}$$

in which γ is the difference between the inclinations of elements i and j:

$$\gamma = \overset{i}{\beta} - \overset{j}{\beta} \tag{5.22}$$

5.4.2 *Displacement Coefficients*

Equations comparable to (5.20) for the displacements at point i can be written using (4.63). They are

$$\begin{aligned}
\overset{i}{u}_{\bar{x}} &= u_x(\overset{j}{P}_{\bar{x}}, \overset{j}{P}_{\bar{y}}) \cos \overset{i}{\beta} + u_y(\overset{j}{P}_{\bar{x}}, \overset{j}{P}_{\bar{y}}) \sin \overset{i}{\beta} \\
\overset{i}{u}_{\bar{y}} &= -u_x(\overset{j}{P}_{\bar{x}}, \overset{j}{P}_{\bar{y}}) \sin \overset{i}{\beta} + u_y(\overset{j}{P}_{\bar{x}}, \overset{j}{P}_{\bar{y}}) \cos \overset{i}{\beta}
\end{aligned} \tag{5.23}$$

The notation $u_x(\overset{j}{P}_{\bar{x}}, \overset{j}{P}_{\bar{y}})$, etc., is used to emphasize that the displacement components depend on the stress discontinuities at element j. The boundary influence coefficients for the displacements—$\overset{ij}{B}_{\bar{x}\bar{x}}$, etc., in (5.15)—can be calculated directly from these equations by using the appropriate expressions for the displacement vector components from either (5.11) or (5.13). The coefficients for isotropic elasticity are

$$
\begin{aligned}
\overset{ij}{B}_{\bar{x}\bar{x}} &= \frac{1}{2\mu}\big[(3-4\nu)\bar{F}_1\cos\gamma - \bar{y}\big(\bar{F}_2\sin\gamma - \bar{F}_3\cos\gamma\big)\big] \\
\overset{ij}{B}_{\bar{x}\bar{y}} &= \frac{1}{2\mu}\big[(3-4\nu)\bar{F}_1\sin\gamma - \bar{y}\big(\bar{F}_2\cos\gamma + \bar{F}_3\sin\gamma\big)\big] \\
\overset{ij}{B}_{\bar{y}\bar{x}} &= \frac{1}{2\mu}\big[-(3-4\nu)\bar{F}_1\sin\gamma - \bar{y}\big(\bar{F}_2\cos\gamma + \bar{F}_3\sin\gamma\big)\big] \\
\overset{ij}{B}_{\bar{y}\bar{y}} &= \frac{1}{2\mu}\big[(3-4\nu)\bar{F}_1\cos\gamma + \bar{y}\big(\bar{F}_2\sin\gamma - \bar{F}_3\cos\gamma\big)\big]
\end{aligned}
\tag{5.24}
$$

in which angle γ is defined by (5.22).

5.5 Element Self Effects

The boundary influence coefficients are well defined for all the boundary elements and can be evaluated directly within a computer program using the preceding equations. When element j coincides with element i the influence coefficients $\overset{ii}{A}_{\bar{x}\bar{x}}$, etc., in (5.2) and $\overset{ii}{B}_{\bar{x}\bar{x}}$, etc., in (5.15) represent element *self effects*, i.e., the tractions and displacements at element i caused by the stress discontinuities $\overset{i}{P}_{\bar{x}}$ and $\overset{i}{P}_{\bar{y}}$ at the same element. We saw in Sect. 4.12 that the displacements are continuous along a line segment subjected to loads $P_{\bar{x}}$ and $P_{\bar{y}}$ but the stress tensor components are discontinuous. As a consequence, the self-effect influence coefficients for the displacements ($\overset{ii}{B}_{\bar{x}\bar{x}}$, etc.,) are unambiguously defined while those for the tractions ($\overset{ii}{A}_{\bar{x}\bar{x}}$, etc.,) require special attention.

To determine the element self-effect coefficients $\overset{ii}{A}_{\bar{x}\bar{x}}$, etc., for an isotropic material we evaluate (5.21) for the case that $\bar{x} = \bar{y} = 0$ and $\gamma = 0$. The resulting expressions are

$$
\begin{aligned}
\overset{ii}{A}_{\bar{x}\bar{x}} &= 2(1-\nu)\bar{F}_3(0,0) \\
\overset{ii}{A}_{\bar{x}\bar{y}} &= (1-2\nu)\bar{F}_2(0,0) \\
\overset{ii}{A}_{\bar{y}\bar{x}} &= -(1-2\nu)\bar{F}_2(0,0) \\
\overset{ii}{A}_{\bar{y}\bar{y}} &= 2(1-\nu)\bar{F}_3(0,0)
\end{aligned}
\tag{5.25}
$$

It follows from (5.8) that $\bar{F}_2(0,0) = 0$ and $\bar{F}_3(0,0_{\pm}) = \mp\big[4(1-\nu)\big]^{-1}$ so that the self-effect terms for the boundary influence coefficients for the tractions are

$$\overset{ii}{A}_{\bar{x}\bar{y}} = \overset{ii}{A}_{\bar{y}\bar{x}} = 0$$
$$\overset{ii}{A}_{\bar{x}\bar{x}} = \overset{ii}{A}_{\bar{y}\bar{y}} = \mp\frac{1}{2} \qquad \text{for } \bar{y} = 0_{\pm} \tag{5.26}$$

A similar process can be used to find the self-effect coefficients for an anisotropic material. It follows from (5.14) that the stresses in this case depend on the functions $I_3(\bar{x}, \bar{y}, \gamma_k)$ and $I_4(\bar{x}, \bar{y}, \gamma_k)$, $k = 1$ or 2, evaluated for $\bar{x} = \bar{y} = 0$ and $\beta = \overset{i}{\beta}$. Referring to equations (4.184) and (4.179), the required values of these functions are

$$I_3(0, 0_{\pm}, \gamma_k) = \mp\frac{\pi\gamma_k \sin\overset{i}{\beta}}{\gamma_k^2\cos^2\overset{i}{\beta} + \sin^2\overset{i}{\beta}}$$
$$I_4(0, 0_{\pm}, \gamma_k) = \pm\frac{\pi\gamma_k^2 \cos\overset{i}{\beta}}{\gamma_k^2\cos^2\overset{i}{\beta} + \sin^2\overset{i}{\beta}} \tag{5.27}$$

When substituted into (5.14) these quantities define the stress components $\sigma_{xx}(\overset{i}{P}_{\bar{x}}, \overset{i}{P}_{\bar{y}})$, etc., which can then be used to compute the tractions $\overset{i}{t}_{\bar{x}}$ and $\overset{i}{t}_{\bar{y}}$ from (5.20). It is left as an exercise for the reader to verify that the self-effect terms for the boundary influence coefficients for the tractions for an orthotropic material are the same as those given above for an isotropic material, (5.26).

Returning now to the situation depicted in Fig. 5.1b, we conclude from (5.26) that the values of the influence coefficients $\overset{ii}{A}_{\bar{x}\bar{x}}$ and $\overset{ii}{A}_{\bar{y}\bar{y}}$ depend on the way in which curve S' is approached. The positive and negative sides of the curve are specified with reference to coordinate $\bar{y}$, or $\boldsymbol{n}$, the local outward normal to the boundary S in Fig. 5.1a. For the situation shown in the figure, the negative side of curve S' represents the boundary S of the cavity, and $\overset{ii}{A}_{\bar{x}\bar{x}}$ and $\overset{ii}{A}_{\bar{y}\bar{y}}$ are both equal to $+\frac{1}{2}$. If we were concerned with the finite region enclosed by the curve instead of the cavity, the values of these coefficients would be $-\frac{1}{2}$. We must therefore use different values for the self-effect coefficients $\overset{ii}{A}_{\bar{x}\bar{x}}$ and $\overset{ii}{A}_{\bar{y}\bar{y}}$ depending upon which region we are considering.

Notice, however, that if we traverse curve S' in the clockwise sense (instead of the counterclockwise sense as in Fig. 5.1b) then the negative side of the curve defines the boundary of a finite region and the values of the self-effect coefficients would be $+\frac{1}{2}$. We can therefore avoid having to use different values for the coefficients $\overset{ii}{A}_{\bar{x}\bar{x}}$ and $\overset{ii}{A}_{\bar{y}\bar{y}}$ if we agree always to traverse the boundary in such a way that the outward normal points *away* from the region of interest. Accordingly, we adopt the following convention: *The boundary of a finite body is traversed in the clockwise sense whereas the boundary of a cavity is traversed in the counterclockwise sense*. This convention leads to some simplification in computer programming because it means that the influence coefficients $\overset{ii}{A}_{\bar{x}\bar{x}}$ and $\overset{ii}{A}_{\bar{y}\bar{y}}$ are equal to $+\frac{1}{2}$ for both types of problem.

5.6 Interior/Exterior Problems

Equations (5.2) and (5.15) form the basis of a general numerical procedure for solving boundary value problems in plane elasticity. In this procedure—the stress discontinuity method—we divide the boundary (or boundaries) of the region to be analyzed into N elements and associate stress discontinuities $P_{\bar{x}}$ and $P_{\bar{y}}$ with each element. We then set up and solve a system of algebraic equations to find the stress discontinuities that produce the prescribed boundary displacements or tractions. The displacements and stresses anywhere in the region of interest can then be computed by summing the effects of the stress discontinuities at the N boundary elements.

As mentioned in Sect. 4.8, boundary value problems in elasticity can be posed for traction, displacement, or mixed boundary conditions. The type of boundary conditions prescribed for any particular problem governs the form of the system of algebraic equations to be solved. For example, if tractions $\overset{i}{t}_{\bar{x}} = (\overset{i}{t}_{\bar{x}})_0$ and $\overset{i}{t}_{\bar{y}} = (\overset{i}{t}_{\bar{y}})_0$ are prescribed at the ith boundary element, then the ith equations of the system are, by (5.2),

$$\begin{aligned}\sum_{j=1}^{N} \overset{ij}{A}_{\bar{x}\bar{x}} \overset{j}{P}_{\bar{x}} + \sum_{j=1}^{N} \overset{ij}{A}_{\bar{x}\bar{y}} \overset{j}{P}_{\bar{y}} &= (\overset{i}{t}_{\bar{x}})_0 \\ \sum_{j=1}^{N} \overset{ij}{A}_{\bar{y}\bar{x}} \overset{j}{P}_{\bar{x}} + \sum_{j=1}^{N} \overset{ij}{A}_{\bar{y}\bar{y}} \overset{j}{P}_{\bar{y}} &= (\overset{i}{t}_{\bar{y}})_0\end{aligned} \tag{5.28}$$

Similarly, if displacements $\overset{i}{u}_{\bar{x}} = (\overset{i}{u}_{\bar{x}})_0$ and $\overset{i}{u}_{\bar{y}} = (\overset{i}{u}_{\bar{y}})_0$ are prescribed, the ith equations are, by (5.15),

$$\begin{aligned}\sum_{j=1}^{N} \overset{ij}{B}_{\bar{x}\bar{x}} \overset{j}{P}_{\bar{x}} + \sum_{j=1}^{N} \overset{ij}{B}_{\bar{x}\bar{y}} \overset{j}{P}_{\bar{y}} &= (\overset{i}{u}_{\bar{x}})_0 \\ \sum_{j=1}^{N} \overset{ij}{B}_{\bar{y}\bar{x}} \overset{j}{P}_{\bar{x}} + \sum_{j=1}^{N} \overset{ij}{B}_{\bar{y}\bar{y}} \overset{j}{P}_{\bar{y}} &= (\overset{i}{u}_{\bar{y}})_0\end{aligned} \tag{5.29}$$

Mixed formulations involving specification of $\overset{i}{u}_{\bar{x}}$ and $\overset{i}{t}_{\bar{y}}$ or $\overset{i}{t}_{\bar{x}}$ and $\overset{i}{u}_{\bar{y}}$ are handled by selecting the appropriate ones of (5.28) and (5.29). Proceeding in this way for

$i = 1$ to N, we obtain a system of $2N$ algebraic equations in $2N$ unknown stress discontinuity components. We will represent this system of equations as follows:

$$\sum_{j=1}^{N} \overset{ij}{C}_{\bar{x}\bar{x}} \overset{j}{P}_{\bar{x}} + \sum_{j=1}^{N} \overset{ij}{C}_{\bar{x}\bar{y}} \overset{j}{P}_{\bar{y}} = \overset{i}{b}_{\bar{x}}$$
$$\sum_{j=1}^{N} \overset{ij}{C}_{\bar{y}\bar{x}} \overset{j}{P}_{\bar{x}} + \sum_{j=1}^{N} \overset{ij}{C}_{\bar{y}\bar{y}} \overset{j}{P}_{\bar{y}} = \overset{i}{b}_{\bar{y}} \tag{5.30}$$

for $i = 1$ to N. The quantities $\overset{i}{b}_{\bar{x}}$ and $\overset{i}{b}_{\bar{y}}$ stand for the known boundary values of traction or displacement, whichever is appropriate, and $\overset{ij}{C}_{\bar{x}\bar{x}}$, etc., are the corresponding influence coefficients from (5.28) or (5.29).

Finally, in respect of interior problems we note that some provision must be made to prevent rigid body movements of any finite region. This can be accomplished by prescribing at least three displacement components at two points along the boundary of the region. Alternatively, one can take advantage of symmetry conditions, if they exist, to achieve the same effect. A procedure for incorporating symmetry conditions in the stress discontinuity method is discussed in the next section. Symmetry conditions often occur in practical problems (both interior and exterior problems) so it is desirable to have a way of handling them.

5.7 Symmetry Conditions

A line of symmetry exists for a problem when the elastic properties of the material, the geometric configuration of the boundaries, and the loading conditions are all symmetrical with respect to the line in question. The first requirement is not needed for a homogeneous, isotropic elastic material because its properties are independent of both position and direction. In the case of an orthotropic or transversely isotropic material with directions of elastic symmetry along the x, y axes, any line of geometric symmetry must be in these directions.

Two physical consequences of a line of symmetry may be noted. First, no normal displacements occur across the line, and, second, no shear tractions act along it.

A line of symmetry can be introduced with respect to a given boundary element by adding an image element at the appropriate location. The line of symmetry acts as a mirror, and the components of the stress discontinuity at the image element are found by reflecting those of the actual element across this line. A consequence of such a reflection is that the shear components of stress discontinuity at the actual and image boundary elements are always *equal in magnitude but opposite in sign*; the normal components are always equal. By including the effects of the image elements in the boundary influence coefficients the effective size of the problem is reduced by

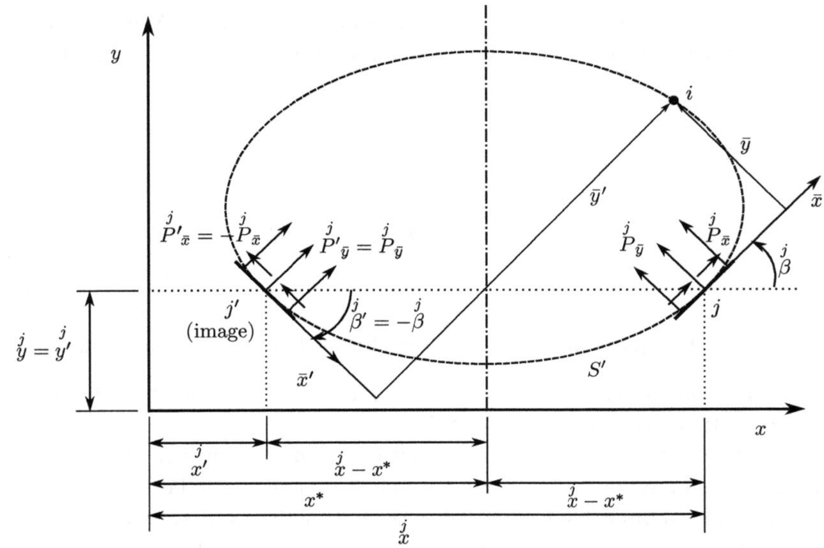

Fig. 5.4 Symmetry condition for the line $x = x^*$

half because the unknowns in the system of algebraic equations are associated with the actual elements.

In order to explain this procedure we will examine in detail the particular case for which $x = x^*$ is a line of symmetry. This case is illustrated in Fig. 5.4. For simplicity, we will discuss only the procedure for computing the boundary influence coefficients. Coefficients for arbitrary points in the region of interest are found in an analogous fashion.

The actual boundary element is assumed to be centered at the point $x = \overset{j}{x} > x^*$, $y = \overset{j}{y}$ and oriented at an angle $\overset{j}{\beta}$ with respect to the positive x direction. The stress discontinuity components at at this element are $\overset{j}{P}_{\bar{x}}$ and $\overset{j}{P}_{\bar{y}}$, as shown in Fig. 5.4. The midpoint of the image element is located at the point $x = \overset{j}{x'}$, $y = \overset{j}{y'}$. The element is oriented at the angle $\overset{j}{\beta'} = -\overset{j}{\beta}$ and it has stress discontinuities $\overset{j}{P'}_{\bar{x}} = -\overset{j}{P}_{\bar{x}}$ and $\overset{j}{P'}_{\bar{y}} = +\overset{j}{P}_{\bar{y}}$. The y coordinates of the two elements are equal, i.e., $\overset{j}{y'} = \overset{j}{y}$. The x coordinate of the image element is found by noting that the actual and image elements are equidistant from the line of symmetry, $x = x^*$. Referring to Fig. 5.4, we write

$$
\begin{aligned}
\overset{j}{x'} &= x^* - (\overset{j}{x} - x^*) \\
&= 2x^* - \overset{j}{x} \\
\overset{j}{y'} &= \overset{j}{y}
\end{aligned}
\tag{5.31}
$$

The boundary displacements ($\overset{i}{u}_{\bar{x}}$ and $\overset{i}{u}_{\bar{y}}$) and stresses ($\overset{i}{\sigma}_{\bar{x}\bar{x}}$, $\overset{i}{\sigma}_{\bar{y}\bar{y}}$, and $\overset{i}{\sigma}_{\bar{x}\bar{y}}$) at the point $x = \overset{i}{x}$, $y = \overset{i}{y}$ (point i in the figure) due to stress discontinuities $\overset{j}{P}{}'_{\bar{x}}$ and $\overset{j}{P}{}'_{\bar{y}}$ at the image element can now be computed in terms of the local coordinates (see (5.16))

$$
\begin{aligned}
\bar{x}' &= (\overset{i}{x} - \overset{j}{x}{}') \cos \overset{j}{\beta}{}' + (\overset{i}{y} - \overset{j}{y}{}') \sin \overset{j}{\beta}{}' \\
&= (\overset{i}{x} - 2x^* + \overset{j}{x}) \cos(-\overset{j}{\beta}) + (\overset{i}{y} - \overset{j}{y}) \sin(-\overset{j}{\beta}) \\
\bar{y}' &= -(\overset{i}{x} - \overset{j}{x}{}') \sin \overset{j}{\beta}{}' + (\overset{i}{y} - \overset{j}{y}{}') \cos \overset{j}{\beta}{}' \\
&= -(\overset{i}{x} - 2x^* + \overset{j}{x}) \sin(-\overset{j}{\beta}) + (\overset{i}{y} - \overset{j}{y}) \cos(-\overset{j}{\beta})
\end{aligned}
\tag{5.32}
$$

and functions $F_k(\bar{x}', \bar{y}')$ from (5.7) and (5.8), or functions $I_k(\bar{x}', \bar{y}', \gamma_m)$ from (4.180).

The local coordinates $\bar{x}'$ and $\bar{y}'$ define the position of point i relative to the midpoint of the image element, j', as shown in Fig. 5.4. In evaluating the displacements and stresses at this point from the results of Sect. 5.4, the sign change for the shear component of the stress discontinuity at the image element must be observed. Influence coefficients $\overset{ij}{B}_{\bar{x}\bar{x}}$, etc., in (5.15) and $\overset{ij}{A}_{\bar{x}\bar{x}}$, etc., in (5.2) that automatically incorporate the effects of the image element can then be computed using the procedure outlined in Sect. 5.4; it is merely necessary to include in the computations the displacements and stresses due to *both* the actual and image boundary elements.

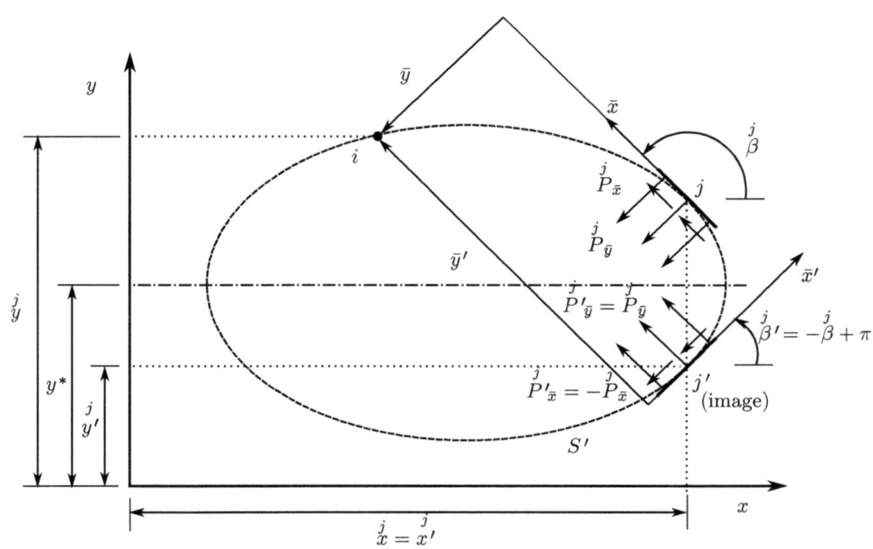

Fig. 5.5 Symmetry condition for the line $y = y^*$

A similar procedure can be followed for the case that $y = y^*$ is a line of symmetry, as illustrated in Fig. 5.5. The x and y coordinates of the midpoint of the image element in this case are

$$\begin{aligned} \overset{j}{x}{}' &= \overset{j}{x} \\ \overset{j}{y}{}' &= y^* - (\overset{j}{y} - y^*) \\ &= 2y^* - \overset{j}{y} \end{aligned} \tag{5.33}$$

The image element is inclined at an angle $\overset{j}{\beta}{}' = -\overset{j}{\beta} + \pi$ from the positive x direction and has stress discontinuities $\overset{j}{P}{}'_{\bar{x}} = -\overset{j}{P}_{\bar{x}}$ and $\overset{j}{P}{}'_{\bar{y}} = +\overset{j}{P}_{\bar{y}}$. The local coordinates of point $x = \overset{i}{x}$, $y = \overset{i}{y}$ with respect to this element are (see (5.16))

$$\begin{aligned} \bar{x}' &= (\overset{i}{x} - \overset{j}{x}) \cos(\pi - \overset{j}{\beta}) + (\overset{i}{y} - 2y^* + \overset{j}{y}) \sin(\pi - \overset{j}{\beta}) \\ \bar{y}' &= -(\overset{i}{x} - \overset{j}{x}) \sin(\pi - \overset{j}{\beta}) + (\overset{i}{y} - 2y^* + \overset{j}{y}) \cos(\pi - \overset{j}{\beta}) \end{aligned} \tag{5.34}$$

The influence coefficients can now be computed using the same process as before.

If both $x = x^*$ and $y = y^*$ are lines of symmetry the influence coefficients are computed by supplying *three* image boundary elements. This situation is shown in Fig. 5.6. The local coordinates associated with each of the image elements can be written in the same way as before. Notice the sign change for the $\bar{x}$ component of stress discontinuity each time a line of symmetry is crossed.

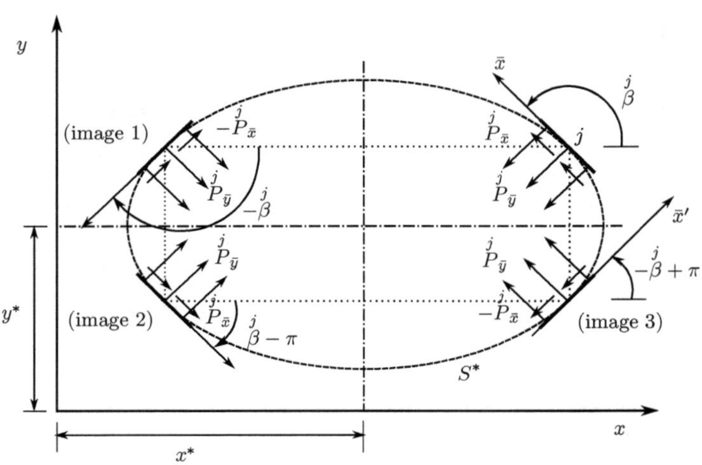

Fig. 5.6 Symmetry conditions for the lines $x = x^*$ and $y = y^*$

Finally, we remark that although Figs. 5.4–5.6 have been drawn assuming counterclockwise traversal of the boundary, the same results are obtained for clockwise traversal. The procedure given above therefore applies to both interior and exterior problems.

5.8 An Algorithm to Identify Points Inside the Region of Interest

When evaluating the displacements and stresses at points inside a region V, it is often useful to perform the computations over a regular grid overlying selected portions of the region. Frequently, however, some points of such a computational grid will fall outside the region of interest and are therefore not of practical importance. It is useful to have a way of automatically identifying these points within a computer program, and we consider this matter next.

The use of straight line elements in a boundary element method means that every closed boundary has a polygonal shape. Depending on the problem being solved, an N-sided polygon will either represent a finite object, as in Fig. 1.6, or the boundary of a hole in an infinite region, as in Fig. 5.3. Moreover, both a finite object and an infinite region might be multiply connected, i.e., contain several holes, and it is necessary to consider these cases, too. The essential question is this: Is a point $(x,\ y)$ inside the region of interest, or not?

Boundary element methods that use straight line elements typically require calculation of the angles subtended by lines drawn from the endpoints of the elements to a field point $(x,\ y)$, and these angles provide a convenient means of determining whether the point is inside or outside the region of interest V. For example, Fig. 5.7a shows that the angle $\Delta\theta = \bar{\theta}_1 - \bar{\theta}_2$ subtended by a typical element on the boundary

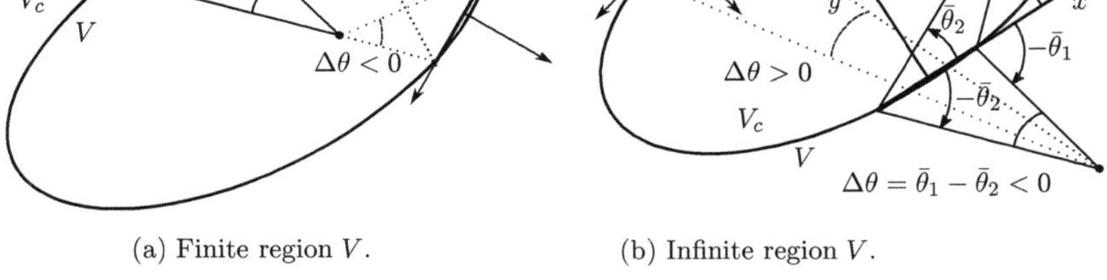

Fig. 5.7 Points inside and outside of region V

of a finite region is negative if point (x, y) is inside the region but can be either positive or negative if the point is outside it. Considering all N elements on the boundary of a finite region V, it is easy to see that the following conditions hold:

$$\sum_{i=1}^{N} (\Delta\theta)_i = \begin{cases} -2\pi & \text{if point } (x, y) \text{ is inside } V \\ 0 & \text{if point } (x, y) \text{ is outside } V \end{cases} \tag{5.35}$$

Similarly, if region V is exterior to a closed boundary with N elements, as depicted in Fig. 5.7b, then

$$\sum_{i=1}^{N} (\Delta\theta)_i = \begin{cases} 0 & \text{if point } (x, y) \text{ is inside } V \\ +2\pi & \text{if point } (x, y) \text{ is outside } V \end{cases} \tag{5.36}$$

For problems involving a single closed contour with N boundary elements conditions (5.35) and (5.36) directly answer the question of whether or not a point (x, y) is inside the region of interest V. It is merely necessary to keep track of the summation of included angles associated with each field point.

The case of an infinite region V with multiple holes can be handled by applying (5.36) to each hole in turn. If point (x, y) is inside one of the holes (that is, if it is outside region V) the summation of included angles for that hole will be equal to 2π and the point should be excluded from consideration.

The case of a finite body containing a hole can be handled by considering a combination of conditions (5.35) and (5.36). If the boundary defining the body has N_1 elements and the one defining the hole has N_2 elements, we have that

$$\sum_{i=1}^{N_1} (\Delta\theta)_i + \sum_{i=1}^{N_2} (\Delta\theta)_i = \begin{cases} -2\pi & \text{if point } (x, y) \text{ is inside } V \\ 0 & \text{if point } (x, y) \text{ is outside } V \end{cases} \tag{5.37}$$

in which the region of interest V is inside the body but outside the hole. The same conditions apply if multiple holes are present.

It is a simple matter to incorporate the preceding logic into a computer program.

5.9 Examples

5.9.1 Exterior Problem

As an example of an exterior problem we consider the case of a circular hole in an infinite plate under uniaxial tension $\sigma_{xx} = p$ $(p > 0)$ at infinity. As explained in Sect. 4.8, we handle the boundary conditions for this problem by subjecting the circular boundary to the negatives of the tractions that would exist if the hole were not

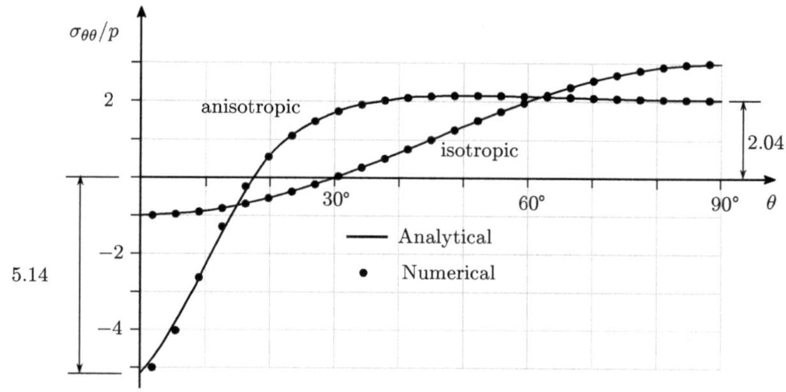

Fig. 5.8 Stress $\sigma_{\theta\theta}$ at boundary of circular hole; isotropic and anisotropic elasticity

present. The complete solution is then given by the sums of the stress and displacement changes (i.e., the induced stresses and displacements) and the original conditions, taking the original displacements to be zero everywhere. This guarantees that the resultant tractions will vanish along the boundary of the hole and that the displacements will be zero at infinity.

Standard coordinate transformation formulas can be used within a computer program to generate the induced traction boundary conditions, given the locations and orientations of the individual boundary elements. For the problem under consideration, a 25 element approximation to one-quarter of the circular boundary was used and separate analyses were made for isotropic and orthotropic elastic materials. Computed values of the circumferential stress $\sigma_{\theta\theta}$ along the boundary of the hole are compared with the exact solutions in Fig. 5.8 and are seen to be in good agreement in both cases.

The analytical solution for the case of anisotropic elasticity is given by Green and Taylor (1945) for the elastic parameters $s_{11} = 15.5 \times 10^{-4}(\text{kPa})^{-1}$, $s_{22} = 0.587 \times 10^{-4}(\text{kPa})^{-1}$, $s_{12} = -0.33 \times 10^{-4}(\text{kPa})^{-1}$, and $s_{66} = 11.5 \times 10^{-4}(\text{kPa})^{-1}$ and plane stress conditions (see (4.31)). These elastic constants are supposed to represent a spruce board with its grain parallel to the y axis, that is, perpendicular to the applied uniaxial tension. The variation of $\sigma_{\theta\theta}$ along the circular boundary for isotropic elasticity is well known (e.g., Sokolnikoff 1956) and is independent of the elastic constants for the material. The latter point is interesting to note in light of the fact that Poisson's ratio ν features prominently in the combinations of the functions used to evaluate the boundary influence coefficients $A^{ij}_{\bar{x}\bar{x}}$, etc., for the stress discontinuity method (see (5.21)). Empirically, it is found that the numerical results for the stresses do not depend on Poisson's ratio over the permissible range of this parameter, $-1 < \nu < 0.5$. It is worth mentioning, however, that the values of the stress discontinuities $P_{\bar{x}}$ and $P_{\bar{y}}$ vary significantly with Poisson's ratio, even though their linear combinations produce stress values that do not.

5.9.2 Interior Problem I

We consider next the interior problem of a circular disk of radius R subjected to a normal compression $\sigma_n = -p$ over two arcs of length $2\alpha R$ centered on the ends of a diameter of the disk. The geometry and loading conditions for the problem are shown in Fig. 5.9, where it can be seen that the lines $x = 0$ and $y = 0$ are both lines of symmetry. By incorporating these lines of symmetry in the boundary element model the disk is automatically held stationary at its center and no additional conditions are required to prevent rigid body movements.

The analytical solution to this problem for isotropic elasticity is given by Hondros (1959). A numerical approximation to the solution was obtained by dividing one-quarter of the disk boundary into 25 straight line elements of length $\pi R/50$ with a normal traction $t_{\bar{y}} = -p$ applied over one element. Figure 5.10 shows comparisons between the approximate and exact values of the stresses σ_{xx} and σ_{yy} along the y axis of the disk. The normalized stresses are within plotting accuracy of the analytical solution near the center of the disk and deviate only slightly from the exact values when y/R is close to the applied load. At $y/R = 0.95$, for example, the computed value of σ_{yy} is in error by about 4.5%.

The computed results are given at regular intervals along the y axis simply for convenience. The results could just as well have been found at other points in the disk, subject only to the restriction that they should not be determined "too close" to the boundary. It is found empirically that numerical solutions based on constant strength stress discontinuities are generally unreliable at points within circles of diameter equal to one element length ($2a$) centered at the midpoint or either endpoint of a boundary element, except, of course, at the midpoint itself. We will see later, however, that this restriction can be relaxed for methods involving more elaborate, "higher order" boundary elements.

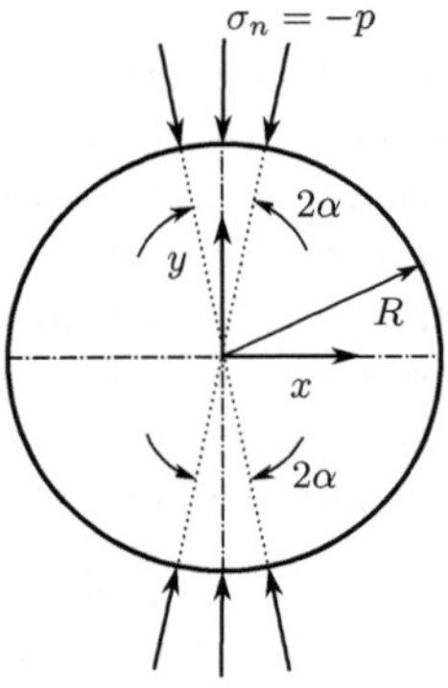

Fig. 5.9 Isotropic circular disk under diametral compression

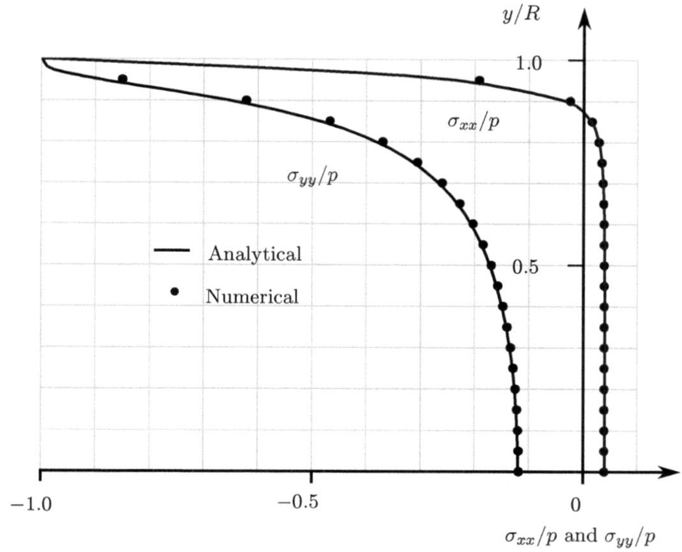

Fig. 5.10 Stresses along y axis of the disk in Fig. 5.9

5.9.3 Interior Problem II

A somewhat more challenging interior problem is depicted in Fig. 5.11. The disk in this case is composed of an anisotropic elastic material with planes of elastic symmetry parallel to the x axis. The disk is compressed by point loads P applied along a diameter that is inclined at an angle θ from the x axis. The analytical solution to this problem was originally obtained by Lekhnitskii (1968); detailed results were recorded more recently by Claesson and Bohloli (2002), including the example problem discussed below.

Fig. 5.11 Anisotropic circular disk under diametral compression

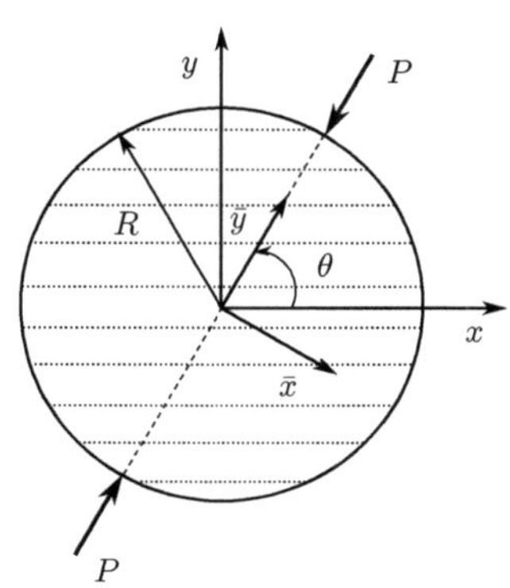

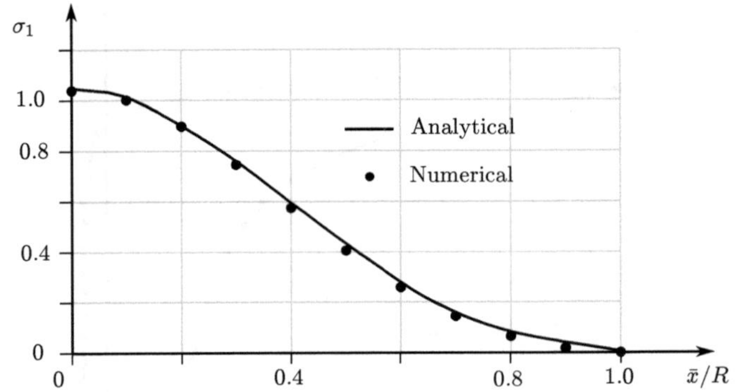

Fig. 5.12 Major principal stress σ_1 along a diameter of the disk perpendicular to the applied load

The absence of symmetry in the problem means that a boundary element analysis must employ additional displacement boundary conditions in order to prevent rigid body movements of the disk. This can be accomplished by setting the normal displacement $u_{\bar{y}}$ equal to zero at two elements straddling one end of the loaded diameter and also fixing $u_{\bar{x}}$ at the two elements. For consistency, the load P is modeled by applying normal tractions $t_{\bar{y}} = -p$ to two elements straddling the other end of the diameter.

Claesson and Bohloli (2002) analyzed this problem for the following parameters: $\theta = \pi/6$, $E_x = 0.507 \times 10^5$ MPa, $E_y = 0.400 \times 10^5$ MPa, $G_{xy} = 0.156 \times 10^5$ MPa, and $\nu_{xy} = \nu_{yx} E_x/E_y = 0.355$. (Only four independent elastic constants occur in the case of plane stress; see (4.31).) In addition, the load P was defined in terms of a scale factor $P/(\pi RL) = 1$, where R is the radius and L is the thickness of the disk. A boundary element representation of this problem used 240 equal length elements along the periphery of the disk. A normal stress p was applied to two elements along an arc of length $Rd\theta = \pi R/60$ (and thickness L) so that $p = 60$ represents an applied load equivalent to that considered by Claesson and Bohloli.

Figure 5.12 shows dimensionless values of the major principal stress σ_1 along a diameter of the disk perpendicular to the applied loading. The numerical results are in good agreement with the analytical solution: the maximum tensile stress at the center of the disk was computed as 1.038 compared to the value 1.044 reported by Claesson and Bohloli. In addition, the ratio $\mid \sigma_1/\sigma_2 \mid$ at the center of the disk was computed as 0.361 compared to 0.363 for the analytical solution.

5.10 Higher Order Elements

A straightforward generalization of the foregoing approach can be devised for straight-line elements with applied tractions (stress discontinuities) that vary along their lengths. The starting point for such an analysis is the solution to the problem of a concentrated force at a point because this solution can be integrated along a line segment to produce arbitrary distributions of loading. We will outline the integration process below for an isotropic material for the case that the loading varies quadratically along the elements. A similar process can be used for an anisotropic material.

The general procedure for variable loading is comparable to that described in Sect. 4.12.1 for constant tractions $t_i = P_i$ applied to the line segment $| \ x \ | \leq a, \ y = 0$ in an infinite elastic solid. Now, however, we consider the arbitrarily oriented line segment $| \ \bar{x} \ | \leq a, \ \bar{y} = 0$ and suppose that the tractions are quadratic functions of $\bar{x}$, i.e., we take

$$
\begin{aligned}
p_{\bar{x}}(\bar{x}) &= \mathsf{a}_{\bar{x}}\bar{x}^2 + \mathsf{b}_{\bar{x}}\bar{x} + \mathsf{c}_{\bar{x}} \\
p_{\bar{y}}(\bar{x}) &= \mathsf{a}_{\bar{y}}\bar{x}^2 + \mathsf{b}_{\bar{y}}\bar{x} + \mathsf{c}_{\bar{y}}
\end{aligned}
\tag{5.38}
$$

where $\mathsf{a}_{\bar{x}}$, $\mathsf{a}_{\bar{y}}$, etc., are constants. For this loading, an infinitesimal element of length $d\xi_{\bar{x}}$ at the point $\bar{x} = \xi_{\bar{x}}, \ \bar{y} = 0$ is subjected to a force with components $F_{\bar{x}} = p_{\bar{x}}(\xi_{\bar{x}}) \, d\xi_{\bar{x}}$ and $F_{\bar{y}} = p_{\bar{y}}(\xi_{\bar{x}}) \, d\xi_{\bar{x}}$. The solution to the problem in question is obtained by substituting this force into versions of (4.143) and (4.144) written in terms of $\bar{x}, \ \bar{y}$ coordinates, replacing $\bar{x}$ by $\bar{x} - \xi_{\bar{x}}$, and integrating the resulting expressions with respect to $\xi_{\bar{x}}$ between the limits $-a$ and $+a$.

It is convenient for these calculations to write the quadratic expressions (5.38) in terms of three element shape functions, defined as follows:

$$
\begin{aligned}
\alpha_1(\bar{x}) &= \frac{1}{2b^2}\bar{x}^2 - \frac{1}{2b}\bar{x} \\
\alpha_2(\bar{x}) &= 1 - \frac{1}{b^2}\bar{x}^2 \\
\alpha_3(\bar{x}) &= \frac{1}{2b^2}\bar{x}^2 + \frac{1}{2b}\bar{x}
\end{aligned}
\tag{5.39}
$$

where $b < a$. These shape functions relate to a discontinuous element with collocation points at $\bar{x} = 0, \ \bar{y} = 0$ and $\bar{x} = \pm b, \ \bar{y} = 0$. They are plotted in Fig. 5.13 for the case that $b = a\sqrt{3}/2$. (The three points $b/a = \pm\sqrt{3}/2$ and $b/a = 0$ are the so-called Gauss-Chebyshev points for the element.) It can be seen from the figure that shape function $\alpha_1(\bar{x})$ is equal to $+1$ at $\bar{x} = -b$ (collocation point 1 in the figure) and is zero at $\bar{x} = 0$ (point 2) and $\bar{x} = +b$ (point 3). Shape functions $\alpha_2(\bar{x})$ and $\alpha_3(\bar{x})$ are similarly equal to +1 at $\bar{x} = 0$ and $\bar{x} = b$, respectively, but are zero at the other collocation points. Equations (5.38) can now be written as

$$
\begin{aligned}
p_{\bar{x}}(\bar{x}) &= p_{\bar{x}}(-b)\alpha_1(\bar{x}) + p_{\bar{x}}(0)\alpha_2(\bar{x}) + p_{\bar{x}}(b)\alpha_3(\bar{x}) \\
p_{\bar{y}}(\bar{x}) &= p_{\bar{y}}(-b)\alpha_1(\bar{x}) + p_{\bar{y}}(0)\alpha_2(\bar{x}) + p_{\bar{y}}(b)\alpha_3(\bar{x})
\end{aligned}
\tag{5.40}
$$

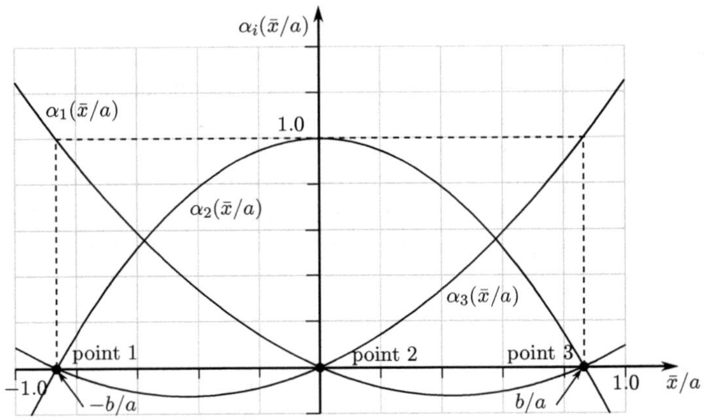

Fig. 5.13 Shape functions $\alpha_i(\bar{x}/a)$, $i = 1$ to 3, for a discontinuous quadratic element

or

$$
\begin{aligned}
p_{\bar{x}}(\bar{x}) &= (p_{\bar{x}})_1\alpha_1(\bar{x}) + (p_{\bar{x}})_2\alpha_2(\bar{x}) + (p_{\bar{x}})_3\alpha_3(\bar{x}) \\
p_{\bar{y}}(\bar{x}) &= (p_{\bar{y}})_1\alpha_1(\bar{x}) + (p_{\bar{y}})_2\alpha_2(\bar{x}) + (p_{\bar{y}})_3\alpha_3(\bar{x})
\end{aligned} \tag{5.41}
$$

The integration process described above involves the quantities $p_{\bar{x}}(\xi_{\bar{x}})$ and $p_{\bar{y}}(\xi_{\bar{x}})$, which in turn involve the shape functions $\alpha_1(\xi_{\bar{x}})$, $\alpha_2(\xi_{\bar{x}})$, and $\alpha_1(\xi_{\bar{x}})$. The integrations can be simplified by rewriting the latter expressions as functions of $\bar{x} - \xi_{\bar{x}}$. To this end, we observe that the quantities $\xi_{\bar{x}}$ and $\xi_{\bar{x}}^2$ can be written as

$$
\begin{aligned}
\xi_{\bar{x}} &= -(\bar{x} - \xi_{\bar{x}}) + \bar{x} \\
\xi_{\bar{x}}^2 &= (\bar{x} - \xi_{\bar{x}})^2 - 2\bar{x}(\bar{x} - \xi_{\bar{x}}) + \bar{x}^2
\end{aligned} \tag{5.42}
$$

Using these results in (5.39) (with $\bar{x}$ replaced by $\xi_{\bar{x}}$) we find

$$
\begin{aligned}
\alpha_1(\xi_{\bar{x}}) &= \frac{1}{2b^2}\left[(\bar{x} - \xi_{\bar{x}})^2 - (2\bar{x} - b)(\bar{x} - \xi_{\bar{x}}) + \bar{x}(\bar{x} - b)\right] \\
\alpha_2(\xi_{\bar{x}}) &= -\frac{1}{b^2}\left[(\bar{x} - \xi_{\bar{x}})^2 - 2\bar{x}(\bar{x} - \xi_{\bar{x}}) + (\bar{x}^2 - b^2)\right] \\
\alpha_3(\xi_{\bar{x}}) &= \frac{1}{2b^2}\left[(\bar{x} - \xi_{\bar{x}})^2 - (2\bar{x} + b)(\bar{x} - \xi_{\bar{x}}) + \bar{x}(\bar{x} + b)\right]
\end{aligned} \tag{5.43}
$$

These equations can be compactly written as

$$
\alpha_k(\xi_{\bar{x}}) = \mathsf{A}_k(\bar{x} - \xi_{\bar{x}})^2 + \mathsf{B}_k(\bar{x} - \xi_{\bar{x}}) + \mathsf{C}_k \tag{5.44}
$$

where $k = 1,\ 2$, or 3 and

$$\begin{aligned}
\mathsf{A}_1 &= \frac{1}{2b^2}, \quad \mathsf{B}_1 = -\mathsf{A}_1(2\bar{x} - b), \quad \mathsf{C}_1 = \mathsf{A}_1\bar{x}(\bar{x} - b) \\
\mathsf{A}_2 &= -\frac{1}{b^2}, \quad \mathsf{B}_2 = -\mathsf{A}_2 2\bar{x}, \quad \mathsf{C}_2 = \mathsf{A}_2(\bar{x}^2 - b^2) \\
\mathsf{A}_3 &= \frac{1}{2b^2}, \quad \mathsf{B}_3 = -\mathsf{A}_3(2\bar{x} + b), \quad \mathsf{C}_3 = \mathsf{A}_3\bar{x}(\bar{x} + b)
\end{aligned} \tag{5.45}$$

The displacements and stresses in the infinite domain due to variable loading along the line segment $|\ \bar{x}\ | \leq a,\ \bar{y} = 0$ can now be obtained by integrating the appropriate forms of (4.143) and (4.144), written in terms of $\bar{x},\ \bar{y}$ coordinates. (Equations comparable to (4.163) can be used to eliminate multiplications by $\bar{x} - \xi_{\bar{x}}$ in these expressions.) The equation for $u_{\bar{x}}$, for example, is

$$\begin{aligned}
u_{\bar{x}} &= \frac{1}{2\mu}\int_{-a}^{a} p_{\bar{x}}(\xi_{\bar{x}})\Big[(3 - 4\nu)g(\bar{x} - \xi_{\bar{x}}, \bar{y}) + \bar{y}\frac{\partial}{\partial \bar{y}}g(\bar{x} - \xi_{\bar{x}}, \bar{y})\Big]\,\mathrm{d}\xi_{\bar{x}} \\
&+ \frac{1}{2\mu}\int_{-a}^{a} p_{\bar{y}}(\xi_{\bar{x}})\Big[-\bar{y}\frac{\partial}{\partial \bar{x}}g(\bar{x} - \xi_{\bar{x}}, \bar{y})\Big]\,\mathrm{d}\xi_{\bar{x}}
\end{aligned} \tag{5.46}$$

or, using (5.41),

$$\begin{aligned}
u_{\bar{x}} &= \frac{1}{2\mu}\sum_{k=1}^{3}(p_{\bar{x}})_k\int_{-a}^{a}\Big[(3 - 4\nu)\alpha_k(\xi_{\bar{x}})g(\bar{x} - \xi_{\bar{x}}, \bar{y}) + \bar{y}\alpha_k(\xi_{\bar{x}})\frac{\partial}{\partial \bar{y}}g(\bar{x} - \xi_{\bar{x}}, \bar{y})\Big]\,\mathrm{d}\xi_{\bar{x}} \\
&+ \frac{1}{2\mu}\sum_{k=1}^{3}(p_{\bar{y}})_k\int_{-a}^{a}\Big[-\bar{y}\alpha_k(\xi_{\bar{x}})\frac{\partial}{\partial \bar{x}}g(\bar{x} - \xi_{\bar{x}}, \bar{y})\Big]\,\mathrm{d}\xi_{\bar{x}}
\end{aligned} \tag{5.47}$$

Similar expressions can be written for $u_{\bar{y}}$ and $\sigma_{\bar{x}\bar{x}}$, $\sigma_{\bar{y}\bar{y}}$, and $\sigma_{\bar{x}\bar{x}}$. As can be seen from (5.47), these quantities involve integrals of the form

$$\int_{-a}^{a} \alpha_k(\xi_{\bar{x}})(\bullet)\,\mathrm{d}\xi_{\bar{x}}$$

where $(\bullet) = g(\bar{x} - \xi_{\bar{x}}, \bar{y})$ or one of its derivatives with respect to $\bar{x}$ and $\bar{y}$. These integrals can be defined as follows:

$$\begin{aligned}
(\bar{G}_1)_k &= \int_{-a}^{a} \alpha_k(\xi_{\bar{x}})g(\bar{x} - \xi_{\bar{x}}, \bar{y})\,\mathrm{d}\xi_{\bar{x}} \\
(\bar{G}_2)_k &= \int_{-a}^{a} \alpha_k(\xi_{\bar{x}})\frac{\partial}{\partial \bar{x}}g(\bar{x} - \xi_{\bar{x}}, \bar{y})\,\mathrm{d}\xi_{\bar{x}} \\
(\bar{G}_3)_k &= \int_{-a}^{a} \alpha_k(\xi_{\bar{x}})\frac{\partial}{\partial \bar{y}}g(\bar{x} - \xi_{\bar{x}}, \bar{y})\,\mathrm{d}\xi_{\bar{x}} \\
(\bar{G}_4)_k &= \int_{-a}^{a} \alpha_k(\xi_{\bar{x}})\frac{\partial^2}{\partial \bar{x}\partial \bar{y}}g(\bar{x} - \xi_{\bar{x}}, \bar{y})\,\mathrm{d}\xi_{\bar{x}} \\
(\bar{G}_5)_k &= \int_{-a}^{a} \alpha_k(\xi_{\bar{x}})\frac{\partial^2}{\partial \bar{x}^2}g(\bar{x} - \xi_{\bar{x}}, \bar{y})\,\mathrm{d}\xi_{\bar{x}}
\end{aligned} \tag{5.48}$$

The reader should confirm that the displacements and stresses can be written in terms of definitions (5.48) as

$$u_{\bar{x}} = \frac{1}{2\mu}\sum_{k=1}^{3}(p_{\bar{x}})_k\big[(3-4\nu)(\bar{G}_1)_k + \bar{y}(\bar{G}_3)_k\big] + \frac{1}{2\mu}\sum_{k=1}^{3}(p_{\bar{y}})_k\big[-\bar{y}(\bar{G}_2)_k\big]$$
$$u_{\bar{y}} = \frac{1}{2\mu}\sum_{k=1}^{3}(p_{\bar{x}})_k\big[-\bar{y}(\bar{G}_2)_k\big] + \frac{1}{2\mu}\sum_{k=1}^{3}(p_{\bar{y}})_k\big[(3-4\nu)(\bar{G}_1)_k - \bar{y}(\bar{G}_3)_k\big] \quad (5.49)$$

and

$$\sigma_{\bar{x}\bar{x}} = \sum_{k=1}^{3}(p_{\bar{x}})_k\big[(3-2\nu)(\bar{G}_2)_k + \bar{y}(\bar{G}_4)_k\big] + \sum_{k=1}^{3}(p_{\bar{y}})_k\big[2\nu(\bar{G}_3)_k - \bar{y}(\bar{G}_5)_k\big]$$
$$\sigma_{\bar{y}\bar{y}} = \sum_{k=1}^{3}(p_{\bar{x}})_k\big[-(1-2\nu)(\bar{G}_2)_k - \bar{y}(\bar{G}_4)_k\big] + \sum_{k=1}^{3}(p_{\bar{y}})_k\big[2(1-\nu)(\bar{G}_3)_k + \bar{y}(\bar{G}_5)_k\big] \quad (5.50)$$
$$\sigma_{\bar{x}\bar{y}} = \sum_{k=1}^{3}(p_{\bar{x}})_k\big[2(1-\nu)(\bar{G}_3)_k - \bar{y}(\bar{G}_5)_k\big] + \sum_{k=1}^{3}(p_{\bar{y}})_k\big[(1-2\nu)(\bar{G}_2)_k - \bar{y}(\bar{G}_4)_k\big]$$

The similarity between these expressions and (5.9) and (5.10) should be noted. Equations (5.49) and (5.50) involve six boundary parameters—$(p_{\bar{x}})_k$ and $(p_{\bar{y}})_k$ for $k = 1$ to 3—while (5.9) and (5.10) only involve two—$P_{\bar{x}}$ and $P_{\bar{y}}$—but the expressions are otherwise essentially the same. The displacements and stresses in the global x, y coordinate system for our higher order elements can be written by using transformation formulas (4.62) and (4.66) to obtain equations comparable to (5.11) and (5.12). This simple task is left as an exercise for the interested reader.

Functions $(\bar{G}_1)_k$ to $(\bar{G}_5)_k$ involve three types of integrals. Using (5.44), function $(G_3)_k$, for example, can be written as

$$\begin{aligned}(\bar{G}_3)_k = \mathsf{A}_k &\int_{-a}^{a}(\bar{x}-\xi_{\bar{x}})^2\frac{\partial}{\partial\bar{y}}g(\bar{x}-\xi_{\bar{x}},\bar{y})\,\mathrm{d}\xi_{\bar{x}} \\ + \mathsf{B}_k &\int_{-a}^{a}(\bar{x}-\xi_{\bar{x}})\frac{\partial}{\partial\bar{y}}g(\bar{x}-\xi_{\bar{x}},\bar{y})\,\mathrm{d}\xi_{\bar{x}} \\ + \mathsf{C}_k &\int_{-a}^{a}\frac{\partial}{\partial\bar{y}}g(\bar{x}-\xi_{\bar{x}},\bar{y})\,\mathrm{d}\xi_{\bar{x}}\end{aligned} \quad (5.51)$$

The last integral is the same as function $F_3(\bar{x}, \bar{y}) = \bar{F}_3$ defined in Sect. 5.3, and we will write this here as $(\bar{F}_3)_C$ to signify that is is associated with the constant part of the shape function $\alpha_k(\xi_{\bar{x}})$ in (5.44). The other two integrals will be denoted as $(\bar{F}_3)_L$ for the linear term and $(\bar{F}_3)_Q$ for the quadratic term. Function $(\bar{G}_3)_k$ can then be written as

$$(\bar{G}_3)_k = \mathsf{A}_k(\bar{F}_3)_Q + \mathsf{B}_k(\bar{F}_3)_L + \mathsf{C}_k(\bar{F}_3)_C \tag{5.52}$$

The five functions defined in (5.48) can all be represented in this way.

The required integrals can be carried out analytically using the integration formulas given in Appendix A. The integrals $(\bar{F}_1)_C$, etc., are the same as $\bar{F}_1(\bar{x}, \bar{y})$, etc., in (5.7) and (5.8). The other integrals are as follows:

$$\begin{aligned}
(\bar{F}_1)_L &= \int_{-a}^{a} (\bar{x} - \xi_{\bar{x}}) g(\bar{x} - \xi_{\bar{x}}, \bar{y})\, \mathrm{d}\xi_{\bar{x}} \\
&= \frac{1}{8\pi(1-\nu)} \left[\bar{r}_1^2 \ln \bar{r}_1 - \bar{r}_2^2 \ln \bar{r}_2 + 2a\bar{x} \right] \\
(\bar{F}_2)_L &= \int_{-a}^{a} (\bar{x} - \xi_{\bar{x}}) \frac{\partial}{\partial \bar{x}} g(\bar{x} - \xi_{\bar{x}}, \bar{y})\, \mathrm{d}\xi_{\bar{x}} \\
&= \frac{1}{4\pi(1-\nu)} \left[\bar{y}(\bar{\theta}_1 - \bar{\theta}_2) - 2a \right] \\
(\bar{F}_3)_L &= \int_{-a}^{a} (\bar{x} - \xi_{\bar{x}}) \frac{\partial}{\partial \bar{y}} g(\bar{x} - \xi_{\bar{x}}, \bar{y})\, \mathrm{d}\xi_{\bar{x}} \\
&= \frac{1}{4\pi(1-\nu)} \bar{y} \ln \frac{\bar{r}_1}{\bar{r}_2} \\
(\bar{F}_4)_L &= \int_{-a}^{a} (\bar{x} - \xi_{\bar{x}}) \frac{\partial^2}{\partial \bar{x} \partial \bar{y}} g(\bar{x} - \xi_{\bar{x}}, \bar{y})\, \mathrm{d}\xi_{\bar{x}} \\
&= \frac{1}{4\pi(1-\nu)} \left[\frac{(\bar{x} - a)\bar{y}}{\bar{r}_1^2} - \frac{(\bar{x} + a)\bar{y}}{\bar{r}_2^2} + (\bar{\theta}_1 - \bar{\theta}_2) \right] \\
(\bar{F}_5)_L &= \int_{-a}^{a} (\bar{x} - \xi_{\bar{x}}) \frac{\partial^2}{\partial \bar{x}^2} g(\bar{x} - \xi_{\bar{x}}, \bar{y})\, \mathrm{d}\xi_{\bar{x}} \\
&= \frac{1}{4\pi(1-\nu)} \left[\frac{(\bar{x} - a)^2}{\bar{r}_1^2} - \frac{(\bar{x} + a)^2}{\bar{r}_2^2} - \ln \frac{\bar{r}_1}{\bar{r}_2} \right]
\end{aligned} \tag{5.53}$$

$$\begin{aligned}
(\bar{F}_1)_Q &= \int_{-a}^{a} (\bar{x} - \xi_{\bar{x}})^2 g(\bar{x} - \xi_{\bar{x}}, \bar{y})\, \mathrm{d}\xi_{\bar{x}} \\
&= \frac{1}{12\pi(1-\nu)} \left[(\bar{x} - a)^3 \ln \bar{r}_1 - (\bar{x} + a)^3 \ln \bar{r}_2 + \bar{y}^3(\bar{\theta}_1 - \bar{\theta}_2) + 2a\left(\bar{x}^2 - \bar{y}^2 + \frac{a^2}{3} \right) \right] \\
(\bar{F}_2)_Q &= \int_{-a}^{a} (\bar{x} - \xi_{\bar{x}})^2 \frac{\partial}{\partial \bar{x}} g(\bar{x} - \xi_{\bar{x}}, \bar{y})\, \mathrm{d}\xi_{\bar{x}} \\
&= \frac{-1}{4\pi(1-\nu)} \left[\bar{y}^2 \ln \frac{\bar{r}_1}{\bar{r}_2} + 2a\bar{x} \right] \\
(\bar{F}_3)_Q &= \int_{-a}^{a} (\bar{x} - \xi_{\bar{x}})^2 \frac{\partial}{\partial \bar{y}} g(\bar{x} - \xi_{\bar{x}}, \bar{y})\, \mathrm{d}\xi_{\bar{x}} \\
&= \frac{\bar{y}}{4\pi(1-\nu)} \left[\bar{y}(\bar{\theta}_1 - \bar{\theta}_2) - 2a \right]
\end{aligned} \tag{5.54}$$

$$
\begin{aligned}
(\bar{F}_4)_Q &= \int_{-a}^{a} (\bar{x} - \xi_{\bar{x}})^2 \frac{\partial^2}{\partial \bar{x} \partial \bar{y}} g(\bar{x} - \xi_{\bar{x}}, \bar{y})\, \mathrm{d}\xi_{\bar{x}} \\
&= \frac{\bar{y}}{4\pi(1-\nu)} \left[\frac{(\bar{x}-a)^2}{\bar{r}_1^2} - \frac{(\bar{x}+a)^2}{\bar{r}_2^2} - 2 \ln \frac{\bar{r}_1}{\bar{r}_2} \right] \\
(\bar{F}_5)_Q &= \int_{-a}^{a} (\bar{x} - \xi_{\bar{x}})^2 \frac{\partial^2}{\partial \bar{x}^2} g(\bar{x} - \xi_{\bar{x}}, \bar{y})\, \mathrm{d}\xi_{\bar{x}} \\
&= \frac{1}{4\pi(1-\nu)} \left[\frac{(\bar{x}-a)^3}{\bar{r}_1^2} - \frac{(\bar{x}+a)^3}{\bar{r}_2^2} - 2\bar{y}(\bar{\theta}_1 - \bar{\theta}_2) + 4a \right]
\end{aligned}
$$

The preceding results can be combined in a similar manner as described for constant strength elements in Sect. 5.2. In the present case, each boundary element i will have six parameters: $(\overset{i}{p}_{\bar{x}})_k$ and $(\overset{i}{p}_{\bar{y}})_k$ for $k = 1$ to 3. The governing system of algebraic equations for N boundary elements will therefore be of size $6N \times 6N$. These equations can be developed from the following expressions, which are comparable to (5.2):

$$
\begin{aligned}
\sum_{l=1}^{3} (\overset{i}{t}_{\bar{x}})_l &= \sum_{k=1}^{3} \left[\sum_{j=1}^{N} (\overset{ij}{A}_{\bar{x}\,\bar{x}})_{lk} (\overset{j}{p}_{\bar{x}})_k + \sum_{j=1}^{N} (\overset{ij}{A}_{\bar{x}\,\bar{y}})_{lk} (\overset{j}{p}_{\bar{y}})_k \right] \\
\sum_{l=1}^{3} (\overset{i}{t}_{\bar{y}})_l &= \sum_{k=1}^{3} \left[\sum_{j=1}^{N} (\overset{ij}{A}_{\bar{y}\,\bar{x}})_{lk} (\overset{j}{p}_{\bar{x}})_k + \sum_{j=1}^{N} (\overset{ij}{A}_{\bar{y}\,\bar{y}})_{lk} (\overset{j}{p}_{\bar{y}})_k \right]
\end{aligned}
\tag{5.55}
$$

The influence coefficients $(\overset{ij}{A}_{\bar{x}\,\bar{x}})_{lk}$, etc., are calculated in essentially the same way as $\overset{ij}{A}_{\bar{x}\,\bar{x}}$, etc., in (5.21), noting that l and k range from 1 to 3 as i and j range from 1 to N. Similar expressions for the boundary displacements can be written by analogy with (5.15). They are:

$$
\begin{aligned}
\sum_{l=1}^{3} (\overset{i}{u}_{\bar{x}})_l &= \sum_{k=1}^{3} \left[\sum_{j=1}^{N} (\overset{ij}{B}_{\bar{x}\,\bar{x}})_{lk} (\overset{j}{p}_{\bar{x}})_k + \sum_{j=1}^{N} (\overset{ij}{B}_{\bar{x}\,\bar{y}})_{lk} (\overset{j}{p}_{\bar{y}})_k \right] \\
\sum_{l=1}^{3} (\overset{i}{u}_{\bar{y}})_l &= \sum_{k=1}^{3} \left[\sum_{j=1}^{N} (\overset{ij}{B}_{\bar{y}\,\bar{x}})_{lk} (\overset{j}{p}_{\bar{x}})_k + \sum_{j=1}^{N} (\overset{ij}{B}_{\bar{y}\,\bar{y}})_{lk} (\overset{j}{p}_{\bar{y}})_k \right]
\end{aligned}
\tag{5.56}
$$

The required system of algebraic equations is formed by considering the boundary conditions at each of the $3N$ collocation points and using the appropriate ones of (5.55) and (5.56). To this end, we number the collocation points consecutively as $m = 3(i-1) + l$ and $n = 3(j-1) + k$ and let $\overset{n}{p}_{\bar{x}}$ and $\overset{n}{p}_{\bar{y}}$ be the applied tractions (the stress discontinuities) at collocation point n and $\overset{m}{b}_{\bar{x}}$ and $\overset{m}{b}_{\bar{y}}$ be the prescribed values of traction or displacement at collocation point m. Following the approach outlined in Sect. 5.6, we then represent the $6N \times 6N$ system of algebraic equations as

$$\sum_{n=1}^{3N} \overset{mn}{C}_{\bar{x}\bar{x}} \overset{n}{p}_{\bar{x}} + \sum_{n=1}^{3N} \overset{mn}{C}_{\bar{x}\bar{y}} \overset{n}{p}_{\bar{y}} = \overset{m}{b}_{\bar{x}}$$
$$\sum_{n=1}^{3N} \overset{mn}{C}_{\bar{y}\bar{x}} \overset{n}{p}_{\bar{x}} + \sum_{n=1}^{3N} \overset{mn}{C}_{\bar{y}\bar{y}} \overset{n}{p}_{\bar{y}} = \overset{m}{b}_{\bar{y}} \quad (5.57)$$

for $m = 1$ to $3N$.

5.10.1 Example

As an example of the use of quadratic stress discontinuity elements we consider the problem of a simply-supported beam subjected to a uniformly distributed load $\sigma_{yy} = -p$ (i.e., compression). The beam has unit thickness and a length to depth ratio of 1 : 1. A numerical model of this problem used 25 equal size quadratic elements on the top and on the bottom of a symmetric half of the beam, with 5 elements on the end. The upper surface of the beam was subjected to a uniform traction $t_y = -p$ and one element on the end of the lower surface was loaded by a traction $t_y = 25p$ to maintain equilibrium. Rigid body translation was prevented by fixing the y component of displacement at an element on the end of the beam; the x component was fixed by symmetry.

The bending stress σ_{xx}/p at the center of the beam is compared with elementary beam theory in Fig. 5.14 for both quadratic elements and constant elements. (The constant element model used three times as many elements to make the number of unknowns the same in both cases.) It can be seen that the results for the quadratic element model agree well with elementary beam theory whereas those for constant

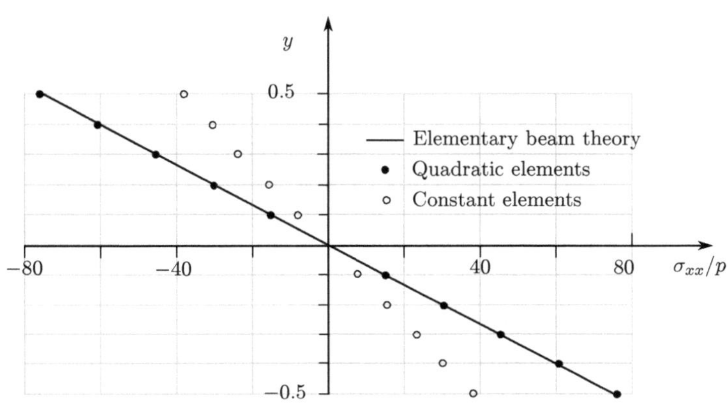

Fig. 5.14 Bending stress in a simply-supported beam

strength elements do not, despite both models having the same number of degrees of freedom.

This example exposes a major deficiency of the stress discontinuity method based on piecewise constant boundary elements: such elements are not capable of accurately representing bending stresses.

5.11 Integral Formulation

The stress discontinuity method was presented in this chapter without explicit reference to boundary integral equations. Instead, an influence function approach was adopted to show how the solutions derived in Sect. 4.12 for constant tractions applied to a finite line segment in an infinite elastic solid could be superimposed to construct a numerical procedure for solving general boundary value problems. Another way of viewing this approach is to consider it as a discretized representation of the integral formulas given in Sect. 4.18 for the displacements, stresses, and tractions for the point force analogue of the single layer potential. We will conclude this discussion of the stress discontinuity method by highlighting the equivalence of the two approaches for calculating the stress components. Similar calculations can be done for the displacements and tractions.

Referring to (4.299), we write the stress components $\sigma_{ij}(\boldsymbol{x})$ at the sample point $\boldsymbol{x}$ in terms of a distribution of forces $\rho_k(\boldsymbol{\xi})\,\mathrm{d}S(\boldsymbol{\xi})$ at the load point $\boldsymbol{\xi}$ as

$$\sigma_{ij}(\boldsymbol{x}) = \int_S \rho_k(\boldsymbol{\xi}) B_{kij}(\boldsymbol{x}, \boldsymbol{\xi})\,\mathrm{d}S(\boldsymbol{\xi}) \tag{5.58}$$

or, in unabridged form,

$$\begin{aligned}
\sigma_{xx}(\boldsymbol{x}) &= \int_S \left[\rho_x(\boldsymbol{\xi}) B_{xxx}(\boldsymbol{x}, \boldsymbol{\xi}) + \rho_y(\boldsymbol{\xi}) B_{yxx}(\boldsymbol{x}, \boldsymbol{\xi})\right] \mathrm{d}S(\boldsymbol{\xi}) \\
\sigma_{yy}(\boldsymbol{x}) &= \int_S \left[\rho_x(\boldsymbol{\xi}) B_{xyy}(\boldsymbol{x}, \boldsymbol{\xi}) + \rho_y(\boldsymbol{\xi}) B_{yyy}(\boldsymbol{x}, \boldsymbol{\xi})\right] \mathrm{d}S(\boldsymbol{\xi}) \\
\sigma_{xy}(\boldsymbol{x}) &= \int_S \left[\rho_x(\boldsymbol{\xi}) B_{xxy}(\boldsymbol{x}, \boldsymbol{\xi}) + \rho_y(\boldsymbol{\xi}) B_{yxy}(\boldsymbol{x}, \boldsymbol{\xi})\right] \mathrm{d}S(\boldsymbol{\xi})
\end{aligned} \tag{5.59}$$

We next suppose that the boundary S is approximated by N straight line elements with the jth element centered at point $\overset{j}{\boldsymbol{\xi}}$ (such that $\xi_x = \overset{j}{x}$ and $\xi_y = \overset{j}{y}$) and inclined from the x axis by an angle $\overset{j}{\beta}$, so that $(\boldsymbol{x}, \boldsymbol{\xi}) = (x - \overset{j}{x}, y - \overset{j}{y})$. Taking the applied loading over the element as constant, i.e., $\rho_x = \overset{j}{P}_x$ and $\rho_y = \overset{j}{P}_y$ for $-\overset{j}{a} \le \bar{x} \le \overset{j}{a}$, $\bar{y} = 0$, where the local coordinates $\bar{x}$ and $\bar{y}$ are given by (5.16), the discretized forms of (5.59) can be written as

$$\sigma_{xx}(x, y) = \sum_{j=1}^{N} \overset{j}{P}_x \int_{-\overset{j}{a}}^{\overset{j}{a}} B_{xxx}(x - \overset{j}{x}, y - \overset{j}{y})\,\mathrm{d}\bar{x} + \sum_{j=1}^{N} \overset{j}{P}_y \int_{-\overset{j}{a}}^{\overset{j}{a}} B_{yxx}(x - \overset{j}{x}, y - \overset{j}{y})\,\mathrm{d}\bar{x}$$

$$\sigma_{yy}(x, y) = \sum_{j=1}^{N} \overset{j}{P}_x \int_{-\overset{j}{a}}^{\overset{j}{a}} B_{xyy}(x - \overset{j}{x}, y - \overset{j}{y})\, \mathrm{d}\bar{x} + \sum_{j=1}^{N} \overset{j}{P}_y \int_{-\overset{j}{a}}^{\overset{j}{a}} B_{yyy}(x - \overset{j}{x}, y - \overset{j}{y})\, \mathrm{d}\bar{x} \tag{5.60}$$

$$\sigma_{xy}(x, y) = \sum_{j=1}^{N} \overset{j}{P}_x \int_{-\overset{j}{a}}^{\overset{j}{a}} B_{xxy}(x - \overset{j}{x}, y - \overset{j}{y})\, \mathrm{d}\bar{x} + \sum_{j=1}^{N} \overset{j}{P}_y \int_{-\overset{j}{a}}^{\overset{j}{a}} B_{yxy}(x - \overset{j}{x}, y - \overset{j}{y})\, \mathrm{d}\bar{x}$$

We have set $\mathrm{d}S(\boldsymbol{\xi}) = \mathrm{d}\bar{x}$ in writing these equations because $\bar{x} = \xi_{\bar{x}}$.

The coefficients $B_{xxx}(x - \overset{j}{x}, y - \overset{j}{y})$, etc., for an isotropic material are given by (4.256), and these can be expressed in terms of $\bar{x}$ and $\bar{y}$ coordinates using the result

$$\begin{aligned} x - \overset{j}{x} &= \bar{x} \cos \overset{j}{\beta} - \bar{y} \sin \overset{j}{\beta} \\ y - \overset{j}{y} &= \bar{x} \sin \overset{j}{\beta} + \bar{y} \cos \overset{j}{\beta} \end{aligned} \tag{5.61}$$

In addition, the applied loads (the stress discontinuities) $\overset{j}{P}_x$ and $\overset{j}{P}_y$ can be expressed in terms of their $\bar{x}$ and $\bar{y}$ components using similar transformation formulas, i.e.,

$$\begin{aligned} \overset{j}{P}_x &= \overset{j}{P}_{\bar{x}} \cos \overset{j}{\beta} - \overset{j}{P}_{\bar{y}} \sin \overset{j}{\beta} \\ \overset{j}{P}_y &= \overset{j}{P}_{\bar{x}} \sin \overset{j}{\beta} + \overset{j}{P}_{\bar{y}} \cos \overset{j}{\beta} \end{aligned} \tag{5.62}$$

Substituting (5.61) and (5.62) into (5.60) we obtain expressions of the form

$$\begin{aligned} \sigma_{yy}(x, y) = &\sum_{j=1}^{N} \overset{j}{P}_{\bar{x}} \int_{-\overset{j}{a}}^{\overset{j}{a}} \left[B_{xyy}(\bar{x}, \bar{y}) \cos \overset{j}{\beta} + B_{yyy}(\bar{x}, \bar{y}) \sin \overset{j}{\beta}\right] \mathrm{d}\bar{x} \\ &+ \sum_{j=1}^{N} \overset{j}{P}_{\bar{y}} \int_{-\overset{j}{a}}^{\overset{j}{a}} \left[-B_{xyy}(\bar{x}, \bar{y}) \sin \overset{j}{\beta} + B_{yyy}(\bar{x}, \bar{y}) \cos \overset{j}{\beta}\right] \mathrm{d}\bar{x} \end{aligned} \tag{5.63}$$

with similar results for $\sigma_{xx}(x, y)$ and $\sigma_{xy}(x, y)$. Using the chain rule for partial differentiation and the definitions (4.256) for $B_{xyy}(\bar{x}, \bar{y})$ and $B_{yyy}(\bar{x}, \bar{y})$ the integrals in (5.63) are found to be

$$\begin{aligned} &\int_{-a}^{a} \left(B_{xyy}(\bar{x}, \bar{y}) \cos \beta + B_{yyy}(\bar{x}, \bar{y}) \sin \beta\right) \mathrm{d}\bar{x} \\ &= \frac{\partial f}{\partial \bar{x}} - 2(1 - \nu)\left(\cos 2\beta \frac{\partial f}{\partial \bar{x}} - \sin 2\beta \frac{\partial f}{\partial \bar{y}}\right) \\ &\quad - \bar{y}\left(\cos 2\beta \frac{\partial^2 f}{\partial \bar{x} \partial \bar{y}} - \sin 2\beta \frac{\partial^2 f}{\partial \bar{y}^2}\right) \end{aligned} \tag{5.64}$$

and

$$\int_{-a}^{a} \big(-B_{xyy}(\bar{x}, \bar{y}) \sin\beta + B_{yyy}(\bar{x}, \bar{y}) \cos\beta\big)\, \mathrm{d}\bar{x}$$
$$= \frac{\partial f}{\partial \bar{y}} + (1 - 2\nu)\big(\sin 2\beta \frac{\partial f}{\partial \bar{x}} + \cos 2\beta \frac{\partial f}{\partial \bar{y}}\big)$$
$$- \bar{y}\big(\sin 2\beta \frac{\partial^2 f}{\partial \bar{x} \partial \bar{y}} + \cos 2\beta \frac{\partial^2 f}{\partial \bar{y}^2}\big) \tag{5.65}$$

in which we have written a and β instead of $\overset{j}{a}$ and $\overset{j}{\beta}$ for notational convenience and where $f(\bar{x}, \bar{y})$ is defined in (5.7). Recalling that $\bar{F}_2 = F(\bar{x}, \bar{y}) = \partial f(\bar{x}, \bar{y})/\partial \bar{x}$, etc., (see (5.8)), it can be seen from these results that a typical set of terms in the summation indicated in (5.63) is the same as the second one of (5.12).

Similar results can be found for $\sigma_{xx}(x, y)$ and $\sigma_{xy}(x, y)$, leading to the conclusion that the influence function approach for the stress discontinuity method is completely equivalent to a comparable discretization of the single layer elastic potential integral formula (5.58) for stresses. This conclusion also holds for the single layer integral formulas (4.299) for the displacements and tractions.

5.11.1 Generalizations of the Stress Discontinuity Method

We noted at the outset that the stress discontinuity method is an indirect boundary element method because the stress discontinuities (i.e., the densities $\rho_k(\boldsymbol{\xi})$ in (5.58)) do not, in general, have physical meaning. Exceptions to this statement occur for some problems, and the interested reader might wish to pursue this topic in more detail.

In one such problem, the region of interest contains a thin, stiff inhomogeneity (a membrane or shell type inhomogeneity) or a so-called interphase (see Benveniste and Miloh (2001) and Mogilevskaya et al. (2021) and the references therein). The boundary conditions for the problem involve continuity of displacements and jumps in stresses that are expressed in terms of various derivatives of the displacements. Another example is the problem of an Eshelby inclusion (Elshelby 1957), in which the region of interest contains a domain of the same properties as the surrounding material but subjected to stress-free transformational strains (eigenstrains). The boundary conditions for the latter problem also involve continuity of displacements and jumps in stresses that are expressed in terms of the prescribed eigenstrains (see Mogilevskaya and Stolarski 2015).

5.12 Exercises/Projects

The reader's mastery of the boundary element approach will be facilitated by writing computer programs to implement the algorithms described in this book. The particular programming language used is not important, and fully functional programs can be written in languages such as Python and Matlab. Several open-source boundary element computer programs are available on the Internet, and some readers might wish to pursue this option. In addition, three simple FORTRAN computer programs are listed in the book by Crouch and Starfield (1983), including one for the stress discontinuity method (called the fictitious stress method by Crouch and Starfield), and some readers might want to use one of these as a staring point because of the close correspondence between the computer code and mathematical formulas presented in this book.

Regardless of which avenue the reader chooses to follow, however, it is important for him or her to gain first-hand programming experience. The exercises, or projects, given below are suggested with this thought in mind.

1. Prepare a computer program for the two-dimensional stress discontinuity method for isotropic elasticity, as outlined in Sect. 5.6. Verify your program by solving the example problems described in Sect. 5.9.
2. Modify your program from Exercise 1 to include the case of orthotropic (transversely isotropic) elasticity.
3. Modify your program from Exercise 1 to use piecewise quadratic elements as described in Sect. 5.10.

References

Benveniste, Y., and T. Miloh. 2001. Imperfect soft and stiff interfaces in two-dimensional elasticity. *Mechanics of Materials* 33: 309–23.

Claesson, J., and B. Bohloli. 2002. Brazilian test: stress field and tensile strength of anisotropic rocks using an analytical solution. *International Journal of Rock Mechanics and Mining Sciences* 39: 991–1004.

Crouch, S.L., and A.M. Starfield. 1983. *Boundary element methods in solid mechanics*. George Allen & Unwin.

Eshelby, J.D. 1957. The determination of the elastic field of an ellipsoidal inclusion, and related problems. *Proceedings of the royal society of London. Series A. Mathematical and Physical Sciences* 241: 376–96.

Green, A.E., and G.I. Taylor. 1945. Stress systems in aeolotropic plates. *Proceedings of the Royal Society of London. Series A. Mathematical and Physical Sciences* 184: 181–95.

Hondros, G. 1959. The evaluation of Poisson's ratio and the modulus of materials of a low tensile resistance by the Brazilian (indirect tensile) test, with particular reference to concrete. *Australian Journal of Applied Science* 10: 243–68.

Lekhnitskii, S. G. 1968. *Anisotropic plates*. Gordon and Breach Scientific Publications.

Mogilevskaya, S.G., and H.K. Stolarski. 2015. On 'strange' properties of some symmetric inhomogeneities. *Proceedings of the Royal Society A: Mathematical, Physical and Engineering Sciences* 471: 20150157.

Mogilevskaya, S.G., A.Y. Zemlyanova, and V.I. Kushch. 2021. Fiber- and particle-reinforced composite materials with the Gurtin-Murdoch and Steigmann-Ogden surface energy endowed interfaces. *Applied Mechanics Reviews* 73 (5): 050801.
Sokolnikoff, I.S. 1956. *Mathematical theory of elasticity*, 2nd ed. McGraw-Hill.

Chapter 6
The Displacement Discontinuity Method

6.1 Introduction

Many practical problems in solid mechanics involve bodies containing thin, slit-like openings or cracks. A crack has two surfaces, or boundaries, one effectively coinciding with the other. The stress discontinuity method breaks down for such problems, because the effects of elements placed along one crack surface are indistinguishable from the effects of those placed along the other. Another boundary element method, however, can be devised to cope directly with problems of this type. The simplest version of this method, the *displacement discontinuity method*, employs the analytical solution to the problem of a constant discontinuity in displacement over a finite line segment in the $x,\ y$ plane of an infinite elastic solid, as presented in Sect. 4.13.

The displacement discontinuity method is based on the notion that one can make a discrete approximation to a continuous distribution of displacement discontinuity along a crack. The method was originally presented (Crouch 1976) as an influence function technique in which a line crack Γ in two dimensions is represented as a number of elemental, constant strength displacement discontinuities over straight-line segments joined end to end. In this version of the approach, the analytical solution to the problem of a constant displacement discontinuity over an arbitrarily oriented, finite line segment is used to express the tractions (t_x, t_y) at one element in terms of the displacement discontinuities (D_x, D_y) at another. By considering the combined effects of N such elements along the crack, the tractions at a typical element i can be expressed in terms of the displacement discontinuities at all N elements, $j = 1$ to N. If the tractions are prescribed for $i = 1$ to N, one then obtains a system of $2N$ algebraic equations in $2N$ unknowns— the values of the displacement discontinuities at all the elements— which can be solved by standard numerical procedures. The displacements and stresses at selected points within the region of interest can subsequently be found by adding the influences of the N elemental displacement discontinuities at these points.

S. L. Crouch and S. G. Mogilevskaya, *A First Course in Boundary Element Methods*, Mathematical Engineering, https://doi.org/10.1007/978-3-031-63341-6_6

This general approach is reminiscent of the numerical procedure discussed in Chap. 5 for the stress discontinuity method. Like the stress discontinuity method, the displacement discontinuity method can, with some restrictions, be applied to general boundary value problems. For these applications the displacement discontinuities do not have physical meaning. For crack problems, the displacement discontinuities represent approximations to the relative movements between the crack surfaces and consequently are physically meaningful. The same is true for problems involving thin, soft inhomogeneities or a spring type interphase (Benveniste and Miloh 2001).

Computational details of the displacement discontinuity method are covered in this chapter. We will show how the method can easily be adapted to enforce special boundary constraints between crack surfaces such as no overlapping and frictional sliding. We will also explain how to include more elaborate—and realistic—distributions of displacement discontinuity along the elements, leading to more accurate higher order implementations of the approach. Finally, we will show that the influence function approach is equivalent to solving numerically the integral equations related to the analogue of the double layer potential, Eq. (4.300).

6.2 Numerical Procedure

The results of Sect. 4.13 form the basis of the displacement discontinuity method. To illustrate the general approach, we consider the problem depicted in Fig. 6.1a for a line crack Γ in an infinite elastic solid. The crack is assumed to be very long in the z direction, perpendicular to the plane of the figure, so that plane strain conditions apply for the x, y plane. The local coordinates s and n are respectively tangent and normal to the crack; they therefore vary from point to point along its length. We will suppose that the surfaces of the crack are subjected to a uniform normal traction $\sigma_n = -p$ (i.e., compression) along its entire length, with the shear traction σ_s equal to zero.

A numerical solution to the problem will be sought by dividing the crack Γ into N straight line segments joined end to end, as shown in Fig. 6.1b. The length of a

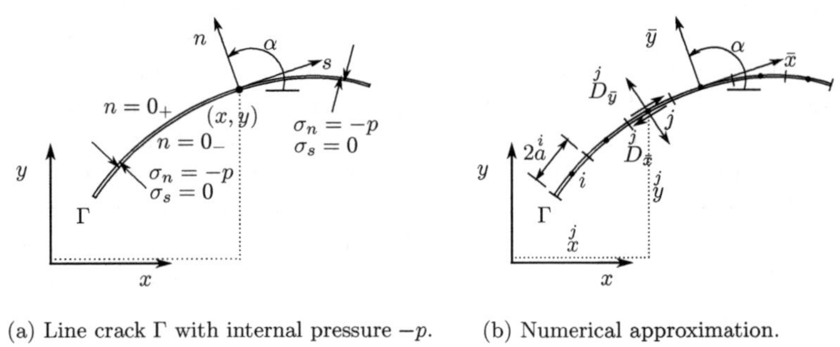

(a) Line crack Γ with internal pressure $-p$. (b) Numerical approximation.

Fig. 6.1 Representation of a crack by displacement discontinuity elements

typical boundary element i is $2\overset{i}{a}$. The boundary conditions for this approximation to the original problem can be written as

$$\overset{i}{\sigma}_n = -p, \qquad \overset{i}{\sigma}_s = 0 \tag{6.1}$$

for $i = 1$ to N. Just as in the case of the stress discontinuity method (see Sect. 5.2), each boundary element has its own local coordinate system, so that the s, n coordinates for element i are different from those for element j. It will again be convenient to refer to these coordinates as $\bar{x}$ (for coordinate s) and $\bar{y}$ (for coordinate n). The positive direction for $\bar{x}$ is taken as the direction in which the crack geometry is specified, element by element, and the positive direction for $\bar{y}$ is chosen such that $\bar{x}$ and $\bar{y}$ form a right-handed system.

The piecewise constant components of displacement discontinuity for a typical element j are the differences in displacement between the two surfaces of the element, $-\overset{j}{a} \le \bar{x} \le \overset{j}{a}$, $\bar{y} = 0_{\pm}$, as follows:

$$\begin{aligned} \overset{j}{D}_{\bar{x}} &= (\overset{j}{u}_{\bar{x}})^{+} - (\overset{j}{u}_{\bar{x}})^{-} \\ \overset{j}{D}_{\bar{y}} &= (\overset{j}{u}_{\bar{y}})^{+} - (\overset{j}{u}_{\bar{y}})^{-} \end{aligned} \tag{6.2}$$

The superscripts "+" and "−" in these expressions denote the positive and negative surfaces of the crack with respect to local coordinate $\bar{y}$. By definition, the displacements are positive in the positive $\bar{x}$ and $\bar{y}$ directions, and this means that the normal component of displacement discontinuity $\overset{j}{D}_{\bar{y}}$ is positive if the two surfaces of the element displace away from one another, that is, if the crack locally opens.[1] The shear component $\overset{j}{D}_{\bar{x}}$ is positive if the positive surface of the element moves to the right with respect to the negative surface (cf. Fig. 6.1b). It should be mentioned that this sign convention is opposite to the one used by Crouch (1976) and later by Crouch and Starfield (1983).

The outward normals of the two surfaces of an element point in opposite directions: $\overset{j}{n}_{\bar{x}} = 0$, $\overset{j}{n}_{\overline{y}} = -1$ for the "upper" surface, $\overline{y} = 0_{+}$, and $\overset{j}{n}_{\bar{x}} = 0$, $\overset{j}{n}_{\overline{y}} = +1$ for the "lower" surface, $\overline{y} = 0_{-}$. The tractions on these surfaces are self-equilibrated, so that $(\overset{j}{t}_{\bar{x}})^{+} = -(\overset{j}{t}_{\bar{x}})^{-}$ and $(\overset{j}{t}_{\bar{y}})^{+} = -(\overset{j}{t}_{\bar{y}})^{-}$. If the normal and shear stresses on the element are negative, as depicted in Fig. 6.2, one has, by virtue of (4.16), $(\overset{j}{t}_{\bar{x}})^{+} > 0$, $(\overset{j}{t}_{\bar{y}})^{+} > 0$ and $(\overset{j}{t}_{\bar{x}})^{-} < 0$, $(\overset{j}{t}_{\bar{y}})^{-} < 0$. In the following discussion, we will use the lower surface of the element as a reference for consistency with the usage adopted for the stress discontinuity method. Negative tractions $(\overset{j}{t}_{\bar{x}})^{-} = \overset{j}{\sigma}_{\overline{x}\overline{y}} < 0$ and $(\overset{j}{t}_{\bar{y}})^{-} =$

[1] We will suppose for the moment that movement of the crack surfaces is not restricted in any way. We will later consider how to impose the physical constraint that the surfaces cannot overlap.

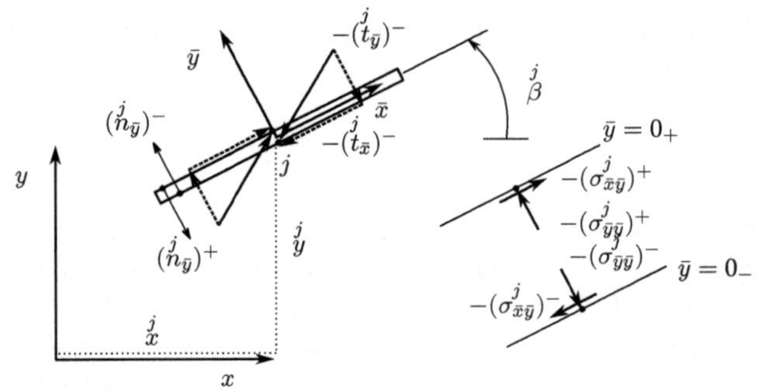

Fig. 6.2 Tractions on surfaces of displacement discontinuity element

$\overset{j}{\sigma}_{\bar{y}\bar{y}} < 0$ on this surface will correspond to positive displacement discontinuities $\overset{j}{D}_{\bar{x}}$ and $\overset{j}{D}_{\bar{y}}$ for the element. For notational convenience, we will in most instances suppress the superscript "−" when referring to these quantities, i.e., we will write $\overset{j}{t}_{\bar{x}}$ and $\overset{j}{t}_{\bar{y}}$ instead of $(\overset{j}{t}_{\bar{x}})^-$ and $(\overset{j}{t}_{\bar{y}})^-$.

Using either of the two solutions given in Sect. 4.13, together with suitable coordinate transformation equations to account for the locations and orientations of all the displacement discontinuity elements, the tractions at the midpoint of element i can be written as

$$
\begin{aligned}
\overset{i}{t}_{\bar{x}} &= \sum_{j=1}^{N} \overset{ij}{A}_{\bar{x}\bar{x}} \overset{j}{D}_{\bar{x}} + \sum_{j=1}^{N} \overset{ij}{A}_{\bar{x}\bar{y}} \overset{j}{D}_{\bar{y}} \\
\overset{i}{t}_{\bar{y}} &= \sum_{j=1}^{N} \overset{ij}{A}_{\bar{y}\bar{x}} \overset{j}{D}_{\bar{x}} + \sum_{j=1}^{N} \overset{ij}{A}_{\bar{y}\bar{y}} \overset{j}{D}_{\bar{y}}
\end{aligned}
\tag{6.3}
$$

for $i = 1$ to N. The quantities $\overset{ij}{A}_{\bar{x}\bar{x}}$, etc., in these equations are the boundary traction influence coefficients for the problem. The coefficient $\overset{ij}{A}_{\bar{x}\bar{y}}$, for example, gives the shear traction $\overset{i}{t}_{\bar{x}}$ at the midpoint of the ith element due to a constant unit normal displacement discontinuity $\overset{j}{D}_{\bar{y}} = 1$ at the jth element. It will be convenient to use the same symbols for these influence coefficients as the ones introduced previously for the stress discontinuity method. The coefficients in the two cases are, of course, different, and it will be clear from context which ones are being used.

A numerical solution to the problem specified by (6.1) can be obtained by setting $\overset{i}{t}_{\bar{x}} = \overset{i}{\sigma}_s = 0$ and $\overset{i}{t}_{\bar{y}} = \overset{i}{\sigma}_n = -p$ and solving the following system of $2N$ simultaneous linear algebraic equations in $2N$ unknowns:

$$\begin{aligned} &\sum_{j=1}^{N} \overset{ij}{A}_{\bar{x}\,\bar{x}} \overset{j}{D}_{\bar{x}} + \sum_{j=1}^{N} \overset{ij}{A}_{\bar{x}\,\bar{y}} \overset{j}{D}_{\bar{y}} = 0 \\ &\sum_{j=1}^{N} \overset{ij}{A}_{\bar{y}\,\bar{x}} \overset{j}{D}_{\bar{x}} + \sum_{j=1}^{N} \overset{ij}{A}_{\bar{y}\,\bar{y}} \overset{j}{D}_{\bar{y}} = -p \end{aligned} \tag{6.4}$$

Once these equations have been solved, the displacements and stresses at any point in the body (except at the element endpoints) can be expressed in terms of the N elemental displacement discontinuities $\overset{j}{D}_{\bar{x}}$ and $\overset{j}{D}_{\bar{y}}$. Detailed computational aspects of the approach are discussed below.

6.3 Coordinate Transformations

The simplest version of the displacement discontinuity method is based on the analytical solution to the problem of a constant strength displacement discontinuity over an arbitrarily oriented, finite line segment in an infinite elastic solid. We obtain below expressions for the displacements and stresses in the global x, y coordinate system due to displacement discontinuities referred to local $\bar{x}$, $\bar{y}$ coordinates for both isotropic and anisotropic materials. Just as in the case of the stress discontinuity method, these results can be used to compute the boundary influence coefficients, this time for the displacement discontinuity method.

6.3.1 Isotropic Material

The displacements and stresses in the $\bar{x}$, $\bar{y}$ coordinate system (see (5.4)) due to a displacement discontinuity $D_{\bar{x}}$ and $D_{\bar{y}}$ along the line segment $|\,\bar{x}\,| \leq a$, $\bar{y} = 0$ in an isotropic material can be written by inspection of (4.192) and (4.193) because there is no preferred direction for an infinite, homogeneous, isotropic elastic body. Thus, we replace all occurrences of x and y in these equations by $\bar{x}$ and $\bar{y}$ and obtain the following results:

$$u_{\bar{x}} = D_{\bar{x}}\big[-2(1-\nu)f_{,\bar{y}} - \bar{y}f_{,\bar{y}\bar{y}}\big] + D_{\bar{y}}\big[(1-2\nu)f_{,\bar{x}} + \bar{y}f_{,\bar{x}\bar{y}}\big]$$
$$u_{\bar{y}} = D_{\bar{x}}\big[-(1-2\nu)f_{,\bar{x}} + \bar{y}f_{,\bar{x}\bar{y}}\big] + D_{\bar{y}}\big[-2(1-\nu)f_{,\bar{y}} + \bar{y}f_{,\bar{y}\bar{y}}\big] \tag{6.5}$$

and

$$\sigma_{\bar{x}\bar{x}} = 2\mu D_{\bar{x}}\big[-2f_{,\bar{x}\bar{y}} - \bar{y}f_{,\bar{x}\bar{y}\bar{y}}\big] + 2\mu D_{\bar{y}}\big[-f_{,\bar{y}\bar{y}} - \bar{y}f_{,\bar{y}\bar{y}\bar{y}}\big]$$
$$\sigma_{\bar{y}\bar{y}} = 2\mu D_{\bar{x}}\big[\bar{y}f_{,\bar{x}\bar{y}\bar{y}}\big] + 2\mu D_{\bar{y}}\big[-f_{,\bar{y}\bar{y}} + \bar{y}f_{,\bar{y}\bar{y}\bar{y}}\big] \tag{6.6}$$
$$\sigma_{\bar{x}\bar{y}} = 2\mu D_{\bar{x}}\big[-f_{,\bar{y}\bar{y}} - \bar{y}f_{,\bar{y}\bar{y}\bar{y}}\big] + 2\mu D_{\bar{y}}\big[\bar{y}f_{,\bar{x}\bar{y}\bar{y}}\big]$$

Function $f(\bar{x}, \bar{y})$ in these expressions and its derivatives through second order are given by (5.7) and (5.8). The two third order derivatives of $f(\bar{x}, \bar{y})$ needed to evaluate (6.6) are

$$f_{,\bar{x}\bar{y}\bar{y}} = F_6(\bar{x}, \bar{y}) = \frac{1}{4\pi(1-\nu)}\left[\frac{(\bar{x}-a)^2 - \bar{y}^2}{\bar{r}_1^4} - \frac{(\bar{x}+a)^2 - \bar{y}^2}{\bar{r}_2^4}\right]$$
$$f_{,\bar{y}\bar{y}\bar{y}} = F_7(\bar{x}, \bar{y}) = \frac{2\bar{y}}{4\pi(1-\nu)}\left[\frac{(\bar{x}-a)}{\bar{r}_1^4} - \frac{(\bar{x}+a)}{\bar{r}_2^4}\right] \tag{6.7}$$

In terms of functions $\bar{F}_1 = F_1(\bar{x}, \bar{y})$ through $\bar{F}_7 = F_7(\bar{x}, \bar{y})$ defined in (5.8) and (6.7), Eqs. (6.5) and (6.6) can be rewritten as

$$u_{\bar{x}} = D_{\bar{x}}\big[-2(1-\nu)\bar{F}_3 + \bar{y}\bar{F}_5\big] + D_{\bar{y}}\big[(1-2\nu)\bar{F}_2 + \bar{y}\bar{F}_4\big]$$
$$u_{\bar{y}} = D_{\bar{x}}\big[-(1-2\nu)\bar{F}_2 + \bar{y}\bar{F}_4\big] + D_{\bar{y}}\big[-2(1-\nu)\bar{F}_3 - \bar{y}\bar{F}_5\big] \tag{6.8}$$

and

$$\sigma_{\bar{x}\bar{x}} = 2\mu D_{\bar{x}}\big[-2\bar{F}_4 - \bar{y}\bar{F}_6\big] + 2\mu D_{\bar{y}}\big[\bar{F}_5 - \bar{y}\bar{F}_7\big]$$
$$\sigma_{\bar{y}\bar{y}} = 2\mu D_{\bar{x}}\big[\bar{y}\bar{F}_6\big] + 2\mu D_{\bar{y}}\big[\bar{F}_5 + \bar{y}\bar{F}_7\big] \tag{6.9}$$
$$\sigma_{\bar{x}\bar{y}} = 2\mu D_{\bar{x}}\big[\bar{F}_5 - \bar{y}\bar{F}_7\big] + 2\mu D_{\bar{y}}\big[\bar{y}\bar{F}_6\big]$$

Finally, the displacements and stresses in the global x, y coordinate system are obtained by using transformation formulas (4.62) and (4.66). Thus, we find

$$\begin{aligned} u_x &= D_{\bar{x}}\big[(1-2\nu)\sin\beta\bar{F}_2 - 2(1-\nu)\cos\beta\bar{F}_3 - \bar{y}(\sin\beta\bar{F}_4 - \cos\beta\bar{F}_5)\big] \\ &\quad + D_{\bar{y}}\big[(1-2\nu)\cos\beta\bar{F}_2 + 2(1-\nu)\sin\beta\bar{F}_3 + \bar{y}(\cos\beta\bar{F}_4 + \sin\beta\bar{F}_5)\big] \\ u_y &= D_{\bar{x}}\big[-(1-2\nu)\cos\beta\bar{F}_2 - 2(1-\nu)\sin\beta\bar{F}_3 + \bar{y}(\cos\beta\bar{F}_4 + \sin\beta\bar{F}_5)\big] \\ &\quad + D_{\bar{y}}\big[(1-2\nu)\sin\beta\bar{F}_2 - 2(1-\nu)\cos\beta\bar{F}_3 + \bar{y}(\sin\beta\bar{F}_4 - \cos\beta\bar{F}_5)\big] \end{aligned} \tag{6.10}$$

and

$$\begin{aligned} \sigma_{xx} &= 2\mu D_{\bar{x}}\big[-\bar{F}_4 - \cos 2\beta\bar{F}_4 - \sin 2\beta\bar{F}_5 - \bar{y}(\cos 2\beta\bar{F}_6 - \sin 2\beta\bar{F}_7)\big] \\ &\quad + 2\mu D_{\bar{y}}\big[\bar{F}_5 - \bar{y}(\sin 2\beta\bar{F}_6 + \cos 2\beta\bar{F}_7)\big] \\ \sigma_{yy} &= 2\mu D_{\bar{x}}\big[-\bar{F}_4 + \cos 2\beta\bar{F}_4 + \sin 2\beta\bar{F}_5 + \bar{y}(\cos 2\beta\bar{F}_6 - \sin 2\beta\bar{F}_7)\big] \\ &\quad + 2\mu D_{\bar{y}}\big[\bar{F}_5 + \bar{y}(\sin 2\beta\bar{F}_6 + \cos 2\beta\bar{F}_7)\big] \\ \sigma_{xy} &= 2\mu D_{\bar{x}}\big[-\sin 2\beta\bar{F}_4 + \cos 2\beta\bar{F}_5 - \bar{y}(\sin 2\beta\bar{F}_6 + \cos 2\beta\bar{F}_7)\big] \\ &\quad + 2\mu D_{\bar{y}}\big[\bar{y}(\cos 2\beta\bar{F}_6 - \sin 2\beta\bar{F}_7)\big] \end{aligned} \tag{6.11}$$

6.3.2 *Orthotropic Material*

The displacements and stresses in the global x, y coordinate system for an orthotropic (transversely isotropic) material are obtained by substituting (4.204) into (4.215) and (4.218). The results can be written as follows:

$$\begin{aligned} u_x &= \frac{q_3}{2\pi}D_{\bar{x}}\left[-\frac{1}{1+q_1}\left\{\gamma_1 J_1(\bar{x},\bar{y},\gamma_1)\sin\beta - J_2(\bar{x},\bar{y},\gamma_1)\cos\beta\right\}\right. \\ &\qquad \left. + \frac{1}{1+q_2}\left\{\gamma_2 J_1(\bar{x},\bar{y},\gamma_2)\sin\beta - J_2(\bar{x},\bar{y},\gamma_2)\cos\beta\right\}\right] \\ &\quad + \frac{q_3}{2\pi}D_{\bar{y}}\left[-\frac{1}{1+q_1}\left\{\gamma_1 J_1(\bar{x},\bar{y},\gamma_1)\cos\beta + J_2(\bar{x},\bar{y},\gamma_1)\sin\beta\right\}\right. \\ &\qquad \left. + \frac{1}{1+q_2}\left\{\gamma_2 J_1(\bar{x},\bar{y},\gamma_2)\cos\beta + J_2(\bar{x},\bar{y},\gamma_2)\sin\beta\right\}\right] \\ u_y &= \frac{q_3}{2\pi}D_{\bar{x}}\left[-\frac{q_1}{1+q_1}\left\{\frac{1}{\gamma_1}J_1(\bar{x},\bar{y},\gamma_1)\cos\beta + J_2(\bar{x},\bar{y},\gamma_1)\sin\beta\right\}\right. \\ &\qquad \left. + \frac{q_2}{1+q_2}\left\{\frac{1}{\gamma_2}J_1(\bar{x},\bar{y},\gamma_2)\cos\beta + J_2(\bar{x},\bar{y},\gamma_2)\sin\beta\right\}\right] \end{aligned} \tag{6.12}$$

$$+\frac{q_3}{2\pi}D_{\bar{y}}\left[\frac{q_1}{1+q_1}\left\{\frac{1}{\gamma_1}J_1(\bar{x},\bar{y},\gamma_1)\sin\beta - J_2(\bar{x},\bar{y},\gamma_1)\cos\beta\right\}\right.$$
$$\left.-\frac{q_2}{1+q_2}\left\{\frac{1}{\gamma_2}J_1(\bar{x},\bar{y},\gamma_2)\sin\beta - J_2(\bar{x},\bar{y},\gamma_2)\cos\beta\right\}\right]$$

and

$$\sigma_{xx} = \frac{c_{66}q_3}{2\pi}D_{\bar{x}}\left[\frac{1}{\gamma_1^2}\left\{J_3(\bar{x},\bar{y},\gamma_1)\cos\beta + \gamma_1 J_4(\bar{x},\bar{y},\gamma_1)\sin\beta\right\}\right.$$
$$\left.-\frac{1}{\gamma_2^2}\left\{J_3(\bar{x},\bar{y},\gamma_2)\cos\beta + \gamma_2 J_4(\bar{x},\bar{y},\gamma_2)\sin\beta\right\}\right]$$
$$+\frac{c_{66}q_3}{2\pi}D_{\bar{y}}\left[-\frac{1}{\gamma_1^2}\left\{J_3(\bar{x},\bar{y},\gamma_1)\sin\beta - \gamma_1 J_4(\bar{x},\bar{y},\gamma_1)\cos\beta\right\}\right.$$
$$\left.+\frac{1}{\gamma_2^2}\left\{J_3(\bar{x},\bar{y},\gamma_2)\sin\beta - \gamma_2 J_4(\bar{x},\bar{y},\gamma_2)\cos\beta\right\}\right]$$

$$\sigma_{yy} = \frac{c_{66}q_3}{2\pi}D_{\bar{x}}\left[-\left\{J_3(\bar{x},\bar{y},\gamma_1)\cos\beta + \gamma_1 J_4(\bar{x},\bar{y},\gamma_1)\sin\beta\right\}\right.$$
$$\left.+\left\{J_3(\bar{x},\bar{y},\gamma_2)\cos\beta + \gamma_2 J_4(\bar{x},\bar{y},\gamma_2)\sin\beta\right\}\right] \tag{6.13}$$
$$+\frac{c_{66}q_3}{2\pi}D_{\bar{y}}\left[\left\{J_3(\bar{x},\bar{y},\gamma_1)\sin\beta - \gamma_1 J_4(\bar{x},\bar{y},\gamma_1)\cos\beta\right\}\right.$$
$$-\left\{J_3(\bar{x},\bar{y},\gamma_2)\sin\beta - \gamma_2 J_4(\bar{x},\bar{y},\gamma_2)\cos\beta\right\}$$

$$\sigma_{xy} = \frac{c_{66}q_3}{2\pi}D_{\bar{x}}\left[-\left\{J_3(\bar{x},\bar{y},\gamma_1)\sin\beta - \frac{1}{\gamma_1}J_4(\bar{x},\bar{y},\gamma_1)\cos\beta\right\}\right.$$
$$\left.+\left\{J_3(\bar{x},\bar{y},\gamma_2)\sin\beta - \frac{1}{\gamma_2}J_4(\bar{x},\bar{y},\gamma_2)\cos\beta\right\}\right]$$
$$+\frac{c_{66}q_3}{2\pi}D_{\bar{y}}\left[-\left\{J_3(\bar{x},\bar{y},\gamma_1)\cos\beta + \frac{1}{\gamma_1}J_4(\bar{x},\bar{y},\gamma_1)\sin\beta\right\}\right.$$
$$\left.+\left\{J_3(\bar{x},\bar{y},\gamma_2)\cos\beta + \frac{1}{\gamma_2}J_4(\bar{x},\bar{y},\gamma_2)\sin\beta\right\}\right]$$

The quantity q_3 in the preceding equations is defined in (4.216).

6.4 Computation of Boundary Influence Coefficients

The procedure for computing the boundary influence coefficients for the displacement discontinuity method exactly parallels that described for the stress discontinuity method in Sect. 5.4. In the present case we use the notation $\sigma_{xx}(\overset{j}{D}_{\bar{x}}, \overset{j}{D}_{\bar{y}})$, etc., to indicate that the stress components depend on the displacement discontinuities for element j, and similarly for the displacements $u_x(\overset{j}{D}_{\bar{x}}, \overset{j}{D}_{\bar{y}})$ and $u_y(\overset{j}{D}_{\bar{x}}, \overset{j}{D}_{\bar{y}})$.

6.4.1 *Traction Coefficients*

Referring to (5.20), we write the following formulas for the tractions at element i due to the displacement discontinuities at element j:

$$\overset{i}{t}_{\bar{x}} = \left[\sigma_{yy}(\overset{j}{D}_{\bar{x}}, \overset{j}{D}_{\bar{y}}) - \sigma_{xx}(\overset{j}{D}_{\bar{x}}, \overset{j}{D}_{\bar{y}})\right] \sin\overset{i}{\beta}\cos\overset{i}{\beta} + \sigma_{xy}(\overset{j}{D}_{\bar{x}}, \overset{j}{D}_{\bar{y}}) \cos 2\overset{i}{\beta}$$

$$\overset{i}{t}_{\bar{y}} = \sigma_{xx}(\overset{j}{D}_{\bar{x}}, \overset{j}{D}_{\bar{y}}) \sin^2\overset{i}{\beta} - \sigma_{xy}(\overset{j}{D}_{\bar{x}}, \overset{j}{D}_{\bar{y}}) \sin 2\overset{i}{\beta} + \sigma_{yy}(\overset{j}{D}_{\bar{x}}, \overset{j}{D}_{\bar{y}}) \cos^2\overset{i}{\beta} \tag{6.14}$$

The boundary influence coefficients for the tractions—$\overset{ij}{A}_{\bar{x}\bar{x}}$, etc., in (6.3)—can be calculated directly from these equations by using the appropriate expressions for the stress tensor components from either (6.11) or (6.13). The coefficients for isotropic elasticity, for example, are

$$\begin{aligned}
\overset{ij}{A}_{\bar{x}\bar{x}} &= 2\mu\big[\big(\bar{F}_4 \sin 2\gamma + \bar{F}_5 \cos 2\gamma\big) + \bar{y}\big(\bar{F}_6 \sin 2\gamma - \bar{F}_7 \cos 2\gamma\big)\big] \\
\overset{ij}{A}_{\bar{x}\bar{y}} &= 2\mu\big[\bar{y}\big(\bar{F}_6 \cos 2\gamma + \bar{F}_7 \sin 2\gamma\big)\big] \\
\overset{ij}{A}_{\bar{y}\bar{x}} &= 2\mu\big[-\bar{F}_4 + \big(\bar{F}_4 \cos 2\gamma - \bar{F}_5 \sin 2\gamma\big) + \bar{y}\big(\bar{F}_6 \cos 2\gamma + \bar{F}_7 \sin 2\gamma\big)\big] \\
\overset{ij}{A}_{\bar{y}\bar{y}} &= 2\mu\big[\bar{F}_5 - \bar{y}\big(\bar{F}_6 \sin 2\gamma - \bar{F}_7 \cos 2\gamma\big)\big]
\end{aligned} \tag{6.15}$$

where $\gamma = \overset{i}{\beta} - \overset{j}{\beta}$.

6.4.2 Displacement Coefficients

Equations comparable to (5.15) for the displacement discontinuity method can be written as

$$\overset{i}{u}_{\bar{x}} = \sum_{j=1}^{N} \overset{ij}{B}_{\bar{x}\bar{x}} \overset{j}{D}_{\bar{x}} + \sum_{j=1}^{N} \overset{ij}{B}_{\bar{x}\bar{y}} \overset{j}{D}_{\bar{y}}$$
$$\overset{i}{u}_{\bar{y}} = \sum_{j=1}^{N} \overset{ij}{B}_{\bar{y}\bar{x}} \overset{j}{D}_{\bar{x}} + \sum_{j=1}^{N} \overset{ij}{B}_{\bar{y}\bar{y}} \overset{j}{D}_{\bar{y}} \tag{6.16}$$

As in the case of the traction coefficients, we use the same symbols for the displacement coefficients as the ones used previously for the stress discontinuity method. The coefficients are of course different. The displacements at element i due to the displacement discontinuities at element j are found using the formulas

$$\overset{i}{u}_{\bar{x}} = u_x(\overset{j}{D}_{\bar{x}}, \overset{j}{D}_{\bar{y}}) \cos \overset{i}{\beta} + u_y(\overset{j}{D}_{\bar{x}}, \overset{j}{D}_{\bar{y}}) \sin \overset{i}{\beta}$$
$$\overset{i}{u}_{\bar{y}} = -u_x(\overset{j}{D}_{\bar{x}}, \overset{j}{D}_{\bar{y}}) \sin \overset{i}{\beta} + u_y(\overset{j}{D}_{\bar{x}}, \overset{j}{D}_{\bar{y}}) \cos \overset{i}{\beta} \tag{6.17}$$

cf. (5.23). The boundary influence coefficients for the displacements—$\overset{ij}{B}_{\bar{x}\bar{x}}$, etc., in (6.16)—can be calculated directly from these equations by using the appropriate expressions for the displacement vector components from either (6.10) or (6.12). The coefficients for isotropic elasticity are

$$\overset{ij}{B}_{\bar{x}\bar{x}} = -(1-2\nu)\bar{F}_2 \sin\gamma - 2(1-\nu)\bar{F}_3 \cos\gamma + \bar{y}\big(\bar{F}_4 \sin\gamma + \bar{F}_5 \cos\gamma\big)$$
$$\overset{ij}{B}_{\bar{x}\bar{y}} = (1-2\nu)\bar{F}_2 \cos\gamma - 2(1-\nu)\bar{F}_3 \sin\gamma + \bar{y}\big(\bar{F}_4 \cos\gamma - \bar{F}_5 \sin\gamma\big)$$
$$\overset{ij}{B}_{\bar{y}\bar{x}} = -(1-2\nu)\bar{F}_2 \cos\gamma + 2(1-\nu)\bar{F}_3 \sin\gamma + \bar{y}\big(\bar{F}_4 \cos\gamma - \bar{F}_5 \sin\gamma\big)$$
$$\overset{ij}{B}_{\bar{y}\bar{y}} = -(1-2\nu)\bar{F}_2 \sin\gamma - 2(1-\nu)\bar{F}_3 \cos\gamma - \bar{y}\big(\bar{F}_4 \sin\gamma + \bar{F}_5 \cos\gamma\big) \tag{6.18}$$

where again $\gamma = \overset{i}{\beta} - \overset{j}{\beta}$.

6.5 Element Self Effects

The self-effect influence coefficients for the displacement discontinuity method can be calculated in a similar way as was done for the stress discontinuity method. In the present case, we are concerned with the the self-effect coefficients for the displacements ($\overset{ii}{B}_{\bar{x}\bar{x}}$, etc.,) because they are discontinuous along the line segment; the self-effect coefficients for the tractions ($\overset{ii}{A}_{\bar{x}\bar{x}}$, etc.,) are unambiguously defined.

To determine coefficients $\overset{ii}{B}_{\bar{x}\bar{x}}$, etc., we evaluate the expressions given in Sect. 6.3 for the displacements u_i for the case that $\bar{x} = \bar{y} = 0$ and $\gamma = 0$. For an isotropic material the resulting expressions are

$$\begin{aligned} \overset{ii}{B}_{\bar{x}\bar{x}} &= -2(1-\nu)\bar{F}_3(0,0) \\ \overset{ii}{B}_{\bar{x}\bar{y}} &= (1-2\nu)\bar{F}_2(0,0) \\ \overset{ii}{B}_{\bar{y}\bar{x}} &= -(1-2\nu)\bar{F}_2(0,0) \\ \overset{ii}{B}_{\bar{y}\bar{y}} &= -2(1-\nu)\bar{F}_3(0,0) \end{aligned} \tag{6.19}$$

Because $\bar{F}_2(0,0) = 0$ and $\bar{F}_3(0,0_\pm) = \pm\left[4(1-\nu)\right]^{-1}$ (see (5.8)) the self-effect terms for the boundary influence coefficients for the displacements are

$$\begin{aligned} \overset{ii}{B}_{\bar{x}\bar{y}} &= \overset{ii}{B}_{\bar{y}\bar{x}} = 0 \\ \overset{ii}{B}_{\bar{x}\bar{x}} &= \overset{ii}{B}_{\bar{y}\bar{y}} = \pm\frac{1}{2} \qquad \text{for } \bar{y} = 0_\pm \end{aligned} \tag{6.20}$$

Similar considerations apply for the case of an anisotropic material, and the reader should verify that the self-effect terms for this case are also given by (6.20).

The ambiguity in sign for $\overset{ii}{B}_{\bar{x}\bar{x}}$ and $\overset{ii}{B}_{\bar{y}\bar{y}}$ in (6.20) is easily resolved once the directions of $\bar{x}$ and $\bar{y}$ for each element are defined. As depicted in Fig. 6.1(b), the direction of traversal of a crack boundary defines the positive direction for $\bar{x}$, and the direction for $\bar{y}$ is such that these coordinates form a right-handed system. In the event that the boundary elements are connected to form a closed contour it is convenient to adopt the same boundary traversal convention as for the stress discontinuity method. Thus, the positive direction for $\bar{y}$ is always taken to point in the direction of the outward normal to the region of interest. The correct values of the self-effect influence coefficients are then those pertaining to $\bar{y} = 0_-$ and the coefficients $\overset{ii}{B}_{\bar{x}\bar{x}}$ and $\overset{ii}{B}_{\bar{y}\bar{y}}$ will have the unique value $-\frac{1}{2}$. After the values of the displacements $(\overset{i}{u}_{\bar{x}})^-$ and $(\overset{i}{u}_{\bar{y}})^-$ have been calculated using (6.16), definitions (6.2) can be used to find $(\overset{i}{u}_{\bar{x}})^+$ and $(\overset{i}{u}_{\bar{y}})^+$.

6.6 Interior/Exterior Problems

The system of algebraic equations for the displacement discontinuity method is formed in the same way as explained in Sect. 5.6 for the stress discontinuity method. Thus, if the tractions $\overset{i}{t}_{\bar{x}} = (\overset{i}{t}_{\bar{x}})_0$ and $\overset{i}{t}_{\bar{y}} = (\overset{i}{t}_{\bar{y}})_0$ are prescribed at the ith element, the ith equations in the system are, by (6.3),

$$\begin{aligned}
\sum_{j=1}^{N} \overset{ij}{A}_{\bar{x}\bar{x}} \overset{j}{D}_{\bar{x}} + \sum_{j=1}^{N} \overset{ij}{A}_{\bar{x}\bar{y}} \overset{j}{D}_{\bar{y}} &= (\overset{i}{t}_{\bar{x}})_0 \\
\sum_{j=1}^{N} \overset{ij}{A}_{\bar{y}\bar{x}} \overset{j}{D}_{\bar{x}} + \sum_{j=1}^{N} \overset{ij}{A}_{\bar{y}\bar{y}} \overset{j}{D}_{\bar{y}} &= (\overset{i}{t}_{\bar{y}})_0
\end{aligned} \tag{6.21}$$

Similarly, if displacements $\overset{i}{u}_{\bar{x}} = (\overset{i}{u}_{\bar{x}})_0$ and $\overset{i}{u}_{\bar{y}} = (\overset{i}{u}_{\bar{y}})_0$ are prescribed, the ith equations are, by (6.16),

$$\begin{aligned}
\sum_{j=1}^{N} \overset{ij}{B}_{\bar{x}\bar{x}} \overset{j}{D}_{\bar{x}} + \sum_{j=1}^{N} \overset{ij}{B}_{\bar{x}\bar{y}} \overset{j}{D}_{\bar{y}} &= (\overset{i}{u}_{\bar{x}})_0 \\
\sum_{j=1}^{N} \overset{ij}{B}_{\bar{y}\bar{x}} \overset{j}{D}_{\bar{x}} + \sum_{j=1}^{N} \overset{ij}{B}_{\bar{y}\bar{y}} \overset{j}{D}_{\bar{y}} &= (\overset{i}{u}_{\bar{y}})_0
\end{aligned} \tag{6.22}$$

Mixed formulations involving specification of $\overset{i}{u}_{\bar{x}}$ and $\overset{i}{t}_{\bar{y}}$ or $\overset{i}{t}_{\bar{x}}$ and $\overset{i}{u}_{\bar{y}}$ are handled by selecting the appropriate ones of (6.21) and (6.22). Proceeding in this way for $i = 1$ to N, we obtain a system of $2N$ algebraic equations in $2N$ unknown displacement discontinuity components. We will represent this system of equations as follows:

$$\begin{aligned}
\sum_{j=1}^{N} \overset{ij}{C}_{\bar{x}\bar{x}} \overset{j}{D}_{\bar{x}} + \sum_{j=1}^{N} \overset{ij}{C}_{\bar{x}\bar{y}} \overset{j}{D}_{\bar{y}} &= \overset{i}{b}_{\bar{x}} \\
\sum_{j=1}^{N} \overset{ij}{C}_{\bar{y}\bar{x}} \overset{j}{D}_{\bar{x}} + \sum_{j=1}^{N} \overset{ij}{C}_{\bar{y}\bar{y}} \overset{j}{D}_{\bar{y}} &= \overset{i}{b}_{\bar{y}}
\end{aligned} \tag{6.23}$$

for $i = 1$ to N. The quantities $\overset{i}{b}_{\bar{x}}$ and $\overset{i}{b}_{\bar{y}}$ stand for the known boundary values of traction or displacement, whichever is appropriate, and $\overset{ij}{C}_{\bar{x}\bar{x}}$, etc., are the corresponding influence coefficients from (6.21) or (6.22).

Problems involving an infinite body with cracks—even curved or intersecting cracks—are especially easy to solve by the displacement discontinuity method. The boundary elements in this case do not form a closed contour but we must nevertheless distinguish between their positive and negative sides in solving the problem. This is necessary in order to interpret the values of the displacements $(\overset{i}{u}_{\bar{x}})^{\pm}$ and $(\overset{i}{u}_{\bar{y}})^{\pm}$ computed for each element. In accordance with the discussion in Sect. 6.2, we will adopt the convention that any prescribed displacement must be given for the negative side of an element, that is, $(\overset{i}{u}_{\bar{x}})^{-} = (\overset{i}{u}_{\bar{x}})_0$ and $(\overset{i}{u}_{\bar{y}})^{-} = (\overset{i}{u}_{\bar{y}})_0$. The tractions $\overset{i}{t}_{\bar{x}}$ and $\overset{i}{t}_{\bar{y}}$ required to produce these displacements will be determined as part of the solution to the problem and the displacements on the positive side of the element can be found from the relations $(\overset{i}{u}_{\bar{x}})^{+} = \overset{i}{D}_{\bar{x}} + (\overset{i}{u}_{\bar{x}})^{-}$ and $(\overset{i}{u}_{\bar{y}})^{+} = \overset{i}{D}_{\bar{y}} + (\overset{i}{u}_{\bar{y}})^{-}$.

If the boundary elements are connected to form a closed contour, we can consider boundary value problems for either the interior or the exterior region of the contour. Solutions to interior (finite body) problems are found in the same way as for the stress discontinuity method; rigid body movements can be prevented by prescribing at least three displacement components at two points along the boundary of the region, or else by using symmetry conditions, if these exist for the problem in question.

Exterior (cavity) problems, on the other hand, must be approached differently when using the displacement discontinuity method. An interior problem is associated with every exterior problem, and the solutions to both are determined together. Provision must therefore be made to prevent rigid body movement of the interior region of a contour, even though we are only interested in the exterior problem. If two lines of symmetry exist for a certain problem, then the interior region will automatically be fixed with respect to these lines. The displacement discontinuity components (which in this case are not physically meaningful) will thus be determined uniquely all along the closed contour, and the displacements and stresses in the exterior region can be expressed as linear combinations of them. If the problem does not involve symmetry it is still possible to prevent rigid body movements by placing two additional displacement discontinuity elements *inside* the closed contour and fixing the displacements of their negative sides (Crouch and Starfield 1983).

All things considered, however, it is probably better to use another method, for example, the stress discontinuity method, for general exterior problems and to restrict the use of the displacement discontinuity method for problems involving only cracks. It should be evident to the reader that the two methods can easily be combined to create a more versatile tool for practical applications. This topic will be considered in Chap. 8.

6.7 Symmetry Conditions

Symmetry conditions can be incorporated in the displacement discontinuity method in exactly the same way as explained in Sect. 5.7 for the stress discontinuity method. That is, we introduce a line of symmetry with respect to a given boundary element by

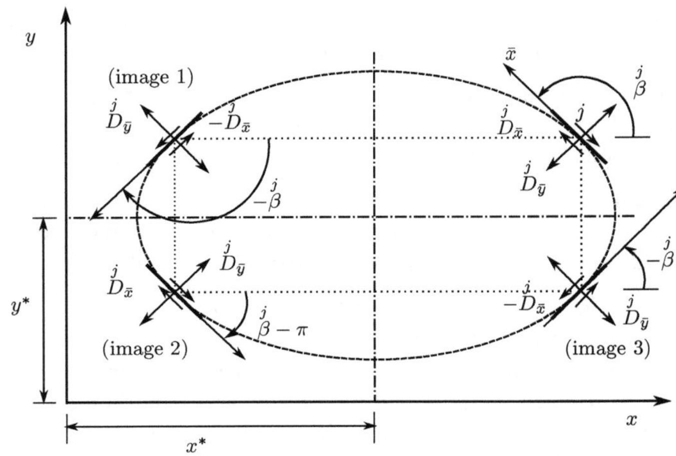

Fig. 6.3 Symmetry conditions for the lines $x = x^*$ and $y = y^*$

adding an image element at the appropriate location. This is depicted in Fig. 6.3 for the case that both $x = x^*$ and $y = y^*$ are lines of symmetry. The normal components for the actual and image elements are equal, but the shear components change sign each time a line of symmetry is crossed.

6.8 Iterative Solution of Systems of Algebraic Equations

Boundary element methods always lead to systems of simultaneous linear algebraic equations of the form $\boldsymbol{AX} = \boldsymbol{b}$. The number of unknowns in the system depends on the type and dimensionality of the problem as well as the complexity of the way that the boundary parameters are assumed to vary over the boundary elements. The simple method described in Chap. 1 has N unknowns (also called "degrees of freedom") $\boldsymbol{X}$, and the coefficient matrix $\boldsymbol{A}$ is $N \times N$. The two-dimensional stress and displacement discontinuity methods with piecewise constant approximations have two unknowns per element so $\boldsymbol{X}$ is a vector of size $2N$ and $\boldsymbol{A}$ is a $2N \times 2N$ matrix.

Simultaneous linear algebraic equations can be solved in several ways (e.g., Carnahan et al. 1969). For example, one could use a standard computer routine to invert matrix $\boldsymbol{A}$ and then obtain values for $\boldsymbol{X}$ by multiplying this inverse, $\boldsymbol{A}^{-1}$, by the known quantities $\boldsymbol{b}$. One could also use the so-called Gaussian elimination scheme in which the system of equations is reduced to triangular form and the unknowns are found by back-substitution from the last equation to the first. These techniques are called direct solution procedures and involve numerical approximations only to the extent of round-off or truncation errors made during arithmetic operations. With a fully populated $N \times N$ coefficient matrix the solution time for a direct solution procedure is proportional to N^3.

An alternative approach is to use an indirect solution procedure, exemplified by an iterative technique called the *method of successive over-relaxation*.[2] Iterative methods involve making arbitrary initial approximations for the values of $\overset{i}{X}$, $i = 1$ to N, and then systematically refining these until they satisfy all the equations to within a desired accuracy. An iterative procedure is particularly useful for solving systems of equations such as (6.23) because, as will be discussed in the next section, this allows for the imposition of constraints such as a no-overlap condition for crack surfaces during the iteration process.

6.8.1 Procedure for an N × N System

To explain the iterative solution process we will first consider an $N \times N$ system, written as

$$\sum_{j=1}^{N} \overset{ij}{A}\overset{j}{X} = \overset{i}{b} \tag{6.24}$$

for $i = 1$ to N. Equivalently, these equations can be written as

$$\sum_{j=1}^{i-1} \overset{ij}{A}\overset{j}{X} + \overset{ii}{A}\overset{i}{X} + \sum_{j=i+1}^{N} \overset{ij}{A}\overset{j}{X} = \overset{i}{b} \tag{6.25}$$

which can be solved for $\overset{i}{X}$ to give

$$\overset{i}{X} = \frac{1}{\overset{ii}{A}}\Big[\overset{i}{b} - \sum_{j=1}^{i-1} \overset{ij}{A}\overset{j}{X} - \sum_{j=i+1}^{N} \overset{ij}{A}\overset{j}{X}\Big] \tag{6.26}$$

If we now let $\overset{j}{X}{}^{(n)}$ be the nth approximation of $\overset{j}{X}$, then (6.26) can be viewed as an iterative formula for computing the next approximation of $\overset{i}{X}$, i.e.,

$$\overset{i}{X}{}^{(n+1)} = \frac{1}{\overset{ii}{A}}\Big[\overset{i}{b} - \sum_{j=1}^{i-1} \overset{ij}{A}\overset{j}{X}{}^{(n+1)} - \sum_{j=i+1}^{N} \overset{ij}{A}\overset{j}{X}{}^{(n)}\Big] \tag{6.27}$$

The first summed term on the right-hand side of this equation contains the $(n + 1)$th approximations of $\overset{j}{X}$ because these approximations have already been made for

[2] The term "relaxation" refers to the way in which the initial errors in the solution vector are reduced, or relaxed, as the calculations proceed.

$j < i$. The iteration process is said to have converged when the magnitude of the difference between two successive iterates for each equation is less than a specified value ε:

$$\mid \overset{i}{X}{}^{(n+1)} - \overset{i}{X}{}^{(n)} \mid < \varepsilon \tag{6.28}$$

for $i = 1$ to N.

Equation (6.27) is called the *Gauss-Seidel method.* Formally, the method can be shown to converge if matrix $\boldsymbol{A}$ is symmetric and positive definite (Golub and Van Loan 1996) but it is not easy to show in practice when this is true, especially when N is large. It is well known, however, that the method converges under less restrictive circumstances, and this has been the authors' experience in the case of boundary element methods.

The method of successive over-relaxation is a simple variant of the Gauss-Seidel method that is designed to accelerate convergence of the iteration process. This is achieved by applying a correction factor at each stage of iteration according to the following algorithm:

$$\overset{i}{X}{}^{(n+1)} = \overset{i}{X}{}^{(n)} + \omega \Delta \overset{i}{X}{}^{(n+1)} \tag{6.29}$$

where ω is the *over-relaxation factor* and where

$$\Delta \overset{i}{X}{}^{(n+1)} = \frac{1}{\underset{A}{ii}} \Big[\overset{i}{b} - \sum_{j=1}^{i-1} \overset{ij}{A} \overset{j}{X}{}^{(n+1)} - \sum_{j=i+1}^{N} \overset{ij}{A} \overset{j}{X}{}^{(n)} \Big] - \overset{i}{X}{}^{(n)}$$

or

$$\Delta \overset{i}{X}{}^{(n+1)} = \frac{1}{\underset{A}{ii}} \Big[\overset{i}{b} - \sum_{j=1}^{i-1} \overset{ij}{A} \overset{j}{X}{}^{(n+1)} - \sum_{j=i}^{N} \overset{ij}{A} \overset{j}{X}{}^{(n)} \Big] \tag{6.30}$$

It can be shown (e.g., Varga 2000) that ω must satisfy the constraint $0 < \omega < 2$. If $\omega = 1$ the method of successive over-relaxation reduces to the Gauss-Seidel method; "under-relaxation" occurs for $\omega < 1$. Theoretically, an optimum value of the over-relaxation factor can be computed for a given matrix $\boldsymbol{A}$ (Varga 2000) but this process involves a substantial amount of computer time and is seldom done in practice. It is usually sufficient to assume a value for ω (say, $\omega = 1.4$), which can then be adjusted for a given class of problem by trial and error.

Taken together, the summed terms on the right-hand side of (6.30) represent the ith term of the product of the matrix $\overset{ij}{A}$ and *current values* of $\overset{j}{X}$, $j = 1$ to N. Because the computer always retains the most recently computed values of $\overset{j}{X}$, these terms will be obtained automatically, without having to distinguish between the approximations $\overset{j}{X}{}^{(n+1)}$ for $j < i$ and $\overset{j}{X}{}^{(n)}$ for $j \geq i$. Furthermore, the absolute value of the quantity $\Delta \overset{i}{X}{}^{(n+1)}$ provides a measure of the convergence of the iteration process, as specified by (6.28).

6.8.2 *Procedure for a 2N × 2N System*

The method of successive over-relaxation presented above for an $N \times N$ system of equations can easily be extended to the current situation. Thus, the counterparts of (6.26) for the displacement discontinuity method can be written as

$$\begin{aligned} \overset{i}{D}_{\bar{x}} &= \frac{1}{\overset{ii}{C}_{\bar{x}\bar{x}}}\left[\overset{i}{b}_{\bar{x}} - \sum_{j=1}^{i-1}\overset{ij}{C}_{\bar{x}\bar{x}}\overset{j}{D}_{\bar{x}} - \sum_{j=i+1}^{N}\overset{ij}{C}_{\bar{x}\bar{x}}\overset{j}{D}_{\bar{x}} - \sum_{j=1}^{N}\overset{ij}{C}_{\bar{x}\bar{y}}\overset{j}{D}_{\bar{y}}\right] \\ \overset{i}{D}_{\bar{y}} &= \frac{1}{\overset{ii}{C}_{\bar{y}\bar{y}}}\left[\overset{i}{b}_{\bar{y}} - \sum_{j=1}^{N}\overset{ij}{C}_{\bar{y}\bar{x}}\overset{j}{D}_{\bar{x}} - \sum_{j=1}^{i-1}\overset{ij}{C}_{\bar{y}\bar{y}}\overset{j}{D}_{\bar{y}} - \sum_{j=i+1}^{N}\overset{ij}{C}_{\bar{y}\bar{y}}\overset{j}{D}_{\bar{y}}\right] \end{aligned} \qquad (6.31)$$

If $\overset{j}{D}_{\bar{x}}{}^{(n)}$ and $\overset{j}{D}_{\bar{y}}{}^{(n)}$ are the nth approximations of $\overset{j}{D}_{\bar{x}}$ and $\overset{j}{D}_{\bar{y}}$, then (6.31) can be viewed as iterative formulas for computing the next approximations of $\overset{i}{D}_{\bar{x}}$ and $\overset{i}{D}_{\bar{y}}$, i.e.,

$$\begin{aligned} \overset{i}{D}_{\bar{x}}{}^{(n+1)} &= \frac{1}{\overset{ii}{C}_{\bar{x}\bar{x}}}\left[\overset{i}{b}_{\bar{x}} - \sum_{j=1}^{i-1}\overset{ij}{C}_{\bar{x}\bar{x}}\overset{j}{D}_{\bar{x}}{}^{(n+1)} - \sum_{j=1}^{i-1}\overset{ij}{C}_{\bar{x}\bar{y}}\overset{j}{D}_{\bar{y}}{}^{(n+1)} \right. \\ &\qquad \left. - \sum_{j=i+1}^{N}\overset{ij}{C}_{\bar{x}\bar{x}}\overset{j}{D}_{\bar{x}}{}^{(n)} - \sum_{j=i}^{N}\overset{ij}{C}_{\bar{x}\bar{y}}\overset{j}{D}_{\bar{y}}{}^{(n)}\right] \\ \overset{i}{D}_{\bar{y}}{}^{(n+1)} &= \frac{1}{\overset{ii}{C}_{\bar{y}\bar{y}}}\left[\overset{i}{b}_{\bar{y}} - \sum_{j=1}^{i-1}\overset{ij}{C}_{\bar{y}\bar{x}}\overset{j}{D}_{\bar{x}}{}^{(n+1)} - \sum_{j=1}^{i-1}\overset{ij}{C}_{\bar{y}\bar{y}}\overset{j}{D}_{\bar{y}}{}^{(n+1)} \right. \\ &\qquad \left. - \sum_{j=i}^{N}\overset{ij}{C}_{\bar{y}\bar{x}}\overset{j}{D}_{\bar{x}}{}^{(n)} - \sum_{j=i+1}^{N}\overset{ij}{C}_{\bar{y}\bar{y}}\overset{j}{D}_{\bar{y}}{}^{(n)}\right] \end{aligned} \qquad (6.32)$$

The first sets of summed terms on the right-hand sides of these equations represent the $(n+1)$th approximations of $\overset{j}{D}_{\bar{x}}$ and $\overset{j}{D}_{\bar{y}}$ because these approximations have already been made for $j < i$.

The differences between two successive iterates are given as

$$\begin{aligned} \Delta\overset{i}{D}_{\bar{x}}{}^{(n+1)} &= \overset{i}{D}_{\bar{x}}{}^{(n+1)} - \overset{i}{D}_{\bar{x}}{}^{(n)} \\ \Delta\overset{i}{D}_{\bar{y}}{}^{(n+1)} &= \overset{i}{D}_{\bar{y}}{}^{(n+1)} - \overset{i}{D}_{\bar{y}}{}^{(n)} \end{aligned} \qquad (6.33)$$

and the iteration process is said to have converged when

$$
\begin{aligned}
&| \Delta \overset{i}{D}_{\bar{x}}{}^{(n+1)} | < \varepsilon \\
&| \Delta \overset{i}{D}_{\bar{y}}{}^{(n+1)} | < \varepsilon
\end{aligned}
\tag{6.34}
$$

for $i = 1$ to N, where ε is a specified tolerance limit. Combining (6.32) and (6.33), it can be seen that all of the summations will involve the current values of the approximations for the displacement discontinuity components. These values are automatically held in computer memory so it is not necessary to keep track of them during the iteration process.

The method of successive over-relaxation applied to (6.33) gives the equations

$$
\begin{aligned}
\overset{i}{D}_{\bar{x}}{}^{(n+1)} &= \overset{i}{D}_{\bar{x}}{}^{(n)} + \omega \, \Delta \, \overset{i}{D}_{\bar{x}}{}^{(n+1)} \\
\overset{i}{D}_{\bar{y}}{}^{(n+1)} &= \overset{i}{D}_{\bar{y}}{}^{(n)} + \omega \, \Delta \, \overset{i}{D}_{\bar{y}}{}^{(n+1)}
\end{aligned}
\tag{6.35}
$$

6.9 Crack Closure and Sliding

It has been tacitly assumed in the discussion so far that all cracks are open along their entire lengths, i.e., it has been assumed that $\overset{i}{D}_{\bar{y}} > 0$ for $i = 1$ to N. Allowance must in general be made for the possibility that parts of the crack surfaces will touch and transmit load at the points of contact. Although the resultant tractions are not known a priori at a point of contact, it is known that the normal component of displacement discontinuity is zero at such a point. If the system of algebraic equations (6.23) is solved by iteration, a no-overlap condition can be enforced simply by prohibiting the occurrence of negative values of $\overset{i}{D}_{\bar{y}}$. This is easily accomplished during the iteration process by setting all negative values of $\overset{i}{D}_{\bar{y}}{}^{(n)}$ equal to zero, where n represents the nth iteration. The associated trial values of the normal traction $\overset{i}{t}_{\bar{y}}{}^{(n)}$ are then given by the second one of (6.3). It will be found that the normal traction $\overset{i}{t}_{\bar{y}}$ is negative at points along a closed crack—i.e., the associated normal stress $(\overset{i}{\sigma}_{\bar{y}\bar{y}})_{total}$ is compressive—as would be expected on physical grounds: the crack surfaces must in effect be pushed away from one another to prevent the occurrence of an overlap.

A boundary condition must also be specified for the $\bar{x}$ direction in the case of crack closure. The difficulty here is that, unlike the normal component of displacement discontinuity, the shear component is not subject to a limiting value based on the geometry of the element. One may postulate, however, that the total shear traction $\overset{i}{t}_{\bar{x}}$ at any element midpoint cannot exceed the value of the total normal traction $\overset{i}{t}_{\bar{y}}$ at that point multiplied by the coefficient of friction, μ_0, for the crack surfaces. This leads to the following algorithm for determining $\overset{i}{D}_{\bar{x}}{}^{(n)}$ for the case that $\overset{i}{D}_{\bar{y}}{}^{(n)}=0$:

If $\mid \overset{i}{t}_{\bar{x}}{}^{(n-1)} \mid \leq \mu_0 \mid \overset{i}{t}_{\bar{y}}{}^{(n-1)} \mid$, set $\overset{i}{D}_{\bar{x}}{}^{(n)} = \overset{i}{D}_{\bar{x}}{}^{(n-1)}$

If $\mid \overset{i}{t}_{\bar{x}}{}^{(n-1)} \mid > \mu_0 \mid \overset{i}{t}_{\bar{y}}{}^{(n-1)} \mid$, set $(\overset{i}{t}_{\bar{x}})_0 = \text{Sign}[\overset{i}{D}_{\bar{x}}{}^{(n-1)}] \times \mu_0 \mid \overset{i}{t}_{\bar{y}}{}^{(n-1)} \mid$ and solve for $\overset{i}{D}_{\bar{x}}{}^{(n)}$.

It should be realized that contact problems with friction are path-dependent and must be solved incrementally. This means that the basic algorithm outlined above is applicable to a single loading (or unloading) increment; the solution at each step must take into account the solution at all preceding steps. A discussion of this methodology is given by Crouch and Starfield (1983) for a related problem.

The above constraints for crack closure with possible frictional sliding can also be viewed as a mixed complementarity problem (DeBremaecker et al. 2000; Ritz and Pollard 2011) whose solution can be obtained using the PATH algorithm developed by Dirkse and Ferris (1995). Details of this implementation are beyond the scope of this discussion and the interested reader is referred to the literature for further information (e.g., DeBremaecker et al. 2000; DeBremaecker and Ferris 2000; Ritz et al. 2012).

6.10 Computation of the Normal Stress Parallel to a Boundary

The shear and normal tractions along a boundary involve only some of the components of the stress tensor. In two dimensions, the traction components $t_{\bar{x}} = \sigma_{\bar{x}\bar{y}}$ and $t_{\bar{y}} = \sigma_{\bar{y}\bar{y}}$ do not involve $\sigma_{\bar{x}\bar{x}}$, the normal stress parallel to the boundary and just inside it, as depicted in Fig. 6.4. This quantity, sometimes called the "hoop" stress, is important because it serves to define stress concentration factors in practical situations, for example, for an array of holes in a plate.

Computation of the normal stress parallel to a boundary was a simple matter for the stress discontinuity method. Influence coefficients for this purpose were determined

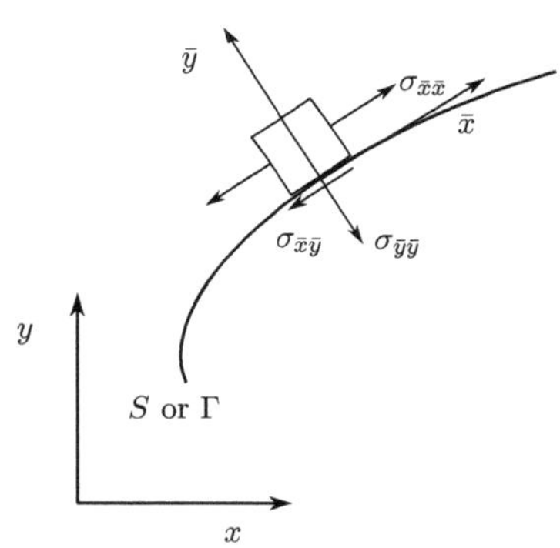

Fig. 6.4 Normal stress $\sigma_{\bar{x}\bar{x}}$ parallel to a boundary

directly from the basic elasticity solution used to construct the numerical method, namely, the solution to the problem of constant tractions applied to a line segment in the x, y plane of an infinite solid. This solution has the characteristic that all of the stress tensor components are *discontinuous* at the midpoint of the line segment in question.

The situation is different for the displacement discontinuity method. In this case, the basic elasticity solution used to construct the numerical method has the characteristic that all of the stress components are *continuous* at the midpoint of a line segment where a constant displacement discontinuity occurs.[3] Not including the endpoints of this line segment, the stresses are continuous everywhere in the x, y plane, and superposition of a number of these solutions will produce continuous stresses at the midpoint of each displacement discontinuity element. On physical grounds, however, one might expect these stresses to be discontinuous in some situations.

We explore below the conditions for which a discontinuity occurs in the normal stress parallel to the surfaces of a displacement discontinuity element. We also describe a numerical procedure for computing the value of the stress discontinuity in terms of the displacements of the two surfaces of the displacement discontinuity element and its neighbors.

We note that for infinite regions the stresses involved in the results of this section refer to induced stresses or stress changes (see Sect. 4.8). The initial stresses at infinity must be added to the stress changes to give the resultant stresses at any boundary element.

6.10.1 Isotropic Material

Because an isotropic elastic material does not have any preferred directions, Hooke's law can be written for the stresses and strains in a local $\bar{x}$, $\bar{y}$ coordinate system directly from (4.40). The first two of (4.40) can be put in the form

$$\begin{aligned}\sigma_{\bar{x}\bar{x}} &= \frac{2\mu}{1-2\nu}\left[(1-\nu)\varepsilon_{\bar{x}\bar{x}} + \nu\varepsilon_{\bar{y}\bar{y}}\right] \\ \sigma_{\bar{y}\bar{y}} &= \frac{2\mu}{1-2\nu}\left[\nu\varepsilon_{\bar{x}\bar{x}} + (1-\nu)\varepsilon_{\bar{y}\bar{y}}\right]\end{aligned} \tag{6.36}$$

After solving the second of these equations for $\varepsilon_{\bar{y}\bar{y}}$ and substituting the resulting expression into the first equation, we find that

$$\sigma_{\bar{x}\bar{x}} = \frac{2\mu}{1-\nu}\varepsilon_{\bar{x}\bar{x}} + \frac{\nu}{1-\nu}\sigma_{\bar{y}\bar{y}}$$

or

[3] But see Sect. 6.12 in this connection.

$$\sigma_{\bar{x}\bar{x}} = \frac{2\mu}{1-\nu}\frac{\partial u_{\bar{x}}}{\partial \bar{x}} + \frac{\nu}{1-\nu}\sigma_{\bar{y}\bar{y}} \tag{6.37}$$

Writing this relation for the two surfaces $\bar{y} = 0_+$ and $\bar{y} = 0_-$ gives

$$\begin{aligned} \sigma^+_{\bar{x}\bar{x}} &= \frac{2\mu}{1-\nu}\frac{\partial u^+_{\bar{x}}}{\partial \bar{x}} + \frac{\nu}{1-\nu}\sigma^+_{\bar{y}\bar{y}} \\ \sigma^-_{\bar{x}\bar{x}} &= \frac{2\mu}{1-\nu}\frac{\partial u^-_{\bar{x}}}{\partial \bar{x}} + \frac{\nu}{1-\nu}\sigma^-_{\bar{y}\bar{y}} \end{aligned} \tag{6.38}$$

The normal stresses $\sigma^+_{\bar{y}\bar{y}}$ and $\sigma^-_{\bar{y}\bar{y}}$ in these expressions are equal, by equilibrium, and it follows that

$$\sigma^+_{\bar{x}\bar{x}} - \sigma^-_{\bar{x}\bar{x}} = \frac{2\mu}{1-\nu}\left[\frac{\partial u^+_{\bar{x}}}{\partial \bar{x}} - \frac{\partial u^-_{\bar{x}}}{\partial \bar{x}}\right] \tag{6.39}$$

or

$$\sigma^+_{\bar{x}\bar{x}} - \sigma^-_{\bar{x}\bar{x}} = \frac{2\mu}{1-\nu}\frac{\partial}{\partial \bar{x}}\Delta u_{\bar{x}} \tag{6.40}$$

We therefore conclude that $\sigma_{\bar{x}\bar{x}}$ is discontinuous whenever the tangential derivative of the shear component of displacement discontinuity is nonzero. For a piecewise constant displacement discontinuity element the tangential derivative must be approximated numerically because $\partial D_{\bar{x}}/\partial \bar{x}$ is always zero.

In order to find the stresses $\sigma^+_{\bar{x}\bar{x}}$ and $\sigma^-_{\bar{x}\bar{x}}$ at the ith displacement discontinuity element using (6.38), we must compute approximations to the derivatives $\partial u^+_{\bar{x}}/\partial \bar{x}$ and $\partial u^-_{\bar{x}}/\partial \bar{x}$ at the midpoint of this element. This is accomplished using standard finite difference formulas (Carnahan et al. 1969). Assuming initially that elements $i-1$, i, and $i+1$ are co-linear, and letting $f(\bar{x})$ stand for either $u^+_{\bar{x}}$ or $u^-_{\bar{x}}$, we note three separate finite difference formulas for $\partial f(\bar{x})/\partial \bar{x}$ at the point $\bar{x} = \overset{i}{\bar{x}}$:

$$\begin{aligned} \left(\frac{\partial f}{\partial \bar{x}}\right)_{\bar{x}=\overset{i}{\bar{x}}} &\doteq \frac{f(\overset{i+1}{\bar{x}}) - f(\overset{i}{\bar{x}})}{\overset{i+1}{\bar{x}} - \overset{i}{\bar{x}}} && \text{forward difference} \\ \left(\frac{\partial f}{\partial \bar{x}}\right)_{\bar{x}=\overset{i}{\bar{x}}} &\doteq \frac{f(\overset{i}{\bar{x}}) - f(\overset{i-1}{\bar{x}})}{\overset{i}{\bar{x}} - \overset{i-1}{\bar{x}}} && \text{backward difference} \\ \left(\frac{\partial f}{\partial \bar{x}}\right)_{\bar{x}=\overset{i}{\bar{x}}} &\doteq \frac{f(\overset{i+1}{\bar{x}}) - f(\overset{i-1}{\bar{x}})}{\overset{i+1}{\bar{x}} - \overset{i-1}{\bar{x}}} && \text{central difference} \end{aligned} \tag{6.41}$$

Usually, however, the elements have different inclinations, as depicted in Fig. 6.5. The displacements shown in the figure can be taken as those for either the positive or negative sides of the elements. The required derivatives at the midpoint of element

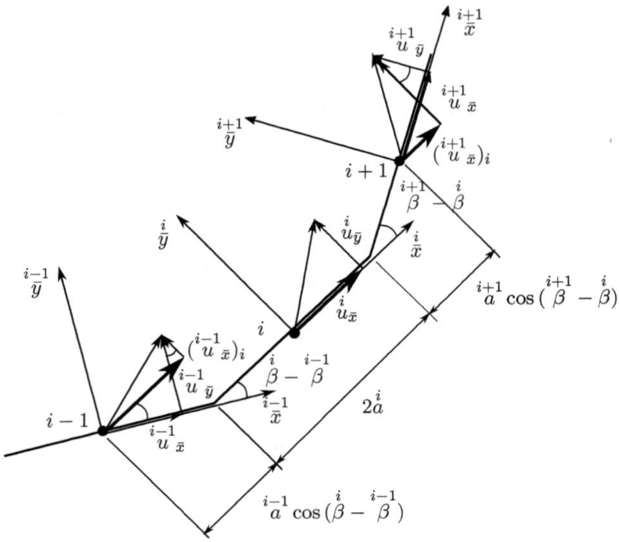

Fig. 6.5 Displacements for elements with different inclinations

i in this case are expressed in terms of the displacements of the adjoining elements *in the* $\overset{i}{\bar{x}}$ *direction*. These displacements are indicated by the bold arrows in Fig. 6.5 and are denoted as $(\overset{i-1}{u}_{\bar{x}})_i$ and $(\overset{i+1}{u}_{\bar{x}})_i$ for elements $i-1$ and $i+1$. We find from the geometry in the figure that

$$
\begin{aligned}
(\overset{i-1}{u}_{\bar{x}})_i &= \overset{i-1}{u}_{\bar{x}}\cos\big(\overset{i}{\beta}-\overset{i-1}{\beta}\big)+\overset{i-1}{u}_{\bar{y}}\sin\big(\overset{i}{\beta}-\overset{i-1}{\beta}\big)\\
(\overset{i+1}{u}_{\bar{x}})_i &= \overset{i+1}{u}_{\bar{x}}\cos\big(\overset{i+1}{\beta}-\overset{i}{\beta}\big)-\overset{i+1}{u}_{\bar{y}}\sin\big(\overset{i}{\beta}-\overset{i-1}{\beta}\big)
\end{aligned}
\tag{6.42}
$$

Applying the central difference formula, for example, we write

$$
\begin{aligned}
\left(\frac{\partial u_{\bar{x}}}{\partial \bar{x}}\right)_{\bar{x}=\overset{i}{\bar{x}}} \doteq \Big[\Big\{&\overset{i+1}{u}_{\bar{x}}\cos\big(\overset{i+1}{\beta}-\overset{i}{\beta}\big)-\overset{i-1}{u}_{\bar{x}}\cos\big(\overset{i}{\beta}-\overset{i-1}{\beta}\big)\Big\}\\
&-\Big\{\overset{i+1}{u}_{\bar{y}}\sin\big(\overset{i+1}{\beta}-\overset{i}{\beta}\big)+\overset{i-1}{u}_{\bar{y}}\sin\big(\overset{i}{\beta}-\overset{i-1}{\beta}\big)\Big\}\Big]\Big/\overset{i}{\Delta s}
\end{aligned}
\tag{6.43}
$$

in which the quantity

$$
\overset{i}{\Delta s}=\overset{i-1}{a}\cos\big(\overset{i}{\beta}-\overset{i-1}{\beta}\big)+2\overset{i}{a}+\overset{i+1}{a}\cos\big(\overset{i+1}{\beta}-\overset{i}{\beta}\big)
\tag{6.44}
$$

is the projected distance in the $\overset{i}{\bar{x}}$ direction between the midpoints of elements $i-1$ and $i+1$. Expressions similar to (6.43) can be written for forward and backward differences and in all cases $u_{\bar{x}}$ and $u_{\bar{y}}$ can be either $u^+_{\bar{x}}$ and $u^+_{\bar{y}}$ or $u^-_{\bar{x}}$ and $u^-_{\bar{y}}$.

Combination of (6.38) and (6.43) enables us to compute the normal stress parallel to a displacement discontinuity element of arbitrary orientation in terms of the normal stress acting on the element (i.e., the normal traction) and the shear and normal displacements at the midpoints of neighboring elements. An alternative way of computing $\sigma^{\pm}_{\bar{x}\bar{x}}$ can be found by expressing the results in terms of the displacement discontinuity components. This approach is often more convenient because it does not require us to find separately the displacements of the two surfaces of the boundary elements.

Using (6.39), definitions (6.2), and a central difference approximation, we have

$$\overset{i+}{\sigma}_{\bar{x}\bar{x}} - \overset{i-}{\sigma}_{\bar{x}\bar{x}} \doteq \frac{2\mu}{1-\nu}\Bigg[\Big\{\overset{i+1}{D}_{\bar{x}}\cos(\overset{i+1}{\beta} - \overset{i}{\beta}) - \overset{i-1}{D}_{\bar{x}}\cos(\overset{i}{\beta} - \overset{i-1}{\beta})\Big\}$$
$$- \Big\{\overset{i+1}{D}_{\bar{y}}\sin(\overset{i+1}{\beta} - \overset{i}{\beta}) + \overset{i-1}{D}_{\bar{y}}\sin(\overset{i}{\beta} - \overset{i-1}{\beta})\Big\}\Bigg]\Big/\overset{i}{\Delta s} \tag{6.45}$$

The normal stresses parallel to the two surfaces of the ith element are obtained from the identities

$$\begin{aligned}\overset{i+}{\sigma}_{\bar{x}\bar{x}} &= \frac{1}{2}\left(\overset{i+}{\sigma}_{\bar{x}\bar{x}} + \overset{i-}{\sigma}_{\bar{x}\bar{x}}\right) + \frac{1}{2}\left(\overset{i+}{\sigma}_{\bar{x}\bar{x}} - \overset{i-}{\sigma}_{\bar{x}\bar{x}}\right)\\ \overset{i-}{\sigma}_{\bar{x}\bar{x}} &= \frac{1}{2}\left(\overset{i+}{\sigma}_{\bar{x}\bar{x}} + \overset{i-}{\sigma}_{\bar{x}\bar{x}}\right) - \frac{1}{2}\left(\overset{i+}{\sigma}_{\bar{x}\bar{x}} - \overset{i-}{\sigma}_{\bar{x}\bar{x}}\right)\end{aligned} \tag{6.46}$$

in which the term $\frac{1}{2}\left(\overset{i+}{\sigma}_{\bar{x}\bar{x}} + \overset{i-}{\sigma}_{\bar{x}\bar{x}}\right)$ is *continuous*. This term represents the combined effects of the N elemental displacement discontinuities along the boundary and is found using (6.11) for element j and the first one of (4.67) for element i:

$$\frac{1}{2}\left(\overset{i+}{\sigma}_{\bar{x}\bar{x}} + \overset{i-}{\sigma}_{\bar{x}\bar{x}}\right) = 2\mu\sum_{j=1}^{N}\Big[-\bar{F}_4 - \bar{F}_4\cos 2\gamma + \bar{F}_5\sin 2\gamma - \bar{y}(\bar{F}_6\cos 2\gamma + \bar{F}_7\sin 2\gamma)\Big]\overset{j}{D}_{\bar{x}}$$
$$+ 2\mu\sum_{j=1}^{N}\Big[\bar{F}_5 + \bar{y}(\bar{F}_6\sin 2\gamma - \bar{F}_7\cos 2\gamma)\Big]\overset{j}{D}_{\bar{y}} \tag{6.47}$$

in which $\gamma = \overset{i}{\beta} - \overset{j}{\beta}$. Combination of (6.45)–(6.47) gives the required result.

6.10.2 Orthotropic Material

We next consider the case of an orthotropic (transversely isotropic) material with directions of elastic symmetry along the x, y coordinate axes. The plane strain version of Hooke's law for such a material is given by (4.34). (These general expressions also apply to plane stress as the inverse of (4.31).) The stresses and strains in these expressions must be transformed to local $\bar{x}$, $\bar{y}$ coordinates in order to obtain the counterpart to (6.38) for orthotropic elasticity, and this is accomplished using (4.67) for the stresses and a similar set of expressions for the strains. After some simple algebra we find that the resulting stress-strain relations can be written as

$$\begin{aligned}
\sigma_{\bar{x}\bar{x}} &= C_{11}\varepsilon_{\bar{x}\bar{x}} + C_{12}\varepsilon_{\bar{y}\bar{y}} + 2C_{13}\varepsilon_{\bar{x}\bar{y}} \\
\sigma_{\bar{y}\bar{y}} &= C_{21}\varepsilon_{\bar{x}\bar{x}} + C_{22}\varepsilon_{\bar{y}\bar{y}} + 2C_{23}\varepsilon_{\bar{x}\bar{y}} \\
\sigma_{\bar{x}\bar{y}} &= C_{31}\varepsilon_{\bar{x}\bar{x}} + C_{32}\varepsilon_{\bar{y}\bar{y}} + 2C_{33}\varepsilon_{\bar{x}\bar{y}}
\end{aligned} \tag{6.48}$$

in which

$$\begin{aligned}
C_{11} &= c_{11}\cos^4\beta + c_{22}\sin^4\beta + 2c_{12}\sin^2\beta\cos^2\beta + c_{66}\sin^2 2\beta \\
C_{12} &= C_{21} = (c_{11}+c_{22})\sin^2\beta\cos^2\beta + c_{12}(\sin^4\beta + \cos^4\beta) - c_{66}\sin^2 2\beta \\
C_{13} &= C_{31} = -(c_{11}-c_{12})\sin\beta\cos^3\beta + (c_{22}-c_{12})\sin^3\beta\cos\beta + c_{66}\sin 2\beta\cos 2\beta \\
C_{22} &= c_{11}\sin^4\beta + c_{22}\cos^4\beta + 2c_{12}\sin^2\beta\cos^2\beta + c_{66}\sin^2 2\beta \\
C_{23} &= C_{32} = -(c_{11}-c_{12})\sin^3\beta\cos\beta + (c_{22}-c_{12})\sin\beta\cos^3\beta - c_{66}\sin 2\beta\cos 2\beta \\
C_{33} &= (c_{11}-c_{12})\sin^2\beta\cos^2\beta + (c_{22}-c_{12})\sin^2\beta\cos^2\beta + c_{66}\cos^2 2\beta
\end{aligned} \tag{6.49}$$

The last two of (6.48) can be put in the form

$$\begin{aligned}
C_{22}\varepsilon_{\bar{y}\bar{y}} + 2C_{23}\varepsilon_{\bar{x}\bar{y}} &= \sigma_{\bar{y}\bar{y}} - C_{21}\varepsilon_{\bar{x}\bar{x}} \\
C_{32}\varepsilon_{\bar{y}\bar{y}} + 2C_{33}\varepsilon_{\bar{x}\bar{y}} &= \sigma_{\bar{x}\bar{y}} - C_{31}\varepsilon_{\bar{x}\bar{x}}
\end{aligned} \tag{6.50}$$

These equations can be solved for $\varepsilon_{\bar{y}y}$ and $2\varepsilon_{\bar{x}\bar{y}}$ and the results substituted into the first one of (6.48) to obtain the following expression for $\sigma_{\bar{x}\bar{x}}$ in terms of the strain $\varepsilon_{\bar{x}\bar{x}} = \partial u_{\bar{x}}/\partial\bar{x}$ and stresses $\sigma_{\bar{y}\bar{y}}$ and $\sigma_{\bar{x}\bar{y}}$:

$$\begin{aligned}
\sigma_{\bar{x}\bar{x}} = {} & \frac{1}{C_{22}C_{33}-C_{23}^2}\left[C_{11}C_{22}C_{33} - C_{11}C_{23}^2 - C_{33}C_{12}^2 + 2C_{12}C_{13}C_{23} - C_{22}C_{13}^2\right]\frac{\partial u_{\bar{x}}}{\partial\bar{x}} \\
& + \frac{1}{C_{22}C_{33}-C_{23}^2}\left[\left(C_{33}C_{12} - C_{13}C_{23}\right)\sigma_{\bar{y}\bar{y}} + \left(C_{22}C_{13} - C_{12}C_{23}\right)\sigma_{\bar{x}\bar{y}}\right]
\end{aligned} \tag{6.51}$$

This expression is applicable to both the positive and negative surfaces of the element and is comparable to (6.38) for the case of orthotropic elasticity. The stresses $\sigma^{+}_{\bar{y}\bar{y}}$, $\sigma^{+}_{\bar{x}\bar{y}}$ and $\sigma^{-}_{\bar{y}\bar{y}}$, $\sigma^{-}_{\bar{x}\bar{y}}$ are equal, by equilibrium, and therefore

$$\sigma^{+}_{\bar{x}\bar{x}} - \sigma^{-}_{\bar{x}\bar{x}} = \frac{1}{C_{22}C_{33} - C_{23}^2}\left[C_{11}C_{22}C_{33} - C_{11}C_{23}^2 - C_{33}C_{12}^2 + 2C_{12}C_{13}C_{23} - C_{22}C_{13}^2\right]\frac{\partial}{\partial \bar{x}}\Delta u_{\bar{x}} \tag{6.52}$$

which again indicates that $\sigma_{\bar{x}\bar{x}}$ is discontinuous whenever the tangential derivative of the shear component of displacement discontinuity is nonzero.

Equations (6.51) and (6.52) can be used to compute values of $\overset{i}{\sigma}{}^{\pm}_{\bar{x}\bar{x}}$ for an orthotropic elastic material in the same way as described for an isotropic material.

6.11 Examples

As illustrations of the displacement discontinuity method we will obtain numerical solutions to two of the problems considered in Section 5.9 for the stress discontinuity method: a circular hole in an infinite plate under uniaxial stress $\sigma_{xx} = p$ $(p > 0)$ at infinity, and a circular disk subjected to diametral compression. More elaborate examples of the displacement discontinuity method are considered in Chap. 8.

6.11.1 Exterior Problem—Circular Hole in a Plate

This problem was solved numerically using a 25 element representation of a symmetric one-quarter of the circular boundary. No special provisions for fixing displacements of the interior region were necessary because the symmetry conditions ensured that this region was held stationary about its center.

The circumferential stress $\sigma_{\theta\theta}$ along the boundary of the hole was computed using the approach described in Sect. 6.10 for both isotropic and orthotropic elastic materials. The elastic constants were taken to be the same as in the examples in Sect. 5.9. The results are compared with the analytical solutions in Fig. 6.6 and are seen to be of similar accuracy to those obtained by the stress discontinuity method (Fig. 5.8).

6.11.2 Interior Problem—Circular Disk

This problem, depicted in Fig. 5.11, was solved using the same boundary discretization, elastic parameters, and loading conditions as described in Sect. 5.9.3. The numerical results for the displacement discontinuity method, shown in Fig. 6.7, are

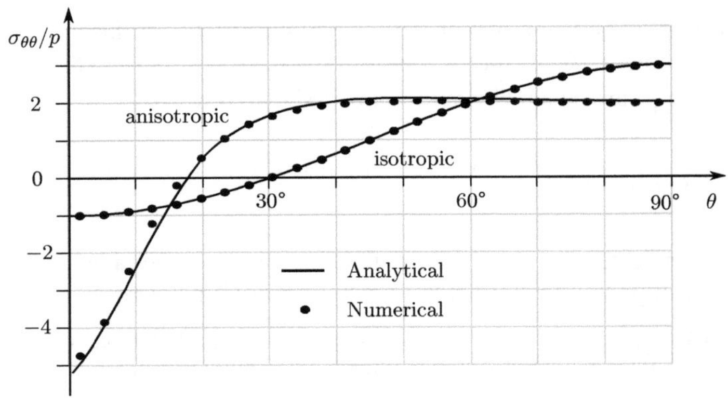

Fig. 6.6 Stress $\sigma_{\theta\theta}$ at the boundary of a circular hole

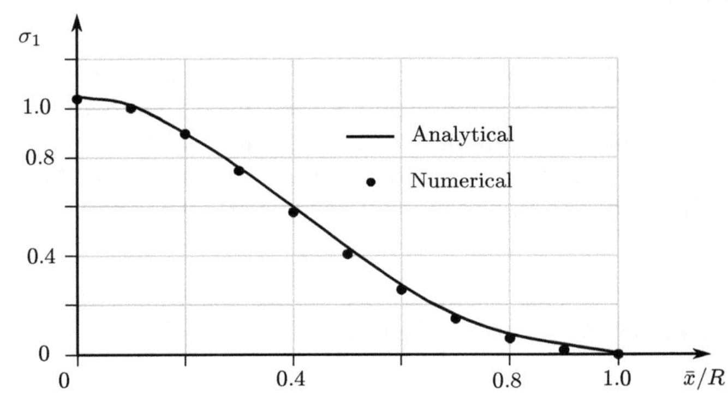

Fig. 6.7 Major principal stress σ_1 along a diameter of the disk (see Fig. 5.11) perpendicular to the applied load

again of comparable accuracy to those obtained previously by the stress discontinuity method (Fig. 5.12).

It should be observed that the same elemental displacement discontinuity components used to compute the results the circular disk also serve to determine a numerical solution to a companion problem for the exterior region. This problem concerns an infinite orthotropic elastic body containing a circular hole that is subjected to normal loads over two small diametrically opposite circular arcs along its boundary and unloaded over the remainder. Stresses and displacements in the infinite region would have been evaluated for this problem if points exterior to the circular boundary had been sampled.

6.12 Higher Order Elements I—Isotropic Elasticity

Higher order elements for the displacement discontinuity method can be developed in the same way as was done for the stress discontinuity method in Sect. 5.10. Results for isotropic elasticity are given in this section; a formulation for anisotropic elasticity is summarized in Sect. 6.13.

By analogy with (5.41), quadratic variations of the displacement discontinuity components over the line segment $|\bar{x}| \leq a$, $\bar{y} = 0$ can be written as

$$\begin{aligned} \Delta u_{\bar{x}}(\bar{x}) &= (\Delta u_{\bar{x}})_1 \alpha_1(\bar{x}) + (\Delta u_{\bar{x}})_2 \alpha_2(\bar{x}) + (\Delta u_{\bar{x}})_3 \alpha_3(\bar{x}) \\ \Delta u_{\bar{y}}(\bar{x}) &= (\Delta u_{\bar{y}})_1 \alpha_1(\bar{x}) + (\Delta u_{\bar{y}})_2 \alpha_2(\bar{x}) + (\Delta u_{\bar{y}})_3 \alpha_3(\bar{x}) \end{aligned} \tag{6.53}$$

where, for $k = 1$ to 3, $(\Delta u_{\bar{x}})_k$ and $(\Delta u_{\bar{y}})_k$ are the displacement discontinuity components at the element collocation points and $\alpha_k(\bar{x})$ are the shape functions given in (5.39).

An infinitesimal element $\mathrm{d}\xi_{\bar{x}}$ of the line segment at the point $\bar{x} = \xi_{\bar{x}}$, $\bar{y} = 0$ can be regarded as a concentrated displacement discontinuity with components $\mathcal{D}_{\bar{x}} = \Delta u_{\bar{x}}(\xi_{\bar{x}})\,\mathrm{d}\xi_{\bar{x}}$ and $\mathcal{D}_{\bar{y}} = \Delta u_{\bar{y}}(\xi_{\bar{x}})\,\mathrm{d}\xi_{\bar{x}}$. Expressions for the displacements and stresses in an infinite, isotropic elastic body due to a quadratically varying displacement discontinuity are then obtained by substituting $\mathcal{D}_{\bar{x}}$ and $\mathcal{D}_{\bar{y}}$ into versions of (4.221) and (4.222) written in terms of $\bar{x}$, $\bar{y}$ coordinates, replacing $\bar{x}$ by $\bar{x} - \xi_{\bar{x}}$, and integrating the resulting expressions with respect to $\xi_{\bar{x}}$ between the limits $-a$ and $+a$.

Proceeding in the same way as in Sect. 5.10, we find, for example, the following expression for $\sigma_{\bar{y}\bar{y}}$:

$$\begin{aligned} \sigma_{\bar{y}\bar{y}} = {} & 2\mu \int_{-a}^{a} \Delta u_{\bar{x}}(\xi_{\bar{x}}) \left[\bar{y} \frac{\partial^3}{\partial \bar{x} \partial \bar{y}^2} g(\bar{x} - \xi_{\bar{x}}, \bar{y}) \right] \mathrm{d}\xi_{\bar{x}} \\ & + 2\mu \int_{-a}^{a} \Delta u_{\bar{y}}(\xi_{\bar{x}}) \left[-\frac{\partial^2}{\partial \bar{y}^2} g(\bar{x} - \xi_{\bar{x}}, \bar{y}) + \bar{y} \frac{\partial^3}{\partial \bar{y}^3} g(\bar{x} - \xi_{\bar{x}}, \bar{y}) \right] \mathrm{d}\xi_{\bar{x}} \end{aligned} \tag{6.54}$$

or, using (6.53),

$$\begin{aligned} \sigma_{\bar{y}\bar{y}} = {} & 2\mu \sum_{k=1}^{3} (\Delta u_{\bar{x}})_k \int_{-a}^{a} \left[\bar{y} \alpha_k(\xi_{\bar{x}}) \frac{\partial^3}{\partial \bar{x} \partial \bar{y}^2} g(\bar{x} - \xi_{\bar{x}}, \bar{y}) \right] \mathrm{d}\xi_{\bar{x}} \\ & + 2\mu \sum_{k=1}^{3} (\Delta u_{\bar{y}})_k \int_{-a}^{a} \left[-\alpha_k(\xi_{\bar{x}}) \frac{\partial^2}{\partial \bar{y}^2} g(\bar{x} - \xi_{\bar{x}}, \bar{y}) + \bar{y} \alpha_k(\xi_{\bar{x}}) \frac{\partial^3}{\partial \bar{y}^3} g(\bar{x} - \xi_{\bar{x}}, \bar{y}) \right] \mathrm{d}\xi_{\bar{x}} \end{aligned} \tag{6.55}$$

Similar expressions can be written for the stress components $\sigma_{\bar{x}\bar{x}}$ and $\sigma_{\bar{x}\bar{y}}$ and the displacements $u_{\bar{x}}$ and $u_{\bar{y}}$. The results involve integrals of the products of the shape functions $\alpha_k(\xi_{\bar{x}})$ and the derivatives through third order of the function $g(\bar{x} - \xi_{\bar{x}}, \bar{y})$. In addition to the quantities $(\bar{G}_2)_k$ through $(\bar{G}_5)_k$ in (5.48), we therefore require the integrals

$$(\bar{G}_6)_k = \int_{-a}^{a} \alpha_k(\xi_{\bar{x}}) \frac{\partial^3}{\partial\bar{x}\partial\bar{y}^2} g(\bar{x} - \xi_{\bar{x}}, \bar{y})\, \mathrm{d}\xi_{\bar{x}}$$

$$(\bar{G}_7)_k = \int_{-a}^{a} \alpha_k(\xi_{\bar{x}}) \frac{\partial^3}{\partial\bar{y}^3} g(\bar{x} - \xi_{\bar{x}}, \bar{y})\, \mathrm{d}\xi_{\bar{x}} \tag{6.56}$$

The reader should confirm that the displacements and stresses can be written in terms of $(\bar{G}_2)_k$ through $(\bar{G}_7)_k$ as follows (cf. (6.8) and (6.9)):

$$u_{\bar{x}} = \sum_{k=1}^{3} (\Delta u_{\bar{x}})_k \big[-2(1-\nu)(\bar{G}_3)_k + \bar{y}(\bar{G}_5)_k\big] + \sum_{k=1}^{3} (\Delta u_{\bar{y}})_k \big[(1-2\nu)(\bar{G}_2)_k + \bar{y}(\bar{G}_4)_k\big]$$

$$u_{\bar{y}} = \sum_{k=1}^{3} (\Delta u_{\bar{x}})_k \big[-(1-2\nu)(\bar{G}_2)_k + \bar{y}(\bar{G}_4)_k\big] + \sum_{k=1}^{3} (\Delta u_{\bar{y}})_k \big[-2(1-\nu)(\bar{G}_3)_k - \bar{y}(\bar{G}_5)_k\big] \tag{6.57}$$

and

$$\sigma_{\bar{x}\bar{x}} = 2\mu \sum_{k=1}^{3} (\Delta u_{\bar{x}})_k \big[-2(\bar{G}_4)_k - \bar{y}(\bar{G}_6)_k\big] + 2\mu \sum_{k=1}^{3} (\Delta u_{\bar{y}})_k \big[(\bar{G}_5)_k - \bar{y}(\bar{G}_7)_k\big]$$

$$\sigma_{\bar{y}\bar{y}} = 2\mu \sum_{k=1}^{3} (\Delta u_{\bar{x}})_k \big[\bar{y}(\bar{G}_6)_k\big] + 2\mu \sum_{k=1}^{3} (\Delta u_{\bar{y}})_k \big[(\bar{G}_5)_k + \bar{y}(\bar{G}_7)_k\big] \tag{6.58}$$

$$\sigma_{\bar{x}\bar{y}} = 2\mu \sum_{k=1}^{3} (\Delta u_{\bar{x}})_k \big[(\bar{G}_5)_k - \bar{y}(\bar{G}_7)_k\big] + 2\mu \sum_{k=1}^{3} (\Delta u_{\bar{y}})_k \big[\bar{y}(\bar{G}_6)_k\big]$$

Notice that when $\bar{y} = 0$ the normal stress $\sigma_{\bar{x}\bar{x}}$ due to the shear component of displacement discontinuity is proportional to the terms $(\Delta u_{\bar{x}})_k(\bar{G}_4)_k$. The function $(\bar{G}_4)_k$ contains the function $(\bar{F}_4)_L$, which, referring to (5.53), involves the jump term $\bar{\theta}_1 - \bar{\theta}_2$. We conclude from this that the finite difference approximations used in Sect. 6.10 to compute a discontinuity in the normal stress parallel to a boundary are not necessary when using quadratic displacement discontinuity elements.

The quantities $(\bar{G}_6)_k$ and $(\bar{G}_7)_k$ in (6.58) are defined in the same way as $(\bar{G}_3)_k$ in (5.51) and (5.52). The integrals $(\bar{F}_6)_C$ and $(\bar{F}_7)_C$ are the same as $F_6(\bar{x}, \bar{y})$ and $F_7(\bar{x}, \bar{y})$ in (6.7). The other integrals are

$$
\begin{aligned}
(\bar{F}_6)_L &= \int_{-a}^{a} (\bar{x}-\xi_{\bar{x}}) \frac{\partial^3}{\partial \bar{x} \partial \bar{y}^2} g(\bar{x}-\xi_{\bar{x}}, \bar{y})\, \mathrm{d}\xi_{\bar{x}} \\
&= \frac{2}{4\pi(1-\nu)} \left[\frac{(\bar{x}-a)^3}{\bar{r}_1^4} - \frac{(\bar{x}+a)^3}{\bar{r}_2^4} \right] \\
(\bar{F}_7)_L &= \int_{-a}^{a} (\bar{x}-\xi_{\bar{x}}) \frac{\partial^3}{\partial \bar{y}^3} g(\bar{x}-\xi_{\bar{x}}, \bar{y})\, \mathrm{d}\xi_{\bar{x}} \\
&= \frac{1}{4\pi(1-\nu)} \bar{y} \left[\frac{1}{\bar{r}_1^2} - \frac{1}{\bar{r}_2^2} + \frac{2(\bar{x}-a)^2}{\bar{r}_1^4} - \frac{2(\bar{x}+a)^2}{\bar{r}_2^4} \right]
\end{aligned}
\tag{6.59}
$$

$$
\begin{aligned}
(\bar{F}_6)_Q &= \int_{-a}^{a} (\bar{x}-\xi_{\bar{x}})^2 \frac{\partial^3}{\partial \bar{x} \partial \bar{y}^2} g(\bar{x}-\xi_{\bar{x}}, \bar{y})\, \mathrm{d}\xi_{\bar{x}} \\
&= \frac{1}{4\pi(1-\nu)} \left[\frac{(\bar{x}-a)^2}{\bar{r}_1^2} - \frac{(\bar{x}+a)^2}{\bar{r}_2^2} + \frac{2(\bar{x}-a)^4}{\bar{r}_1^4} - \frac{2(\bar{x}+a)^4}{\bar{r}_2^4} - 2 \ln \frac{\bar{r}_1}{\bar{r}_2} \right] \\
(\bar{F}_7)_Q &= \int_{-a}^{a} (\bar{x}-\xi_{\bar{x}})^2 \frac{\partial^3}{\partial \bar{y}^3} g(\bar{x}-\xi_{\bar{x}}, \bar{y})\, \mathrm{d}\xi_{\bar{x}} \\
&= \frac{2}{4\pi(1-\nu)} \left[\frac{(\bar{x}-a)\bar{y}}{\bar{r}_1^2} - \frac{(\bar{x}+a)\bar{y}}{\bar{r}_2^2} + \frac{(\bar{x}-a)^3\bar{y}}{\bar{r}_1^4} - \frac{(\bar{x}+a)^3\bar{y}}{\bar{r}_2^4} + (\bar{\theta}_1 - \bar{\theta}_2) \right]
\end{aligned}
\tag{6.60}
$$

Boundary influence coefficients for quadratically varying displacement discontinuities can be developed from the preceding results by essentially the same procedure as was used for constant strength elements. Equations comparable to (5.55) and (5.56) for the displacement discontinuity method are

$$
\begin{aligned}
\sum_{l=1}^{3} (\overset{i}{t}_{\bar{x}})_l &= \sum_{k=1}^{3} \left[\sum_{j=1}^{N} (\overset{ij}{A}_{\bar{x}\bar{x}})_{lk} (\Delta \overset{j}{u}_{\bar{x}})_k + \sum_{j=1}^{N} (\overset{ij}{A}_{\bar{x}\bar{y}})_{lk} (\Delta \overset{j}{u}_{\bar{y}})_k \right] \\
\sum_{l=1}^{3} (\overset{i}{t}_{\bar{y}})_l &= \sum_{k=1}^{3} \left[\sum_{j=1}^{N} (\overset{ij}{A}_{\bar{y}\bar{x}})_{lk} (\Delta \overset{j}{u}_{\bar{x}})_k + \sum_{j=1}^{N} (\overset{ij}{A}_{\bar{y}\bar{y}})_{lk} (\Delta \overset{j}{u}_{\bar{y}})_k \right]
\end{aligned}
\tag{6.61}
$$

and

$$
\begin{aligned}
\sum_{l=1}^{3} (\overset{i}{u}_{\bar{x}})_l &= \sum_{k=1}^{3} \left[\sum_{j=1}^{N} (\overset{ij}{B}_{\bar{x}\bar{x}})_{lk} (\Delta \overset{j}{u}_{\bar{x}})_k + \sum_{j=1}^{N} (\overset{ij}{B}_{\bar{x}\bar{y}})_{lk} (\Delta \overset{j}{u}_{\bar{y}})_k \right] \\
\sum_{l=1}^{3} (\overset{i}{u}_{\bar{y}})_l &= \sum_{k=1}^{3} \left[\sum_{j=1}^{N} (\overset{ij}{B}_{\bar{y}\bar{x}})_{lk} (\Delta \overset{j}{u}_{\bar{x}})_k + \sum_{j=1}^{N} (\overset{ij}{B}_{\bar{y}\bar{y}})_{lk} (\Delta \overset{j}{u}_{\bar{y}})_k \right]
\end{aligned}
\tag{6.62}
$$

The required system of algebraic equations for the displacement discontinuity method is formed by considering the boundary conditions at each of the $3N$ collocation points and using the appropriate ones of (6.61) and (6.62), just as in the process described for the stress discontinuity method. If we number the collocation points consecutively as $m = 3(i-1)+l$ and $n = 3(j-1)+k$, let $\overset{m}{b}_{\bar{x}}$ and $\overset{m}{b}_{\bar{y}}$ be the prescribed values of traction or displacement at collocation point m, and let $\Delta\overset{n}{u}_{\bar{x}}$ and $\Delta\overset{n}{u}_{\bar{y}}$ be the displacement discontinuities at collocation point n, we represent the $6N \times 6N$ system of algebraic equations as

$$\sum_{n=1}^{3N} \overset{mn}{C}_{\bar{x}\bar{x}} \Delta\overset{n}{u}_{\bar{x}} + \sum_{n=1}^{3N} \overset{mn}{C}_{\bar{x}\bar{y}} \Delta\overset{n}{u}_{\bar{y}} = \overset{m}{b}_{\bar{x}}$$
$$\sum_{n=1}^{3N} \overset{mn}{C}_{\bar{y}\bar{x}} \Delta\overset{n}{u}_{\bar{x}} + \sum_{n=1}^{3N} \overset{mn}{C}_{\bar{y}\bar{y}} \Delta\overset{n}{u}_{\bar{y}} = \overset{m}{b}_{\bar{y}} \tag{6.63}$$

where m ranges from 1 to $3N$.

After (6.63) have been solved for a given problem the displacements and stresses at specified points in the region of interest can be calculated by summing the effects of the displacement discontinuities at the N boundary elements. For these calculations it is convenient to use formulas (4.62) and (4.66) to transform displacements (6.57) and stresses (6.58) to the global x, y coordinate system.

6.13 Higher Order Elements II—Anisotropic Elasticity

The displacements u_i and stresses σ_{ij} in an infinite, orthotropic (or transversely isotropic) elastic material due to a concentrated displacement discontinuity with components $\mathcal{D}_x$ and $\mathcal{D}_y$ are given in Sect. 4.14.2. For present purposes we need these results expressed in terms of the local components of the concentrated displacement discontinuity, $\mathcal{D}_{\bar{x}}$ and $\mathcal{D}_{\bar{y}}$, which is achieved by substituting (4.204) into (4.226) and (4.227). The displacement component u_x, for example, is given as

$$\begin{aligned} u_x = {} & \frac{q_3}{2\pi}\mathcal{D}_{\bar{x}}\left[-\frac{1}{1+q_1}\left\{\gamma_1\frac{\partial\Omega(x,y_1)}{\partial x}\sin\beta - \frac{\partial\Omega(x,y_1)}{\partial y_i}\cos\beta\right\}\right. \\ & \left. + \frac{1}{1+q_2}\left\{\gamma_2\frac{\partial\Omega(x,y_2)}{\partial x}\sin\beta - \frac{\partial\Omega(x,y_2)}{\partial y_i}\cos\beta\right\}\right] \\ & + \frac{q_3}{2\pi}\mathcal{D}_{\bar{y}}\left[-\frac{1}{1+q_1}\left\{\gamma_1\frac{\partial\Omega(x,y_1)}{\partial x}\cos\beta + \frac{\partial\Omega(x,y_1)}{\partial y_i}\sin\beta\right\}\right. \\ & \left. + \frac{1}{1+q_2}\left\{\gamma_2\frac{\partial\Omega(x,y_2)}{\partial x}\cos\beta + \frac{\partial\Omega(x,y_2)}{\partial y_i}\sin\beta\right\}\right] \end{aligned} \tag{6.64}$$

The results for a quadratically varying displacement discontinuity over the line segment $-a \leq \overline{x} \leq a$, $\overline{y} = 0$ are obtained by replacing $\mathcal{D}_{\bar{x}}$ and $\mathcal{D}_{\bar{y}}$ in (6.64) (and other similar equations) by $\Delta u_{\bar{x}}(\xi_{\bar{x}})\,\mathrm{d}\xi_{\bar{x}}$ and $\Delta u_{\bar{y}}(\xi_{\bar{x}})\,\mathrm{d}\xi_{\bar{x}}$, with $\Delta u_{\overline{x}}(\overline{x})$ and $\Delta u_{\overline{y}}(\overline{x})$ defined as in (6.53), replacing $\bar{x}$ by $\bar{x} - \xi_{\bar{x}}$, and integrating with respect to $\xi_{\bar{x}}$ between the limits $-a$ and $+a$. These calculations will involve integrals such as

$$\int_{-a}^{a} \frac{\alpha_k(\xi_{\bar{x}})}{A_i(\bar{x}-\xi_{\bar{x}})^2 + B_i(\bar{x}-\xi_{\bar{x}})\bar{y} + C_i\bar{y}^2}\,\mathrm{d}\xi_{\bar{x}}$$

$$\int_{-a}^{a} \frac{(\bar{x}-\xi_{\bar{x}})\alpha_k(\xi_{\bar{x}})}{A_i(\bar{x}-\xi_{\bar{x}})^2 + B_i(\bar{x}-\xi_{\bar{x}})\bar{y} + C_i\bar{y}^2}\,\mathrm{d}\xi_{\bar{x}}$$

$$\int_{-a}^{a} \frac{(\bar{x}-\xi_{\bar{x}})^2\alpha_k(\xi_{\bar{x}})}{\left[A_i(\bar{x}-\xi_{\bar{x}})^2 + B_i(\bar{x}-\xi_{\bar{x}})\bar{y} + C_i\bar{y}^2\right]^2}\,\mathrm{d}\xi_{\bar{x}}$$

and so forth, where $k = 1$, 2, or 3 and where $\alpha_k(\xi_{\bar{x}})$ are given in (5.44). If we set $u = \overline{x} - \xi_{\bar{x}}$ and $A = A_i$, $B = B_i\bar{y}$, and $C = C_i\bar{y}^2$, a typical integral will have the form

$$\int_{-a}^{a} \frac{(\bar{x}-\xi_{\bar{x}})}{A_i(\bar{x}-\xi_{\bar{x}})^2 + B_i(\bar{x}-\xi_{\bar{x}})\bar{y} + C_i\bar{y}^2}\,\mathrm{d}\xi_{\bar{x}} = -\int_{\bar{x}+a}^{\bar{x}-a} \frac{u}{Au^2 + Bu + C}\,\mathrm{d}u \tag{6.65}$$

This and other integrals needed to complete the analysis can be evaluated using standard integration formulas (e.g., Petit Bois 1961). The required formulas are summarized in Appendix A for the case that $B^2 - 4AC < 0$. It is easy to show from (4.179) that $B^2 - 4AC = -4\bar{y}^2/\gamma_i^2 < 0$ for our problem.[4]

6.14 Integral Formulation

We conclude this discussion of the displacement discontinuity method by showing that the method can be viewed as a discretized representation of the integral formulas given in Sect. 4.18.2 for the displacements, stresses, and tractions for the concentrated displacement discontinuity analogue of the double layer potential. As was done for the stress discontinuity method, we will demonstrate the equivalence of the two approaches for calculation of the stress components.

The stresses $\sigma_{ij}(\boldsymbol{x})$ at a sample point $\boldsymbol{x}$ due to a distribution of concentrated displacement discontinuities $\eta(\boldsymbol{\xi})\,\mathrm{d}S(\boldsymbol{\xi})$ at a load point $\boldsymbol{\xi}$ along a smooth curve S

---

[4] As usual, the case for which $\bar{y} \to 0_{\pm}$ is handled by a limiting procedure.

are given by (4.300). In unabridged form comparable to (5.59) for the single layer formulation they are:

$$
\begin{aligned}
\sigma_{xx}(\boldsymbol{x}) &= \int_S \left[\eta_x(\boldsymbol{\xi})C_{xxx}(\boldsymbol{x},\boldsymbol{\xi}) + \eta_y(\boldsymbol{\xi})C_{yxx}(\boldsymbol{x},\boldsymbol{\xi})\right] \mathrm{d}S(\boldsymbol{\xi}) \\
\sigma_{yy}(\boldsymbol{x}) &= \int_S \left[\eta_x(\boldsymbol{\xi})C_{xyy}(\boldsymbol{x},\boldsymbol{\xi}) + \eta_y(\boldsymbol{\xi})C_{yyy}(\boldsymbol{x},\boldsymbol{\xi})\right] \mathrm{d}S(\boldsymbol{\xi}) \\
\sigma_{xy}(\boldsymbol{x}) &= \int_S \left[\eta_x(\boldsymbol{\xi})C_{xxy}(\boldsymbol{x},\boldsymbol{\xi}) + \eta_y(\boldsymbol{\xi})C_{yxy}(\boldsymbol{x},\boldsymbol{\xi})\right] \mathrm{d}S(\boldsymbol{\xi})
\end{aligned}
\tag{6.66}
$$

We next suppose that the boundary S is approximated by N straight line elements with the jth element centered at point $\boldsymbol{\xi}$ (such that $\xi_x = \overset{j}{x}$ and $\xi_y = \overset{j}{y}$) and inclined from the x axis by an angle $\overset{j}{\beta}$, so that $(\boldsymbol{x},\boldsymbol{\xi}) = (x - \overset{j}{x}, y - \overset{j}{y})$. Taking the displacement discontinuity over the element as constant, i.e., $\eta_x = \overset{j}{D}_x$ and $\eta_y = \overset{j}{D}_y$ for $-\overset{j}{a} \le \bar{x} \le \overset{j}{a}$, $\bar{y} = 0$, where the local coordinates $\bar{x}$ and $\bar{y}$ are given by (5.16), the discretized forms of (6.66) can be written as

$$
\begin{aligned}
\sigma_{xx}(x,y) &= \sum_{j=1}^{N} \overset{j}{D}_x \int_{-\overset{j}{a}}^{\overset{j}{a}} C_{xxx}(x-\overset{j}{x}, y-\overset{j}{y})\,\mathrm{d}\bar{x} + \sum_{j=1}^{N} \overset{j}{D}_y \int_{-\overset{j}{a}}^{\overset{j}{a}} C_{yxx}(x-\overset{j}{x}, y-\overset{j}{y})\,\mathrm{d}\bar{x} \\
\sigma_{yy}(x,y) &= \sum_{j=1}^{N} \overset{j}{D}_x \int_{-\overset{j}{a}}^{\overset{j}{a}} C_{xyy}(x-\overset{j}{x}, y-\overset{j}{y})\,\mathrm{d}\bar{x} + \sum_{j=1}^{N} \overset{j}{D}_y \int_{-\overset{j}{a}}^{\overset{j}{a}} C_{yyy}(x-\overset{j}{x}, y-\overset{j}{y})\,\mathrm{d}\bar{x} \\
\sigma_{xy}(x,y) &= \sum_{j=1}^{N} \overset{j}{D}_x \int_{-\overset{j}{a}}^{\overset{j}{a}} C_{xxy}(x-\overset{j}{x}, y-\overset{j}{y})\,\mathrm{d}\bar{x} + \sum_{j=1}^{N} \overset{j}{D}_y \int_{-\overset{j}{a}}^{\overset{j}{a}} C_{yxy}(x-\overset{j}{x}, y-\overset{j}{y})\,\mathrm{d}\bar{x}
\end{aligned}
\tag{6.67}
$$

As in Sect. 5.11 we note that $\bar{x} = \xi_{\bar{x}}$ so that and $\mathrm{d}S(\boldsymbol{\xi}) = \mathrm{d}\bar{x}$. The coefficients $C_{xxx}(x - \overset{j}{x}, y - \overset{j}{y})$, etc., for an isotropic material can be recovered by elaboration of (4.291).

As in the case of the single layer potential analogue considered in Sect. 5.11, we seek to express all the quantities on the right-hand sides of (6.67) in terms of $\bar{x}$ and $\bar{y}$ coordinates. For the component $\sigma_{yy}(x,y)$, for example, we write

$$
\begin{aligned}
\sigma_{yy}(x,y) = &\sum_{j=1}^{N} \overset{j}{D}_{\bar{x}} \int_{-\overset{j}{a}}^{\overset{j}{a}} \left[C_{xyy}(\bar{x},\bar{y})\cos\overset{j}{\beta} + C_{yyy}(\bar{x},\bar{y})\sin\overset{j}{\beta}\right] \mathrm{d}\bar{x} \\
&+ \sum_{j=1}^{N} \overset{j}{D}_{\bar{y}} \int_{-\overset{j}{a}}^{\overset{j}{a}} \left[-C_{xyy}(\bar{x},\bar{y})\sin\overset{j}{\beta} + C_{yyy}(\bar{x},\bar{y})\cos\overset{j}{\beta}\right] \mathrm{d}\bar{x}
\end{aligned}
\tag{6.68}
$$

with similar expressions for $\sigma_{xx}(x, y)$ and $\sigma_{xy}(x, y)$.

Referring to (4.291), the coefficients $C_{xyy}(x - \overset{j}{x}, y - \overset{j}{y})$ and $C_{yyy}(x - \overset{j}{x}, y - \overset{j}{y})$ in (6.67) can be written as

$$C_{xyy}(x - \overset{j}{x}, y - \overset{j}{y}) = \frac{2\mu}{4\pi(1-\nu)}\frac{1}{r^2}\left[-n_x + 2r_{,y}r_{,y}n_x + 2r_{,x}(1 - 4r_{,y}r_{,y})\frac{\partial r}{\partial \bar{y}}\right]$$

$$C_{yyy}(x - \overset{j}{x}, y - \overset{j}{y}) = \frac{2\mu}{4\pi(1-\nu)}\frac{1}{r^2}\left[n_y + 2r_{,y}r_{,y}n_y + 2r_{,y}(1 - 4r_{,y}r_{,y})\frac{\partial r}{\partial \bar{y}}\right] \tag{6.69}$$

where $r = \left[(x - \overset{j}{x})^2 + (y - \overset{j}{y})^2\right]^{1/2} = (\bar{x}^2 + \bar{y}^2)^{1/2}$ (see (5.61)), and where we have written $\partial r/\partial \xi_{\bar{y}} = -\partial r/\partial \bar{y}$ for $\partial r/\partial \boldsymbol{n}(\boldsymbol{\xi})$ in (4.291) in accordance with (4.282). For the element $-\overset{j}{a} \le \bar{x} \le \overset{j}{a}$, $\bar{y} = 0$, $n_x = -\sin\overset{j}{\beta}$ and $n_y = \cos\overset{j}{\beta}$ and (6.69) become

$$C_{xyy}(x - \overset{j}{x}, y - \overset{j}{y}) = \frac{2\mu}{4\pi(1-\nu)}\frac{1}{r^2}\left[\sin\overset{j}{\beta}(1 - 2r_{,y}r_{,y}) + 2r_{,x}(1 - 4r_{,y}r_{,y})\frac{\partial r}{\partial \bar{y}}\right]$$

$$C_{yyy}(x - \overset{j}{x}, y - \overset{j}{y}) = \frac{2\mu}{4\pi(1-\nu)}\frac{1}{r^2}\left[\cos\overset{j}{\beta}(1 + 2r_{,y}r_{,y}) + 2r_{,y}(1 - 4r_{,y}r_{,y})\frac{\partial r}{\partial \bar{y}}\right] \tag{6.70}$$

For comparison with the results of the influence coefficient formulation of the displacement discontinuity method given earlier in this chapter it is convenient to write these formulas in terms of derivatives of function $g(x, y)$ (see (4.252)) rather than of r. To this end, we note the following relationships, which can be verified by direct differentiation:

$$\begin{aligned}
\frac{\partial^2 \ln r}{\partial y^2} &= \frac{1}{r^2}(1 - 2r_{,y}r_{,y}) \\
\frac{\partial^2 \ln r}{\partial x \partial y} &= -\frac{2}{r^2}r_{,x}r_{,y} \\
y\frac{\partial^3 \ln r}{\partial y^3} &= -\frac{4}{r^2}r_{,y}r_{,y} - \frac{2}{r^2}r_{,y}r_{,y}(1 - 4r_{,y}r_{,y}) \\
y\frac{\partial^3 \ln r}{\partial x \partial y^2} &= -\frac{2}{r^2}r_{,x}r_{,y}(1 - 4r_{,y}r_{,y})
\end{aligned} \tag{6.71}$$

Substituting these results into (6.70), using the chain rule for partial differentiation, and performing some algebraic manipulations[5] leads to the following expressions for the quantities in the integrands of (6.68):

$$
\begin{aligned}
& C_{xyy}(\bar{x},\bar{y})\cos\overset{j}{\beta} + C_{yyy}(\bar{x},\bar{y})\sin\overset{j}{\beta} \\
&= 2\mu\Big[-\frac{\partial^2 g}{\partial\bar{x}\partial\bar{y}} + \cos 2\overset{j}{\beta}\frac{\partial^2 g}{\partial\bar{x}\partial\bar{y}} - \sin 2\overset{j}{\beta}\frac{\partial^2 g}{\partial\bar{y}^2} \\
&\quad + \bar{y}\Big(\cos 2\overset{j}{\beta}\frac{\partial^3 g}{\partial\bar{x}\partial\bar{y}^2} - \sin 2\overset{j}{\beta}\frac{\partial^3 g}{\partial\bar{y}^3}\Big)\Big] \\
& -C_{xyy}(\bar{x},\bar{y})\sin\overset{j}{\beta} + C_{yyy}(\bar{x},\bar{y})\cos\overset{j}{\beta} \\
&= 2\mu\Big[-\frac{\partial^2 g}{\partial\bar{y}^2} + \bar{y}\Big(\sin 2\overset{j}{\beta}\frac{\partial^3 g}{\partial\bar{x}\partial\bar{y}^2} + \cos 2\overset{j}{\beta}\frac{\partial^3 g}{\partial\bar{y}^3}\Big)\Big]
\end{aligned}
\tag{6.72}
$$

The integrals in (6.68) are then given as follows:

$$
\begin{aligned}
& \int_{-a}^{a}\Big[C_{xyy}(\bar{x},\bar{y})\cos\beta + C_{yyy}(\bar{x},\bar{y})\sin\beta\Big]\,\mathrm{d}\bar{x} \\
&= 2\mu\Big[-\frac{\partial^2 f}{\partial\bar{x}\partial\bar{y}} + \cos 2\beta\frac{\partial^2 f}{\partial\bar{x}\partial\bar{y}} - \sin 2\beta\frac{\partial^2 f}{\partial\bar{y}^2} \\
&\quad + \bar{y}\Big(\cos 2\beta\frac{\partial^3 f}{\partial\bar{x}\partial\bar{y}^2} - \sin 2\beta\frac{\partial^3 f}{\partial\bar{y}^3}\Big)\Big] \\
& \int_{-a}^{a}\Big[-C_{xyy}(\bar{x},\bar{y})\sin\beta + C_{yyy}(\bar{x},\bar{y})\cos\beta\Big]\,\mathrm{d}\bar{x} \\
&= 2\mu\Big[-\frac{\partial^2 f}{\partial\bar{y}^2} + \bar{y}\Big(\sin 2\beta\frac{\partial^3 f}{\partial\bar{x}\partial\bar{y}^2} + \cos 2\beta\frac{\partial^3 f}{\partial\bar{y}^3}\Big)\Big]
\end{aligned}
\tag{6.73}
$$

in which have written a and β instead of $\overset{j}{a}$ and $\overset{j}{\beta}$ for notational convenience and where $f(\bar{x},\bar{y})$ is defined in (5.7). Writing $\bar{F}_4 = F(\bar{x},\bar{y}) = \partial^2 f(\bar{x},\bar{y})/\partial\bar{x}\partial\bar{y}$, etc. (see (5.8) and (6.7)), Eq. (6.73) become

[5] Results comparable to (4.163) for $\bar{x}$, $\bar{y}$ coordinates are used to eliminate terms involving derivatives of function $g(\bar{x},\bar{y})$ multiplied by coordinate $\bar{x}$.

$$\int_{-a}^{a}\left[C_{xyy}(\bar{x},\bar{y})\cos\beta + C_{yyy}(\bar{x},\bar{y})\sin\beta\right]\mathrm{d}\bar{x}$$
$$= 2\mu\big[-\bar{F}_4 + \cos 2\beta\bar{F}_4 + \sin 2\beta\bar{F}_5 + \bar{y}\big(\cos 2\beta\bar{F}_6 - \sin 2\beta\bar{F}_7\big)\big]$$
$$\int_{-a}^{a}\left[-C_{xyy}(\bar{x},\bar{y})\sin\beta + C_{yyy}(\bar{x},\bar{y})\cos\beta\right]\mathrm{d}\bar{x}$$
$$= 2\mu\big[\bar{F}_5 + \bar{y}\big(\sin 2\beta\bar{F}_6 + \cos 2\beta\bar{F}_7\big)\big] \tag{6.74}$$

A typical set of terms in the summations indicated in (6.68) is the same as the second one of (6.11) and we accordingly conclude that the influence function approach for the displacement discontinuity method is completely equivalent to a comparable discretization of the double layer elastic potential integral formulas (4.300).

6.15 Exercises/Projects

1. Prepare a computer program for the two-dimensional displacement discontinuity method for isotropic elasticity, as outlined in Sect. 6.6. Verify your program by solving the example problems described in Sect. 6.11.
2. Modify your program from Exercise 1 to include the case of orthotropic (transversely isotropic) elasticity.
3. Modify your program from Exercise 1 to use piecewise quadratic elements as described in Sect. 6.12.

References

Benveniste, Y., and T. Miloh. 2001. Imperfect soft and stiff interfaces in two-dimensional elasticity. *Mechanics & Materials* 33: 309–23.

Carnahan, B., H.A. Luther, and J.O. Wilkes. 1969. *Applied numerical methods*. Wiley.

Crouch, S.L. 1976. Solution of plane elasticity problems by the displacement discontinuity method. *International Journal for Numerical Methods in Engineering* 10: 301–43.

Crouch, S.L., and A.M. Starfield. 1983. *Boundary element methods in solid mechanics*. George Allen & Unwin.

DeBremaecker, J.-Cl., M. C. Ferris, and D. Ralph. 2000. Compressional fractures considered as contact problems and mixed complementarity problems. *Engineering Fracture Mechanics* 66: 287–303.

DeBremaecker, J.-Cl., and M. C. Ferris. 2000. A comparison of two algorithms for solving closed crack problems. *Engineering Fracture Mechanics* 66: 601–05.

Dirkse, S.P., and M.C. Ferris. 1995. The PATH solver: A non-monotonic stabilization scheme for mixed complementarity problems. *Optimization methods and software* 5: 123–56.

Golub, G.H., and C.F. Van Loan. 1996. *Matrix computations*, 3rd ed. The Johns Hopkins University Press.

Petit Bois, G. 1961. *Tables of indefinite integrals*. Dover.

Ritz, E., and D.D. Pollard. 2011. Closure of circular arc cracks under general loading: effects on stress intensity factors. *International Journal of Fracture* 167: 3–14.

Ritz, E., O. Mutlo, and D.D. Pollard. 2012. Integrating complementarity into the 2D displacement discontinuity boundary element method to model faults and fractures with frictional contact properties. *Computers and Geosciences* 45: 304–12.

Varga, R. S. 2000. *Matrix iterative analysis*. 2nd edn. Springer.

Chapter 7
The Direct Boundary Integral Method for Elasticity

7.1 Introduction

Somigliana's formula for the displacements (4.328) is the starting point for our treatment of the direct boundary integral method for elasticity. This formula expresses the displacements $u_i(\boldsymbol{\xi})$ at a point $\boldsymbol{\xi}$ inside an elastic region V in terms of integrals of the tractions t_j and displacements u_j over the boundary S of this region, as follows:

$$u_i(\boldsymbol{\xi}) = \int_S t_j(\boldsymbol{x}) U_{ij}(\boldsymbol{x}, \boldsymbol{\xi}) \, \mathrm{d}S(\boldsymbol{x}) - \int_S u_j(\boldsymbol{x}) T_{ij}(\boldsymbol{x}, \boldsymbol{\xi}) \, \mathrm{d}S(\boldsymbol{x}) \tag{7.1}$$

where $\boldsymbol{\xi} \neq \boldsymbol{x}$. Recall that $U_{ij}(\boldsymbol{x}, \boldsymbol{\xi})$ and $T_{ij}(\boldsymbol{x}, \boldsymbol{\xi})$ in this expression are tensor fields that give, respectively, the displacements and tractions in the j direction at point $\boldsymbol{x}$ (the sample point) due to a unit concentrated force in the i direction at point $\boldsymbol{\xi}$ (the load point). Recall also that the components of *both* the tractions $t_j(\boldsymbol{x})$ and the displacements $u_j(\boldsymbol{x})$ along the boundary S are only known *after* a particular boundary value problem has been solved; prior to this, only one-half of the required information is available, through definition of the boundary conditions for the problem (see Sect. 4.8). As we will presently see, the remaining information can be obtained numerically by transforming (7.1) into a system of simultaneous linear algebraic equations in which the unknowns are components of either displacement or traction along discrete elements of boundary S.

It is worth mentioning again the similarity between Somigliana's formula for elasticity and Green's representation formula for potential theory. Green's formula, (2.62),

$$\phi(\boldsymbol{\xi}) = \int_S q(\boldsymbol{x}) G(\boldsymbol{x}, \boldsymbol{\xi}) \, \mathrm{d}S(\boldsymbol{x}) - \int_S \phi(\boldsymbol{x}) Q(\boldsymbol{x}, \boldsymbol{\xi}) \, \mathrm{d}S(\boldsymbol{x}) \tag{7.2}$$

S. L. Crouch and S. G. Mogilevskaya, *A First Course in Boundary Element Methods*, Mathematical Engineering, https://doi.org/10.1007/978-3-031-63341-6_7

gives the potential $\phi(\boldsymbol{\xi})$ at a source point $\boldsymbol{\xi}$ inside region V in terms of two boundary integrals, one involving a single layer potential with density function $q(\boldsymbol{x})$, and the other a double layer potential with density function $\phi(\boldsymbol{x})$. Somigliana's formula (7.1) can be regarded as a vectorial version of this result. It gives the displacement components $u_i(\boldsymbol{\xi})$ at a source (or load) point $\boldsymbol{\xi}$ in terms of two boundary integrals, one involving the vector analogue of the single layer potential with density function $t_j(\boldsymbol{x})$, and the other with the vector analogue of the double layer potential with density function $u_j(\boldsymbol{x})$.

In this chapter we will develop two versions of the direct boundary integral method for isotropic elasticity: one for an assumed piecewise constant approximation of the boundary displacements and tractions and the other for an assumed piecewise linear approximation of these parameters. We again make the additional assumption that the boundary S can be represented with sufficient accuracy by N straight line segments joined end to end. As in our previous analyses, this approximation allows us to evaluate the integrals over the individual boundary elements analytically. The modifications required to develop comparable versions of direct boundary integral methods for anisotropic elasticity are straightforward but we will not pursue this matter in detail. Instead, we will give a brief summary of the necessary changes and leave the full development of the methods to the interested reader.

Another formulation of the direct boundary integral method for elasticity can be developed using the special version of Somigliana's displacement equation that has been modified to account for the special terms that arise as the source point $\boldsymbol{\xi}$ in region V is allowed to coincide with the sample point $\boldsymbol{x}$ on the boundary S of this region *before* any integrations have been performed. This approach is only subtly different from the other one and both give exactly the same results for cases in which the integrations can be done analytically, which generally requires the boundary to be approximated in piecewise fashion by straight line segments. If the integrations are done numerically then the second approach must be used.

Much of the literature on boundary element methods is based on the use of curvilinear elements with assumed quadratic variations (for two-dimensional problems) of the boundary displacements and tractions and numerical integration of the underlying boundary integral equations. We will conclude our presentation of the direct boundary integral method by explaining the basic steps needed to implement this formulation of the method.

7.2 Numerical Procedure I—Piecewise Constant Boundary Parameters

Equation (7.1) can be written in unabridged form for plane problems as

$$\begin{aligned} u_x(\boldsymbol{\xi}) &= \int_S \left[t_x(\boldsymbol{x})U_{xx}(\boldsymbol{x},\boldsymbol{\xi}) + t_y(\boldsymbol{x})U_{xy}(\boldsymbol{x},\boldsymbol{\xi})\right] \mathrm{d}S(\boldsymbol{x}) \\ &\quad - \int_S \left[u_x(\boldsymbol{x})T_{xx}(\boldsymbol{x},\boldsymbol{\xi}) + u_y(\boldsymbol{x})T_{xy}(\boldsymbol{x},\boldsymbol{\xi})\right] \mathrm{d}S(\boldsymbol{x}) \\ u_y(\boldsymbol{\xi}) &= \int_S \left[t_x(\boldsymbol{x})U_{yx}(\boldsymbol{x},\boldsymbol{\xi}) + t_y(\boldsymbol{x})U_{yy}(\boldsymbol{x},\boldsymbol{\xi})\right] \mathrm{d}S(\boldsymbol{x}) \\ &\quad - \int_S \left[u_x(\boldsymbol{x})T_{yx}(\boldsymbol{x},\boldsymbol{\xi}) + u_y(\boldsymbol{x})T_{yy}(\boldsymbol{x},\boldsymbol{\xi})\right] \mathrm{d}S(\boldsymbol{x}) \end{aligned} \tag{7.3}$$

where U_{xx}, etc., and T_{xx}, etc., for isotropic elasticity are given in (4.251) and (4.265) or, equivalently, by (4.269) and (4.272).

In practice, the integrals in these expressions are evaluated by dividing the boundary contour S into discrete elements and approximating $u_j(\boldsymbol{x})$ and $t_j(\boldsymbol{x})$ over each element by polynomials. The polynomials are fitted to a specified number of collocation points along the elements (typically two or three points per element), so that the displacements and tractions are approximated in piecewise fashion along the entire boundary.

In order to focus attention on the basic structure of the numerical approach, we will initially assume that the displacements and tractions are constant over each element; in subsequent sections of this chapter we will consider more elaborate (and more accurate) higher order approximations of the boundary displacements and tractions.

As shown in Fig. 7.1, we take the sample point $\boldsymbol{x}$ at the center of a typical element j on the discretized boundary contour. The local $\overline{x}$, $\overline{y}$ coordinate system originating at this point is chosen such that $\overline{y}$ points in the direction of the local outward normal to

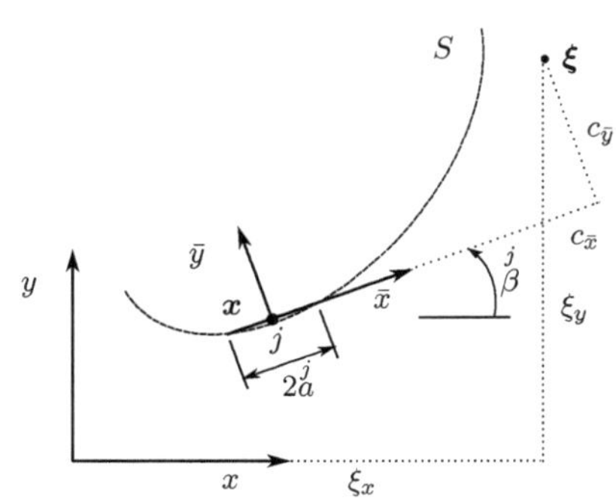

Fig. 7.1 Definition of local coordinates for element j

the boundary. The local coordinates are then defined by the transformation formulas (3.4) through (3.7). In particular, the load point $\boldsymbol{\xi}$ has coordinates $\bar{x} = c_{\bar{x}}$ and $\bar{y} = c_{\bar{y}}$.

Using the above definitions, we write (7.3) for the discretized boundary contour as

$$\begin{aligned}
u_x(\boldsymbol{\xi}) &= \sum_{j=1}^{N} \int_{-\overset{j}{a}}^{\overset{j}{a}} \Big[\overset{j}{t}_x(\bar{x}) U_{xx}(\bar{x} - c_{\bar{x}}, c_{\bar{y}}) + \overset{j}{t}_y(\bar{x}) U_{xy}(\bar{x} - c_{\bar{x}}, c_{\bar{y}})\Big] \,\mathrm{d}\bar{x} \\
&\quad - \sum_{j=1}^{N} \int_{-\overset{j}{a}}^{\overset{j}{a}} \Big[\overset{j}{u}_x(\bar{x}) T_{xx}(\bar{x} - c_{\bar{x}}, c_{\bar{y}}) + \overset{j}{u}_y(\bar{x}) T_{xy}(\bar{x} - c_{\bar{x}}, c_{\bar{y}})\Big] \,\mathrm{d}\bar{x} \\
u_y(\boldsymbol{\xi}) &= \sum_{j=1}^{N} \int_{-\overset{j}{a}}^{\overset{j}{a}} \Big[\overset{j}{t}_x(\bar{x}) U_{yx}(\bar{x} - c_{\bar{x}}, c_{\bar{y}}) + \overset{j}{t}_y(\bar{x}) U_{yy}(\bar{x} - c_{\bar{x}}, c_{\bar{y}})\Big] \,\mathrm{d}\bar{x} \\
&\quad - \sum_{j=1}^{N} \int_{-\overset{j}{a}}^{\overset{j}{a}} \Big[\overset{j}{u}_x(\bar{x}) T_{yx}(\bar{x} - c_{\bar{x}}, c_{\bar{y}}) + \overset{j}{u}_y(\bar{x}) T_{yy}(\bar{x} - c_{\bar{x}}, c_{\bar{y}})\Big] \,\mathrm{d}\bar{x}
\end{aligned} \tag{7.4}$$

Expressions for tensor fields U_{ij} and T_{ij} in terms of local coordinates $\bar{x}$ and $\bar{y}$ are given in Sect. 7.3.

We next make the assumption that the displacements and tractions are constant along the elements. Letting $\overset{j}{u}_x$, $\overset{j}{u}_y$ and $\overset{j}{t}_x$, $\overset{j}{t}_y$ be the values of the displacements and tractions for element j, Eqs. (7.4) can be represented as

$$\begin{aligned}
u_x(\boldsymbol{\xi}) &= \sum_{j=1}^{N} \Big[\overset{j}{t}_x \int_{-\overset{j}{a}}^{\overset{j}{a}} U_{xx}(\bar{x} - c_{\bar{x}}, c_{\bar{y}}) \,\mathrm{d}\bar{x} + \overset{j}{t}_y \int_{-\overset{j}{a}}^{\overset{j}{a}} U_{xy}(\bar{x} - c_{\bar{x}}, c_{\bar{y}}) \,\mathrm{d}\bar{x}\Big] \\
&\quad - \sum_{j=1}^{N} \Big[\overset{j}{u}_x \int_{-\overset{j}{a}}^{\overset{j}{a}} T_{xx}(\bar{x} - c_{\bar{x}}, c_{\bar{y}}) \,\mathrm{d}\bar{x} + \overset{j}{u}_y \int_{-\overset{j}{a}}^{\overset{j}{a}} T_{xy}(\bar{x} - c_{\bar{x}}, c_{\bar{y}}) \,\mathrm{d}\bar{x}\Big] \\
u_y(\boldsymbol{\xi}) &= \sum_{j=1}^{N} \Big[\overset{j}{t}_x \int_{-\overset{j}{a}}^{\overset{j}{a}} U_{yx}(\bar{x} - c_{\bar{x}}, c_{\bar{y}}) \,\mathrm{d}\bar{x} + \overset{j}{t}_y \int_{-\overset{j}{a}}^{\overset{j}{a}} U_{yy}(\bar{x} - c_{\bar{x}}, c_{\bar{y}}) \,\mathrm{d}\bar{x}\Big] \\
&\quad - \sum_{j=1}^{N} \Big[\overset{j}{u}_x \int_{-\overset{j}{a}}^{\overset{j}{a}} T_{yx}(\bar{x} - c_{\bar{x}}, c_{\bar{y}}) \,\mathrm{d}\bar{x} + \overset{j}{u}_y \int_{-\overset{j}{a}}^{\overset{j}{a}} T_{yy}(\bar{x} - c_{\bar{x}}, c_{\bar{y}}) \,\mathrm{d}\bar{x}\Big]
\end{aligned} \tag{7.5}$$

The indicated integrations can be performed analytically and the equations written as a set of algebraic expressions of the following form:

$$u_x(\boldsymbol{\xi}) = \sum_{j=1}^{N}\Big[\mathbb{U}_{xx}(\overset{j}{\boldsymbol{x}}, \boldsymbol{\xi})\overset{j}{t}_x + \mathbb{U}_{xy}(\overset{j}{\boldsymbol{x}}, \boldsymbol{\xi})\overset{j}{t}_y\Big] - \sum_{j=1}^{N}\Big[\mathbb{T}_{xx}(\overset{j}{\boldsymbol{x}}, \boldsymbol{\xi})\overset{j}{u}_x + \mathbb{T}_{xy}(\overset{j}{\boldsymbol{x}}, \boldsymbol{\xi})\overset{j}{u}_y\Big]$$

$$(7.6)$$

$$u_y(\boldsymbol{\xi}) = \sum_{j=1}^{N}\Big[\mathbb{U}_{yx}(\overset{j}{\boldsymbol{x}}, \boldsymbol{\xi})\overset{j}{t}_x + \mathbb{U}_{yy}(\overset{j}{\boldsymbol{x}}, \boldsymbol{\xi})\overset{j}{t}_y\Big] - \sum_{j=1}^{N}\Big[\mathbb{T}_{yx}(\overset{j}{\boldsymbol{x}}, \boldsymbol{\xi})\overset{j}{u}_x + \mathbb{T}_{yy}(\overset{j}{\boldsymbol{x}}, \boldsymbol{\xi})\overset{j}{u}_y\Big]$$

in which we have used the notation

$$\mathbb{U}_{pq}(\overset{j}{\boldsymbol{x}}, \boldsymbol{\xi}) = \int_{-\overset{j}{a}}^{\overset{j}{a}} U_{pq}(\bar{x} - c_{\overline{x}}, c_{\overline{y}})\,\mathrm{d}\bar{x}$$

$$(7.7)$$

$$\mathbb{T}_{pq}(\overset{j}{\boldsymbol{x}}, \boldsymbol{\xi}) = \int_{-\overset{j}{a}}^{\overset{j}{a}} T_{pq}(\bar{x} - c_{\overline{x}}, c_{\overline{y}})\,\mathrm{d}\bar{x}$$

where p and q stand for either x or y.

Equations (7.6) have two distinct applications. First, after a boundary value problem has been solved and the components of displacement and traction are known for all N boundary elements, the equations can be evaluated directly to give the displacements at an arbitrary point $\boldsymbol{\xi}$ within the region of interest. Similar equations can be developed from (4.332) to compute the stress components at point $\boldsymbol{\xi}$. These are presented in Sect. 7.4.

The other application for equations (7.6) is to consider the limiting case in which point $\boldsymbol{\xi}$ is taken at the midpoint of a boundary element, say point $\overset{i}{\boldsymbol{x}}$.[1] We then have

$$u_x(\overset{i}{\boldsymbol{x}}) = \sum_{j=1}^{N}\Big[\mathbb{U}_{xx}(\overset{j}{\boldsymbol{x}}, \overset{i}{\boldsymbol{x}})\overset{j}{t}_x + \mathbb{U}_{xy}(\overset{j}{\boldsymbol{x}}, \overset{i}{\boldsymbol{x}})\overset{j}{t}_y\Big] - \sum_{j=1}^{N}\Big[\mathbb{T}_{xx}(\overset{j}{\boldsymbol{x}}, \overset{i}{\boldsymbol{x}})\overset{j}{u}_x + \mathbb{T}_{xy}(\overset{j}{\boldsymbol{x}}, \overset{i}{\boldsymbol{x}})\overset{j}{u}_y\Big]$$

$$(7.8)$$

$$u_y(\overset{i}{\boldsymbol{x}}) = \sum_{j=1}^{N}\Big[\mathbb{U}_{yx}(\overset{j}{\boldsymbol{x}}, \overset{i}{\boldsymbol{x}})\overset{j}{t}_x + \mathbb{U}_{yy}(\overset{j}{\boldsymbol{x}}, \overset{i}{\boldsymbol{x}})\overset{j}{t}_y\Big] - \sum_{j=1}^{N}\Big[\mathbb{T}_{yx}(\overset{j}{\boldsymbol{x}}, \overset{i}{\boldsymbol{x}})\overset{j}{u}_x + \mathbb{T}_{yy}(\overset{j}{\boldsymbol{x}}, \overset{i}{\boldsymbol{x}})\overset{j}{u}_y\Big]$$

or, setting $u_x(\overset{i}{\boldsymbol{x}}) = \overset{i}{u}_x$, $u_y(\overset{i}{\boldsymbol{x}}) = \overset{i}{u}_y$, and $\mathbb{T}_{xx}(\overset{j}{\boldsymbol{x}}, \overset{i}{\boldsymbol{x}}) = \overset{ij}{\mathbb{T}}_{xx}$, etc., these equations can be written as

[1] Because limits are taken of the integrated expressions (7.7), the jump terms associated with kernels $T_{pq}(\bar{x} - c_{\bar{x}}, c_{\bar{y}})$ are accounted for directly.

$$\begin{aligned}\overset{i}{u}_x &= \sum_{j=1}^{N}\left(\overset{ij}{\mathbb{U}}_{xx}\overset{j}{t}_x + \overset{ij}{\mathbb{U}}_{xy}\overset{j}{t}_y\right) - \sum_{j=1}^{N}\left(\overset{ij}{\mathbb{T}}_{xx}\overset{j}{u}_x + \overset{ij}{\mathbb{T}}_{xy}\overset{j}{u}_y\right)\\ \overset{i}{u}_y &= \sum_{j=1}^{N}\left(\overset{ij}{\mathbb{U}}_{yx}\overset{j}{t}_x + \overset{ij}{\mathbb{U}}_{yy}\overset{j}{t}_y\right) - \sum_{j=1}^{N}\left(\overset{ij}{\mathbb{T}}_{yx}\overset{j}{u}_x + \overset{ij}{\mathbb{T}}_{yy}\overset{j}{u}_y\right)\end{aligned} \tag{7.9}$$

or

$$\begin{aligned}\sum_{j=1}^{N}\left((\overset{ij}{\mathbb{T}}_{xx} + \overset{ij}{\mathbb{I}})\overset{j}{u}_x + \overset{ij}{\mathbb{T}}_{xy}\overset{j}{u}_y\right) &= \sum_{j=1}^{N}\left(\overset{ij}{\mathbb{U}}_{xx}\overset{j}{t}_x + \overset{ij}{\mathbb{U}}_{xy}\overset{j}{t}_y\right)\\ \sum_{j=1}^{N}\left(\overset{ij}{\mathbb{T}}_{yx}\overset{j}{u}_x + (\overset{ij}{\mathbb{T}}_{yy} + \overset{ij}{\mathbb{I}})\overset{j}{u}_y\right) &= \sum_{j=1}^{N}\left(\overset{ij}{\mathbb{U}}_{yx}\overset{j}{t}_x + \overset{ij}{\mathbb{U}}_{yy}\overset{j}{t}_y\right)\end{aligned} \tag{7.10}$$

where $\overset{ij}{\mathbb{I}}$ is the $N \times N$ identity matrix: $\overset{ij}{\mathbb{I}} = 1$ if $i = j$; $\overset{ij}{\mathbb{I}} = 0$ if $i \neq j$.

If we consider N possible positions for point $\overset{i}{\boldsymbol{x}}$ on the discretized boundary contour, Eqs. (7.10) represent a system of $2N$ simultaneous linear algebraic equations in $2N$ unknowns—the unspecified values of displacement or traction at the midpoints of the N boundary elements. The form of this system of equations depends on the nature of the boundary conditions in a particular problem. For example, suppose that the quantities $\overset{j}{t}_x = (\overset{j}{t}_x)_o$ and $\overset{j}{t}_y = (\overset{j}{t}_y)_o$ are known for all N boundary elements. The right-hand sides of the equations

$$\begin{aligned}\sum_{j=1}^{N}\left((\overset{ij}{\mathbb{T}}_{xx} + \overset{ij}{\mathbb{I}})\overset{j}{u}_x + \overset{ij}{\mathbb{T}}_{xy}\overset{j}{u}_y\right) &= \sum_{j=1}^{N}\left(\overset{ij}{\mathbb{U}}_{xx}(\overset{j}{t}_x)_o + \overset{ij}{\mathbb{U}}_{xy}(\overset{j}{t}_y)_o\right)\\ \sum_{j=1}^{N}\left(\overset{ij}{\mathbb{T}}_{yx}\overset{j}{u}_x + (\overset{ij}{\mathbb{T}}_{yy} + \overset{ij}{\mathbb{I}})\overset{j}{u}_y\right) &= \sum_{j=1}^{N}\left(\overset{ij}{\mathbb{U}}_{yx}(\overset{j}{t}_x)_o + \overset{ij}{\mathbb{U}}_{yy}(\overset{j}{t}_y)_o\right)\end{aligned} \tag{7.11}$$

can then be evaluated directly, leading to a set of algebraic equations of the form $\mathbf{AX} = \mathbf{b}$, which can be solved by standard numerical techniques to obtain the values of the boundary displacements $\overset{i}{u}_x$ and $\overset{i}{u}_y$ for $i = 1$ to N. A similar process is used to set up the system of algebraic equations for the case that the displacements $\overset{i}{u}_x = (\overset{i}{u}_x)_o$ and $\overset{i}{u}_y = (\overset{i}{u}_y)_o$ are prescribed for all N boundary elements; the tractions $\overset{i}{t}_x$ and $\overset{i}{t}_y$, $i = 1$ to N, are then the unknowns in the system.

For a mixed boundary value problem, the system of equations is formed by rearranging (7.10) so that the $2N$ known boundary parameters are on the right-hand sides of the equals signs and the $2N$ unknown parameters are on the left. If, for example, the x component of traction $\overset{j}{t}_x = (\overset{j}{t}_x)_o$ and the y component of displacement $\overset{j}{u}_y = (\overset{j}{u}_y)_o$ are specified for all N boundary elements, the system of equations can be written as

$$
\begin{aligned}
\sum_{j=1}^{N}\Big((\overset{ij}{\mathbb{T}}_{xx} + \overset{ij}{\mathbb{I}})\overset{j}{u}_x - \overset{ij}{\mathbb{U}}_{xy}\overset{j}{t}_y\Big) &= \sum_{j=1}^{N}\Big(\overset{ij}{\mathbb{U}}_{xx}(\overset{j}{t}_x)_o - \overset{ij}{\mathbb{T}}_{xy}(\overset{j}{u}_y)_o\Big) \\
\sum_{j=1}^{N}\Big(\overset{ij}{\mathbb{T}}_{yx}\overset{j}{u}_x - \overset{ij}{\mathbb{U}}_{yy}\overset{j}{t}_y\Big) &= \sum_{j=1}^{N}\Big(\overset{ij}{\mathbb{U}}_{yx}(\overset{j}{t}_x)_o - (\overset{ij}{\mathbb{T}}_{yy} + \overset{ij}{\mathbb{I}})(\overset{j}{u}_y)_o\Big)
\end{aligned}
\tag{7.12}
$$

The unknowns in the system are now the displacements $\overset{i}{u}_x$ and tractions $\overset{i}{t}_y$ for $i = 1$ to N.

Once the appropriate system of algebraic equations has been set up and solved, all of the boundary displacements and tractions will have been determined, and the displacements and stresses anywhere in the region of interest can be computed using (7.6) and comparable expressions for the stresses, as mentioned above.

7.3 Coordinate Transformations; Evaluation of Integrals

The tensor fields U_{ij} and T_{ij} in (4.251) and (4.265) can be written in terms of local $\bar{x}$, $\bar{y}$ coordinates by using transformation formulas (3.6) and (3.7), and the chain rule for partial differentiation:

$$
\begin{aligned}
\frac{\partial}{\partial x} &= \frac{\partial \bar{x}}{\partial x}\frac{\partial}{\partial \bar{x}} + \frac{\partial \bar{y}}{\partial x}\frac{\partial}{\partial \bar{y}} = \cos\overset{j}{\beta}\frac{\partial}{\partial \bar{x}} - \sin\overset{j}{\beta}\frac{\partial}{\partial \bar{y}} \\
\frac{\partial}{\partial y} &= \frac{\partial \bar{x}}{\partial y}\frac{\partial}{\partial \bar{x}} + \frac{\partial \bar{y}}{\partial y}\frac{\partial}{\partial \bar{y}} = \sin\overset{j}{\beta}\frac{\partial}{\partial \bar{x}} + \cos\overset{j}{\beta}\frac{\partial}{\partial \bar{y}}
\end{aligned}
\tag{7.13}
$$

After some algebraic manipulation we obtain the following expressions for the terms of $U_{pq}(\overline{x} - c_{\overline{x}}, \overline{y} - c_{\overline{y}})$:

$$
\begin{aligned}
U_{xx} &= \frac{1}{2\mu}\Big[(3-4\nu)g + (\bar{y}-c_{\bar{y}})(\sin 2\,{}^{j}\beta \frac{\partial g}{\partial \bar{x}} + \cos 2\,{}^{j}\beta \frac{\partial g}{\partial \bar{y}})\Big] \\
U_{yx} &= \frac{1}{2\mu}\Big[-(\bar{y}-c_{\bar{y}})(\cos 2\,{}^{j}\beta \frac{\partial g}{\partial \bar{x}} - \sin 2\,{}^{j}\beta \frac{\partial g}{\partial \bar{y}})\Big] \\
U_{xy} &= U_{yx} = \frac{1}{2\mu}\Big[-(\bar{y}-c_{\bar{y}})(\cos 2\,{}^{j}\beta \frac{\partial g}{\partial \bar{x}} - \sin 2\,{}^{j}\beta \frac{\partial g}{\partial \bar{y}})\Big] \\
U_{yy} &= \frac{1}{2\mu}\Big[(3-4\nu)g - (\bar{y}-c_{\bar{y}})(\sin 2\,{}^{j}\beta \frac{\partial g}{\partial \bar{x}} + \cos 2\,{}^{j}\beta \frac{\partial g}{\partial \bar{y}})\Big]
\end{aligned} \tag{7.14}
$$

where function g is

$$
g(\bar{x}-c_{\bar{x}}, \bar{y}-c_{\bar{y}}) = -\frac{1}{4\pi(1-\nu)} \ln\Big[(\bar{x}-c_{\bar{x}})^2 + (\bar{y}-c_{\bar{y}})^2\Big]^{1/2} \tag{7.15}
$$

We note that the following identities were used in writing (7.14), to eliminate the occurrence of terms multiplied by $(\bar{x}-c_{\bar{x}})$:

$$
\begin{aligned}
(\bar{x}-c_{\bar{x}})\frac{\partial g}{\partial \bar{y}} &= (\bar{y}-c_{\bar{y}})\frac{\partial g}{\partial \bar{x}} \\
(\bar{x}-c_{\bar{x}})\frac{\partial g}{\partial \bar{x}} &= -\frac{1}{4\pi(1-\nu)} - (\bar{y}-c_{\bar{y}})\frac{\partial g}{\partial \bar{y}}
\end{aligned} \tag{7.16}
$$

This measure was taken to simplify the integrations specified in (7.7). The constant term in the second one of (7.16) corresponds to rigid body displacements and is disregarded in the subsequent analysis.

Similarly, the values of $T_{pq}(\bar{x}-c_{\bar{x}}, \bar{y}-c_{\bar{y}})$ are found to be

$$
\begin{aligned}
T_{xx} &= 2(1-\nu)\frac{\partial g}{\partial \bar{y}} + (\bar{y}-c_{\bar{y}})(\sin 2\,{}^{j}\beta \frac{\partial^2 g}{\partial \bar{x}\partial \bar{y}} - \cos 2\,{}^{j}\beta \frac{\partial^2 g}{\partial \bar{x}^2}) \\
T_{yx} &= (1-2\nu)\frac{\partial g}{\partial \bar{x}} - (\bar{y}-c_{\bar{y}})(\cos 2\,{}^{j}\beta \frac{\partial^2 g}{\partial \bar{x}\partial \bar{y}} + \sin 2\,{}^{j}\beta \frac{\partial^2 g}{\partial \bar{x}^2}) \\
T_{xy} &= -(1-2\nu)\frac{\partial g}{\partial \bar{x}} - (\bar{y}-c_{\bar{y}})(\cos 2\,{}^{j}\beta \frac{\partial^2 g}{\partial \bar{x}\partial \bar{y}} + \sin 2\,{}^{j}\beta \frac{\partial^2 g}{\partial \bar{x}^2}) \\
T_{yy} &= 2(1-\nu)\frac{\partial g}{\partial \bar{y}} - (\bar{y}-c_{\bar{y}})(\sin 2\,{}^{j}\beta \frac{\partial^2 g}{\partial \bar{x}\partial \bar{y}} - \cos 2\,{}^{j}\beta \frac{\partial^2 g}{\partial \bar{x}^2})
\end{aligned} \tag{7.17}
$$

In writing these equations, use was made of the following identities, which were obtained by differentiating (7.16) with respect to $\bar{x}$:

$$
\begin{aligned}
(\bar{x} - c_{\bar{x}})\frac{\partial^2 g}{\partial \bar{x}\partial \bar{y}} &= -\frac{\partial g}{\partial \bar{y}} + (\bar{y} - c_{\bar{y}})\frac{\partial^2 g}{\partial \bar{x}^2} \\
(\bar{x} - c_{\bar{x}})\frac{\partial^2 g}{\partial \bar{x}^2} &= -\frac{\partial g}{\partial \bar{x}} - (\bar{y} - c_{\bar{y}})\frac{\partial^2 g}{\partial \bar{x}\partial \bar{y}}
\end{aligned}
\tag{7.18}
$$

It is easy to show that the required integrals of function $g(\bar{x} - c_{\bar{x}}, \bar{y} - c_{\bar{y}})$ and its derivatives over the line segment $-\overset{j}{a} \leq \bar{x} \leq \overset{j}{a},\ \bar{y} = 0$ can be expressed in terms of the following functions:

$$
\begin{aligned}
\bar{T}_1 &= \int_{-\overset{j}{a}}^{\overset{j}{a}} g(\bar{x} - c_{\bar{x}}, \bar{y} - c_{\bar{y}})\,\mathrm{d}\bar{x}\ |_{\bar{y}=0} \\
&= -\frac{1}{4\pi(1-\nu)}\Big[c_{\bar{y}}(\bar{\theta}_1 - \bar{\theta}_2) - (c_{\bar{x}} - \overset{j}{a})\ln \bar{r}_1 + (c_{\bar{x}} + \overset{j}{a})\ln \bar{r}_2 - 2\overset{j}{a}\Big] \\
\bar{T}_2 &= \int_{-\overset{j}{a}}^{\overset{j}{a}} \frac{\partial}{\partial \bar{x}} g(\bar{x} - c_{\bar{x}}, \bar{y} - c_{\bar{y}})\,\mathrm{d}\bar{x}\ |_{\bar{y}=0} = -\frac{1}{4\pi(1-\nu)}\ln\frac{\bar{r}_1}{\bar{r}_2} \\
\bar{T}_3 &= \int_{-\overset{j}{a}}^{\overset{j}{a}} \frac{\partial}{\partial \bar{y}} g(\bar{x} - c_{\bar{x}}, \bar{y} - c_{\bar{y}})\,\mathrm{d}\bar{x}\ |_{\bar{y}=0} = \frac{1}{4\pi(1-\nu)}(\bar{\theta}_1 - \bar{\theta}_2) \\
\bar{T}_4 &= \int_{-\overset{j}{a}}^{\overset{j}{a}} \frac{\partial^2}{\partial \bar{x}\partial \bar{y}} g(\bar{x} - c_{\bar{x}}, \bar{y} - c_{\bar{y}})\,d\bar{x}\ |_{\bar{y}=0} = \frac{1}{4\pi(1-\nu)}\Big[\frac{c_{\bar{y}}}{\bar{r}_1^2} - \frac{c_{\bar{y}}}{\bar{r}_2^2}\Big] \\
\bar{T}_5 &= \int_{-\overset{j}{a}}^{\overset{j}{a}} \frac{\partial^2}{\partial \bar{x}^2} g(\bar{x} - c_{\bar{x}}, \bar{y} - c_{\bar{y}})\,\mathrm{d}\bar{x}\ |_{\bar{y}=0} = \frac{1}{4\pi(1-\nu)}\Big[\frac{(c_{\bar{x}} - \overset{j}{a})}{\bar{r}_1^2} - \frac{(c_{\bar{x}} + \overset{j}{a})}{\bar{r}_2^2}\Big] \\
\bar{T}_6 &= \int_{-\overset{j}{a}}^{\overset{j}{a}} \frac{\partial^3}{\partial \bar{x}\partial \bar{y}^2} g(\bar{x} - c_{\bar{x}}, \bar{y} - c_{\bar{y}})\,\mathrm{d}\bar{x}\ |_{\bar{y}=0} \\
&= -\frac{1}{4\pi(1-\nu)}\Big[\frac{((c_{\bar{x}} - \overset{j}{a})^2 - c_{\bar{y}}^2)}{\bar{r}_1^4} - \frac{((c_{\bar{x}} + \overset{j}{a})^2 - c_{\bar{y}}^2)}{\bar{r}_2^4}\Big] \\
\bar{T}_7 &= \int_{-\overset{j}{a}}^{\overset{j}{a}} \frac{\partial^3}{\partial \bar{x}^2\partial \bar{y}} g(\bar{x} - c_{\bar{x}}, \bar{y} - c_{\bar{y}})\,\mathrm{d}\bar{x}\ |_{\bar{y}=0} = \frac{2}{4\pi(1-\nu)}\Big[\frac{(c_{\bar{x}} - \overset{j}{a})c_{\bar{y}}}{\bar{r}_1^4} - \frac{(c_{\bar{x}} + \overset{j}{a})c_{\bar{y}}}{\bar{r}_2^4}\Big]
\end{aligned}
\tag{7.19}
$$

In these equations, $\bar{r}_1$, $\bar{r}_2$ and $\bar{\theta}_1$, $\bar{\theta}_2$ are the distances and angles from the end points of element j to the load point $\boldsymbol{\xi}$. The last two functions in (7.19) will be needed in the next section to evaluate the stress components $\sigma_{ij}(\boldsymbol{\xi})$ and we record them here for completeness.

It instructive to compare functions (7.19) with similar functions used for the stress and displacement discontinuity methods, Eqs. (5.7), (5.8), and (6.7). It can be seen that the "$\bar{T}$ functions" can be recovered from the "$\bar{F}$ functions" by replacing $\bar{x}$ and $\bar{y}$ in the latter with $c_{\bar{x}}$ and $c_{\bar{y}}$ and changing the signs of the expressions with odd order derivatives. This is a consequence of the reversal of the load point $\boldsymbol{\xi}$ and the sample point $\boldsymbol{x}$ in the two sets of expressions, as depicted in Figs. 5.2 and 7.1.

The integrated coefficients $\mathbb{U}_{pq}(\overset{j}{\boldsymbol{x}}, \boldsymbol{\xi})$ and $\mathbb{T}_{pq}(\overset{j}{\boldsymbol{x}}, \boldsymbol{\xi})$ in (7.7) can now be explicitly written out by combining (7.14), (7.17) and (7.19) to get

$$
\begin{aligned}
\mathbb{U}_{xx}(\overset{j}{\boldsymbol{x}}, \boldsymbol{\xi}) &= \int_{-\overset{j}{a}}^{\overset{j}{a}} U_{xx}(\bar{x} - c_{\bar{x}}, c_{\bar{y}})\,\mathrm{d}\bar{x} \\
&= \frac{1}{2\mu}\Big[(3-4\nu)\bar{T}_1 - c_{\bar{y}}(\sin 2\overset{j}{\beta}\bar{T}_2 + \cos 2\overset{j}{\beta}\bar{T}_3)\Big] \\
\mathbb{U}_{yx}(\overset{j}{\boldsymbol{x}}, \boldsymbol{\xi}) &= \int_{-\overset{j}{a}}^{\overset{j}{a}} U_{yx}(\bar{x} - c_{\bar{x}}, c_{\bar{y}})\,\mathrm{d}\bar{x} \\
&= \frac{1}{2\mu}\Big[c_{\bar{y}}(\cos 2\overset{j}{\beta}\bar{T}_2 - \sin 2\overset{j}{\beta}\bar{T}_3)\Big] \\
\mathbb{U}_{xy}(\overset{j}{\boldsymbol{x}}, \boldsymbol{\xi}) &= \int_{-\overset{j}{a}}^{\overset{j}{a}} U_{xy}(\bar{x} - c_{\bar{x}}, c_{\bar{y}})\,\mathrm{d}\bar{x} \\
&= \mathbb{U}_{yx}(\overset{j}{\boldsymbol{x}}, \boldsymbol{\xi}) \\
\mathbb{U}_{yy}(\overset{j}{\boldsymbol{x}}, \boldsymbol{\xi}) &= \int_{-\overset{j}{a}}^{\overset{j}{a}} U_{yy}(\bar{x} - c_{\bar{x}}, c_{\bar{y}})\,\mathrm{d}\bar{x} \\
&= \frac{1}{2\mu}\Big[(3-4\nu)\bar{T}_1 + c_{\bar{y}}(\sin 2\overset{j}{\beta}\bar{T}_2 + \cos 2\overset{j}{\beta}\bar{T}_3)\Big]
\end{aligned}
\tag{7.20}
$$

and

$$
\begin{aligned}
\mathbb{T}_{xx}(\overset{j}{\boldsymbol{x}}, \boldsymbol{\xi}) &= \int_{-\overset{j}{a}}^{\overset{j}{a}} T_{xx}(\bar{x} - c_{\bar{x}}, c_{\bar{y}})\,\mathrm{d}\bar{x} \\
&= 2(1-\nu)\bar{T}_3 - c_{\bar{y}}(\sin 2\overset{j}{\beta}\bar{T}_4 - \cos 2\overset{j}{\beta}\bar{T}_5) \\
\mathbb{T}_{yx}(\overset{j}{\boldsymbol{x}}, \boldsymbol{\xi}) &= \int_{-\overset{j}{a}}^{\overset{j}{a}} T_{yx}(\bar{x} - c_{\bar{x}}, c_{\bar{y}})\,\mathrm{d}\bar{x} \\
&= (1-2\nu)\bar{T}_2 + c_{\bar{y}}(\cos 2\overset{j}{\beta}\bar{T}_4 + \sin 2\overset{j}{\beta}\bar{T}_5) \\
\mathbb{T}_{xy}(\overset{j}{\boldsymbol{x}}, \boldsymbol{\xi}) &= \int_{-\overset{j}{a}}^{\overset{j}{a}} T_{xy}(\bar{x} - c_{\bar{x}}, c_{\bar{y}})\,\mathrm{d}\bar{x} \\
&= -(1-2\nu)\bar{T}_2 + c_{\bar{y}}(\cos 2\overset{j}{\beta}\bar{T}_4 + \sin 2\overset{j}{\beta}\bar{T}_5) \\
\mathbb{T}_{yy}(\overset{j}{\boldsymbol{x}}, \boldsymbol{\xi}) &= \int_{-\overset{j}{a}}^{\overset{j}{a}} T_{yy}(\bar{x} - c_{\bar{x}}, c_{\bar{y}})\,\mathrm{d}\bar{x} \\
&= 2(1-\nu)\bar{T}_3 + c_{\bar{y}}(\sin 2\overset{j}{\beta}\bar{T}_4 - \cos 2\overset{j}{\beta}\bar{T}_5).
\end{aligned}
\tag{7.21}
$$

7.4 Approximations of Somigliana's Formulas for Stresses

Expressions for the stress components at point $\boldsymbol{\xi}$ can be developed in an analogous way as was done for the displacements. Equation (4.332)—repeated here—is the basis for this approach:

$$\sigma_{ij}(\boldsymbol{\xi}) = \int_S t_k(\boldsymbol{x}) D_{kij}(\boldsymbol{x}, \boldsymbol{\xi})\, \mathrm{d}S(\boldsymbol{x}) - \int_S u_k(\boldsymbol{x}) S_{kij}(\boldsymbol{x}, \boldsymbol{\xi})\, \mathrm{d}S(\boldsymbol{x}) \tag{7.22}$$

Proceeding as in Sect. 7.2, we write this equation in unabridged form for a boundary contour divided into N straight line elements with piecewise constant displacements and tractions as follows:

$$\begin{aligned}
\sigma_{xx}(\boldsymbol{\xi}) &= \sum_{j=1}^{N} \overset{j}{t}_x \int_{-\overset{j}{a}}^{\overset{j}{a}} D_{xxx}(\overline{x} - c_{\overline{x}}, c_{\overline{y}})\, \mathrm{d}\bar{x} + \sum_{j=1}^{N} \overset{j}{t}_y \int_{-\overset{j}{a}}^{\overset{j}{a}} D_{yxx}(\overline{x} - c_{\overline{x}}, c_{\overline{y}})\, \mathrm{d}\overline{x} \\
&- \sum_{j=1}^{N} \overset{j}{u}_x \int_{-\overset{j}{a}}^{\overset{j}{a}} S_{xxx}(\overline{x} - c_{\overline{x}}, c_{\overline{y}})\, \mathrm{d}\bar{x} - \sum_{j=1}^{N} \overset{j}{u}_y \int_{-\overset{j}{a}}^{\overset{j}{a}} S_{yxx}(\overline{x} - c_{\overline{x}}, c_{\overline{y}})\, \mathrm{d}\overline{x} \\
\sigma_{yy}(\boldsymbol{\xi}) &= \sum_{j=1}^{N} \overset{j}{t}_x \int_{-\overset{j}{a}}^{\overset{j}{a}} D_{xyy}(\overline{x} - c_{\overline{x}}, c_{\overline{y}})\, \mathrm{d}\bar{x} + \sum_{j=1}^{N} \overset{j}{t}_y \int_{-\overset{j}{a}}^{\overset{j}{a}} D_{yyy}(\overline{x} - c_{\overline{x}}, c_{\overline{y}})\, \mathrm{d}\overline{x} \\
&- \sum_{j=1}^{N} \overset{j}{u}_x \int_{-\overset{j}{a}}^{\overset{j}{a}} S_{xyy}(\overline{x} - c_{\overline{x}}, c_{\overline{y}})\, \mathrm{d}\bar{x} - \sum_{j=1}^{N} \overset{j}{u}_y \int_{-\overset{j}{a}}^{\overset{j}{a}} S_{yyy}(\overline{x} - c_{\overline{x}}, c_{\overline{y}})\, \mathrm{d}\overline{x} \qquad (7.23) \\
\sigma_{xy}(\boldsymbol{\xi}) &= \sum_{j=1}^{N} \overset{j}{t}_x \int_{-\overset{j}{a}}^{\overset{j}{a}} D_{xxy}(\overline{x} - c_{\overline{x}}, c_{\overline{y}})\, \mathrm{d}\bar{x} + \sum_{j=1}^{N} \overset{j}{t}_y \int_{-\overset{j}{a}}^{\overset{j}{a}} D_{yxy}(\overline{x} - c_{\overline{x}}, c_{\overline{y}})\, \mathrm{d}\overline{x} \\
&- \sum_{j=1}^{N} \overset{j}{u}_x \int_{-\overset{j}{a}}^{\overset{j}{a}} S_{xxy}(\overline{x} - c_{\overline{x}}, c_{\overline{y}})\, \mathrm{d}\bar{x} - \sum_{j=1}^{N} \overset{j}{u}_y \int_{-\overset{j}{a}}^{\overset{j}{a}} S_{yxy}(\overline{x} - c_{\overline{x}}, c_{\overline{y}})\, \mathrm{d}\overline{x}
\end{aligned}$$

It can be seen that the results involve integrals of the form

$$\begin{aligned}
\mathbb{D}_{pqr}(\overset{j}{\boldsymbol{x}}, \boldsymbol{\xi}) &= \int_{-\overset{j}{a}}^{\overset{j}{a}} D_{pqr}(\overline{x} - c_{\overline{x}}, c_{\overline{y}})\, \mathrm{d}\overline{x} \\
\mathbb{S}_{pqr}(\overset{j}{\boldsymbol{x}}, \boldsymbol{\xi}) &= \int_{-\overset{j}{a}}^{\overset{j}{a}} S_{pqr}(\overline{x} - c_{\overline{x}}, c_{\overline{y}})\, \mathrm{d}\overline{x}
\end{aligned} \tag{7.24}$$

where p, q, and r are either x or y. As before, all of these integrals can be evaluated analytically.

The first step in evaluating integrals (7.24) is to write the tensor fields D_{kij} and S_{kij} in terms of local coordinates.[2] This is accomplished by following the same approach used in Sect. 7.3 to express U_{ij} and T_{ij} in terms of $\overline{x} - c_{\overline{x}}$ and $\overline{y} - c_{\overline{y}}$. As before, we obtain the expressions for general values of $\overline{y}$, recognizing that the required integrals pertain to the special case $\overline{y} = 0$.

The individual terms of $D_{pqr}(\overline{x} - c_{\overline{x}}, \overline{y} - c_{\overline{y}})$ can be written as

$$
\begin{aligned}
D_{xxx} &= -2(1-\nu)\big(\cos \overset{j}{\beta}\frac{\partial g}{\partial \overline{x}} - \sin \overset{j}{\beta}\frac{\partial g}{\partial \overline{y}}\big) - \cos \overset{j}{\beta}(\cos 2\overset{j}{\beta}\frac{\partial g}{\partial \overline{x}} - \sin 2\overset{j}{\beta}\frac{\partial g}{\partial \overline{y}}) \\
&\quad - (\overline{y} - c_{\overline{y}})(\cos 3\overset{j}{\beta}\frac{\partial^2 g}{\partial \overline{x}\partial \overline{y}} + \sin 3\overset{j}{\beta}\frac{\partial^2 g}{\partial \overline{x}^2}) \\
D_{xyy} &= -2\nu\big(\cos \overset{j}{\beta}\frac{\partial g}{\partial \overline{x}} - \sin \overset{j}{\beta}\frac{\partial g}{\partial \overline{y}}\big) + \cos \overset{j}{\beta}(\cos 2\overset{j}{\beta}\frac{\partial g}{\partial \overline{x}} - \sin 2\overset{j}{\beta}\frac{\partial g}{\partial \overline{y}}) \\
&\quad + (\overline{y} - c_{\overline{y}})(\cos 3\overset{j}{\beta}\frac{\partial^2 g}{\partial \overline{x}\partial \overline{y}} + \sin 3\overset{j}{\beta}\frac{\partial^2 g}{\partial \overline{x}^2}) \\
D_{xxy} &= -(1-2\nu)\big(\sin \overset{j}{\beta}\frac{\partial g}{\partial \overline{x}} + \cos \overset{j}{\beta}\frac{\partial g}{\partial \overline{y}}\big) - \cos \overset{j}{\beta}(\sin 2\overset{j}{\beta}\frac{\partial g}{\partial \overline{x}} + \cos 2\overset{j}{\beta}\frac{\partial g}{\partial \overline{y}}) \\
&\quad - (\overline{y} - c_{\overline{y}})(\sin 3\overset{j}{\beta}\frac{\partial^2 g}{\partial \overline{x}\partial \overline{y}} - \cos 3\overset{j}{\beta}\frac{\partial^2 g}{\partial \overline{x}^2}) \\
D_{yxx} &= -2\nu\big(\sin \overset{j}{\beta}\frac{\partial g}{\partial \overline{x}} + \cos \overset{j}{\beta}\frac{\partial g}{\partial \overline{y}}\big) - \sin \overset{j}{\beta}(\cos 2\overset{j}{\beta}\frac{\partial g}{\partial \overline{x}} - \sin 2\overset{j}{\beta}\frac{\partial g}{\partial \overline{y}}) \\
&\quad - (\overline{y} - c_{\overline{y}})(\sin 3\overset{j}{\beta}\frac{\partial^2 g}{\partial \overline{x}\partial \overline{y}} - \cos 3\overset{j}{\beta}\frac{\partial^2 g}{\partial \overline{x}^2}) \\
D_{yyy} &= -2(1-\nu)\big(\sin \overset{j}{\beta}\frac{\partial g}{\partial \overline{x}} + \cos \overset{j}{\beta}\frac{\partial g}{\partial \overline{y}}\big) + \sin \overset{j}{\beta}(\cos 2\overset{j}{\beta}\frac{\partial g}{\partial \overline{x}} - \sin 2\overset{j}{\beta}\frac{\partial g}{\partial \overline{y}}) \\
&\quad + (\overline{y} - c_{\overline{y}})(\sin 3\overset{j}{\beta}\frac{\partial^2 g}{\partial \overline{x}\partial \overline{y}} - \cos 3\overset{j}{\beta}\frac{\partial^2 g}{\partial \overline{x}^2}) \\
D_{yxy} &= -(1-2\nu)\big(\cos \overset{j}{\beta}\frac{\partial g}{\partial \overline{x}} - \sin \overset{j}{\beta}\frac{\partial g}{\partial \overline{y}}\big) - \sin \overset{j}{\beta}(\sin 2\overset{j}{\beta}\frac{\partial g}{\partial \overline{x}} + \cos 2\overset{j}{\beta}\frac{\partial g}{\partial \overline{y}}) \\
&\quad + (\overline{y} - c_{\overline{y}})(\cos 3\overset{j}{\beta}\frac{\partial^2 g}{\partial \overline{x}\partial \overline{y}} + \sin 3\overset{j}{\beta}\frac{\partial^2 g}{\partial \overline{x}^2})
\end{aligned}
\tag{7.25}
$$

Similarly, the individual terms of $S_{pqr}(\overline{x} - c_{\overline{x}}, \overline{y} - c_{\overline{y}})$ can be written as

$$
\begin{aligned}
S_{xxx} &= 2\mu\Big[-\cos \overset{j}{\beta}\frac{\partial^2 g}{\partial \overline{x}\partial \overline{y}} - \sin \overset{j}{\beta}\frac{\partial^2 g}{\partial \overline{x}^2} - \cos \overset{j}{\beta}(\cos 2\overset{j}{\beta}\frac{\partial^2 g}{\partial \overline{x}\partial \overline{y}} + \sin 2\overset{j}{\beta}\frac{\partial^2 g}{\partial \overline{x}^2}) \\
&\quad - (\overline{y} - c_{\overline{y}})(\cos 3\overset{j}{\beta}\frac{\partial^3 g}{\partial \overline{x}\partial \overline{y}^2} + \sin 3\overset{j}{\beta}\frac{\partial^3 g}{\partial \overline{x}^2\partial \overline{y}})\Big]
\end{aligned}
$$

[2] Expressions for D_{kij} and S_{kij} in terms of function $g(\boldsymbol{x}, \boldsymbol{\xi})$ are given in Appendix B.

$$S_{xyy} = 2\mu\Big[-\sin\overset{j}{\beta}(\sin 2\overset{j}{\beta}\frac{\partial^2 g}{\partial\overline{x}\partial\overline{y}} - \cos 2\overset{j}{\beta}\frac{\partial^2 g}{\partial\overline{x}^2}) + (\overline{y} - c_{\overline{y}})(\cos 3\overset{j}{\beta}\frac{\partial^3 g}{\partial\overline{x}\partial\overline{y}^2} + \sin 3\overset{j}{\beta}\frac{\partial^3 g}{\partial\overline{x}^2\partial\overline{y}})\Big]$$

$$S_{xxy} = 2\mu\Big[-\cos\overset{j}{\beta}(\sin 2\overset{j}{\beta}\frac{\partial^2 g}{\partial\overline{x}\partial\overline{y}} - \cos 2\overset{j}{\beta}\frac{\partial^2 g}{\partial\overline{x}^2}) - (\overline{y} - c_{\overline{y}})(\sin 3\overset{j}{\beta}\frac{\partial^3 g}{\partial\overline{x}\partial\overline{y}^2} - \cos 3\overset{j}{\beta}\frac{\partial^3 g}{\partial\overline{x}^2\partial\overline{y}})\Big] \tag{7.26}$$

$$S_{yxx} = S_{xxy}$$

$$S_{yyy} = 2\mu\Big[-\sin\overset{j}{\beta}\frac{\partial^2 g}{\partial\overline{x}\partial\overline{y}} + \cos\overset{j}{\beta}\frac{\partial^2 g}{\partial\overline{x}^2} + \sin\overset{j}{\beta}(\cos 2\overset{j}{\beta}\frac{\partial^2 g}{\partial\overline{x}\partial\overline{y}} + \sin 2\overset{j}{\beta}\frac{\partial^2 g}{\partial\overline{x}^2})$$
$$+ (\overline{y} - c_{\overline{y}})(\sin 3\overset{j}{\beta}\frac{\partial^3 g}{\partial\overline{x}\partial\overline{y}^2} - \cos 3\overset{j}{\beta}\frac{\partial^3 g}{\partial\overline{x}^2\partial\overline{y}})\Big]$$

$$S_{yxy} = S_{xyy}$$

The integrated coefficients $\mathbb{D}_{pqr}(\overset{j}{\boldsymbol{x}}, \boldsymbol{\xi})$ and $\mathbb{S}_{pqr}(\overset{j}{\boldsymbol{x}}, \boldsymbol{\xi})$ in (7.24) are now found by combining the above expressions with (7.19) to get

$$\begin{aligned}
\mathbb{D}_{xxx}(\overset{j}{\boldsymbol{x}}, \boldsymbol{\xi}) &= \int_{-\overset{j}{a}}^{\overset{j}{a}} D_{xxx}(\overline{x} - c_{\overline{x}}, c_{\overline{y}})\,\mathrm{d}\overline{x} \\
&= -2(1-\nu)\big(\cos\overset{j}{\beta}\bar{T}_2 - \sin\overset{j}{\beta}\bar{T}_3\big) - \cos\overset{j}{\beta}(\cos 2\overset{j}{\beta}\bar{T}_2 - \sin 2\overset{j}{\beta}\bar{T}_3) \\
&\quad + c_{\bar{y}}(\cos 3\overset{j}{\beta}\bar{T}_4 + \sin 3\overset{j}{\beta}\bar{T}_5) \\
\mathbb{D}_{xyy}(\overset{j}{\boldsymbol{x}}, \boldsymbol{\xi}) &= \int_{-\overset{j}{a}}^{\overset{j}{a}} D_{xyy}(\overline{x} - c_{\overline{x}}, c_{\overline{y}})\,\mathrm{d}\overline{x} \\
&= -2\nu\big(\cos\overset{j}{\beta}\bar{T}_2 - \sin\overset{j}{\beta}\bar{T}_3\big) + \cos\overset{j}{\beta}(\cos 2\overset{j}{\beta}\bar{T}_2 - \sin 2\overset{j}{\beta}\bar{T}_3) \\
&\quad - c_{\bar{y}}(\cos 3\overset{j}{\beta}\bar{T}_4 + \sin 3\overset{j}{\beta}\bar{T}_5) \\
\mathbb{D}_{xxy}(\overset{j}{\boldsymbol{x}}, \boldsymbol{\xi}) &= \int_{-\overset{j}{a}}^{\overset{j}{a}} D_{xxy}(\overline{x} - c_{\overline{x}}, c_{\overline{y}})\,\mathrm{d}\overline{x} \\
&= -(1-2\nu)\big(\sin\overset{j}{\beta}\bar{T}_2 + \cos\overset{j}{\beta}\bar{T}_3\big) - \cos\overset{j}{\beta}(\sin 2\overset{j}{\beta}\bar{T}_2 + \cos 2\overset{j}{\beta}\bar{T}_3) \\
&\quad + c_{\bar{y}}(\sin 3\overset{j}{\beta}\bar{T}_4 - \cos 3\overset{j}{\beta}\bar{T}_5) \qquad (7.27) \\
\mathbb{D}_{yxx}(\overset{j}{\boldsymbol{x}}, \boldsymbol{\xi}) &= \int_{-\overset{j}{a}}^{\overset{j}{a}} D_{yxx}(\overline{x} - c_{\overline{x}}, c_{\overline{y}})\,\mathrm{d}\overline{x} \\
&= -2\nu\big(\sin\overset{j}{\beta}\bar{T}_2 + \cos\overset{j}{\beta}\bar{T}_3\big) - \sin\overset{j}{\beta}(\cos 2\overset{j}{\beta}\bar{T}_2 - \sin 2\overset{j}{\beta}\bar{T}_3) \\
&\quad + c_{\bar{y}}(\sin 3\overset{j}{\beta}\bar{T}_4 - \cos 3\overset{j}{\beta}\bar{T}_5)
\end{aligned}$$

$$\mathbb{D}_{yyy}(\overset{j}{\boldsymbol{x}}, \boldsymbol{\xi}) = \int_{-\overset{j}{a}}^{\overset{j}{a}} D_{yyy}(\overline{x} - c_{\overline{x}}, c_{\overline{y}})\, \mathrm{d}\overline{x}$$
$$= -2(1-\nu)\big(\sin\overset{j}{\beta}\bar{T}_2 + \cos\overset{j}{\beta}\bar{T}_3\big) + \sin\overset{j}{\beta}(\cos 2\overset{j}{\beta}\bar{T}_2 - \sin 2\overset{j}{\beta}\bar{T}_3)$$
$$- c_{\bar{y}}(\sin 3\overset{j}{\beta}\bar{T}_4 - \cos 3\overset{j}{\beta}\bar{T}_5)$$

$$\mathbb{D}_{yxy}(\overset{j}{\boldsymbol{x}}, \boldsymbol{\xi}) = \int_{-\overset{j}{a}}^{\overset{j}{a}} D_{yxy}(\overline{x} - c_{\overline{x}}, c_{\overline{y}})\, \mathrm{d}\overline{x}$$
$$= -(1-2\nu)\big(\cos\overset{j}{\beta}\bar{T}_2 - \sin\overset{j}{\beta}\bar{T}_3\big) - \sin\overset{j}{\beta}(\sin 2\overset{j}{\beta}\bar{T}_2 + \cos 2\overset{j}{\beta}\bar{T}_3)$$
$$- c_{\bar{y}}(\cos 3\overset{j}{\beta}\bar{T}_4 + \sin 3\overset{j}{\beta}\bar{T}_5)$$

and

$$\mathbb{S}_{xxx}(\overset{j}{\boldsymbol{x}}, \boldsymbol{\xi}) = \int_{-\overset{j}{a}}^{\overset{j}{a}} S_{xxx}(\overline{x} - c_{\overline{x}}, c_{\overline{y}})\, \mathrm{d}\overline{x}$$
$$= 2\mu\Big[-\cos\overset{j}{\beta}\bar{T}_4 - \sin\overset{j}{\beta}\bar{T}_5 - \cos\overset{j}{\beta}(\cos 2\overset{j}{\beta}\bar{T}_4 + \sin 2\overset{j}{\beta}\bar{T}_5) + c_{\bar{y}}(\cos 3\overset{j}{\beta}\bar{T}_6 + \sin 3\overset{j}{\beta}\bar{T}_7)\Big]$$
$$\mathbb{S}_{xyy}(\overset{j}{\boldsymbol{x}}, \boldsymbol{\xi}) = \int_{-\overset{j}{a}}^{\overset{j}{a}} S_{xyy}(\overline{x} - c_{\overline{x}}, c_{\overline{y}})\, \mathrm{d}\overline{x}$$
$$= 2\mu\Big[-\sin\overset{j}{\beta}(\sin 2\overset{j}{\beta}\bar{T}_4 - \cos 2\overset{j}{\beta}\bar{T}_5) - c_{\bar{y}}(\cos 3\overset{j}{\beta}\bar{T}_6 + \sin 3\overset{j}{\beta}\bar{T}_7)\Big]$$
$$\mathbb{S}_{xxy}(\overset{j}{\boldsymbol{x}}, \boldsymbol{\xi}) = \int_{-\overset{j}{a}}^{\overset{j}{a}} S_{xxy}(\overline{x} - c_{\overline{x}}, c_{\overline{y}})\, \mathrm{d}\overline{x} \tag{7.28}$$
$$= 2\mu\Big[-\cos\overset{j}{\beta}(\sin 2\overset{j}{\beta}\bar{T}_4 - \cos 2\overset{j}{\beta}\bar{T}_5) + c_{\bar{y}}(\sin 3\overset{j}{\beta}\bar{T}_6 - \cos 3\overset{j}{\beta}\bar{T}_7)\Big]$$
$$\mathbb{S}_{yxx}(\overset{j}{\boldsymbol{x}}, \boldsymbol{\xi}) = \mathbb{S}_{xxy}(\overset{j}{\boldsymbol{x}}, \boldsymbol{\xi})$$
$$\mathbb{S}_{yyy}(\overset{j}{\boldsymbol{x}}, \boldsymbol{\xi}) = \int_{-\overset{j}{a}}^{\overset{j}{a}} S_{yyy}(\overline{x} - c_{\overline{x}}, c_{\overline{y}})\, \mathrm{d}\overline{x}$$
$$= 2\mu\Big[-\sin\overset{j}{\beta}\bar{T}_4 + \cos\overset{j}{\beta}\bar{T}_5 + \sin\overset{j}{\beta}(\cos 2\overset{j}{\beta}\bar{T}_4 + \sin 2\overset{j}{\beta}\bar{T}_5) - c_{\bar{y}}(\sin 3\overset{j}{\beta}\bar{T}_6 - \cos 3\overset{j}{\beta}\bar{T}_7)\Big]$$
$$\mathbb{S}_{yxy}(\overset{j}{\boldsymbol{x}}, \boldsymbol{\xi}) = \mathbb{S}_{xyy}(\overset{j}{\boldsymbol{x}}, \boldsymbol{\xi}).$$

7.5 Example

As an illustration of the direct boundary element method for elasticity we will consider again the problem of a circular hole of radius R in an plate under uniaxial tension $\sigma_{xx} = p$ at infinity (see Sects. 5.9 and 6.11). Further examples of the method are given later in this chapter.

The circular hole problem was solved assuming plane strain conditions and using a 25 element approximation of a symmetric one-quarter of the circular boundary. Because the problem at hand is a traction boundary value problem, the unknowns in the system (7.11) are the actual displacements $\overset{i}{u}_x$ and $\overset{i}{u}_y$ of the boundary. These can be translated into local displacements $\overset{i}{u}_{\bar{x}}$ and $\overset{i}{u}_{\bar{y}}$ using (4.63) and compared with the analytical solution for the radial and tangential displacements at the boundary $r = R$ (Sokolnikoff 1956). The *changes* of displacement (i.e., the displacements due to the presence of the hole) are

$$
\begin{aligned}
u_r &= \frac{Rp}{4\mu}\left[1 + (3 - 4\nu)\cos 2\theta\right] \\
u_\theta &= -\frac{Rp}{4\mu}(3 - 4\nu)\sin 2\theta
\end{aligned}
\tag{7.29}
$$

The shear component of displacement $\overset{i}{u}_{\bar{x}}$ is the same as $\overset{i}{u}_\theta$ while the normal component $\overset{i}{u}_{\bar{y}}$ is the negative of $\overset{i}{u}_r$ because the local $\bar{x}$ coordinate points in the direction of increasing angle θ and the local $\bar{y}$ coordinate points toward the origin, opposite to the direction of r.

Computed values of the radial and tangential displacements are compared with the exact solution (7.29) in Fig. 7.2 for the case that $\nu = 0.1$ and $p/\mu = 10^{-3}$. These displacements can be used as described in Sect. 6.10 to compute the circumferential stress along the boundary; the results are found to be within plotting accuracy of the analytical solution.

7.6 Numerical Procedure II—Piecewise Linear Boundary Parameters

In presenting a more elaborate treatment of the direct boundary integral method we will assume that the displacements and tractions vary linearly between the endpoints, or nodal points, of the boundary elements. A closed boundary contour divided into N straight line elements has N connection points, or nodes, and the problem is formulated in terms of the displacements and tractions at these points rather than at a set of intermediate collocation points as was done for the stress and displacement

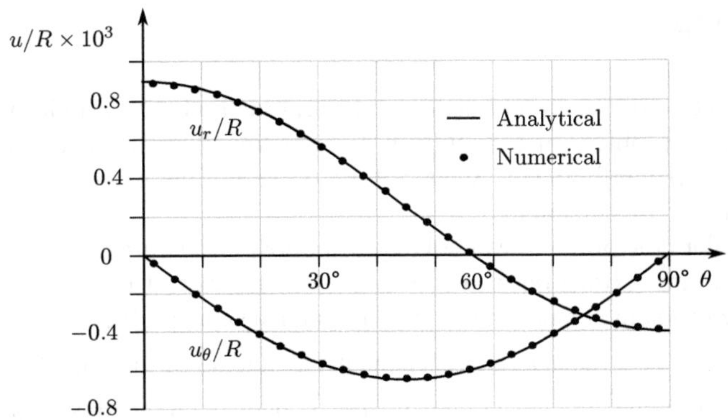

Fig. 7.2 Radial and tangential displacements on the wall of a circular hole

discontinuity methods with piecewise quadratic elements. If the boundary parameters were allowed to vary quadratically along the elements the closed boundary contour would have $2N$ nodal points.

The geometry for the problem is given in Fig. 7.3. As shown in the figure, we take the sample point $\boldsymbol{x}$ at the midpoint of element J, which lies between nodes j and $j+1$. Notice that the capital letter J refers to the element while the lower case letter j refers to one of its two nodal points. In the same way, element $J-1$ lies between nodes $j-1$ and j. After integrating over the elements the sample points will be referred to the nodes. The sample point at node j, for example, will have contributions from both element J and element $J-1$.

The coordinate transformation formulas for this problem are the same as those given in Sect. 3.9 for a similar geometric configuration for the direct method for Laplace's equation. Using (3.46)–(3.48), we write (7.3) as

$$
\begin{aligned}
u_x(\boldsymbol{\xi}) &= \sum_{J=1}^{N} \int_{-\overset{J}{a}}^{\overset{J}{a}} \left[\overset{J}{t}_x(\bar{x}) U_{xx}(\bar{x} - c_{\overline{x}}, c_{\overline{y}}) + \overset{J}{t}_y(\bar{x}) U_{xy}(\bar{x} - c_{\overline{x}}, c_{\overline{y}}) \right] \mathrm{d}\bar{x} \\
&\quad - \sum_{J=1}^{N} \int_{-\overset{J}{a}}^{\overset{J}{a}} \left[\overset{J}{u}_x(\bar{x}) T_{xx}(\bar{x} - c_{\overline{x}}, c_{\overline{y}}) + \overset{J}{u}_y(\bar{x}) T_{xy}(\bar{x} - c_{\overline{x}}, c_{\overline{y}}) \right] \mathrm{d}\bar{x} \\
u_y(\boldsymbol{\xi}) &= \sum_{J=1}^{N} \int_{-\overset{J}{a}}^{\overset{J}{a}} \left[\overset{J}{t}_x(\bar{x}) U_{yx}(\bar{x} - c_{\overline{x}}, c_{\overline{y}}) + \overset{J}{t}_y(\bar{x}) U_{yy}(\bar{x} - c_{\overline{x}}, c_{\overline{y}}) \right] \mathrm{d}\bar{x} \\
&\quad - \sum_{J=1}^{N} \int_{-\overset{J}{a}}^{\overset{J}{a}} \left[\overset{J}{u}_x(\bar{x}) T_{yx}(\bar{x} - c_{\overline{x}}, c_{\overline{y}}) + \overset{J}{u}_y(\bar{x}) T_{yy}(\bar{x} - c_{\overline{x}}, c_{\overline{y}}) \right] \mathrm{d}\bar{x}
\end{aligned}
\tag{7.30}
$$

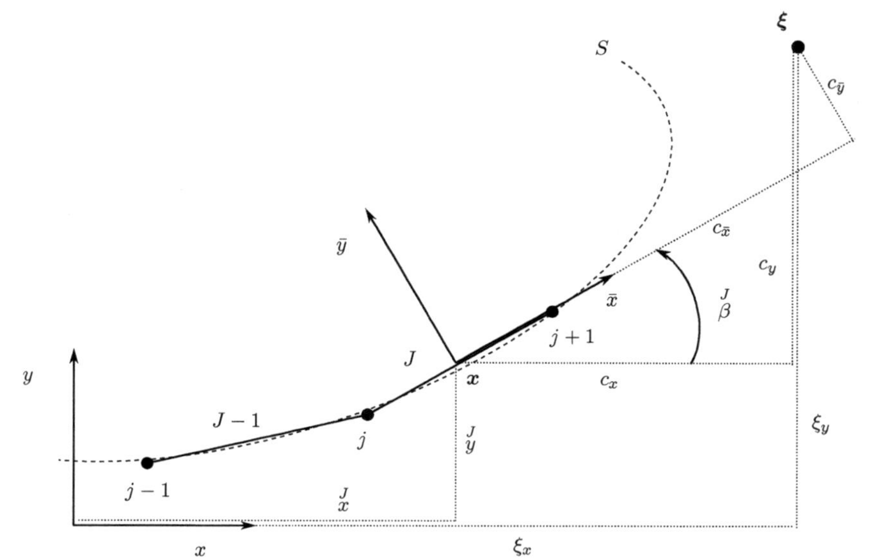

Fig. 7.3 Discretized boundary contour

in which we have again written $c_{\overline{x}}$ and $c_{\overline{y}}$ instead of $\overset{J}{c}_{\bar{x}}$ and $\overset{J}{c}_{\bar{y}}$ for notational convenience. Expressions for the quantities $T_{xx}(\overline{x} - c_{\overline{x}}, c_{\overline{y}})$, etc., in (7.30) are given in Sect. 7.3. In the present case, however, the angle $\overset{j}{\beta}$ appearing in those expressions is designated as $\overset{J}{\beta}$.

We next assume that the displacements $\overset{J}{u}_x(\overline{x})$ and $\overset{J}{u}_y(\overline{x})$ in (7.30) can be expressed as linear functions of $\overline{x}$:

$$
\begin{aligned}
\overset{J}{u}_x(\bar{x}) &= \frac{1}{2}\left(1 - \frac{\bar{x}}{\overset{J}{a}}\right)(\overset{j}{u}_x)_J + \frac{1}{2}\left(1 + \frac{\bar{x}}{\overset{J}{a}}\right)(\overset{j+1}{u}_x)_J \\
\overset{J}{u}_y(\bar{x}) &= \frac{1}{2}\left(1 - \frac{\bar{x}}{\overset{J}{a}}\right)(\overset{j}{u}_y)_J + \frac{1}{2}\left(1 + \frac{\bar{x}}{\overset{J}{a}}\right)(\overset{j+1}{u}_y)_J
\end{aligned}
\qquad (7.31)
$$

or

$$
\begin{aligned}
\overset{J}{u}_x(\bar{x}) &= \overset{J}{\psi}_1(\bar{x})(\overset{j}{u}_x)_J + \overset{J}{\psi}_2(\bar{x})(\overset{j+1}{u}_x)_J \\
\overset{J}{u}_y(\bar{x}) &= \overset{J}{\psi}_1(\bar{x})(\overset{j}{u}_y)_J + \overset{J}{\psi}_2(\bar{x})(\overset{j+1}{u}_y)_J
\end{aligned}
\qquad (7.32)
$$

where $\overset{J}{\psi}_1(\bar{x}) = \frac{1}{2}(1 - \bar{x}/\overset{J}{a})$ and $\overset{J}{\psi}_2(\bar{x}) = \frac{1}{2}(1 + \bar{x}/\overset{J}{a})$ are the element shape functions. These quantities differ from the linear variation shape functions used previously for discontinuous elements (Sect. 3.9) in that they are expressed in terms of the element endpoints $\pm\overset{J}{a}$ rather than two points away from the ends. Note that $\overset{J}{\psi}_1 = 1$ and $\overset{J}{\psi}_2 = 0$ at $\bar{x} = -\overset{J}{a}$, and that $\overset{J}{\psi}_1 = 0$ and $\overset{J}{\psi}_2 = 1$ at $\bar{x} = +\overset{J}{a}$. Similarly, we assume that the tractions $\overset{J}{t}_x(\bar{x})$ and $\overset{J}{t}_y(\bar{x})$ in (7.30) can be written as

$$
\begin{aligned}
\overset{J}{t}_x(\bar{x}) &= \overset{J}{\psi}_1(\bar{x})(\overset{j}{t}_x)_J + \overset{J}{\psi}_2(\bar{x})(\overset{j+1}{t_x})_J \\
\overset{J}{t}_y(\bar{x}) &= \overset{J}{\psi}_1(\bar{x})(\overset{j}{t}_y)_J + \overset{J}{\psi}_2(\bar{x})(\overset{j+1}{t_y})_J
\end{aligned}
\tag{7.33}
$$

The notation $(\overset{j}{\bullet})_J$ and $(\overset{j+1}{\bullet})_J$ means that we are considering the values of displacement and traction at nodes j and $j+1$ of element J. This notation is required because each node j is shared by two elements, J and $J-1$ (see Fig. 7.3). In general, due to different element inclinations, the tractions at node j of element $J-1$ are not the same as those at node j of element J, i.e.,

$$
\begin{aligned}
(\overset{j}{t}_x)_{J-1} &\neq (\overset{j}{t}_x)_J \\
(\overset{j}{t}_y)_{J-1} &\neq (\overset{j}{t}_y)_J
\end{aligned}
\tag{7.34}
$$

The displacements at node j, on the other hand, are single-valued:

$$
\begin{aligned}
(\overset{j}{u}_x)_{J-1} &= (\overset{j}{u}_x)_J \\
(\overset{j}{u}_y)_{J-1} &= (\overset{j}{u}_y)_J
\end{aligned}
\tag{7.35}
$$

We could therefore suppress the element identifiers J and $J-1$ and simply refer to the nodal displacements as $\overset{j}{u}_x$ and $\overset{j}{u}_y$. For now, however, it is useful to retain the node/element notation in order to distinguish between the effects of neighboring elements.

If expressions (7.32) and (7.33) are substituted into (7.30), one obtains discretized versions of Somigliana's formulas involving integrals of kernel–shape function products of the form

$$
\int_{-\overset{J}{a}}^{\overset{J}{a}} U_{pq}(\bar{x} - c_{\bar{x}}, c_{\bar{y}})\overset{J}{\psi}_m(\bar{x})\,\mathrm{d}\bar{x}
\tag{7.36}
$$

and

$$\int_{-\overset{J}{a}}^{\overset{J}{a}} T_{pq}(\bar{x} - c_{\bar{x}}, c_{\bar{y}})\overset{J}{\psi}_m(\bar{x})\,\mathrm{d}\bar{x} \tag{7.37}$$

where p and q are either x or y, and m is 1 or 2. All of these integrals can be evaluated analytically, as described below.

7.7 Evaluation of Integrals

It is easy to see from (7.14) and (7.17) that expressions (7.36) and (7.37) can be evaluated in terms of integrals of the function $g(\overline{x} - c_{\overline{x}}, \overline{y} - c_{\overline{y}})$ and its derivatives multiplied by shape functions $\overset{J}{\psi}_1(\bar{x}) = \frac{1}{2}(1 - \bar{x}/\overset{J}{a})$ and $\overset{J}{\psi}_2(\bar{x}) = \frac{1}{2}(1 + \bar{x}/\overset{J}{a})$, with $\overline{y}$ then set equal to 0. The integration process can be simplified somewhat by writing the shape functions in the form

$$\begin{aligned} \overset{J}{\psi}_1(\bar{x}) &= -\frac{1}{2\overset{J}{a}}[(c_{\bar{x}} - \overset{J}{a}) + (\bar{x} - c_{\bar{x}})] \\ \overset{J}{\psi}_2(\bar{x}) &= \frac{1}{2\overset{J}{a}}[(c_{\bar{x}} + \overset{J}{a}) + (\bar{x} - c_{\bar{x}})] \end{aligned} \tag{7.38}$$

Then, letting $(\bullet)$ stand for $g(\bar{x}, \bar{y})$ or one of its derivatives, we have

$$\begin{aligned} \int_{-\overset{J}{a}}^{\overset{J}{a}} (\bullet)\,\overset{J}{\psi}_1(\bar{x})\,\mathrm{d}\bar{x} &= -\frac{1}{2\overset{J}{a}}(c_{\bar{x}} - \overset{J}{a})\int_{-\overset{J}{a}}^{\overset{J}{a}} (\bullet)\,\mathrm{d}\bar{x} - \frac{1}{2\overset{J}{a}}\int_{-\overset{J}{a}}^{\overset{J}{a}} (\bar{x} - c_{\bar{x}})\,(\bullet)\,\mathrm{d}\bar{x} \\ \int_{-\overset{J}{a}}^{\overset{J}{a}} (\bullet)\,\overset{J}{\psi}_2(\bar{x})\,\mathrm{d}\bar{x} &= \frac{1}{2\overset{J}{a}}(c_{\bar{x}} + \overset{J}{a})\int_{-\overset{J}{a}}^{\overset{J}{a}} (\bullet)\,\mathrm{d}\bar{x} + \frac{1}{2\overset{J}{a}}\int_{-\overset{J}{a}}^{\overset{J}{a}} (\bar{x} - c_{\bar{x}})\,(\bullet)\,\mathrm{d}\bar{x} \end{aligned} \tag{7.39}$$

Setting $u = \bar{x} - c_{\overline{x}}$ and $v = \bar{y} - c_{\overline{y}}$, the required integrals can easily be obtained from the indefinite integration formulas tabulated in Appendix A for function $g(u, v) = -[4\pi(1 - \nu]^{-1} \ln \sqrt{u^2 + v^2}$ and its derivatives.

Proceeding in this way, we obtain the following results for the terms involving shape function $\overset{J}{\psi}_1(\overline{x})$:

$$
\begin{aligned}
\overset{\psi_1}{(I_1)}_J &= \int_{-\overset{J}{a}}^{\overset{J}{a}} g(\bar{x}-c_{\bar{x}}, \bar{y}-c_{\bar{y}})\overset{J}{\psi}_1(\bar{x})\, \mathrm{d}\bar{x}\ |_{\bar{y}=0} \\
&= \frac{1}{8\pi(1-\nu)}\Big[\frac{1}{\overset{J}{a}}(c_{\bar{x}}-\overset{J}{a})c_{\bar{y}}(\bar{\theta}_1-\bar{\theta}_2) + (c_{\bar{x}}-\overset{J}{a})\ln\bar{r}_1 - (c_{\bar{x}}+\overset{J}{a})\ln\bar{r}_2 \\
&\quad + \frac{1}{2\overset{J}{a}}\Big\{c_{\bar{y}}^2 - (c_{\bar{x}}-\overset{J}{a})(c_{\bar{x}}+\overset{J}{a})\Big\}\ln(\frac{\bar{r}_1}{\bar{r}_2}) - (c_{\bar{x}}-\overset{J}{a}) + \overset{J}{a}\Big] \\
\overset{\psi_1}{(I_2)}_J &= \int_{-\overset{J}{a}}^{\overset{J}{a}} \frac{\partial}{\partial\bar{x}} g(\bar{x}-c_{\bar{x}}, \bar{y}-c_{\bar{y}})\overset{J}{\psi}_1(\bar{x})\, \mathrm{d}\bar{x}\ |_{\bar{y}=0} \\
&= \frac{1}{8\pi(1-\nu)}\Big[-\frac{1}{\overset{J}{a}}c_{\bar{y}}(\bar{\theta}_1-\bar{\theta}_2) + \frac{1}{\overset{J}{a}}(c_{\bar{x}}-\overset{J}{a})\ln(\frac{\bar{r}_1}{\bar{r}_2}) + 2\Big] \\
\overset{\psi_1}{(I_3)}_J &= \int_{-\overset{J}{a}}^{\overset{J}{a}} \frac{\partial}{\partial\bar{y}} g(\bar{x}-c_{\bar{x}}, \bar{y}-c_{\bar{y}})\overset{J}{\psi}_1(\bar{x})\, \mathrm{d}\bar{x}\ |_{\bar{y}=0} \\
&= -\frac{1}{8\pi(1-\nu)}\Big[\frac{1}{\overset{J}{a}}(c_{\bar{x}}-\overset{J}{a})(\bar{\theta}_1-\bar{\theta}_2) + \frac{1}{\overset{J}{a}}c_{\bar{y}}\ln(\frac{\bar{r}_1}{\bar{r}_2})\Big] \\
\overset{\psi_1}{(I_4)}_J &= \int_{-\overset{J}{a}}^{\overset{J}{a}} \frac{\partial^2}{\partial\bar{x}\partial\bar{y}} g(\bar{x}-c_{\bar{x}}, \bar{y}-c_{\bar{y}})\overset{J}{\psi}_1(\bar{x})\, \mathrm{d}\bar{x}\ |_{\bar{y}=0} \qquad (7.40) \\
&= \frac{1}{8\pi 1-\nu)}\Big[\frac{1}{\overset{J}{a}}(\bar{\theta}_1-\bar{\theta}_2) - \frac{2c_{\bar{y}}}{\bar{r}_2^2}\Big] \\
\overset{\psi_1}{(I_5)}_J &= \int_{-\overset{J}{a}}^{\overset{J}{a}} \frac{\partial^2}{\partial\bar{x}^2} g(\bar{x}-c_{\bar{x}}, \bar{y}-c_{\bar{y}})\overset{J}{\psi}_1(\bar{x})\, \mathrm{d}\bar{x}\ |_{\bar{y}=0} \\
&= -\frac{1}{8\pi(1-\nu)}\Big[\frac{1}{\overset{J}{a}}\ln(\frac{\bar{r}_1}{\bar{r}_2}) + \frac{2(c_{\bar{x}}+\overset{J}{a})}{\bar{r}_2^2}\Big] \\
\overset{\psi_1}{(I_6)}_J &= \int_{-\overset{J}{a}}^{\overset{J}{a}} \frac{\partial^3}{\partial\bar{x}\partial\bar{y}^2} g(\bar{x}-c_{\bar{x}}, \bar{y}-c_{\bar{y}})\overset{J}{\psi}_1(\bar{x})\, \mathrm{d}\bar{x}\ |_{\bar{y}=0} \\
&= -\frac{1}{8\pi(1-\nu)}\Big[\frac{1}{\overset{J}{a}}(c_{\bar{x}}-\overset{J}{a})(\frac{1}{\bar{r}_1^2} - \frac{1}{\bar{r}_2^2}) - \frac{4(c_{\bar{x}}+\overset{J}{a})^2}{\bar{r}_2^4}\Big] \\
\overset{\psi_1}{(I_7)}_J &= \int_{-\overset{J}{a}}^{\overset{J}{a}} \frac{\partial^3}{\partial\bar{x}^2\partial\bar{y}} g(\bar{x}-c_{\bar{x}}, \bar{y}-c_{\bar{y}})\overset{J}{\psi}_1(\bar{x})\, \mathrm{d}\bar{x}\ |_{\bar{y}=0} \\
&= \frac{1}{8\pi(1-\nu)}\Big[\frac{1}{\overset{J}{a}}c_{\bar{y}}(\frac{1}{\bar{r}_1^2} - \frac{1}{\bar{r}_2^2}) - \frac{4(c_{\bar{x}}+\overset{J}{a})c_{\bar{y}}}{\bar{r}_2^4}\Big]
\end{aligned}
$$

The last two terms will be needed later (Sect. 7.11) to evaluate the stress components $\sigma_{ij}(\boldsymbol{\xi})$ and we record them here for completeness.

Similar results for the terms involving shape function $\overset{J}{\psi}_2(\bar{x})$ are

$$
\begin{aligned}
(\overset{\psi_2}{I_1})_J &= \int_{-\overset{J}{a}}^{\overset{J}{a}} g(\bar{x}-c_{\bar{x}},\, \bar{y}-c_{\bar{y}})\overset{J}{\psi}_2(\bar{x})\,\mathrm{d}\bar{x}\ |_{\bar{y}=0} \\
&= -\frac{1}{8\pi(1-\nu)}\Big[\frac{1}{\overset{J}{a}}(c_{\bar{x}}+\overset{J}{a})c_{\bar{y}}(\bar{\theta}_1-\bar{\theta}_2) - (c_{\bar{x}}-\overset{J}{a})\ln\bar{r}_1 + (c_{\bar{x}}+\overset{J}{a})\ln\bar{r}_2 \\
&\quad + \frac{1}{2\overset{J}{a}}\Big\{c_{\bar{y}}^2 - (c_{\bar{x}}-\overset{J}{a})(c_{\bar{x}}+\overset{J}{a})\Big\}\ln(\frac{\bar{r}_1}{\bar{r}_2}) - (c_{\bar{x}}+\overset{J}{a}) - \overset{J}{a}\Big] \\
(\overset{\psi_2}{I_2})_J &= \int_{-\overset{J}{a}}^{\overset{J}{a}} \frac{\partial}{\partial\bar{x}} g(\bar{x}-c_{\bar{x}},\, \bar{y}-c_{\bar{y}})\overset{J}{\psi}_2(\bar{x})\,\mathrm{d}\bar{x}\ |_{\bar{y}=0} \\
&= -\frac{1}{8\pi(1-\nu)}\Big[-\frac{1}{\overset{J}{a}}c_{\bar{y}}(\bar{\theta}_1-\bar{\theta}_2) + \frac{1}{\overset{J}{a}}(c_{\bar{x}}+\overset{J}{a})\ln(\frac{\bar{r}_1}{\bar{r}_2}) + 2\Big] \\
(\overset{\psi_2}{I_3})_J &= \int_{-\overset{J}{a}}^{\overset{J}{a}} \frac{\partial}{\partial\bar{y}} g(\bar{x}-c_{\bar{x}},\, \bar{y}-c_{\bar{y}})\overset{J}{\psi}_2(\bar{x})\,\mathrm{d}\bar{x}\ |_{\bar{y}=0} \\
&= \frac{1}{8\pi(1-\nu)}\Big[\frac{1}{\overset{J}{a}}(c_{\bar{x}}+\overset{J}{a})(\bar{\theta}_1-\bar{\theta}_2) + \frac{1}{\overset{J}{a}}c_{\bar{y}}\ln(\frac{\bar{r}_1}{\bar{r}_2})\Big] \\
(\overset{\psi_2}{I_4})_J &= \int_{-\overset{J}{a}}^{\overset{J}{a}} \frac{\partial^2}{\partial\bar{x}\partial\bar{y}} g(\bar{x}-c_{\bar{x}},\, \bar{y}-c_{\bar{y}})\overset{J}{\psi}_2(\bar{x})\,\mathrm{d}\bar{x}\ |_{\bar{y}=0} \\
&= -\frac{1}{8\pi(1-\nu)}\Big[\frac{1}{\overset{J}{a}}(\bar{\theta}_1-\bar{\theta}_2) - \frac{2c_{\bar{y}}}{\bar{r}_1^2}\Big] \\
(\overset{\psi_2}{I_5})_J &= \int_{-\overset{J}{a}}^{\overset{J}{a}} \frac{\partial^2}{\partial\bar{x}^2} g(\bar{x}-c_{\bar{x}},\, \bar{y}-c_{\bar{y}})\overset{J}{\psi}_2(\bar{x})\,\mathrm{d}\bar{x}\ |_{\bar{y}=0} \\
&= \frac{1}{8\pi(1-\nu)}\Big[\frac{1}{\overset{J}{a}}\ln(\frac{\bar{r}_1}{\bar{r}_2}) + \frac{2(c_{\bar{x}}-\overset{J}{a})}{\bar{r}_1^2}\Big] \\
(\overset{\psi_2}{I_6})_J &= \int_{-\overset{J}{a}}^{\overset{J}{a}} \frac{\partial^3}{\partial\bar{x}\partial\bar{y}^2} g(\bar{x}-c_{\bar{x}},\, \bar{y}-c_{\bar{y}})\overset{J}{\psi}_2(\bar{x})\,\mathrm{d}\bar{x}\ |_{\bar{y}=0} \\
&= \frac{1}{8\pi(1-\nu)}\Big[\frac{1}{\overset{J}{a}}(c_{\bar{x}}+\overset{J}{a})(\frac{1}{\bar{r}_1^2}-\frac{1}{\bar{r}_2^2}) - \frac{4(c_{\bar{x}}-\overset{J}{a})^2}{\bar{r}_1^4}\Big] \\
(\overset{\psi_2}{I_7})_J &= \int_{-\overset{J}{a}}^{\overset{J}{a}} \frac{\partial^3}{\partial\bar{x}^2\partial\bar{y}} g(\bar{x}-c_{\bar{x}},\, \bar{y}-c_{\bar{y}})\overset{J}{\psi}_2(\bar{x})\,\mathrm{d}\bar{x}\ |_{\bar{y}=0} \\
&= -\frac{1}{8\pi(1-\nu)}\Big[\frac{1}{\overset{J}{a}}c_{\bar{y}}(\frac{1}{\bar{r}_1^2}-\frac{1}{\bar{r}_2^2}) - \frac{4(c_{\bar{x}}-\overset{J}{a})c_{\bar{y}}}{\bar{r}_1^4}\Big]
\end{aligned} \tag{7.41}
$$

Finally, the integrals of the kernel–shape function products in (7.36) and (7.37) can be written in terms of the integrated expressions $(\overset{\psi_m}{I_1})_J$ through $(\overset{\psi_m}{I_5})_J$ in (7.40) and (7.41) as follows:

$$
\begin{aligned}
\overset{\psi_m}{\mathbb{U}}_{xx}(\overset{J}{\boldsymbol{x}},\boldsymbol{\xi}) &= \int_{-\overset{J}{a}}^{\overset{J}{a}} U_{xx}(\bar{x}-c_{\bar{x}},c_{\bar{y}})\overset{J}{\psi}_m(\bar{x})\,\mathrm{d}\bar{x}\\
&= \frac{1}{2\mu}\left[(3-4\nu)(\overset{\psi_m}{I_1})_J - c_{\bar{y}}\left(\sin 2\overset{J}{\beta}(\overset{\psi_m}{I_2})_J + \cos 2\overset{J}{\beta}(\overset{\psi_m}{I_3})_J\right)\right]\\
\overset{\psi_m}{\mathbb{U}}_{yx}(\overset{J}{\boldsymbol{x}},\boldsymbol{\xi}) &= \int_{-\overset{J}{a}}^{\overset{J}{a}} U_{yx}(\bar{x}-c_{\bar{x}},c_{\bar{y}})\overset{J}{\psi}_m(\bar{x})\,\mathrm{d}\bar{x}\\
&= \frac{1}{2\mu}\left[c_{\bar{y}}\left(\cos 2\overset{J}{\beta}(\overset{\psi_m}{I_2})_J - \sin 2\overset{J}{\beta}(\overset{\psi_m}{I_3})_J\right)\right]\\
\overset{\psi_m}{\mathbb{U}}_{xy}(\overset{J}{\boldsymbol{x}},\boldsymbol{\xi}) &= \int_{-\overset{J}{a}}^{\overset{J}{a}} U_{xy}(\bar{x}-c_{\bar{x}},c_{\bar{y}})\overset{J}{\psi}_m(\bar{x})\,\mathrm{d}\bar{x} = \overset{\psi_m}{\mathbb{U}}_{yx}(\overset{J}{\boldsymbol{x}},\boldsymbol{\xi})\\
\overset{\psi_m}{\mathbb{U}}_{yy}(\overset{J}{\boldsymbol{x}},\boldsymbol{\xi}) &= \int_{-\overset{J}{a}}^{\overset{J}{a}} U_{yy}(\bar{x}-c_{\bar{x}},c_{\bar{y}})\overset{J}{\psi}_m(\bar{x})\,\mathrm{d}\bar{x}\\
&= \frac{1}{2\mu}\left[(3-4\nu)(\overset{\psi_m}{I_1})_J + c_{\bar{y}}\left(\sin 2\overset{J}{\beta}(\overset{\psi_m}{I_2})_J + \cos 2\overset{J}{\beta}(\overset{\psi_m}{I_3})_J\right)\right]
\end{aligned}
\tag{7.42}
$$

and

$$
\begin{aligned}
\overset{\psi_m}{\mathbb{T}}_{xx}(\overset{J}{\boldsymbol{x}},\boldsymbol{\xi}) &= \int_{-\overset{J}{a}}^{\overset{J}{a}} T_{xx}(\bar{x}-c_{\bar{x}},c_{\bar{y}})\overset{J}{\psi}_m(\bar{x})\,\mathrm{d}\bar{x}\\
&= 2(1-\nu)(\overset{\psi_m}{I_3})_J - c_{\bar{y}}\left(\sin 2\overset{J}{\beta}(\overset{\psi_m}{I_4})_J - \cos 2\overset{J}{\beta}(\overset{\psi_m}{I_5})_J\right)\\
\overset{\psi_m}{\mathbb{T}}_{yx}(\overset{J}{\boldsymbol{x}},\boldsymbol{\xi}) &= \int_{-\overset{J}{a}}^{\overset{J}{a}} T_{yx}(\bar{x}-c_{\bar{x}},c_{\bar{y}})\overset{J}{\psi}_m(\bar{x})\,\mathrm{d}\bar{x}\\
&= (1-2\nu)(\overset{\psi_m}{I_2})_J + c_{\bar{y}}\left(\cos 2\overset{J}{\beta}(\overset{\psi_m}{I_4})_J + \sin 2\overset{J}{\beta}(\overset{\psi_m}{I_5})_J\right)\\
\overset{\psi_m}{\mathbb{T}}_{xy}(\overset{J}{\boldsymbol{x}},\boldsymbol{\xi}) &= \int_{-\overset{J}{a}}^{\overset{J}{a}} T_{xy}(\bar{x}-c_{\bar{x}},c_{\bar{y}})\overset{J}{\psi}_m(\bar{x})\,\mathrm{d}\bar{x}\\
&= -(1-2\nu)(\overset{\psi_m}{I_2})_J + c_{\bar{y}}\left(\cos 2\overset{J}{\beta}(\overset{\psi_m}{I_4})_J + \sin 2\overset{J}{\beta}(\overset{\psi_m}{I_5})_J\right)\\
\overset{\psi_m}{\mathbb{T}}_{yy}(\overset{J}{\boldsymbol{x}},\boldsymbol{\xi}) &= \int_{-\overset{J}{a}}^{\overset{J}{a}} T_{yy}(\bar{x}-c_{\bar{x}},c_{\bar{y}})\overset{J}{\psi}_m(\bar{x})\,\mathrm{d}\bar{x}\\
&= 2(1-\nu)(\overset{\psi_m}{I_3})_J + c_{\bar{y}}\left(\sin 2\overset{J}{\beta}(\overset{\psi_m}{I_4})_J - \cos 2\overset{J}{\beta}(\overset{\psi_m}{I_5})_J\right).
\end{aligned}
\tag{7.43}
$$

7.8 Approximation of Somigliana's Formulas for Displacements

Combining Eqs. (7.30), (7.32), (7.33), (7.42), and (7.43), we obtain the following numerical representation of Somigliana's formulas for the case in which the displacements and tractions vary linearly over N straight line elements of the discretized boundary contour S:

$$
\begin{aligned}
u_x(\boldsymbol{\xi}) = {} & \sum_{J=1}^{N}\left[\overset{\psi_1}{\mathbb{U}}_{xx}(\overset{J}{\boldsymbol{x}},\boldsymbol{\xi})(\overset{j}{t_x})_J + \overset{\psi_2}{\mathbb{U}}_{xx}(\overset{J}{\boldsymbol{x}},\boldsymbol{\xi})(\overset{j+1}{t_x})_J\right] + \sum_{J=1}^{N}\left[\overset{\psi_1}{\mathbb{U}}_{xy}(\overset{J}{\boldsymbol{x}},\boldsymbol{\xi})(\overset{j}{t_y})_J + \overset{\psi_2}{\mathbb{U}}_{xy}(\overset{J}{\boldsymbol{x}},\boldsymbol{\xi})(\overset{j+1}{t_y})_J\right] \\
& - \sum_{J=1}^{N}\left[\overset{\psi_1}{\mathbb{T}}_{xx}(\overset{J}{\boldsymbol{x}},\boldsymbol{\xi})(\overset{j}{u_x})_J + \overset{\psi_2}{\mathbb{T}}_{xx}(\overset{J}{\boldsymbol{x}},\boldsymbol{\xi})(\overset{j+1}{u_x})_J\right] - \sum_{J=1}^{N}\left[\overset{\psi_1}{\mathbb{T}}_{xy}(\overset{J}{\boldsymbol{x}},\boldsymbol{\xi})(\overset{j}{u_y})_J + \overset{\psi_2}{\mathbb{T}}_{xy}(\overset{J}{\boldsymbol{x}},\boldsymbol{\xi})(\overset{j+1}{u_y})_J\right]
\end{aligned}
\tag{7.44}
$$

$$
\begin{aligned}
u_y(\boldsymbol{\xi}) = {} & \sum_{J=1}^{N}\left[\overset{\psi_1}{\mathbb{U}}_{yx}(\overset{J}{\boldsymbol{x}},\boldsymbol{\xi})(\overset{j}{t_x})_J + \overset{\psi_2}{\mathbb{U}}_{yx}(\overset{J}{\boldsymbol{x}},\boldsymbol{\xi})(\overset{j+1}{t_x})_J\right] + \sum_{J=1}^{N}\left[\overset{\psi_1}{\mathbb{U}}_{yy}(\overset{J}{\boldsymbol{x}},\boldsymbol{\xi})(\overset{j}{t_y})_J + \overset{\psi_2}{\mathbb{U}}_{yy}(\overset{J}{\boldsymbol{x}},\boldsymbol{\xi})(\overset{j+1}{t_y})_J\right] \\
& - \sum_{J=1}^{N}\left[\overset{\psi_1}{\mathbb{T}}_{yx}(\overset{J}{\boldsymbol{x}},\boldsymbol{\xi})(\overset{j}{u_x})_J + \overset{\psi_2}{\mathbb{T}}_{yx}(\overset{J}{\boldsymbol{x}},\boldsymbol{\xi})(\overset{j+1}{u_x})_J\right] - \sum_{J=1}^{N}\left[\overset{\psi_1}{\mathbb{T}}_{yy}(\overset{J}{\boldsymbol{x}},\boldsymbol{\xi})(\overset{j}{u_y})_J + \overset{\psi_2}{\mathbb{T}}_{yy}(\overset{J}{\boldsymbol{x}},\boldsymbol{\xi})(\overset{j+1}{u_y})_J\right]
\end{aligned}
$$

These equations play the same role for our higher order formulation of the direct boundary integral method as (7.6) played for the simplified treatment of the method discussed previously. That is, if all of the nodal point displacements and tractions are known, Eqs. (7.44) can be used to compute the values of the displacements at the point $\boldsymbol{\xi} = (\xi_x, \xi_y)$. Comparable formulas for the stress components at point $\boldsymbol{\xi}$ are presented in Sect. 7.11.

7.9 Formation of the System of Equations

The next step in our treatment of the direct boundary integral method is to set up a system of algebraic equations from which to determine the unknown values of the boundary displacements and tractions at the element nodal points. For this, it is convenient to use a form of Somigliana's formulas for the displacements with the sample point taken at a node, say the point $\overset{j}{\boldsymbol{x}}$, rather than at the center of an element. As shown in Fig. 7.4, the integrated kernel-shape function products now have to be evaluated for two separate local coordinate systems, one originating at the midpoint of element J, and the other at the midpoint of element $J-1$. The coordinates of load point $\boldsymbol{\xi}$ for the two systems are, for element J,

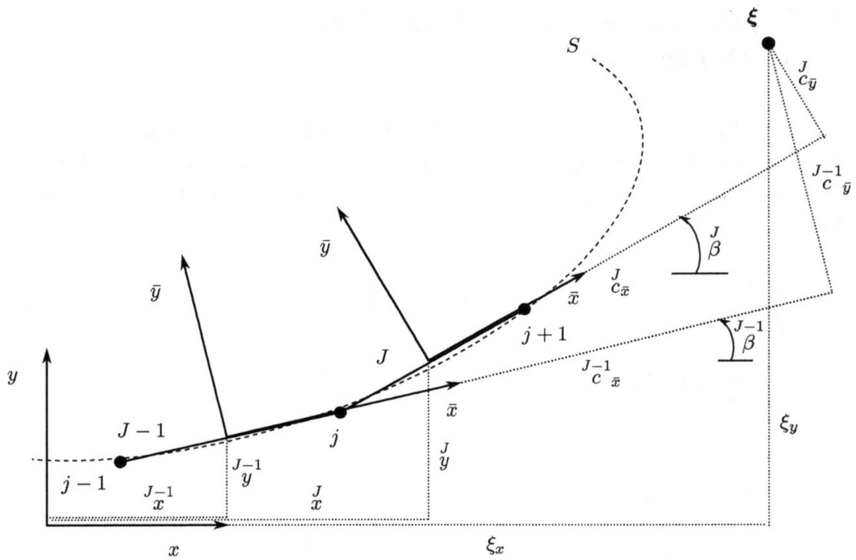

Fig. 7.4 Geometry for boundary elements $J-1$ and J

$$
\begin{aligned}
\overset{J}{c}_{\overline{x}} &= (\xi_x - \overset{J}{x})\cos\overset{J}{\beta} + (\xi_y - \overset{J}{y})\sin\overset{J}{\beta} \\
\overset{J}{c}_{\overline{y}} &= -(\xi_x - \overset{J}{x})\sin\overset{J}{\beta} + (\xi_y - \overset{J}{y})\cos\overset{J}{\beta}
\end{aligned}
\tag{7.45}
$$

and, for element $J-1$,

$$
\begin{aligned}
\overset{J-1}{c}_{\overline{x}} &= (\xi_x - \overset{J-1}{x})\cos\overset{J-1}{\beta} + (\xi_y - \overset{J-1}{y})\sin\overset{J-1}{\beta} \\
\overset{J-1}{c}_{\overline{y}} &= -(\xi_x - \overset{J-1}{x})\sin\overset{J-1}{\beta} + (\xi_y - \overset{J-1}{y})\cos\overset{J-1}{\beta}
\end{aligned}
\tag{7.46}
$$

Also, we denote the half-lengths of the two elements as $\overset{J}{a}$ and $\overset{J-1}{a}$.

The discretized version of (7.1) will now involve integrals of the form (see (7.42))

$$
\begin{aligned}
\left(\mathbb{U}_{xx}(\overset{j}{\boldsymbol{x}},\boldsymbol{\xi})\right)_J &= \int_{\overset{J}{a}}^{\overset{J}{a}} U_{xx}(\bar{x}-\overset{J}{c}_{\bar{x}},\overset{J}{c}_{\bar{y}})\psi_1(\bar{x})\,\mathrm{d}\bar{x} \\
&= \frac{1}{2\mu}\left[(3-4\nu)(\overset{\psi_1}{I_1})_J - \overset{J}{c}_{\bar{y}}\left(\sin 2\overset{J}{\beta}(\overset{\psi_1}{I_2})_J + \cos 2\overset{J}{\beta}(\overset{\psi_1}{I_3})_J\right)\right]
\end{aligned}
\tag{7.47}
$$

for element J, and

$$\left(\mathbb{U}_{xx}(\overset{j}{\boldsymbol{x}},\boldsymbol{\xi})\right)_{J-1} = \int_{\overset{J-1}{-a}}^{\overset{J-1}{a}} U_{xx}(\bar{x} - \overset{J-1}{c_{\bar{x}}}, \overset{J-1}{c_{\bar{y}}}) \overset{J-1}{\psi_2}(\bar{x})\,\mathrm{d}\bar{x}$$
$$= \frac{1}{2\mu}\left[(3-4\nu)(\overset{\psi_2}{I_1})_{J-1} - \overset{J-1}{c}_{\bar{y}}\left(\sin 2\overset{J-1}{\beta}(\overset{\psi_2}{I_2})_{J-1} + \cos 2\overset{J-1}{\beta}(\overset{\psi_2}{I_3})_{J-1}\right)\right] \tag{7.48}$$

for element $J-1$. The reader should confirm that the required expressions are, using (7.30), (7.32), and (7.33), together with similar equations for element $J-1$,

$$\begin{aligned} u_x(\boldsymbol{\xi}) = & \sum_{j=1}^{N}\left[\left(\mathbb{U}_{xx}(\overset{j}{\boldsymbol{x}},\boldsymbol{\xi})\right)_J(\overset{j}{t_x})_J + \left(\mathbb{U}_{xy}(\overset{j}{\boldsymbol{x}},\boldsymbol{\xi})\right)_J(\overset{j}{t_y})_J\right] \\ & - \sum_{j=1}^{N}\left[\left(\mathbb{T}_{xx}(\overset{j}{\boldsymbol{x}},\boldsymbol{\xi})\right)_J(\overset{j}{u_x})_J + \left(\mathbb{T}_{xy}(\overset{j}{\boldsymbol{x}},\boldsymbol{\xi})\right)_J(\overset{j}{u_y})_J\right] \\ & + \sum_{j=1}^{N}\left[\left(\mathbb{U}_{xx}(\overset{j}{\boldsymbol{x}},\boldsymbol{\xi})\right)_{J-1}(\overset{j}{t_x})_{J-1} + \left(\mathbb{U}_{xy}(\overset{j}{\boldsymbol{x}},\boldsymbol{\xi})\right)_{J-1}(\overset{j}{t_y})_{J-1}\right] \\ & - \sum_{j=1}^{N}\left[\left(\mathbb{T}_{xx}(\overset{j}{\boldsymbol{x}},\boldsymbol{\xi})\right)_{J-1}(\overset{j}{u_x})_{J-1} + \left(\mathbb{T}_{xy}(\overset{j}{\boldsymbol{x}},\boldsymbol{\xi})\right)_{J-1}(\overset{j}{u_y})_{J-1}\right] \end{aligned} \tag{7.49}$$

$$\begin{aligned} u_y(\boldsymbol{\xi}) = & \sum_{j=1}^{N}\left[\left(\mathbb{U}_{yx}(\overset{j}{\boldsymbol{x}},\boldsymbol{\xi})\right)_J(\overset{j}{t_x})_J + \left(\mathbb{U}_{yy}(\overset{j}{\boldsymbol{x}},\boldsymbol{\xi})\right)_J(\overset{j}{t_y})_J\right] \\ & - \sum_{j=1}^{N}\left[\left(\mathbb{T}_{yx}(\overset{j}{\boldsymbol{x}},\boldsymbol{\xi})\right)_J(\overset{j}{u_x})_J + \left(\mathbb{T}_{yy}(\overset{j}{\boldsymbol{x}},\boldsymbol{\xi})\right)_J(\overset{j}{u_y})_J\right] \\ & + \sum_{j=1}^{N}\left[\left(\mathbb{U}_{yx}(\overset{j}{\boldsymbol{x}},\boldsymbol{\xi})\right)_{J-1}(\overset{j}{t_x})_{J-1} + \left(\mathbb{U}_{yy}(\overset{j}{\boldsymbol{x}},\boldsymbol{\xi})\right)_{J-1}(\overset{j}{t_y})_{J-1}\right] \\ & - \sum_{j=1}^{N}\left[\left(\mathbb{T}_{yx}(\overset{j}{\boldsymbol{x}},\boldsymbol{\xi})\right)_{J-1}(\overset{j}{u_x})_{J-1} + \left(\mathbb{T}_{yy}(\overset{j}{\boldsymbol{x}},\boldsymbol{\xi})\right)_{J-1}(\overset{j}{u_y})_{J-1}\right] \end{aligned}$$

Following the general approach described in Sect. 7.2, we let point $\boldsymbol{\xi}$ be one of the nodal points on the discretized boundary contour, say point $\overset{i}{\boldsymbol{x}}$. (We suppose for the moment that $\overset{i}{\boldsymbol{x}} \neq \overset{j-1}{\boldsymbol{x}}, \overset{j}{\boldsymbol{x}}$, or $\overset{j+1}{\boldsymbol{x}}$.) Then, starting from (7.49), letting $u_x(\overset{i}{\boldsymbol{x}}) = \overset{i}{u_x}$ and $u_y(\overset{i}{\boldsymbol{x}}) = \overset{i}{u_y}$, and using (7.35), we obtain

$$
\begin{aligned}
\overset{i}{u}_x = & \sum_{j=1}^{N}\Big[\big(\mathbb{U}_{xx}(\overset{j}{\boldsymbol{x}},\overset{i}{\boldsymbol{x}})\big)_J(\overset{j}{t}_x)_J + \big(\mathbb{U}_{xy}(\overset{j}{\boldsymbol{x}},\overset{i}{\boldsymbol{x}})\big)_J(\overset{j}{t}_y)_J\Big] - \sum_{j=1}^{N}\Big[\big(\mathbb{T}_{xx}(\overset{j}{\boldsymbol{x}},\overset{i}{\boldsymbol{x}})_J\big)\overset{j}{u}_x + \big(\mathbb{T}_{xy}(\overset{j}{\boldsymbol{x}},\overset{i}{\boldsymbol{x}})\big)_J\overset{j}{u}_y\Big] \\
& + \sum_{j=1}^{N}\Big[\big(\mathbb{U}_{xx}(\overset{j}{\boldsymbol{x}},\overset{i}{\boldsymbol{x}})\big)_{J-1}(\overset{j}{t}_x)_{J-1} + \big(\mathbb{U}_{xy}(\overset{j}{\boldsymbol{x}},\overset{i}{\boldsymbol{x}})\big)_{J-1}(\overset{j}{t}_y)_{J-1}\Big] \\
& - \sum_{j=1}^{N}\Big[\big(\mathbb{T}_{xx}(\overset{j}{\boldsymbol{x}},\overset{i}{\boldsymbol{x}})\big)_{J-1}\overset{j}{u}_x + \big(\mathbb{T}_{xy}(\overset{j}{\boldsymbol{x}},\overset{i}{\boldsymbol{x}})\big)_{J-1}\overset{j}{u}_y\Big]
\end{aligned}
\tag{7.50}
$$

$$
\begin{aligned}
\overset{i}{u}_y = & +\sum_{j=1}^{N}\Big[\big(\mathbb{U}_{yx}(\overset{j}{\boldsymbol{x}},\overset{i}{\boldsymbol{x}})\big)_J(\overset{j}{t}_x)_J + \big(\mathbb{U}_{yy}(\overset{j}{\boldsymbol{x}},\overset{i}{\boldsymbol{x}})\big)_J(\overset{j}{t}_y)_J\Big] - \sum_{j=1}^{N}\Big[\big(\mathbb{T}_{yx}(\overset{j}{\boldsymbol{x}},\overset{i}{\boldsymbol{x}})\big)_J\overset{j}{u}_x + \big(\mathbb{T}_{yy}(\overset{j}{\boldsymbol{x}},\overset{i}{\boldsymbol{x}})\big)_J\overset{j}{u}_y\Big] \\
& + \sum_{j=1}^{N}\Big[\big(\mathbb{U}_{yx}(\overset{j}{\boldsymbol{x}},\overset{i}{\boldsymbol{x}})\big)_{J-1}(\overset{j}{t}_x)_{J-1} + \big(\mathbb{U}_{yy}(\overset{j}{\boldsymbol{x}},\overset{i}{\boldsymbol{x}})\big)_{J-1}(\overset{j}{t}_y)_{J-1}\Big] \\
& - \sum_{j=1}^{N}\Big[\big(\mathbb{T}_{yx}(\overset{j}{\boldsymbol{x}},\overset{i}{\boldsymbol{x}})\big)_{J-1}\overset{j}{u}_x + \big(\mathbb{T}_{yy}(\overset{j}{\boldsymbol{x}},\overset{i}{\boldsymbol{x}})\big)_{J-1}\overset{j}{u}_y\Big]
\end{aligned}
$$

Writing $(\overset{ij}{\mathbb{U}}_{xx})_J$, etc., for $(\mathbb{U}_{xx}(\overset{j}{\boldsymbol{x}},\overset{i}{\boldsymbol{x}}))_J$, etc., these equations can be put in the form

$$
\begin{aligned}
& \sum_{j=1}^{N}\Big[(\overset{ij}{\mathbb{T}}_{xx})_J + (\overset{ij}{\mathbb{T}}_{xx})_{J-1} + \overset{ij}{\mathbb{I}}\Big]\overset{j}{u}_x + \sum_{j=1}^{N}\Big[(\overset{ij}{\mathbb{T}}_{xy})_J + (\overset{ij}{\mathbb{T}}_{xy})_{J-1}\Big]\overset{j}{u}_y \\
& \quad = \sum_{j=1}^{N}\Big[(\overset{ij}{\mathbb{U}}_{xx})_J(\overset{j}{t}_x)_J + (\overset{ij}{\mathbb{U}}_{xx})_{J-1}(\overset{j}{t}_x)_{J-1}\Big] + \sum_{j=1}^{N}\Big[(\overset{ij}{\mathbb{U}}_{xy})_J(\overset{j}{t}_y)_J + (\overset{ij}{\mathbb{U}}_{xy})_{J-1}(\overset{j}{t}_y)_{J-1}\Big]
\end{aligned}
\tag{7.51}
$$

$$
\begin{aligned}
& \sum_{j=1}^{N}\Big[(\overset{ij}{\mathbb{T}}_{yx})_J + (\overset{ij}{\mathbb{T}}_{yx})_{J-1}\Big]\overset{j}{u}_x + \sum_{j=1}^{N}\Big[(\overset{ij}{\mathbb{T}}_{yy})_J + (\overset{ij}{\mathbb{T}}_{yy})_{J-1} + \overset{ij}{\mathbb{I}}\Big]\overset{j}{u}_y \\
& \quad = \sum_{j=1}^{N}\Big[(\overset{ij}{\mathbb{U}}_{yx})_J(\overset{j}{t}_x)_J + (\overset{ij}{\mathbb{U}}_{yx})_{J-1}(\overset{j}{t}_x)_{J-1}\Big] + \sum_{j=1}^{N}\Big[(\overset{ij}{\mathbb{U}}_{yy})_J(\overset{j}{t}_y)_J + (\overset{ij}{\mathbb{U}}_{yy})_{J-1}(\overset{j}{t}_y)_{J-1}\Big].
\end{aligned}
$$

Structurally, Eqs. (7.51) are comparable to (7.10), and a numerical solution to a given boundary value problem proceeds in essentially the same manner as summarized in Sect. 7.2. Notice, however, that in the present case we have associated *six* boundary parameters with every nodal point: two displacement components and four traction components. In order to obtain a system of $2N$ algebraic equations in $2N$ unknowns from (7.51), we must therefore prescribe *four* parameters at each node. The simplest situation occurs for a traction boundary value problem, in which the tractions on both "sides" of every nodal point are given. Equations (7.51) can then be solved directly for the $2N$ unknown displacement components, $\overset{i}{u}_x$ and $\overset{i}{u}_y$ for $i = 1$ to N.

Mixed boundary conditions for any particular nodal point can be handled in a similar fashion. If, for example, the displacements $\overset{i}{u}_x$ and $\overset{i}{u}_y$ at nodal point i are prescribed, and if the tractions $(\overset{i}{t}_x)_I$ and $(\overset{i}{t}_y)_I$ on one side of this node are known, then the tractions $(\overset{i}{t}_x)_{I-1}$ and $(\overset{i}{t}_y)_{I-1}$ on the other side of the node are the unknowns for that point. If none of the tractions associated with nodal point i are known, a solution can still be found by assuming that these tractions are continuous, i.e., $(\overset{i}{t}_x)_{I-1} = (\overset{i}{t}_x)_I$ and $(\overset{i}{t}_y)_{I-1} = (\overset{i}{t}_y)_I$. Although this assumption is good in many cases, it is open to question when elements I and $I-1$ have different inclinations. It should be noted, however, that many applications—especially in geomechanics—involve traction boundary value problems for which the tractions on both sides of the nodes are known.

7.10 Special Cases for Integrals of Kernel-Shape Function Products

The integrated kernel-shape function products $(\overset{ij}{\mathbb{U}}_{xx})_J$, etc., in Eqs. (7.51) can be evaluated directly within a computer program by using definitions (7.42) and (7.43), together with Eqs. (7.45) and (7.46), noting that ξ_x and ξ_y in the latter equations are to be replaced by $\overset{i}{x}$ and $\overset{i}{y}$, respectively. Special attention must be paid, however, to situations in which the load point $\overset{i}{\boldsymbol{x}}$ coincides with one of the three nodal points involved in the integrals over elements $J-1$ and J (i.e., the sample point $\overset{j}{\boldsymbol{x}}$ and neighboring points $\overset{j-1}{\boldsymbol{x}}$ and $\overset{j+1}{\boldsymbol{x}}$), as depicted in Fig. 7.5. At these three locations, we use Eqs. (7.49) and consider the limiting cases for which the load point $\boldsymbol{\xi}$ approaches the boundary from *inside* the region of interest. (A comparable calculation is given by Crouch and Starfield (1983) for the case that $\boldsymbol{\xi} = \overset{i}{\boldsymbol{x}}$ approaches the boundary from *outside* the region of interest.)

Before examining these limiting cases, it is helpful to note from Fig. 7.5 that for both element J and element $J-1$ the values of $c_{\overline{x}}$ and $c_{\overline{y}}$ can be expressed in terms of $\bar{r}_1$ and $\bar{\theta}_1$ as follows:

$$
\begin{aligned}
c_{\overline{x}} &= a + \bar{r}_1 \cos \bar{\theta}_1 \\
c_{\overline{y}} &= \bar{r}_1 \sin \bar{\theta}_1
\end{aligned}
\tag{7.52}
$$

in which a is one-half the length of the element. Equations (7.52) are used when evaluating integrals (7.40) in the near vicinity of the right endpoint of an element, where $\bar{r}_2 \approx 2a$ and $\bar{\theta}_2 \approx 0$.

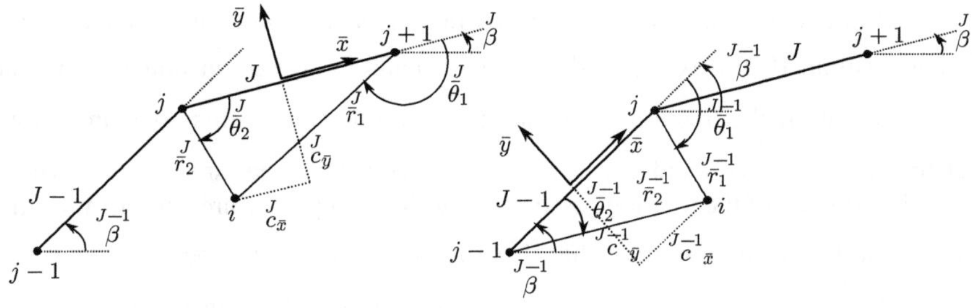

(a) Element J; node $i \rightarrow$ node $j+1$ (b) Element $J-1$; node $i \rightarrow$ node j

Fig. 7.5 Parameters for integration formulas, elements J and $J-1$

Similarly, $c_{\bar{x}}$ and $c_{\bar{y}}$ can be expressed in terms of $\bar{r}_2$ and $\bar{\theta}_2$ as

$$\begin{aligned} c_{\bar{x}} &= -a + \bar{r}_2 \cos \bar{\theta}_2 \\ c_{\bar{y}} &= \bar{r}_2 \sin \bar{\theta}_2 \end{aligned} \tag{7.53}$$

Equations (7.53) are used in the near vicinity of the left endpoint of an element, where $\bar{r}_1 \approx 2a$ and $\bar{\theta}_1 \approx -\pi$. (Recall that $-\pi \leq \bar{\theta}_1 \leq +\pi$. A value of $+\pi$ would be used for $\bar{\theta}_1$ if we wanted to approach the element endpoint from *outside* the region of interest.)

Considering first the limiting case in which $\boldsymbol{\xi} \rightarrow \overset{j+1}{\boldsymbol{x}}$ (or $i = j+1$), Fig. 7.5a, we use (7.52) and write $\overset{J}{c}_{\bar{x}} = \overset{J}{a} + \varepsilon \cos \overset{J}{\bar{\theta}}_1$ and $\overset{J}{c}_{\bar{y}} = \varepsilon \sin \overset{J}{\bar{\theta}}_1$, where $\overset{J}{\bar{r}}_1 = \varepsilon$. In the limit, as $\varepsilon \rightarrow 0$, we have $\overset{J}{c}_{\bar{x}} = \overset{J}{a}$, $\overset{J}{c}_{\bar{y}} = 0$, $\overset{J}{\bar{\theta}}_2 = 0$, and $\overset{J}{\bar{r}}_2 = 2\overset{J}{a}$. Substituting these limiting values into (7.40), we obtain the following results for the integrals $(\overset{\psi_1}{I_1})_J$ through $(\overset{\psi_1}{I_5})_J$:

$$\begin{aligned} (\overset{\psi_1}{I_1})_J &= -\frac{\overset{J}{a}}{4\pi(1-\nu)}\left[\ln(2\overset{J}{a}) - \frac{1}{2}\right] \\ (\overset{\psi_1}{I_2})_J &= \frac{1}{4\pi(1-\nu)} \\ (\overset{\psi_1}{I_3})_J &= 0 \\ \overset{J}{c}_{\bar{y}}\,(\overset{\psi_1}{I_4})_J &= 0 \\ \overset{J}{c}_{\bar{y}}\,(\overset{\psi_1}{I_5})_J &= 0. \end{aligned} \tag{7.54}$$

Notice that we have given limiting values of the *products* $\overset{J}{c}_{\overline{y}}(\overset{\psi_1}{I_4})_J$ and $\overset{J}{c}_{\overline{y}}(\overset{\psi_1}{I_5})_J$. This is done because, as shown in (7.43), the terms $(\overset{\psi_1}{I_4})_J$ and $(\overset{\psi_1}{I_5})_J$ always appear in the analysis as products with $\overset{J}{c}_{\overline{y}}$. Notice also that we have used the relation $\varepsilon \ln \varepsilon = 0$ as $\varepsilon \to 0$ in treating limits of expressions involving the quantity $\ln \overset{J}{\bar{r}}_1 = \ln \varepsilon$.

Similarly, when $\boldsymbol{\xi} \to \overset{j-1}{\boldsymbol{x}}$ (or $i = j - 1$), Fig. 7.5b, we use (7.53) to obtain $\overset{J-1}{c}_{\overline{x}} = -\overset{J-1}{a} + \varepsilon \cos \overset{J-1}{\bar{\theta}_2}$ and $\overset{J-1}{c}_{\overline{y}} = \varepsilon \sin \overset{J-1}{\bar{\theta}_2}$, where now $\overset{J-1}{\bar{r}_2} = \varepsilon$. In the limit, as $\varepsilon \to 0$, we have $\overset{J-1}{c}_{\overline{x}} = -\overset{J-1}{a}$, $\overset{J-1}{c}_{\overline{y}} = 0$, $\overset{J-1}{\bar{\theta}}_1 = -\pi$, and $\overset{J-1}{\bar{r}_1} = 2\overset{J-1}{a}$. Using these limiting values in (7.41) with J replaced by $J - 1$, we obtain the following results for the integrals $\overset{\psi_2}{I_1}$ through $\overset{\psi_2}{I_5}$:

$$
\begin{aligned}
(\overset{\psi_2}{I_1})_{J-1} &= -\frac{\overset{J-1}{a}}{4\pi(1-\nu)}\left[\ln(2\overset{J-1}{a}) - \frac{1}{2}\right] \\
(\overset{\psi_2}{I_2})_{J-1} &= -\frac{1}{4\pi(1-\nu)} \\
(\overset{\psi_2}{I_3})_{J-1} &= 0 \\
\overset{J-1}{c}_{\overline{y}}\,(\overset{\psi_2}{I_4})_{J-1} &= 0 \\
\overset{J-1}{c}_{\overline{y}}\,(\overset{\psi_2}{I_5})_{J-1} &= 0
\end{aligned}
\tag{7.55}
$$

When $\boldsymbol{\xi} \to \overset{j}{\boldsymbol{x}}$ (i.e., when nodes i and j are coincident), we consider contributions from both elements $J - 1$ and J. This is depicted in Fig. 7.6 where, for convenience, we approach node j along a line of constant inclination bisecting the angle between elements $J - 1$ and J. This line makes an angle of $\frac{1}{2}(\pi - \gamma)$ with both element I and element $J - 1$, where $\gamma = \overset{J-1}{\beta} - \overset{J}{\beta}$ is the difference of the inclinations of the two elements. The distance from the load point $\boldsymbol{\xi}$ to the sample point $\overset{j}{\boldsymbol{x}}$ is ε, and we seek the limiting values of the first five expressions in (7.40) and (7.41) as $\varepsilon \to 0$.

To evaluate the contributions from element J, Fig. 7.6a, we use (7.53) and write $\overset{J}{c}_{\overline{x}} = -\overset{J}{a} + \varepsilon \cos \overset{J}{\bar{\theta}}_2$ and $\overset{J}{c}_{\overline{y}} = \varepsilon \sin \overset{J}{\bar{\theta}}_2$, where $\overset{J}{\bar{r}}_2 = \varepsilon$. In addition, when $\varepsilon \to 0$ we have that $\overset{J}{\bar{r}}_1 = 2\overset{J}{a}$ and $\overset{J}{\bar{\theta}}_1 = -\pi$. Using these values in (7.40), setting $\overset{J}{\bar{\theta}}_2 = -\frac{1}{2}(\pi - \gamma)$, and taking limits as $\varepsilon \to 0$, we obtain the following results:

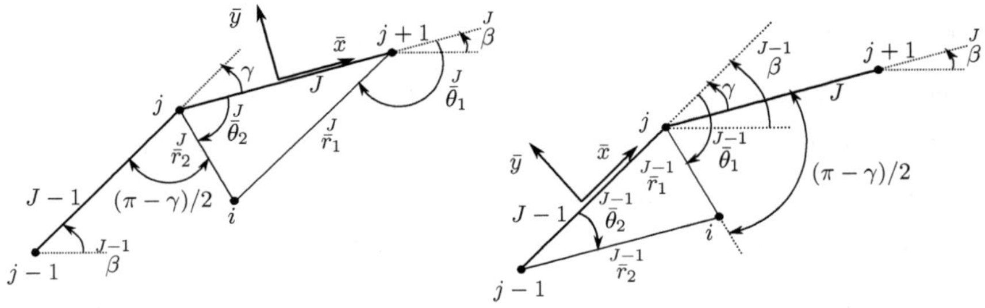

(a) Element J; node $i \to$ node j (b) Element J–1; node $i \to$ node j

Fig. 7.6 Geometry used to evaluate self-effect terms for nodal point j

$$\begin{aligned}
\overset{\psi_1}{(I_1)}_J &= -\frac{\overset{J}{a}}{4\pi(1-\nu)}\left(\ln(2\overset{J}{a}) - \frac{3}{2}\right)\\
\overset{\psi_1}{(I_2)}_J &= -\frac{1}{4\pi(1-\nu)}\left(\ln(2\overset{J}{a}) - 1\right) + \frac{1}{4\pi(1-\nu)}\ln\varepsilon\\
\overset{\psi_1}{(I_3)}_J &= -\frac{1}{8\pi(1-\nu)}(\pi+\gamma)\\
\overset{J}{c_{\bar{y}}}\,\overset{\psi_1}{(I_4)}_J &= -\frac{1}{8\pi(1-\nu)}(1+\cos\gamma)\\
\overset{J}{c_{\bar{y}}}\,\overset{\psi_1}{(I_5)}_J &= \frac{1}{8\pi(1-\nu)}\sin\gamma
\end{aligned} \tag{7.56}$$

With the exception of the singular term $\ln\varepsilon$ in the expression for $\overset{\psi_1}{(I_2)}_J$, all of these values are finite. It can be shown from (7.42), however, that coefficients $\overset{jj}{(\mathbb{U}_{xx})}_J$, etc., contain the quantity $\overset{\psi_1}{(I_2)}_J$ only as a product with $\overset{J}{c_{\bar{y}}}$ $(= \varepsilon \sin \overset{J}{\bar{\theta}_2})$, and, as noted above, $\varepsilon \ln \varepsilon = 0$ as $\varepsilon \to 0$. It can also be shown from (7.43) that $\overset{\psi_1}{(I_2)}_J$ appears alone in the coefficients $\overset{jj}{(\mathbb{T}_{yx})}_J$ and $\overset{jj}{(\mathbb{T}_{xy})}_J$. This does not cause any computational difficulties either, because, as we will soon see, the singular term contributed by element J is canceled by a corresponding singular term from element $J-1$.

Similarly, the contributions from element $J-1$, Fig. 7.6b, are found by using (7.52) and setting $\overset{J-1}{c_{\bar{x}}} = \overset{J-1}{a} + \varepsilon \cos \overset{J-1}{\bar{\theta}}_1$ and $\overset{J-1}{c_{\bar{y}}} = \varepsilon \sin \overset{J-1}{\bar{\theta}}_1$, where now $\overset{J-1}{\bar{r}_1} = \varepsilon$. In the limit, as $\varepsilon \to 0$, we have $\overset{J-1}{\bar{r}_2} = 2a$ and $\overset{J-1}{\bar{\theta}}_2 = 0$. Using these values in (7.41), again with J replaced by $J-1$, setting $\overset{J-1}{\bar{\theta}}_1 = -\frac{1}{2}(\pi+\gamma)$, and taking limits as $\varepsilon \to 0$, we obtain the following results:

$$\begin{aligned}
\overset{\psi_2}{(I_1)}_{J-1} &= -\frac{\overset{J-1}{a}}{4\pi(1-\nu)}\left(\ln(2\overset{J-1}{a}) - \frac{3}{2}\right)\\
\overset{\psi_2}{(I_2)}_{J-1} &= \frac{1}{4\pi(1-\nu)}\left(\ln(2\overset{J-1}{a}) - 1\right) - \frac{1}{4\pi(1-\nu)}\ln\varepsilon\\
\overset{\psi_2}{(I_3)}_{J-1} &= -\frac{1}{8\pi(1-\nu)}(\pi+\gamma)\\
\overset{J-1}{c_{\bar{y}}}\overset{\psi_2}{(I_4)}_{J-1} &= \frac{1}{8\pi(1-\nu)}(1+\cos\gamma)\\
\overset{J-1}{c_{\bar{y}}}\overset{\psi_2}{(I_5)}_{J-1} &= \frac{1}{8\pi(1-\nu)}\sin\gamma.
\end{aligned} \tag{7.57}$$

7.10.1 Self-effect Coefficients

The self-effect influence coefficients $\overset{jj}{(\mathbb{U}_{xx})}_J$, etc., and $\overset{jj}{(\mathbb{T}_{xx})}_J$, etc., can now be obtained by using (7.56) and (7.57) in (7.42) and (7.43), with appropriate notational changes. The reader should verify that the results are as follows:

$$\begin{aligned}
\overset{jj}{(\mathbb{U}_{xx})}_J &= -\frac{(3-4\nu)\overset{J}{a}}{8\pi\mu(1-\nu)}\left(\ln(2\overset{J}{a}) - \frac{3}{2}\right)\\
\overset{jj}{(\mathbb{U}_{yx})}_J &= \overset{jj}{(\mathbb{U}_{xy})}_J = 0\\
\overset{jj}{(\mathbb{U}_{yy})}_J &= -\frac{(3-4\nu)\overset{J}{a}}{8\pi\mu(1-\nu)}\left(\ln(2\overset{J}{a}) - \frac{3}{2}\right)\\
\overset{jj}{(\mathbb{U}_{xx})}_{J-1} &= -\frac{(3-4\nu)\overset{J-1}{a}}{8\pi\mu(1-\nu)}\left(\ln(2\overset{J-1}{a}) - \frac{3}{2}\right)\\
\overset{jj}{(\mathbb{U}_{yx})}_{J-1} &= \overset{jj}{(\mathbb{U}_{xy})}_{J-1} = 0\\
\overset{jj}{(\mathbb{U}_{yy})}_{J-1} &= -\frac{(3-4\nu)\overset{J-1}{a}}{8\pi\mu(1-\nu)}\left(\ln(2\overset{J-1}{a}) - \frac{3}{2}\right)
\end{aligned} \tag{7.58}$$

and

$$
\begin{aligned}
\overset{jj}{\mathbb{T}}_{xx} &= (\overset{jj}{\mathbb{T}}_{xx})_J + (\overset{jj}{\mathbb{T}}_{xx})_{J-1} + \overset{jj}{\mathbb{I}} = \frac{1}{2} - \frac{\gamma}{2\pi} - \frac{1}{4\pi(1-\nu)} \sin\gamma \cos(\overset{J-1}{\beta} + \overset{J}{\beta}) \\
\overset{jj}{\mathbb{T}}_{yx} &= (\overset{jj}{\mathbb{T}}_{yx})_J + (\overset{jj}{\mathbb{T}}_{yx})_{J-1} = \frac{(1-2\nu)}{4\pi(1-\nu)} \ln \frac{\overset{J-1}{a}}{\overset{J}{a}} - \frac{1}{4\pi(1-\nu)} \sin\gamma \sin(\overset{J-1}{\beta} + \overset{J}{\beta})
\end{aligned}
\tag{7.59}
$$

$$
\begin{aligned}
\overset{jj}{\mathbb{T}}_{xy} &= (\overset{jj}{\mathbb{T}}_{xy})_J + (\overset{jj}{\mathbb{T}}_{xy})_{J-1} = -\frac{(1-2\nu)}{4\pi(1-\nu)} \ln \frac{\overset{J-1}{a}}{\overset{J}{a}} - \frac{1}{4\pi(1-\nu)} \sin\gamma \sin(\overset{J-1}{\beta} + \overset{J}{\beta}) \\
\overset{jj}{\mathbb{T}}_{yy} &= (\overset{jj}{\mathbb{T}}_{yy})_J + (\overset{jj}{\mathbb{T}}_{yy})_{J-1} + \overset{jj}{\mathbb{I}} = \frac{1}{2} - \frac{\gamma}{2\pi} + \frac{1}{4\pi(1-\nu)} \sin\gamma \cos(\overset{J-1}{\beta} + \overset{J}{\beta}).
\end{aligned}
$$

7.11 Approximation of Somigliana's Formulas for Stresses

Expressions for the stress components at point $\boldsymbol{\xi}$ can be developed in an analogous way as was done for the displacements in Sect. 7.8. Equation (7.22) is the basis for this approach.

Proceeding as in Sect. 7.6, we write this equation in unabridged form for a boundary contour divided into N straight line elements as follows:

$$
\begin{aligned}
\sigma_{xx}(\boldsymbol{\xi}) &= \sum_{J=1}^{N} \int_{-\overset{J}{a}}^{\overset{J}{a}} \left[\overset{J}{t}_x(\overline{x}) D_{xxx}(\overline{x} - c_{\overline{x}}, c_{\overline{y}}) + \overset{J}{t}_y(\overline{x}) D_{yxx}(\overline{x} - c_{\overline{x}}, c_{\overline{y}}) \right] \mathrm{d}\overline{x} \\
&\quad - \sum_{J=1}^{N} \int_{-\overset{J}{a}}^{\overset{J}{a}} \left[\overset{J}{u}_x(\overline{x}) S_{xxx}(\overline{x} - c_{\overline{x}}, c_{\overline{y}}) + \overset{J}{u}_y(\overline{x}) S_{yxx}(\overline{x} - c_{\overline{x}}, c_{\overline{y}}) \right] \mathrm{d}\overline{x} \\
\sigma_{yy}(\boldsymbol{\xi}) &= \sum_{J=1}^{N} \int_{-\overset{J}{a}}^{\overset{J}{a}} \left[\overset{J}{t}_x(\overline{x}) D_{xyy}(\overline{x} - c_{\overline{x}}, c_{\overline{y}}) + \overset{J}{t}_y(\overline{x}) D_{yyy}(\overline{x} - c_{\overline{x}}, c_{\overline{y}}) \right] \mathrm{d}\overline{x} \\
&\quad - \sum_{J=1}^{N} \int_{-\overset{J}{a}}^{\overset{J}{a}} \left[\overset{J}{u}_x(\overline{x}) S_{xyy}(\overline{x} - c_{\overline{x}}, c_{\overline{y}}) + \overset{J}{u}_y(\overline{x}) S_{yyy}(\overline{x} - c_{\overline{x}}, c_{\overline{y}}) \right] \mathrm{d}\overline{x} \qquad (7.60) \\
\sigma_{xy}(\boldsymbol{\xi}) &= \sum_{J=1}^{N} \int_{-\overset{J}{a}}^{\overset{J}{a}} \left[\overset{J}{t}_x(\overline{x}) D_{xxy}(\overline{x} - c_{\overline{x}}, c_{\overline{y}}) + \overset{J}{t}_y(\overline{x}) D_{yxy}(\overline{x} - c_{\overline{x}}, c_{\overline{y}}) \right] \mathrm{d}\overline{x} \\
&\quad - \sum_{J=1}^{N} \int_{-\overset{J}{a}}^{\overset{J}{a}} \left[\overset{J}{u}_x(\overline{x}) S_{xxy}(\overline{x} - c_{\overline{x}}, c_{\overline{y}}) + \overset{J}{u}_y(\overline{x}) S_{yxy}(\overline{x} - c_{\overline{x}}, c_{\overline{y}}) \right] \mathrm{d}\overline{x}
\end{aligned}
$$

Substituting definitions (7.32) and (7.33) into these expressions, it can be seen that the results involve integrals of the form

$$\int_{-\overset{J}{a}}^{\overset{J}{a}} D_{pqr}(\overline{x} - c_{\overline{x}}, c_{\overline{y}})\overset{J}{\psi}_m(\overline{x})\,\mathrm{d}\overline{x} \tag{7.61}$$

and

$$\int_{-\overset{J}{a}}^{\overset{J}{a}} S_{pqr}(\overline{x} - c_{\overline{x}}, c_{\overline{y}})\overset{J}{\psi}_m(\overline{x})\,\mathrm{d}\overline{x} \tag{7.62}$$

where p, q, and r are either x or y, and m is 1 or 2. As before, all of these integrals can be evaluated analytically. The individual terms of $D_{pqr}(\overline{x} - c_{\overline{x}}, \overline{y} - c_{\overline{y}})$ and $S_{pqr}(\overline{x} - c_{\overline{x}}, \overline{y} - c_{\overline{y}})$ are given by (7.25) and (7.26) with $\overset{j}{\beta}$ replaced by $\overset{J}{\beta}$.

Combining (7.39), (7.61), and (7.25), we obtain the following expressions:

$$\begin{aligned}
\overset{\psi_m}{\mathbb{D}}_{xxx}(\overset{J}{\boldsymbol{x}}, \boldsymbol{\xi}) &= \int_{-\overset{J}{a}}^{\overset{J}{a}} D_{xxx}(\overline{x} - c_{\overline{x}}, c_{\overline{y}})\overset{J}{\psi}_m(\overline{x})\,\mathrm{d}\overline{x} \\
&= -2(1-\nu)\Big(\cos\overset{J}{\beta}(\overset{\psi_m}{I_2})_J - \sin\overset{J}{\beta}(\overset{\psi_m}{I_3})_J\Big) - \cos\overset{J}{\beta}\Big(\cos 2\overset{J}{\beta}(\overset{\psi_m}{I_2})_J - \sin 2\overset{J}{\beta}(\overset{\psi_m}{I_3})_J\Big) \\
&\quad + c_{\overline{y}}\Big(\cos 3\overset{J}{\beta}(\overset{\psi_m}{I_4})_J + \sin 3\overset{J}{\beta}(\overset{\psi_m}{I_5})_J\Big) \\
\overset{\psi_m}{\mathbb{D}}_{xyy}(\overset{J}{\boldsymbol{x}}, \boldsymbol{\xi}) &= \int_{-\overset{J}{a}}^{\overset{J}{a}} D_{xyy}(\overline{x} - c_{\overline{x}}, c_{\overline{y}})\overset{J}{\psi}_m(\overline{x})\,\mathrm{d}\overline{x} \\
&= -2\nu\Big(\cos\overset{J}{\beta}(\overset{\psi_m}{I_2})_J - \sin\overset{J}{\beta}(\overset{\psi_m}{I_3})_J\Big) + \cos\overset{J}{\beta}\Big(\cos 2\overset{J}{\beta}(\overset{\psi_m}{I_2})_J - \sin 2\overset{J}{\beta}(\overset{\psi_m}{I_3})_J\Big) \\
&\quad - c_{\overline{y}}\Big(\cos 3\overset{J}{\beta}(\overset{\psi_m}{I_4})_J + \sin 3\overset{J}{\beta}(\overset{\psi_m}{I_5})_J\Big) \\
\overset{\psi_m}{\mathbb{D}}_{xxy}(\overset{J}{\boldsymbol{x}}, \boldsymbol{\xi}) &= \int_{-\overset{J}{a}}^{\overset{J}{a}} D_{xxy}(\overline{x} - c_{\overline{x}}, c_{\overline{y}})\overset{J}{\psi}_m(\overline{x})\,\mathrm{d}\overline{x} \\
&= -(1-2\nu)\Big(\sin\overset{J}{\beta}(\overset{\psi_m}{I_2})_J + \cos\overset{J}{\beta}(\overset{\psi_m}{I_3})_J\Big) - \cos\overset{J}{\beta}\Big(\sin 2\overset{J}{\beta}(\overset{\psi_m}{I_2})_J + \cos 2\overset{J}{\beta}(\overset{\psi_m}{I_3})_J\Big) \\
&\quad + c_{\overline{y}}\Big(\sin 3\overset{J}{\beta}(\overset{\psi_m}{I_4})_J - \cos 3\overset{J}{\beta}(\overset{\psi_m}{I_5})_J\Big)
\end{aligned} \tag{7.63}$$

$$\begin{aligned}
\overset{\psi_m}{\mathbb{D}}_{yxx}(\overset{J}{\boldsymbol{x}}, \boldsymbol{\xi}) &= \int_{-\overset{J}{a}}^{\overset{J}{a}} D_{yxx}(\overline{x} - c_{\overline{x}}, c_{\overline{y}})\overset{J}{\psi}_m(\overline{x})\,\mathrm{d}\overline{x} \\
&= -2\nu\Big(\sin\overset{J}{\beta}(\overset{\psi_m}{I_2})_J + \cos\overset{J}{\beta}(\overset{\psi_m}{I_3})_J\Big) - \sin\overset{J}{\beta}\Big(\cos 2\overset{J}{\beta}(\overset{\psi_m}{I_2})_J - \sin 2\overset{J}{\beta}(\overset{\psi_m}{I_3})_J\Big) \\
&\quad + c_{\overline{y}}\Big(\sin 3\overset{J}{\beta}(\overset{\psi_m}{I_4})_J - \cos 3\overset{J}{\beta}(\overset{\psi_m}{I_5})_J\Big) \\
\overset{\psi_m}{\mathbb{D}}_{yyy}(\overset{J}{\boldsymbol{x}}, \boldsymbol{\xi}) &= \int_{-\overset{J}{a}}^{\overset{J}{a}} D_{yyy}(\overline{x} - c_{\overline{x}}, c_{\overline{y}})\overset{J}{\psi}_m(\overline{x})\,\mathrm{d}\overline{x} \\
&= -2(1-\nu)\Big(\sin\overset{J}{\beta}(\overset{\psi_m}{I_2})_J + \cos\overset{J}{\beta}(\overset{\psi_m}{I_3})_J\Big) + \sin\overset{J}{\beta}\Big(\cos 2\overset{J}{\beta}(\overset{\psi_m}{I_2})_J - \sin 2\overset{J}{\beta}(\overset{\psi_m}{I_3})_J\Big) \\
&\quad - c_{\overline{y}}\Big(\sin 3\overset{J}{\beta}(\overset{\psi_m}{I_4})_J - \cos 3\overset{J}{\beta}(\overset{\psi_m}{I_5})_J\Big)
\end{aligned}$$

$$\begin{aligned}
\overset{\psi_m}{\mathbb{D}}_{yxy}(\overset{J}{\boldsymbol{x}},\boldsymbol{\xi}) &= \int_{-\overset{J}{a}}^{\overset{J}{a}} D_{yxy}(\overline{x}-c_{\overline{x}}, c_{\overline{y}})\overset{J}{\psi}_m(\overline{x})\,\mathrm{d}\overline{x} \\
&= -(1-2\nu)\Big(\cos\overset{J}{\beta}(\overset{\psi_m}{I_2})_J - \sin\overset{J}{\beta}(\overset{\psi_m}{I_3})_J\Big) - \sin\overset{J}{\beta}\Big(\sin 2\overset{J}{\beta}(\overset{\psi_m}{I_2})_J + \cos 2\overset{J}{\beta}(\overset{\psi_m}{I_3})_J\Big) \\
&\quad - c_{\overline{y}}\Big(\cos 3\overset{J}{\beta}(\overset{\psi_m}{I_4})_J + \sin 3\overset{J}{\beta}(\overset{\psi_m}{I_5})_J\Big)
\end{aligned}$$

Likewise, the integrated quantities specified in (7.62) can be written as (see (7.26))

$$\begin{aligned}
\overset{\psi_m}{\mathbb{S}}_{xxx}(\overset{J}{\boldsymbol{x}},\boldsymbol{\xi}) &= \int_{-\overset{J}{a}}^{\overset{J}{a}} S_{xxx}(\overline{x}-c_{\overline{x}}, c_{\overline{y}})\overset{J}{\psi}_m(\overline{x})\,\mathrm{d}\overline{x} \\
&= 2\mu\Big[-\cos\overset{J}{\beta}(\overset{\psi_m}{I_4})_J - \sin\overset{J}{\beta}(\overset{\psi_m}{I_5})_J \\
&\quad - \cos\overset{J}{\beta}\Big(\cos 2\overset{J}{\beta}(\overset{\psi_m}{I_4})_J + \sin 2\overset{J}{\beta}(\overset{\psi_m}{I_5})_J\Big) + c_{\overline{y}}\Big(\cos 3\overset{J}{\beta}(\overset{\psi_m}{I_6})_J + \sin 3\overset{J}{\beta}(\overset{\psi_m}{I_7})_J\Big)\Big] \\
\overset{\psi_m}{\mathbb{S}}_{xyy}(\overset{J}{\boldsymbol{x}},\boldsymbol{\xi}) &= \int_{-\overset{J}{a}}^{\overset{J}{a}} S_{xyy}(\overline{x}-c_{\overline{x}}, c_{\overline{y}})\overset{J}{\psi}_m(\overline{x})\,\mathrm{d}\overline{x} \\
&= 2\mu\Big[-\sin\overset{J}{\beta}\Big(\sin 2\overset{J}{\beta}(\overset{\psi_m}{I_4})_J - \cos 2\overset{J}{\beta}(\overset{\psi_m}{I_5})_J\Big) - c_{\overline{y}}\Big(\cos 3\overset{J}{\beta}(\overset{\psi_m}{I_6})_J + \sin 3\overset{J}{\beta}(\overset{\psi_m}{I_7})_J\Big)\Big] \\
\overset{\psi_m}{\mathbb{S}}_{xxy}(\overset{J}{\boldsymbol{x}},\boldsymbol{\xi}) &= \int_{-\overset{J}{a}}^{\overset{J}{a}} S_{xxy}(\overline{x}-c_{\overline{x}}, c_{\overline{y}})\overset{J}{\psi}_m(\overline{x})\,\mathrm{d}\overline{x} \\
&= 2\mu\Big[-\cos\overset{J}{\beta}\Big(\sin 2\overset{J}{\beta}(\overset{\psi_m}{I_4})_J - \cos 2\overset{J}{\beta}(\overset{\psi_m}{I_5})_J\Big) + c_{\overline{y}}\Big(\sin 3\overset{J}{\beta}(\overset{\psi_m}{I_6})_J - \cos 3\overset{J}{\beta}(\overset{\psi_m}{I_7})_J\Big)\Big]
\end{aligned} \tag{7.64}$$

$$\begin{aligned}
\overset{\psi_m}{\mathbb{S}}_{yxx}(\overset{J}{\boldsymbol{x}},\boldsymbol{\xi}) &= \overset{\psi_m}{\mathbb{S}}_{xxy}(\overset{J}{\boldsymbol{x}},\boldsymbol{\xi}) \\
\overset{\psi_m}{\mathbb{S}}_{yyy}(\overset{J}{\boldsymbol{x}},\boldsymbol{\xi}) &= \int_{-\overset{J}{a}}^{\overset{J}{a}} S_{yyy}(\overline{x}-c_{\overline{x}}, c_{\overline{y}})\overset{J}{\psi}_m(\overline{x})\,\mathrm{d}\overline{x} \\
&= 2\mu\Big[-\sin\overset{J}{\beta}(\overset{\psi_m}{I_4})_J + \cos\overset{J}{\beta}(\overset{\psi_m}{I_5})_J \\
&\quad + \sin\overset{J}{\beta}\Big(\cos 2\overset{J}{\beta}(\overset{\psi_m}{I_4})_J + \sin 2\overset{J}{\beta}(\overset{\psi_m}{I_5})_J\Big) - c_{\overline{y}}\Big(\sin 3\overset{J}{\beta}(\overset{\psi_m}{I_6})_J\Big) - \cos 3\overset{J}{\beta}(\overset{\psi_m}{I_7})_J\Big] \\
\overset{\psi_m}{\mathbb{S}}_{yxy}(\overset{J}{\boldsymbol{x}},\boldsymbol{\xi}) &= \overset{\psi_m}{\mathbb{S}}_{xyy}(\overset{J}{\boldsymbol{x}},\boldsymbol{\xi})
\end{aligned}$$

Functions $\overset{\psi_m}{I_1}$ through $\overset{\psi_m}{I_7}$ in (7.63) and (7.64) are given by (7.40) and (7.41).

Finally, combining Eqs. (7.60), (7.32), (7.33), (7.63), and (7.64), we obtain the following expressions for the stress components at an arbitrary point $\boldsymbol{\xi}$ in region V for the case that the displacements and tractions are assumed to vary linearly over N straight line elements of the discretized boundary contour S:

$$
\begin{aligned}
\sigma_{xx}(\boldsymbol{\xi}) = {} & \sum_{J=1}^{N}\Big[\overset{\psi_1}{\mathbb{D}}_{xxx}(\overset{J}{\boldsymbol{x}},\boldsymbol{\xi})(\overset{j}{t}_x)_J + \overset{\psi_2}{\mathbb{D}}_{xxx}(\overset{J}{\boldsymbol{x}},\boldsymbol{\xi})(\overset{j+1}{t_x})_J\Big] + \sum_{J=1}^{N}\Big[\overset{\psi_1}{\mathbb{D}}_{yxx}(\overset{J}{\boldsymbol{x}},\boldsymbol{\xi})(\overset{j}{t}_y)_J + \overset{\psi_2}{\mathbb{D}}_{yxx}(\overset{J}{\boldsymbol{x}},\boldsymbol{\xi})(\overset{j+1}{t_y})_J\Big] \\
& - \sum_{J=1}^{N}\Big[\overset{\psi_1}{\mathbb{S}}_{xxx}(\overset{J}{\boldsymbol{x}},\boldsymbol{\xi})(\overset{j}{u}_x)_J + \overset{\psi_2}{\mathbb{S}}_{xxx}(\overset{J}{\boldsymbol{x}},\boldsymbol{\xi})(\overset{j+1}{u_x})_J\Big] - \sum_{J=1}^{N}\Big[\overset{\psi_1}{\mathbb{S}}_{yxx}(\overset{J}{\boldsymbol{x}},\boldsymbol{\xi})(\overset{j}{u}_y)_J + \overset{\psi_2}{\mathbb{S}}_{yxx}(\overset{J}{\boldsymbol{x}},\boldsymbol{\xi})(\overset{j+1}{u_y})_J\Big] \\[2ex]
\sigma_{yy}(\boldsymbol{\xi}) = {} & \sum_{J=1}^{N}\Big[\overset{\psi_1}{\mathbb{D}}_{xyy}(\overset{J}{\boldsymbol{x}},\boldsymbol{\xi})(\overset{j}{t}_x)_J + \overset{\psi_2}{\mathbb{D}}_{xyy}(\overset{J}{\boldsymbol{x}},\boldsymbol{\xi})(\overset{j+1}{t_x})_J\Big] + \sum_{J=1}^{N}\Big[\overset{\psi_1}{\mathbb{D}}_{yyy}(\overset{J}{\boldsymbol{x}},\boldsymbol{\xi})(\overset{j}{t}_y)_J + \overset{\psi_2}{\mathbb{D}}_{yyy}(\overset{J}{\boldsymbol{x}},\boldsymbol{\xi})(\overset{j+1}{t_y})_J\Big] \\
& - \sum_{J=1}^{N}\Big[\overset{\psi_1}{\mathbb{S}}_{xyy}(\overset{J}{\boldsymbol{x}},\boldsymbol{\xi})(\overset{j}{u}_x)_J + \overset{\psi_2}{\mathbb{S}}_{xyy}(\overset{J}{\boldsymbol{x}},\boldsymbol{\xi})(\overset{j+1}{u_x})_J\Big] - \sum_{J=1}^{N}\Big[\overset{\psi_1}{\mathbb{S}}_{yyy}(\overset{J}{\boldsymbol{x}},\boldsymbol{\xi})(\overset{j}{u}_y)_J + \overset{\psi_2}{\mathbb{S}}_{yyy}(\overset{J}{\boldsymbol{x}},\boldsymbol{\xi})(\overset{j+1}{u_y})_J\Big] \qquad (7.65) \\[2ex]
\sigma_{xy}(\boldsymbol{\xi}) = {} & \sum_{J=1}^{N}\Big[\overset{\psi_1}{\mathbb{D}}_{xxy}(\overset{J}{\boldsymbol{x}},\boldsymbol{\xi})(\overset{j}{t}_x)_J + \overset{\psi_2}{\mathbb{D}}_{xxy}(\overset{J}{\boldsymbol{x}},\boldsymbol{\xi})(\overset{j+1}{t_x})_J\Big] + \sum_{J=1}^{N}\Big[\overset{\psi_1}{\mathbb{D}}_{yxy}(\overset{J}{\boldsymbol{x}},\boldsymbol{\xi})(\overset{j}{t}_y)_J + \overset{\psi_2}{\mathbb{D}}_{yxy}(\overset{J}{\boldsymbol{x}},\boldsymbol{\xi})(\overset{j+1}{t_y})_J\Big] \\
& - \sum_{J=1}^{N}\Big[\overset{\psi_1}{\mathbb{S}}_{xxy}(\overset{J}{\boldsymbol{x}},\boldsymbol{\xi})(\overset{j}{u}_x)_J + \overset{\psi_2}{\mathbb{S}}_{xxy}(\overset{J}{\boldsymbol{x}},\boldsymbol{\xi})(\overset{j+1}{u_x})_J\Big] - \sum_{J=1}^{N}\Big[\overset{\psi_1}{\mathbb{S}}_{yxy}(\overset{J}{\boldsymbol{x}},\boldsymbol{\xi})(\overset{j}{u}_y)_J + \overset{\psi_2}{\mathbb{S}}_{yxy}(\overset{J}{\boldsymbol{x}},\boldsymbol{\xi})(\overset{j+1}{u_y})_J\Big]
\end{aligned}
$$

These equations can be evaluated directly within a computer program once the boundary displacements and tractions have been determined. The only restriction to keep in mind is that point $\boldsymbol{\xi}$ should not be taken at, or very close to, any of the nodal points of the discretized boundary contour, because the quantities $\overset{\psi_m}{\mathbb{S}}_{pqr}(\overset{J}{\boldsymbol{x}},\boldsymbol{\xi})$ and $\overset{\psi_m}{\mathbb{D}}_{pqr}(\overset{J}{\boldsymbol{x}},\boldsymbol{\xi})$ contain strong singularities at these points.

7.12 Symmetry Conditions

Symmetry conditions can be incorporated in the direct boundary integral method in essentially the same way as explained in Sect. 5.7 for the stress discontinuity method. In the present case, all of the boundary elements—including the image elements—are subjected to loading (displacements and tractions) in the global x, y coordinate directions rather than in local $\bar{x}$, $\bar{y}$ directions. The influence coefficients are still calculated in terms of local coordinates, however, so that if $x = x^*$ is a line of symmetry, for example, Eqs. (5.32) define the local coordinates of the image for element j. The y components of displacement and traction for the image element are the same as those for the actual element, that is, $\overset{j}{u'}_y = \overset{j}{u}_y$ and $\overset{j}{t'}_y = \overset{j}{t}_y$, but the x components are of opposite sign: $\overset{j}{u'}_x = -\overset{j}{u}_x$ and $\overset{j}{t'}_x = -\overset{j}{t}_x$. As before, the contributions from the image elements are added to those from the actual elements to calculate sets of influence coefficients for a given problem.

A line of symmetry $y = y^*$ is handled in the same way and, again, lines of symmetry for both the x and y directions can be incorporated if desired.

7.13 Example

The solution to the problem of an elliptical hole in a large isotropic, linearly elastic plate under uniaxial loading (e.g., Muskhelishvili 1963; Timoshenko and Goodier 1970) provides a useful test of the direct boundary integral method with linear variations of the boundary parameters. An elliptical boundary can be conveniently divided into elements of differing sizes, and this gives a good way to demonstrate the accuracy of the method for nonuniform element spacing.

Cartesian coordinates (x, y) are related to elliptical coordinates (u, v) by the transformation

$$x = c \cosh u \cos v$$
$$y = c \sinh u \sin v \tag{7.66}$$

The coordinate $u = u_0$ defines an ellipse with semiaxes $a = c \cosh u_0$ and $b = c \sinh u_0$ and foci at $x = \pm c$; different values of u_0 define different ellipses with the same foci, known as confocal ellipses. Values of v ranging from 0 to 2π define hyperbolas orthogonal to the ellipses. The coordinates u and v for elliptical boundaries are comparable to r and θ for circular boundaries. Thus, the hoop stress for an elliptical boundary is denoted as σ_{vv}. For a uniaxial stress p applied at an angle ϑ from the x axis, σ_{vv} can be written as (Timoshenko and Goodier 1970)

$$(\sigma_{vv})_{u=u_0} = p \frac{\sinh 2u_0 + \cos 2\vartheta - e^{2u_0} \cos 2(\vartheta - v)}{\cosh 2u_0 - \cos 2v} \tag{7.67}$$

As a specific example we consider the case that $u_0 = 0.20273$ and $c = 2.4495$ (arbitrary length units). These parameters define an ellipse with the ratio of the major to minor axes a/b approximately equal to 5; the value selected for c makes the length of the major axis equal to 5. In addition, we take the stress at infinity as $\sigma_{yy} = p$ so that $\vartheta = \pi/2$. The stress σ_{vv} along a symmetric one-quarter of the elliptical hole is plotted in Fig. 7.7, together with numerical results for the case that one-quarter of the boundary was divided into 15 elements with lengths ranging from 0.054 to 0.2613, a 48-fold variation. The largest error is approximately 6%, and this is reduced to 2% by doubling the number of elements.

7.14 Anisotropy

The tensor fields $U_{ij}(\boldsymbol{x}, \boldsymbol{\xi})$, $T_{ij}(\boldsymbol{x}, \boldsymbol{\xi})$, $D_{kij}(\boldsymbol{x}, \boldsymbol{\xi})$, and $S_{kij}(\boldsymbol{x}, \boldsymbol{\xi})$ needed to develop the direct boundary integral method for plane problems in anisotropic elasticity are given in Appendix B. These quantities can be expressed in terms of local $\bar{x}$, $\bar{y}$ coordinates as was done for isotropic elasticity by using coordinate transformation formulas

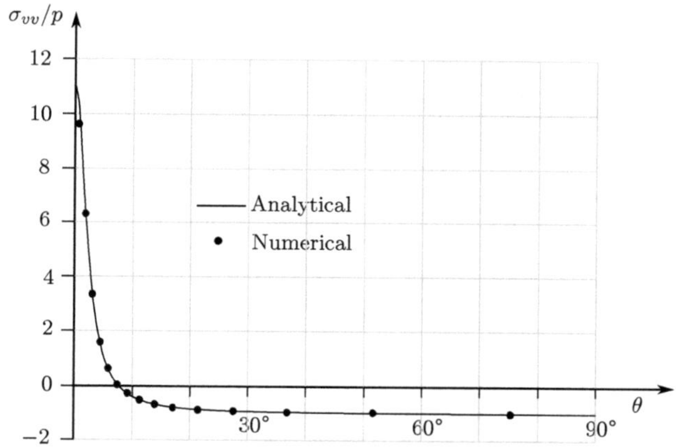

Fig. 7.7 Hoop stress σ_{vv} at boundary of elliptical hole

(3.6) and (3.7) and the chain rule for partial differentiation. Referring to Appendix B, the components $U_{xx}(\bar{x} - c_{\bar{x}}, \bar{y} - c_{\bar{y}})$, etc., evaluated for $\bar{y} = 0$, for example, can be written as

$$
\begin{aligned}
U_{xx} &= \frac{1}{2\pi(q_1 - q_2)c_{66}} \left[\frac{\gamma_1}{q_1} \ln \bar{r}(\gamma_1) - \frac{\gamma_2}{q_2} \ln \bar{r}(\gamma_2) \right] \\
U_{yx} &= -\frac{1}{2\pi(q_1 - q_2)c_{66}} \left[\bar{\theta}(\gamma_1) - \bar{\theta}(\gamma_2) \right] \\
U_{xy} &= U_{yx} \\
U_{yy} &= -\frac{1}{2\pi(q_1 - q_2)c_{66}} \left[\frac{q_1}{\gamma_1} \ln \bar{r}(\gamma_1) - \frac{q_2}{\gamma_2} \ln \bar{r}(\gamma_2) \right]
\end{aligned}
\tag{7.68}
$$

in which $\bar{r}(\gamma_i)$ and $\bar{\theta}(\gamma_i)$, for $i = 1$ or 2, are the *scaled* distances and angles from the load point $\boldsymbol{\xi}$ to the sample point $\boldsymbol{x}$:

$$
\begin{aligned}
\bar{r}(\gamma_i) &= \left[A_i(\bar{x} - c_{\bar{x}})^2 - B_i(\bar{x} - c_{\bar{x}})c_{\bar{y}} + C_i c_{\bar{y}}^2 \right]^{1/2} \\
\bar{\theta}(\gamma_i) &= \arctan \frac{1}{\gamma_i} \frac{(\bar{x} - c_{\bar{x}}) \sin\beta - c_{\bar{y}} \cos\beta}{(\bar{x} - c_{\bar{x}}) \cos\beta + c_{\bar{y}} \sin\beta}
\end{aligned}
\tag{7.69}
$$

where A_i, B_i, and C_i are given as in (4.179), i.e.,

$$
\begin{aligned}
A_i &= (\gamma_i^2 \cos^2\beta + \sin^2\beta)/\gamma_i^2 \\
B_i &= (1 - \gamma_i^2) \sin 2\beta/\gamma_i^2 \\
C_i &= (\gamma_i^2 \sin^2\beta + \cos^2\beta)/\gamma_i^2
\end{aligned}
\tag{7.70}
$$

The components $T_{xx}(\bar{x} - c_{\bar{x}}, \bar{y} - c_{\bar{y}})$, etc., for $\bar{y} = 0$ can similarly be written as

$$
\begin{aligned}
T_{xx} &= -\frac{1}{2\pi(q_1 - q_2)}\left[\frac{(1+q_1)}{\gamma_1 q_1}\frac{c_{\bar{y}}}{\bar{r}^2(\gamma_1)} - \frac{(1+q_2)}{\gamma_2 q_2}\frac{c_{\bar{y}}}{\bar{r}^2(\gamma_2)}\right] \\
T_{yx} &= -\frac{1}{2\pi(q_1 - q_2)}\left[\frac{(1+q_1)}{\gamma_1}\frac{A_1(\bar{x} - c_{\bar{x}}) - B_1 c_{\bar{y}}/2}{\bar{r}^2(\gamma_1)} - \frac{(1+q_2)}{\gamma_2}\frac{A_2(\bar{x} - c_{\bar{x}}) - B_2 c_{\bar{y}}/2}{\bar{r}^2(\gamma_2)}\right] \\
T_{xy} &= -\frac{1}{2\pi(q_1 - q_2)}\left[\frac{\gamma_1(1+q_1)}{q_1}\frac{A_1(\bar{x} - c_{\bar{x}}) - B_1 c_{\bar{y}}/2}{\bar{r}^2(\gamma_1)} - \frac{\gamma_2(1+q_2)}{q_2}\frac{A_2(\bar{x} - c_{\bar{x}}) - B_2 c_{\bar{y}}/2}{\bar{r}^2(\gamma_2)}\right] \\
T_{yy} &= -\frac{1}{2\pi(q_1 - q_2)}\left[\frac{(1+q_1)}{\gamma_1}\frac{c_{\bar{y}}}{\bar{r}^2(\gamma_1)} - \frac{(1+q_2)}{\gamma_2}\frac{c_{\bar{y}}}{\bar{r}^2(\gamma_2)}\right]
\end{aligned}
\tag{7.71}
$$

Equations (7.68) and (7.71) can be used in place of (7.14) and (7.17) to develop boundary influence coefficients for anisotropic elasticity comparable to (7.20) and (7.21), or (7.42) and (7.43). Once again, all of the required integrations can be done analytically.

Finally, the influence coefficients needed to calculate the stress components can be obtained by suitably integrating the quantities $D_{kij}(\boldsymbol{x}, \boldsymbol{\xi})$, and $S_{kij}(\boldsymbol{x}, \boldsymbol{\xi})$ listed in Appendix B.

7.15 Another Formulation of the Method

We saw in Chap. 3 that the direct boundary integral method for Laplace's equation could be formulated in two ways, with both formulations leading to the same final result. In both cases the boundary S of region V was approximated by N straight line elements and the boundary parameters (the potential and its normal derivative) were assumed to be piecewise constant along these elements. In the first formulation the source point $\boldsymbol{\xi}$ was assumed to be inside the region of interest (i.e., not on the boundary), the elemental boundary integrals were evaluated analytically, and the system of algebraic equations needed to find the unknown boundary parameters was obtained by considering the limiting case in which the source point was moved to the boundary. The second formulation used a special version of Green's representation formula in which the source point was assumed to lie on the boundary and the jump term in the boundary integral formula was extracted by a limiting process prior to any boundary discretization and integration.

A similar treatment is possible for elasticity using the special version of Somigliana's displacement formula developed in Sect. 4.23 for the limiting case that $\boldsymbol{\xi} \to \boldsymbol{x}$. The revised formula (4.353) can be discretized using the same procedure described in Sects. 7.8 and 7.9, the only difference being the way in which the self-effect terms for the boundary influence coefficients are calculated. Previously, these terms were found by substituting (7.58) and (7.59) into (7.51), noting the presence of the identity matrix $\overset{ij}{\mathbb{I}}$ in the latter equations. Now, the jump terms for the self-effect coefficients for the displacements are given directly by (4.354) while the remaining terms for these coefficients must be evaluated as Cauchy principal value integrals. We leave it as an exercise for the reader to confirm that this calculation accounts for the presence of the logarithm terms in the coefficients $\overset{jj}{\mathbb{T}}_{yx}$ and $\overset{jj}{\mathbb{T}}_{xy}$ in (7.59).

Finally, we note that the relations given previously for the displacements $u_i(\boldsymbol{\xi})$ and stresses $\sigma_{ij}(\boldsymbol{\xi})$ inside the region of interest (see (7.1) and (7.60)) do not depend on which approach is used to develop the system of algebraic equations from which to find the unknown boundary parameters.

7.16 Alternative Calculation of Self-effect Coefficients

The self-effect coefficients for the displacements can also be determined indirectly by imagining that the boundary S is traction-free but is subjected to a rigid body displacement. If the boundary tractions are zero, (4.353) reduces to

$$c_{ij}u_j(\boldsymbol{\xi}) = -\!\!\!\!\!\!\int_S u_j(\boldsymbol{x})T_{ij}(\boldsymbol{x},\boldsymbol{\xi})\,\mathrm{d}S(\boldsymbol{x}) \tag{7.72}$$

or, in discretized form for piecewise linear boundary parameters,

$$\begin{aligned}
c_{xx}\overset{i}{u}_x + c_{xy}\overset{i}{u}_y &= -\sum_{j=1}^{N}\Big[(\overset{ij}{\mathbb{T}}_{xx})_J + (\overset{ij}{\mathbb{T}}_{xx})_{J-1}\Big]\overset{j}{u}_x - \sum_{j=1}^{N}\Big[(\overset{ij}{\mathbb{T}}_{xy})_J + (\overset{ij}{\mathbb{T}}_{xy})_{J-1}\Big]\overset{j}{u}_y \\
c_{yx}\overset{i}{u}_x + c_{yy}\overset{i}{u}_y &= -\sum_{j=1}^{N}\Big[(\overset{ij}{\mathbb{T}}_{yx})_J + (\overset{ij}{\mathbb{T}}_{yx})_{J-1}\Big]\overset{j}{u}_x - \sum_{j=1}^{N}\Big[(\overset{ij}{\mathbb{T}}_{yy})_J + (\overset{ij}{\mathbb{T}}_{yy})_{J-1}\Big]\overset{j}{u}_y
\end{aligned} \tag{7.73}$$

Now, suppose that the entire boundary is displaced by a constant amount in the x direction, e.g., $\overset{j}{u}_x = (u_x)_0$ and $\overset{j}{u}_y = 0$ for $j = 1$ to N. We then have

$$\begin{aligned}
c_{xx}(u_x)_0 &= -\sum_{j=1}^{N}\Big[(\overset{ij}{\mathbb{T}}_{xx})_J + (\overset{ij}{\mathbb{T}}_{xx})_{J-1}\Big](u_x)_0 \\
c_{yx}(u_x)_0 &= -\sum_{j=1}^{N}\Big[(\overset{ij}{\mathbb{T}}_{yx})_J + (\overset{ij}{\mathbb{T}}_{yx})_{J-1}\Big](u_x)_0
\end{aligned} \tag{7.74}$$

so that

$$c_{xx} + (\overset{ii}{\mathbb{T}}_{xx})_J + (\overset{ii}{\mathbb{T}}_{xx})_{J-1} = - \sum_{j=1;\ j\neq i}^{N} \left[(\overset{ij}{\mathbb{T}}_{xx})_J + (\overset{ij}{\mathbb{T}}_{xx})_{J-1} \right]$$

$$c_{yx} + (\overset{ii}{\mathbb{T}}_{yx})_J + (\overset{ii}{\mathbb{T}}_{yx})_{J-1} = - \sum_{j=1;\ j\neq i}^{N} \left[(\overset{ij}{\mathbb{T}}_{yx})_J + (\overset{ij}{\mathbb{T}}_{yx})_{J-1} \right] \tag{7.75}$$

This result indirectly determines the diagonal terms of the influence coefficients for row i of the matrix (including the jump terms c_{xx} and c_{yx}) as the negative sums of the other coefficients in that same row.

A similar calculation for the case of a rigid body displacement in the y direction gives

$$c_{xy} + (\overset{ii}{\mathbb{T}}_{xy})_J + (\overset{ii}{\mathbb{T}}_{xy})_{J-1} = - \sum_{j=1;\ j\neq i}^{N} \left[(\overset{ij}{\mathbb{T}}_{xy})_J + (\overset{ij}{\mathbb{T}}_{xy})_{J-1} \right]$$

$$c_{yy} + (\overset{ii}{\mathbb{T}}_{yy})_J + (\overset{ii}{\mathbb{T}}_{yy})_{J-1} = - \sum_{j=1;\ j\neq i}^{N} \left[(\overset{ij}{\mathbb{T}}_{yy})_J + (\overset{ij}{\mathbb{T}}_{yy})_{J-1} \right] \tag{7.76}$$

which indirectly determines the other diagonal terms of the boundary influence coefficient matrix for the displacements.

7.17 Curvilinear Elements

In our analyses so far we have assumed that a boundary can be approximated by a number of straight line elements joined end to end. This assumption is reasonable in many cases and has the advantage that all integrals along the boundary elements can be evaluated analytically. If portions of the boundary have sharp curvature, however, a large number of straight line boundary elements may be required to adequately represent the geometry. Such situations can be better handled using curvilinear elements, a topic we consider next.

7.17.1 Element Shape Functions

We have previously used the concept of shape functions to specify the variation of boundary parameters such as displacements and tractions along straight line elements in terms of their values at discrete points along the elements (see, for example, (7.32) and (7.33)). The same idea can be used to define the geometry of an element, in which case the values of the x, y coordinates along the element are specified in terms of their values at discrete points. This geometric variation—which can include straight lines but is not restricted to them—is customarily expressed in terms of a so-called intrinsic (dimensionless) coordinate η, where $-1 \leq \eta \leq 1$. The associated shape functions for a quadratic (i.e., parabolic) variation of the boundary geometry are

$$\begin{aligned}\psi_1(\eta) &= \frac{1}{2}\eta(\eta - 1)\\ \psi_2(\eta) &= 1 - \eta^2 \\ \psi_3(\eta) &= \frac{1}{2}\eta(\eta + 1)\end{aligned} \tag{7.77}$$

in which the subscripts refer to the three reference points $\eta = -1$ (point 1), $\eta = 0$ (point 2), and $\eta = 1$ (point 3), as depicted in Fig. 7.8.

If the $(x,\ y)$ coordinates of the three reference points are $(x_1,\ y_1)$, $(x_2,\ y_2)$, and $(x_3,\ y_3)$ the values of x and y at any point along the element are given as

$$\begin{aligned}x(\eta) &= \psi_1(\eta)x_1 + \psi_2(\eta)x_2 + \psi_3(\eta)x_3\\ y(\eta) &= \psi_1(\eta)y_1 + \psi_2(\eta)y_2 + \psi_3(\eta)y_3\end{aligned} \tag{7.78}$$

or, in vector form,

$$\boldsymbol{x}(\eta) = \psi_1(\eta)\boldsymbol{x}_1 + \psi_2(\eta)\boldsymbol{x}_2 + \psi_3(\eta)\boldsymbol{x}_3 \tag{7.79}$$

This formula expresses the $(x,\ y)$ coordinates of the sample point $\boldsymbol{x}$ in terms of the intrinsic coordinate η.

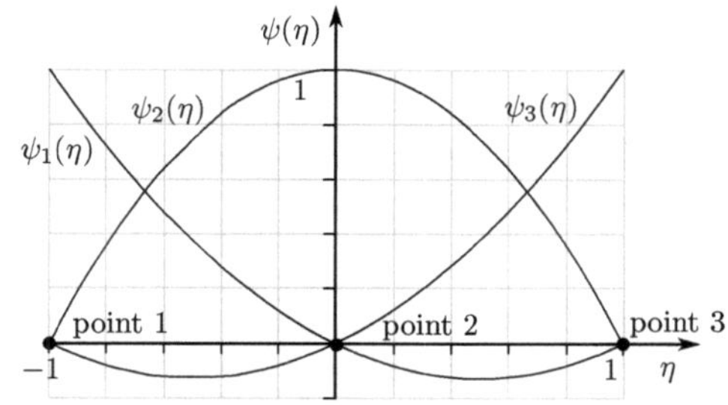

Fig. 7.8 Quadratic shape functions in terms of intrinsic coordinate η

Using (7.77), (7.79) can be alternatively written as

$$\boldsymbol{x}(\eta) = \frac{1}{2}(\boldsymbol{x}_1 - 2\boldsymbol{x}_2 + \boldsymbol{x}_3)\eta^2 + \frac{1}{2}(\boldsymbol{x}_3 - \boldsymbol{x}_1)\eta + \boldsymbol{x}_2 \tag{7.80}$$

If the three reference points lie on a straight line and if point 2 is midway between points 1 and 3, then $\boldsymbol{x}_2 = (\boldsymbol{x}_1 + \boldsymbol{x}_3)/2$ and (7.80) becomes

$$\boldsymbol{x}(\eta) = \frac{1}{2}(1 - \eta)\boldsymbol{x}_1 + \frac{1}{2}(1 + \eta)\boldsymbol{x}_3 \tag{7.81}$$

In this case, the boundary geometry is linear and is defined by two shape functions, $(1 - \eta)/2$ and $(1 + \eta)/2$.

If the shape functions used to define the element geometry are the same as those used to specify the boundary parameters they are called *isoparametric*. The displacements and tractions for an isoparametric quadratic element can therefore be written as

$$\begin{aligned} u_x(\eta) &= \psi_1(\eta)(u_x)_1 + \psi_2(\eta)(u_x)_2 + \psi_3(\eta)(u_x)_3 \\ u_y(\eta) &= \psi_1(\eta)(u_y)_1 + \psi_2(\eta)(u_y)_2 + \psi_3(\eta)(u_y)_3 \end{aligned} \tag{7.82}$$

and

$$\begin{aligned} t_x(\eta) &= \psi_1(\eta)(t_x)_1 + \psi_2(\eta)(t_x)_2 + \psi_3(\eta)(t_x)_3 \\ t_y(\eta) &= \psi_1(\eta)(t_y)_1 + \psi_2(\eta)(t_y)_2 + \psi_3(\eta)(t_y)_3 \end{aligned} \tag{7.83}$$

The reference points 1, 2, and 3 in these expressions are the same as those used to specify the element geometry in (7.78).

The reader should recognize that it is possible to have a situation in which the displacements and tractions vary quadratically along a straight line element. The element geometry is then specified by (7.81) while the boundary parameters are approximated as in (7.82) and (7.83). Such an element is useful in practice but should not technically be referred to as isoparametric.

7.17.2 Discretization of Integrals; Load Point ξ not on Boundary

To see how the preceding results can be implemented numerically, consider the following discretization of (7.3) for N boundary elements:

$$u_x(\boldsymbol{\xi}) = \sum_{J=1}^{N} \int_{\overset{J}{\Delta S}} \left[\overset{J}{t}_x(\boldsymbol{x}) U_{xx}(\boldsymbol{x}, \boldsymbol{\xi}) + \overset{J}{t}_y(\boldsymbol{x}) U_{xy}(\boldsymbol{x}, \boldsymbol{\xi}) \right] \mathrm{d}S(\boldsymbol{x}) - \sum_{J=1}^{N} \int_{\overset{J}{\Delta S}} \left[\overset{J}{u}_x(\boldsymbol{x}) T_{xx}(\boldsymbol{x}, \boldsymbol{\xi}) + \overset{J}{u}_y(\boldsymbol{x}) T_{xy}(\boldsymbol{x}, \boldsymbol{\xi}) \right] \mathrm{d}S(\boldsymbol{x})$$

$$u_y(\boldsymbol{\xi}) = \sum_{J=1}^{N} \int_{\overset{J}{\Delta S}} \left[\overset{J}{t}_x(\boldsymbol{x}) U_{yx}(\boldsymbol{x}, \boldsymbol{\xi}) + \overset{J}{t}_y(\boldsymbol{x}) U_{yy}(\boldsymbol{x}, \boldsymbol{\xi}) \right] \mathrm{d}S(\boldsymbol{x}) - \sum_{J=1}^{N} \int_{\overset{J}{\Delta S}} \left[\overset{J}{u}_x(\boldsymbol{x}) T_{yx}(\boldsymbol{x}, \boldsymbol{\xi}) + \overset{J}{u}_y(\boldsymbol{x}) T_{yy}(\boldsymbol{x}, \boldsymbol{\xi}) \right] \mathrm{d}S(\boldsymbol{x}) \tag{7.84}$$

Just as with (7.3), Eqs. (7.84) are written for the case that the load point $\boldsymbol{\xi}$ is not on the boundary S. We will assume for now that the displacements $\overset{J}{u}_x(\boldsymbol{x})$ and $\overset{J}{u}_y(\boldsymbol{x})$ and the tractions $\overset{J}{t}_x(\boldsymbol{x})$ and $\overset{J}{t}_y(\boldsymbol{x})$ are known for all the elements, $J = 1$ to N; that is, we will assume that a numerical solution to a boundary value problem has already been found. We will later return to the case that the load point $\boldsymbol{\xi}$ is on the boundary S and perform a similar discretization of (4.353).

The geometry for a typical quadratic element J is shown in Fig. 7.9. Reference point 1 is at the left end of the element (relative to the direction of boundary traversal) and point 3 is at the right end; point 2 is at an intermediate point chosen to define the appropriate parabolic arc representing the boundary along the element. These three points have coordinates $\overset{J}{\boldsymbol{x}}_1 = (\overset{J}{x}_1, \overset{J}{y}_1)$, $\overset{J}{\boldsymbol{x}}_2 = (\overset{J}{x}_2, \overset{J}{y}_2)$, and $\overset{J}{\boldsymbol{x}}_3 = (\overset{J}{x}_3, \overset{J}{y}_3)$. Notice that, in general, point 2 does not need to be midway between points 1 and 3. Using (7.79) the value of $\boldsymbol{x}$ in (7.84) is

$$\boldsymbol{x} = \psi_1(\eta)\overset{J}{\boldsymbol{x}}_1 + \psi_2(\eta)\overset{J}{\boldsymbol{x}}_2 + \psi_3(\eta)\overset{J}{\boldsymbol{x}}_3 \tag{7.85}$$

Fig. 7.9 Geometry for quadratic element J in global x, y coordinates

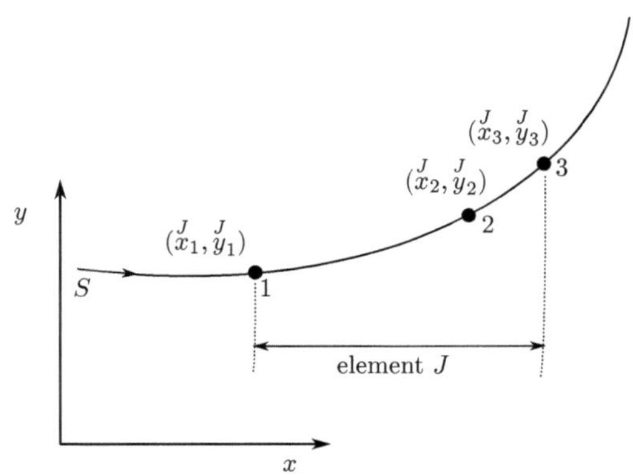

The differential of arc length at point $\boldsymbol{x}$ of the boundary S is defined as $\mathrm{d}S(\boldsymbol{x}) = \left[\mathrm{d}x^2 + \mathrm{d}y^2\right]^{1/2}$, or, because x and y are functions of the intrinsic coordinate η,

$$\mathrm{d}S(\boldsymbol{x}) = \left[\left(\frac{\mathrm{d}x(\eta)}{\mathrm{d}\eta}\right)^2 + \left(\frac{\mathrm{d}y(\eta)}{\mathrm{d}\eta}\right)^2\right]^{1/2} \mathrm{d}\eta \tag{7.86}$$

This expression can also be written as

$$\mathrm{d}S(\boldsymbol{x}) = \mathcal{J}(\eta)\,\mathrm{d}\eta \tag{7.87}$$

in which $\mathcal{J}(\eta)$ is called the Jacobian of transformation (7.78). The transformation from the global physical coordinates $\boldsymbol{x}$ to the local intrinsic coordinate η is then achieved by substituting (7.87) in (7.84) and using (7.85). For the quadratic element defined by (7.77) and (7.78) the Jacobian is

$$\begin{aligned}\mathcal{J}(\eta) &= \left[\left(\frac{\mathrm{d}x(\eta)}{\mathrm{d}\eta}\right)^2 + \left(\frac{\mathrm{d}y(\eta)}{\mathrm{d}\eta}\right)^2\right]^{1/2} \\ &= \left[\left\{(\overset{J}{x}_1 - 2\overset{J}{x}_2 + \overset{J}{x}_3)\eta + \frac{\overset{J}{x}_3 - \overset{J}{x}_1}{2}\right\}^2 + \left\{(\overset{J}{y}_1 - 2\overset{J}{y}_2 + \overset{J}{y}_3)\eta + \frac{\overset{J}{y}_3 - \overset{J}{y}_1}{2}\right\}^2\right]^{1/2}\end{aligned} \tag{7.88}$$

Consider now a typical term of (7.84) such as, for example,

$$\int_{\overset{J}{\Delta S}} \overset{J}{t}_x(\boldsymbol{x}) U_{xx}(\boldsymbol{x}, \boldsymbol{\xi})\,\mathrm{d}S(\boldsymbol{x}) \tag{7.89}$$

Referring to (4.251), $U_{xx}(\boldsymbol{x}, \boldsymbol{\xi})$ can be written as

$$U_{xx}(\boldsymbol{x}, \boldsymbol{\xi}) = -\frac{1}{8\mu\pi(1-\nu)}\left[(3-4\nu)\ln\sqrt{(x-\xi_x)^2 + (y-\xi_y)^2} - \frac{(x-\xi_x)^2}{(x-\xi_x)^2 + (y-\xi_y)^2}\right] \tag{7.90}$$

and, using (7.85), we have that

$$\begin{aligned} x - \xi_x &= \psi_1(\eta)\overset{J}{x}_1 + \psi_2(\eta)\overset{J}{x}_2 + \psi_3(\eta)\overset{J}{x}_3 - \xi_x \\ y - \xi_y &= \psi_1(\eta)\overset{J}{y}_1 + \psi_2(\eta)\overset{J}{y}_2 + \psi_3(\eta)\overset{J}{y}_3 - \xi_y \end{aligned} \tag{7.91}$$

For a specified location of the load point $\boldsymbol{\xi} = (\xi_x, \xi_y)$, the quantity $U_{xx}(\boldsymbol{x}, \boldsymbol{\xi})$ is therefore a function of η, i.e.,

$$U_{xx}(\boldsymbol{x}, \boldsymbol{\xi}) = U_{xx}(\eta) \tag{7.92}$$

Using (7.83), the traction $\overset{J}{t}_x(\boldsymbol{x})$ can be written as

$$\overset{J}{t}_x(\boldsymbol{x}) = \psi_1(\eta)(\overset{J}{t}_x)_1 + \psi_2(\eta)(\overset{J}{t}_x)_2 + \psi_3(\eta)(\overset{J}{t}_x)_3 \tag{7.93}$$

which is also a function of η:

$$\overset{J}{t}_x(\boldsymbol{x}) = \overset{J}{t}_x(\eta) \tag{7.94}$$

Collecting the foregoing results we have that

$$\int_{\overset{J}{\Delta S}} \overset{J}{t}_x(\boldsymbol{x}) U_{xx}(\boldsymbol{x}, \boldsymbol{\xi})\, \mathrm{d}S(\boldsymbol{x}) = \sum_{k=1}^{3} (\overset{J}{t}_x)_k \int_{-1}^{1} U_{xx}(\eta)\psi_k(\eta)\mathcal{J}(\eta)\, \mathrm{d}\eta \tag{7.95}$$

Integrals of this form can easily be evaluated using Gauss-Legendre quadrature (see, for example, Abramowitz and Stegun 1965). The standard formula for this is

$$\int_{-1}^{1} f(x)\, \mathrm{d}x \cong \sum_{i=1}^{n} w_i f(x_i) \tag{7.96}$$

in which the abscissae x_i are roots of the Legendre polynomial of order n and w_i are the associated weight factors. The abscissae and weight factors for various degrees of approximation are tabulated in numerous references. Most researchers suggest using $n = 10$ with quadratic isoparametric elements. If the load point $\boldsymbol{\xi}$ is far enough away from a boundary a lower number of points might suffice, but this should be established empirically for specific problems.

A similar procedure is used to evaluate terms such as, for example,

$$\int_{\overset{J}{\Delta S}} \overset{J}{u}_y(\boldsymbol{x}) T_{yy}(\boldsymbol{x}, \boldsymbol{\xi})\, \mathrm{d}S(\boldsymbol{x}) \tag{7.97}$$

in (7.84), leading to the result

$$\int_{\overset{J}{\Delta S}} \overset{J}{u}_y(\boldsymbol{x}) T_{yy}(\boldsymbol{x}, \boldsymbol{\xi})\, \mathrm{d}S(\boldsymbol{x}) = \sum_{k=1}^{3} (\overset{J}{u}_y)_k \int_{-1}^{1} T_{yy}(\eta)\psi_k(\eta)\mathcal{J}(\eta)\, \mathrm{d}\eta \tag{7.98}$$

Using (4.265), $T_{yy}(\boldsymbol{x}, \boldsymbol{\xi})$ can be written as

$$\begin{aligned} T_{yy}(\boldsymbol{x}, \boldsymbol{\xi}) = & -\frac{1}{4\pi(1-\nu)}\left[(1-2\nu)\frac{x-\xi_x}{(x-\xi_x)^2+(y-\xi_y)^2} + \frac{2(x-\xi_x)(y-\xi_y)^2}{((x-\xi_x)^2+(y-\xi_y)^2)^2}\right] n_x(\boldsymbol{x}) \\ & - \frac{1}{4\pi(1-\nu)}\left[2(1-\nu)\frac{y-\xi_y}{(x-\xi_x)^2+(y-\xi_y)^2} - (y-\xi_y)\frac{(x-\xi_x)^2-(y-\xi_y)^2}{((x-\xi_x)^2+(y-\xi_y)^2)^2}\right] n_y(\boldsymbol{x}) \end{aligned} \tag{7.99}$$

where $n_x(\boldsymbol{x})$ and $n_y(\boldsymbol{x})$ are the x and y components of the unit outward normal to the boundary at the sample point $\boldsymbol{x}$. These quantities represent the components of a unit vector perpendicular to the parabola passing through the three nodal point locations $(\overset{J}{x}_1,\ \overset{J}{y}_1)$, $(\overset{J}{x}_2,\ \overset{J}{y}_2)$, and $(\overset{J}{x}_3,\ \overset{J}{y}_3)$. If this parabola is defined by the equation

$$y = Ax^2 + Bx + C \tag{7.100}$$

the slope is

$$\frac{dy}{dx} = 2Ax + B \tag{7.101}$$

in which

$$\begin{aligned} A &= \frac{-\overset{J}{x}_1(\overset{J}{y}_3 - \overset{J}{y}_2) + \overset{J}{x}_2(\overset{J}{y}_3 - \overset{J}{y}_1) - \overset{J}{x}_3(\overset{J}{y}_2 - \overset{J}{y}_1)}{(\overset{J}{x}_3 - \overset{J}{x}_1)(\overset{J}{x}_3 - \overset{J}{x}_2)(\overset{J}{x}_2 - \overset{J}{x}_1)} \\ B &= -(\overset{J}{x}_1 + \overset{J}{x}_3)A + \frac{\overset{J}{y}_3 - \overset{J}{y}_1}{\overset{J}{x}_3 - \overset{J}{x}_1} \end{aligned} \tag{7.102}$$

The tangent angles at the three reference points $(\overset{J}{x}_k,\ \overset{J}{y}_k)$, $k = 1$ to 3, are therefore

$$\overset{J}{\beta}_k = \arctan(2A\overset{J}{x}_k + B) \tag{7.103}$$

and the components of the unit normals at these points are

$$\begin{aligned} (\overset{J}{n}_x)_k &= \cos(\overset{J}{\beta}_k + \pi/2) = -\sin \overset{J}{\beta}_k \\ (\overset{J}{n}_y)_k &= \sin(\overset{J}{\beta}_k + \pi/2) = \cos \overset{J}{\beta}_k \end{aligned} \tag{7.104}$$

The other integrals in (7.84) can be treated in the same way, and similar expressions for the stresses at point $\boldsymbol{\xi}$ can be developed from (4.332).

7.17.3 Element and Nodal Point Numbering

A closed contour represented by N quadratic boundary elements has $2N$ nodal points, N of which are at the connecting points between the individual elements. There are also N nodal points along the elements and away from the ends. These nodes are typically taken at the element midpoints but this is a matter of computational

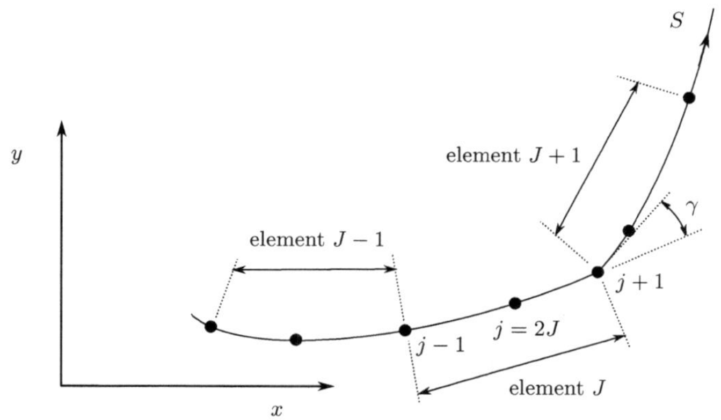

Fig. 7.10 Element and nodal point numbering for three connecting quadratic elements

convenience and not a requirement. In what follows we will refer to them as "midpoint nodes" simply to distinguish them from "endpoint nodes".

Figure 7.10 shows three connecting quadratic elements numbered $J-1$, J, and $J+1$. We will adopt the convention that midpoint nodes have even numbers such that, for example, the midpoint node for element J is designated as $j=2J$ (as in Sect. 7.6 we use capital letters for elements and lower case letters for nodes). Each element has three reference points, and the endpoint nodes of adjacent elements coincide. It can be seen from Fig. 7.10 that reference point 3 for element $J-1$ is the same as reference point 1 for element J and that reference point 1 for element $J+1$ is the same as reference point 3 for element J, i.e.,

$$
\begin{aligned}
(J-1)_3 &= (J)_1 \\
(J+1)_1 &= (J)_3
\end{aligned}
$$

With the understanding that the midpoint node j is an even number, these endpoint nodes have numbers $j-1$ and $j+1$, respectively, as shown in the figure.

7.17.4 Discretization of Integrals for Load Point $\boldsymbol{\xi}$ on Boundary

Discretization of (4.353) proceeds in a similar manner as discussed previously for discretization of (7.3). Now, the load point $\boldsymbol{\xi}$ is located at a boundary node, say $\overset{i}{\boldsymbol{x}}$, so that the discretized boundary integral equations are, for example,

$$c_{xx}u_x(\overset{i}{\boldsymbol{x}}) + c_{xy}u_y(\overset{i}{\boldsymbol{x}}) = \sum_{J=1}^{N} \int_{\overset{J}{\Delta S}} \left[\overset{J}{t}_x(\boldsymbol{x}) U_{xx}(\boldsymbol{x}, \overset{i}{\boldsymbol{x}}) + \overset{J}{t}_y(\boldsymbol{x}) U_{xy}(\boldsymbol{x}, \overset{i}{\boldsymbol{x}}) \right] \mathrm{d}S(\boldsymbol{x})$$
$$- \sum_{J=1}^{N} \fint_{\overset{J}{\Delta S}} \left[\overset{J}{u}_x(\boldsymbol{x}) T_{xx}(\boldsymbol{x}, \overset{i}{\boldsymbol{x}}) + \overset{J}{u}_y(\boldsymbol{x}) T_{xy}(\boldsymbol{x}, \overset{i}{\boldsymbol{x}}) \right] \mathrm{d}S(\boldsymbol{x})$$

(7.105)

$$c_{yx}u_x(\overset{i}{\boldsymbol{x}}) + c_{yy}u_y(\overset{i}{\boldsymbol{x}}) = \sum_{J=1}^{N} \int_{\overset{J}{\Delta S}} \left[\overset{J}{t}_x(\boldsymbol{x}) U_{yx}(\boldsymbol{x}, \overset{i}{\boldsymbol{x}}) + \overset{J}{t}_y(\boldsymbol{x}) U_{yy}(\boldsymbol{x}, \overset{i}{\boldsymbol{x}}) \right] \mathrm{d}S(\boldsymbol{x})$$
$$- \sum_{J=1}^{N} \fint_{\overset{J}{\Delta S}} \left[\overset{J}{u}_x(\boldsymbol{x}) T_{yx}(\boldsymbol{x}, \overset{i}{\boldsymbol{x}}) + \overset{J}{u}_y(\boldsymbol{x}) T_{yy}(\boldsymbol{x}, \overset{i}{\boldsymbol{x}}) \right] \mathrm{d}S(\boldsymbol{x})$$

where $\boldsymbol{x}$ is given by (7.85). We will assume in this section that the load point $\overset{i}{\boldsymbol{x}}$ is not located at one of the three nodes of element J (the sample points for the element) or at an endpoint of a connecting element $J-1$ or $J+1$. In this case all of the other integrals in (7.105) are nonsingular and can be evaluated in the same way as before, with $\xi_x = \overset{i}{x}$ and $\xi_y = \overset{i}{y}$ representing one of the three nodal points on element J. We will consider in the next section an indirect procedure for evaluating the Cauchy principal value integrals in (7.105).

Integrals such as

$$\int_{\overset{J}{\Delta S}} \overset{J}{t}_x(\boldsymbol{x}) U_{xx}(\boldsymbol{x}, \overset{i}{\boldsymbol{x}})\, \mathrm{d}S(\boldsymbol{x}) \tag{7.106}$$

and

$$\int_{\overset{J}{\Delta S}} \overset{J}{u}_y(\boldsymbol{x}) T_{yy}(\boldsymbol{x}, \overset{i}{\boldsymbol{x}})\, \mathrm{d}S(\boldsymbol{x}) \tag{7.107}$$

in (7.105) can be evaluated in an analogous way to that described previously. Instead of (7.90) we now have

$$U_{xx}(\boldsymbol{x}, \overset{i}{\boldsymbol{x}}) = -\frac{1}{8\mu\pi(1-\nu)} \left[(3-4\nu) \ln \sqrt{(x-\overset{i}{x})^2 + (y-\overset{i}{y})^2} - \frac{(x-\overset{i}{x})^2}{(x-\overset{i}{x})^2 + (y-\overset{i}{y})^2} \right] \tag{7.108}$$

in which

$$x - \overset{i}{x} = \psi_1(\eta)\overset{J}{x}_1 + \psi_2(\eta)\overset{J}{x}_2 + \psi_3(\eta)\overset{J}{x}_3 - \overset{i}{x}$$

(7.109)

$$y - \overset{i}{y} = \psi_1(\eta)\overset{J}{y}_1 + \psi_2(\eta)\overset{J}{y}_2 + \psi_3(\eta)\overset{J}{y}_3 - \overset{i}{y}$$

Also, using (4.265), $T_{yy}(\boldsymbol{x}, \overset{i}{\boldsymbol{x}})$ can be written as

$$T_{yy}(\boldsymbol{x}, \overset{i}{\boldsymbol{x}}) = -\frac{1}{4\pi(1-\nu)}\Big[(1-2\nu)\frac{x-\overset{i}{x}}{(x-\overset{i}{x})^2+(y-\overset{i}{y})^2} + \frac{2(x-\overset{i}{x})(y-\overset{i}{y})^2}{\big((x-\overset{i}{x})^2+(y-\overset{i}{y})^2\big)^2}\Big]n_x(\boldsymbol{x})$$
$$-\frac{1}{4\pi(1-\nu)}\Big[2(1-\nu)\frac{y-\overset{i}{y}}{(x-\overset{i}{x})^2+(y-\overset{i}{y})^2} - (y-\overset{i}{y})\frac{(x-\overset{i}{x})^2-(y-\overset{i}{y})^2}{\big((x-\overset{i}{x})^2+(y-\overset{i}{y})^2\big)^2}\Big]n_y(\boldsymbol{x}) \tag{7.110}$$

where the components of the unit outward normal $n_x(\boldsymbol{x})$ and $n_y(\boldsymbol{x})$ are given by (7.104).

Substitution of (7.109) into (7.108) and (7.110) shows that $U_{xx}(\boldsymbol{x}, \overset{i}{\boldsymbol{x}})$ and $T_{yy}(\boldsymbol{x}, \overset{i}{\boldsymbol{x}})$ are functions of the intrinsic coordinate η, i.e.,

$$U_{xx}(\boldsymbol{x}, \overset{i}{\boldsymbol{x}}) = U_{xx}(\eta) \tag{7.111}$$

and

$$T_{yy}(\boldsymbol{x}, \overset{i}{\boldsymbol{x}}) = T_{yy}(\eta) \tag{7.112}$$

The other components of the tensor fields U_{pq} and T_{pq} in (7.105) can be written in a similar way by referring to (4.251) and (4.265).

Integrals (7.106) and (7.107) can be expressed as

$$\int_{\overset{J}{\Delta S}} \overset{J}{t}_x(\boldsymbol{x}) U_{xx}(\boldsymbol{x}, \overset{i}{\boldsymbol{x}})\,\mathrm{d}S(\boldsymbol{x}) = \sum_{k=1}^{3} (\overset{J}{t}_x)_k \int_{-1}^{1} U_{xx}(\eta)\psi_k(\eta)\mathcal{J}(\eta)\,\mathrm{d}\eta \tag{7.113}$$

and

$$\int_{\overset{J}{\Delta S}} \overset{J}{u}_y(\boldsymbol{x}) T_{yy}(\boldsymbol{x}, \overset{i}{\boldsymbol{x}})\,\mathrm{d}S(\boldsymbol{x}) = \sum_{k=1}^{3} (\overset{J}{u}_y)_k \int_{-1}^{1} T_{yy}(\eta)\psi_k(\eta)\mathcal{J}(\eta)\,\mathrm{d}\eta \tag{7.114}$$

and similarly for the other discretized integrals in (7.105). The indicated integrations can be carried out numerically using (7.96).

7.17.5 Indirect Evaluation of Cauchy Principal Value Integrals

The terms in (7.105) involving Cauchy principal value integrals can be evaluated indirectly using the procedure outlined in Sect. 7.16. These terms arise when the load point $\overset{i}{\boldsymbol{x}}$ is at either the midpoint of element I, $\overset{I}{\boldsymbol{x}}_2$, or one of its two endpoints,

$\overset{I}{\boldsymbol{x}}_1$ or $\overset{I}{\boldsymbol{x}}_3$. In the latter case, two connecting elements, $I-1$ and $I+1$ are also involved because they each share an endpoint with element I. In particular, referring to Figs. 7.9 and 7.10 with I in place of J, $\overset{I-1}{\boldsymbol{x}}_3 = \overset{I}{\boldsymbol{x}}_1$ and $\overset{I+1}{\boldsymbol{x}}_1 = \overset{I}{\boldsymbol{x}}_3$. Therefore, the integrations for elements $I-1$ and $I+1$ also require special treatment.

We will consider below the case that $\overset{i}{\boldsymbol{x}}$ is at the midpoint of element I, i.e., $\overset{i}{\boldsymbol{x}} = \overset{I}{\boldsymbol{x}}_2$; the analyses for the other situations are similar. We note that this general approach was originally devised by Lachat and Watson 1976 and is in widespread use today.

Recall that (7.72) is written for the case that the boundary tractions are zero. This equation can be discretized for the case that $\overset{i}{\boldsymbol{x}} = \overset{I}{\boldsymbol{x}}_2$, as follows:

$$\begin{aligned} c_{xx}(\overset{I}{u}_x)_2 + c_{xy}(\overset{I}{u}_y)_2 &= -\fint_{\overset{I}{\Delta S}} \overset{I}{u}_x(\boldsymbol{x}) T_{xx}(\boldsymbol{x}, \overset{I}{\boldsymbol{x}}_2) \mathrm{d}S(\boldsymbol{x}) - \fint_{\overset{I}{\Delta S}} \overset{I}{u}_y(\boldsymbol{x}) T_{xy}(\boldsymbol{x}, \overset{I}{\boldsymbol{x}}_2) \mathrm{d}S(\boldsymbol{x}) \\ &- \sum_{J=1;\,J\neq I}^{N} \left\{ \int_{\overset{J}{\Delta S}} \overset{J}{u}_x(\boldsymbol{x}) T_{xx}(\boldsymbol{x}, \overset{I}{\boldsymbol{x}}_2) \mathrm{d}S(\boldsymbol{x}) + \int_{\overset{J}{\Delta S}} \overset{J}{u}_y(\boldsymbol{x}) T_{xy}(\boldsymbol{x}, \overset{I}{\boldsymbol{x}}_2) \mathrm{d}S(\boldsymbol{x}) \right\} \end{aligned} \tag{7.115}$$

$$\begin{aligned} c_{yx}(\overset{I}{u}_x)_2 + c_{yy}(\overset{I}{u}_y)_2 &= -\fint_{\overset{I}{\Delta S}} \overset{I}{u}_x(\boldsymbol{x}) T_{yx}(\boldsymbol{x}, \overset{I}{\boldsymbol{x}}_2) \mathrm{d}S(\boldsymbol{x}) - \fint_{\overset{I}{\Delta S}} \overset{I}{u}_y(\boldsymbol{x}) T_{yy}(\boldsymbol{x}, \overset{I}{\boldsymbol{x}}_2) \mathrm{d}S(\boldsymbol{x}) \\ &- \sum_{J=1;\,J\neq I}^{N} \left\{ \int_{\overset{J}{\Delta S}} \overset{J}{u}_x(\boldsymbol{x}) T_{yx}(\boldsymbol{x}, \overset{I}{\boldsymbol{x}}_2) \mathrm{d}S(\boldsymbol{x}) + \int_{\overset{J}{\Delta S}} \overset{J}{u}_y(\boldsymbol{x}) T_{yy}(\boldsymbol{x}, \overset{I}{\boldsymbol{x}}_2) \mathrm{d}S(\boldsymbol{x}) \right\} \end{aligned}$$

in which we have written $(\overset{I}{u}_x)_2$ and $(\overset{I}{u}_y)_2$ for $u_x(\overset{I}{\boldsymbol{x}}_2)$ and $u_y(\overset{I}{\boldsymbol{x}}_2)$ (cf. Fig. 7.9).

The summed terms on the right-hand side of (7.115) do not contain any singularities and can be expressed as functions of the intrinsic coordinate η for element J by using (7.85) and the relations

$$\begin{aligned} \overset{J}{u}_x &= \psi_1(\eta)(\overset{J}{u}_x)_1 + \psi_2(\eta)(\overset{J}{u}_x)_2 + \psi_3(\eta)(\overset{J}{u}_x)_3 \\ \overset{J}{u}_y &= \psi_1(\eta)(\overset{J}{u}_y)_1 + \psi_2(\eta)(\overset{J}{u}_y)_2 + \psi_3(\eta)(\overset{J}{u}_y)_3 \end{aligned} \tag{7.116}$$

This gives, for example,

$$\int_{\overset{J}{\Delta S}} \overset{J}{u}_x(\boldsymbol{x}) T_{xx}(\boldsymbol{x}, \overset{I}{\boldsymbol{x}}_2) \mathrm{d}S(\boldsymbol{x}) = \sum_{k=1}^{3} (\overset{J}{u}_x)_k \int_{-1}^{1} \overset{J}{T}_{xx}(\eta) \psi_k(\eta) \mathcal{J}(\eta) \mathrm{d}\eta \tag{7.117}$$

in which we have introduced the notation $\overset{J}{T}_{xx}(\eta)$ to emphasize that $T_{xx}(\eta)$ is calculated using the global coordinates

$$\begin{aligned} x - \overset{I}{x}_2 &= \psi_1(\eta)\overset{J}{x}_1 + \psi_2(\eta)\overset{J}{x}_2 + \psi_3(\eta)\overset{J}{x}_3 - \overset{I}{x}_2 \\ y - \overset{I}{y}_2 &= \psi_1(\eta)\overset{J}{y}_1 + \psi_2(\eta)\overset{J}{y}_2 + \psi_3(\eta)\overset{J}{y}_3 - \overset{I}{y}_2 \end{aligned} \tag{7.118}$$

cf. (7.91). The values of $\overset{J}{T}_{xx}(\eta)$, etc., are obtained using (7.118) and (4.265).

A similar calculation can be made for the terms in (7.115) involving element I. In this case we write

$$\fint_{\overset{I}{\Delta S}} \overset{I}{u}_x(\boldsymbol{x}) T_{xx}(\boldsymbol{x}, \overset{I}{\boldsymbol{x}}_2) \mathrm{d}S(\boldsymbol{x}) = (\overset{I}{u}_x)_1 \int_{-1}^{1} \overset{I}{T}_{xx}(\eta)\psi_1(\eta)\mathcal{J}(\eta)\mathrm{d}\eta + (\overset{I}{u}_x)_2 \fint_{-1}^{1} \overset{I}{T}_{xx}(\eta)\psi_2(\eta)\mathcal{J}(\eta)\mathrm{d}\eta$$
$$+ (\overset{I}{u}_x)_3 \int_{-1}^{1} \overset{I}{T}_{xx}(\eta)\psi_3(\eta)\mathcal{J}(\eta)\mathrm{d}\eta \qquad (7.119)$$

with comparable expressions for the other Cauchy principal value integrals in (7.115). Because we have taken the load point $\overset{i}{\boldsymbol{x}}$ at the midpoint node of element I, the integrals involving $\psi_1(\eta)$ and $\psi_3(\eta)$ in (7.119) are nonsingular.

Collecting results, we write

$$c_{xx}(\overset{I}{u}_x)_2 + c_{xy}(\overset{I}{u}_y)_2 = -(\overset{I}{u}_x)_1 \int_{-1}^{1} \overset{I}{T}_{xx}(\eta)\psi_1(\eta)\mathcal{J}(\eta)\mathrm{d}\eta - (\overset{I}{u}_x)_2 \fint_{-1}^{1} \overset{I}{T}_{xx}(\eta)\psi_2(\eta)\mathcal{J}(\eta)\mathrm{d}\eta$$
$$- (\overset{I}{u}_x)_3 \int_{-1}^{1} \overset{I}{T}_{xx}(\eta)\psi_3(\eta)\mathcal{J}(\eta)\mathrm{d}\eta - (\overset{I}{u}_y)_1 \int_{-1}^{1} \overset{I}{T}_{xx}(\eta)\psi_1(\eta)\mathcal{J}(\eta)\mathrm{d}\eta$$
$$- (\overset{I}{u}_y)_2 \fint_{-1}^{1} \overset{I}{T}_{xy}(\eta)\psi_2(\eta)\mathcal{J}(\eta)\mathrm{d}\eta - (\overset{I}{u}_y)_3 \int_{-1}^{1} \overset{I}{T}_{xy}(\eta)\psi_3(\eta)\mathcal{J}(\eta)\mathrm{d}\eta \qquad (7.120)$$
$$- \sum_{J=1;J\neq I}^{N} \left\{ \sum_{k=1}^{3} (\overset{J}{u}_x)_k \int_{-1}^{1} \overset{J}{T}_{xx}(\eta)\psi_k(\eta)\mathcal{J}(\eta)\mathrm{d}\eta + \sum_{k=1}^{3} (\overset{J}{u}_y)_k \int_{-1}^{1} \overset{J}{T}_{xy}(\eta)\psi_k(\eta)\mathcal{J}(\eta)\mathrm{d}\eta \right\}$$

with a similar expression for the second equation of (7.115).

If we now assume that the entire boundary is displaced by a constant amount $(u_x)_0$ in the x direction, with $u_y = 0$, (7.120) can be written as

$$c_{xx} + \fint_{-1}^{1} \overset{I}{T}_{xx}(\eta)\psi_2(\eta)\mathcal{J}(\eta)\mathrm{d}\eta = -\int_{-1}^{1} \overset{I}{T}_{xx}(\eta)\psi_1(\eta)\mathcal{J}(\eta)\mathrm{d}\eta - \int_{-1}^{1} \overset{I}{T}_{xx}(\eta)\psi_3(\eta)\mathcal{J}(\eta)\mathrm{d}\eta$$
$$- \sum_{J=1;J\neq I}^{N} \sum_{k=1}^{3} \int_{-1}^{1} \overset{J}{T}_{xx}(\eta)\psi_k(\eta)\mathcal{J}(\eta)\mathrm{d}\eta \qquad (7.121)$$

This result shows that the combination of the jump term c_{xx} and the Cauchy principal value integral involving T_{xx} for a midpoint node can be calculated indirectly by summing the integrals involving T_{xx} for all the other nodes. This result can easily be extended to give similar results for the other Cauchy principal value integrals, e.g.,

$$c_{xy} + \fint_{-1}^{1} \overset{I}{T}_{xy}(\eta)\psi_2(\eta)\mathcal{J}(\eta)\mathrm{d}\eta = -\int_{-1}^{1} \overset{I}{T}_{xy}(\eta)\psi_1(\eta)\mathcal{J}(\eta)\mathrm{d}\eta - \int_{-1}^{1} \overset{I}{T}_{xy}(\eta)\psi_3(\eta)\mathcal{J}(\eta)\mathrm{d}\eta$$
$$- \sum_{J=1;J\neq I}^{N} \sum_{k=1}^{3} \int_{-1}^{1} \overset{J}{T}_{xy}(\eta)\psi_k(\eta)\mathcal{J}(\eta)\mathrm{d}\eta \qquad (7.122)$$

and so forth.

Finally, when the load point $\overset{i}{\boldsymbol{x}}$ is at an endpoint node, say the node at $\overset{I}{\boldsymbol{x}}_1$ (or, equivalently, the node at $\overset{I-1}{\boldsymbol{x}}_3$), the left-hand side of (7.121) will be

$$c_{xx} + \fint_{-1}^{1} \overset{I-1}{T}_{xx}(\eta)\psi_3(\eta)\mathcal{J}(\eta)\mathrm{d}\eta + \fint_{-1}^{1} \overset{I}{T}_{xx}(\eta)\psi_1(\eta)\mathcal{J}(\eta)\mathrm{d}\eta \tag{7.123}$$

and terms on the right-hand side of (7.121) will involve all the other boundary points. The two integrals in this expression are identified as Hadamard finite part integrals because the load point is located at an endpoint node. This does not allow for the selection of a symmetric interval around the load point, as required by the definition of a Cauchy principal value integral (see, for example, (2.19)). Of course, these integrals are not explicitly evaluated in our indirect solution procedure.

7.17.6 Integration of Logarithm Terms

The logarithm term in $U_{xx}(\boldsymbol{x}, \overset{i}{\boldsymbol{x}})$ in (7.108) (and also in a similar expression for $U_{yy}(\boldsymbol{x}, \overset{i}{\boldsymbol{x}})$) needs separate treatment for the case that $\overset{i}{\boldsymbol{x}}$ is at one of the three reference points of element J because the logarithm is singular at these points. We consider this circumstance next. It is important to point out, however, that $U_{xx}(\boldsymbol{x}, \overset{i}{\boldsymbol{x}})$ is well defined for all other elements and standard Gaussian quadrature can be used to evaluate the integrals.

Special quadrature formulas have been developed to handle integrals of the form

$$\begin{aligned}\int_0^1 \ln\frac{1}{x} f(x)\,\mathrm{d}x &= -\int_0^1 \ln x f(x)\,\mathrm{d}x \\ &\cong \sum_{i=1}^{n} w_i f(x_i)\end{aligned} \tag{7.124}$$

e.g., Stroud and Secrest (1966). The abscissae x_i and weight factors w_i in this formula automatically account for the presence of the singularity at the lower limit of integration, $x = 0$. (The values of x_i and w_i in (7.124) are clearly not the same as those in (7.96); it is standard practice in numerical analysis to use the same symbols for different quadrature formulas.)

To use (7.124) for integration of $U_{xx}(\boldsymbol{x}, \overset{i}{\boldsymbol{x}})$ it is convenient to separate this quantity into two parts: one part with the singular term and the other without. Thus, we write

$$U_{xx}(\boldsymbol{x}, \overset{i}{\boldsymbol{x}}) = U^{*}_{xx}(\boldsymbol{x}, \overset{i}{\boldsymbol{x}}) + U^{**}_{xx}(\boldsymbol{x}, \overset{i}{\boldsymbol{x}}) \tag{7.125}$$

in which

$$U^{*}_{xx}(\boldsymbol{x}, \overset{i}{\boldsymbol{x}}) = -\frac{(3-4\nu)}{8\mu\pi(1-\nu)} \ln\sqrt{(x-\overset{i}{x})^2 + (y-\overset{i}{y})^2} \tag{7.126}$$

and

$$U^{**}_{xx}(\boldsymbol{x}, \overset{i}{\boldsymbol{x}}) = +\frac{1}{8\mu\pi(1-\nu)} \frac{(x-\overset{i}{x})^2}{(x-\overset{i}{x})^2 + (y-\overset{i}{y})^2} \tag{7.127}$$

Equation (7.127) can be integrated using ordinary Gaussian quadrature and does not require special attention. The same decomposition as in (7.125) shows that $U^{*}_{yy}(\boldsymbol{x}, \overset{i}{\boldsymbol{x}}) = U^{*}_{xx}(\boldsymbol{x}, \overset{i}{\boldsymbol{x}})$ so it is sufficient to consider (7.126) for both cases. We therefore seek to evaluate

$$\int_{\overset{J}{\Delta S}} \overset{J}{t}_x(\boldsymbol{x}) U^{*}_{xx}(\boldsymbol{x}, \overset{i}{\boldsymbol{x}})\, \mathrm{d}S(\boldsymbol{x}) \tag{7.128}$$

for the case that the load point $\overset{i}{\boldsymbol{x}}$ is, in turn, located at each of the three reference points of element J: $\overset{J}{\boldsymbol{x}}_1 = (\overset{J}{x}_1, \overset{J}{y}_1)$, $\overset{J}{\boldsymbol{x}}_2 = (\overset{J}{x}_2, \overset{J}{y}_2)$, and $\overset{J}{\boldsymbol{x}}_3 = (\overset{J}{x}_3, \overset{J}{y}_3)$. A similar calculation is needed when $\overset{i}{\boldsymbol{x}}$ is one of the endpoint nodes of the connecting elements, $J-1$ and $J+1$, but the details for this are the same as discussed below.

Considering first the case that $\overset{i}{\boldsymbol{x}} = \overset{J}{\boldsymbol{x}}_1$, we introduce a change of variable $\eta = 2t - 1$ and write the shape functions (7.77) as

$$\begin{aligned}\psi_1(t) &= (t-1)(2t-1) \\ \psi_2(t) &= 4t(1-t) \\ \psi_3(t) &= t(2t-1)\end{aligned} \tag{7.129}$$

The new intrinsic coordinate t is defined for the range $0 \le t \le 1$ specified in quadrature formula (7.124) and the reference points 1, 2, and 3 are at $t = 0$, $t = 1/2$, and $t = 1$, respectively. The x, y coordinates of a point along element J are given as

$$\begin{aligned}x - \overset{J}{x}_1 &= \psi_1(t)\overset{J}{x}_1 + \psi_2(t)\overset{J}{x}_2 + \psi_3(t)\overset{J}{x}_3 - \overset{J}{x}_1 \\ y - \overset{J}{y}_1 &= \psi_1(t)\overset{J}{y}_1 + \psi_2(t)\overset{J}{y}_2 + \psi_3(t)\overset{J}{y}_3 - \overset{J}{y}_1\end{aligned} \tag{7.130}$$

Combining (7.129) and (7.130) gives

$$\begin{aligned}x - \overset{J}{x}_1 &= t\Big[2t\big(\overset{J}{x}_1 - 2\overset{J}{x}_2 + \overset{J}{x}_3\big) - \big(3\overset{J}{x}_1 - 4\overset{J}{x}_2 + \overset{J}{x}_3\big)\Big] \\ &= t\big(A_x t + B_x\big) \\ y - \overset{J}{y}_1 &= t\Big[2t\big(\overset{J}{y}_1 - 2\overset{J}{y}_2 + \overset{J}{y}_3\big) - \big(3\overset{J}{y}_1 - 4\overset{J}{y}_2 + \overset{J}{y}_3\big)\Big] \\ &= t\big(A_y t + B_y\big)\end{aligned} \tag{7.131}$$

We then have that

$$
\begin{aligned}
U^*_{xx}(\boldsymbol{x}, \overset{J}{\boldsymbol{x}}_1) &= -\frac{(3-4\nu)}{8\mu\pi(1-\nu)} \ln\Big[t^2\Big\{\big(A_x t + B_x\big)^2 + \big(A_y t + B_y\big)^2\Big\}\Big]^{1/2} \\
&= -\frac{(3-4\nu)}{8\mu\pi(1-\nu)} \ln t - \frac{(3-4\nu)}{8\mu\pi(1-\nu)} \ln\Big[\big(A_x t + B_x\big)^2 + \big(A_y t + B_y\big)^2\Big]^{1/2}
\end{aligned}
\tag{7.132}
$$

The second term in this expression is nonsingular and can be combined with the quantity $U^{**}_{xx}(\boldsymbol{x}, \overset{J}{\boldsymbol{x}}_1)$ and integrated using (7.96). Denoting the first term as $\hat{U}^*_{xx}(\boldsymbol{x}, \overset{J}{\boldsymbol{x}}_1)$ and proceeding as before we find

$$
\int_{\overset{J}{\Delta S}} \overset{J}{t}_x(\boldsymbol{x})\hat{U}^*_{xx}(\boldsymbol{x}, \overset{J}{\boldsymbol{x}}_1)\,\mathrm{d}S(\boldsymbol{x}) = -\frac{(3-4\nu)}{8\mu\pi(1-\nu)} \sum_{k=1}^{3} (\overset{J}{t}_x)_k \int_0^1 \ln(t)\psi_k(t)\mathcal{J}(t)\,\mathrm{d}t
\tag{7.133}
$$

in which the Jacobian is

$$
\begin{aligned}
\mathcal{J}(t) &= \left[\left(\frac{\mathrm{d}x(t)}{\mathrm{d}t}\right)^2 + \left(\frac{\mathrm{d}y(t)}{\mathrm{d}t}\right)^2\right]^{1/2} \\
&= \left[\left\{(4t-3)\overset{J}{x}_1 + 4(1-2t)\overset{J}{x}_2 + (4t-1)\overset{J}{x}_3\right\}^2 + \left\{(4t-3)\overset{J}{y}_1 + 4(1-2t)\overset{J}{y}_2 + (4t-1)\overset{J}{y}_3\right\}^2\right]
\end{aligned}
\tag{7.134}
$$

Quadrature formula (7.124) can be applied directly to (7.133).

A similar process is followed for the case that the load point is at the other end of the element, i.e., at $\overset{i}{\boldsymbol{x}} = \overset{J}{\boldsymbol{x}}_3$. In order to have the singularity at this point, however, we use the change of variable $\eta = 1 - 2t$ and write the shape functions (7.77) as

$$
\begin{aligned}
\psi_1(t) &= t(2t-1) \\
\psi_2(t) &= 4t(1-t) \\
\psi_3(t) &= (t-1)(2t-1)
\end{aligned}
\tag{7.135}
$$

Following the same procedure as above, we find that

$$
\int_{\overset{J}{\Delta S}} \overset{J}{t}_x(\boldsymbol{x})\hat{U}^*_{xx}(\boldsymbol{x}, \overset{J}{\boldsymbol{x}}_3)\,\mathrm{d}S(\boldsymbol{x}) = -\frac{(3-4\nu)}{8\mu\pi(1-\nu)} \sum_{k=1}^{3} (\overset{J}{t}_x)_k \int_0^1 \ln(t)\psi_k(t)\mathcal{J}(t)\,\mathrm{d}t
\tag{7.136}
$$

in which now

$$
\mathcal{J}(t) = \left[\left\{(4t-1)\overset{J}{x}_1 + 4(1-2t)\overset{J}{x}_2 + (4t-3)\overset{J}{x}_3\right\}^2 + \left\{(4t-1)\overset{J}{y}_1 + 4(1-2t)\overset{J}{y}_2 + (4t-3)\overset{J}{y}_3\right\}^2\right]
\tag{7.137}
$$

Finally, when the singularity is at reference point 2, i.e., when $\overset{i}{\boldsymbol{x}} = \overset{J}{\boldsymbol{x}}_2$, it is necessary to perform the integation over the element in two parts, first from reference point 1 to point 2, and then from point 2 to point 3. For the first part we set $\eta = -t$ in (7.77) and for the second we use $\eta = t$. Both integrals are then in the form required to use the special integration formula (7.124). Details of these calculations are left as an exercise for the interested reader.

7.17.7 Setting Up the System of Algebraic Equations

For the discretization of boundary integrals (7.84) we did not need to identify individual nodal point numbers and could simply refer to the three reference points for integration along element J. This was possible because we assumed that the load point $\boldsymbol{\xi}$ was not on the boundary and that the boundary displacements and tractions were known for all the elements, $J = 1$ to N. The situation is different, however, when we discretize (4.353). In this case, point $\boldsymbol{\xi}$ is on the boundary and only one-half of the boundary data is known. In order to keep track of the unknown boundary displacements and/or tractions, and then set up a system of algebraic equations from which to find them, it is useful to label them according to their nodal point numbers.

After discretizing all the terms in (7.105) a system of algebraic equations can be formed to find the boundary displacement and traction components that were not prescribed at the outset. The displacements are single-valued at all the boundary nodal points but the tractions at the element endpoints are not necessarily the same, just as in the case of linear variations of the boundary parameters considered in Sect. 7.6. To simplify the following discussion we will suppose that the tractions are specified at all the nodal points so that the boundary displacements are the unknowns for the problem. A similar, but somewhat more elaborate, process can be followed for mixed boundary value problems.

The required system of equations will be of the form

$$\begin{aligned} \sum_{j=1}^{2N} \overset{ij}{A}_{xx} \overset{j}{u}_x + \sum_{j=1}^{2N} \overset{ij}{A}_{xy} \overset{j}{u}_y &= \overset{i}{b}_x \\ \sum_{j=1}^{2N} \overset{ij}{A}_{yx} \overset{j}{u}_x + \sum_{j=1}^{2N} \overset{ij}{A}_{yy} \overset{j}{u}_y &= \overset{i}{b}_y \end{aligned} \tag{7.138}$$

for $i = 1$ to $2N$. These equations are assembled, as usual, by letting points $\overset{i}{\boldsymbol{x}}$ and $\overset{j}{\boldsymbol{x}}$ each occupy all of the $2N$ nodal points along the boundary S. Thus, for a fixed value of $\overset{i}{\boldsymbol{x}}$ we let $\overset{j}{\boldsymbol{x}}$ range from 1 to $2N$ to calculate the influence coefficients $\overset{ij}{A}_{xx}$, etc., a process that is then repeated for the next value of $\overset{i}{\boldsymbol{x}}$. The coefficients $\overset{ij}{A}_{xx}$, etc., in (7.138) involve integrals with T_{ij} on the right-hand side of (7.105) while the quantities $\overset{i}{b}_x$ and $\overset{i}{b}_y$ represent the sums of the of integrals the integrals with U_{ij}.

When the sample point $\overset{j}{\boldsymbol{x}}$ is at an element endpoint the coefficients $\overset{ij}{A}_{xx}$, etc., have contributions from two elements, e.g.,

$$\overset{ij}{A}_{xx} = (\overset{ij}{A}_{xx})_{J-1} + (\overset{ij}{A}_{xx})_{J} \tag{7.139}$$

and so forth. The logic within a computer program can easily accommodate this requirement. The components of the unit outward normals at the endpoints of two connecting elements will in general be different but this is of no consequence because these components are associated with the individual elements, as indicated, for example, by (7.110). The quantities $\overset{i}{b}_x$ and $\overset{i}{b}_y$ in (7.138) can be calculated directly for these points.

7.18 Exercises/Projects

1. Prepare a computer program for the two-dimensional direct boundary integral method for isotropic elasticity, as outlined in Sects. 7.2–7.4. Verify your program by solving the example problem described in Sect. 7.5 and use the approach described in Sect. 6.10 to calculate the hoop stress for the problem.
2. Prepare a computer program comparable to that from Exercise 1 but for the case that the displacements and tractions vary linearly over straight line elements, as in Sects. 7.6–7.12.
3. Modify your program from Exercise 2 to follow the approach explained in Sect. 7.15 for calculating the self-effect terms for the boundary influence coefficients.
4. Rewrite your program from Exercise 3 using numerical integration to evaluate all of the boundary integrals.
5. Modify your program from Exercise 4 for isoparametric quadratic elements as described in Sect. 7.17.
6. Prepare a computer program for the two-dimensional direct boundary integral method for anisotropic elasticity using quadratic isoparametric elements.

References

Abramowitz, M., and I.S. Stegun. 1965. *Handbook of mathematical functions*. Dover.

Crouch, S.L., and A.M. Starfield. 1983. *Boundary element methods in solid mechanics*. George Allen & Unwin.

Lachat, J.C., and J.O. Watson. 1976. Effective numerical treatment of boundary integral equations: A formulation for three-dimensional elastostatics. *The International Journal for Numerical Methods in Engineering* 10: 991–1005.

Muskhelishvili, N. I. 1963. *Some basic problems of the mathematical theory of elasticity*, 2nd (English) edn. P. Noordhoff.

Sokolnikoff, I.S. 1956. *Mathematical theory of elasticity*, 2nd ed. McGraw-Hill.

Stroud, A.H., and D. Secrest. 1966. *Gaussian quadrature formulas*. Prentice-Hall.

Timoshenko, S. P., and J. N. Goodier. 1970. *Theory of elasticity*, 3rd. edn. McGraw-Hill.

References

Chapter 8
Extensions and Refinements

8.1 Introduction

Three different two-dimensional boundary element methods for elasticity were introduced in the previous three chapters of this book. These methods can all be used to solve general boundary value problems in elasticity and they have in common the important characteristic of being able to treat infinite regions without introducing any additional approximations. As noted in Chap. 4, this is because elasticity problems can be formulated in terms of the stress changes that occur when a cavity is introduced in an infinite region; the resulting stress and displacement *changes* are zero at infinity when the boundary loading is in equilibrium. We will show in this chapter that boundary element methods can be extended to treat other types of elasticity problems, including those involving semi-infinite and piecewise homogeneous regions.

Before discussing these and other extensions of the basic approach, it is worth summarizing the basic steps needed to write a computer program for a boundary element method, irrespective of the computer programming language used. In addition to graphics pre- and post-processing packages, a typical boundary element computer program consists of five parts, or modules (Crouch and Starfield 1983):

- Module 1 defines the locations of the boundary elements and specifies the boundary conditions for each one;
- Module 2 computes the boundary influence coefficients and sets up the system of algebraic equations in terms of the specified boundary conditions;
- Module 3 solves the system of algebraic equations;
- Module 4 computes the unknown displacement and/or stress components at each boundary element; and
- Module 5 provides the option to compute displacements and stresses at specified points inside the region of interest.

This program structure is independent of the boundary element method being used. In fact, as one changes from one boundary element method to another, only modules

S. L. Crouch and S. G. Mogilevskaya, *A First Course in Boundary Element Methods*,
Mathematical Engineering, https://doi.org/10.1007/978-3-031-63341-6_8

2 and 5 need to be changed. The modular character remains intact for all of the extensions and refinements described in this chapter. It will be seen that the necessary modifications within modules can always be made in a straightforward manner, even when the derivation of the appropriate formulas is not especially simple.

It follows that the boundary element technique provides a highly flexible approach to the numerical solution of problems, one that enables a user to employ a level of sophistication appropriate to a particular situation. In some cases, mathematical solutions to certain singular problems in elasticity can be exploited to provide a more satisfactory numerical solution for special classes of problems. Examples of this are the half-plane and bonded half-plane solutions of Sects. 8.3 and 8.5. In other cases, tailor-made elements can be developed to solve a particular problem. An example associated with calculation of the stress intensity factors at a crack tip is given in Sect. 8.11.

But with three methods to choose from, the reader might well ask: "Which method should I use?" Although there is not a simple answer to this question, a few observations about the methods can be made. The displacement discontinuity method is specifically designed for problems involving cracks or crack-like features such as rock joints, faults, and bedding planes whereas the other two methods are less suitable for such problems. The stress discontinuity and direct boundary integral methods are better choices for problems involving multiply-connected regions (i.e., those with holes) than the displacement discontinuity method because no special arrangements are required to prevent rigid body movements of the interiors of the holes. In addition, the stress discontinuity method has the advantage that it provides a value for the normal stress parallel to an element (the so-called hoop stress) whereas the other methods require additional numerical approximations to find this quantity. Versions of all three methods can be developed with higher order (e.g., linear or quadratic) variations of the boundary parameters. Sometimes, it might be beneficial to combine the methods to take advantage their individual strengths, a topic considered in Sect. 8.9.

The perceptive reader will have recognized that there are important differences between direct and indirect formulations of the boundary element approach. An indirect formulation such as the stress discontinuity method can be developed using a singular solution integrated over a straight line segment as a basic building block and then employing the principle of superposition to form a system of algebraic equations from which to find the unknown boundary parameters. This "influence function" process is more physical and intuitive than mathematical, and first-time users often find it easier to understand than a formulation developed by discretizing boundary integral equations. Of course, as we saw in Sect. 5.11, the stress discontinuity method can also be developed from a boundary integral formula (the analogue of the single layer potential) but this is not essential to the basic method. The same can be said for the displacement discontinuity method (the analogue of the double layer potential).

In the direct boundary integral method the roles of the source and sample points are reversed in comparison to an indirect formulation. This reversal is a consequence of the use of the reciprocal theorem to derive the underlying boundary integral equations for the method and, it can be argued, makes the direct formulation less amenable to a simple physical interpretation than an indirect one. This notwithstanding, the direct boundary integral approach is by far the method of choice of researchers in the field.

8.2 Anti-Plane Strain; Generalized Plane Strain

A state of *anti-plane strain* for the x, y plane is defined by the conditions $u_x = u_y = 0$ and $u_z = u_z(x, y)$. It can be shown from the Navier-Cauchy equations (4.81) that, if the body forces are zero, these conditions mean that $\nabla^2 u_z(x, y) = 0$. Using (4.12) and (4.30) it follows that the only nonzero stress components for an anti-plane strain problem are $\sigma_{xz} = \mu \partial u_z(x, y)/\partial x$ and $\sigma_{yz} = \mu \partial u_z(x, y)/\partial y$.

Generalized plane strain represents a combination of "ordinary" plane strain for the x, y plane (see Sect. 4.6) and anti-plane strain as defined above. The displacements and stresses in both cases are independent of z. The nonzero displacement and stress components for plane strain are u_x, u_y and σ_{xx}, σ_{yy}, σ_{xy}, $\sigma_{zz} = \nu(\sigma_{xx} + \sigma_{yy})$, while for anti-plane strain they are the out of plane displacement u_z and the shear stresses σ_{xz} and σ_{yz}.

Generalized plane strain is applicable to specialized three-dimensional problems in which the geometry is two-dimensional but the stress field has nonzero out of plane shear components. A long, straight tunnel in rock provides a good example of such a problem. If the tunnel is uniform along its length, a representative cross section can be treated as a plane strain problem for the x, y plane. If in addition the principal directions of the initial stress field are not aligned with the x, y, z coordinate axes the shear stresses σ_{xz} and σ_{yz} at infinity must be taken into account. (The normal stress σ_{zz} along the axis of the tunnel does not contribute to the displacements and stresses in the x, y plane.)

We have already introduced several boundary element methods for handling the plane strain part of a generalized plane strain problem so this aspect does not require further discussion. The anti-plane strain problem has not yet been addressed, however, and we turn our attention to this next.

Because the out of plane displacement $u_z(x, y)$ satisfies Laplace's equation in two dimensions, any of the boundary element methods discussed in Chaps. 1 and 3 can be used to solve an anti-plane strain problem. Using the direct method with Green's representation formula (2.60), and with the displacement $u_z(x, y)$ playing the part of the potential $\phi(x, y)$, we write

$$u_z(\boldsymbol{\xi}) = \int_S \frac{\partial u_z(\boldsymbol{x})}{\partial \boldsymbol{n}(\boldsymbol{x})} G(\boldsymbol{x}, \boldsymbol{\xi})\, \mathrm{d}S(\boldsymbol{x}) - \int_S u_z(\boldsymbol{x}) \frac{\partial G(\boldsymbol{x}, \boldsymbol{\xi})}{\partial \boldsymbol{n}(\boldsymbol{x})}\, \mathrm{d}S(\boldsymbol{x}) \tag{8.1}$$

in which $G(\boldsymbol{x}, \boldsymbol{\xi})$ and $\partial G(\boldsymbol{x}, \boldsymbol{\xi})/\partial \boldsymbol{n}(\boldsymbol{x})$ are given by (2.41) and (2.57), and where $\boldsymbol{x}$ is a sample point on the boundary S of region V and $\boldsymbol{\xi}$ is the load point inside this region. The quantity $\partial u_z(\boldsymbol{x})/\partial \boldsymbol{n}(\boldsymbol{x})$ can be written as

$$\begin{aligned}\frac{\partial u_z(\boldsymbol{x})}{\partial \boldsymbol{n}(\boldsymbol{x})} &= \frac{\partial}{\partial x_i} u_z(\boldsymbol{x}) n_i(\boldsymbol{x}) \\ &= \frac{\partial}{\partial x} u_z(\boldsymbol{x}) n_x(\boldsymbol{x}) + \frac{\partial}{\partial y} u_z(\boldsymbol{x}) n_y(\boldsymbol{x})\end{aligned} \tag{8.2}$$

or, using (4.12), (4.30), and (4.7),

$$\frac{\partial u_z(\boldsymbol{x})}{\partial \boldsymbol{n}(\boldsymbol{x})} = \frac{1}{\mu} t_z(\boldsymbol{x}) \tag{8.3}$$

We therefore have that

$$u_z(\boldsymbol{\xi}) = \int_S \frac{1}{\mu} t_z(\boldsymbol{x}) G(\boldsymbol{x}, \boldsymbol{\xi})\, \mathrm{d}S(\boldsymbol{x}) - \int_S u_z(\boldsymbol{x}) \frac{\partial G(\boldsymbol{x}, \boldsymbol{\xi})}{\partial \boldsymbol{n}(\boldsymbol{x})}\, \mathrm{d}S(\boldsymbol{\xi}) \tag{8.4}$$

This equation can be discretized in the same way as described in Chap. 3 and used to find an approximation to the displacements $u_z(\boldsymbol{x})$ or the tractions $t_z(\boldsymbol{x})$ along the boundary. In this regard, there is a formal equivalence between the quantities $\phi(\boldsymbol{x})$ and $q(\boldsymbol{x})$ in (3.1) and $u_z(\boldsymbol{x})$ and $t_z(\boldsymbol{x})/\mu$ in (8.4).

As an example, consider the problem of a circular tunnel of radius A in an infinite rock mass with shear modulus μ and shear stresses $\sigma_{xz} = p_{xz}$ and $\sigma_{yz} = p_{yz}$ at infinity (all other far-field stress components are taken as zero). The tractions on the boundary of the tunnel are zero. The analytical solution to this problem is recorded by Brady and Bray (1978) and, in the present notation, is given as

$$u_z = -\frac{1}{\mu}(x p_{xz} + y p_{yz}) \frac{A^2}{r^2} \tag{8.5}$$

and

$$\begin{aligned}\sigma_{rz} &= \left(\frac{x}{r} p_{xz} + \frac{y}{r} p_{yz}\right)\left(1 - \frac{A^2}{r^2}\right) \\ \sigma_{\theta z} &= -\left(\frac{y}{r} p_{xz} - \frac{x}{r} p_{yz}\right)\left(1 + \frac{A^2}{r^2}\right)\end{aligned} \tag{8.6}$$

A numerical solution to the problem was found using the equivalent of algebraic system (3.22) with 100 equal size boundary elements around the periphery of the hole. The stresses at infinity were taken as $p_{xz}/\mu = 1$ and $p_{yz}/\mu = 1/2$ and induced tractions $t_z(\boldsymbol{x})$ due to shear stresses $-p_{xz}/\mu$ and $-p_{yz}/\mu$ were applied to the boundary of the hole according to the procedure explained in Sect. 4.8. Representative values of the boundary displacement u_z (i.e., values of u_z for $r = A$) are compared with the analytical solution (8.5) in Fig. 8.1. Similarly, values of the stresses σ_{xz} and σ_{yz} along the x axis are compared with (8.6) in Fig. 8.2. In both cases, the numerical results are within plotting accuracy of the exact solution.

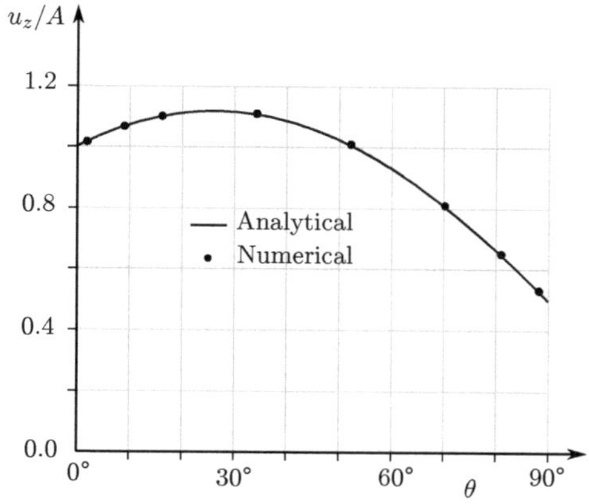

Fig. 8.1 Displacement u_z along the boundary of the tunnel for anti-plane strain problem

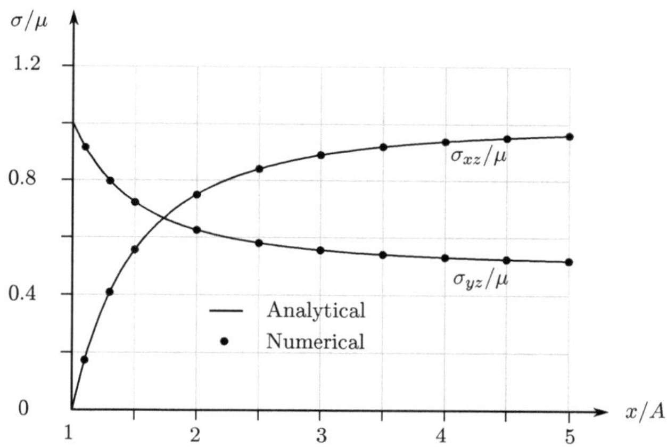

Fig. 8.2 Stresses σ_{xz} and σ_{yz} along x axis for anti-plane strain problem

8.3 Half-Plane Solutions

Many practical problems in solid mechanics involve bodies that can be idealized as elastic half-planes. The problem of an edge crack in a large elastic sheet under tension, for example, can be treated in this way. So too can certain problems in rock mechanics and geological engineering, where the boundary of the half-plane usually represents the surface of the earth, which is typically taken to be traction-free. A long, shallow tunnel parallel to the earth's surface, for instance, can often be represented by the plane strain (or generalized plane strain) problem of a hole near the boundary of a half-plane with a traction-free surface.

Half-plane problems can be solved using any of the boundary element methods for elasticity described in this book. The surface of the half-plane can be treated like any other boundary and divided into elements in the usual way, and the relevant equations can then be constructed so as to satisfy the required boundary conditions at these elements. Of course, it would be impossible to divide the entire surface into boundary elements, and so a "sufficiently long" portion of the surface must be treated in this manner. In most cases a rather large number of boundary elements is needed to give a satisfactory representation of the surface effects.

A more accurate and efficient way of solving half-plane problems can be developed by using special singular solutions that automatically meet the prescribed boundary conditions along the surface. Two such solutions for an elastic half-plane with a traction-free boundary are especially useful for our purposes: one for a concentrated force and the other for a concentrated displacement discontinuity in the interior of the half-plane. Using these solutions, boundary element methods can be developed that satisfy exactly the boundary conditions along the entire surface of the half-plane, so that boundary elements need only be placed along internal contours (e.g., holes or cavities within the half-plane). The related problem of a concentrated force applied directly to the surface of an elastic half-plane is also of interest because this solution can be integrated along the surface to provide general loading conditions instead of zero tractions.

In this section we will give derivations of the solutions for concentrated forces and concentrated displacement discontinuities for both isotropic and orthotropic (transversely isotropic) elastic half-planes. We will also show how the basic results can be used in practice but will leave their detailed implementations to the interested reader.

8.3.1 Concentrated Force in an Isotropic Elastic Half-Plane

The solution to the problem of a concentrated force in an isotropic, linearly elastic half-plane with a traction-free boundary was originally given by Melan (1932). We will present an alternative derivation of this solution below in order to maintain a consistent notation and also to document the basic procedure for later reference.

The subject problem is illustrated in Fig. 8.3. We suppose that the force has components $F_x > 0$ and $F_y > 0$ and is situated at the point $x = 0$, $y = -\ell$ in the half-plane $y \le 0$, as shown in the figure. We also suppose that the surface of the half-plane is traction-free, i.e., $t_x = \sigma_{xy} = 0$ and $t_y = \sigma_{yy} = 0$ on $y = 0$ for $-\infty < x < \infty$.

The solution to this problem is obtained in two stages using the method of images (cf. Sect. 1.10). In the first stage of the analysis an infinite body containing two concentrated forces is considered, as shown in Fig. 8.4. One of these, in $y < 0$, represents the actual force and the other, in $y > 0$, represents its image, reflected about the line $y = 0$. As shown in the figure, the image force has components $+F_x$ and $-F_y$. By symmetry, this arrangement makes the shear traction $t_x = \sigma_{xy}$ vanish all along $y = 0$. The normal traction $t_y = \sigma_{yy}$ on $y = 0$, however, is not zero, and in the second stage of the analysis a supplemental solution is obtained to make it vanish. The supplemental solution, therefore, relates to the elastic half-plane $y \le 0$ with prescribed tractions $t_x = \sigma_{xy} = 0$, $t_y = \sigma_{yy} \ne 0$ along its surface.

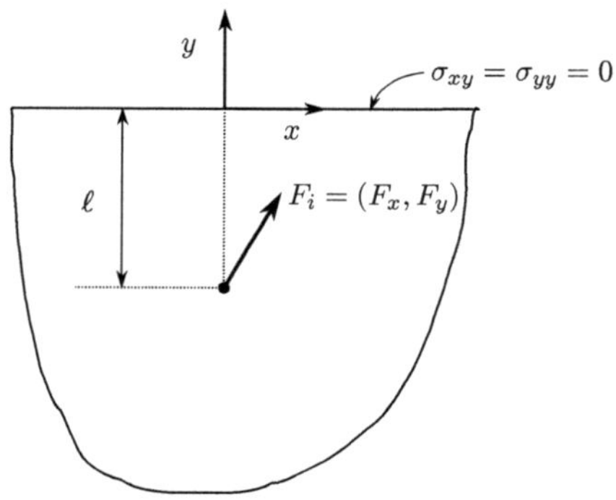

Fig. 8.3 Concentrated force in an elastic half-plane

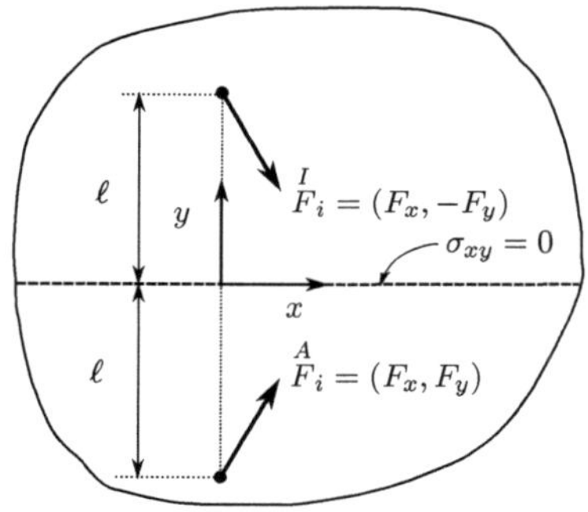

Fig. 8.4 Arrangement of actual and image forces

Denoting the displacements and stresses due to the actual force as $\overset{A}{u}_i$ and $\overset{A}{\sigma}_{ij}$, those due to its image as $\overset{I}{u}_i$ and $\overset{I}{\sigma}_{ij}$, and those resulting from the supplemental solution as $\overset{S}{u}_i$ and $\overset{S}{\sigma}_{ij}$, the complete solution for the half-plane $y \leq 0$ can be written as

$$u_i = \overset{A}{u}_i + \overset{I}{u}_i + \overset{S}{u}_i$$
$$\sigma_{ij} = \overset{A}{\sigma}_{ij} + \overset{I}{\sigma}_{ij} + \overset{S}{\sigma}_{ij} \tag{8.7}$$

The values of $\overset{A}{u}_i$ and $\overset{A}{\sigma}_{ij}$ for the actual force are found by replacing y by $y + \ell$ in (4.143) and (4.144) to get

$$\overset{A}{u}_x = \frac{F_x}{2\mu}\big[(3-4\nu)\overset{A}{g} + (y+\ell)\overset{A}{g}_{,y}\big] + \frac{F_y}{2\mu}\big[-(y+\ell)\overset{A}{g}_{,x}\big]$$
$$\overset{A}{u}_y = \frac{F_x}{2\mu}\big[-(y+\ell)\overset{A}{g}_{,x}\big] + \frac{F_y}{2\mu}\big[(3-4\nu)\overset{A}{g} - (y+\ell)\overset{A}{g}_{,y}\big] \tag{8.8}$$

and

$$\overset{A}{\sigma}_{xx} = F_x\big[(3-2\nu)\overset{A}{g}_{,x} + (y+\ell)\overset{A}{g}_{,xy}\big] + F_y\big[2\nu\overset{A}{g}_{,y} + (y+\ell)\overset{A}{g}_{,yy}\big]$$

$$\overset{A}{\sigma}_{yy} = F_x\big[-(1-2\nu)\overset{A}{g}_{,x} - (y+\ell)\overset{A}{g}_{,xy}\big] + F_y\big[2(1-\nu)\overset{A}{g}_{,y} - (y+\ell)\overset{A}{g}_{,yy}\big] \quad (8.9)$$

$$\overset{A}{\sigma}_{xy} = F_x\big[2(1-\nu)\overset{A}{g}_{,y} + (y+\ell)\overset{A}{g}_{,yy}\big] + F_y\big[(1-2\nu)\overset{A}{g}_{,x} - (y+\ell)\overset{A}{g}_{,xy}\big]$$

The function $\overset{A}{g}$ in these equations is

$$\overset{A}{g}(x,y) = -\frac{1}{4\pi(1-\nu)}\ln[x^2+(y+\ell)^2]^{1/2} \quad (8.10)$$

In writing (8.8) and (8.9) we have made use of the relations

$$\begin{aligned} x\overset{A}{g}_{,y} &= (y+\ell)\overset{A}{g}_{,x} \\ x\overset{A}{g}_{,x} &= -(y+\ell)\overset{A}{g}_{,y} - \frac{1}{4\pi(1-v)} \\ x\overset{A}{g}_{,xy} &= -\overset{A}{g}_{,y} + (y+\ell)\overset{A}{g}_{,xx} \\ x\overset{A}{g}_{,xx} &= -\overset{A}{g}_{,x} - (y+\ell)\overset{A}{g}_{,xy} \end{aligned} \quad (8.11)$$

to eliminate terms involving derivatives of function $\overset{A}{g}(x,y)$ multiplied by coordinate x. The constant term from the second one of (8.11) has been omitted in the above expression for $\overset{A}{u}_x$.

Similarly, $\overset{I}{u}_i$ and $\overset{I}{\sigma}_{ij}$ for the image force are found by replacing y by $y-\ell$ in (4.143) and (4.144) and changing the sign of the terms involving F_y, thus

$$\begin{aligned} \overset{I}{u}_x &= \frac{F_x}{2\mu}\big[(3-4\nu)\overset{I}{g} + (y-\ell)\overset{I}{g}_{,y}\big] + \frac{F_y}{2\mu}\big[(y-\ell)\overset{I}{g}_{,x}\big] \\ \overset{I}{u}_y &= \frac{F_x}{2\mu}\big[-(y-\ell)\overset{I}{g}_{,x}\big] + \frac{F_y}{2\mu}\big[-(3-4\nu)\overset{I}{g} + (y-\ell)\overset{I}{g}_{,y}\big] \end{aligned} \quad (8.12)$$

$$\overset{I}{\sigma}_{xx} = F_x\big[(3-2\nu)\overset{I}{g}_{,x} + (y-\ell)\overset{I}{g}_{,xy}\big] + F_y\big[-2\nu\overset{I}{g}_{,y} - (y-\ell)\overset{I}{g}_{,yy}\big]$$

$$\overset{I}{\sigma}_{yy} = F_x\big[-(1-2\nu)\overset{I}{g}_{,x} - (y-\ell)\overset{I}{g}_{,xy}\big] + F_y\big[-2(1-\nu)\overset{I}{g}_{,y} + (y-\ell)\overset{I}{g}_{,yy}\big] \quad (8.13)$$

$$\overset{I}{\sigma}_{xy} = F_x\big[2(1-\nu)\overset{I}{g}_{,y} + (y-\ell)\overset{I}{g}_{,yy}\big] + F_y\big[-(1-2\nu)\overset{I}{g}_{,x} + (y-\ell)\overset{I}{g}_{,xy}\big]$$

where

$$\overset{I}{g}(x,y) = -\frac{1}{4\pi(1-\nu)}\ln[x^2+(y-\ell)^2]^{1/2} \tag{8.14}$$

It is easy to show from these results that, on $y=0$, $\overset{A}{\sigma}_{xy} = -\overset{I}{\sigma}_{xy}$ and $\overset{A}{\sigma}_{yy} = \overset{I}{\sigma}_{yy}$. In order to remove the tractions from the surface of the half-plane the following conditions must hold for the supplemental solution: $\overset{S}{\sigma}_{xy} = 0$ and $\overset{S}{\sigma}_{yy} = -2\overset{I}{\sigma}_{yy}$ on $y=0$ for $-\infty < x < \infty$. (The latter condition is written in terms of $\overset{I}{\sigma}_{yy}$ rather than $\overset{A}{\sigma}_{yy}$ because the function $\overset{I}{g}$ does not contain any singularities in $y \le 0$.)

We saw in Sect. 4.9.3 that the solution to an elasticity problem containing a shear-free plane $y=0$ can be expressed in terms of a special Papkovitch-Neuber function $\Phi(x,y)$, which we will denote here as $\overset{S}{\Phi}(x,y)$ to identify it with our supplemental solution. The displacements $\overset{S}{u}_i$ and stresses $\overset{S}{\sigma}_{ij}$ associated with this function are given by (4.99) and (4.100), i.e.,

$$\begin{aligned}\overset{S}{u}_x &= -(1-2\nu)\overset{S}{\Phi}_{,x} - y\overset{S}{\Phi}_{,xy}\\ \overset{S}{u}_y &= 2(1-\nu)\overset{S}{\Phi}_{,y} - y\overset{S}{\Phi}_{,yy}\end{aligned} \tag{8.15}$$

and

$$\begin{aligned}\overset{S}{\sigma}_{xx} &= 2\mu(\overset{S}{\Phi}_{,yy} + y\overset{S}{\Phi}_{,yyy})\\ \overset{S}{\sigma}_{yy} &= 2\mu(\overset{S}{\Phi}_{,yy} - y\overset{S}{\Phi}_{,yyy})\\ \overset{S}{\sigma}_{xy} &= -2\mu y\overset{S}{\Phi}_{,xyy}\end{aligned} \tag{8.16}$$

The last one of (8.16) shows that $\overset{S}{\sigma}_{xy} = 0$ on $y=0$, as required.

Using the second one of (8.16) and the expression for $\overset{I}{\sigma}_{yy}$ from (8.13), the requirement that $\overset{S}{\sigma}_{yy} = -2\overset{I}{\sigma}_{yy}$ on $y=0$ means that the boundary value of $\overset{S}{\Phi}_{,yy}$ must satisfy the condition

$$\overset{S}{\Phi}_{,yy} = \frac{1}{\mu}F_x\big[(1-2\nu)\overset{I}{g}_{,x} - \ell\overset{I}{g}_{,xy}\big] + \frac{1}{\mu}F_y\big[2(1-\nu)\overset{I}{g}_{,y} + \ell\overset{I}{g}_{,yy}\big] \quad \text{on } y=0 \tag{8.17}$$

or

$$\overset{S}{\Phi}_{,yy} = p(x,y) \qquad \text{on } y = 0 \tag{8.18}$$

Function $p(x,y)$ is harmonic and regular (i.e., it is continuously differentiable and has no singularities) in $y \le 0$ and has the same boundary values as those required for $\overset{S}{\Phi}_{,yy}$. Since $\overset{S}{\Phi}_{,yy}$ must also be a harmonic function it follows from the uniqueness theorem for the Dirichlet problem (e.g., Sokolnikoff and Redheffer 1966) that (8.18) holds everywhere in $y \le 0$ and not just on the boundary, that is,

$$\overset{S}{\Phi}_{,yy} = \frac{1}{\mu}F_x\big[(1-2\nu)\overset{I}{g}_{,x} - \ell\overset{I}{g}_{,xy}\big] + \frac{1}{\mu}F_y\big[2(1-\nu)\overset{I}{g}_{,y} + \ell\overset{I}{g}_{,yy}\big] \qquad y \le 0 \tag{8.19}$$

Substitution of this result into (8.16) gives the following expressions for the stresses $\overset{S}{\sigma}_{ij}$:

$$
\begin{aligned}
\overset{S}{\sigma}_{xx} &= 2F_x\big[(1-2\nu)(\overset{I}{g}_{,x} + y\overset{I}{g}_{,xy}) - \ell(\overset{I}{g}_{,xy} + y\overset{I}{g}_{,xyy})\big] \\
&\quad + 2F_y\big[2(1-\nu)(\overset{I}{g}_{,y} + y\overset{I}{g}_{,yy}) + \ell(\overset{I}{g}_{,yy} + y\overset{I}{g}_{,yyy})\big] \\[1ex]
\overset{S}{\sigma}_{yy} &= 2F_x\big[(1-2\nu)(\overset{I}{g}_{,x} - y\overset{I}{g}_{,xy}) - \ell(\overset{I}{g}_{,xy} - y\overset{I}{g}_{,xyy})\big] \\
&\quad + 2F_y\big[2(1-\nu)(\overset{I}{g}_{,y} - y\overset{I}{g}_{,yy}) + \ell(\overset{I}{g}_{,yy} - y\overset{I}{g}_{,yyy})\big] \\[1ex]
\overset{S}{\sigma}_{xy} &= 2F_x\big[(1-2\nu)y\overset{I}{g}_{,yy} - \ell y\overset{I}{g}_{;yyy}\big] + 2F_y\big[-2(1-\nu)y\overset{I}{g}_{,xy} - \ell y\overset{I}{g}_{,xyy}\big]
\end{aligned}
\tag{8.20}
$$

The displacements $\overset{S}{u}_i$ associated with these stresses are found using (8.15) after computing $\overset{S}{\Phi}_{,x}$ and $\overset{S}{\Phi}_{,y}$ by integrating $\overset{S}{\Phi}_{,xx} = -\overset{S}{\Phi}_{,yy}$ and $\overset{S}{\Phi}_{,yy}$. The results are

$$
\begin{aligned}
\overset{S}{u}_x &= \frac{1}{\mu}F_x\big[(1-2\nu)^2\overset{I}{g} + (1-2\nu)(y-\ell)\overset{I}{g}_{,y} - \ell y\overset{I}{g}_{,yy}\big] \\
&\quad + \frac{1}{\mu}F_y\Big[2(1-\nu)(1-2\nu)\int \overset{I}{g}_{,y}\,\mathrm{d}x - 2(1-\nu)(y+\ell)\overset{I}{g}_{,x} + \ell(\overset{I}{g}_{,x} - y\overset{I}{g}_{,xy})\Big] \\[1ex]
\overset{S}{u}_y &= \frac{1}{\mu}F_x\Big[2(1-\nu)(1-2\nu)\int \overset{I}{g}_{,x}\,\mathrm{d}y - (1-2\nu)(y+\ell)\overset{I}{g}_{,x} - \ell(\overset{I}{g}_{,x} - y\overset{I}{g}_{,xy})\Big] \\
&\quad + \frac{1}{\mu}F_y\big[4(1-\nu)^2\overset{I}{g} - 2(1-\nu)(y-\ell)\overset{I}{g}_{,y} - \ell y\overset{I}{g}_{,yy}\big]
\end{aligned}
\tag{8.21}
$$

8.3.2 Concentrated Force on the Surface of an Isotropic Elastic Half-Plane

The problem of a concentrated force F_y applied to the surface of a linearly elastic half-plane is known as Flamant's problem. The solution to this problem can be found in several ways (see, for example, Timoshenko and Goodier 1970), including use of the special Papkovitch-Neuber functions $\Phi(x, y)$ and $\Psi(x, y)$ introduced in Sect. 4.9.3, an approach we will use here. The representation in terms of function $\Phi(x, y)$ relates to the case of normal loading of the surface because this function automatically satisfies the condition $\sigma_{xy} = 0$ on $y = 0$ for $-\infty < x < \infty$; the representation in terms of function $\Psi(x, y)$ is appropriate to the case of shear loading because it automatically satisfies the condition $\sigma_{yy} = 0$ on $y = 0$ for $-\infty < x < \infty$.

Considering first the case of normal loading, the second one of (4.100) shows that

$$\sigma_{yy} = 2\mu \frac{\partial^2}{\partial y^2}\Phi(x, y) \qquad \text{on } y = 0 \tag{8.22}$$

We have previously seen (see (1.19)) that the harmonic function $\arctan y/(x-a) - \arctan y/(x+a)$ is equal to zero on $y = 0$ for $|\, x \,| > a$ and undergoes a step change equal to $-\pi$ for $-a \le x \le a$ and $y = 0_-$. The problem of a constant normal stress $\sigma_{yy} = p_y$ applied to the line segment $-a \le x \le a$ on the surface of an otherwise unloaded half-plane $y \le 0$ is therefore determined by the function

$$\frac{\partial^2}{\partial y^2}\Phi(x, y) = -\frac{p_y}{2\pi\mu}\left(\arctan \frac{y}{x-a} - \arctan \frac{y}{x+a}\right) \tag{8.23}$$

A concentrated force F_y at the origin is the limiting case of this loading for which $p_y = F_y/(2a)$ and $a \to 0$. An application of L'Hopital's rule to (8.23) then gives

$$\frac{\partial^2}{\partial y^2}\Phi(x, y) = -\frac{F_y}{2\pi\mu}\frac{y}{x^2 + y^2} \tag{8.24}$$

Using (4.100) we find that the stresses in $y \le 0$ are given as

$$\begin{aligned}
\sigma_{xx} &= -\frac{2F_y}{\pi}\frac{x^2 y}{(x^2+y^2)^2} \\
\sigma_{yy} &= -\frac{2F_y}{\pi}\frac{y^3}{(x^2+y^2)^2} \\
\sigma_{xy} &= -\frac{2F_y}{\pi}\frac{x y^2}{(x^2+y^2)^2}
\end{aligned} \tag{8.25}$$

or, setting $x = r\cos\theta$ and $y = r\sin\theta$,

$$\sigma_{xx} = -\frac{2F_y}{\pi}\frac{\cos^2\theta\sin\theta}{r}$$

$$\sigma_{yy} = -\frac{2F_y}{\pi}\frac{\sin^3\theta}{r} \tag{8.26}$$

$$\sigma_{xy} = -\frac{2F_y}{\pi}\frac{\cos\theta\sin^2\theta}{r}$$

The stresses are defined everywhere except at the origin, $r = 0$. Similar to the case of a concentrated force in an infinite region (Sect. 4.11) we avoid this conceptional difficulty by imagining that a small semi-circular hole of radius $r = \varepsilon$ is excluded from the analysis. It is easy to show that the resultant traction on the boundary of this hole is statically equivalent to the force F_y. It can be seen from (8.26) that the stresses go to zero as $r \to \infty$.

The displacements due to the concentrated force can be obtained by integrating $\partial^2\Phi/\partial y^2$ and $\partial^2\Phi/\partial x^2 = -\partial^2\Phi/\partial y^2$ to find $\partial\Phi/\partial y$ and $\partial\Phi/\partial x$ and using these in (4.99). The results, however, are not especially useful in practice and we will not present them. The y component of displacement, for example, tends to plus infinity at the point of application of the force $F_y > 0$, and to minus infinity at large distances from this point. These infinite displacements are due to the nature of the applied boundary loading. Although we have adopted a plane strain condition for the x, y plane, it is implicit that the force F_y is applied along the entire z axis. (Otherwise, the assumed plane strain condition would not occur.) The total applied force (i.e., force/length times length) is therefore infinite, and the unbounded displacements are needed to accommodate it. This circumstance also occurs for the case of a concentrated force in a plane (or a half-plane), but in applying this solution in a boundary element method the resultant of all the applied forces vanishes for equilibrated boundary loading and the associated displacements are well defined. Loading of the surface of a half-plane (to represent an initial stress state in the earth due to hilly terrain, for example) will not satisfy this condition and hence the associated displacements are not meaningful.

The problem of a concentrated force F_x is treated similarly using the third one of (4.108) to find

$$\frac{\partial^2}{\partial y^2}\Psi(x, y) = -\frac{F_x}{2\pi\mu}\frac{y}{x^2 + y^2} \tag{8.27}$$

It can then be shown that the stresses in $y \leq 0$ are given as

$$\sigma_{xx} = -\frac{2F_x}{\pi}\frac{\cos^3\theta}{r}$$

$$\sigma_{yy} = -\frac{2F_x}{\pi}\frac{\cos\theta\sin^2\theta}{r} \tag{8.28}$$

$$\sigma_{xy} = -\frac{2F_x}{\pi}\frac{\cos^2\theta\sin\theta}{r}$$

8.3.3 *Concentrated Displacement Discontinuity in an Isotropic Elastic Half-Plane*

The problem of a concentrated displacement discontinuity in an isotropic, linearly elastic half-plane with a traction-free boundary is illustrated in Fig. 8.5. The discontinuity is located at the point $x = 0,\ y = -\ell$ in the half-plane $y \leq 0$, and its components $\mathcal{D}_{\bar{x}}$ and $\mathcal{D}_{\bar{y}}$ are referred to the local coordinate system specified by the conditions

$$\bar{x} = x\cos\beta + (y + \ell)\sin\beta$$
$$\bar{y} = -x\sin\beta + (y + \ell)\cos\beta \tag{8.29}$$

The solution to this problem is also obtained using the method of images with the resulting displacements and stresses given by (8.7). The image discontinuity is located at the point $x = 0,\ y = +\ell$ has components $-\mathcal{D}_{\bar{x}}$ and $+\mathcal{D}_{\bar{y}}$, as depicted

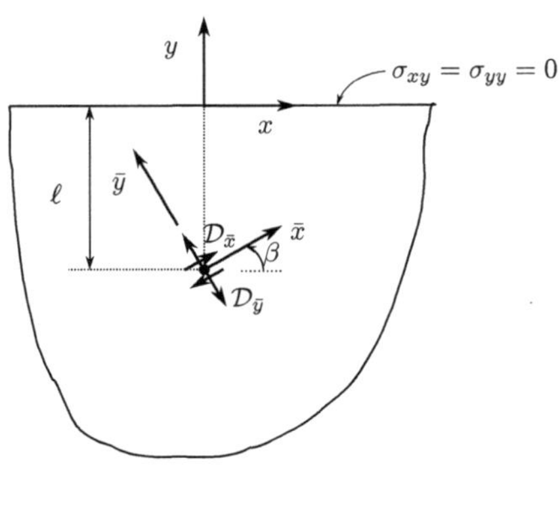

Fig. 8.5 Concentrated displacement discontinuity in an elastic half-plane

Fig. 8.6 Arrangement of actual and image displacement discontinuities

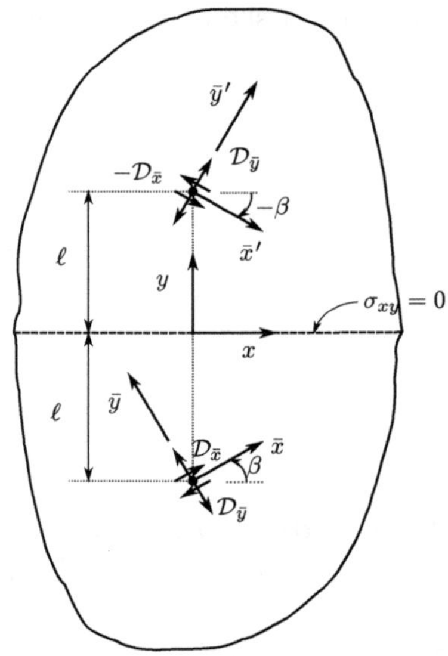

in Fig. 8.6. This combination of displacement discontinuities does not produce any shear stress on $y = 0$. The local $\bar{x}'$ coordinate axis for the image displacement discontinuity is inclined at an angle $-\beta$ from the x direction. The local coordinates for the image discontinuity are defined as

$$\bar{x}' = x \cos\beta - (y - \ell) \sin\beta$$
$$\bar{y}' = x \sin\beta + (y - \ell) \cos\beta \tag{8.30}$$

The displacements $\overset{A}{u}_i$ and stresses $\overset{A}{\sigma}_{ij}$ for the actual discontinuity can be expressed in terms of the function

$$\overset{A}{g}(\bar{x}, \bar{y}) = -\frac{1}{4\pi(1-\nu)} \ln[\bar{x}^2 + \bar{y}^2]^{1/2} \tag{8.31}$$

using (4.221) and (4.222) and coordinate transformation formulas (4.62) and (4.66). (The process is the same as that used in Sect. 6.3.1 for a related problem.) Omitting the simple details of the computation, we obtain the following results:

$$\begin{aligned}\overset{A}{u}_x &= \mathcal{D}_{\bar{x}}\big[(1-2\nu)\sin\beta \overset{A}{g}_{,\bar{x}} - 2(1-\nu)\cos\beta \overset{A}{g}_{,\bar{y}} - \bar{y}(\sin\beta \overset{A}{g}_{,\bar{x}\bar{y}} - \cos\beta \overset{A}{g}_{,\bar{x}\bar{x}})\big] \\ &\quad + \mathcal{D}_{\bar{y}}\big[(1-2\nu)\cos\beta \overset{A}{g}_{,\bar{x}} + 2(1-\nu)\sin \overset{A}{g}_{,\bar{y}} + \bar{y}(\cos\beta \overset{A}{g}_{,\bar{x}\bar{y}} + \sin\beta \overset{A}{g}_{,\bar{x}\bar{x}})\big] \\ \overset{A}{u}_y &= \mathcal{D}_{\bar{x}}\big[-(1-2\nu)\cos\beta \overset{A}{g}_{,\bar{x}} - 2(1-\nu)\sin\beta \overset{A}{g}_{,\bar{y}} + \bar{y}(\cos\beta \overset{A}{g}_{,\bar{x}\bar{y}} + \sin\beta \overset{A}{g}_{,\bar{x}\bar{x}})\big]\end{aligned} \tag{8.32}$$

$$+\,\mathcal{D}_{\bar{y}}\big[(1-2\nu)\sin\beta\overset{A}{g}_{,\bar{x}}-2(1-\nu)\cos\beta\overset{A}{g}_{,\bar{y}}+\bar{y}(\sin\beta\overset{A}{g}_{,\bar{x}\,\bar{y}}-\cos\beta\overset{A}{g}_{,\bar{x}\,\bar{x}})\big]$$

$$\begin{aligned}
\overset{A}{\sigma}_{xx}&=2\mu\mathcal{D}_{\bar{x}}\big[-2\cos^2\beta\overset{A}{g}_{,\bar{x}\,\bar{y}}-\sin 2\beta\overset{A}{g}_{,\bar{x}\,\bar{x}}-\bar{y}(\cos 2\beta\overset{A}{g}_{,\bar{x}\,\bar{y}\,\bar{y}}-\sin 2\beta\overset{A}{g}_{,\bar{y}\,\bar{y}\,\bar{y}})\big]\\
&\quad+2\mu\mathcal{D}_{\bar{y}}\big[\overset{A}{g}_{,\bar{x}\,\bar{x}}-\bar{y}(\sin 2\beta\overset{A}{g}_{,\bar{x}\,\bar{y}\,\bar{y}}+\cos 2\beta\overset{A}{g}_{,\bar{y}\,\bar{y}\,\bar{y}})\big]\\
\overset{A}{\sigma}_{yy}&=2\mu\mathcal{D}_{\bar{x}}\big[-2\sin^2\beta\overset{A}{g}_{,\bar{x}\,\bar{y}}+\sin 2\beta\overset{A}{g}_{,\bar{x}\,\bar{x}}+\bar{y}(\cos 2\beta\overset{A}{g}_{,\bar{x}\,\bar{y}\,\bar{y}}-\sin 2\beta\overset{A}{g}_{,\bar{y}\,\bar{y}\,\bar{y}})\big]\\
&\quad+2\mu\mathcal{D}_{\bar{y}}\big[\overset{A}{g}_{,\bar{x}\,\bar{x}}+\bar{y}(\sin 2\beta\overset{A}{g}_{,\bar{x}\,\bar{y}\,\bar{y}}+\cos 2\beta\overset{A}{g}_{,\bar{y}\,\bar{y}\,\bar{y}})\big] \qquad (8.33)\\
\overset{A}{\sigma}_{xy}&=2\mu\mathcal{D}_{\bar{x}}\big[-\sin 2\beta\overset{A}{g}_{,\bar{x}\,\bar{y}}+\cos 2\beta\overset{A}{g}_{,\bar{x}\,\bar{x}}-\bar{y}(\sin 2\beta\overset{A}{g}_{,\bar{x}\,\bar{y}\,\bar{y}}+\cos 2\beta\overset{A}{g}_{,\bar{y}\,\bar{y}\,\bar{y}})\big]\\
&\quad+2\mu\mathcal{D}_{\bar{y}}\big[\bar{y}(\cos 2\beta\overset{A}{g}_{,\bar{x}\,\bar{y}\,\bar{y}}-\sin 2\beta\overset{A}{g}_{,\bar{y}\,\bar{y}\,\bar{y}})\big]
\end{aligned}$$

In the same way, the displacements $\overset{I}{u}_i$ and stresses $\overset{I}{\sigma}_{ij}$ for the image discontinuity can be written as

$$\begin{aligned}
\overset{I}{u}_x&=\mathcal{D}_{\bar{x}}\big[(1-2\nu)\sin\beta\overset{I}{g}_{,\bar{x}'}+2(1-\nu)\cos\beta\overset{I}{g}_{,\bar{y}'}-\bar{y}'(\sin\beta\overset{I}{g}_{,\bar{x}'\,\bar{y}'}+\cos\beta\overset{I}{g}_{,\bar{x}'\,\bar{x}'})\big]\\
&\quad+\mathcal{D}_{\bar{y}}\big[(1-2\nu)\cos\beta\overset{I}{g}_{,\bar{x}'}-2(1-\nu)\sin\beta\overset{I}{g}_{,\bar{y}'}+\bar{y}'(\cos\beta\overset{I}{g}_{,\bar{x}'\,\bar{y}'}-\sin\beta\overset{I}{g}_{,\bar{x}'\,\bar{x}'})\big] \qquad (8.34)\\
\overset{I}{u}_y&=\mathcal{D}_{\bar{x}}\big[(1-2\nu)\cos\beta\overset{I}{g}_{,\bar{x}'}-2(1-\nu)\sin\beta\overset{I}{g}_{,\bar{y}'}-\bar{y}'(\cos\beta\overset{I}{g}_{,\bar{x}'\,\bar{y}'}-\sin\beta\overset{I}{g}_{,\bar{x}'\,\bar{x}'})\big]\\
&\quad+\mathcal{D}_{\bar{y}}\big[-(1-2\nu)\sin\beta\overset{I}{g}_{,\bar{x}'}-2(1-\nu)\cos\beta\overset{I}{g}_{,\bar{y}'}-\bar{y}'(\sin\beta\overset{I}{g}_{,\bar{x}'\,\bar{y}'}+\cos\beta\overset{I}{g}_{,\bar{x}'\,\bar{x}'})\big]
\end{aligned}$$

$$\begin{aligned}
\overset{I}{\sigma}_{xx}&=2\mu\mathcal{D}_{\bar{x}}\big[2\cos^2\beta\overset{I}{g}_{,\bar{x}'\,\bar{y}'}-\sin 2\beta\overset{I}{g}_{,\bar{x}'\,\bar{x}'}+\bar{y}'(\cos 2\beta\overset{I}{g}_{,\bar{x}'\,\bar{y}'\,\bar{y}'}+\sin 2\beta\overset{I}{g}_{,\bar{y}'\,\bar{y}'\,\bar{y}'})\big]\\
&\quad+2\mu\mathcal{D}_{\bar{y}}\big[\overset{I}{g}_{,\bar{x}'\,\bar{x}'}+\bar{y}'(\sin 2\beta\overset{I}{g}_{,\bar{x}'\,\bar{y}'\,\bar{y}'}-\cos 2\beta\overset{I}{g}_{,\bar{y}'\,\bar{y}'\,\bar{y}'})\big]\\
\overset{I}{\sigma}_{yy}&=2\mu\mathcal{D}_{\bar{x}}\big[2\sin^2\beta\overset{I}{g}_{,\bar{x}'\,\bar{y}'}+\sin 2\beta\overset{I}{g}_{,\bar{x}'\,\bar{x}'}-\bar{y}'(\cos 2\beta\overset{I}{g}_{,\bar{x}'\,\bar{y}'\,\bar{y}'}+\sin 2\beta\overset{I}{g}_{,\bar{y}'\,\bar{y}'\,\bar{y}'})\big]\\
&\quad+2\mu\mathcal{D}_{\bar{y}}\big[\overset{I}{g}_{,\bar{x}'\,\bar{x}'}-\bar{y}'(\sin 2\beta\overset{I}{g}_{,\bar{x}'\,\bar{y}'\,\bar{y}'}-\cos 2\beta\overset{I}{g}_{,\bar{y}'\,\bar{y}'\,\bar{y}'})\big] \qquad (8.35)\\
\overset{I}{\sigma}_{xy}&=2\mu\mathcal{D}_{\bar{x}}\big[-\sin 2\beta\overset{I}{g}_{,\bar{x}'\,\bar{y}'}-\cos 2\beta\overset{I}{g}_{,\bar{x}'\,\bar{x}'}-\bar{y}'(\sin 2\beta\overset{I}{g}_{,\bar{x}'\,\bar{y}'\,\bar{y}'}-\cos 2\beta\overset{I}{g}_{,\bar{y}'\,\bar{y}'\,\bar{y}'})\big]\\
&\quad+2\mu\mathcal{D}_{\bar{y}}\big[\bar{y}'(\cos 2\beta\overset{I}{g}_{,\bar{x}'\,\bar{y}'\,\bar{y}'}+\sin 2\beta\overset{I}{g}_{,\bar{y}'\,\bar{y}'\,\bar{y}'})\big]
\end{aligned}$$

where

$$\overset{I}{g}(\bar{x}', \bar{y}') = -\frac{1}{4\pi(1-\nu)} \ln[\bar{x}'^2 + \bar{y}'^2]^{1/2} \tag{8.36}$$

It is easy to verify that $\overset{A}{\sigma}_{xy} + \overset{I}{\sigma}_{xy} = 0$ and $\overset{A}{\sigma}_{yy} = \overset{I}{\sigma}_{yy}$ on $y = 0$ and hence that the displacements $\overset{S}{u}_i$ and stresses $\overset{S}{\sigma}_{ij}$ for the supplemental solution can be written in terms of a single harmonic function $\overset{S}{\Phi}(x, y)$, as in (8.15) and (8.16). This representation ensures that $\overset{S}{\sigma}_{xy} = 0$ on $y = 0$ for $-\infty < x < \infty$ and in order to make the resultant normal stress σ_{yy} also vanish along this line we require that

$$\begin{aligned} \overset{S}{\Phi}_{,yy} = {} & 2\mathcal{D}_{\bar{x}}\big[-2\sin^2\beta \overset{I}{g}_{,\bar{x}'\bar{y}'} - \sin 2\beta \overset{I}{g}_{,\bar{x}'\bar{x}'} + \bar{y}'(\cos 2\beta \overset{I}{g}_{,\bar{x}'\bar{y}'\bar{y}'} + \sin 2\beta \overset{I}{g}_{,\bar{y}'\bar{y}'\bar{y}'})\big] \\ & + 2\mathcal{D}_{\bar{y}}\big[-\overset{I}{g}_{,\bar{x}'\bar{x}'} + \bar{y}'(\sin 2\beta \overset{I}{g}_{,\bar{x}'\bar{y}'\bar{y}'} - \cos 2\beta \overset{I}{g}_{,\bar{y}'\bar{y}'\bar{y}'})\big] \quad \text{on } y = 0 \end{aligned} \tag{8.37}$$

The derivatives with respect to $\bar{x}'$ and $\bar{y}'$ in this expression can be written in terms of derivatives with respect to x and y by using the chain rule for partial differentiation. The reader should confirm that this calculation gives the following results:

$$\begin{aligned} \overset{I}{g}_{,\bar{x}'\bar{y}'} &= \cos 2\beta \overset{I}{g}_{,xy} - \sin 2\beta \overset{I}{g}_{,yy} \\ \overset{I}{g}_{,\bar{x}'\bar{x}'} &= \cos 2\beta \overset{I}{g}_{,xx} - \sin 2\beta \overset{I}{g}_{,xy} \\ \overset{I}{g}_{,\bar{x}'\bar{y}'\bar{y}'} &= \cos 3\beta \overset{I}{g}_{,xyy} - \sin 3\beta \overset{I}{g}_{,yyy} \\ \overset{I}{g}_{,\bar{y}'\bar{y}'\bar{y}'} &= \cos 3\beta \overset{I}{g}_{,yyy} + \sin 3\beta \overset{I}{g}_{,xyy} \end{aligned} \tag{8.38}$$

Noting that $\bar{y}' = x \sin\beta - \ell\cos\beta$ on $y = 0$ and using (8.38) in (8.37) we find that, on $y = 0$,

$$\begin{aligned} \overset{S}{\Phi}_{,yy} = {} & \mathcal{D}_{\bar{x}}\big[4\sin^2\beta \overset{I}{g}_{,xy} + 2\sin 2\beta \overset{I}{g}_{,yy} + 2(x\sin\beta - \ell\cos\beta)(\cos\beta \overset{I}{g}_{,xyy} - \sin\beta \overset{I}{g}_{,yyy})\big] \\ & + \mathcal{D}_{\bar{y}}\big[2\sin 2\beta \overset{I}{g}_{,xy} + 2\cos 2\beta \overset{I}{g}_{,yy} - 2(x\sin\beta - \ell\cos\beta)(\sin\beta \overset{I}{g}_{,xyy} + \cos\beta \overset{I}{g}_{,yyy})\big] \end{aligned} \tag{8.39}$$

If this function were harmonic the uniqueness theorem for the Dirichlet problem would assure that (8.39) holds everywhere in $y \le 0$ and not just on the boundary. But this function is not harmonic due to the presence of the terms multiplied by x so the uniqueness theorem is not directly applicable. Function $\overset{S}{\Phi}_{,yy}$ can, however, be made harmonic without changing its value on $y = 0$ by using a simple device described by Rongved and Frasier (1958): Given a harmonic function $\Theta(x, y)$ it can be verified that the function $x\partial\Theta/\partial y - y\partial\Theta/\partial x$ is also harmonic. It follows that the function $x\partial\Theta/\partial y$, which is not harmonic, can be made harmonic *without changing its value on* $y = 0$ by adding the function $-y\partial\Theta/\partial x$.

Following this procedure, addition of the following terms will make $\overset{S}{\Phi}_{,yy}$ harmonic and regular in $y \leq 0$ without changing its value on $y = 0$:

$$\mathcal{D}_{\bar{x}}\big[-\sin 2\beta y \overset{I}{g}_{,xxy} + 2\sin^2\beta y \overset{I}{g}_{,xyy})\big] + \mathcal{D}_{\bar{y}}\big[2\sin^2\beta y \overset{I}{g}_{,xxy} + \sin 2\beta y \overset{I}{g}_{,xyy}\big]$$

We therefore obtain the following expression for $\overset{S}{\Phi}_{,yy}$, valid for $y \leq 0$:

$$\begin{aligned}
\overset{S}{\Phi}_{,yy} = \mathcal{D}_{\bar{x}}\big[&4\sin^2\beta \overset{I}{g}_{,xy} + 2\sin 2\beta \overset{I}{g}_{,yy} + 2(x\sin\beta - \ell\cos\beta)(\cos\beta \overset{I}{g}_{,xyy} - \sin\beta \overset{I}{g}_{,yyy}) \\
&+ 2y\sin\beta(\sin\beta \overset{I}{g}_{,xyy} + \cos\beta \overset{I}{g}_{,yyy})\big] \\
+ \mathcal{D}_{\bar{y}}\big[&2\sin 2\beta \overset{I}{g}_{,xy} + 2\cos 2\beta \overset{I}{g}_{,yy} - 2(x\sin\beta - \ell\cos\beta)(\sin\beta \overset{I}{g}_{,xyy} + \cos\beta \overset{I}{g}_{,yyy}) \\
&+ 2y\sin\beta(\cos\beta \overset{I}{g}_{,xyy} - \sin\beta \overset{I}{g}_{,yyy})\big]
\end{aligned} \tag{8.40}$$

The stresses $\overset{S}{\sigma}_{ij}$ for the supplemental solution can now be obtained by using (8.40) in (4.100). The associated displacements $\overset{S}{u}_i$ are obtained by integrating $\overset{S}{\Phi}_{,yy}$ and $\overset{S}{\Phi}_{,xx} = -\overset{S}{\Phi}_{,yy}$ to find $\overset{S}{\Phi}_{,y}$ and $\overset{S}{\Phi}_{,x}$ and using (4.99) while imposing the condition that the displacements vanish at infinity. It is convenient to write these stresses and displacements in terms of derivatives of function $\overset{I}{g}$ with respect to the image coordinates, $\bar{x}'$ and $\bar{y}'$, and this is accomplished by again using the chain rule for partial differentiation. The calculations are straightforward but rather tedious and we present here only the final results. The displacements are

$$\begin{aligned}
\overset{S}{u}_x = 2\mathcal{D}_{\bar{x}}\Big[&-(1-2\nu)\sin\beta \overset{I}{g}_{,\bar{x}'} - \big\{(1-2\nu)\bar{y}'\sin\beta + y\sin 2\beta\big\}\overset{I}{g}_{,\bar{x}'\bar{y}'} \\
&+ \big\{(1-2\nu)(\bar{y}'\cos\beta - y) - 2y\sin^2\beta\big\}\overset{I}{g}_{,\bar{y}'\bar{y}'} - y(\bar{y}'\sin 2\beta - y\sin\beta)\overset{I}{g}_{,\bar{x}'\bar{y}'\bar{y}'} \\
&+ y(\bar{y}'\cos 2\beta - y\cos\beta)\overset{I}{g}_{,\bar{y}'\bar{y}'\bar{y}'}\Big] \\
+ 2\mathcal{D}_{\bar{y}}\Big[&-(1-2\nu)\cos\beta \overset{I}{g}_{,\bar{x}'} + \big\{(1-2\nu)(\bar{y}'\cos\beta - y) - y\big\}\overset{I}{g}_{,\bar{x}'\bar{y}'} + (1-2\nu)\bar{y}'\sin\beta \overset{I}{g}_{,\bar{y}'\bar{y}'} \\
&+ y(\bar{y}'\cos 2\beta - y\cos\beta)\overset{I}{g}_{,\bar{x}'\bar{y}'\bar{y}'} + y(\bar{y}'\sin 2\beta - y\sin\beta)\overset{I}{g}_{,\bar{y}'\bar{y}'\bar{y}'}\Big]
\end{aligned}$$

(8.41)

$$\begin{aligned}
\overset{S}{u}_y = 2\mathcal{D}_{\bar{x}}\Big[&2(1-\nu)\sin\beta \overset{I}{g}_{,\bar{y}'} + \big\{2(1-\nu)(\bar{y}'\cos\beta - y) + 2y\sin^2\beta\big\}\overset{I}{g}_{,\bar{x}'\bar{y}'} \\
&+ \big\{2(1-\nu)\bar{y}'\sin\beta - y\sin 2\beta\big\}\overset{I}{g}_{,\bar{y}'\bar{y}'} - y(\bar{y}'\cos 2\beta - y\cos\beta)\overset{I}{g}_{,\bar{x}'\bar{y}'\bar{y}'} \\
&- y(\bar{y}'\sin 2\beta - y\sin\beta)\overset{I}{g}_{,\bar{y}'\bar{y}'\bar{y}'}\Big]
\end{aligned}$$

$$
\begin{aligned}
&+ 2\mathcal{D}_{\bar{y}}\Big[2(1-\nu)\cos\beta \overset{I}{g}_{,\bar{y}'} + 2(1-\nu)\bar{y}'\sin\beta \overset{I}{g}_{,\bar{x}'\bar{y}'} - \big\{2(1-\nu)(\bar{y}'\cos\beta - y) + y\big\}\overset{I}{g}_{,\bar{y}'\bar{y}'} \\
&\quad - y(\bar{y}'\sin 2\beta - y\sin\beta)\overset{I}{g}_{,\bar{x}'\bar{y}'\bar{y}'} + y(\bar{y}'\cos 2\beta - y\cos\beta)\overset{I}{g}_{,\bar{y}'\bar{y}'\bar{y}'}\Big]
\end{aligned}
$$

while the stresses are

$$
\begin{aligned}
\overset{S}{\sigma}_{xx} &= 4\mu\mathcal{D}_{\bar{x}}\Big[-2\sin^2\beta\overset{I}{g}_{,\bar{x}'\bar{y}'} + \sin 2\beta\overset{I}{g}_{,\bar{y}'\bar{y}'} - 3y\sin\beta\big(\sin 2\beta\overset{I}{g}_{,\bar{x}'\bar{y}'\bar{y}'} - \cos 2\beta\overset{I}{g}_{,\bar{y}'\bar{y}'\bar{y}'}\big) \\
&\qquad + (\bar{y}'\cos 2\beta - y\cos\beta)\overset{I}{g}_{,\bar{x}'\bar{y}'\bar{y}'} + (\bar{y}'\sin 2\beta - y\sin\beta)\overset{I}{g}_{,\bar{y}'\bar{y}'\bar{y}'} \\
&\qquad + y(\bar{y}'\cos 3\beta - y\cos 2\beta)\overset{I}{g}_{,\bar{x}'\bar{y}'\bar{y}'\bar{y}'} + y(\bar{y}'\sin 3\beta - y\sin 2\beta)\overset{I}{g}_{,\bar{y}'\bar{y}'\bar{y}'\bar{y}'}\Big] \\
&\quad + 4\mu\mathcal{D}_{\bar{y}}\Big[\overset{I}{g}_{,\bar{y}'\bar{y}'} - 2y\sin^3\beta\overset{I}{g}_{,\bar{x}'\bar{y}'\bar{y}'} + y\cos\beta(1+2\sin^2\beta)\overset{I}{g}_{,\bar{y}'\bar{y}'\bar{y}'} \\
&\qquad + (\bar{y}'\sin 2\beta - y\sin\beta)\overset{I}{g}_{,\bar{x}'\bar{y}'\bar{y}'} - (\bar{y}'\cos 2\beta - y\cos\beta)\overset{I}{g}_{,\bar{y}'\bar{y}'\bar{y}'} \\
&\qquad + y(\bar{y}'\sin 3\beta - y\sin 2\beta)\overset{I}{g}_{,\bar{x}'\bar{y}'\bar{y}'\bar{y}'} - y(\bar{y}'\cos 3\beta - y\cos 2\beta)\overset{I}{g}_{,\bar{y}'\bar{y}'\bar{y}'\bar{y}'}\Big]
\end{aligned}
$$

$$
\begin{aligned}
\overset{S}{\sigma}_{yy} &= 4\mu\mathcal{D}_{\bar{x}}\Big[-2\sin^2\beta\overset{I}{g}_{,\bar{x}'\bar{y}'} + \sin 2\beta\overset{I}{g}_{,\bar{y}'\bar{y}'} + 3y\sin\beta\big(\sin 2\beta\overset{I}{g}_{,\bar{x}'\bar{y}'\bar{y}'} - \cos 2\beta\overset{I}{g}_{,\bar{y}'\bar{y}'\bar{y}'}\big) \\
&\qquad + (\bar{y}'\cos 2\beta - y\cos\beta)\overset{I}{g}_{,\bar{x}'\bar{y}'\bar{y}'} + (\bar{y}'\sin 2\beta - y\sin\beta)\overset{I}{g}_{,\bar{y}'\bar{y}'\bar{y}'} \\
&\qquad - y(\bar{y}'\cos 3\beta - y\cos 2\beta)\overset{I}{g}_{,\bar{x}'\bar{y}'\bar{y}'\bar{y}'} - y(\bar{y}'\sin 3\beta - y\sin 2\beta)\overset{I}{g}_{,\bar{y}'\bar{y}'\bar{y}'\bar{y}'}\Big] \\
&\quad + 4\mu\mathcal{D}_{\bar{y}}\Big[\overset{I}{g}_{,\bar{y}'\bar{y}'} + 2y\sin^3\beta\overset{I}{g}_{,\bar{x}'\bar{y}'\bar{y}'} - y\cos\beta(1+2\sin^2\beta)\overset{I}{g}_{,\bar{y}'\bar{y}'\bar{y}'} \qquad (8.42) \\
&\qquad + (\bar{y}'\sin 2\beta - y\sin\beta)\overset{I}{g}_{,\bar{x}'\bar{y}'\bar{y}'} - (\bar{y}'\cos 2\beta - y\cos\beta)\overset{I}{g}_{,\bar{y}'\bar{y}'\bar{y}'} \\
&\qquad - y(\bar{y}'\sin 3\beta - y\sin 2\beta)\overset{I}{g}_{,\bar{x}'\bar{y}'\bar{y}'\bar{y}'} + y(\bar{y}'\cos 3\beta - y\cos 2\beta)\overset{I}{g}_{,\bar{y}'\bar{y}'\bar{y}'\bar{y}'}\Big]
\end{aligned}
$$

$$
\begin{aligned}
\overset{S}{\sigma}_{xy} &= 4\mu\mathcal{D}_{\bar{x}}\Big[-3y\sin\beta\big(\cos 2\beta\overset{I}{g}_{,\bar{x}'\bar{y}'\bar{y}'} + \sin 2\beta\overset{I}{g}_{,\bar{y}'\bar{y}'\bar{y}'}\big) \\
&\qquad - y(\bar{y}'\sin 3\beta - y\sin 2\beta)\overset{I}{g}_{,\bar{x}'\bar{y}'\bar{y}'\bar{y}'} + y(\bar{y}'\cos 3\beta - y\cos 2\beta)\overset{I}{g}_{,\bar{y}'\bar{y}'\bar{y}'\bar{y}'}\Big] \\
&\quad + 4\mu\mathcal{D}_{\bar{y}}\Big[-y\cos\beta\big(1+2\sin^2\beta\big)\overset{I}{g}_{,\bar{x}'\bar{y}'\bar{y}'} - 2y\sin^3\beta\overset{I}{g}_{,\bar{y}'\bar{y}'\bar{y}'} \\
&\qquad + y(\bar{y}'\cos 3\beta - y\cos 2\beta)\overset{I}{g}_{,\bar{x}'\bar{y}'\bar{y}'\bar{y}'} + y(\bar{y}'\sin 3\beta - y\sin 2\beta)\overset{I}{g}_{,\bar{y}'\bar{y}'\bar{y}'\bar{y}'}\Big]
\end{aligned}
$$

Because the stresses and displacements for the image displacement discontinuity and the supplemental solution are expressed in terms the same function, these quantities could, if desired, be combined.

Finally, we note that the solution for a constant displacement discontinuity $D_{\bar{x}},\ D_{\bar{y}}$ over the line segment $-a \leq \bar{x} \leq a,\ \bar{y} = 0$ in the half-plane can be recovered from the preceding results by replacing $\overset{A}{g}$ and $\overset{I}{g}$ by $\overset{A}{f}$ and $\overset{I}{f}$, where

$$
\overset{A}{f}(\bar{x}, \bar{y}) = \frac{-1}{4\pi(1-\nu)}\Big[\bar{y}\big(\arctan\frac{\bar{x}-a}{\bar{y}} - \arctan\frac{\bar{x}+a}{\bar{y}}\big)
$$

$$- (\bar{x} - a) \ln \sqrt{(\bar{x} - a)^2 + \bar{y}^2} + (\bar{x} + a) \ln \sqrt{(\bar{x} + a)^2 + \bar{y}^2} - 2a \Big] \tag{8.43}$$

and

$$\overset{I}{f}(\bar{x}', \bar{y}') = \frac{-1}{4\pi(1 - \nu)} \Big[\bar{y}' (\arctan \frac{\bar{x}' - a}{\bar{y}'} - \arctan \frac{\bar{x}' + a}{\bar{y}'}) - (\bar{x}' - a) \ln \sqrt{(\bar{x}' - a)^2 + \bar{y}'^2} + (\bar{x}' + a) \ln \sqrt{(\bar{x}' + a)^2 + \bar{y}'^2} - 2a \Big] \tag{8.44}$$

8.3.4 *Concentrated Force in an Orthotropic Elastic Half-Plane*

We consider next the case of a concentrated force in an orthotropic (transversely isotropic) elastic half-plane with a traction-free boundary. The directions of elastic symmetry for the half-plane are in the x and y directions and the force $F_x > 0$, $F_y > 0$ is located at the point $x = 0$, $y = -\ell$ $(\ell > 0)$ in the half-plane $y \leq 0$. The solution to this problem again uses the method of images, with the results in the form (8.7).

The displacements and stresses due to the actual force can be written from (4.160) and (4.161) by replacing y by $y + \ell$, that is, by replacing y_i by $y_i + \ell_i$, where $\ell_i = \ell/\gamma_i$ for $i = 1$ and 2. It is convenient for present purposes, however, to express the solution in terms of a function $\overset{A}{h}(x, y_i + \ell_i)$, defined as

$$\overset{A}{h}(x, y_i + \ell_i) = \frac{1}{2\pi(q_1 - q_2)} \ln \sqrt{x^2 + (y_i + \ell_i)^2} \tag{8.45}$$

It is easy to show from (4.160) and (4.161) that the displacements and stresses can then be written as

$$\begin{aligned}
\overset{A}{u}_x &= \frac{1}{c_{66}} F_x \Big[\frac{\gamma_1}{q_1} \overset{A}{h}(x, y_1 + \ell_1) - \frac{\gamma_2}{q_2} \overset{A}{h}(x, y_2 + \ell_2) \Big] \\
&\quad - \frac{1}{c_{66}} F_y \Big[\int \frac{\partial}{\partial x} \overset{A}{h}(x, y_1 + \ell_1)\, \mathrm{d}y_1 - \int \frac{\partial}{\partial x} \overset{A}{h}(x, y_2 + \ell_2)\, \mathrm{d}y_2 \Big] \\
\overset{A}{u}_y &= -\frac{1}{c_{66}} F_x \Big[\int \frac{\partial}{\partial x} \overset{A}{h}(x, y_1 + \ell_1)\, \mathrm{d}y_1 - \int \frac{\partial}{\partial x} \overset{A}{h}(x, y_2 + \ell_2)\, \mathrm{d}y_2 \Big] \\
&\quad - \frac{1}{c_{66}} F_y \Big[\frac{q_1}{\gamma_1} \overset{A}{h}(x, y_1 + \ell_1) - \frac{q_2}{\gamma_2} \overset{A}{h}(x, y_2 + \ell_2) \Big]
\end{aligned} \tag{8.46}$$

and

$$\overset{A}{\sigma}_{xx} = F_x\Big[\frac{(1+q_1)}{\gamma_1 q_1}\frac{\partial}{\partial x}\overset{A}{h}(x, y_1+\ell_1) - \frac{(1+q_2)}{\gamma_2 q_2}\frac{\partial}{\partial x}\overset{A}{h}(x, y_2+\ell_2)\Big] + F_y\Big[\frac{(1+q_1)}{\gamma_1^2}\frac{\partial}{\partial y_1}\overset{A}{h}(x, y_1+\ell_1) - \frac{(1+q_2)}{\gamma_2^2}\frac{\partial}{\partial y_2}\overset{A}{h}(x, y_2+\ell_2)\Big]$$

$$\overset{A}{\sigma}_{yy} = -F_x\Big[\frac{\gamma_1(1+q_1)}{q_1}\frac{\partial}{\partial x}\overset{A}{h}(x, y_1+\ell_1) - \frac{\gamma_2(1+q_2)}{q_2}\frac{\partial}{\partial x}\overset{A}{h}(x, y_2+\ell_2)\Big] - F_y\Big[(1+q_1)\frac{\partial}{\partial y_1}\overset{A}{h}(x, y_1+\ell_1) - (1+q_2)\frac{\partial}{\partial y_2}\overset{A}{h}(x, y_2+\ell_2)\Big] \tag{8.47}$$

$$\overset{A}{\sigma}_{xy} = F_x\Big[\frac{(1+q_1)}{q_1}\frac{\partial}{\partial y_1}\overset{A}{h}(x, y_1+\ell_1) - \frac{(1+q_2)}{q_2}\frac{\partial}{\partial y_2}\overset{A}{h}(x, y_2+\ell_2)\Big] - F_y\Big[\frac{(1+q_1)}{\gamma_1}\frac{\partial}{\partial x}\overset{A}{h}(x, y_1+\ell_1) - \frac{(1+q_2)}{\gamma_2}\frac{\partial}{\partial x}\overset{A}{h}(x, y_2+\ell_2)\Big]$$

The image force solution is defined similarly in terms of a function $\overset{I}{h}(x, y_i - \ell_i)$, defined as

$$\overset{I}{h}(x, y_i - \ell_i) = \frac{1}{2\pi(q_1 - q_2)} \ln\sqrt{x^2 + (y_i - \ell_i)^2} \tag{8.48}$$

The displacements and stresses associated with this function are

$$\begin{aligned} \overset{I}{u}_x &= \frac{1}{c_{66}}F_x\Big[\frac{\gamma_1}{q_1}\overset{I}{h}(x, y_1-\ell_1) - \frac{\gamma_2}{q_2}\overset{I}{h}(x, y_2-\ell_2)\Big] + \frac{1}{c_{66}}F_y\Big[\int\frac{\partial}{\partial x}\overset{I}{h}(x, y_1-\ell_1)\,\mathrm{d}y_1 - \int\frac{\partial}{\partial x}\overset{I}{h}(x, y_2-\ell_2)\,\mathrm{d}y_2\Big] \\ \overset{I}{u}_y &= -\frac{1}{c_{66}}F_x\Big[\int\frac{\partial}{\partial x}\overset{I}{h}(x, y_1-\ell_1)\,\mathrm{d}y_1 - \int\frac{\partial}{\partial x}\overset{I}{h}(x, y_2-\ell_2)\,\mathrm{d}y_2\Big] + \frac{1}{c_{66}}F_y\Big[\frac{q_1}{\gamma_1}\overset{I}{h}(x, y_1-\ell_1) - \frac{q_2}{\gamma_2}\overset{I}{h}(x, y_2-\ell_2)\Big] \end{aligned} \tag{8.49}$$

and

$$\overset{I}{\sigma}_{xx} = F_x\Big[\frac{(1+q_1)}{\gamma_1 q_1}\frac{\partial}{\partial x}\overset{I}{h}(x, y_1-\ell_1) - \frac{(1+q_2)}{\gamma_2 q_2}\frac{\partial}{\partial x}\overset{I}{h}(x, y_2-\ell_2)\Big] - F_y\Big[\frac{(1+q_1)}{\gamma_1^2}\frac{\partial}{\partial y_1}\overset{I}{h}(x, y_1-\ell_1) - \frac{(1+q_2)}{\gamma_2^2}\frac{\partial}{\partial y_2}\overset{I}{h}(x, y_2-\ell_2)\Big]$$

$$\begin{aligned}\overset{I}{\sigma}_{yy} &= -F_x\Big[\frac{\gamma_1(1+q_1)}{q_1}\frac{\partial}{\partial x}\overset{I}{h}(x, y_1-\ell_1) - \frac{\gamma_2(1+q_2)}{q_2}\frac{\partial}{\partial x}\overset{I}{h}(x, y_2-\ell_2)\Big] \\ &\quad + F_y\Big[(1+q_1)\frac{\partial}{\partial y_1}\overset{I}{h}(x, y_1-\ell_1) - (1+q_2)\frac{\partial}{\partial y_2}\overset{I}{h}(x, y_2-\ell_2)\Big] \end{aligned} \tag{8.50}$$

$$\begin{aligned}\overset{I}{\sigma}_{xy} &= F_x\Big[\frac{(1+q_1)}{q_1}\frac{\partial}{\partial y_1}\overset{I}{h}(x, y_1-\ell_1) - \frac{(1+q_2)}{q_2}\frac{\partial}{\partial y_2}\overset{I}{h}(x, y_2-\ell_2)\Big] \\ &\quad + F_y\Big[\frac{(1+q_1)}{\gamma_1}\frac{\partial}{\partial x}\overset{I}{h}(x, y_1-\ell_1) - \frac{(1+q_2)}{\gamma_2}\frac{\partial}{\partial x}\overset{I}{h}(x, y_2-\ell_2)\Big]\end{aligned}$$

It should be noted that the signs of the terms involving the force component F_y have been changed in (8.49) and (8.50) in order to make $y = 0$ a line of symmetry for the actual and image solutions.

It is easy to show from these results that $\overset{A}{\sigma}_{xy} = -\overset{I}{\sigma}_{xy}$ and $\overset{A}{\sigma}_{yy} = \overset{I}{\sigma}_{yy}$ on $y = 0$. In order to remove the tractions from the surface of the half-plane the supplemental solution must therefore satisfy the following conditions: $\overset{S}{\sigma}_{xy} = 0$ and $\overset{S}{\sigma}_{yy} = -2\overset{I}{\sigma}_{yy}$ on $y = 0$. As discussed in Sect. 8.3.1, we express the boundary values of $\overset{S}{\sigma}_{yy}$ in terms of the image force solution because this solution does not introduce additional singularities in $y \le 0$.

Referring to (4.135) and (4.136), we express the required supplemental solution in terms of a harmonic function $\overset{S}{\Phi}(x, y)$ as

$$\begin{aligned}\overset{S}{u}_x &= \frac{\gamma_1}{1+q_1}\frac{\partial}{\partial x}\overset{S}{\Phi}(x, y_1) - \frac{\gamma_2}{1+q_2}\frac{\partial}{\partial x}\overset{S}{\Phi}(x, y_2) \\ \overset{S}{u}_y &= \frac{q_1}{1+q_1}\frac{\partial}{\partial y_1}\overset{S}{\Phi}(x, y_1) - \frac{q_2}{1+q_2}\frac{\partial}{\partial y_2}\overset{S}{\Phi}(x, y_2)\end{aligned} \tag{8.51}$$

and

$$\begin{aligned}\overset{S}{\sigma}_{xx} &= -c_{66}\Big[\frac{1}{\gamma_1}\frac{\partial^2}{\partial y_1^2}\overset{S}{\Phi}(x, y_1) - \frac{1}{\gamma_2}\frac{\partial^2}{\partial y_2^2}\overset{S}{\Phi}(x, y_2)\Big] \\ \overset{S}{\sigma}_{yy} &= c_{66}\Big[\gamma_1\frac{\partial^2}{\partial y_1^2}\overset{S}{\Phi}(x, y_1) - \gamma_2\frac{\partial^2}{\partial y_2^2}\overset{S}{\Phi}(x, y_2)\Big] \\ \overset{S}{\sigma}_{xy} &= c_{66}\Big[\frac{\partial^2}{\partial x\partial y_1}\overset{S}{\Phi}(x, y_1) - \frac{\partial^2}{\partial x\partial y_2}\overset{S}{\Phi}(x, y_2)\Big]\end{aligned} \tag{8.52}$$

The last one of (8.52) shows that $\overset{S}{\sigma}_{xy} = 0$ on $y = 0$ because functions $\overset{S}{\Phi}(x, y_1)$ and $\overset{S}{\Phi}(x, y_2)$ and all their derivatives are equal when $y = 0$, i.e., when $y_i = 0$ for $i = 1$ and 2 (see (4.134) and the ensuing discussion). The second one of (8.52) gives

$$\begin{aligned}\left(\overset{S}{\sigma}_{yy}\right)_{y=0} &= c_{66}\Big[\gamma_1 \frac{\partial^2}{\partial y_1^2}\overset{S}{\Phi}(x, y_1) - \gamma_2 \frac{\partial^2}{\partial y_2^2}\overset{S}{\Phi}(x, y_2)\Big]_{y=0} \\ &= c_{66}(\gamma_1 - \gamma_2)\Big[\frac{\partial^2}{\partial y^2}\overset{S}{\Phi}(x, y)\Big]_{y=0}\end{aligned} \tag{8.53}$$

and similarly the second one of (8.50) can be used to write

$$\begin{aligned}\left(\overset{I}{\sigma}_{yy}\right)_{y=0} = &-F_x\Big[\frac{\gamma_1(1+q_1)}{q_1}\frac{\partial}{\partial x}\overset{I}{h}(x, y-\ell_1) - \frac{\gamma_2(1+q_2)}{q_2}\frac{\partial}{\partial x}\overset{I}{h}(x, y-\ell_2)\Big]_{y=0} \\ &+ F_y\Big[(1+q_1)\frac{\partial}{\partial y}\overset{I}{h}(x, y-\ell_1) - (1+q_2)\frac{\partial}{\partial y}\overset{I}{h}(x, y-\ell_2)\Big]_{y=0}\end{aligned} \tag{8.54}$$

Imposing the condition $\overset{S}{\sigma}_{yy} = -2\overset{I}{\sigma}_{yy}$ on $y = 0$, we find that

$$\begin{aligned}c_{66}(\gamma_1 - \gamma_2)\frac{\partial^2}{\partial y^2}\overset{S}{\Phi}(x, y) = 2F_x\Big[&\frac{\gamma_1(1+q_1)}{q_1}\frac{\partial}{\partial x}\overset{I}{h}(x, y-\ell_1) - \frac{\gamma_2(1+q_2)}{q_2}\frac{\partial}{\partial x}\overset{I}{h}(x, y-\ell_2)\Big] \\ &- 2F_y\Big[(1+q_1)\frac{\partial}{\partial y}\overset{I}{h}(x, y-\ell_1) - (1+q_2)\frac{\partial}{\partial y}\overset{I}{h}(x, y-\ell_2)\Big]_{y=0}\end{aligned} \tag{8.55}$$

The right-hand side of (8.55) defines a function that contains no singular points in $y \leq 0$ and is continuously differentiable in the same domain. This function is also harmonic in x, y coordinates and so, by the uniqueness theorem for the Dirichlet problem, the equation holds everywhere in $y \leq 0$ and not just on the boundary.

We therefore have that the potential $\overset{S}{\Phi}(x, y)$ for the supplemental solution is determined by the following equation:

$$\begin{aligned}\frac{\partial^2}{\partial y^2}\overset{S}{\Phi}(x, y) = &\frac{2}{c_{66}(\gamma_1 - \gamma_2)}F_x\Big[\frac{\gamma_1(1+q_1)}{q_1}\frac{\partial}{\partial x}\overset{I}{h}(x, y-\ell_1) - \frac{\gamma_2(1+q_2)}{q_2}\frac{\partial}{\partial x}\overset{I}{h}(x, y-\ell_2)\Big] \\ &- \frac{2}{c_{66}(\gamma_1 - \gamma_2)}F_y\Big[(1+q_1)\frac{\partial}{\partial y}\overset{I}{h}(x, y-\ell_1) - (1+q_2)\frac{\partial}{\partial y}\overset{I}{h}(x, y-\ell_2)\Big]\end{aligned} \tag{8.56}$$

Expressions for $\partial\overset{S}{\Phi}(x, y)/\partial y$, $\partial\overset{S}{\Phi}(x, y)/\partial x$, and $\partial^2\overset{S}{\Phi}(x, y)/\partial x\partial y$ can be calculated directly from this relation and used in (8.51) and (8.52) to obtain the following values for the displacements and stresses for the supplemental solution:

$$
\begin{aligned}
\overset{S}{u}_x = &-\frac{2}{c_{66}(\gamma_1-\gamma_2)} F_x\Big[\frac{\gamma_1^2}{q_1}\overset{I}{h}(x, y_1-\ell_1)+\frac{\gamma_2^2}{q_2}\overset{I}{h}(x, y_2-\ell_2) \\
&\quad -\frac{\gamma_1\gamma_2(1+q_2)}{q_2(1+q_1)}\overset{I}{h}(x, y_1-\ell_2)-\frac{\gamma_1\gamma_2(1+q_1)}{q_1(1+q_2)}\overset{I}{h}(x, y_2-\ell_1)\Big] \\
&-\frac{2}{c_{66}(\gamma_1-\gamma_2)} F_y\Big[\gamma_1\int\frac{\partial}{\partial x}\overset{I}{h}(x, y_1-\ell_1)\,\mathrm{d}y_1+\gamma_2\int\frac{\partial}{\partial x}\overset{I}{h}(x, y_2-\ell_2)\,\mathrm{d}y_2 \\
&\quad -\frac{\gamma_1(1+q_2)}{(1+q_1)}\int\frac{\partial}{\partial x}\overset{I}{h}(x, y_1-\ell_2)\,\mathrm{d}y_1-\frac{\gamma_2(1+q_1)}{(1+q_2)}\int\frac{\partial}{\partial x}\overset{I}{h}(x, y_2-\ell_1)\,\mathrm{d}y_2\Big]
\end{aligned}
\tag{8.57}
$$

$$
\begin{aligned}
\overset{S}{u}_y = &\frac{2}{c_{66}(\gamma_1-\gamma_2)} F_x\Big[\gamma_1\int\frac{\partial}{\partial x}\overset{I}{h}(x, y_1-\ell_1)\,\mathrm{d}y_1+\gamma_2\int\frac{\partial}{\partial x}\overset{I}{h}(x, y_2-\ell_2)\,\mathrm{d}y_2 \\
&\quad -\frac{\gamma_2 q_1(1+q_2)}{q_2(1+q_1)}\int\frac{\partial}{\partial x}\overset{I}{h}(x, y_1-\ell_2)\,\mathrm{d}y_1-\frac{\gamma_1 q_2(1+q_1)}{q_1(1+q_2)}\int\frac{\partial}{\partial x}\overset{I}{h}(x, y_2-\ell_1)\,\mathrm{d}y_2\Big] \\
&-\frac{2}{c_{66}(\gamma_1-\gamma_2)} F_y\Big[q_1\overset{I}{h}(x, y_1-\ell_1)+q_2\overset{I}{h}(x, y_2-\ell_2) \\
&\quad -\frac{q_1(1+q_2)}{(1+q_1)}\overset{I}{h}(x, y_1-\ell_2)-\frac{q_2(1+q_1)}{(1+q_2)}\overset{I}{h}(x, y_2-\ell_1)\Big]
\end{aligned}
$$

and

$$
\begin{aligned}
\overset{S}{\sigma}_{xx} = &-\frac{2}{(\gamma_1-\gamma_2)} F_x\Big[\frac{(1+q_1)}{q_1}\frac{\partial}{\partial x}\overset{I}{h}(x, y_1-\ell_1)+\frac{(1+q_2)}{q_2}\frac{\partial}{\partial x}\overset{I}{h}(x, y_2-\ell_2) \\
&\quad -\frac{\gamma_2(1+q_2)}{\gamma_1 q_2}\frac{\partial}{\partial x}\overset{I}{h}(x, y_1-\ell_2)-\frac{\gamma_1(1+q_1)}{\gamma_2 q_1}\frac{\partial}{\partial x}\overset{I}{h}(x, y_2-\ell_1)\Big] \\
&+\frac{2}{(\gamma_1-\gamma_2)} F_y\Big[\frac{(1+q_1)}{\gamma_1}\frac{\partial}{\partial y_1}\overset{I}{h}(x, y_1-\ell_1)+\frac{(1+q_2)}{\gamma_2}\frac{\partial}{\partial y_2}\overset{I}{h}(x, y_2-\ell_2) \\
&\quad -\frac{(1+q_2)}{\gamma_1}\frac{\partial}{\partial y_1}\overset{I}{h}(x, y_1-\ell_2)-\frac{(1+q_1)}{\gamma_2}\frac{\partial}{\partial y_2}\overset{I}{h}(x, y_2-\ell_1)\Big]
\end{aligned}
$$

$$
\begin{aligned}
\overset{S}{\sigma}_{yy} = &\frac{2}{(\gamma_1-\gamma_2)} F_x\Big[\frac{\gamma_1^2(1+q_1)}{q_1}\frac{\partial}{\partial x}\overset{I}{h}(x, y_1-\ell_1)+\frac{\gamma_2^2(1+q_2)}{q_2}\frac{\partial}{\partial x}\overset{I}{h}(x, y_2-\ell_2) \\
&\quad -\frac{\gamma_1\gamma_2(1+q_2)}{q_2}\frac{\partial}{\partial x}\overset{I}{h}(x, y_1-\ell_2)-\frac{\gamma_1\gamma_2(1+q_1)}{q_1}\frac{\partial}{\partial x}\overset{I}{h}(x, y_2-\ell_1)\Big] \\
&-\frac{2}{(\gamma_1-\gamma_2)} F_y\Big[\gamma_1(1+q_1)\frac{\partial}{\partial y_1}\overset{I}{h}(x, y_1-\ell_1)+\gamma_2(1+q_2)\frac{\partial}{\partial y_2}\overset{I}{h}(x, y_2-\ell_2) \\
&\quad -\gamma_1(1+q_2)\frac{\partial}{\partial y_1}\overset{I}{h}(x, y_1-\ell_2)-\gamma_2(1+q_1)\frac{\partial}{\partial y_2}\overset{I}{h}(x, y_2-\ell_1)\Big]
\end{aligned}
\tag{8.58}
$$

$$
\overset{S}{\sigma}_{xy} = -\frac{2}{(\gamma_1-\gamma_2)} F_x\Big[\frac{\gamma_1(1+q_1)}{q_1}\Big\{\frac{\partial}{\partial y_1}\overset{I}{h}(x, y_1-\ell_1)-\frac{\partial}{\partial y_2}\overset{I}{h}(x, y_2-\ell_1)\Big\}
$$

$$+\frac{\gamma_2(1+q_2)}{q_2}\Big\{\frac{\partial}{\partial y_2}\overset{I}{h}(x,y_2-\ell_2)-\frac{\partial}{\partial y_1}\overset{I}{h}(x,y_1-\ell_2)\Big\}\Big]$$

$$-\frac{2}{(\gamma_1-\gamma_2)}F_y\Big[(1+q_1)\Big\{\frac{\partial}{\partial x}\overset{I}{h}(x,y_1-\ell_1)-\frac{\partial}{\partial x}\overset{I}{h}(x,y_2-\ell_1)\Big\}$$

$$+(1+q_2)\Big\{\frac{\partial}{\partial x}\overset{I}{h}(x,y_2-\ell_2)-\frac{\partial}{\partial x}\overset{I}{h}(x,y_1-\ell_2)\Big\}\Big]$$

8.3.5 *Concentrated Force on the Surface of an Orthotropic Elastic Half-Plane*

We saw in the case of an isotropic elastic material that a concentrated force $F_x,\ F_y$ on the surface of the half-plane $y \le 0$ can be treated as the limiting case of the problem of constant stresses $\sigma_{xy} = p_x$ and $\sigma_{yy} = p_y$ applied over the line segment $-a \le x \le a$ on $y = 0$ with the resultant loads $2ap_x\ (= F_x)$ and $2ap_y\ (= F_y)$ kept constant as $a \to 0$. The same approach can be used for an orthotropic elastic half-plane, again using harmonic functions $\Phi(x, y)$ and $\Psi(x, y)$, but now specialized for an orthotropic (or transversely isotropic) material as explained in Sect. 4.10.4. We again assume that the directions of elastic symmetry are in the $x,\ y$ coordinate directions and, at least temporarily, that $\gamma_1 \ne \gamma_2$.

For the case of normal loading, representation (4.136) in terms of a harmonic function $\Phi(x, y)$ automatically satisfies the condition that the shear stress σ_{xy} is zero on $y = 0$ for all values of x. Guided by the solution for the isotropic case, (8.23), we take

$$\frac{\partial^2}{\partial y_i^2}\Phi(x,y_i) = A\Big(\arctan\frac{y_i}{x-a}-\arctan\frac{y_i}{x+a}\Big) \tag{8.59}$$

where A is a constant. The second one of (4.136) gives

$$\sigma_{yy} = c_{66}A\Big[\gamma_1\big(\arctan\frac{y_1}{x-a}-\arctan\frac{y_1}{x+a}\big)-\gamma_2\big(\arctan\frac{y_2}{x-a}-\arctan\frac{y_2}{x+a}\big)\Big] \tag{8.60}$$

On the boundary $y = 0_-$ the functions $\arctan y_i/(x-a) - \arctan y_i/(x+a)$ are equal to zero if $\mid x \mid > a$ and are equal to $-\pi$ if $\mid x \mid \le a$. The requirement that $\sigma_{yy} = p_y$ for $\mid x \mid \le a$ on $y = 0$ determines constant A as

$$A = -\frac{p_y}{\pi c_{66}(\gamma_1-\gamma_2)} \tag{8.61}$$

so that

$$\frac{\partial^2}{\partial y_i^2}\Phi(x, y_i) = -\frac{p_y}{\pi c_{66}(\gamma_1 - \gamma_2)}\left(\arctan\frac{y_i}{x-a} - \arctan\frac{y_i}{x+a}\right) \tag{8.62}$$

a result that should be compared with (8.23).

As in the case of isotropic elasticity, a concentrated force F_y at the origin is the limiting case for which $p_y = F_y/(2a)$ and is kept constant as $a \to 0$. An application of L'Hopital's rule to (8.62) then gives

$$\frac{\partial^2}{\partial y_i^2}\Phi(x, y_i) = -\frac{F_y}{\pi c_{66}(\gamma_1 - \gamma_2)}\frac{y_i}{x^2 + y_i^2} \tag{8.63}$$

Using this result in (4.136) gives the following stress components due to the concentrated force F_y:

$$\sigma_{xx} = \frac{F_y}{\pi(\gamma_1 - \gamma_2)}\left[\frac{1}{\gamma_1}\frac{y_1}{x^2 + y_1^2} - \frac{1}{\gamma_2}\frac{y_2}{x^2 + y_2^2}\right]$$

$$\sigma_{yy} = -\frac{F_y}{\pi(\gamma_1 - \gamma_2)}\left[\gamma_1\frac{y_1}{x^2 + y_1^2} - \gamma_2\frac{y_2}{x^2 + y_2^2}\right] \tag{8.64}$$

$$\sigma_{xy} = -\frac{F_y}{\pi(\gamma_1 - \gamma_2)}\left[\frac{x}{x^2 + y_1^2} - \frac{x}{x^2 + y_2^2}\right]$$

Expressions for the displacements could be obtained by using (8.62) in (4.135) but we will omit these because, as noted for the isotropic case, the displacements for this problem are not meaningful for practical applications.

A similar analysis for a concentrated force F_x at the origin leads to the results

$$\frac{\partial^2}{\partial y_i^2}\Psi(x, y_i) = -\frac{F_x}{\pi(\gamma_1 - \gamma_2)}\frac{y_i}{x^2 + y_i^2} \tag{8.65}$$

and

$$\sigma_{xx} = -\frac{F_x}{\pi\gamma_1 - \gamma_2)}\left[\frac{1}{\gamma_1^2}\frac{x}{x^2 + y_1^2} - \frac{1}{\gamma_2^2}\frac{x}{x^2 + y_2^2}\right]$$

$$\sigma_{yy} = \frac{F_x}{\pi(\gamma_1 - \gamma_2)}\left[\frac{x}{x^2 + y_1^2} - \frac{x}{x^2 + y_2^2}\right] \tag{8.66}$$

$$\sigma_{xy} = -\frac{F_x}{\pi(\gamma_1 - \gamma_2)}\Big[\frac{1}{\gamma_1}\frac{y_1}{x^2 + y_1^2} - \frac{1}{\gamma_2}\frac{y_2}{x^2 + y_2^2}\Big]$$

We conclude this discussion by demonstrating that the above solution reduces to that given previously for isotropic elasticity when $\gamma_1 = \gamma_2 = 1$. The stresses (8.64) and (8.66) are all of the form $0/0$ when $\gamma_1 = \gamma_2$ because the scaled coordinates y_1 and y_2 are equal in this case. Limiting expressions for these quantities can be found by fixing one of the parameters, $\gamma_2 = \gamma_0$ say, and then applying L'Hopital's rule for $\gamma_1 \to \gamma_0$. Evaluation of σ_{yy} in (8.64) in this way, for example, is done as follows:

$$\begin{aligned}\sigma_{yy} &= -\frac{F_y}{\pi}\lim_{\gamma_1\to\gamma_0}\left\{\frac{\partial}{\partial\gamma_1}\Big[\frac{y}{x^2 + y^2/\gamma_1^2} - \frac{y}{x^2 + y^2/\gamma_0^2}\Big]\Big/\frac{\partial}{\partial\gamma_1}(\gamma_1 - \gamma_0)\right\}\\ &= -\frac{2F_y}{\pi}\frac{y^3/\gamma_0^3}{(x^2 + y^2/\gamma_0^2)^2}\end{aligned} \tag{8.67}$$

If $\gamma_0 = 1$ this expression reduces to the second one of (8.25).

8.3.6 Concentrated Displacement Discontinuity in an Orthotropic Elastic Half-Plane

The problem of a concentrated displacement discontinuity in an orthotropic (or transversely isotropic) elastic half-plane $y \le 0$ with a traction-free boundary is depicted in Fig. 8.5. As in the corresponding problem for isotropic elasticity the displacement discontinuity is located at the point $x = 0,\ y = -\ell$ and has components $\mathcal{D}_{\bar{x}}$ and $\mathcal{D}_{\bar{y}}$ relative to a local $\bar{x},\ \bar{y}$ coordinate system. The local $\bar{x}$ coordinate axis is inclined at an angle β from the global x direction and the x and y axes are directions of elastic symmetry.

In using the method of images to solve this problem it is convenient to write the solution in terms of $x,\ y$ coordinates instead of $\bar{x},\ \bar{y}$ and $\bar{x}',\ \bar{y}'$ coordinates (as was done for the isotropic case) because this allows for the direct use of the results of Sect. 4.14.2. The displacement discontinuity components $\mathcal{D}_x$ and $\mathcal{D}_y$ are related to the components $\mathcal{D}_{\bar{x}}$ and $\mathcal{D}_{\bar{y}}$ by Eqs. (4.204). The values of these components for the image discontinuity are $-\mathcal{D}_x$ and $+\mathcal{D}_y$.

The displacements and stresses due to the actual displacement discontinuity can be written from (4.226) and (4.227) by replacing y_i by $y_i + \ell_i$, where $\ell_i = \ell/\gamma_i$ for $i = 1$ and 2. For notational convenience, we use (4.224), (4.226), and (4.227) to define a function $\overset{A}{V}(x, y_i + \ell_i)$ as

$$\overset{A}{V}(x, y_i + \ell_i) = -\frac{q_3}{2\pi}\Big[\cos\beta\ln\sqrt{x^2 + (y_i + \ell_i)^2} - \frac{\sin\beta}{\gamma_i}\arctan\frac{y_i + \ell_i}{x}\Big] \tag{8.68}$$

in which the quantity q_3 is defined in (4.216), and write

$$\begin{aligned}\overset{A}{u}_x &= \mathcal{D}_x\Big[\frac{1}{1+q_1}\frac{\partial}{\partial y_1}\overset{A}{V}(x, y_1+\ell_1) - \frac{1}{1+q_2}\frac{\partial}{\partial y_2}\overset{A}{V}(x, y_2+\ell_2)\Big]\\ &\quad - \mathcal{D}_y\Big[\frac{\gamma_1}{1+q_1}\frac{\partial}{\partial x}\overset{A}{V}(x, y_1+\ell_1) - \frac{\gamma_2}{1+q_2}\frac{\partial}{\partial x}\overset{A}{V}(x, y_2+\ell_2)\Big]\\ \overset{A}{u}_y &= -\mathcal{D}_x\Big[\frac{q_1}{\gamma_1(1+q_1)}\frac{\partial}{\partial x}\overset{A}{V}(x, y_1+\ell_1) - \frac{q_2}{\gamma_2(1+q_2)}\frac{\partial}{\partial x}\overset{A}{V}(x, y_2+\ell_2)\Big]\\ &\quad - \mathcal{D}_y\Big[\frac{q_1}{1+q_1}\frac{\partial}{\partial y_1}\overset{A}{V}(x, y_1+\ell_1) - \frac{q_2}{(1+q_2)}\frac{\partial}{\partial y_2}\overset{A}{V}(x, y_2+\ell_2)\Big]\end{aligned} \tag{8.69}$$

and

$$\begin{aligned}\overset{A}{\sigma}_{xx} &= c_{66}\mathcal{D}_x\Big[\frac{1}{\gamma_1^2}\frac{\partial^2}{\partial x\partial y_1}\overset{A}{V}(x, y_1+\ell_1) - \frac{1}{\gamma_2^2}\frac{\partial^2}{\partial x\partial y_2}\overset{A}{V}(x, y_2+\ell_2)\Big]\\ &\quad + c_{66}\mathcal{D}_y\Big[\frac{1}{\gamma_1}\frac{\partial^2}{\partial y_1^2}\overset{A}{V}(x, y_1+\ell_1) - \frac{1}{\gamma_2}\frac{\partial^2}{\partial y_2^2}\overset{A}{V}(x, y_2+\ell_2)\Big]\\ \overset{A}{\sigma}_{yy} &= -c_{66}\mathcal{D}_x\Big[\frac{\partial^2}{\partial x\partial y_1}\overset{A}{V}(x, y_1+\ell_1) - \frac{\partial^2}{\partial x\partial y_2}\overset{A}{V}(x, y_2+\ell_2)\Big]\\ &\quad - c_{66}\mathcal{D}_y\Big[\gamma_1\frac{\partial^2}{\partial y_1^2}\overset{A}{V}(x, y_1+\ell_1) - \gamma_2\frac{\partial^2}{\partial y_2^2}\overset{A}{V}(x, y_2+\ell_2)\Big]\\ \overset{A}{\sigma}_{xy} &= c_{66}\mathcal{D}_x\Big[\frac{1}{\gamma_1}\frac{\partial^2}{\partial y_1^2}\overset{A}{V}(x, y_1+\ell_1) - \frac{1}{\gamma_2}\frac{\partial^2}{\partial y_2^2}\overset{A}{V}(x, y_2+\ell_2)\Big]\\ &\quad - c_{66}\mathcal{D}_y\Big[\frac{\partial^2}{\partial x\partial y_1}\overset{A}{V}(x, y_1+\ell_1) - \frac{\partial^2}{\partial x\partial y_2}\overset{A}{V}(x, y_2+\ell_2)\Big]\end{aligned} \tag{8.70}$$

The solution for the image displacement discontinuity is defined similarly in terms of a function $\overset{I}{V}(x, y_i - \ell_i)$, i.e.,

$$\overset{I}{V}(x, y_i-\ell_i) = -\frac{q_3}{2\pi}\Big[\cos\beta\ln\sqrt{x^2+(y_i-\ell_i)^2} + \frac{\sin\beta}{\gamma_i}\arctan\frac{y_i-\ell_i}{x}\Big] \tag{8.71}$$

Note that the sign of the second term in this expression has been changed to reflect the fact that the orientation of the image displacement discontinuity is the negative of that of the actual discontinuity.

The displacements and stresses associated with function (8.71) are

$$
\begin{aligned}
\overset{I}{u}_x &= -\mathcal{D}_x\Big[\frac{1}{1+q_1}\frac{\partial}{\partial y_1}\overset{I}{V}(x, y_1-\ell_1) - \frac{1}{1+q_2}\frac{\partial}{\partial y_2}\overset{I}{V}(x, y_2-\ell_2)\Big] \\
&\quad - \mathcal{D}_y\Big[\frac{\gamma_1}{1+q_1}\frac{\partial}{\partial x}\overset{I}{V}(x, y_1-\ell_1) - \frac{\gamma_2}{1+q_2}\frac{\partial}{\partial x}\overset{I}{V}(x, y_2-\ell_2)\Big] \\
\overset{I}{u}_y &= \mathcal{D}_x\Big[\frac{q_1}{\gamma_1(1+q_1)}\frac{\partial}{\partial x}\overset{I}{V}(x, y_1-\ell_1) - \frac{q_2}{\gamma_2(1+q_2)}\frac{\partial}{\partial x}\overset{I}{V}(x, y_2-\ell_2)\Big] \\
&\quad - \mathcal{D}_y\Big[\frac{q_1}{1+q_1}\frac{\partial}{\partial y_1}\overset{I}{V}(x, y_1-\ell_1) - \frac{q_2}{(1+q_2)}\frac{\partial}{\partial y_2}\overset{I}{V}(x, y_2-\ell_2)\Big]
\end{aligned}
\tag{8.72}
$$

and

$$
\begin{aligned}
\overset{I}{\sigma}_{xx} &= -c_{66}\mathcal{D}_x\Big[\frac{1}{\gamma_1^2}\frac{\partial^2}{\partial x\partial y_1}\overset{I}{V}(x, y_1-\ell_1) - \frac{1}{\gamma_2^2}\frac{\partial^2}{\partial x\partial y_2}\overset{I}{V}(x, y_2-\ell_2)\Big] \\
&\quad + c_{66}\mathcal{D}_y\Big[\frac{1}{\gamma_1}\frac{\partial^2}{\partial y_1^2}\overset{I}{V}(x, y_1-\ell_1) - \frac{1}{\gamma_2}\frac{\partial^2}{\partial y_2^2}\overset{I}{V}(x, y_2-\ell_2)\Big] \\
\overset{I}{\sigma}_{yy} &= c_{66}\mathcal{D}_x\Big[\frac{\partial^2}{\partial x\partial y_1}\overset{I}{V}(x, y_1-\ell_1) - \frac{\partial^2}{\partial x\partial y_2}\overset{I}{V}(x, y_2-\ell_2)\Big] \\
&\quad - c_{66}\mathcal{D}_y\Big[\gamma_1\frac{\partial^2}{\partial y_1^2}\overset{I}{V}(x, y_1-\ell_1) - \gamma_2\frac{\partial^2}{\partial y_2^2}\overset{I}{V}(x, y_2-\ell_2)\Big] \\
\overset{I}{\sigma}_{xy} &= -c_{66}\mathcal{D}_x\Big[\frac{1}{\gamma_1}\frac{\partial^2}{\partial y_1^2}\overset{I}{V}(x, y_1-\ell_1) - \frac{1}{\gamma_2}\frac{\partial^2}{\partial y_2^2}\overset{I}{V}(x, y_2-\ell_2)\Big] \\
&\quad - c_{66}\mathcal{D}_y\Big[\frac{\partial^2}{\partial x\partial y_1}\overset{I}{V}(x, y_1-\ell_1) - \frac{\partial^2}{\partial x\partial y_2}\overset{I}{V}(x, y_2-\ell_2)\Big]
\end{aligned}
\tag{8.73}
$$

It is easy to show that the following relations hold on $y = 0$: $\partial^2\overset{A}{V}(x, y_i + \ell_i)/\partial x\partial y_i = -\partial^2\overset{I}{V}(x, y_i - \ell_i)/\partial x\partial y_i$ and $\partial^2\overset{A}{V}(x, y_i + \ell_i)/\partial y_i^2 = \partial^2\overset{I}{V}(x, y_i - \ell_i)/\partial y_i^2$. It then follows from (8.70) and (8.73) that $\overset{A}{\sigma}_{xy} = -\overset{I}{\sigma}_{xy}$ and $\overset{A}{\sigma}_{yy} = \overset{I}{\sigma}_{yy}$ on $y = 0$. We therefore again conclude that the supplemental solution must satisfy the conditions $\overset{S}{\sigma}_{xy} = 0$ and $\overset{S}{\sigma}_{yy} = -2\overset{I}{\sigma}_{yy}$ on $y = 0$.

Using the same approach as that used for a concentrated force in an orthotropic elastic half-plane, we find that the potential $\overset{S}{\Phi}(x, y)$ for the supplemental solution for the present problem is determined by the following equation:

$$\frac{\partial^2}{\partial y^2}\overset{S}{\Phi}(x,y) = -\frac{2}{(\gamma_1-\gamma_2)}\mathcal{D}_x\Big[\frac{\partial^2}{\partial x\partial y}\overset{I}{V}(x,y-\ell_1) - \frac{\partial^2}{\partial x\partial y}\overset{I}{V}(x,y-\ell_2)\Big] + \frac{2}{(\gamma_1-\gamma_2)}\mathcal{D}_y\Big[\gamma_1\frac{\partial^2}{\partial y^2}\overset{I}{V}(x,y-\ell_1) - \gamma_2\frac{\partial^2}{\partial y^2}\overset{I}{V}(x,y-\ell_2)\Big] \tag{8.74}$$

It can then be shown from (8.51) and (8.52) that the displacements and stresses for the supplemental solution are given as

$$\begin{aligned}\overset{S}{u}_x = {} & \frac{2}{(\gamma_1-\gamma_2)}\mathcal{D}_x\Big[\frac{\gamma_1}{(1+q_1)}\frac{\partial}{\partial y_1}\overset{I}{V}(x,y_1-\ell_1) + \frac{\gamma_2}{(1+q_2)}\frac{\partial}{\partial y_2}\overset{I}{V}(x,y_2-\ell_2) \\ & - \frac{\gamma_1}{(1+q_1)}\frac{\partial}{\partial y_1}\overset{I}{V}(x,y_1-\ell_2) - \frac{\gamma_2}{(1+q_2)}\frac{\partial}{\partial y_2}\overset{I}{V}(x,y_2-\ell_1)\Big] \\ & + \frac{2}{(\gamma_1-\gamma_2)}\mathcal{D}_y\Big[\frac{\gamma_1^2}{(1+q_1)}\frac{\partial}{\partial x}\overset{I}{V}(x,y_1-\ell_1) + \frac{\gamma_2^2}{(1+q_2)}\frac{\partial}{\partial x}\overset{I}{V}(x,y_2-\ell_2) \\ & - \frac{\gamma_1\gamma_2}{(1+q_1)}\frac{\partial}{\partial x}\overset{I}{V}(x,y_1-\ell_2) - \frac{\gamma_1\gamma_2}{(1+q_2)}\frac{\partial}{\partial x}\overset{I}{V}(x,y_2-\ell_1)\Big]\end{aligned} \tag{8.75}$$

$$\begin{aligned}\overset{S}{u}_y = {} & -\frac{2}{(\gamma_1-\gamma_2)}\mathcal{D}_x\Big[\frac{q_1}{(1+q_1)}\frac{\partial}{\partial x}\overset{I}{V}(x,y_1-\ell_1) + \frac{q_2}{(1+q_2)}\frac{\partial}{\partial x}\overset{I}{V}(x,y_2-\ell_2) \\ & - \frac{q_1}{(1+q_1)}\frac{\partial}{\partial x}\overset{I}{V}(x,y_1-\ell_2) - \frac{q_2}{(1+q_2)}\frac{\partial}{\partial x}\overset{I}{V}(x,y_2-\ell_1)\Big] \\ & + \frac{2}{(\gamma_1-\gamma_2)}\mathcal{D}_y\Big[\frac{\gamma_1 q_1}{(1+q_1)}\frac{\partial}{\partial y_1}\overset{I}{V}(x,y_1-\ell_1) + \frac{\gamma_2 q_2}{(1+q_2)}\frac{\partial}{\partial y_2}\overset{I}{V}(x,y_2-\ell_2) \\ & - \frac{\gamma_2 q_1}{(1+q_1)}\frac{\partial}{\partial y_1}\overset{I}{V}(x,y_1-\ell_2) - \frac{\gamma_1 q_2}{(1+q_2)}\frac{\partial}{\partial y_2}\overset{I}{V}(x,y_2-\ell_1)\Big]\end{aligned}$$

$$\begin{aligned}\overset{S}{\sigma}_{xx} = {} & \frac{2c_{66}}{(\gamma_1-\gamma_2)}\mathcal{D}_x\Big[\frac{1}{\gamma_1}\frac{\partial^2}{\partial x\partial y_1}\overset{I}{V}(x,y_1-\ell_1) + \frac{1}{\gamma_2}\frac{\partial^2}{\partial x\partial y_2}\overset{I}{V}(x,y_2-\ell_2) \\ & - \frac{1}{\gamma_1}\frac{\partial^2}{\partial x\partial y_1}\overset{I}{V}(x,y_1-\ell_2) - \frac{1}{\gamma_2}\frac{\partial^2}{\partial x\partial y_2}\overset{I}{V}(x,y_2-\ell_1)\Big] \\ & - \frac{2c_{66}}{(\gamma_1-\gamma_2)}\mathcal{D}_y\Big[\frac{\partial^2}{\partial y_1^2}\overset{I}{V}(x,y_1-\ell_1) + \frac{\partial^2}{\partial y_2^2}\overset{I}{V}(x,y_2-\ell_2) \\ & - \frac{\gamma_2}{\gamma_1}\frac{\partial^2}{\partial y_1^2}\overset{I}{V}(x,y_1-\ell_2) - \frac{\gamma_1}{\gamma_2}\frac{\partial^2}{\partial y_2^2}\overset{I}{V}(x,y_2-\ell_1)\Big]\end{aligned}$$

$$\overset{S}{\sigma}_{yy} = -\frac{2c_{66}}{(\gamma_1-\gamma_2)}\mathcal{D}_x\Big[\gamma_1\frac{\partial^2}{\partial x\partial y_1}\overset{I}{V}(x,y_1-\ell_1) + \gamma_2\frac{\partial^2}{\partial x\partial y_2}\overset{I}{V}(x,y_2-\ell_2)$$

$$
\begin{aligned}
&- \gamma_1 \frac{\partial^2}{\partial x \partial y_1} \overset{I}{V}(x, y_1 - \ell_2) - \gamma_2 \frac{\partial^2}{\partial x \partial y_2} \overset{I}{V}(x, y_2 - \ell_1) \Big] \\
&+ \frac{2c_{66}}{(\gamma_1 - \gamma_2)} \mathcal{D}_y \Big[\gamma_1^2 \frac{\partial^2}{\partial y_1^2} \overset{I}{V}(x, y_1 - \ell_1) + \gamma_2^2 \frac{\partial^2}{\partial y_2^2} \overset{I}{V}(x, y_2 - \ell_2) \\
&- \gamma_1 \gamma_2 \frac{\partial^2}{\partial y_1^2} \overset{I}{V}(x, y_1 - \ell_2) - \gamma_1 \gamma_2 \frac{\partial^2}{\partial y_2^2} \overset{I}{V}(x, y_2 - \ell_1) \Big]
\end{aligned}
\tag{8.76}
$$

$$
\begin{aligned}
\overset{S}{\sigma}_{xy} = {} & \frac{2c_{66}}{(\gamma_1 - \gamma_2)} \mathcal{D}_x \Big[\Big\{ \frac{\partial^2}{\partial y_1^2} \overset{I}{V}(x, y_1 - \ell_1) - \frac{\partial^2}{\partial y_2^2} \overset{I}{V}(x, y_2 - \ell_1) \Big\} \\
&+ \Big\{ \frac{\partial^2}{\partial y_2^2} \overset{I}{V}(x, y_2 - \ell_2) - \frac{\partial^2}{\partial y_1^2} \overset{I}{V}(x, y_1 - \ell_2) \Big\} \Big] \\
&+ \frac{2c_{66}}{(\gamma_1 - \gamma_2)} \mathcal{D}_y \Big[\gamma_1 \Big\{ \frac{\partial^2}{\partial x \partial y_1} \overset{I}{V}(x, y_1 - \ell_1) - \frac{\partial^2}{\partial x \partial y_2} \overset{I}{V}(x, y_2 - \ell_1) \Big\} \\
&+ \gamma_2 \Big\{ \frac{\partial^2}{\partial x \partial y_2} \overset{I}{V}(x, y_2 - \ell_2) - \frac{\partial^2}{\partial x \partial y_1} \overset{I}{V}(x, y_1 - \ell_2) \Big\} \Big]
\end{aligned}
$$

8.4 Piecewise Homogeneous Regions I

We saw in Sect. 3.12 that the direct boundary integral method for potential problems could be extended to solve problems involving *piecewise homogeneous* regions. A similar approach can be applied to elasticity problems and we consider this topic next.

We will suppose that the region of interest consists of two finite subregions, as illustrated in Fig. 8.7. The two subregions, labeled V_1 and V_2 in the figure, are each assumed to be homogeneous, isotropic, and linearly elastic, with elastic constants ν_1, μ_1 and ν_2, μ_2. (The same approach applies to anisotropic materials as well as to

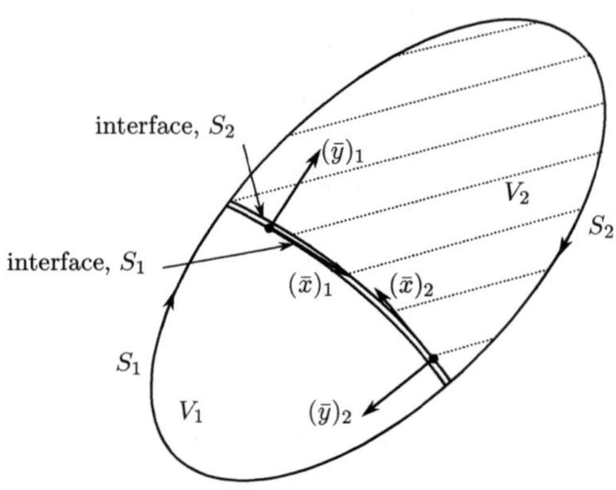

Fig. 8.7 Piecewise homogeneous body consisting of two isotropic, linearly elastic subregions

multiple subregions but we will limit our discussion to isotropic materials and two subregions.) The boundary contours, S_1 of subregion V_1 and S_2 of subregion V_2, are traversed in a clockwise sense, in accordance with the convention adopted previously. The local $\bar{x},\ \bar{y}$ coordinates associated with the two contours then appear as shown in the figure. The common portions of the two contours define the *interface* between the two subregions. The local coordinates $(\bar{x})_1,\ (\bar{y})_1$ and $(\bar{x})_2,\ (\bar{y})_2$ are oppositely directed all along the interface, i.e., $(\bar{x})_1 = -(\bar{x})_2$ and $(\bar{y})_1 = -(\bar{y})_2$.

A boundary value problem for the piecewise homogeneous body of Fig. 8.7 is defined by the usual displacement and traction boundary conditions along the "free" portions of the boundary contours S_1 and S_2, as well as by *continuity conditions* for the components of displacement and traction along the interface between the two subregions. The latter conditions assume that the interface is perfectly bonded; other types of interface conditions are considered in Sect. 8.8.

The continuity conditions for a point $\boldsymbol{x}$ on the interface can be written in terms of global $x,\ y$ coordinates as

$$\begin{aligned} t_x^{[1]}(\boldsymbol{x}) &= -t_x^{[2]}(\boldsymbol{x}) \\ t_y^{[1]}(\boldsymbol{x}) &= -t_y^{[2]}(\boldsymbol{x}) \end{aligned} \tag{8.77}$$

for the tractions, and

$$\begin{aligned} u_x^{[1]}(\boldsymbol{x}) &= u_x^{[2]}(\boldsymbol{x}) \\ u_y^{[1]}(\boldsymbol{x}) &= u_y^{[2]}(\boldsymbol{x}) \end{aligned} \tag{8.78}$$

for the displacements. The minus signs in (8.77) are a consequence of equilibrium: the sum of the tractions at points along the interface are zero. In terms of the local coordinates shown in Fig. 8.7 the continuity conditions are

$$\begin{aligned} t_{\bar{x}}^{[1]}(\boldsymbol{x}) &= t_{\bar{x}}^{[2]}(\boldsymbol{x}) \\ t_{\bar{y}}^{[1]}(\boldsymbol{x}) &= t_{\bar{y}}^{[2]}(\boldsymbol{x}) \end{aligned} \tag{8.79}$$

and

$$\begin{aligned} u_{\bar{x}}^{[1]}(\boldsymbol{x}) &= -u_{\bar{x}}^{[2]}(\boldsymbol{x}) \\ u_{\bar{y}}^{[1]}(\boldsymbol{x}) &= -u_{\bar{y}}^{[2]}(\boldsymbol{x}) \end{aligned} \tag{8.80}$$

The signs in these expressions are a consequence of the opposite directions for the local coordinates $(\bar{x})_1$, $(\bar{y})_1$ and $(\bar{x})_2$, $(\bar{y})_2$ along the interface.

A boundary element solution to the problem under consideration can be found by dividing the contours S_1 and S_2 into a number of straight line segments, joined end to end, and making an assumption about the way in which the boundary parameters vary over these segments. For simplicity we will assume in the following discussion that the boundary parameters are piecewise constant. Including the portions of the contours that represent the interface, there will be N_1 boundary elements along S_1 and N_2 boundary elements along S_2. Elements along the two sides of the interface must match exactly.[1] There will then be four unknowns associated with each interface element: two displacement components $u_{\bar{x}}$ and $u_{\bar{y}}$ and two traction components $t_{\bar{x}}$ and $t_{\bar{y}}$. Continuity conditions (8.79) and (8.80) give four relations to be satisfied at each interface element, so the problem is solvable.

More elaborate variations of the boundary parameters can also be assumed—piecewise linear, for example—but the numerical procedure for these is essentially the same as for piecewise constant variations. The stress and displacement discontinuity methods require the use of discontinuous elements to avoid strong singularities at the element endpoints, and such elements simply have two or three collocation points inside an element instead of one. In addition, the direct boundary integral method requires the use of either discontinuous elements or double nodes for the common situation in which adjoining elements have different inclinations because the tractions are in general different on either side of an element junction. For these reasons, it is sufficient to consider only the case for which the displacements and tractions are assumed to be piecewise constant from element to element.

Any of the boundary element methods discussed previously can be used to solve boundary value problems for piecewise homogeneous elastic bodies. The numerical procedures are slightly different for the indirect and direct formulations, however, so we will describe these separately below.

8.4.1 Procedure for the Stress (or Displacement) Discontinuity Method

We consider first the numerical procedure for the stress discontinuity method. The procedure for the displacement discontinuity method is in all respects analogous to this and we will not discuss it here.

The problem depicted in Fig. 8.7 can be treated as two separate boundary value problems, one for subregion V_1 and the other for subregion V_2. These two problems, of course, are linked by the interface continuity conditions (8.79) and (8.80). Proceeding in essentially the same way as in Chap. 5, we associate stress discontinuities $P_{\bar{x}}$ and $P_{\bar{y}}$ with each element of the two boundary contours. The displacements and stresses in subregion V_1 depend only on the stress discontinuities $P_{\bar{x}}^{[1]}$ and $P_{\bar{y}}^{[1]}$ at

[1] Matching elements will have the same x, y coordinates but their inclinations will differ by π.

the N_1 elements of contour S_1, and the displacements and stresses in subregion V_2 depend only on the applied tractions $P_{\bar{x}}^{[2]}$ and $P_{\bar{y}}^{[2]}$ at the N_2 elements of contour S_2. Our problem, therefore, is to find $P_{\bar{x}}$ and $P_{\bar{y}}$ at each of the $N_1 + N_2 = N$ boundary elements such that the boundary conditions and interface continuity conditions are met.

In setting up the system of algebraic equations for the composite problem, it will be convenient to number the boundary elements consecutively for the two subregions, starting with subregion V_1. Elements 1 to N_1 then lie along contour S_1 and elements $N_1 + 1$ to $N_1 + N_2 = N$ lie along contour S_2. Referring to equations (5.2), the tractions on the boundary of subregion V_1 can be written as

$$
\begin{aligned}
(\overset{i}{t}_{\bar{x}})^{[1]} &= \sum_{j=1}^{N_1} (\overset{ij}{A}_{\bar{x}\,\bar{x}})^{[1]} (\overset{j}{P}_{\bar{x}})^{[1]} + \sum_{j=1}^{N_1} (\overset{ij}{A}_{\bar{x}\,\bar{y}})^{[1]} (\overset{j}{P}_{\bar{y}})^{[1]} \\
(\overset{i}{t}_{\bar{y}})^{[1]} &= \sum_{j=1}^{N_1} (\overset{ij}{A}_{\bar{y}\,\bar{x}})^{[1]} (\overset{j}{P}_{\bar{x}})^{[1]} + \sum_{j=1}^{N_1} (\overset{ij}{A}_{\bar{y}\,\bar{y}})^{[1]} (\overset{j}{P}_{\bar{y}})^{[1]}
\end{aligned}
\tag{8.81}
$$

for $i = 1$ to N_1, while the tractions on the boundary of subregion V_2 are

$$
\begin{aligned}
(\overset{i}{t}_{\bar{x}})^{[2]} &= \sum_{j=N_1+1}^{N} (\overset{ij}{A}_{\bar{x}\,\bar{x}})^{[2]} (\overset{j}{P}_{\bar{x}})^{[2]} + \sum_{j=N_1+1}^{N} (\overset{ij}{A}_{\bar{x}\,\bar{y}})^{[2]} (\overset{j}{P}_{\bar{y}})^{[2]} \\
(\overset{i}{t}_{\bar{y}})^{[2]} &= \sum_{j=N_1+1}^{N} (\overset{ij}{A}_{\bar{y}\,\bar{x}})^{[2]} (\overset{j}{P}_{\bar{x}})^{[2]} + \sum_{j=N_1+1}^{N} (\overset{ij}{A}_{\bar{y}\,\bar{y}})^{[2]} (\overset{j}{P}_{\bar{y}})^{[2]}
\end{aligned}
\tag{8.82}
$$

for $i = N_1 + 1$ to N.

Similar sets of equations can be written for the displacements on the boundaries of the two subregions. Using (5.15), we represent these as

$$
\begin{aligned}
(\overset{i}{u}_{\bar{x}})^{[1]} &= \sum_{j=1}^{N_1} (\overset{ij}{B}_{\bar{x}\,\bar{x}})^{[1]} (\overset{j}{P}_{\bar{x}})^{[1]} + \sum_{j=1}^{N_1} (\overset{ij}{B}_{\bar{x}\,\bar{y}})^{[1]} (\overset{j}{P}_{\bar{y}})^{[1]} \\
(\overset{i}{u}_{\bar{y}})^{[1]} &= \sum_{j=1}^{N_1} (\overset{ij}{B}_{\bar{y}\,\bar{x}})^{[1]} (\overset{j}{P}_{\bar{x}})^{[1]} + \sum_{j=1}^{N_1} (\overset{ij}{B}_{\bar{y}\,\bar{y}})^{[1]} (\overset{j}{P}_{\bar{y}})^{[1]}
\end{aligned}
\tag{8.83}
$$

for $i = 1$ to N_1, and

$$
\begin{aligned}
(\overset{i}{u}_{\bar{x}})^{[2]} &= \sum_{j=N_1+1}^{N} (\overset{ij}{B}_{\bar{x}\bar{x}})^{[2]}(\overset{j}{P}_{\bar{x}})^{[2]} + \sum_{j=N_1+1}^{N} (\overset{ij}{B}_{\bar{x}\bar{y}})^{[2]}(\overset{j}{P}_{\bar{y}})^{[2]} \\
(\overset{i}{u}_{\bar{y}})^{[2]} &= \sum_{j=N_1+1}^{N} (\overset{ij}{B}_{\bar{y}\bar{x}})^{[2]}(\overset{j}{P}_{\bar{x}})^{[2]} + \sum_{j=N_1+1}^{N} (\overset{ij}{B}_{\bar{y}\bar{y}})^{[2]}(\overset{j}{P}_{\bar{y}})^{[2]}
\end{aligned}
\tag{8.84}
$$

for $i = N_1 + 1$ to N. The boundary influence coefficients in (8.81)–(8.84) are computed exactly as explained in Sect. 5.4. It is only necessary to ensure that the appropriate elastic constants are used for each subregion.

Equations (8.81)–(8.84) can be used to set up a system of $2N$ algebraic equations in $2N$ unknown stress discontinuity components $P_{\bar{x}}$ and $P_{\bar{y}}$. We will represent this system as

$$
\begin{aligned}
\sum_{j=1}^{N} \overset{ij}{C}_{\bar{x}\bar{x}} \overset{j}{P}_{\bar{x}} + \sum_{j=1}^{N} \overset{ij}{C}_{\bar{x}\bar{y}} \overset{j}{P}_{\bar{y}} &= \overset{i}{b}_{\bar{x}} \\
\sum_{j=1}^{N} \overset{ij}{C}_{\bar{y}\bar{x}} \overset{j}{P}_{\bar{x}} + \sum_{j=1}^{N} \overset{ij}{C}_{\bar{y}\bar{y}} \overset{j}{P}_{\bar{y}} &= \overset{i}{b}_{\bar{y}}
\end{aligned}
\tag{8.85}
$$

cf. (5.30), where the first N_1 components of $\overset{j}{P}_{\bar{x}}$ and $\overset{j}{P}_{\bar{y}}$ represent the N_1 values of $(\overset{j}{P}_{\bar{x}})^{[1]}$ and $(\overset{j}{P}_{\bar{y}})^{[1]}$, and the remaining N_2 components represent the N_2 values of $(\overset{j}{P}_{\bar{x}})^{[2]}$ and $(\overset{j}{P}_{\bar{y}})^{[2]}$. The other quantities in (8.85) are obtained by using (8.81)–(8.84) and considering the boundary conditions or the interface continuity conditions for each element in turn. To this end, we note that that a typical boundary element i must either be on a free portion of one of the boundary contours or else along the interface between the two subregions. In the former case the ith ones of (8.85) are established by considering the boundary conditions for the element in question, just as in Sect. 5.6, whereas in the latter case they are found by employing the interface continuity conditions.

For purposes of illustration we will assume that the tractions are prescribed along boundary contour S_1 in Fig. 8.7 while the displacements are prescribed along S_2. We will further assume that overall equilibrium is maintained for these loading conditions.

We will suppose first that the ith element is on a free portion of one of the boundary contours, say S_1. If we suppose further that the tractions $(\overset{i}{t}_{\bar{x}})^{[1]} = (\overset{i}{t}_{\bar{x}})_0$ and $(\overset{i}{t}_{\bar{y}})^{[1]} = (\overset{i}{t}_{\bar{y}})_0$ are prescribed at this element, then (8.81) and (8.82) can be used to set the following values for the quantities $\overset{i}{b}_{\bar{x}}$, $\overset{i}{b}_{\bar{y}}$, and $\overset{ij}{C}_{\bar{x}\bar{x}}$, etc., in (8.85):

$$
\begin{aligned}
\overset{i}{b}_{\bar{x}} &= (\overset{i}{t}_{\bar{x}})_0 \\
\overset{i}{b}_{\bar{y}} &= (\overset{i}{t}_{\bar{y}})_0 \\
\overset{ij}{C}_{\bar{x}\bar{x}} &= \begin{cases} (\overset{ij}{A}_{\bar{x}\bar{x}})^{[1]} & j \leq N_1 \\ 0 & N_1 + 1 \leq j \leq N \end{cases}
\end{aligned}
\tag{8.86}
$$

with similar relations for the coefficients $\overset{ij}{C}_{\bar{x}\bar{y}}$, $\overset{ij}{C}_{\bar{y}\bar{x}}$, and $\overset{ij}{C}_{\bar{y}\bar{y}}$. On the other hand, if the ith element is on the free portion of contour S_2, and if the displacements $(\overset{i}{u}_{\bar{x}})^{[2]} = (\overset{i}{u}_{\bar{x}})_0$ and $(\overset{i}{u}_{\bar{y}})^{[2]} = (\overset{i}{u}_{\bar{y}})_0$ are prescribed, then (8.83) and (8.84) can be used to write

$$
\begin{aligned}
\overset{i}{b}_{\bar{x}} &= (\overset{i}{u}_{\bar{x}})_0 \\
\overset{i}{b}_{\bar{y}} &= (\overset{i}{u}_{\bar{y}})_0 \\
\overset{ij}{C}_{\bar{x}\bar{x}} &= \begin{cases} 0 & j \leq N_1 \\ (\overset{ij}{B}_{\bar{x}\bar{x}})^{[2]} & N_1 + 1 \leq j \leq N \end{cases}
\end{aligned}
\tag{8.87}
$$

with similar relations for the other coefficients in (8.85). Mixed boundary conditions for either contour can be taken into account by making simple adjustments to the parameters in (8.86) and (8.87) (cf. Sect. 5.6).

We consider finally the case that the ith element lies along the interface between the two subregions. As stated previously, an interface element actually consists of two coincident elements, one for contour S_1 and the other for contour S_2. Thus, if the ith element is on one side of the interface there will be a matching element, say the i^*th, on the other side. Also, four continuity conditions must be satisfied at each interface element. Two of these can be used to set up the equations pertaining to the element on one side of the interface, and the other two can be used similarly for the element on the other side. For example, if the ith element is on the S_1 side of the interface, the two conditions (8.79) for traction continuity can be written as

$$
(\overset{i}{t}_{\bar{x}})^{[1]} - (\overset{i^*}{t}_{\bar{x}})^{[2]} = 0
\tag{8.88}
$$

$$(\overset{i}{t}_{\bar{y}})^{[1]} - (\overset{i^*}{t}_{\bar{y}})^{[2]} = 0$$

It then follows from (8.81) and (8.82) that the quantities $\overset{i}{b}_{\bar{x}}$, $\overset{i}{b}_{\bar{y}}$, and $\overset{ij}{C}_{\bar{x}\bar{x}}$, etc., in (8.85) are

$$\overset{i}{b}_{\bar{x}} = (\overset{i}{t}_{\bar{x}})^{[1]} - (\overset{i^*}{t}_{\bar{x}})^{[2]} = 0$$

$$\overset{i}{b}_{\bar{y}} = (\overset{i}{t}_{\bar{y}})^{[1]} - (\overset{i^*}{t}_{\bar{y}})^{[2]} = 0 \tag{8.89}$$

$$\overset{ij}{C}_{\bar{x}\bar{x}} = \begin{cases} (\overset{ij}{A}_{\bar{x}\bar{x}})^{[1]} & j \le N_1 \\ -(\overset{i^*j}{A}_{\bar{x}\bar{x}})^{[2]} & N_1 + 1 \le j \le N \end{cases}$$

and so forth for the coefficients $\overset{ij}{C}_{\bar{x}\bar{y}}$, $\overset{ij}{C}_{\bar{y}\bar{x}}$, and $\overset{ij}{C}_{\bar{y}\bar{y}}$.

Interface continuity conditions for the displacements are enforced by considering the corresponding circumstance that the ith element is on the S_2 side of the interface (because we are determining the coefficients for the ith equations) while the i^*th element is on the S_1 side. Conditions (8.80) can then be expressed as

$$(\overset{i}{u}_{\bar{x}})^{[2]} + (\overset{i^*}{u}_{\bar{x}})^{[1]} = 0$$

$$(\overset{i}{u}_{\bar{y}})^{[2]} + (\overset{i^*}{u}_{\bar{y}})^{[1]} = 0 \tag{8.90}$$

and (8.83) and (8.84) can be combined to give

$$\overset{i}{b}_{\bar{x}} = (\overset{i}{u}_{\bar{x}})^{[2]} + (\overset{i^*}{u}_{\bar{x}})^{[1]} = 0$$

$$\overset{i}{b}_{\bar{y}} = (\overset{i}{u}_{\bar{y}})^{[2]} + (\overset{i^*}{u}_{\bar{y}})^{[1]} = 0 \tag{8.91}$$

$$\overset{ij}{C}_{\bar{x}\bar{x}} = \begin{cases} (\overset{i^*j}{B}_{\bar{x}\bar{x}})^{[1]} & j \le N_1 \\ (\overset{ij}{B}_{\bar{x}\bar{x}})^{[2]} & N_1 + 1 \le j \le N \end{cases}$$

again with similar relations for the other coefficients in (8.85).

The reader should be able to recognize that the above algorithms can be introduced rather easily into a basic computer program for the stress discontinuity method (or, for that matter, the displacement discontinuity method). A more substantial restructuring of the program would be required if it were desired to store only the non-zero terms of the coefficients $\overset{ij}{C}_{\bar{x}\bar{x}}$, etc., in (8.85), but this is essentially a file-handling problem and we will not explore it here.

8.4.2 *Procedure for the Direct Boundary Integral Method*

A numerical procedure for solving the problem of Fig. 8.7 by the direct boundary integral method can be developed in a similar way (see also Rizzo and Shippy 1968). We again consider the composite problem as two separate boundary value problems, linked by the interface continuity conditions (8.77) and (8.78). If boundary elements 1 to N_1 lie along contour S_1 and elements $N_1 + 1$ to $N_1 + N_2 = N$ lie along contour S_2, then (7.10) can be used to form the following two sets of equations for subregions V_1 and V_2:

$$\sum_{j=1}^{N_1}\left[(\overset{ij}{\mathbb{T}}_{xx} + \overset{ij}{\mathbb{I}})^{[1]}(\overset{j}{u}_x)^{[1]} + (\overset{ij}{\mathbb{T}}_{xy})^{[1]}(\overset{j}{u}_y)^{[1]}\right] = \sum_{j=1}^{N_1}\left[(\overset{ij}{\mathbb{U}}_{xx})^{[1]}(\overset{j}{t}_x)^{[1]} + (\overset{ij}{\mathbb{U}}_{xy})^{[1]}(\overset{j}{t}_y)^{[1]}\right]$$
$$\sum_{j=1}^{N_1}\left[(\overset{ij}{\mathbb{T}}_{yx})^{[1]}(\overset{j}{u}_x)^{[1]} + (\overset{ij}{\mathbb{T}}_{yy} + \overset{ij}{\mathbb{I}})^{[1]}(\overset{j}{u}_y)^{[1]}\right] = \sum_{j=1}^{N_1}\left[(\overset{ij}{\mathbb{U}}_{yx})^{[1]}(\overset{j}{t}_x)^{[1]} + (\overset{ij}{\mathbb{U}}_{yy})^{[1]}(\overset{j}{t}_y)^{[1]}\right] \tag{8.92}$$

where i ranges from 1 to N_1, and

$$\sum_{j=N_1+1}^{N}\left[(\overset{ij}{\mathbb{T}}_{xx} + \overset{ij}{\mathbb{I}})^{[2]}(\overset{j}{u}_x)^{[2]} + (\overset{ij}{\mathbb{T}}_{xy})^{[2]}(\overset{j}{u}_y)^{[2]}\right] = \sum_{j=N_1+1}^{N}\left[(\overset{ij}{\mathbb{U}}_{xx})^{[2]}(\overset{j}{t}_x)^{[2]} + (\overset{ij}{\mathbb{U}}_{xy})^{[2]}(\overset{j}{t}_y)^{[2]}\right]$$
$$\sum_{j=N_1+1}^{N}\left[(\overset{ij}{\mathbb{T}}_{yx})^{[2]}(\overset{j}{u}_x)^{[2]} + (\overset{ij}{\mathbb{T}}_{yy} + \overset{ij}{\mathbb{I}})^{[2]}(\overset{j}{u}_y)^{[2]}\right] = \sum_{j=N_1+1}^{N}\left[(\overset{ij}{\mathbb{U}}_{yx})^{[2]}(\overset{j}{t}_x)^{[2]} + (\overset{ij}{\mathbb{U}}_{yy})^{[2]}(\overset{j}{t}_y)^{[2]}\right] \tag{8.93}$$

where i ranges from $N_1 + 1$ to N. The boundary influence coefficients in these equations are obtained by using (7.20) and (7.21), with elastic constants ν_1, μ_1 for subregion V_1 and ν_2, μ_2 for subregion V_2.

Equations (8.92) and (8.93) are $2N$ relations involving $4N$ boundary parameters: the x and y components of the displacements (u_x and u_y) and tractions (t_x and t_y) at each of the N boundary elements. For an element on a free portion of one of the boundary contours two of the four parameters will be specified as boundary conditions, leaving the other two parameters as unknowns. All four of the boundary parameters are unknown for an element on the interface, but by using continuity conditions (8.77) and (8.78) we can associate two of the parameters with an element on one side of the interface and the other two with the matching element on the other side. Equations (8.92) and (8.93) can therefore be used to set up the following system of $2N$ algebraic equations in $2N$ unknowns:

$$\begin{aligned} \sum_{j=1}^{N} \overset{ij}{A}_{xx} \overset{j}{X}_x + \sum_{j=1}^{N} \overset{ij}{A}_{xy} \overset{j}{X}_y &= \overset{i}{b}_x \\ \sum_{j=1}^{N} \overset{ij}{A}_{yx} \overset{j}{X}_x + \sum_{j=1}^{N} \overset{ij}{A}_{yy} \overset{j}{X}_y &= \overset{i}{b}_y \end{aligned} \tag{8.94}$$

where $i = 1$ to N. Just as explained in Sect. 7.2, the quantities $\overset{i}{b}_x$ and $\overset{i}{b}_y$ in these equations are certain linear combinations of the known boundary parameters, and $\overset{ij}{A}_{xx}$, etc., are the appropriate boundary influence coefficients associated with the unknown boundary parameters $\overset{j}{X}_x$ and $\overset{j}{X}_y$.

As an illustration of the procedure used to construct the system (8.94) from equations (8.92) and (8.93), we will again suppose that the tractions $(\overset{i}{t}_x)^{[1]} = (\overset{i}{t}_x)_0$ and $(\overset{i}{t}_y)^{[1]} = (\overset{i}{t}_y)_0$ are specified on the free portion of contour S_1, and that the displacements $(\overset{i}{u}_x)^{[2]} = (\overset{i}{u}_x)_0$ and $(\overset{i}{u}_y)^{[2]} = (\overset{i}{u}_y)_0$ are specified on the free portion of contour S_2. Then, if we treat the displacements $(\overset{j}{u}_x)^{[1]}$ and $(\overset{j}{u}_y)^{[1]}$ as the unknown boundary parameters on the S_1 side of the interface, and the tractions $(\overset{j}{t}_x)^{[2]}$ and $(\overset{j}{t}_y)^{[2]}$ as the unknowns on the S_2 side, the quantities $\overset{j}{X}_x$ and $\overset{j}{X}_y$ will be defined as follows:

$$\begin{aligned} \overset{j}{X}_x &= \begin{cases} (\overset{j}{u}_x)^{[1]} & j = 1 \text{ to } N_1 \\ (\overset{j}{t}_x)^{[2]} & j = N_1 + 1 \text{ to } N \end{cases} \\ \overset{j}{X}_y &= \begin{cases} (\overset{j}{u}_y)^{[1]} & j = 1 \text{ to } N_1 \\ (\overset{j}{t}_y)^{[2]} & j = N_1 + 1 \text{ to } N \end{cases} \end{aligned} \tag{8.95}$$

Again, it is tacitly assumed that overall equilibrium is maintained for the specified boundary loading conditions.

If there are M boundary elements on each side of the interface, then, by virtue of our assumptions about the boundary conditions for the other elements, the tractions $(\overset{j}{t}_x)^{[1]}$ and $(\overset{j}{t}_y)^{[1]}$ are known for $j = 1$ to $N_1 - M$, and the displacements $(\overset{j}{u}_x)^{[2]}$ and $(\overset{j}{u}_y)^{[2]}$ are known for $j = N_1 + 1 + M$ to N. Using (8.92) for all elements of contour S_1 ($i = 1$ to N_1) and (8.93) for all elements of contour S_2 ($i = N_1 + 1$ to N), the quantities $\overset{i}{b}_x$ and $\overset{i}{b}_y$ in (8.94) can be written as

$$\begin{aligned}\overset{i}{b}_x &= \sum_{j=1}^{N_1-M}\left((\overset{ij}{\mathbb{U}}_{xx})^{[1]}(\overset{j}{t}_x)_0 + (\overset{ij}{\mathbb{U}}_{xy})^{[1]}(\overset{j}{t}_y)_0\right)\\ \overset{i}{b}_y &= \sum_{j=1}^{N_1-M}\left((\overset{ij}{\mathbb{U}}_{yx})^{[1]}(\overset{j}{t}_x)_0 + (\overset{ij}{\mathbb{U}}_{yy})^{[1]}(\overset{j}{t}_y)_0\right)\end{aligned} \tag{8.96}$$

for $i = 1$ to $N_1 - M$ and

$$\begin{aligned}\overset{i}{b}_x &= \sum_{j=N_1+1+M}^{N}\left(-(\overset{ij}{\mathbb{T}}_{xx} + \overset{ij}{\mathbb{I}})^{[2]}(\overset{j}{u}_x)_0 - (\overset{ij}{\mathbb{T}}_{xy})^{[2]}(\overset{j}{u}_y)_0\right)\\ \overset{i}{b}_y &= \sum_{j=N_1+1+M}^{N}\left(-(\overset{ij}{\mathbb{T}}_{yx})^{[2]}(\overset{j}{u}_x)_0 - (\overset{ij}{\mathbb{T}}_{yy} + \overset{ij}{\mathbb{I}})^{[2]}(\overset{j}{u}_y)_0\right)\end{aligned} \tag{8.97}$$

for $i = N_1 + M + 1$ to N. For elements along the interface, $i = N_1 - M + 1$ to $N_1 + M$, the values of $\overset{i}{b}_x$ and $\overset{i}{b}_y$ are zero because the interface tractions and displacements are unknown (they are included in (8.95)).

The coefficients $\overset{ij}{A}_{xx}$, etc., for the ith equations of the system (8.94) are found by considering the boundary conditions and interface continuity conditions along the contours S_1 and S_2. According to definitions (8.95), the displacements $(\overset{j}{u}_x)^{[1]}$ and $(\overset{j}{u}_y)^{[1]}$ are the unknown boundary parameters along contour S_1 (where $i \le N_1$) and the tractions $(\overset{j}{t}_x)^{[2]}$ and $(\overset{j}{t}_y)^{[2]}$ are the unknown parameters along contour S_2 (where $N_1 + 1 \le i \le N$). In setting up the system (8.94) from (8.92) and (8.93), therefore, we must eliminate reference to the tractions $(\overset{j}{t}_x)^{[1]}$ and $(\overset{j}{t}_y)^{[1]}$ along the S_1 side of the interface (where $j = N_1 - M + 1$ to N_1), and also to the displacements $(\overset{j}{u}_x)^{[2]}$ and $(\overset{j}{u}_y)^{[2]}$ along the S_2 side (where $j = N_1 + 1$ to $N_1 + M$). This is achieved by using the interface continuity conditions.

As noted previously, an element j on the S_1 side of the interface is necessarily connected to another element j^* on the S_2 side. Continuity conditions (8.77) for the tractions can then be written as

$$\begin{aligned}(\overset{j}{t}_x)^{[1]} &= -(\overset{j^*}{t}_x)^{[2]}\\ (\overset{j}{t}_y)^{[1]} &= -(\overset{j^*}{t}_y)^{[2]}\end{aligned} \tag{8.98}$$

Similarly, if element j is on the S_2 side of the interface and the matching element j^* is on the S_1 side, continuity conditions (8.78) for the displacements become

$$\begin{aligned}(\overset{j}{u}_x)^{[2]} &= (\overset{j^*}{u}_x)^{[1]}\\ (\overset{j}{u}_y)^{[2]} &= (\overset{j^*}{u}_y)^{[1]}\end{aligned} \tag{8.99}$$

It is now a simple matter to specify the coefficients $\overset{ij}{A}_{xx}$, etc., in (8.94). Combining (8.92), (8.95), and (8.98) we have, for $i \leq N_1$,

$$\overset{ij}{A}_{xx} = \begin{cases} (\overset{ij}{\mathbb{T}}_{xx} + \overset{ij}{\mathbb{I}})^{[1]} & j \leq N_1 \\ -(\overset{ij^*}{\mathbb{U}}_{xx})^{[1]} & N_1 + 1 \leq j \leq N_1 + M \\ 0 & N_1 + M + 1 \leq j \leq N \end{cases} \tag{8.100}$$

with similar expressions for the coefficients $\overset{ij}{A}_{xy}$, $\overset{ij}{A}_{yx}$, and $\overset{ij}{A}_{yy}$. The values of the coefficients for $N_1 + 1 \leq i \leq N$ are found by combining (8.93), (8.95), and (8.99):

$$\overset{ij}{A}_{xx} = \begin{cases} 0 & j \leq N_1 - M \\ -(\overset{ij^*}{\mathbb{T}}_{xx} + \overset{ij^*}{\mathbb{I}})^{[2]} & N_1 - M + 1 \leq j \leq N_1 \\ -(\overset{ij}{\mathbb{U}}_{xx})^{[2]} & N_1 + 1 \leq j \leq N \end{cases} \tag{8.101}$$

and so forth for the other coefficients in (8.94).

Finally, we note that the displacements and stresses at points inside either of the two subregions can be obtained by using Somigliana's formulas (7.5) and (7.23) with elastic constants and number of boundary elements appropriate to the subregion in question.

8.4.3 Examples

A simple example of a boundary value problem for a piecewise homgeneous elastic body is shown in Fig. 8.8. The region of interest consists of a large elastic plate with a circular hole of radius $r = b$ containing an annulus $a \le r \le b$ that is perfectly bonded along the interface. The plate and the annulus have elastic constants $\nu_2,\ \mu_2$ and $\nu_1,\ \mu_1$, respectively. The inside wall of the annulus is subjected to a normal stress $\sigma_{rr} = -p$ and the plate is unstressed at infinity. The solution to this problem, satisfying continuity of radial stress and displacement at the interface $r = b$, can be constructed from standard formulas for thick-walled cylinders (see, for example, Jaeger 1962). The radial and hoop stresses are

$$\sigma_{rr} = \begin{cases} (1 - a^2/b^2)^{-1}\Big[(pa^2/b^2 - p') - (p - p')a^2/r^2\Big] & a \le r \le b \\ -p'b^2/r^2 & r \ge b \end{cases}$$
$$\sigma_{\theta\theta} = \begin{cases} (1 - a^2/b^2)^{-1}\Big[(pa^2/b^2 - p') + (p - p')a^2/r^2\Big] & a \le r \le b \\ p'b^2/r^2 & r \ge b \end{cases} \tag{8.102}$$

in which

$$p' = \frac{2(1-\nu_1)pa^2/b^2}{2(1-\nu_1) + (\mu_1/\mu_2 - 1)(1 - a^2/b^2)} \tag{8.103}$$

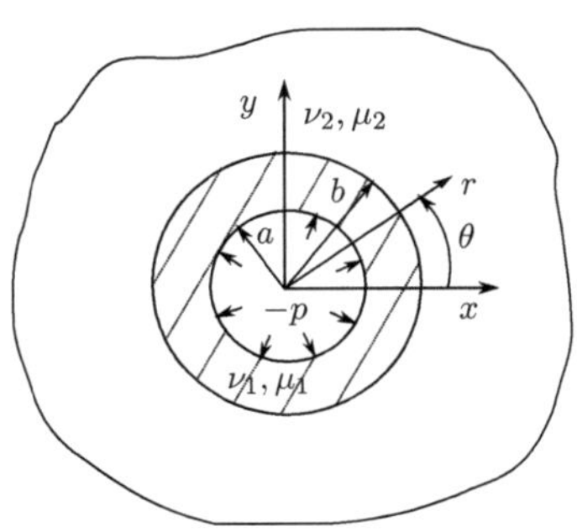

Fig. 8.8 A large plate with a circular hole containing an annulus

A numerical solution to this problem was obtained using the stress discontinuity method for the case that $a/b = 1/2$, $\nu_1 = \nu_2 = 0.25$, $\mu_1/\mu_2 = 2$, and $p/\mu_1 = 10^{-3}$. Symmetry conditions were employed and one-quarter of the circular boundary $r = a$ and the two sides of the interface $r = b$ between $\theta = 0$ and $\theta = \pi/2$ (see Fig. 8.8) were each divided into 25 elements. The results are compared with the analytical solution in Fig. 8.9.

A related problem for the same region of Fig. 8.8 can be defined for the case that the outer radius of the annulus is initially larger than b, say $b + \delta_r$, so the annulus has to be compressed to fit inside the hole in the plate. The stresses for this problem are again given by (8.102), but p' now has the value

$$p' = \frac{2(1-\nu_1)pa^2/b^2 + 2\mu_1(1-a^2/b^2)\delta_r/b}{2(1-\nu_1) + (\mu_1/\mu_2 - 1)(1-a^2/b^2)} \tag{8.104}$$

If the internal pressure p is zero, this expression gives the radial stress at the interface due to the radial mismatch δ_r.

For a numerical solution to this problem we replace the interface continuity condition for the normal displacements by the condition

$$u_{\bar{y}}^{[1]}(\boldsymbol{x}) + \delta_r = -u_{\bar{y}}^{[2]}(\boldsymbol{x}) \tag{8.105}$$

In the stress discontinuity method, for example, this means that the constant $\overset{i}{b}_{\bar{y}}$ in (8.91) now has the value $+\delta_r$. Taking $\delta_r/b = 0.5 \times 10^{-3}$ and leaving all the other parameters from the previous example unchanged, we obtain the results shown in Fig. 8.10.

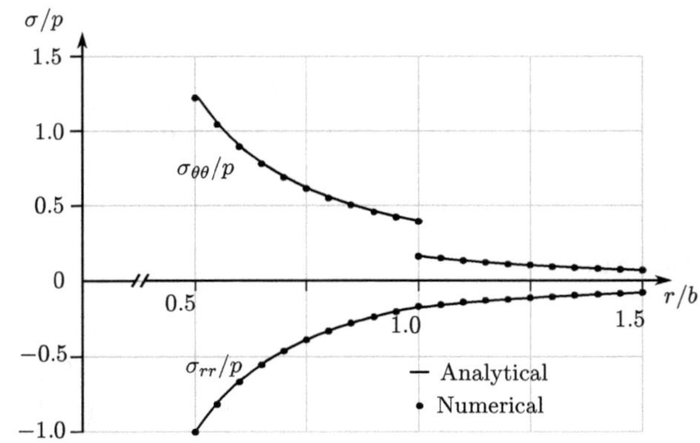

Fig. 8.9 Stresses in annulus ($0.5 \leq r/b \leq 1.0$) and plate ($r/b \geq 1.0$)

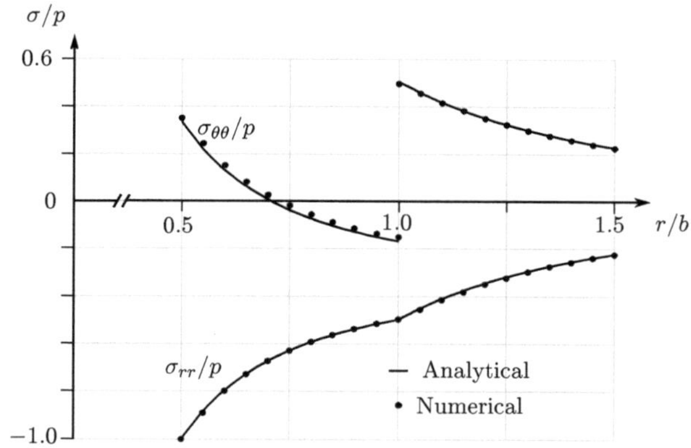

Fig. 8.10 Stresses in annulus ($0.5 \leq r/b \leq 1.0$) and plate ($r/b \geq 1.0$) for radial mismatch of the annulus $\delta_r/b = 0.5 \times 10^{-3}$

8.5 Bonded Half-Planes

The numerical procedures explained in the previous section can be applied to piecewise homogeneous bodies of arbitrary shape. When the interface between two subregions is a straight line, however, another approach to the problem is possible. In this case a special boundary element method can be developed using the analytical solution to the problem of a concentrated force in one of two *bonded half-planes*. When using this solution in a boundary element method the interface continuity conditions are satisfied exactly and boundary elements need only be placed along the boundaries of holes within the half-planes.[2]

The solution to the problem of a concentrated force in one of two isotropic, linearly elastic bonded half-planes was originally given by Frasier and Rongved (1957). The derivation of this solution presented below has some features in common with Frasier and Rongved's approach but makes use of the results from Sect. 8.3.1.

The geometry used for the problem is shown in Fig. 8.11. The lower half-plane, $y \leq 0$, has elastic constants ν and μ and has a concentrated force with components $F_x > 0$ and $F_y > 0$ located at the point $x = 0,\ y = -\ell$. The upper half-plane, $y \geq 0$, has elastic constants ν^* and μ^* and is joined to the lower half-plane along the common boundary, $y = 0$. We will denote the displacements and stresses in the lower half-plane as u_i and σ_{ij}, and those in the upper half-plane as u_i^* and σ_{ij}^*.

We will assume that the stresses are zero at infinity. If these stresses are not zero a relationship between $\sigma_{xx}^{(\infty)}$, $\sigma_{xx}^{*(\infty)}$, and $\sigma_{yy}^{(\infty)}$ can be established to ensure continuity of displacements and tractions across $y = 0$ (see exercise (6)).

[2] The solution for the problem of a displacement discontinuity in one of two bonded half-planes is given by Crouch and Starfield (1983).

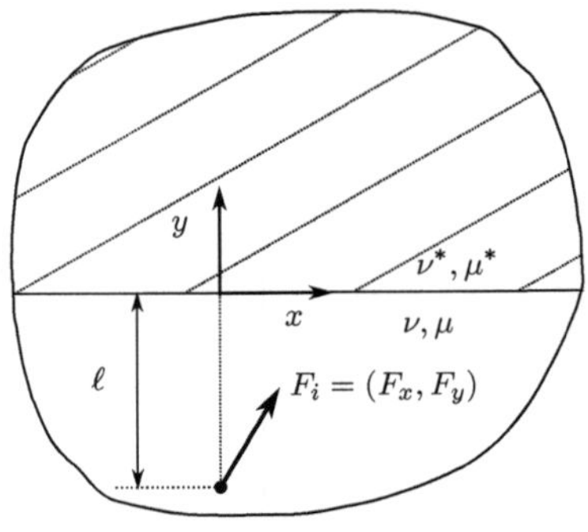

Fig. 8.11 Concentrated force in one of two bonded half-planes

The components of the unit normal to the boundaries of the lower ($y \leq 0$) and upper ($y \geq 0$) half-planes are $n_i = (0, 1)$ and $n_i^* = (0, -1)$. By (4.6), therefore, the corresponding tractions are $t_x = \sigma_{xy}$, $t_y = \sigma_{yy}$ and $t_x^* = \sigma_{xy}^*$, $t_y^* = -\sigma_{yy}^*$, respectively. Interface continuity conditions comparable to (8.77) and (8.78) can then be written as

$$\begin{aligned} \sigma_{xy}^* &= \sigma_{xy} \\ \sigma_{yy}^* &= \sigma_{yy} \end{aligned} \tag{8.106}$$

for the tractions and

$$\begin{aligned} u_x^* &= u_x \\ u_y^* &= u_y \end{aligned} \tag{8.107}$$

for the displacements. In both cases the conditions apply along the entire interface, $-\infty < x < \infty$, $y = 0$.

The displacements and stresses for the lower half-plane can be written as

$$\begin{aligned} u_i &= (\overset{A}{u}_i + \overset{I}{u}_i + \overset{S}{u}_i) + u_i^{**} \\ \sigma_{ij} &= (\overset{A}{\sigma}_{ij} + \overset{I}{\sigma}_{ij} + \overset{S}{\sigma}_{ij}) + \sigma_{ij}^{**} \end{aligned} \tag{8.108}$$

The terms in parentheses represent the solution given in Sect. 8.3.1 for a concentrated force in an elastic half-plane with a traction-free boundary. By definition, this solution has continuous (in fact zero) normal and shear stresses along the interface. Continuity of the displacements is achieved by superimposing normal and shear stresses along $y = 0$. The *same* stresses must be applied to both sides of the interface, $y = 0_-$ and $y = 0_+$, in order to preserve stress (traction) continuity. The quantities u_i^{**} and σ_{ij}^{**}

represent the displacement and stress components in the lower half-plane due to the stresses applied to the negative side of the interface, $y = 0_-$.

The displacements u_i^* and stresses σ_{ij}^* represent the complete solution for the upper half-plane. These quantities depend only on the normal and shear stresses applied to the positive side of the interface, $y = 0_+$.

The problem is reduced to that of finding the normal and shear stresses σ_{yy}^{**} and σ_{xy}^{**} that must be applied to the surfaces of both half-planes in order to achieve continuity of the displacements on $y = 0$. The steps needed to accomplish this are summarized below.

8.5.1 General Expressions for the Lower Half-Plane, $y \leq 0$

The displacements and stresses in an isotropic elastic solid due to arbitrary distributions of normal and shear stresses (tractions) along the line $y = 0$ can be expressed in terms of the two special Papkovitch-Neuber functions $\Phi(x, y)$ and $\Psi(x, y)$ introduced in Sect. 4.9.3 (see (4.109) and (4.110)). For present purposes we denote these functions as Φ^{**} and Ψ^{**} and write

$$\begin{aligned} u_x^{**} &= -(1-2\nu)\Phi_{,x}^{**} + 2(1-\nu)\Psi_{,y}^{**} - y\big(\Psi_{,xx}^{**} + \Phi_{,xy}^{**}\big) \\ u_y^{**} &= 2(1-\nu)\Phi_{,y}^{**} + (1-2\nu)\Psi_{,x}^{**} - y\big(\Psi_{,xy}^{**} + \Phi_{,yy}^{**}\big) \end{aligned} \tag{8.109}$$

and

$$\begin{aligned} \sigma_{xx}^{**} &= 2\mu\big[\Phi_{,yy}^{**} + 2\Psi_{,xy}^{**} + y\big(\Psi_{,xyy}^{**} + \Phi_{,yyy}^{**}\big)\big] \\ \sigma_{yy}^{**} &= 2\mu\big[\Phi_{,yy}^{**} - y\big(\Psi_{,xyy}^{**} + \Phi_{,yyy}^{**}\big)\big] \\ \sigma_{xy}^{**} &= 2\mu\big[\Psi_{,yy}^{**} - y\big(\Psi_{,xxy}^{**} + \Phi_{,xyy}^{**}\big)\big] \end{aligned} \tag{8.110}$$

We require Φ^{**} and Ψ^{**} to be harmonic and regular in $y \leq 0$ and hence seek to express these quantities in terms of the function (8.14), i.e., in terms of

$$\overset{I}{g}(x, y) = -\frac{1}{4\pi(1-\nu)} \ln[x^2 + (y-\ell)^2]^{1/2} \tag{8.111}$$

The boundary values of the stresses and displacements for the lower half-plane are found using (8.108)–(8.110) and referring to the solution presented in Sect. 8.3.1 for a concentrated force in an elastic half-plane. In this way we obtain the following results, all for the case that $y = 0$:

$$\sigma_{yy} = 2\mu\Phi^{**}_{,yy}$$
$$\sigma_{xy} = 2\mu\Psi^{**}_{,yy} \tag{8.112}$$

$$u_x = -(1-2\nu)\Phi^{**}_{,x} + 2(1-\nu)\Psi^{**}_{,y} + \frac{2(1-\nu)}{\mu}\left[F_x\{2(1-\nu)\overset{I}{g} - \ell\overset{I}{g}_{,y}\} + F_y\{(1-2\nu)\int \overset{I}{g}_{,y}\,\mathrm{d}x - \ell\overset{I}{g}_{,x}\}\right]$$
$$u_y = 2(1-\nu)\Phi^{**}_{,y} + (1-2\nu)\Psi^{**}_{,x} + \frac{2(1-\nu)}{\mu}\left[F_x\{(1-2\nu)\int \overset{I}{g}_{,x}\,\mathrm{d}y - \ell\overset{I}{g}_{,x}\} + F_y\{2(1-\nu)\overset{I}{g} + \ell\overset{I}{g}_{,y}\}\right] \tag{8.113}$$

Equations (8.113) were obtained using (8.108), (8.8), (8.12), and (8.21), noting that the following relations hold *for* $y = 0$: $\overset{A}{g} = \overset{I}{g}$, $\overset{A}{g}_{,x} = \overset{I}{g}_{,x}$, and $\overset{A}{g}_{,y} = -\overset{I}{g}_{,y}$.

8.5.2 General Expressions for the Upper Half-Plane, y ≥ 0

Equations comparable to (8.109) and (8.110) for the upper half-plane are

$$u^*_x = -(1-2\nu^*)\Phi^*_{,x} + 2(1-\nu^*)\Psi^*_{,y} - y\left(\Psi^*_{,xx} + \Phi^*_{,xy}\right)$$
$$u^*_y = 2(1-\nu^*)\Phi^*_{,y} + (1-2\nu^*)\Psi^*_{,x} - y\left(\Psi^*_{,xy} + \Phi^*_{,yy}\right) \tag{8.114}$$

and

$$\sigma^*_{xx} = 2\mu^*\left[\Phi^*_{,yy} + 2\Psi^*_{,xy} + y\left(\Psi^*_{,xyy} + \Phi^*_{,yyy}\right)\right]$$
$$\sigma^*_{yy} = 2\mu^*\left[\Phi^*_{,yy} - y\left(\Psi^*_{,xyy} + \Phi^*_{,yyy}\right)\right] \tag{8.115}$$
$$\sigma^*_{xy} = 2\mu^*\left[\Psi^*_{,yy} - y\left(\Psi^*_{,xxy} + \Phi^*_{,xyy}\right)\right]$$

Functions Φ^* and Ψ^* are harmonic and regular in $y \geq 0$ and are therefore expressible in terms of the function (8.10), i.e., in terms of

$$\overset{A}{g}(x, y) = -\frac{1}{4\pi(1-\nu)}\ln[x^2 + (y+\ell)^2]^{1/2} \tag{8.116}$$

On the surface $y = 0$ we have that

$$\begin{aligned} \sigma^*_{yy} &= 2\mu^* \Phi^*_{,yy} \\ \sigma^*_{xy} &= 2\mu^* \Psi^*_{,yy} \end{aligned} \tag{8.117}$$

$$\begin{aligned} u^*_x &= -(1-2\nu^*)\Phi^*_{,x} + 2(1-\nu^*)\Psi^*_{,y} \\ u^*_y &= 2(1-\nu^*)\Phi^*_{,y} + (1-2\nu^*)\Psi^*_{,x} \end{aligned} \tag{8.118}$$

8.5.3 Interface Continuity Conditions, y = 0

Continuity conditions for the stresses ($\sigma^*_{yy} = \sigma_{yy}$ and $\sigma^*_{xy} = \sigma_{xy}$) and displacements ($u^*_x = u_x$ and $u^*_y = u_y$) on $y = 0$ give directly the following relationships between functions Φ^{**}, Ψ^{**} and Φ^*, Ψ^*:

$$\begin{aligned} \Phi^*_{,yy} &= \frac{\mu}{\mu^*}\Phi^{**}_{,yy} \\ \Psi^*_{,yy} &= \frac{\mu}{\mu^*}\Psi^{**}_{,yy} \end{aligned} \tag{8.119}$$

and

$$\begin{aligned} -(1-2\nu^*)\Phi^*_{,x} + 2(1-\nu^*)\Psi^*_{,y} &= -(1-2\nu)\Phi^{**}_{,x} + 2(1-\nu)\Psi^{**}_{,y} \\ &\quad + \frac{2(1-\nu)}{\mu}\Big[F_x\big\{2(1-\nu)\overset{I}{g} - \ell\overset{I}{g}_{,y}\big\} \\ &\quad + F_y\big\{(1-2\nu)\int \overset{I}{g}_{,y}\,\mathrm{d}x - \ell\overset{I}{g}_{,x}\big\}\Big] \\ 2(1-\nu^*)\Phi^*_{,y} + (1-2\nu^*)\Psi^*_{,x} &= 2(1-\nu)\Phi^{**}_{,y} + (1-2\nu)\Psi^{**}_{,x} \\ &\quad + \frac{2(1-\nu)}{\mu}\Big[F_x\big\{(1-2\nu)\int \overset{I}{g}_{,x}\,\mathrm{d}y - \ell\overset{I}{g}_{,x}\big\} \\ &\quad + F_y\big\{2(1-\nu)\overset{I}{g} + \ell\overset{I}{g}_{,y}\big\}\Big] \end{aligned} \tag{8.120}$$

Equations (8.119) can be used to eliminate functions Φ^* and Ψ^* from (8.120) by observing that the solutions for $y \le 0$ and $y \ge 0$ are symmetric with respect to $y = 0$, which means that $\Phi^*_{,x} = \mu\Phi^{**}_{,x}/\mu^*$, $\Phi^*_{,y} = -\mu\Phi^{**}_{,y}/\mu^*$, $\Psi^*_{,x} = \mu\Psi^{**}_{,x}/\mu^*$, and $\Psi^{*}_{,y} = -\mu\Psi^{**}_{,y}/\mu^{*}$ (Frasier and Rongved 1957). Substituting these relations into (8.120), we find that

$$L\Phi^{**}_{,x} - K\Psi^{**}_{,y} = \frac{2(1-\nu)}{\mu}\left[F_x\left\{2(1-\nu)\overset{I}{g} - \ell\overset{I}{g}_{,y}\right\} + F_y\left\{(1-2\nu)\int \overset{I}{g}_{,y}\,\mathrm{d}x - \ell\overset{I}{g}_{,x}\right\}\right]$$

$$L\Psi^{**}_{,x} + K\Phi^{**}_{,y} = -\frac{2(1-\nu)}{\mu}\left[F_x\left\{(1-2\nu)\int \overset{I}{g}_{,x}\,\mathrm{d}y - \ell\overset{I}{g}_{,x}\right\} + F_y\left\{2(1-\nu)\overset{I}{g} + \ell\overset{I}{g}_{,y}\right\}\right] \tag{8.121}$$

where constants K and L are defined as

$$K = 2(1-\nu) + 2(1-\nu^{*})\frac{\mu}{\mu^{*}}$$

$$L = (1-2\nu) - (1-2\nu^{*})\frac{\mu}{\mu^{*}} \tag{8.122}$$

For notational convenience we let

$$\Lambda_1 = \frac{2(1-\nu)}{\mu}F_x\left[2(1-\nu)\overset{I}{g} - \ell\overset{I}{g}_{,y}\right] + \frac{2(1-\nu)}{\mu}F_y\left[(1-2\nu)\int \overset{I}{g}_{,y}\,\mathrm{d}x - \ell\overset{I}{g}_{,x}\right]$$

$$\Lambda_2 = -\frac{2(1-\nu)}{\mu}F_x\left[(1-2\nu)\int \overset{I}{g}_{,x}\,\mathrm{d}y - \ell\overset{I}{g}_{,x}\right] - \frac{2(1-\nu)}{\mu}F_y\left[2(1-\nu)\overset{I}{g} + \ell\overset{I}{g}_{,y}\right] \tag{8.123}$$

so that equations (8.121) can be written as

$$L\Phi^{**}_{,x} - K\Psi^{**}_{,y} = \Lambda_1$$

$$L\Psi^{**}_{,x} + K\Phi^{**}_{,y} = \Lambda_2 \tag{8.124}$$

8.5.4 *Solution for the Lower Half-Plane, y ≤ 0*

The function $\Phi^{**}_{,yy}$ can be found by differentiating the first one of (8.124) with respect to x, multiplying the resulting equation by L and then adding to this the second one of (8.124) differentiated with respect to y and multiplied by K. This gives

$$(K^2 - L^2)\Phi^{**}_{,yy} = L\frac{\partial\Lambda_1}{\partial x} + K\frac{\partial\Lambda_2}{\partial y} \tag{8.125}$$

so that

$$\Phi^{**}_{,yy} = \frac{2(1-\nu)}{\mu(K^2-L^2)}\Big[F_x\{N\overset{I}{g}_{,x} + \ell(K-L)\overset{I}{g}_{,xy}\} + F_y\{-M\overset{I}{g}_{,y} - \ell(K-L)\overset{I}{g}_{,yy}\}\Big] \tag{8.126}$$

in which

$$M = 2(1-\nu)K - (1-2\nu)L$$
$$N = 2(1-\nu)L - (1-2\nu)K \tag{8.127}$$

A similar calculation for function $\Psi^{**}_{,yy}$ gives the result

$$(K^2 - L^2)\Psi^{**}_{,yy} = -K\frac{\partial \Lambda_1}{\partial y} + L\frac{\partial \Lambda_2}{\partial x} \tag{8.128}$$

and

$$\Psi^{**}_{,yy} = \frac{2(1-\nu)}{\mu(K^2-L^2)}\Big[F_x\{-M\overset{I}{g}_{,y} + \ell(K-L)\overset{I}{g}_{,yy}\} + F_y\{-N\overset{I}{g}_{,x} + \ell(K-L)\overset{I}{g}_{,xy}\}\Big] \tag{8.129}$$

The displacements u_i^{**} and stresses σ_{ij}^{**} for $y \le 0$ can now be found by using (8.126) and (8.129) in (8.109) and (8.110). The calculations are straightforward, but tedious, and we will quote only the final results. The displacements are

$$u_x^{**} = \frac{4(1-\nu)}{K+L}\Big[-(\overset{I}{u}_x + \overset{S}{u}_x) - \frac{L}{K-L}\frac{1}{2\mu}\big(F_x\overset{I}{g} - F_y\int \overset{I}{g}_{,y}\,\mathrm{d}x\big)\Big]$$
$$u_y^{**} = \frac{4(1-\nu)}{K+L}\Big[-(\overset{I}{u}_y + \overset{S}{u}_y) - \frac{L}{K-L}\frac{1}{2\mu}\big(F_x\int \overset{I}{g}_{,y}\,\mathrm{d}x + F_y\overset{I}{g}\big)\Big] \tag{8.130}$$

and the stresses are

$$\sigma_{xx}^{**} = \frac{4(1-\nu)}{K+L}\Big[-(\overset{I}{\sigma}_{xx} + \overset{S}{\sigma}_{xx}) - \frac{L}{K-L}\big(F_x\overset{I}{g}_{,x} - F_y\overset{I}{g}_{,y}\big)\Big]$$
$$\sigma_{yy}^{**} = \frac{4(1-\nu)}{K+L}\Big[-(\overset{I}{\sigma}_{yy} + \overset{S}{\sigma}_{yy}) + \frac{L}{K-L}\big(F_x\overset{I}{g}_{,x} - F_y\overset{I}{g}_{,y}\big)\Big] \tag{8.131}$$
$$\sigma_{xy}^{**} = \frac{4(1-\nu)}{K+L}\Big[-(\overset{I}{\sigma}_{xy} + \overset{S}{\sigma}_{xy}) - \frac{L}{K-L}\big(F_x\overset{I}{g}_{,y} + F_y\overset{I}{g}_{,x}\big)\Big]$$

The displacements $\overset{I}{u}_x + \overset{S}{u}_x$ and $\overset{I}{u}_y + \overset{S}{u}_y$ are given by (8.12) and (8.21), and the stresses $\overset{I}{\sigma}_{xx} + \overset{S}{\sigma}_{xx}$, etc., by (8.13) and (8.20). Also, the complete solution for the lower half-plane is given by (8.108).

Three special cases of the above solution are worth mentioning. First, if the two half-planes have the same elastic constants, i.e., if $\nu^* = \nu$ and $\mu^* = \mu$, then the parameters K and L are equal to $4(1-\nu)$ and 0, respectively, and we find from (8.130) and (8.131) that $u_i^{**} = -(\overset{I}{u}_i + \overset{S}{u}_i)$ and $\sigma_{ij}^{**} = -(\overset{I}{\sigma}_{ij} + \overset{S}{\sigma}_{ij})$. The solution then reduces to that of concentrated force in an infinite, homogeneous elastic solid. A second special case of the general solution for the bonded half-planes occurs when the shear modulus of the upper half-plane is zero, i.e., $\mu^* = 0$. Equations (8.122) in this case yield the result $(K+L)^{-1} = 0$, so that $u_i^{**} = 0$ and $\sigma_{ij}^{**} = 0$. The solution therefore reduces to that given in Sect. 8.3 for a concentrated force in an elastic half-plane with a traction-free boundary. Finally, if the upper half-plane is rigid, i.e., if $\mu^* = \infty$, then $K = 2(1-\nu)$ and $L = 1 - 2\nu$ and it can be verified that resultant displacements u_i of the interface are zero. We obtain in this way the solution for a concentrated force in an elastic half-plane whose surface is prevented from deforming.

8.5.5 *Solution for the Upper Half-Plane, y ≥ 0*

Computation of u_i^* and σ_{ij}^* proceeds in a similar way, using (8.119) in (8.114) and (8.115). First, however, we note that the boundary values of $\Phi^{**}_{,yy}$ and $\Psi^{**}_{,yy}$ can be written in terms of the function $\overset{A}{g}(x,y)$ instead of $\overset{I}{g}(x,y)$ by using the following relations, valid for $y = 0$:

$$\begin{aligned} \overset{I}{g}_{,x} &= \overset{A}{g}_{,x} \\ \overset{I}{g}_{,y} &= -\overset{A}{g}_{,y} \end{aligned} \tag{8.132}$$

We may then write, for example,

$$\begin{aligned} \Phi^*_{,yy} &= \frac{\mu}{\mu^*}\left(\Phi^{**}_{,yy}\right)_{y=0} \\ &= \frac{2(1-\nu)}{\mu^*(K^2-L^2)}\left[F_x\{N\overset{A}{g}_{,x} - \ell(K-L)\overset{A}{g}_{,xy}\} + F_y\{M\overset{A}{g}_{,y} - \ell(K-L)\overset{A}{g}_{,yy}\}\right] \end{aligned} \tag{8.133}$$

The right-hand side of this equation defines a function that is harmonic and regular in $y \geq 0$ and we therefore conclude from the uniqueness theorem for the Dirichlet problem that the expression holds everywhere in the upper half-plane and not just on its boundary. Likewise, the function $\Psi^*_{,yy}$, harmonic and regular in $y \geq 0$, is

$$\Psi^*_{,yy} = \frac{2(1-\nu)}{\mu^*(K^2-L^2)}\Big[F_x\big\{M\overset{A}{g}_{,y} - \ell(K-L)\overset{A}{g}_{,yy}\big\} + F_y\big\{-N\overset{A}{g}_{,x} + \ell(K-L)\overset{A}{g}_{,xy}\big\}\Big] \tag{8.134}$$

Using (8.133) and (8.134) in (8.114) and (8.115) it can be shown that the complete solution for the upper half-plane is given as

$$\begin{aligned} u_x^* = \frac{2(1-\nu)}{\mu^*(K^2-L^2)}\Big[&F_x\big\{((3-2\nu)K + 2\nu L - 2\nu^*(K+L))\overset{A}{g} + Y(y)\overset{A}{g}_{,y}\big\} \\ &+ F_y\big\{((3-2\nu)L + 2\nu K - 2\nu^*(K+L))\int \overset{A}{g}_{,y}\,\mathrm{d}x - Y(y)\overset{A}{g}_{,x}\big\}\Big] \\ u_y^* = \frac{2(1-\nu)}{\mu^*(K^2-L^2)}\Big[&F_x\big\{-((3-2\nu)L + 2\nu K - 2\nu^*(K+L))\int \overset{A}{g}_{,y}\,\mathrm{d}x - Y(y)\overset{A}{g}_{,x}\big\} \\ &+ F_y\big\{((3-2\nu)K + 2\nu L - 2\nu^*(K+L))\overset{A}{g} - Y(y)\overset{A}{g}_{,y}\big\}\Big] \end{aligned} \tag{8.135}$$

and

$$\begin{aligned} \sigma_{xx}^* &= \frac{4(1-\nu)}{K^2-L^2}\Big[F_x\big\{(2\nu L + (3-2\nu)K)\overset{A}{g}_{,x} + Y(y)\overset{A}{g}_{,xy}\big\} \\ &\qquad + F_y\big\{(2\nu K + (3-2\nu)L)\overset{A}{g}_{,y} + Y(y)\overset{A}{g}_{,yy}\big\}\Big] \\ \sigma_{yy}^* &= \frac{4(1-\nu)}{K^2-L^2}\Big[F_x\big\{N\overset{A}{g}_{,x} - Y(y)\overset{A}{g}_{,xy}\big\} + F_y\big\{M\overset{A}{g}_{,y} - Y(y)\overset{A}{g}_{,yy}\big\}\Big] \\ \sigma_{xy}^* &= \frac{4(1-\nu)}{K^2-L^2}\Big[F_x\big\{M\overset{A}{g}_{,y} + Y(y)\overset{A}{g}_{,yy}\big\} + F_y\big\{-N\overset{A}{g}_{,x} - Y(y)\overset{A}{g}_{,xy}\big\}\Big] \end{aligned} \tag{8.136}$$

in which function $Y(y)$ is defined as

$$Y(y) = K(y+\ell) + L(y-\ell) \tag{8.137}$$

It is easy to verify that if $\nu^* = \nu$ and $\mu^* = \mu$ the solution reduces to that of a concentrated force in an infinite, homogeneous elastic solid, i.e., $u_i^* = \overset{A}{u}_i = u_i$ and $\sigma_{ij}^* = \overset{A}{\sigma}_{ij} = \sigma_{ij}$, as required.

8.6 Volterra's Formulas and the Displacement Discontinuity Method

We noted in Chap. 6 that the displacement discontinuity method is often referred to in the literature as an indirect boundary element method, even though for crack problems the unknowns in the system—the displacement discontinuities—have physical meaning, just as do the boundary displacements and tractions in the direct boundary integral method. Historically, this terminology probably arose because the basic method can be developed without reference to Somigliana's formulas, which are regarded by most authors as the touchstone for a direct boundary element formulation.

We will demonstrate below that an alternate development of the displacement discontinuity method is possible through the use of Volterra's formulas (Sect. 4.24). While this version of the method is completely equivalent to that obtained by the influence function approach described in Chap. 6, it is useful to show that the method can in fact be developed through a direct boundary integral formulation, just as was done for the method for general boundary value problems using Somigliana's formulas, as discussed in Chap. 7. We will use this observation later (Sect. 8.9) to construct a direct boundary element formulation that combines the attributes of both approaches.

Volterra's formulas (see (4.372)–(4.374), reproduced here for convenience) give the displacements, stresses, and tractions at a point $\boldsymbol{\xi}$ inside a region V in terms of the following integrals of displacement discontinuities $\Delta u_j(\boldsymbol{x}) = u_j^+(\boldsymbol{x}) - u_j^-(\boldsymbol{x})$ (in which the signs are defined as in (4.362)) along a crack boundary Γ:

$$u_i(\boldsymbol{\xi}) = \int_{\Gamma} \Delta u_j(\boldsymbol{x}) T_{ij}(\boldsymbol{x}, \boldsymbol{\xi})\, \mathrm{d}S(\boldsymbol{x}) \tag{8.138}$$

$$\sigma_{ij}(\boldsymbol{\xi}) = \int_{\Gamma} \Delta u_k(\boldsymbol{x}) S_{kij}(\boldsymbol{x}, \boldsymbol{\xi})\, \mathrm{d}S(\boldsymbol{x}) \tag{8.139}$$

$$t_i(\boldsymbol{\xi}) = \int_{\Gamma} \Delta u_j(\boldsymbol{x}) H_{ij}(\boldsymbol{x}, \boldsymbol{\xi})\, \mathrm{d}S(\boldsymbol{x}) \tag{8.140}$$

If the crack is divided into N straight line elements, as depicted in Fig. 8.12, these equations can be discretized in the usual way to give, for example,

$$\begin{aligned} u_x(\boldsymbol{\xi}) &= \sum_{j=1}^{N} \Big(\int_{-\overset{j}{a}}^{\overset{j}{a}} T_{xx}(\bar{x} - c_{\bar{x}}, c_{\bar{y}}) \Delta u_x(\bar{x})\, \mathrm{d}\bar{x} + \int_{-\overset{j}{a}}^{\overset{j}{a}} T_{xy}(\bar{x} - c_{\bar{x}}, c_{\bar{y}}) \Delta u_y(\bar{x})\, \mathrm{d}\bar{x} \Big) \\ u_y(\boldsymbol{\xi}) &= \sum_{j=1}^{N} \Big(\int_{-\overset{j}{a}}^{\overset{j}{a}} T_{yx}(\bar{x} - c_{\bar{x}}, c_{\bar{y}}) \Delta u_x(\bar{x})\, \mathrm{d}\bar{x} + \int_{-\overset{j}{a}}^{\overset{j}{a}} T_{yy}(\bar{x} - c_{\bar{x}}, c_{\bar{y}}) \Delta u_y(\bar{x})\, \mathrm{d}\bar{x} \Big) \end{aligned} \tag{8.141}$$

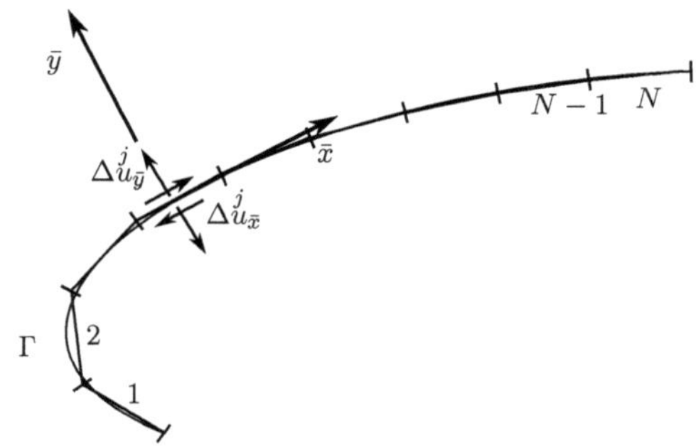

Fig. 8.12 Crack Γ divided into N displacement discontinuity elements

with similar expressions for the individual terms of $\sigma_{ij}(\boldsymbol{\xi})$ and $t_i(\boldsymbol{\xi})$.

Without loss of generality we take the displacement discontinuities as piecewise constant along the elements, i.e., $\Delta u_x(\bar{x}) = \Delta \overset{j}{u}_x$ and $\Delta u_y(\bar{x}) = \Delta \overset{j}{u}_y$ for $-\overset{j}{a} \leq \bar{x} \leq \overset{j}{a}$, $\bar{y} = 0$. Then, using (7.21), (8.141) become

$$
\begin{aligned}
u_x(\boldsymbol{\xi}) &= \sum_{j=1}^{N} \Big(2(1-\nu)\bar{T}_3 - c_{\bar{y}}(\sin 2\overset{j}{\beta}\bar{T}_4 - \cos 2\overset{j}{\beta}\bar{T}_5) \Big) \Delta \overset{j}{u}_x \\
&+ \sum_{j=1}^{N} \Big(-(1-2\nu)\bar{T}_2 + c_{\bar{y}}(\cos 2\overset{j}{\beta}\bar{T}_4 + \sin 2\overset{j}{\beta}\bar{T}_5) \Big) \Delta \overset{j}{u}_y \\
u_y(\boldsymbol{\xi}) &= \sum_{j=1}^{N} \Big((1-2\nu)\bar{T}_2 + c_{\bar{y}}(\cos 2\overset{j}{\beta}\bar{T}_4 + \sin 2\overset{j}{\beta}\bar{T}_5) \Big) \Delta \overset{j}{u}_x \\
&+ \sum_{j=1}^{N} \Big(2(1-\nu)\bar{T}_3 + c_{\bar{y}}(\sin 2\overset{j}{\beta}\bar{T}_4 - \cos 2\overset{j}{\beta}\bar{T}_5) \Big) \Delta \overset{j}{u}_y
\end{aligned}
\tag{8.142}
$$

These equations can be written in terms of the displacement discontinuity components in local $\bar{x}$, $\bar{y}$ coordinates by using the transformation formulas of the type (4.62)

$$
\begin{aligned}
\Delta \overset{j}{u}_x &= \Delta \overset{j}{u}_{\bar{x}} \cos \overset{j}{\beta} - \Delta \overset{j}{u}_{\bar{y}} \sin \overset{j}{\beta} \\
\Delta \overset{j}{u}_y &= \Delta \overset{j}{u}_{\bar{x}} \sin \overset{j}{\beta} + \Delta \overset{j}{u}_{\bar{y}} \cos \overset{j}{\beta}
\end{aligned}
\tag{8.143}
$$

The resulting expressions are

$$
\begin{aligned}
u_x(\boldsymbol{\xi}) = &\sum_{j=1}^{N}\Big(-(1-2\nu)\sin\overset{j}{\beta}\bar{T}_2 + 2(1-\nu)\cos\overset{j}{\beta}\bar{T}_3 - c_{\bar{y}}(\sin\overset{j}{\beta}\bar{T}_4 - \cos\overset{j}{\beta}\bar{T}_5)\Big)\Delta\overset{j}{u}_{\bar{x}} \\
&+\sum_{j=1}^{N}\Big(-(1-2\nu)\cos\overset{j}{\beta}\bar{T}_2 - 2(1-\nu)\sin\overset{j}{\beta}\bar{T}_3 + c_{\bar{y}}(\cos\overset{j}{\beta}\bar{T}_4 + \sin\overset{j}{\beta}\bar{T}_5)\Big)\Delta\overset{j}{u}_{\bar{y}}
\end{aligned}
\tag{8.144}
$$

$$
\begin{aligned}
u_y(\boldsymbol{\xi}) = &\sum_{j=1}^{N}\Big((1-2\nu)\cos\overset{j}{\beta}\bar{T}_2 + 2(1-\nu)\sin\overset{j}{\beta}\bar{T}_3 + c_{\bar{y}}(\cos\overset{j}{\beta}\bar{T}_4 + \sin\overset{j}{\beta}\bar{T}_5)\Big)\Delta\overset{j}{u}_{\bar{x}} \\
&+\sum_{j=1}^{N}\Big(-(1-2\nu)\sin\overset{j}{\beta}\bar{T}_2 + 2(1-\nu)\cos\overset{j}{\beta}\bar{T}_3 + c_{\bar{y}}(\sin\overset{j}{\beta}\bar{T}_4 - \cos\overset{j}{\beta}\bar{T}_5)\Big)\Delta\overset{j}{u}_{\bar{y}}
\end{aligned}
$$

The terms for an individual element j in (8.144) can be compared with the corresponding expressions in (6.10). To this end, we note that $c_{\bar{x}}$ and $c_{\bar{y}}$ play the same role in the present case as do $\bar{x}$ and $\bar{y}$ in (6.10). Also, by comparison of (7.19) and (5.8) it can be seen that $\bar{T}_2 = -\bar{F}_2$ and $\bar{T}_3 = -\bar{F}_3$, but that $\bar{T}_4 = \bar{F}_4$ and $\bar{T}_5 = \bar{F}_5$. Apart from these minor notational differences, the displacements derived using Volterra's formula (8.138) are exactly the same as those obtained for the "indirect" version of the displacement discontinuity method, as asserted. Similar considerations apply for the stresses and tractions in (8.139) and (8.140).

8.7 A Direct Method Based on Somigliana's Formula for Tractions

The version of the direct boundary integral method discussed in Chap. 7 is based on Somigliana's formula for the displacements, (4.328). A comparable boundary integral equation for the tractions was also derived in Sect. 4.22.4 and we next examine how this result can be used to develop another direct boundary integral method. We will again restrict our attention to the case of isotropic elasticity and two dimensions.

8.7.1 Basic Formulation

Somigliana's formula for the tractions, (4.338), repeated here for reference, is

$$
t_i(\boldsymbol{\xi}) = \int_S t_k(\boldsymbol{x})D_{kij}(\boldsymbol{x},\boldsymbol{\xi})n_j(\boldsymbol{\xi})\,\mathrm{d}S(\boldsymbol{x}) - \int_S u_k(\boldsymbol{x})S_{kij}(\boldsymbol{x},\boldsymbol{\xi})n_j(\boldsymbol{\xi})\,\mathrm{d}S(\boldsymbol{x}) \tag{8.145}
$$

The kernel functions $D_{kij}(\boldsymbol{x}, \boldsymbol{\xi})$ and $S_{kij}(\boldsymbol{x}, \boldsymbol{\xi})$ in this expression are defined as in (4.333) and (4.334) (or, alternatively, as in Appendix B). We will suppose that point $\boldsymbol{\xi}$ does not initially lie on the boundary S and imagine that $n_j(\boldsymbol{\xi})$ defines a direction that will become the unit outward normal to the boundary in the limit as $\boldsymbol{\xi}$ goes to S from inside region V.

Equation (8.145) can alternatively be written as (see (4.343))

$$t_i(\boldsymbol{\xi}) = \int_S t_j(\boldsymbol{x}) K_{ij}(\boldsymbol{x}, \boldsymbol{\xi})\, \mathrm{d}S(\boldsymbol{x}) - \int_S u_j(\boldsymbol{x}) H_{ij}(\boldsymbol{x}, \boldsymbol{\xi})\, \mathrm{d}S(\boldsymbol{x}) \tag{8.146}$$

in which the tensor fields $K_{ij}(\boldsymbol{x}, \boldsymbol{\xi})$ and $H_{ij}(\boldsymbol{x}, \boldsymbol{\xi})$ are given as

$$K_{ij}(\boldsymbol{x}, \boldsymbol{\xi}) = -\frac{1}{4\pi(1-\nu)} \frac{1}{r} \left[(1-2\nu) \left(r_{,j} n_i(\boldsymbol{\xi}) - r_{,i} n_j(\boldsymbol{\xi}) + \delta_{ij} \frac{\partial r}{\partial \boldsymbol{n}(\boldsymbol{\xi})} \right) + 2 r_{,i} r_{,j} \frac{\partial r}{\partial \boldsymbol{n}(\boldsymbol{\xi})} \right] \tag{8.147}$$

and

$$\begin{aligned} H_{ij}(\boldsymbol{x}, \boldsymbol{\xi}) = \frac{2\mu}{4\pi(1-\nu)} \frac{1}{r^2} \Big[& n_m(\boldsymbol{x}) n_m(\boldsymbol{\xi}) \delta_{ij} + n_i(\boldsymbol{x}) n_j(\boldsymbol{\xi}) - n_j(\boldsymbol{x}) n_i(\boldsymbol{\xi}) \\ & + 2 r_{,j} n_i(\boldsymbol{\xi}) \frac{\partial r}{\partial \boldsymbol{n}(\boldsymbol{x})} - 2 r_{,i} n_j(\boldsymbol{x}) \frac{\partial r}{\partial \boldsymbol{n}(\boldsymbol{\xi})} + 8 r_{,i} r_{,j} \frac{\partial r}{\partial \boldsymbol{n}(\boldsymbol{x})} \frac{\partial r}{\partial \boldsymbol{n}(\boldsymbol{\xi})} \Big] \end{aligned} \tag{8.148}$$

These expressions were obtained by interchanging subscripts i and j in (4.284) and (4.297), and can be interpreted as follows: $K_{ij}(\boldsymbol{x}, \boldsymbol{\xi})$ and $H_{ij}(\boldsymbol{x}, \boldsymbol{\xi})$ give, respectively, the displacements and tractions in the j direction at point $\boldsymbol{x}$ (the sample point) due to a unit concentrated displacement discontinuity in the i direction at point $\boldsymbol{\xi}$ (the load point). This interpretation will be used later to show the relationship between the present analysis and the displacement discontinuity method.

Equation (8.146) can be written in unabridged form for plane problems as

$$\begin{aligned} t_x(\boldsymbol{\xi}) &= \int_S \Big[t_x(\boldsymbol{x}) K_{xx}(\boldsymbol{x}, \boldsymbol{\xi}) + t_y(\boldsymbol{x}) K_{xy}(\boldsymbol{x}, \boldsymbol{\xi}) \Big] \mathrm{d}S(\boldsymbol{x}) \\ &\quad - \int_S \Big[u_x(\boldsymbol{x}) H_{xx}(\boldsymbol{x}, \boldsymbol{\xi}) + u_y(\boldsymbol{x}) H_{xy}(\boldsymbol{x}, \boldsymbol{\xi}) \Big] \mathrm{d}S(\boldsymbol{x}) \\ t_y(\boldsymbol{\xi}) &= \int_S \Big[t_x(\boldsymbol{x}) K_{yx}(\boldsymbol{x}, \boldsymbol{\xi}) + t_y(\boldsymbol{x}) K_{yy}(\boldsymbol{x}, \boldsymbol{\xi}) \Big] \mathrm{d}S(\boldsymbol{x}) \\ &\quad - \int_S \Big[u_x(\boldsymbol{x}) H_{yx}(\boldsymbol{x}, \boldsymbol{\xi}) + u_y(\boldsymbol{x}) H_{yy}(\boldsymbol{x}, \boldsymbol{\xi}) \Big] \mathrm{d}S(\boldsymbol{x}) \end{aligned} \tag{8.149}$$

Instead of using (8.147) and (8.148), we refer to (8.145) and write $K_{ij}(\boldsymbol{x},\boldsymbol{\xi})$ and $H_{ij}(\boldsymbol{x},\boldsymbol{\xi})$ as follows:

$$
\begin{aligned}
K_{xx}(\boldsymbol{x},\boldsymbol{\xi}) &= D_{xxx}(\boldsymbol{x},\boldsymbol{\xi})n_x(\boldsymbol{\xi}) + D_{xxy}(\boldsymbol{x},\boldsymbol{\xi})n_y(\boldsymbol{\xi}) \\
K_{xy}(\boldsymbol{x},\boldsymbol{\xi}) &= D_{yxx}(\boldsymbol{x},\boldsymbol{\xi})n_x(\boldsymbol{\xi}) + D_{yxy}(\boldsymbol{x},\boldsymbol{\xi})n_y(\boldsymbol{\xi}) \\
K_{yx}(\boldsymbol{x},\boldsymbol{\xi}) &= D_{xyx}(\boldsymbol{x},\boldsymbol{\xi})n_x(\boldsymbol{\xi}) + D_{xyy}(\boldsymbol{x},\boldsymbol{\xi})n_y(\boldsymbol{\xi}) \\
K_{yy}(\boldsymbol{x},\boldsymbol{\xi}) &= D_{yyx}(\boldsymbol{x},\boldsymbol{\xi})n_x(\boldsymbol{\xi}) + D_{yyy}(\boldsymbol{x},\boldsymbol{\xi})n_y(\boldsymbol{\xi})
\end{aligned}
\tag{8.150}
$$

and

$$
\begin{aligned}
H_{xx}(\boldsymbol{x},\boldsymbol{\xi}) &= S_{xxx}(\boldsymbol{x},\boldsymbol{\xi})n_x(\boldsymbol{\xi}) + S_{xxy}(\boldsymbol{x},\boldsymbol{\xi})n_y(\boldsymbol{\xi}) \\
H_{xy}(\boldsymbol{x},\boldsymbol{\xi}) &= S_{yxx}(\boldsymbol{x},\boldsymbol{\xi})n_x(\boldsymbol{\xi}) + S_{yxy}(\boldsymbol{x},\boldsymbol{\xi})n_y(\boldsymbol{\xi}) \\
H_{yx}(\boldsymbol{x},\boldsymbol{\xi}) &= S_{xyx}(\boldsymbol{x},\boldsymbol{\xi})n_x(\boldsymbol{\xi}) + S_{xyy}(\boldsymbol{x},\boldsymbol{\xi})n_y(\boldsymbol{\xi}) \\
H_{yy}(\boldsymbol{x},\boldsymbol{\xi}) &= S_{yyx}(\boldsymbol{x},\boldsymbol{\xi})n_x(\boldsymbol{\xi}) + S_{yyy}(\boldsymbol{x},\boldsymbol{\xi})n_y(\boldsymbol{\xi})
\end{aligned}
\tag{8.151}
$$

Assuming that the displacements and tractions are piecewise constant over N straight line elements of boundary S, (8.149) can be written in discretized form comparable to (7.5) as follows:

$$
\begin{aligned}
t_x(\boldsymbol{\xi}) = {} & \sum_{j=1}^{N}\left[\overset{j}{t}_x \int_{-\overset{j}{a}}^{\overset{j}{a}} K_{xx}(\bar{x}-c_{\bar{x}}, c_{\bar{y}})\,\mathrm{d}\bar{x} + \overset{j}{t}_y \int_{-\overset{j}{a}}^{\overset{j}{a}} K_{xy}(\bar{x}-c_{\bar{x}}, c_{\bar{y}})\,\mathrm{d}\bar{x}\right] \\
& - \sum_{j=1}^{N}\left[\overset{j}{u}_x \int_{-\overset{j}{a}}^{\overset{j}{a}} H_{xx}(\bar{x}-c_{\bar{x}}, c_{\bar{y}})\,\mathrm{d}\bar{x} + \overset{j}{u}_y \int_{-\overset{j}{a}}^{\overset{j}{a}} H_{xy}(\bar{x}-c_{\bar{x}}, c_{\bar{y}})\,\mathrm{d}\bar{x}\right] \\
t_y(\boldsymbol{\xi}) = {} & \sum_{j=1}^{N}\left[\overset{j}{t}_x \int_{-\overset{j}{a}}^{\overset{j}{a}} K_{yx}(\bar{x}-c_{\bar{x}}, c_{\bar{y}})\,\mathrm{d}\bar{x} + \overset{j}{t}_y \int_{-\overset{j}{a}}^{\overset{j}{a}} K_{yy}(\bar{x}-c_{\bar{x}}, c_{\bar{y}})\,\mathrm{d}\bar{x}\right] \\
& - \sum_{j=1}^{N}\left[\overset{j}{u}_x \int_{-\overset{j}{a}}^{\overset{j}{a}} H_{yx}(\bar{x}-c_{\bar{x}}, c_{\bar{y}})\,\mathrm{d}\bar{x} + \overset{j}{u}_y \int_{-\overset{j}{a}}^{\overset{j}{a}} H_{yy}(\bar{x}-c_{\bar{x}}, c_{\bar{y}})\,\mathrm{d}\bar{x}\right]
\end{aligned}
\tag{8.152}
$$

Expressions for the tensor fields $K_{ij}(\boldsymbol{x},\boldsymbol{\xi})$ and $H_{ij}(\boldsymbol{x},\boldsymbol{\xi})$ in terms of local $\bar{x}$, $\bar{y}$ coordinates originating at the midpoint of element j are found by substituting (7.25) and

(7.26) into (8.150) and (8.151), recalling that the quantities $D_{kij}(\boldsymbol{x},\boldsymbol{\xi})$ and $S_{kij}(\boldsymbol{x},\boldsymbol{\xi})$ are symmetric with respect the last two indices. After carrying out the indicated analytical integrations in (8.152), taking the limit, and setting $\boldsymbol{\xi}=\overset{i}{\boldsymbol{x}}$ we have that

$$t_x(\overset{i}{\boldsymbol{x}}) = \sum_{j=1}^{N}\Big[\mathbb{K}_{xx}(\overset{j}{\boldsymbol{x}},\overset{i}{\boldsymbol{x}})\overset{j}{t}_x + \mathbb{K}_{xy}(\overset{j}{\boldsymbol{x}},\overset{i}{\boldsymbol{x}})\overset{j}{t}_y\Big] - \sum_{j=1}^{N}\Big[\mathbb{H}_{xx}(\overset{j}{\boldsymbol{x}},\overset{i}{\boldsymbol{x}})\overset{j}{u}_x + \mathbb{H}_{xy}(\overset{j}{\boldsymbol{x}},\overset{i}{\boldsymbol{x}})\overset{j}{u}_y\Big]$$

$$t_y(\overset{i}{\boldsymbol{x}}) = \sum_{j=1}^{N}\Big[\mathbb{K}_{yx}(\overset{j}{\boldsymbol{x}},\overset{i}{\boldsymbol{x}})\overset{j}{t}_x + \mathbb{K}_{yy}(\overset{j}{\boldsymbol{x}},\overset{i}{\boldsymbol{x}})\overset{j}{t}_y\Big] - \sum_{j=1}^{N}\Big[\mathbb{H}_{yx}(\overset{j}{\boldsymbol{x}},\overset{i}{\boldsymbol{x}})\overset{j}{u}_x + \mathbb{H}_{yy}(\overset{j}{\boldsymbol{x}},\overset{i}{\boldsymbol{x}})\overset{j}{u}_y\Big] \tag{8.153}$$

The values of $\mathbb{K}_{xx}(\overset{j}{\boldsymbol{x}},\overset{i}{\boldsymbol{x}})$, etc., and $\mathbb{H}_{xx}(\overset{j}{\boldsymbol{x}},\overset{i}{\boldsymbol{x}})$, etc., are found by using (7.27) and (7.28) in (8.150) and (8.151), noting that the components of the unit outward normal to the boundary S at point $\boldsymbol{\xi}\to\overset{i}{\boldsymbol{x}}$ are $n_x(\overset{i}{\boldsymbol{x}}) = -\sin\overset{i}{\beta}$ and $n_y(\overset{i}{\boldsymbol{x}}) = \cos\overset{i}{\beta}$, where $\overset{i}{\beta}$ is the inclination of element i.

Setting $t_x(\overset{i}{\boldsymbol{x}}) = \overset{i}{t}_x$, $t_y(\overset{i}{\boldsymbol{x}}) = \overset{i}{t}_y$, and $\mathbb{K}_{xx}(\overset{j}{\boldsymbol{x}},\overset{i}{\boldsymbol{x}}) = \overset{ij}{\mathbb{K}}_{xx}$, etc., equations (8.153) can be written as

$$\sum_{j=1}^{N}\Big[\overset{ij}{\mathbb{H}}_{xx}\overset{j}{u}_x + \overset{ij}{\mathbb{H}}_{xy}\overset{j}{u}_y\Big] = \sum_{j=1}^{N}\Big[(\overset{ij}{\mathbb{K}}_{xx} - \overset{ij}{\mathbb{I}})\overset{j}{t}_x + \overset{ij}{\mathbb{K}}_{xy}\overset{j}{t}_y\Big]$$

$$\sum_{j=1}^{N}\Big[\overset{ij}{\mathbb{H}}_{yx}\overset{j}{u}_x + \overset{ij}{\mathbb{H}}_{yy}\overset{j}{u}_y\Big] = \sum_{j=1}^{N}\Big[\overset{ij}{\mathbb{K}}_{yx}\overset{j}{t}_x + (\overset{ij}{\mathbb{K}}_{yy} - \overset{ij}{\mathbb{I}})\overset{j}{t}_y\Big] \tag{8.154}$$

By considering the boundary conditions at all N elements, that is, letting i range from 1 to N, we obtain a set of $2N$ algebraic equations in $2N$ unknowns of the form $\boldsymbol{AX}=\boldsymbol{b}$, which can be solved by standard numerical techniques. Once the boundary displacements and tractions have been determined the displacements $u_i(\boldsymbol{\xi})$ and stresses $\sigma_{ij}(\boldsymbol{\xi})$ inside the region of interest can be found as explained in Sects. 7.2 and 7.4.

8.7.2 Examples

The foregoing numerical procedure gives results comparable to those obtained by the other boundary element methods for elasticity discussed in this book. Figure 8.13, for example, shows a comparison of the hoop stress with the analytical solution for the problem of a circular hole in a plate under a far-field uniaxial tension $\sigma_{xx}=-p$. As in previous illustrations for this same problem (see Figs. 5.5 and 6.6), the numerical

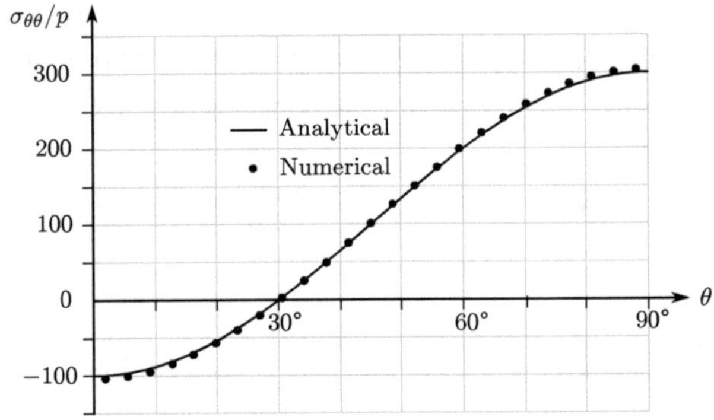

Fig. 8.13 Stress $\sigma_{\theta\theta}$ at boundary of circular hole; direct method based on Somigliana's traction equation

model used full symmetry and one-quarter of the circular boundary was divided into 25 equal size elements. In the present case, the numerical results are somewhat less accurate than those obtained by the stress and displacement discontinuity methods but are still in reasonable agreement with the exact solution.

An interesting aspect of the direct method based Somigliana's traction equation reveals itself when modeling problems for multiply connected regions without employing symmetry conditions. Curiously, for such problems the interiors of cavities appear to undergo rigid body translations, although the displacements and stresses inside the region of interest are unaffected by this circumstance. In particular, the displacements in the region of interest V *do not* contain rigid body movements. By trial, it is found that $\overset{j}{u}_x$ and $\overset{j}{u}_y$ in (8.154) actually represent the displacements on the negative sides of the elements, $(\overset{j}{u}_x)_{\bar{y}=0_-}$ and $(\overset{j}{u}_y)_{\bar{y}=0_-}$, minus constant terms from the rigid body translations, $(\overset{j}{u}_x)_{\bar{y}=0_+}$ and $(\overset{j}{u}_x)_{\bar{y}=0_+}$:

$$\begin{aligned} \overset{j}{u}_x &= (\overset{j}{u}_x)_{\bar{y}=0_-} - (\overset{j}{u}_x)_{\bar{y}=0_+} \\ \overset{j}{u}_y &= (\overset{j}{u}_y)_{\bar{y}=0_-} - (\overset{j}{u}_x)_{\bar{y}=0_+} \end{aligned} \tag{8.155}$$

Again, it is found by trial that the values of $(\overset{j}{u}_x)_{\bar{y}=0_+}$ and $(\overset{j}{u}_x)_{\bar{y}=0_+}$ are same for all N elements. As a result, finite difference approximations of the derivatives of $\overset{j}{u}_x$ and $\overset{j}{u}_y$ used to calculate the hoop stress from the boundary displacements (see Sect. 6.10.1) are unaffected by these constant terms. The resemblance of conditions (8.155) to the definition (4.362) of a displacement discontinuity is worth noting.

As an illustration, consider again the circular hole problem but now with the entire boundary divided into 100 equal size elements. The displacement field plotted in

Fig. 8.14 clearly shows that the interior of the hole undergoes a uniform translation. By substituting the applied tractions $\overset{j}{t}_x$, $\overset{j}{t}_y$ and the calculated boundary displacements $\overset{j}{u}_x$, $\overset{j}{u}_y$ into Somigliana's displacement formulas (7.6) for a single point inside the hole, the components of translation were found to be $0.00633R$ and $0.00074R$ (arbitrary length units) in the x and y directions, respectively, where R is the radius of the hole. This translation can be removed by adding these displacements to each of the $N = 100$ boundary elements, which then gives the boundary displacements $\overset{j}{u}_x = (\overset{j}{u}_x)_{\bar{y}=0_-}$ and $\overset{j}{u}_y = (\overset{j}{u}_y)_{\bar{y}=0_-}$ for $j = 1$ to N. With this simple correction, the displacement field for the subject problem appears as shown in Fig. 8.15.

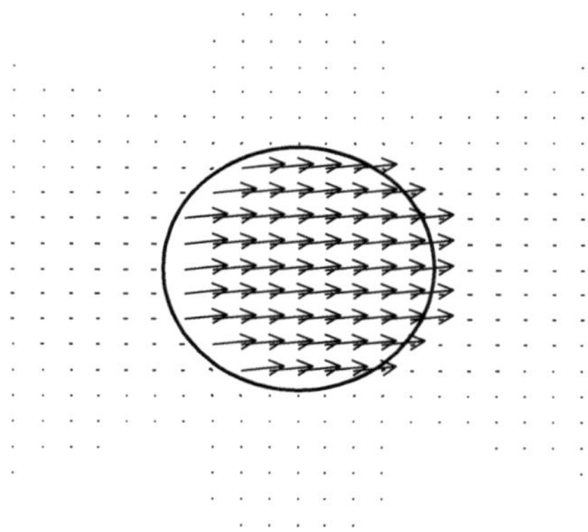

Fig. 8.14 Rigid body translation of inside portion of circular hole; displacement magnitude $6.33R \times 10^{-3}$

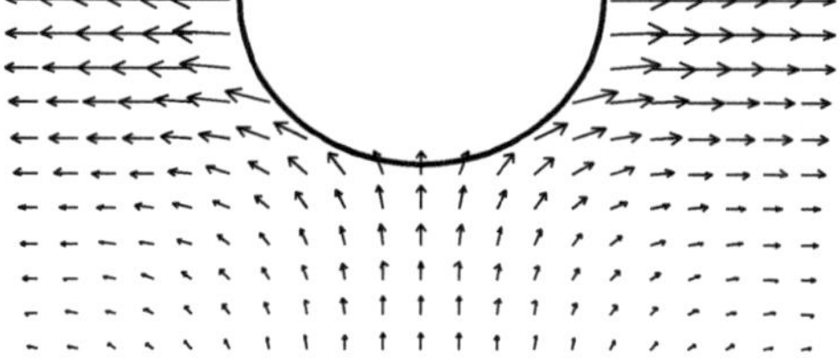

Fig. 8.15 Corrected displacements around circular hole; magnitude of maximum displacement is 0.883×10^{-3}

8.7.3 Alternative Formulation

We next observe that the boundary element method based on Somigliana's formula for the tractions can also be developed by using a concentrated displacement discontinuity in the reciprocal theorem instead of a concentrated force. It is only necessary to consider that the source point $\boldsymbol{\xi}$ is *outside* the region of interest, rather than inside it as in our previous calculations.

Referring to Sect. 4.21, we take problem (2) in the reciprocal theorem as the problem we wish to solve (i.e., a problem with displacements u_i and tractions t_i on the boundary S of region V) and problem (1) as the solution to the problem of a concentrated displacement discontinuity $\mathcal{D}_j(\boldsymbol{\xi})$ at the point $\boldsymbol{\xi}$ on a line outside region V, as shown in Fig. 8.16. We also suppose that the body forces are zero for both problems so the volume integrals in (4.315) can be disregarded. We thus have from (4.285) and (4.295) that

$$
\begin{aligned}
u_j^{(1)} &= K_{kj}(\boldsymbol{x}, \boldsymbol{\xi})\mathcal{D}_k(\boldsymbol{\xi}) \\
t_j^{(1)} &= H_{kj}(\boldsymbol{x}, \boldsymbol{\xi})\mathcal{D}_k(\boldsymbol{\xi}) \\
B_j^{(1)} &= 0
\end{aligned}
\tag{8.156}
$$

and

$$
\begin{aligned}
u_j^{(2)} &= u_j(\boldsymbol{x}) \\
t_j^{(2)} &= t_j(\boldsymbol{x}) \\
B_j^{(2)} &= 0
\end{aligned}
\tag{8.157}
$$

Substitution of these quantities into (4.315) then gives

$$
\int_S u_j(\boldsymbol{x}) H_{kj}(\boldsymbol{x}, \boldsymbol{\xi})\mathcal{D}_k(\boldsymbol{\xi})\, \mathrm{d}S(\boldsymbol{x}) = \int_S t_j(\boldsymbol{x}) K_{kj}(\boldsymbol{x}, \boldsymbol{\xi})\mathcal{D}_k(\boldsymbol{\xi})\, \mathrm{d}S(\boldsymbol{x})
$$

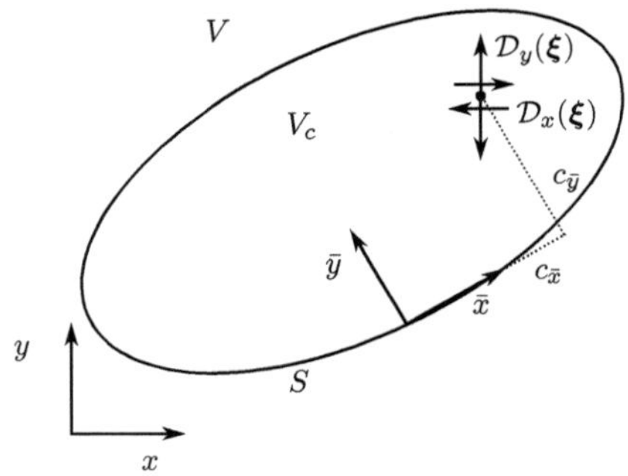

Fig. 8.16 Concentrated displacement discontinuity at a point $\boldsymbol{\xi}$ *on a line outside* the region of interest V

or

$$\mathcal{D}_k(\boldsymbol{\xi}) \int_S u_j(\boldsymbol{x}) H_{kj}(\boldsymbol{x}, \boldsymbol{\xi}) \, \mathrm{d}S(\boldsymbol{x}) = \mathcal{D}_k(\boldsymbol{\xi}) \int_S t_j(\boldsymbol{x}) K_{kj}(\boldsymbol{x}, \boldsymbol{\xi}) \, \mathrm{d}S(\boldsymbol{x}) \tag{8.158}$$

Considering separately the two cases $\mathcal{D}_1(\boldsymbol{\xi}) \neq 0$, $\mathcal{D}_2(\boldsymbol{\xi}) = 0$ and $\mathcal{D}_1(\boldsymbol{\xi}) = 0$, $\mathcal{D}_2(\boldsymbol{\xi}) \neq 0$ we arrive at the result

$$0 = \int_S t_j(\boldsymbol{x}) K_{ij}(\boldsymbol{x}, \boldsymbol{\xi}) \, \mathrm{d}S(\boldsymbol{x}) - \int_S u_j(\boldsymbol{x}) H_{ij}(\boldsymbol{x}, \boldsymbol{\xi}) \, \mathrm{d}S(\boldsymbol{x}) \tag{8.159}$$

Discretizing (8.159) and taking limits in the same way as before gives the following equations:

$$\begin{aligned} \sum_{j=1}^{N} \left(\overset{ij}{\mathbb{H}}_{xx} \overset{j}{u}_x + \overset{ij}{\mathbb{H}}_{xy} \overset{j}{u}_y \right) &= \sum_{j=1}^{N} \left(\overset{ij}{\mathbb{K}}_{xx} \overset{j}{t}_x + \overset{ij}{\mathbb{K}}_{xy} \overset{j}{t}_y \right) \\ \sum_{j=1}^{N} \left(\overset{ij}{\mathbb{H}}_{yx} \overset{j}{u}_x + \overset{ij}{\mathbb{H}}_{yy} \overset{j}{u}_y \right) &= \sum_{j=1}^{N} \left(\overset{ij}{\mathbb{K}}_{yx} \overset{j}{t}_x + \overset{ij}{\mathbb{K}}_{yy} \overset{j}{t}_y \right) \end{aligned} \tag{8.160}$$

Numerically, these equations are the same as (8.154) because the diagonal coefficients $\overset{ii}{\mathbb{K}}_{xx}$ and $\overset{ii}{\mathbb{K}}_{yy}$ must be evaluated differently in the two cases. It can be shown from (8.150) that, for $c_{\bar{x}} = 0$,

$$\begin{aligned} \overset{ii}{\mathbb{K}}_{xx} = \overset{ii}{\mathbb{K}}_{yy} &= -2(1-\nu) \lim_{c_{\bar{y}} \to 0_\pm} \bar{T}_3 \\ &= \begin{cases} -\frac{1}{2} & \text{on } c_{\bar{y}} = 0_+ \\ +\frac{1}{2} & \text{on } c_{\bar{y}} = 0_- \end{cases} \end{aligned} \tag{8.161}$$

According to our boundary traversal convention $c_{\bar{y}}$ is positive in the direction of the outward normal to the boundary (see Fig. 8.16). When point $\boldsymbol{\xi}$ is inside the region of interest we have that $c_{\bar{y}} \to 0_-$; when it is outside then $c_{\bar{y}} \to 0_+$. Thus, the self-effect coefficients in (8.154) are $(\overset{ii}{\mathbb{K}}_{xx})_{0_-} = (\overset{ii}{\mathbb{K}}_{xx})_{0_-} = +1/2$ while those in (8.160) are $(\overset{ii}{\mathbb{K}}_{xx})_{0_+} = (\overset{ii}{\mathbb{K}}_{xx})_{0_+} = -1/2$. Taking into account the $\overset{ii}{\mathbb{I}} = 1$ terms in (8.154), we see that the two sets of expressions are identical.

The result (8.161) follows the limit properties of an adjoint double layer elastic potential, see Sect. 4.18.2 and Eq. (4.388) of Sect. 4.27.

8.7.4 Relationship to the Displacement Discontinuity Method

An interesting observation can be made by comparing Somigliana's traction formula (8.146) with Volterra's (8.140). The latter formula is written for a crack Γ, but, as we noted earlier (see Sect. 6.6), its numerical representation as the displacement discontinuity method is also applicable to a closed contour S, provided provisions are made to prevent movement of the interior region.

As we just noted, the boundary displacements in traction equation (8.146) can be interpreted as pertaining to the negative side of boundary S. We could therefore write this equation as

$$t_i(\boldsymbol{\xi}) = \int_S t_j(\boldsymbol{x})K_{ij}(\boldsymbol{x},\boldsymbol{\xi})\,\mathrm{d}S(\boldsymbol{x}) - \int_S u_j^-(\boldsymbol{x})H_{ij}(\boldsymbol{x},\boldsymbol{\xi})\,\mathrm{d}S(\boldsymbol{x}) \tag{8.162}$$

By adding and subtracting a term with the displacements on the positive side of S, i.e., *outside* region V, and using the definition of the displacement discontinuity, $\Delta u_j(\boldsymbol{x}) = u_j^+(\boldsymbol{x}) - u_j^-(\boldsymbol{x})$, we write

$$t_i(\boldsymbol{\xi}) = \int_S t_j(\boldsymbol{x})K_{ij}(\boldsymbol{x},\boldsymbol{\xi})\,\mathrm{d}S(\boldsymbol{x}) - \int_S u_j^+(\boldsymbol{x})H_{ij}(\boldsymbol{x},\boldsymbol{\xi})\,\mathrm{d}S(\boldsymbol{x}) + \int_S \Delta u_j(\boldsymbol{x})H_{ij}(\boldsymbol{x},\boldsymbol{\xi})\,\mathrm{d}S(\boldsymbol{x}) \tag{8.163}$$

Referring to (8.140) with Γ replaced by S, we see that

$$0 = \int_S t_j(\boldsymbol{x})K_{ij}(\boldsymbol{x},\boldsymbol{\xi})\,\mathrm{d}S(\boldsymbol{x}) - \int_S u_j^+(\boldsymbol{x})H_{ij}(\boldsymbol{x},\boldsymbol{\xi})\,\mathrm{d}S(\boldsymbol{x}) \tag{8.164}$$

which is the same as (8.159) because $u_j(\boldsymbol{x})$ in that equation refers to the displacements of the positive side of boundary S.

8.8 Piecewise Homogeneous Regions II

Somigliana's traction formula (8.146) can be used to develop another numerical procedure for piecewise homogeneous regions. This approach is limited to two dimensions and the case of isotropic elasticity but has the advantage that it allows for the presence of displacement discontinuities (i.e., cracks) along the interfaces between subregions. The tractions can also be discontinuous along the interface, but we will not include this possibility in our analysis. We will consider below the case of two subregions, V_1 and V_2, with elastic constants ν_1, μ_1 and ν_2, μ_2, and with boundaries $S_1 + L_1$ and $S_2 + L_2$, where boundaries L_1 and L_2 define the interface between the subregions, as depicted in Fig. 8.17.[3] For reasons that will become clear in due course, the direction of boundary traversal is taken to be the same along L_1 and

[3] Note that these definitions of S_1 and S_2 differ from those used for the methods discussed in Sect. 8.4; the interfaces are designated separately in the present case.

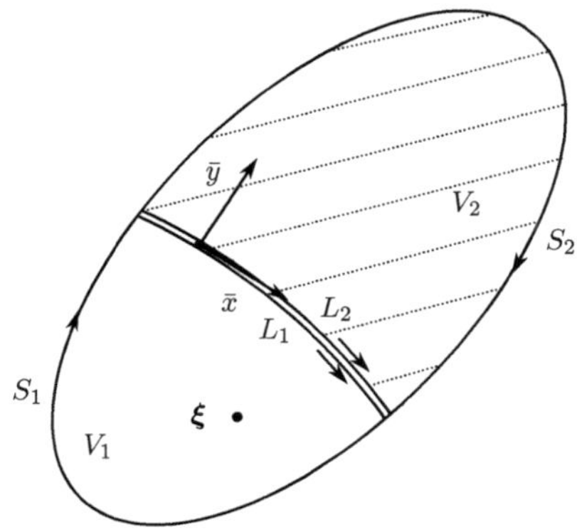

Fig. 8.17 Piecewise homogeneous body consisting of two isotropic, linearly elastic subregions

L_2. (A similar convention was used in Sect. 4.24 to derive Volterra's formulas from Somigliana's displacement equation.) More than two subregions can be treated using the same ideas but we will not consider this possibility.

8.8.1 *Problem Formulation*

Equation (8.146) can be written for each subregion V_k ($k = 1$ or 2) as

$$t_i(\boldsymbol{\xi}) = \int_{S_k+L_k} t_j(\boldsymbol{x})K_{ij}(\boldsymbol{x},\boldsymbol{\xi})\,\mathrm{d}S(\boldsymbol{x}) - \int_{S_k+L_k} u_j(\boldsymbol{x})H_{ij}(\boldsymbol{x},\boldsymbol{\xi})\,\mathrm{d}S(\boldsymbol{x})\,; \qquad \boldsymbol{\xi} \in V_k \tag{8.165}$$

in which $t_i(\boldsymbol{\xi})$ is associated with a normal $\boldsymbol{n}(\boldsymbol{\xi})$ that defines a direction that will become the unit outward normal to the boundary in the limit as $\boldsymbol{\xi}$ goes to $S_k + L_k$ from inside V_k. If the load point $\boldsymbol{\xi}$ is not inside subregion region V_k nor on its boundary $S_k + L_k$ we have the additional relation

$$0 = \int_{S_k+L_k} t_j(\boldsymbol{x})K_{ij}(\boldsymbol{x},\boldsymbol{\xi})\,\mathrm{d}S(\boldsymbol{x}) - \int_{S_k+L_k} u_j(\boldsymbol{x})H_{ij}(\boldsymbol{x},\boldsymbol{\xi})\,\mathrm{d}S(\boldsymbol{x})\,; \qquad \boldsymbol{\xi} \notin V_k \tag{8.166}$$

The kernels $K_{ij}(\boldsymbol{x},\boldsymbol{\xi})$ and $H_{ij}(\boldsymbol{x},\boldsymbol{\xi})$ in these expressions are given in (8.147) and (8.148). The kernel $H_{ij}(\boldsymbol{x},\boldsymbol{\xi})$ can be put in the form

$$H_{ij}(\boldsymbol{x},\boldsymbol{\xi}) = \frac{2\mu_k}{4\pi(1-\nu_k)}Q_{ij}(\boldsymbol{x},\boldsymbol{\xi}) \tag{8.167}$$

in which the *property-independent* sub-kernel $Q_{ij}(\boldsymbol{x},\boldsymbol{\xi})$ is

$$Q_{ij}(\boldsymbol{x},\boldsymbol{\xi}) = \frac{1}{r^2}\Big[n_m(\boldsymbol{x})n_m(\boldsymbol{\xi})\delta_{ij} + n_i(\boldsymbol{x})n_j(\boldsymbol{\xi}) - n_j(\boldsymbol{x})n_i(\boldsymbol{\xi}) + 2r_{,j}n_i(\boldsymbol{\xi})\frac{\partial r}{\partial \boldsymbol{n}(\boldsymbol{x})} - 2r_{,i}n_j(\boldsymbol{x})\frac{\partial r}{\partial \boldsymbol{n}(\boldsymbol{\xi})} + 8r_{,i}r_{,j}\frac{\partial r}{\partial \boldsymbol{n}(\boldsymbol{x})}\frac{\partial r}{\partial \boldsymbol{n}(\boldsymbol{\xi})}\Big] \tag{8.168}$$

If we multiply (8.165) and (8.166) by $4\pi(1-\nu_k)/(2\mu_k)$ and use (8.167), these equations can be written as

$$\frac{4\pi(1-\nu_k)}{2\mu_k}t_i(\boldsymbol{\xi}) = \frac{4\pi(1-\nu_k)}{2\mu_k}\int_{S_k+L_k} t_j(\boldsymbol{x})K_{ij}(\boldsymbol{x},\boldsymbol{\xi})\,\mathrm{d}S(\boldsymbol{x}) - \int_{S_k+L_k} u_j(\boldsymbol{x})Q_{ij}(\boldsymbol{x},\boldsymbol{\xi})\,\mathrm{d}S(\boldsymbol{x})\,; \qquad \boldsymbol{\xi}\in V_k \tag{8.169}$$

and

$$0 = \frac{4\pi(1-\nu_k)}{2\mu_k}\int_{S_k+L_k} t_j(\boldsymbol{x})K_{ij}(\boldsymbol{x},\boldsymbol{\xi})\,\mathrm{d}S(\boldsymbol{x}) - \int_{S_k+L_k} u_j(\boldsymbol{x})Q_{ij}(\boldsymbol{x},\boldsymbol{\xi})\,\mathrm{d}S(\boldsymbol{x})\,; \qquad \boldsymbol{\xi}\notin V_k \tag{8.170}$$

where $k = 1$ or 2. We note that the integrals in (8.169) and (8.170) containing the boundary displacements $u_j(\boldsymbol{x})$ do not involve an elastic property-dependent multiplier while the ones containing the tractions do. This fact forms the basis for the succeeding analysis.

We now consider the problem depicted in Fig. 8.17 for subregions 1 and 2. If load point $\boldsymbol{\xi}$ is in subregion 1, i.e., if $\boldsymbol{\xi}\in V_1$, we have from (8.169) that

$$\frac{4\pi(1-\nu_1)}{2\mu_1}t_i(\boldsymbol{\xi}) = \frac{4\pi(1-\nu_1)}{2\mu_1}\int_{S_1+L_1} t_j(\boldsymbol{x})K_{ij}(\boldsymbol{x},\boldsymbol{\xi})\,\mathrm{d}S(\boldsymbol{x}) - \int_{S_1+L_1} u_j(\boldsymbol{x})Q_{ij}(\boldsymbol{x},\boldsymbol{\xi})\,\mathrm{d}S(\boldsymbol{x}) \tag{8.171}$$

and, from (8.170),

$$0 = \frac{4\pi(1-\nu_2)}{2\mu_2}\int_{S_2+L_2} t_j(\boldsymbol{x})K_{ij}(\boldsymbol{x},\boldsymbol{\xi})\,\mathrm{d}S(\boldsymbol{x}) - \int_{S_2+L_2} u_j(\boldsymbol{x})Q_{ij}(\boldsymbol{x},\boldsymbol{\xi})\,\mathrm{d}S(\boldsymbol{x}) \tag{8.172}$$

Writing the integrals over S_k and L_k separately and adding (8.171) and (8.172) gives

$$\frac{4\pi(1-\nu_1)}{2\mu_1}t_i(\boldsymbol{\xi}) = \sum_{k=1}^{2}\left[\frac{4\pi(1-\nu_k)}{2\mu_k}\int_{S_k} t_j(\boldsymbol{x})K_{ij}^{(k)}(\boldsymbol{x},\boldsymbol{\xi})t_j(\boldsymbol{x})\,\mathrm{d}S(\boldsymbol{x}) - \int_{S_k} u_j(\boldsymbol{x})Q_{ij}(\boldsymbol{x},\boldsymbol{\xi})\,\mathrm{d}S(\boldsymbol{x})\right]$$
$$+\sum_{k=1}^{2}\left[\frac{4\pi(1-\nu_k)}{2\mu_k}\int_{L_k} t_j(\boldsymbol{x})K_{ij}^{(k)}(\boldsymbol{x},\boldsymbol{\xi})\,\mathrm{d}S(\boldsymbol{x}) - \int_{L_k} u_j(\boldsymbol{x})Q_{ij}(\boldsymbol{x},\boldsymbol{\xi})\,\mathrm{d}S(\boldsymbol{x})\right] \tag{8.173}$$

where we have written $K_{ij}^{(k)}(\boldsymbol{x},\boldsymbol{\xi})$ for the kernel $K_{ij}(\boldsymbol{x},\boldsymbol{\xi})$ to signify that it depends on the elastic properties of subregion k; the kernel $Q_{ij}(\boldsymbol{x},\boldsymbol{\xi})$ does not depend on these properties.

The integrals over the interface boundaries L_1 and L_2 in (8.173) can be combined by writing them in terms of a single interface contour, L_{int}, which we will choose to be L_1, i.e., $L_{int} = L_1$. If $n_j(\boldsymbol{x})$ are the components of the outward normal to the boundary of region V_1 at sample point $\boldsymbol{x}$, as shown in Fig. 8.18, then the components of the outward normal to the boundary of region V_2 at this point are $-n_j(\boldsymbol{x})$. By inspection of (8.168), we see that the value of $Q_{ij}(\boldsymbol{x},\boldsymbol{\xi})$ on L_2 is the negative of its value on $L_1 = L_{int}$. Letting $u_j^-(\boldsymbol{x})$ be the displacements on L_1 and $u_j^+(\boldsymbol{x})$ be the displacements on L_2, we then have that

$$\begin{aligned}\sum_{k=1}^{2}\int_{L_k} u_j(\boldsymbol{x})Q_{ij}(\boldsymbol{x},\boldsymbol{\xi})\,\mathrm{d}S(\boldsymbol{x}) &= \int_{L_1} u_j(\boldsymbol{x})Q_{ij}(\boldsymbol{x},\boldsymbol{\xi})\,\mathrm{d}S(\boldsymbol{x}) + \int_{L_2} u_j(\boldsymbol{x})Q_{ij}(\boldsymbol{x},\boldsymbol{\xi})\,\mathrm{d}S(\boldsymbol{x})\\ &= \int_{L_{int}} u_j^-(\boldsymbol{x})Q_{ij}(\boldsymbol{x},\boldsymbol{\xi})\,\mathrm{d}S(\boldsymbol{x}) - \int_{L_{int}} u_j^+(\boldsymbol{x})Q_{ij}(\boldsymbol{x},\boldsymbol{\xi})\,\mathrm{d}S(\boldsymbol{x})\\ &= -\int_{L_{int}} \Delta u_j(\boldsymbol{x})Q_{ij}(\boldsymbol{x},\boldsymbol{\xi})\,\mathrm{d}S(\boldsymbol{x})\end{aligned} \tag{8.174}$$

which is equal to zero if the displacements are continuous, i.e., if $\Delta u_j(\boldsymbol{x}) = u_j^+(\boldsymbol{x}) - u_j^-(\boldsymbol{x}) = 0$.

Similarly, letting $t_j^-(\boldsymbol{x})$ be the tractions on L_1 and $t_j^+(\boldsymbol{x})$ be the tractions on L_2, and noting that, because the tractions are continuous, $t_j^+(\boldsymbol{x}) + t_j^-(\boldsymbol{x}) = 0$, the terms

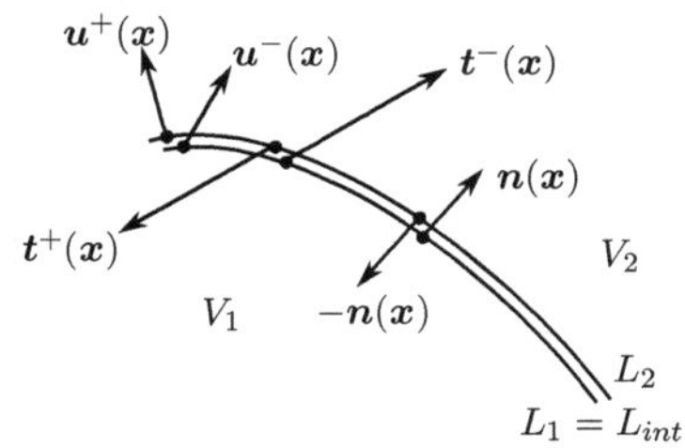

Fig. 8.18 Boundary parameters for point $\boldsymbol{x}$ on interface between subregions V_1 and V_2

$$\sum_{k=1}^{2}\frac{4\pi(1-\nu_k)}{2\mu_k}\int_{L_k} t_j(\boldsymbol{x})K_{ij}^{(k)}(\boldsymbol{x},\boldsymbol{\xi})\,\mathrm{d}S(\boldsymbol{x}) = \frac{4\pi(1-\nu_1)}{2\mu_1}\int_{L_1} t_j^{-}(\boldsymbol{x})K_{ij}^{(1)}(\boldsymbol{x},\boldsymbol{\xi})\,\mathrm{d}S(\boldsymbol{x}) + \frac{4\pi(1-\nu_2)}{2\mu_2}\int_{L_2} t_j^{+}(\boldsymbol{x})K_{ij}^{(2)}(\boldsymbol{x},\boldsymbol{\xi})\,\mathrm{d}S(\boldsymbol{x}) \tag{8.175}$$

can be written as

$$\begin{aligned}\sum_{k=1}^{2}\frac{4\pi(1-\nu_k)}{2\mu_k}\int_{L_k} t_j(\boldsymbol{x})K_{ij}^{(k)}(\boldsymbol{x},\boldsymbol{\xi})\,\mathrm{d}S(\boldsymbol{x}) &= \frac{4\pi(1-\nu_1)}{2\mu_1}\int_{L_{int}} t_j^{-}(\boldsymbol{x})K_{ij}^{(1)}(\boldsymbol{x},\boldsymbol{\xi})\,\mathrm{d}S(\boldsymbol{x}) \\ &\quad - \frac{4\pi(1-\nu_2)}{2\mu_2}\int_{L_{int}} t_j^{-}(\boldsymbol{x})K_{ij}^{(2)}(\boldsymbol{x},\boldsymbol{\xi})\,\mathrm{d}S(\boldsymbol{x}) \\ &= \int_{L_{int}} t_j(\boldsymbol{x})P_{ij}(\boldsymbol{x},\boldsymbol{\xi})\,\mathrm{d}S(\boldsymbol{x})\end{aligned} \tag{8.176}$$

where

$$P_{ij}(\boldsymbol{x},\boldsymbol{\xi}) = \frac{4\pi(1-\nu_1)}{2\mu_1}K_{ij}^{(1)}(\boldsymbol{x},\boldsymbol{\xi}) - \frac{4\pi(1-\nu_2)}{2\mu_2}K_{ij}^{(2)}(\boldsymbol{x},\boldsymbol{\xi}) \tag{8.177}$$

and where we have written $t_j(\boldsymbol{x})$ for $t_j^{-}(\boldsymbol{x})$ as was done previously for the displacement discontinuity method.

Using the preceding results together with (8.167) we write (8.173) as

$$\begin{aligned}\frac{4\pi(1-\nu_1)}{2\mu_1}t_i(\boldsymbol{\xi}) &= \sum_{k=1}^{2}\frac{4\pi(1-\nu_k)}{2\mu_k}\left[\int_{S_k} t_j(\boldsymbol{x})K_{ij}^{(k)}(\boldsymbol{x},\boldsymbol{\xi})\,\mathrm{d}S(\boldsymbol{x}) - \int_{S_k} u_j(\boldsymbol{x})H_{ij}^{(k)}(\boldsymbol{x},\boldsymbol{\xi})\,\mathrm{d}S(\boldsymbol{x})\right] \\ &\quad + \int_{L_{int}} t_j(\boldsymbol{x})P_{ij}(\boldsymbol{x},\boldsymbol{\xi})\,\mathrm{d}S(\boldsymbol{x}) + \int_{L_{int}} \Delta u_j(\boldsymbol{x})Q_{ij}(\boldsymbol{x},\boldsymbol{\xi})\,\mathrm{d}S(\boldsymbol{x})\end{aligned} \tag{8.178}$$

in which $H_{ij}^{(k)}$ denotes the values of kernel function $H_{ij}(\boldsymbol{x},\boldsymbol{\xi})$ for subregion k.

The case for which the load point $\boldsymbol{\xi}$ is in subregion 2 can be treated similarly. If L_{int} is again chosen as L_1 it can be shown that the resulting equation is the same as (8.178), except that the multiplier for $t_i(\boldsymbol{\xi})$ is $4\pi(1-\nu_2)/(2\mu_2)$. If we set

$$A(\boldsymbol{\xi}) = \begin{cases} 4\pi(1-\nu_1)(2\mu_1)^{-1} & \boldsymbol{\xi}\in V_1 \\ 4\pi(1-\nu_2)(2\mu_2)^{-1} & \boldsymbol{\xi}\in V_2 \end{cases} \tag{8.179}$$

both equations can be recovered from the equation

$$\begin{aligned}A(\boldsymbol{\xi})t_i(\boldsymbol{\xi}) &= \sum_{k=1}^{2}\frac{4\pi(1-\nu_k)}{2\mu_k}\left[\int_{S_k} t_j(\boldsymbol{x})K_{ij}^{(k)}(\boldsymbol{x},\boldsymbol{\xi})\,\mathrm{d}S(\boldsymbol{x}) - \int_{S_k} u_j(\boldsymbol{x})H_{ij}^{(k)}(\boldsymbol{x},\boldsymbol{\xi})\,\mathrm{d}S(\boldsymbol{x})\right] \\ &\quad + \int_{L_{int}} t_j(\boldsymbol{x})P_{ij}(\boldsymbol{x},\boldsymbol{\xi})\,\mathrm{d}S(\boldsymbol{x}) + \int_{L_{int}} \Delta u_j(\boldsymbol{x})Q_{ij}(\boldsymbol{x},\boldsymbol{\xi})\,\mathrm{d}S(\boldsymbol{x})\end{aligned} \tag{8.180}$$

8.8.2 Some Observations

Several observations can be made about equation (8.180).

First, if unknown displacement discontinuities exist along the interface then either the tractions must be specified at these locations or else a relationship between tractions and displacement discontinuities must be provided, e.g., a spring-type crack (see Crouch and Starfield 1983). An example is the case of an interface crack subjected to an internal pressure $-p$. The normal stress on the crack surfaces is then equal to $-p$ and the shear stress is zero; the x and y components of applied traction t_x and t_y can be calculated from these using coordinate transformation formulas (4.64).

Second, if the two subregions have the same elastic properties, i.e., if $\nu_1 = \nu_2 = \nu$ and $\mu_1 = \mu_2 = \mu$ then $A(\boldsymbol{\xi}) = 4\pi(1-\nu)/(2\mu)$, $P_{ij} = 0$, and the kernels K_{ij} and H_{ij} are the same for both subregions. If furthermore L_{int} represents a line crack Γ inside region $V = V_1 + V_2$ with boundary $S = S_1 + S_2$, (8.180) can be written as

$$t_i(\boldsymbol{\xi}) = \int_S t_j(\boldsymbol{x}) K_{ij}(\boldsymbol{x}, \boldsymbol{\xi})\, \mathrm{d}S(\boldsymbol{x}) - \int_S u_j(\boldsymbol{x}) H_{ij}(\boldsymbol{x}, \boldsymbol{\xi})\, \mathrm{d}S(\boldsymbol{x}) + \int_\Gamma \Delta u_j(\boldsymbol{x}) H_{ij}(\boldsymbol{x}, \boldsymbol{\xi})\, \mathrm{d}S(\boldsymbol{x}) \tag{8.181}$$

which is the sum of Somigliana's traction equation (8.146) and Volterra's formula (8.140) for the tractions. This equation can be discretized and used to solve boundary value problems for elastic regions containing cracks. (See also Sect. 8.9 in this connection.)

Third, the stresses at points $\boldsymbol{\xi}$ inside subregions V_1 or V_2 can be found by computing the tractions $t_i(\boldsymbol{\xi})$ at these points from (8.173) and then using definitions (4.16), written here as $t_i(\boldsymbol{\xi}) = \sigma_{ij}(\boldsymbol{\xi}) n_j(\boldsymbol{\xi})$. Taking $n_1(\boldsymbol{\xi}) = 1$ and $n_2(\boldsymbol{\xi}) = 0$ gives values for $\sigma_{11}(\boldsymbol{\xi})$ and $\sigma_{21}(\boldsymbol{\xi}) = \sigma_{12}(\boldsymbol{\xi})$; a value for $\sigma_{22}(\boldsymbol{\xi})$ is found by taking $n_1(\boldsymbol{\xi}) = 0$ and $n_2(\boldsymbol{\xi}) = 1$.

The final important observation about (8.180) is that the integrals along the interface L_{int} do not involve the individual displacement components on L_1 and L_2, only their differences. Discretization of (8.180) (discussed below) will therefore lead to a system of algebraic equations in which the tractions and (if applicable) the displacement discontinuities are the unknowns along the interface. (Additionally, of course, the displacements, the tractions, or some combination of these quantities will initially be unknown along boundaries S_1 and S_2.) Once the system of equations has been solved, the displacements and the tractions will be known on boundaries S_1 and S_2 and the tractions will be known on the interface. One can therefore define a boundary value problem for either subregion V_1 or V_2 from which to find the interface displacements. The direct boundary integral method of Chap. 7 can be used for this purpose or, alternatively, the method discussed in Sect. 8.7. The displacements in either subregion can then be computed using suitably discretized versions of Somigliana's formula (4.328).

8.8.3 Incorporation of Stresses at Infinity

If one of the subregions is infinite it is necessary to have a way of including nonzero initial stresses $\sigma_{ij}^{(\infty)}$ at infinity in the analysis. The stresses are nonuniform in the vicinity of a finite subregion that has different elastic properties than those of the surrounding material so the simple superposition procedure used in Sect. 4.8 for an infinite homogeneous region is not applicable. In what follows we will take V_1 as the finite subregion and V_2 as the infinite plane. The interface between these two subregions is again $L_{int} = L_1$.

Proceeding as before, we write (8.165) for a point $\boldsymbol{\xi}$ inside subregion 1 as

$$t_i(\boldsymbol{\xi}) = \int_{S_1+L_1} t_j(\boldsymbol{x}) K_{ij}(\boldsymbol{x}, \boldsymbol{\xi}) \, \mathrm{d}S(\boldsymbol{x}) - \int_{S_1+L_1} u_j(\boldsymbol{x}) H_{ij}(\boldsymbol{x}, \boldsymbol{\xi}) \, \mathrm{d}S(\boldsymbol{x}) \tag{8.182}$$

where $\boldsymbol{\xi} \in V_1$. However, in place of (8.166) we write

$$0 = \int_{S_2+L_2} t_j(\boldsymbol{x}) K_{ij}(\boldsymbol{x}, \boldsymbol{\xi}) \, \mathrm{d}S(\boldsymbol{x}) - \int_{S_2+L_2} u_j(\boldsymbol{x}) H_{ij}(\boldsymbol{x}, \boldsymbol{\xi}) \, \mathrm{d}S(\boldsymbol{x}) + \sigma_{ij}^{(\infty)} n_j(\boldsymbol{\xi}) \tag{8.183}$$

in which $\boldsymbol{\xi} \in V_1$ and $n_j(\boldsymbol{\xi})$ are the components of the normal vector associated with the line on which point $\boldsymbol{\xi}$ is located.

Letting

$$t_i^{(\infty)}(\boldsymbol{\xi}) = \sigma_{ij}^{(\infty)} n_j(\boldsymbol{\xi}) \tag{8.184}$$

the same procedure that was used to obtain (8.178) can be followed to write

$$\begin{aligned} \frac{4\pi(1-\nu_1)}{2\mu_1} t_i(\boldsymbol{\xi}) &= \sum_{k=1}^{2} \frac{4\pi(1-\nu_k)}{2\mu_k} \left[\int_{S_k} t_j(\boldsymbol{x}) K_{ij}^{(k)}(\boldsymbol{x}, \boldsymbol{\xi}) \, \mathrm{d}S(\boldsymbol{x}) - \int_{S_k} u_j(\boldsymbol{x}) H_{ij}^{(k)}(\boldsymbol{x}, \boldsymbol{\xi}) \, \mathrm{d}S(\boldsymbol{x}) \right] \\ &+ \int_{L_{int}} t_j(\boldsymbol{x}) P_{ij}(\boldsymbol{x}, \boldsymbol{\xi}) \, \mathrm{d}S(\boldsymbol{x}) + \int_{L_{int}} \Delta u_j(\boldsymbol{x}) Q_{ij}(\boldsymbol{x}, \boldsymbol{\xi}) \, \mathrm{d}S(\boldsymbol{x}) + \frac{4\pi(1-\nu_2)}{2\mu_2} t_i^{(\infty)}(\boldsymbol{\xi}) \end{aligned} \tag{8.185}$$

A similar equation can be derived for the case that the load point $\boldsymbol{\xi}$ is inside subregion 2 (the infinite plane). The modified version of (8.180) is:

$$\begin{aligned} A(\boldsymbol{\xi}) t_i(\boldsymbol{\xi}) &= \sum_{k=1}^{2} \frac{4\pi(1-\nu_k)}{2\mu_k} \left[\int_{S_k} t_j(\boldsymbol{x}) K_{ij}^{(k)}(\boldsymbol{x}, \boldsymbol{\xi}) \, \mathrm{d}S(\boldsymbol{x}) - \int_{S_k} u_j(\boldsymbol{x}) H_{ij}^{(k)}(\boldsymbol{x}, \boldsymbol{\xi}) \, \mathrm{d}S(\boldsymbol{x}) \right] \\ &+ \int_{L_{int}} t_j(\boldsymbol{x}) P_{ij}(\boldsymbol{x}, \boldsymbol{\xi}) \, \mathrm{d}S(\boldsymbol{x}) + \int_{L_{int}} \Delta u_j(\boldsymbol{x}) Q_{ij}(\boldsymbol{x}, \boldsymbol{\xi}) \, \mathrm{d}S(\boldsymbol{x}) \\ &+ \frac{4\pi(1-\nu_2)}{2\mu_2} t_i^{(\infty)}(\boldsymbol{\xi}) \end{aligned} \tag{8.186}$$

in which $A(\boldsymbol{\xi})$ is again given by (8.179).

8.8.4 Numerical Implementation

Discretization of (8.180) and (8.186) is accomplished in the same way as was done for the other boundary element methods discussed in this book. The calculations are closely similar to those in Sect. 8.7 because the same kernels $K_{ij}(\boldsymbol{x},\boldsymbol{\xi})$ and $H_{ij}(\boldsymbol{x},\boldsymbol{\xi})$ are involved; the kernel $Q_{ij}(\boldsymbol{x},\boldsymbol{\xi})$ is a constant multiple of $H_{ij}(\boldsymbol{x},\boldsymbol{\xi})$ so it is essentially the same. It therefore follows that a numerical procedure for piecewise homogeneous regions can developed in a straightforward manner from the direct boundary integral method based on Somigliana's formula for the tractions.

As an example of a numerical implementation of (8.186) we will assume that the boundary and interface parameters are piecewise constant and that there are N_1 elements along boundary S_1, and N_{int} elements along the interface between subregions V_1 and V_2. Subregion V_2 is infinite and we assume without loss of generality that there is no boundary S_2 for this problem.

The same approach used to write (8.153) can be followed to obtain the discretized form of (8.186). After discretization has been done and the integrations over the elements have been performed, point $\boldsymbol{\xi}$ is moved from inside region V_1 in the direction of the normal to boundary S_1 or L_{int}. Setting $\boldsymbol{x}=\overset{j}{\boldsymbol{x}}$ and $\boldsymbol{\xi}=\overset{i}{\boldsymbol{x}}$ we have

$$
\begin{aligned}
A(\overset{i}{\boldsymbol{x}})t_x(\overset{i}{\boldsymbol{x}}) ={}& \frac{4\pi(1-\nu_1)}{2\mu_1}\left\{\sum_{j=1}^{N_1}\Big[\mathbb{K}^{(1)}_{xx}(\overset{j}{\boldsymbol{x}},\overset{i}{\boldsymbol{x}})\overset{j}{t}_x+\mathbb{K}^{(1)}_{xy}(\overset{j}{\boldsymbol{x}},\overset{i}{\boldsymbol{x}})\overset{j}{t}_y\Big]-\sum_{j=1}^{N_1}\Big[\mathbb{H}^{(1)}_{xx}(\overset{j}{\boldsymbol{x}},\overset{i}{\boldsymbol{x}})\overset{j}{u}_x+\mathbb{H}^{(1)}_{xy}(\overset{j}{\boldsymbol{x}},\overset{i}{\boldsymbol{x}})\overset{j}{u}_y\Big]\right\}\\
&+\sum_{j=N_1+1}^{N}\Big[\mathbb{P}_{xx}(\overset{j}{\boldsymbol{x}},\overset{i}{\boldsymbol{x}})\overset{j}{t}_x+\mathbb{P}_{xy}(\overset{j}{\boldsymbol{x}},\overset{i}{\boldsymbol{x}})\overset{j}{t}_y\Big]+\sum_{j=N_1+1}^{N}\Big[\mathbb{Q}_{xx}(\overset{j}{\boldsymbol{x}},\overset{i}{\boldsymbol{x}})\Delta\overset{j}{u}_x+\mathbb{Q}_{xy}(\overset{j}{\boldsymbol{x}},\overset{i}{\boldsymbol{x}})\Delta\overset{j}{u}_y\Big]\\
&+\frac{4\pi(1-\nu_2)}{2\mu_2}t_x^{(\infty)}(\overset{i}{\boldsymbol{x}})
\end{aligned}
\tag{8.187}
$$

$$
\begin{aligned}
A(\overset{i}{\boldsymbol{x}})t_y(\overset{i}{\boldsymbol{x}}) ={}& \frac{4\pi(1-\nu_1)}{2\mu_1}\left\{\sum_{j=1}^{N_1}\Big[\mathbb{K}^{(1)}_{yx}(\overset{j}{\boldsymbol{x}},\overset{i}{\boldsymbol{x}})\overset{j}{t}_x+\mathbb{K}^{(1)}_{yy}(\overset{j}{\boldsymbol{x}},\overset{i}{\boldsymbol{x}})\overset{j}{t}_y\Big]-\sum_{j=1}^{N_1}\Big[\mathbb{H}^{(1)}_{yx}(\overset{j}{\boldsymbol{x}},\overset{i}{\boldsymbol{x}})\overset{j}{u}_x+\mathbb{H}^{(1)}_{yy}(\overset{j}{\boldsymbol{x}},\overset{i}{\boldsymbol{x}})\overset{j}{u}_y\Big]\right\}\\
&+\sum_{j=N_1+1}^{N}\Big[\mathbb{P}_{yx}(\overset{j}{\boldsymbol{x}},\overset{i}{\boldsymbol{x}})\overset{j}{t}_x+\mathbb{P}_{yy}(\overset{j}{\boldsymbol{x}},\overset{i}{\boldsymbol{x}})\overset{j}{t}_y\Big]+\sum_{j=N_1+1}^{N}\Big[\mathbb{Q}_{yx}(\overset{j}{\boldsymbol{x}},\overset{i}{\boldsymbol{x}})\Delta\overset{j}{u}_x+\mathbb{Q}_{yy}(\overset{j}{\boldsymbol{x}},\overset{i}{\boldsymbol{x}})\Delta\overset{j}{u}_y\Big]\\
&+\frac{4\pi(1-\nu_2)}{2\mu_2}t_y^{(\infty)}(\overset{i}{\boldsymbol{x}})
\end{aligned}
$$

in which $N=N_1+N_{int}$. Because the load point $\boldsymbol{\xi}$ was moved to the boundary from inside region V_1, $A(\overset{i}{\boldsymbol{x}})=4\pi(1-\nu_1)/(2\mu_1)$ so (8.187) can also be written as

$$
\begin{aligned}
t_x(\overset{i}{\boldsymbol{x}}) ={}& \sum_{j=1}^{N_1}\Big[\mathbb{K}^{(1)}_{xx}(\overset{j}{\boldsymbol{x}},\overset{i}{\boldsymbol{x}})\overset{j}{t}_x+\mathbb{K}^{(1)}_{xy}(\overset{j}{\boldsymbol{x}},\overset{i}{\boldsymbol{x}})\overset{j}{t}_y\Big]-\sum_{j=1}^{N_1}\Big[\mathbb{H}^{(1)}_{xx}(\overset{j}{\boldsymbol{x}},\overset{i}{\boldsymbol{x}})\overset{j}{u}_x+\mathbb{H}^{(1)}_{xy}(\overset{j}{\boldsymbol{x}},\overset{i}{\boldsymbol{x}})\overset{j}{u}_y\Big]\\
&+\frac{2\mu_1}{4\pi(1-\nu_1)}\sum_{j=N_1+1}^{N}\Big[\mathbb{P}_{xx}(\overset{j}{\boldsymbol{x}},\overset{i}{\boldsymbol{x}})\overset{j}{t}_x+\mathbb{P}_{xy}(\overset{j}{\boldsymbol{x}},\overset{i}{\boldsymbol{x}})\overset{j}{t}_y\Big]+\sum_{j=N_1+1}^{N}\Big[\mathbb{H}^{(1)}_{xx}(\overset{j}{\boldsymbol{x}},\overset{i}{\boldsymbol{x}})\Delta\overset{j}{u}_x+\mathbb{H}^{(1)}_{xy}(\overset{j}{\boldsymbol{x}},\overset{i}{\boldsymbol{x}})\Delta\overset{j}{u}_y\Big]
\end{aligned}
$$

$$
+\frac{2\mu_1}{4\pi(1-\nu_1)}\frac{4\pi(1-\nu_2)}{2\mu_2}t_x^{(\infty)}(\overset{i}{\boldsymbol{x}})
$$

(8.188)

$$
\begin{aligned}
t_y(\overset{i}{\boldsymbol{x}}) =& \sum_{j=1}^{N_1}\Big[\mathbb{K}_{yx}^{(1)}(\overset{j}{\boldsymbol{x}},\overset{i}{\boldsymbol{x}})\overset{j}{t}_x+\mathbb{K}_{yy}^{(1)}(\overset{j}{\boldsymbol{x}},\overset{i}{\boldsymbol{x}})\overset{j}{t}_y\Big]-\sum_{j=1}^{N_1}\Big[\mathbb{H}_{yx}^{(1)}(\overset{j}{\boldsymbol{x}},\overset{i}{\boldsymbol{x}})\overset{j}{u}_x+\mathbb{H}_{yy}^{(1)}(\overset{j}{\boldsymbol{x}},\overset{i}{\boldsymbol{x}})\overset{j}{u}_y\Big]\\
&+\frac{2\mu_1}{4\pi(1-\nu_1)}\sum_{j=N_1+1}^{N}\Big[\mathbb{P}_{yx}(\overset{j}{\boldsymbol{x}},\overset{i}{\boldsymbol{x}})\overset{j}{t}_x+\mathbb{P}_{yy}(\overset{j}{\boldsymbol{x}},\overset{i}{\boldsymbol{x}})\overset{j}{t}_y\Big]+\sum_{j=N_1+1}^{N}\Big[\mathbb{H}_{yx}^{(1)}(\overset{j}{\boldsymbol{x}},\overset{i}{\boldsymbol{x}})\Delta\overset{j}{u}_x+\mathbb{H}_{yy}^{(1)}(\overset{j}{\boldsymbol{x}},\overset{i}{\boldsymbol{x}})\Delta\overset{j}{u}_y\Big]\\
&+\frac{2\mu_1}{4\pi(1-\nu_1)}\frac{4\pi(1-\nu_2)}{2\mu_2}t_y^{(\infty)}(\overset{i}{\boldsymbol{x}})
\end{aligned}
$$

The components of the unit normal at point $\overset{i}{\boldsymbol{x}}$ can be written in terms of the angle of inclination of element i as $n_x(\overset{i}{\boldsymbol{x}})=-\sin\overset{i}{\beta}$ and $n_y(\overset{i}{\boldsymbol{x}})=\cos\overset{i}{\beta}$, so that $t_x^{(\infty)}(\overset{i}{\boldsymbol{x}})$ and $t_y^{(\infty)}(\overset{i}{\boldsymbol{x}})$ can be written as

$$
t_x^{(\infty)}(\overset{i}{\boldsymbol{x}})=-\sigma_{xx}^{(\infty)}\sin\overset{i}{\beta}+\sigma_{xy}^{(\infty)}\cos\overset{i}{\beta}
$$

(8.189)

$$
t_y^{(\infty)}(\overset{i}{\boldsymbol{x}})=-\sigma_{xy}^{(\infty)}\sin\overset{i}{\beta}+\sigma_{yy}^{(\infty)}\cos\overset{i}{\beta}
$$

Setting $t_x(\overset{i}{\boldsymbol{x}})=\overset{i}{t}_x$, $t_y(\overset{i}{\boldsymbol{x}})=\overset{i}{t}_y$, writing $\mathbb{K}_{xx}^{(1)}(\overset{j}{\boldsymbol{x}},\overset{i}{\boldsymbol{x}})=\overset{ij}{\mathbb{K}}_{xx}$, etc., and $\mathbb{H}_{xx}^{(1)}(\overset{j}{\boldsymbol{x}},\overset{i}{\boldsymbol{x}})=\overset{ij}{\mathbb{H}}_{xx}$, etc., and using (8.189), equations (8.188) become

$$
\begin{aligned}
\overset{i}{t}_x =& \sum_{j=1}^{N_1}\Big[\overset{ij}{\mathbb{K}}_{xx}\overset{j}{t}_x+\overset{ij}{\mathbb{K}}_{xy}\overset{j}{t}_y\Big]-\sum_{j=1}^{N_1}\Big[\overset{ij}{\mathbb{H}}_{xx}\overset{j}{u}_x+\overset{ij}{\mathbb{H}}_{xy}\overset{j}{u}_y\Big]\\
&+\frac{\mu_1}{2\pi(1-\nu_1)}\sum_{j=N_1+1}^{N}\Big[\overset{ij}{\mathbb{P}}_{xx}\overset{j}{t}_x+\overset{ij}{\mathbb{P}}_{xy}\overset{j}{t}_y\Big]+\sum_{j=N_1+1}^{N}\Big[\overset{ij}{\mathbb{H}}_{xx}\Delta\overset{j}{u}_x+\overset{ij}{\mathbb{H}}_{xy}\Delta\overset{j}{u}_y\Big]\\
&+\frac{\mu_1(1-\nu_2)}{\mu_2(1-\nu_1)}\Big[-\sigma_{xx}^{(\infty)}\sin\overset{i}{\beta}+\sigma_{xy}^{(\infty)}\cos\overset{i}{\beta}\Big]
\end{aligned}
$$

(8.190)

$$
\begin{aligned}
\overset{i}{t}_y =& \sum_{j=1}^{N_1}\Big[\overset{ij}{\mathbb{K}}_{yx}\overset{j}{t}_x+\overset{ij}{\mathbb{K}}_{yy}\overset{j}{t}_y\Big]-\sum_{j=1}^{N_1}\Big[\overset{ij}{\mathbb{H}}_{yx}\overset{j}{u}_x+\overset{ij}{\mathbb{H}}_{yy}\overset{j}{u}_y\Big]\\
&+\frac{\mu_1}{2\pi(1-\nu_1)}\sum_{j=N_1+1}^{N}\Big[\overset{ij}{\mathbb{P}}_{yx}\overset{j}{t}_x+\overset{ij}{\mathbb{P}}_{yy}\overset{j}{t}_y\Big]+\sum_{j=N_1+1}^{N}\Big[\overset{ij}{\mathbb{H}}_{yx}\Delta\overset{j}{u}_x+\overset{ij}{\mathbb{H}}_{yy}\Delta\overset{j}{u}_y\Big]\\
&+\frac{\mu_1(1-\nu_2)}{\mu_2(1-\nu_1)}\Big[-\sigma_{xy}^{(\infty)}\sin\overset{i}{\beta}+\sigma_{yy}^{(\infty)}\cos\overset{i}{\beta}\Big]
\end{aligned}
$$

which, with $i = 1$ to N, represent a system of $2N$ algebraic equations in $2N$ unknowns:

$$\sum_{j=1}^{N} \overset{ij}{A}_{xx} \overset{j}{X}_x + \sum_{j=1}^{N} \overset{ij}{A}_{xy} \overset{j}{X}_y = \overset{i}{b}_x$$

$$\sum_{j=1}^{N} \overset{ij}{A}_{yx} \overset{j}{X}_x + \sum_{j=1}^{N} \overset{ij}{A}_{yy} \overset{j}{X}_y = \overset{i}{b}_y \tag{8.191}$$

The quantities $\overset{i}{b}_x$ and $\overset{i}{b}_y$ in these equations are the sums of the terms involving the components of stress at infinity and certain linear combinations of the known boundary parameters, and $\overset{ij}{A}_{xx}$, etc., are the boundary influence coefficients.

If we suppose that the tractions are prescribed on S_1, i.e., if

$$\overset{j}{t}_x = (\overset{j}{t}_x)_0$$

$$\overset{j}{t}_y = (\overset{j}{t}_y)_0 \tag{8.192}$$

then the unknown boundary parameters $\overset{j}{X}_x$ and $\overset{j}{X}_y$ are the displacements $\overset{j}{u}_x$, $\overset{j}{u}_y$ on S_1 and the tractions $\overset{j}{t}_x$, $\overset{j}{t}_y$ and displacement discontinuities $\Delta\overset{j}{u}_x$, $\Delta\overset{j}{u}_y$ on L_{int}. As a specific example, we let N_{dd} of the N_{int} interface elements represent displacement discontinuities. The parameters $\overset{j}{X}_x$ and $\overset{j}{X}_y$ are then:

$$\overset{j}{X}_x = \begin{cases} \overset{j}{u}_x & j = 1 \text{ to } N_1 \\ \overset{j}{t}_x & j = N_1 + 1 \text{ to } M \\ \Delta\overset{j}{u}_x & j = M + 1 \text{ to } N \end{cases}$$

$$\overset{j}{X}_y = \begin{cases} \overset{j}{u}_y & j = 1 \text{ to } N_1 \\ \overset{j}{t}_y & j = N_1 + 1 \text{ to } M \\ \Delta\overset{j}{u}_y & j = M + 1 \text{ to } N \end{cases} \tag{8.193}$$

in which we have set $M = N_1 + N_{int} - N_{dd}$ and $N = N_1 + N_{int}$.

The values of $\overset{i}{b}_x$ and $\overset{i}{b}_y$ in (8.191) are

$$\overset{i}{b}_x = \sum_{j=1}^{N_1}\Big[\big(\overset{ij}{\mathbb{K}}_{xx} - \overset{ij}{\mathbb{I}}\big)(\overset{j}{t}_x)_0 + \overset{ij}{\mathbb{K}}_{xy}(\overset{j}{t}_y)_0\Big] + \frac{\mu_1(1-\nu_2)}{\mu_2(1-\nu_1)}\Big[-\sigma_{xx}^{(\infty)}\sin\overset{i}{\beta} + \sigma_{xy}^{(\infty)}\cos\overset{i}{\beta}\Big]$$

$$\overset{i}{b}_y = \sum_{j=1}^{N_1}\Big[\overset{ij}{\mathbb{K}}_{yx}(\overset{j}{t}_x)_0 + \big(\overset{ij}{\mathbb{K}}_{yy} - \overset{ij}{\mathbb{I}}\big)(\overset{j}{t}_y)_0\Big] + \frac{\mu_1(1-\nu_2)}{\mu_2(1-\nu_1)}\Big[-\sigma_{xy}^{(\infty)}\sin\overset{i}{\beta} + \sigma_{yy}^{(\infty)}\cos\overset{i}{\beta}\Big] \tag{8.194}$$

for $i = 1$ to N_1 and, assuming that the displacement discontinuity elements are traction-free,

$$\overset{i}{b}_x = \frac{\mu_1(1-\nu_2)}{\mu_2(1-\nu_1)}\Big[-\sigma_{xx}^{(\infty)}\sin\overset{i}{\beta} + \sigma_{xy}^{(\infty)}\cos\overset{i}{\beta}\Big]$$

$$\overset{i}{b}_y = \frac{\mu_1(1-\nu_2)}{\mu_2(1-\nu_1)}\Big[-\sigma_{xy}^{(\infty)}\sin\overset{i}{\beta} + \sigma_{yy}^{(\infty)}\cos\overset{i}{\beta}\Big] \tag{8.195}$$

for $i = N_1 + 1$ to N.[4]

Finally, the influence coefficients in (8.191) are, for $i = 1$ to N_1,

$$\overset{ij}{A}_{xx} = \begin{cases} \overset{ij}{\mathbb{H}}_{xx} & j = 1 \text{ to } N_1 \\ -\dfrac{\mu_1}{2\pi(1-\nu_1)}\overset{ij}{\mathbb{P}}_{xx} & j = N_1 + 1 \text{ to } M \\ -\overset{ij}{\mathbb{H}}_{xx} & j = M + 1 \text{ to } N \end{cases} \tag{8.196}$$

with similar expressions for $\overset{ij}{A}_{xy}$, $\overset{ij}{A}_{yx}$, and $\overset{ij}{A}_{yy}$. For elements on the interface, $i = N_1 + 1$ to N,

$$\overset{ij}{A}_{xx} = \begin{cases} \overset{ij}{\mathbb{H}}_{xx} & j = 1 \text{ to } N_1 \\ -\dfrac{\mu_1}{2\pi(1-\nu_1)}\overset{ij}{\mathbb{P}}_{xx} + \overset{ij}{\mathbb{I}} & j = N_1 + 1 \text{ to } M \\ -\overset{ij}{\mathbb{H}}_{xx} & j = M + 1 \text{ to } N \end{cases} \tag{8.197}$$

again with similar expressions for the other coefficients. The value of $\overset{ij}{A}_{xx}$ in (8.197) takes into account that the jth element is located on a bonded portion of the interface.

---

[4] One could also assume a linear spring-type relationship between displacement discontinuities and tractions, in which case $\Delta\overset{j}{u}_x$ and $\Delta\overset{j}{u}_y$ in (8.193) would be replaced by $\Delta\overset{j}{u}_x = \overset{j}{t}_x/\overset{j}{k}_x$ and $\Delta\overset{j}{u}_y = \overset{j}{t}_y/\overset{j}{k}_y$, where $\overset{j}{k}_x$ and $\overset{j}{k}_y$ are the spring constants for element j.

8.8.5 *Example 1: A Circular Elastic Disk in an Elastic Plane with Different Material Properties*

As a simple example of the foregoing numerical procedure we consider the case of a circular disk of radius $r = a$ and elastic constants ν_1, μ_1 in an infinite plane with elastic constants ν_2, μ_2 under an applied stress $\sigma_{xx}^{(\infty)} = p$, $\sigma_{yy}^{(\infty)} = \sigma_{xy}^{(\infty)} = 0$ at infinity. The analytical solution for the tractions σ_{rr} and $\sigma_{r\theta}$ on the periphery of the disk is given as (Muskhelishvili 1963)

$$\sigma_{rr} = p\Big[\frac{\mu_1(1-\nu_2)}{\mu_1 + (1-2\nu_1)\mu_2} + \frac{2\mu_1(1-\nu_2)}{(3-4\nu_2)\mu_1 + \mu_2}\cos 2\theta\Big]$$

$$\sigma_{r\theta} = -p\frac{2\mu_1(1-\nu_2)}{(3-4\nu_2)\mu_1 + \mu_2}\sin 2\theta \tag{8.198}$$

These expressions are plotted in Fig. 8.19 for the case that $\nu_1 = \nu_2 = 0.25$ and $\mu_1/\mu_2 = 5$. The numerical results shown in the figure were obtained by approximating a symmetric one-quarter of the circular boundary by 30 straight line elements. The numerical values of the tractions can be seen to be in good agreement with the exact solution.

An interesting and well known (Eshelby 1957) aspect of this simple problem is that the stress state is constant inside the disk, i.e., for $r < a$

$$\sigma_{xx} = p\Big[\frac{\mu_1(1-\nu_2)}{\mu_1 + (1-2\nu_1)\mu_2} + \frac{2\mu_1(1-\nu_2)}{(3-4\nu_2)\mu_1 + \mu_2}\Big]$$

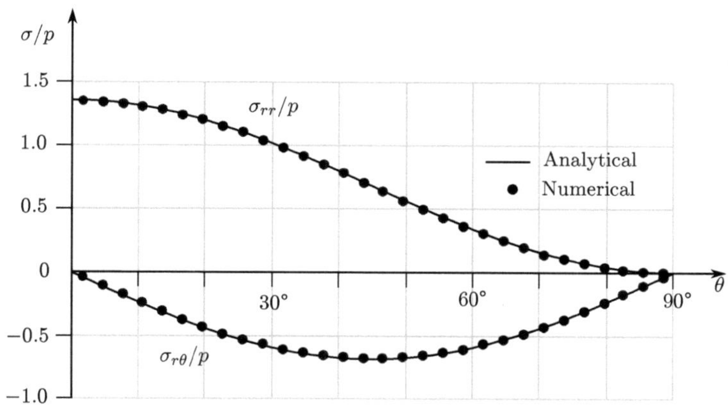

Fig. 8.19 Boundary tractions for an elastic disk in an elastic plane under a stress $\sigma_{xx} = p$ at infinity

$$\sigma_{yy} = p\Big[\frac{\mu_1(1-\nu_2)}{\mu_1 + (1-2\nu_1)\mu_2} - \frac{2\mu_1(1-\nu_2)}{(3-4\nu_2)\mu_1 + \mu_2}\Big] \tag{8.199}$$

$$\sigma_{xy} = 0$$

For the elastic parameters given above, $\sigma_{xx}/p = 1.3636$ and $\sigma_{yy} = 0$. The boundary element solution gives values of σ_{xx}/p slightly less than this, ranging from 1.352 to 1.362 for several dozen sample points inside the disk.

8.8.6 Example 2: A Circular Elastic Disk with an Interface Crack

A problem for a circular elastic disk with an interface crack is depicted in Fig. 8.20. The disk has a traction-free crack along its interface, $-30^\circ \le \theta \le 30^\circ$, $r = a$, but is otherwise perfectly bonded to the surrounding matrix. A uniaxial stress σ_0 is applied at infinity in a direction of 30° from the x axis. We seek results for the r, θ components of displacement discontinuity along the crack and the tractions along the bonded portion of the interface.

This problem is a special case of a problem considered by Toya (1974) for a glass inclusion[5] ($\nu_1 = 0.22$, $\mu_1 = 44.2$ GN/m^2) in an epoxy matrix ($\nu_2 = 0.35$, $\mu_2 = 2.39$ GN/m^2) subjected to a uniaxial tensile stress σ_0 applied at an angle φ from the x axis. The uniaxial tension at infinity is achieved by taking the field stresses as $\sigma_{xx} = 0.75\sigma_0$, $\sigma_{yy} = 0.25\sigma_0$, and $\sigma_{xy} = 0.433\sigma_0$. A numerical solution for the problem was found by approximating the circular interface by 120 straight line elements, 20 of which represented the crack. The computed displacement discontinuities Δu_r and Δu_θ are plotted in Fig. 8.21 for $-30^\circ < \theta < 30^\circ$ and the tractions σ_{rr} and $\sigma_{r\theta}$ in

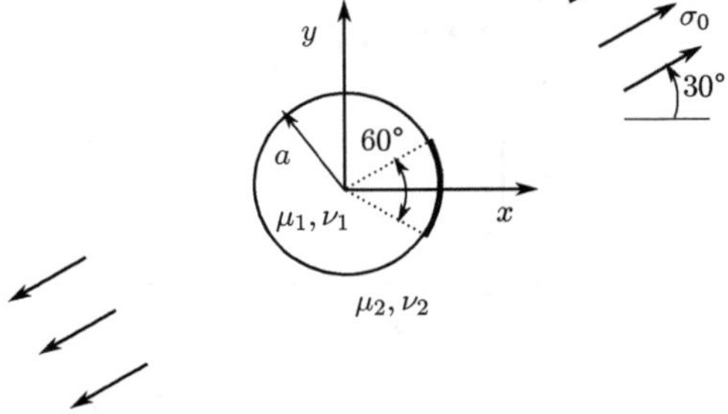

Fig. 8.20 Elastic disk with interface crack

[5] We preserve Toya's terminology but note that in the literature "inclusion" usually refers to an embedded object that is subjected to a deformation (e.g., a thermal expansion or a so-called eigenstrain), in the absence of any external loading. This usage was originally proposed by Eshelby (1957) and is described in detail by Mura (1987).

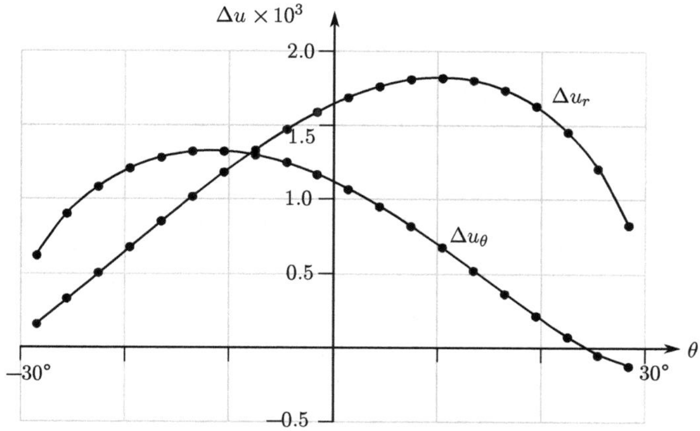

Fig. 8.21 Radial and shear displacement discontinuities for $-30^\circ < \theta < 30^\circ$

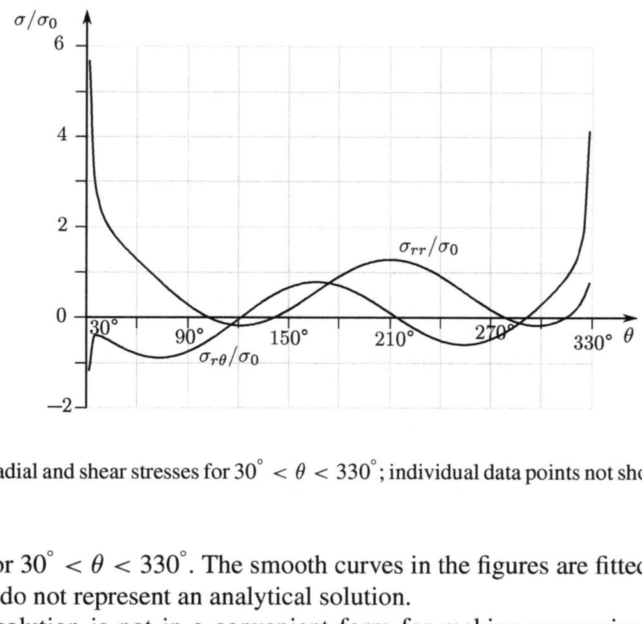

Fig. 8.22 Radial and shear stresses for $30^\circ < \theta < 330^\circ$; individual data points not shown for clarity

Fig. 8.22 for $30^\circ < \theta < 330^\circ$. The smooth curves in the figures are fitted to the data points and do not represent an analytical solution.

Toya's solution is not in a convenient form for making comparisons with our numerical results but a comparison can be made to another numerical solution obtained by the authors (Crouch and Mogilevskaya 2006) using Fourier series approximations to the interface tractions and displacement discontinuities. The results plotted in Figs. 8.21 and 8.22 are in good agreement with that numerical solution.

8.9 Combined Methods

Many practical problems in solid mechanics involve loading of bodies containing cracks. The displacement discontinuity method is capable of modeling such problems but, by itself, is not ideally suited for the purpose. As noted previously, if the body in question contains one or more holes (i.e., if the region of interest is multiply connected), some provision must be made to prevent movement of the interior(s) of the hole(s). Although this can be accomplished by artificially introducing additional displacement discontinuity elements inside a hole boundary and fixing their displacements, this measure can lead to poorly conditioned systems of algebraic equations and can not always be relied upon to produce satisfactory results. In particular, an iterative solution of the system of equations may not be possible and, as a consequence, problems involving crack closure and frictional sliding become difficult to treat. A better approach is to combine the displacement discontinuity method with either the stress discontinuity method or the direct boundary integral method. In this way, displacement discontinuity elements can be used to their best advantage—the modeling of cracks—while the other boundary elements are used for the remainder of the problem.

We will describe two implementations of this "hybrid" approach in this section, one an indirect method and the other a direct method. For simplicity, we will assume in the following discussion that all of the boundary parameters are piecewise constant along the elements. Higher order implementations (e.g., piecewise quadratic variations) can be developed in a similar manner, and are especially straightforward in the case of the combined stress and displacement discontinuity methods because these methods both require the use of discontinuous elements. In the case of the direct boundary integral method discontinuous elements are only needed along the crack and at those places where a crack meets another boundary; standard higher order elements (as in Sects. 7.6 to 7.11) can be used elsewhere.

To explain the hybrid approaches we will consider the specific two-dimensional problem of a hole with boundary S in an infinite, isotropic elastic region V subjected to a normal stress $\sigma_n = -p$. The hole is intersected by a crack Γ, which is also subjected to a normal stress $\sigma_n = -p$. We seek approximations to the distribution of displacement discontinuity along the crack as well as to the displacements and stresses in the vicinity of the hole.

8.9.1 Combined Stress Discontinuity and Displacement Discontinuity Methods

For a numerical solution to this problem we will suppose that the boundary S is approximated by N_S straight line stress discontinuity elements, joined end to end in the usual way. In addition, the crack Γ is approximated by N_Γ displacement discontinuity elements. A typical element i has coordinates $\overset{i}{x}$, $\overset{i}{y}$, length $2\overset{i}{a}$, and inclination $\overset{i}{\beta}$ relative to the global x, y coordinate system. Because we are assuming that the boundary parameters are piecewise constant, we will also suppose that all of the elements have approximately the same length. With reference to a local $\bar{x}$, $\bar{y}$ coordinate system centered at the midpoint of the ith element, the boundary conditions for the problem can be written as follows:

$$\overset{i}{t}_{\bar{x}} = 0$$
$$\overset{i}{t}_{\bar{y}} = -p \tag{8.200}$$

for $i = 1$ to $N_S + N_\Gamma$. The reader should recall that the negative surface of the crack, $\bar{y} = 0_-$, is taken as a reference for the displacement discontinuity elements. Also, boundary S is traversed in a counterclockwise sense with $\bar{x}$ pointing in the direction of traversal and $\bar{y}$ pointing in the direction of the local outward normal.

By combining (5.2) and (6.3) we write the tractions $\overset{i}{t}_{\bar{x}}$ and $\overset{i}{t}_{\bar{y}}$ at the midpoint of element i in terms of the N_S applied tractions $\overset{j}{P}_{\bar{x}}$ and $\overset{j}{P}_{\bar{y}}$, and the N_Γ displacement discontinuities $\overset{j}{D}_{\bar{x}}$ and $\overset{j}{D}_{\bar{y}}$, as follows:

$$\overset{i}{t}_{\bar{x}} = \left(\sum_{j=1}^{N_S} \overset{ij}{A}_{\bar{x}\bar{x}} \overset{j}{P}_{\bar{x}} + \sum_{j=1}^{N_S} \overset{ij}{A}_{\bar{x}\bar{y}} \overset{j}{P}_{\bar{y}}\right) + \left(\sum_{j=N_S+1}^{N} \overset{ij}{A}_{\bar{x}\bar{x}} \overset{j}{D}_{\bar{x}} + \sum_{j=N_S+1}^{N} \overset{ij}{A}_{\bar{x}\bar{y}} \overset{j}{D}_{\bar{y}}\right)$$
$$\overset{i}{t}_{\bar{y}} = \left(\sum_{j=1}^{N_S} \overset{ij}{A}_{\bar{y}\bar{x}} \overset{j}{P}_{\bar{x}} + \sum_{j=1}^{N_S} \overset{ij}{A}_{\bar{y}\bar{y}} \overset{j}{P}_{\bar{y}}\right) + \left(\sum_{j=N_S+1}^{N} \overset{ij}{A}_{\bar{y}\bar{x}} \overset{j}{D}_{\bar{x}} + \sum_{j=N_S+1}^{N} \overset{ij}{A}_{\bar{y}\bar{y}} \overset{j}{D}_{\bar{y}}\right) \tag{8.201}$$

where $N = N_S + N_\Gamma$. The influence coefficients $\overset{ij}{A}_{\bar{x}\bar{x}}$, etc., in these expressions are given by (5.21) for the stress discontinuity elements and by (6.15) for the displacement discontinuity elements.

Letting $\overset{j}{X}_{\bar{x}}$ and $\overset{j}{X}_{\bar{y}}$ stand for either $\overset{j}{P}_{\bar{x}}$ or $\overset{j}{D}_{\bar{x}}$ and either $\overset{j}{P}_{\bar{y}}$ or $\overset{j}{D}_{\bar{y}}$, as appropriate, and using (8.200), we obtain the following system of $2N \times 2N$ algebraic equations

$$\sum_{j=1}^{N} \overset{ij}{A}_{\bar{x}\bar{x}} \overset{j}{X}_{\bar{x}} + \sum_{j=1}^{N} \overset{ij}{A}_{\bar{x}\bar{y}} \overset{j}{X}_{\bar{y}} = 0$$
$$\sum_{j=1}^{N} \overset{ij}{A}_{\bar{y}\bar{x}} \overset{j}{X}_{\bar{x}} + \sum_{j=1}^{N} \overset{ij}{A}_{\bar{y}\bar{y}} \overset{j}{X}_{\bar{y}} = -p \tag{8.202}$$

for $i = 1$ to N.

More generally, if a problem involves both traction and displacement boundary conditions a numerical solution can be obtained from a system of algebraic equations of the form

$$\sum_{j=1}^{N} \overset{ij}{C}_{\bar{x}\bar{x}} \overset{j}{X}_{\bar{x}} + \sum_{j=1}^{N} \overset{ij}{C}_{\bar{x}\bar{y}} \overset{j}{X}_{\bar{y}} = \overset{i}{b}_{\bar{x}}$$

$$\sum_{j=1}^{N} \overset{ij}{C}_{\bar{y}\bar{x}} \overset{j}{X}_{\bar{x}} + \sum_{j=1}^{N} \overset{ij}{C}_{\bar{y}\bar{y}} \overset{j}{X}_{\bar{y}} = \overset{i}{b}_{\bar{y}} \tag{8.203}$$

Now, $\overset{ij}{C}_{\bar{x}\bar{x}}$, etc., represent either the traction coefficients $\overset{ij}{A}_{\bar{x}\bar{x}}$, etc., from (5.21) or (6.15), or the displacement coefficients $\overset{ij}{B}_{\bar{x}\bar{x}}$, etc., from (5.24) or (6.18).

8.9.2 *Combined Direct Boundary Integral and Displacement Discontinuity Methods*

The displacements $u_i(\boldsymbol{\xi})$ at a point $\boldsymbol{\xi}$ that is not initially on either boundary S or Γ of region V are given by combining (7.1) and (8.138) to write

$$u_i(\boldsymbol{\xi}) = \int_S t_j(\boldsymbol{x}) U_{ij}(\boldsymbol{x}, \boldsymbol{\xi})\, \mathrm{d}S(\boldsymbol{x}) - \int_S u_j(\boldsymbol{x}) T_{ij}(\boldsymbol{x}, \boldsymbol{\xi})\, \mathrm{d}S(\boldsymbol{x}) + \int_\Gamma \Delta u_j(\boldsymbol{x}) T_{ij}(\boldsymbol{x}, \boldsymbol{\xi})\, \mathrm{d}S(\boldsymbol{x}) \tag{8.204}$$

Similarly, (8.146) and (8.140) can be combined to give a formula for the tractions, i.e.,

$$t_i(\boldsymbol{\xi}) = \int_S t_j(\boldsymbol{x}) K_{ij}(\boldsymbol{x}, \boldsymbol{\xi})\, \mathrm{d}S(\boldsymbol{x}) - \int_S u_j(\boldsymbol{x}) H_{ij}(\boldsymbol{x}, \boldsymbol{\xi})\, \mathrm{d}S(\boldsymbol{x}) + \int_\Gamma \Delta u_j(\boldsymbol{x}) H_{ij}(\boldsymbol{x}, \boldsymbol{\xi})\, \mathrm{d}S(\boldsymbol{x}) \tag{8.205}$$

(Recall that $t_i(\boldsymbol{\xi})$ represents the tractions along an imaginary line parallel to a local portion of boundary S or Γ. After discretizing the boundaries and carrying out the integrations pertaining to each boundary element, point $\boldsymbol{\xi}$ can be moved to the boundary).

The above equations can be discretized as explained in Sects. 7.2 and 8.6, rearranged, and put in the following forms:

$$\overset{i}{u}_x + \sum_{j=1}^{N_S} \left(\overset{ij}{\mathbb{T}}_{xx} \overset{j}{u}_x + \overset{ij}{\mathbb{T}}_{xy} \overset{j}{u}_y \right) - \sum_{j=N_S+1}^{N} \left(\overset{ij}{\mathbb{T}}_{xx} \Delta \overset{j}{u}_x + \overset{ij}{\mathbb{T}}_{xy} \Delta \overset{j}{u}_y \right) = \sum_{j=1}^{N} \left(\overset{ij}{\mathbb{U}}_{xx} \overset{j}{t}_x + \overset{ij}{\mathbb{U}}_{xy} \overset{j}{t}_y \right) \tag{8.206}$$

$$\overset{i}{u}_y + \sum_{j=1}^{N_S}\left(\overset{ij}{\mathbb{T}}_{yx}\overset{j}{u}_x + \overset{ij}{\mathbb{T}}_{yy}\overset{j}{u}_y\right) - \sum_{j=N_S+1}^{N}\left(\overset{ij}{\mathbb{T}}_{yx}\Delta\overset{j}{u}_x + \overset{ij}{\mathbb{T}}_{yy}\Delta\overset{j}{u}_y\right) = \sum_{j=1}^{N}\left(\overset{ij}{\mathbb{U}}_{yx}\overset{j}{t}_x + \overset{ij}{\mathbb{U}}_{yy}\overset{j}{t}_y\right)$$

and

$$\sum_{j=1}^{N_S}\left(\overset{ij}{\mathbb{H}}_{xx}\overset{j}{u}_x + \overset{ij}{\mathbb{H}}_{xy}\overset{j}{u}_y\right) - \sum_{j=N_S+1}^{N}\left(\overset{ij}{\mathbb{H}}_{xx}\Delta\overset{j}{u}_x + \overset{ij}{\mathbb{H}}_{xy}\Delta\overset{j}{u}_y\right) = -\overset{i}{t}_x + \sum_{j=1}^{N}\left(\overset{ij}{\mathbb{K}}_{xx}\overset{j}{t}_x + \overset{ij}{\mathbb{K}}_{xy}\overset{j}{t}_y\right)$$

$$\sum_{j=1}^{N_S}\left(\overset{ij}{\mathbb{H}}_{yx}\overset{j}{u}_x + \overset{ij}{\mathbb{H}}_{yy}\overset{j}{u}_y\right) - \sum_{j=N_S+1}^{N}\left(\overset{ij}{\mathbb{H}}_{yx}\Delta\overset{j}{u}_x + \overset{ij}{\mathbb{H}}_{yy}\Delta\overset{j}{u}_y\right) = -\overset{i}{t}_y + \sum_{j=1}^{N}\left(\overset{ij}{\mathbb{K}}_{yx}\overset{j}{t}_x + \overset{ij}{\mathbb{K}}_{yy}\overset{j}{t}_y\right) \tag{8.207}$$

where again $N = N_S + N_\Gamma$. The values of the coefficients $\overset{ij}{\mathbb{U}}_{xx}$, etc., and $\overset{ij}{\mathbb{T}}_{xx}$, etc., in (8.206) are obtained by setting $\boldsymbol{\xi} = \overset{i}{\boldsymbol{x}}$ in (7.20) and (7.21). The values of the coefficients $\overset{ij}{\mathbb{K}}_{xx}$, etc., and $\overset{ij}{\mathbb{H}}_{xx}$, etc., are similarly obtained from (7.27) and (7.28), using definitions (8.150) and (8.151).

Equations (8.206) are used when element i is located on boundary S, whereas (8.207) are used when it is located on the crack Γ. The right-hand sides of these equations involve known quantities for the problem with tractions (8.200); in terms of global coordinates these tractions are

$$\overset{i}{t}_x = -\overset{i}{t}_{\bar{y}}\sin\overset{i}{\beta} = +p\sin\overset{i}{\beta}$$

$$\overset{i}{t}_y = \overset{i}{t}_{\bar{y}}\cos\overset{i}{\beta} = -p\cos\overset{i}{\beta} \tag{8.208}$$

for $i = 1$ to N. The unknowns in the system are the displacements $\overset{i}{u}_x$ and $\overset{i}{u}_y$ for elements on the boundary S and the displacement discontinuities $\Delta\overset{i}{u}_x$ and $\Delta\overset{i}{u}_y$ for elements on the crack boundary Γ. The resulting equations can be put in the form

$$\sum_{j=1}^{N}\overset{ij}{C}_{xx}\overset{j}{X}_x + \sum_{j=1}^{N}\overset{ij}{C}_{xy}\overset{j}{X}_y = \overset{i}{b}_x$$

$$\sum_{j=1}^{N}\overset{ij}{C}_{yx}\overset{j}{X}_x + \sum_{j=1}^{N}\overset{ij}{C}_{yy}\overset{j}{X}_y = \overset{i}{b}_y \tag{8.209}$$

for $i = 1$ to $N = N_S + N_\Gamma$, where the coefficients $\overset{ij}{C}_{xx}$, etc., are the values of $\overset{ij}{\mathbb{T}}_{xx}$, etc., for $i \le N_S$, with $j = 1$ to N, and $\overset{ij}{\mathbb{H}}_{yx}$, etc., for $N_S + 1 < i \le N_S + N_\Gamma$, again with $j = 1$ to N. Also, $\overset{i}{b}_x$ and $\overset{i}{b}_y$ are the sums indicated on the right-hand sides of (8.206) and (8.207), taking note of (8.208). Equations (8.209) also apply to situations in which displacements are prescribed over some portions of S instead of tractions, provided (8.206) are suitably rearranged.

8.10 The Dual Boundary Element Method

The so-called dual boundary element method is another technique for modeling crack problems. A theoretical basis for the method was given by Hong and Chen (1988) and numerical implementations were presented by Portela et al. (1992) and Mi and Alibadi (1992). Unlike the displacement discontinuity method, the dual boundary element method discretizes *both* surfaces of a crack. One surface is modeled using Somigliana's formula for the displacements (4.328) and the other using the companion formula for the tractions (4.343). The use of two different (but related) integral representations accounts for the word "dual" in the name of the numerical method. For situations in which non-crack boundaries are also present, the Somigliana displacement equation is used in the usual way. The dual boundary element method can therefore be regarded as another "combined method" in addition to those described in Sect. 8.9. The displacements are taken as the unknowns for both surfaces of the crack, an arrangement that involves twice as many elements for modeling a crack as compared to the displacement discontinuity method, which considers the difference between these displacements.

In describing the dual boundary element method we will initially consider an infinite, linearly elastic solid containing a single crack Γ. One surface of the crack, Γ^-, will be modeled using the displacement equation (4.328)

$$u_i(\boldsymbol{\xi}) = \int_\Gamma t_j(\boldsymbol{x}) U_{ij}(\boldsymbol{x}, \boldsymbol{\xi})\, \mathrm{d}S(\boldsymbol{x}) - \int_\Gamma u_j(\boldsymbol{x}) T_{ij}(\boldsymbol{x}, \boldsymbol{\xi})\, \mathrm{d}S(\boldsymbol{x}) \tag{8.210}$$

and the other, Γ^+, using the traction equation (4.343)

$$t_i(\boldsymbol{\xi}) = \int_\Gamma t_j(\boldsymbol{x}) K_{ij}(\boldsymbol{x}, \boldsymbol{\xi})\, \mathrm{d}S(\boldsymbol{x}) - \int_\Gamma u_j(\boldsymbol{x}) H_{ij}(\boldsymbol{x}, \boldsymbol{\xi})\, \mathrm{d}S(\boldsymbol{x}) \tag{8.211}$$

We emphasize again that both (8.210) and (8.211) are written for the case that point $\boldsymbol{\xi}$ is not at point $\boldsymbol{x}$ on the boundary Γ. Also, $t_i(\boldsymbol{\xi})$ represents the tractions along an imaginary line parallel to a local portion of the boundary. After discretizing the crack surfaces and carrying out the integrations pertaining to each boundary element, point $\boldsymbol{\xi}$ can be moved to the boundary, just as we have done for the other boundary element methods described in this book.

Notice that the integrals in (8.210) and (8.211) extend over both crack surfaces, i.e., $\Gamma = \Gamma^- + \Gamma^+$. The key point here is that Somigliana's formulas relate to a closed boundary and a crack only meets this requirement when both of its surfaces are considered. The choice between using (8.210) or (8.211) depends on which surface is used to define the limit for $\boldsymbol{\xi} \to \Gamma$.

It should be mentioned that the formulation of the dual boundary element method presented here differs from that found in the literature. The difference arises because the alternative formulation of the method uses special versions of Somigliana's formulas that are valid when point $\boldsymbol{\xi}$ is located *on* a boundary (a crack surface in the present case), as in (4.353) for the displacement problem, for example. The resulting integrals containing $T_{ij}(\boldsymbol{x}, \boldsymbol{\xi})$ and $K_{ij}(\boldsymbol{x}, \boldsymbol{\xi})$ are strongly singular and must be evaluated in the sense of the Cauchy principal value; the integral containing $H_{ij}(\boldsymbol{x}, \boldsymbol{\xi})$ is hypersingular and requires special treatment as a Hadamard finite part integral. Our approach of taking limits after the integrals have been evaluated avoids these issues.

Assuming the simple case of N_{Γ^-} and N_{Γ^+} constant strength elements along the crack surfaces[6] and letting $N = N_{\Gamma^-} + N_{\Gamma^+}$, the discretized versions of (8.210) and (8.211) can be written as

$$\begin{aligned} \overset{i}{u}_x + \sum_{j=1}^{N}\left(\overset{ij}{\mathbb{T}}_{xx}\overset{j}{u}_x + \overset{ij}{\mathbb{T}}_{xy}\overset{j}{u}_y\right) &= \sum_{j=1}^{N}\left(\overset{ij}{\mathbb{U}}_{xx}\overset{j}{t}_x + \overset{ij}{\mathbb{U}}_{xy}\overset{j}{t}_y\right) \\ \overset{i}{u}_y + \sum_{j=1}^{N}\left(\overset{ij}{\mathbb{T}}_{yx}\overset{j}{u}_x + \overset{ij}{\mathbb{T}}_{yy}\overset{j}{u}_y\right) &= \sum_{j=1}^{N}\left(\overset{ij}{\mathbb{U}}_{yx}\overset{j}{t}_x + \overset{ij}{\mathbb{U}}_{yy}\overset{j}{t}_y\right) \end{aligned} \tag{8.212}$$

and

$$\begin{aligned} \sum_{j=1}^{N}\left(\overset{ij}{\mathbb{H}}_{xx}\overset{j}{u}_x + \overset{ij}{\mathbb{H}}_{xy}\overset{j}{u}_y\right) &= -\overset{i}{t}_x + \sum_{j=1}^{N}\left(\overset{ij}{\mathbb{K}}_{xx}\overset{j}{t}_x + \overset{ij}{\mathbb{K}}_{xy}\overset{j}{t}_y\right) \\ \sum_{j=1}^{N}\left(\overset{ij}{\mathbb{H}}_{yx}\overset{j}{u}_x + \overset{ij}{\mathbb{H}}_{yy}\overset{j}{u}_y\right) &= -\overset{i}{t}_y + \sum_{j=1}^{N}\left(\overset{ij}{\mathbb{K}}_{yx}\overset{j}{t}_x + \overset{ij}{\mathbb{K}}_{yy}\overset{j}{t}_y\right) \end{aligned} \tag{8.213}$$

see (7.9) and (8.154). In calculating the limiting values of the diagonal terms ($i = j$) of the boundary influence coefficients, $\overset{ii}{\mathbb{T}}_{xx}$, $\overset{ii}{\mathbb{T}}_{yy}$, $\overset{ii}{\mathbb{K}}_{xx}$, $\overset{ii}{\mathbb{K}}_{yy}$, care must be taken to observe the direction in which the local value of $c_{\bar{y}}$ approaches the boundary. If the convention is followed that the positive direction for $c_{\bar{y}}$ always points away from the crack surface, the limiting values of the coefficients are obtained from the conditions $c_{\bar{x}} = 0$ and $c_{\bar{y}} \to 0_-$. A similar situation applies when element j is the "matching element" for element i on the opposite surface of the crack. The midpoint coordinates

[6] The elements must match exactly.

of the two elements are again equal but the limiting values of the coefficients $\overset{ij}{\mathbb{T}}_{xx}$, $\overset{ij}{\mathbb{T}}_{yy}$, $\overset{ij}{\mathbb{K}}_{xx}$, $\overset{ij}{\mathbb{K}}_{yy}$ are now obtained from the conditions $c_{\bar{x}} = 0$ and $c_{\bar{y}} \to 0_+$.

We next suppose that the tractions $\overset{i}{t}_x = \overset{i}{t}_{\bar{x}} \cos\beta - \overset{i}{t}_{\bar{y}} \sin\beta$ and $\overset{i}{t}_y = \overset{i}{t}_{\bar{x}} \sin\beta + \overset{i}{t}_{\bar{y}} \cos\beta$ are known for all elements of the crack surfaces, i.e., for $i = 1$ to $N = N_{\Gamma^-} + N_{\Gamma^+}$. The following system of algebraic equations can then be formed using (8.212) and (8.213):

$$\sum_{j=1}^{N} \overset{ij}{C}_{xx} \overset{j}{u}_x + \sum_{j=1}^{N} \overset{ij}{C}_{xy} \overset{j}{u}_y = \overset{i}{b}_x$$

$$\sum_{j=1}^{N} \overset{ij}{C}_{yx} \overset{j}{u}_x + \sum_{j=1}^{N} \overset{ij}{C}_{yy} \overset{j}{u}_y = \overset{i}{b}_y \tag{8.214}$$

The values of $\overset{i}{b}_x$ and $\overset{i}{b}_y$ are obtained by evaluating the summations on the right-hand sides of (8.212) and (8.213); equations (8.212) are used for $1 \leq j \leq N_{\Gamma^-}$ while (8.213) are used for $N_{\Gamma^-} + 1 \leq j \leq N$. The coefficients $\overset{ij}{C}_{xx}$, etc., are found by considering the locations of elements i and j. If element i is on Γ^- we use (8.212) and write $\overset{ij}{C}_{xx} = \overset{ij}{\mathbb{T}}_{xx} + \overset{ij}{\mathbb{I}}$, $\overset{ij}{C}_{xy} = \overset{ij}{\mathbb{T}}_{xy}$, and so forth, irrespective of the location of element j. Similarly, if element i is on Γ^+ we use (8.213) with $\overset{ij}{C}_{xx} = \overset{ij}{\mathbb{H}}_{xx}$, $\overset{ij}{C}_{xy} = \overset{ij}{\mathbb{H}}_{xy}$, etc., again irrespective of the location of element j.

The modifications needed for the above procedure when the infinite region contains a closed boundary S in addition to the crack are straightforward. The integrals in (8.210) and (8.211) now extend over the entire boundary, $S + \Gamma$, and the summations in (8.212) and (8.213) range from $j = 1$ to $N_{\Gamma^-} + N_{\Gamma^+} + N_S$. The coefficients $\overset{ij}{C}_{xx}$, etc., in (8.214) are assembled in the same way as before, but where now equations (8.212) are used whenever element i is either on the negative crack surface or the boundary S. Situations involving a crack within a finite region are handled similarly; displacement boundary conditions for selected portions of boundary S can be applied by appropriate rearrangement of (8.212) and (8.213).

8.10.1 Example

We have introduced two different numerical methods in this book for analyzing problems containing cracks: the displacement discontinuity method (Chap. 6 and Sect. 8.6) and the dual boundary element method (Sect. 8.10). Both methods represent a crack as two coincident boundaries whose outward normals point in opposite directions. The displacement discontinuity method treats the difference between the displacements on the two crack surfaces as a fundamental boundary parameter while the dual boundary element method considers the displacements of the two surfaces separately. Apart from this distinction the two methods are found to produce virtually the same results for comparable problems.

As a simple example, we consider the case of a pressurized straight line crack $-\ell \leq x \leq \ell$, $y = 0$ in an infinite, isotropic, linearly elastic solid. The analytical solution for the displacement discontinuity along the crack is (Sneddon 1951)

$$\Delta u_y = \frac{2(1-\nu)p\ell}{\mu}\left(1 - \frac{x^2}{\ell^2}\right)^{1/2} \tag{8.215}$$

where p is the magnitude of the pressure in the crack. The dual boundary element method was used to solve this problem for two cases, both with $\nu = 0.1$ and $p/\mu = 10^{-3}$. One case used 20 equal size piecewise constant elements along each crack surface (40 elements in total) and the other 40 elements along each surface (80 in total). The numerical results shown in Fig. 8.23 were obtained by subtracting the displacements for matching elements on the two crack surfaces according to the definition $\Delta u_y = u_y^+ - u_y^-$. A separate calculation was made for the same problem using the displacement discontinuity method. The results (not shown) were the same as those for the dual boundary element method, but using half as many elements.

Similar numerical results could be obtained for the pressurized crack problem by using fewer elements of higher order, but the near-tip solution will always be subject to error unless the crack tip asymptotics are taken into account. Two ways of doing this for the displacement discontinuity method are presented in the next two sections. These approaches are not directly applicable to the dual boundary element method because the solutions are expressed in terms of the near-tip displacement discontinuities, not the individual displacements.

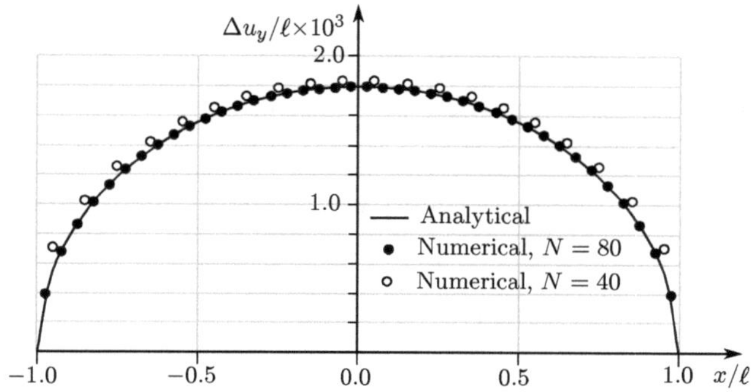

Fig. 8.23 Dual boundary element solution of pressurized crack problem, $N = 40$ and $N = 80$

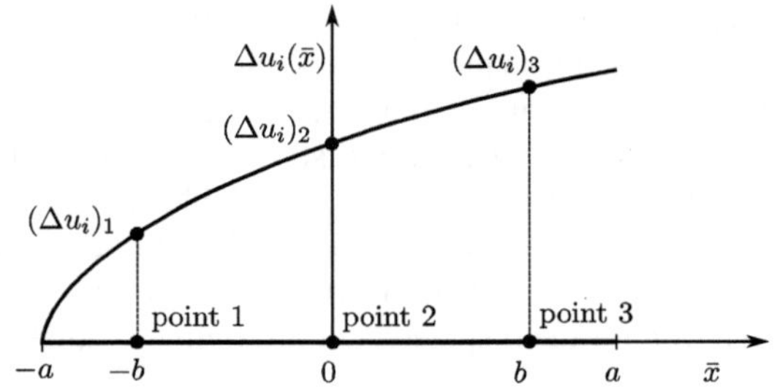

Fig. 8.24 Crack tip element, $-a \leq \bar{x} \leq a,\ \bar{y} = 0$

8.11 Crack Tip Elements I—Square Root Variation Displacement Discontinuity

For linearly elastic materials it is well known that the components of displacement discontinuity in the vicinity of a crack tip are proportional to the square root of the distance from the tip. For accurate modeling of cracks we should therefore use a special representation of $\Delta u_i(\bar{x})$ that reflects this characteristic feature. We will consider the case of a crack tip at the left-hand end of the element $|\ \bar{x}\ |\leq a,\ \bar{y} = 0$, i.e., at the point $\bar{x} = -a,\ \bar{y} = 0$, as shown in Fig. 8.24. The corresponding solution for an element with a crack tip at its right-hand end can be obtained by simply reversing the directions of $\bar{x}$ and $\bar{y}$ for that element.

The displacement discontinuity representation

$$\Delta u_i(\bar{x}) = \mathsf{a}_i(\bar{x} + a)^2 + \mathsf{b}_i(\bar{x} + a)^{3/2} + \mathsf{c}_i(\bar{x} + a)^{1/2} \tag{8.216}$$

where a_i, b_i, and c_i are constants and $i = \bar{x}$ or $\bar{y}$ has the desired square root variation near the crack tip $\bar{x} = -a$ and is also compatible with an assumed piecewise quadratic variation of displacement discontinuity over the other elements of a crack.[7] It should be noted that there is no linear term in this expression, i.e., no term involving $\bar{x} + a$ alone. This is in accord with a general result proved by Gray and Paulino (1998). Equation (8.216) can also be written in terms of element shape functions, defined for the three element collocation points $\bar{x} = -b$ (point 1), $\bar{x} = 0$ (point 2), and $\bar{x} = b$ (point 3) as in (6.53), as follows:

$$\begin{aligned} \Delta u_{\bar{x}}(\bar{x}) &= (\Delta u_{\bar{x}})_1\alpha_1^*(\bar{x}) + (\Delta u_{\bar{x}})_2\alpha_2^*(\bar{x}) + (\Delta u_{\bar{x}})_3\alpha_3^*(\bar{x}) \\ \Delta u_{\bar{y}}(\bar{x}) &= (\Delta u_{\bar{y}})_1\alpha_1^*(\bar{x}) + (\Delta u_{\bar{y}})_2\alpha_2^*(\bar{x}) + (\Delta u_{\bar{y}})_3\alpha_3^*(\bar{x}) \end{aligned} \tag{8.217}$$

[7] Another approach is to use the product of a square root and a polynomial shape function (e.g., Linkov and Mogilevskaya 1998). We note that (8.216) would be of this form if the first term in the expression were $\mathsf{a}_i(\bar{x} + a)^{5/2}$.

It is left as an exercise for the reader to show that the shape functions $\alpha_k^*(\bar{x})$ in these expressions are

$$\alpha_1^*(\bar{x}) = \frac{1}{\sqrt{a-b}K}(\bar{x}+a)^2 - \frac{1}{b\sqrt{a-b}K}\left[(a+b)^{3/2}\bar{x} - a^{3/2}(\bar{x}-b)\right]\sqrt{\bar{x}+a}$$

$$\alpha_2^*(\bar{x}) = -\frac{2}{\sqrt{a}K}(\bar{x}+a)^2 + \frac{1}{b\sqrt{a}K}\left[(a+b)^{3/2}(\bar{x}+b) - (a-b)^{3/2}(\bar{x}-b)\right]\sqrt{\bar{x}+a} \tag{8.218}$$

$$\alpha_3^*(\bar{x}) = \frac{1}{\sqrt{a+b}K}(\bar{x}+a)^2 + \frac{1}{b\sqrt{a+b}K}\left[(a-b)^{3/2}\bar{x} - a^{3/2}(\bar{x}+b)\right]\sqrt{\bar{x}+a}$$

where

$$K = (a+b)^{3/2} - 2a^{3/2} + (a-b)^{3/2} \tag{8.219}$$

The integration process for a crack tip element is the same as described in Sect. 6.12. Thus, we integrate the effects of concentrated displacement discontinuities $\mathcal{D}_{\bar{x}} = \Delta u_{\bar{x}}(\xi_{\bar{x}})\mathrm{d}\xi_{\bar{x}}$ and $\mathcal{D}_{\bar{y}} = \Delta u_{\bar{y}}(\xi_{\bar{x}})\mathrm{d}\xi_{\bar{x}}$ at the point $\bar{x} = \xi_{\bar{x}},\ \bar{y} = 0$ over the line segment $-a \le \bar{x} \le a,\ \bar{y} = 0$. This is achieved by substituting the above values of $\mathcal{D}_{\bar{x}}$ and $\mathcal{D}_{\bar{y}}$ into versions of (4.221) and (4.222) written in terms of $\bar{x}$, $\bar{y}$ coordinates, replacing $\bar{x}$ by $\bar{x} - \xi_{\bar{x}}$, and integrating the resulting expressions with respect to $\xi_{\bar{x}}$ between the limits $-a$ and $+a$.

The concentrated displacement discontinuities $\mathcal{D}_{\bar{x}}$ and $\mathcal{D}_{\bar{y}}$ involve the shape functions $\alpha_1^*(\xi_{\bar{x}})$, $\alpha_2^*(\xi_{\bar{x}})$, and $\alpha_3^*(\xi_{\bar{x}})$, which, using (5.42) in (8.218), can be written as

$$\alpha_k^*(\xi_{\bar{x}}) = \mathsf{A}_k^*(\bar{x}-\xi_{\bar{x}})^2 + \mathsf{B}_k^*(\bar{x}-\xi_{\bar{x}}) + \mathsf{C}_k^* + \left[\mathsf{D}_k^*(\bar{x}-\xi_{\bar{x}}) + \mathsf{E}_k^*\right]\sqrt{(\bar{x}+a)-(\bar{x}-\xi_{\bar{x}})} \tag{8.220}$$

where $k = 1, 2$, or 3 and

$$\mathsf{A}_1^* = \frac{1}{\sqrt{a-b}K},\quad \mathsf{B}_1^* = -2(\bar{x}+a)\mathsf{A}_1^*,\quad \mathsf{C}_1^* = (\bar{x}+a)^2\mathsf{A}_1^*$$

$$\mathsf{A}_2^* = -\frac{2}{\sqrt{a}K},\quad \mathsf{B}_2^* = -2(\bar{x}+a)\mathsf{A}_2^*,\quad \mathsf{C}_2^* = (\bar{x}+a)^2\mathsf{A}_2^*$$

$$\mathsf{A}_3^* = \frac{1}{\sqrt{a+b}K},\quad \mathsf{B}_3^* = -2(\bar{x}+a)\mathsf{A}_3^*,\quad \mathsf{C}_3^* = (\bar{x}+a)^2\mathsf{A}_3^* \tag{8.221}$$

$$\mathsf{D}_1^* = \frac{\left[(a+b)^{3/2} - a^{3/2}\right]}{b\sqrt{a-b}K}, \quad \mathsf{E}_1^* = -\bar{x}\mathsf{D}_1^* + \frac{a^{3/2}}{\sqrt{a-b}K}$$

$$\mathsf{D}_2^* = -\frac{\left[(a+b)^{3/2} - (a-b)^{3/2}\right]}{b\sqrt{a}K}, \quad \mathsf{E}_2^* = -\bar{x}\mathsf{D}_2^* - \frac{(a+b)^{3/2} + (a-b)^{3/2}}{\sqrt{a}K}$$

$$\mathsf{D}_3^* = -\frac{\left[(a-b)^{3/2} - a^{3/2}\right]}{b\sqrt{a+b}K}, \quad \mathsf{E}_3^* = -\bar{x}\mathsf{D}_3^* - \frac{a^{3/2}}{\sqrt{a+b}K}$$

Following the same reasoning as in Sects. 5.10 and 6.12, expressions for the displacements and stresses associated with a crack tip element involve integrals of the form

$$\int_{-a}^{a} \alpha_k^*(\xi_{\bar{x}})(\bullet)\,\mathrm{d}\xi_{\bar{x}}$$

where $(\bullet)$ is a derivative, through third order, of the function $g(\bar{x} - \xi_{\bar{x}}, \bar{y})$ with respect to $\bar{x}$ and $\bar{y}$. Using (8.220) we have that

$$\int_{-a}^{a} \alpha_k^*(\xi_{\bar{x}})(\bullet)\,\mathrm{d}\xi_{\bar{x}} = \mathsf{A}_k^* \int_{-a}^{a} (\bar{x} - \xi_{\bar{x}})^2(\bullet)\,\mathrm{d}\xi_{\bar{x}} + \mathsf{B}_k^* \int_{-a}^{a} (\bar{x} - \xi_{\bar{x}})(\bullet)\,\mathrm{d}\xi_{\bar{x}} + \mathsf{C}_k^* \int_{-a}^{a} (\bullet)\,\mathrm{d}\xi_{\bar{x}} \tag{8.222}$$

$$+ \mathsf{D}_k^* \int_{-a}^{a} (\bar{x} - \xi_{\bar{x}})\sqrt{(\bar{x} + a) - (\bar{x} - \xi_{\bar{x}})}(\bullet)\,\mathrm{d}\xi_{\bar{x}}$$

$$+ \mathsf{E}_k^* \int_{-a}^{a} \sqrt{(\bar{x} + a) - (\bar{x} - \xi_{\bar{x}})}(\bullet)\,\mathrm{d}\xi_{\bar{x}}$$

The integrals multiplied by constants A_k^*, B_k^*, and C_k^* have already been evaluated (see Sects. 5.10 and 6.12) so it only remains to evaluate the integrals multiplied by D_k^* and E_k^* in (8.222). For these we let $u = \bar{x} - \xi_{\bar{x}}$ and $v = \bar{y}$ and write, for example,

$$\int_{-a}^{a} (\bar{x} - \xi_{\bar{x}})\sqrt{(\bar{x} + a) - (\bar{x} - \xi_{\bar{x}})}\frac{\partial}{\partial \bar{x}} g(\bar{x} - \xi_{\bar{x}}, \bar{y})\,\mathrm{d}\xi_{\bar{x}} = -\int_{\bar{x}+a}^{\bar{x}-a} u\sqrt{(\bar{x} + a) - u}\frac{\partial}{\partial u} g(u, v)\,\mathrm{d}u \tag{8.223}$$

$$\int_{-a}^{a} \sqrt{(\bar{x} + a) - (\bar{x} - \xi_{\bar{x}})}\frac{\partial}{\partial \bar{x}} g(\bar{x} - \xi_{\bar{x}}, \bar{y})\,\mathrm{d}\xi_{\bar{x}} = -\int_{\bar{x}+a}^{\bar{x}-a} \sqrt{(\bar{x} + a) - u}\frac{\partial}{\partial u} g(u, v)\,\mathrm{d}u$$

These and similar integrals can be evaluated using the substitution $z^2 = (\bar{x} + a) - u$. Setting $g(u, v) = -C \ln \sqrt{u^2 + v^2}$ as in Appendix A, the second one of (8.223) can be written as

$$\int_{-a}^{a} \sqrt{(\bar{x}+a)-(\bar{x}-\xi_{\bar{x}})}\frac{\partial}{\partial \bar{x}} g(\bar{x}-\xi_{\bar{x}}, \bar{y})\, \mathrm{d}\xi_{\bar{x}} = -2(\bar{x}+a)C \int_{0}^{\sqrt{2a}} \frac{z^2}{z^4 - 2(\bar{x}+a)+\bar{r}_2^2}\, \mathrm{d}z + 2C \int_{0}^{\sqrt{2a}} \frac{z^4}{z^4 - 2(\bar{x}+a)+\bar{r}_2^2}\, \mathrm{d}z \quad (8.224)$$

where $\bar{r}_2^2 = (\bar{x}+a)^2 + \bar{y}^2$ is the distance from the crack tip to an arbitrary point $(\bar{x}, \bar{y})$.

All of the integrals in the last two terms of (8.222) can be evaluated using the preceding substitutions, integration by parts, and standard integration formulas (e.g., Petit Bois 1961). The results are summarized in Appendix A.

The displacements $u_{\bar{x}}$ and $u_{\bar{y}}$ and stresses $\sigma_{\bar{x}\bar{x}}$, $\sigma_{\bar{y}\bar{y}}$, and $\sigma_{\bar{x}\bar{y}}$ associated with the crack tip element can now be found by using (6.57) and (6.58), with $(\bar{G}_3)_k$, etc., replaced by $(\bar{G}_3^*)_k$, etc., where

$$\begin{aligned}(\bar{G}_3^*)_k &= \int_{-a}^{a} \alpha_k^*(\xi_{\bar{x}}) \frac{\partial}{\partial \bar{y}} g(\bar{x}-\xi_{\bar{x}}, \bar{y})\, \mathrm{d}\xi_{\bar{x}} \\ &= \mathsf{A}_k^*(\bar{F}_3)_Q + \mathsf{B}_k^*(\bar{F}_3)_L + \mathsf{C}_k^*(\bar{F}_3)_C \\ &+ \mathsf{D}_k^* \int_{-a}^{a} (\bar{x}-\xi_{\bar{x}})\sqrt{(\bar{x}+a)-(\bar{x}-\xi_{\bar{x}})}\frac{\partial}{\partial \bar{y}} g(\bar{x}-\xi_{\bar{x}}, \bar{y})\, \mathrm{d}\xi_{\bar{x}} \\ &+ \mathsf{E}_k^* \int_{-a}^{a} \sqrt{(\bar{x}+a)-(\bar{x}-\xi_{\bar{x}})}\frac{\partial}{\partial \bar{y}} g(\bar{x}-\xi_{\bar{x}}, \bar{y})\, \mathrm{d}\xi_{\bar{x}}\end{aligned} \quad (8.225)$$

cf. (5.51) and (5.52). The two integrals in (8.225) multiplied by D_k^* and E_k^* will be denoted as $(\bar{F}_3)_{LSR}$ and $(\bar{F}_3)_{CSR}$ so that

$$(\bar{G}_3^*)_k = \mathsf{A}_k^*(\bar{F}_3)_Q + \mathsf{B}_k^*(\bar{F}_3)_L + \mathsf{C}_k^*(\bar{F}_3)_C + \mathsf{D}_k^*(\bar{F}_3)_{LSR} + \mathsf{E}_k^*(\bar{F}_3)_{CSR} \quad (8.226)$$

The quantities $(\bar{G}_2^*)_k$, etc., can be represented similarly.

The values of functions $(\bar{F}_m)_{CSR}$ and $(\bar{F}_m)_{CSR}$ for $m = 2$ to 7 can be evaluated using the crack tip integration formulas in Appendix A. Using the abbreviations

$$\begin{aligned}\Omega_1 &= \ln\left[\frac{2a - \sqrt{2a\bar{r}_2}\cos\bar{\theta}_2/2 + \bar{r}_2}{2a + \sqrt{2a\bar{r}_2}\cos\bar{\theta}_2/2 + \bar{r}_2}\right]^{1/2} \\ \Omega_2 &= \arctan\frac{2\sqrt{2a\bar{r}_2}\sin\bar{\theta}_2/2}{\bar{r}_2 - 2a}\end{aligned} \quad (8.227)$$

and

$$Z_1 = \Omega_1 \sin\frac{\bar{\theta}_2}{2} + \Omega_2 \cos\frac{\bar{\theta}_2}{2}$$

$$Z_2 = \Omega_1 \sin\frac{\bar{\theta}_2}{2} - \Omega_2 \cos\frac{\bar{\theta}_2}{2}$$
$$Z_3 = \Omega_1 \cos\frac{\bar{\theta}_2}{2} + \Omega_2 \sin\frac{\bar{\theta}_2}{2} \tag{8.228}$$
$$Z_4 = \Omega_1 \cos\frac{\bar{\theta}_2}{2} - \Omega_2 \sin\frac{\bar{\theta}_2}{2}$$

the results are

$$\begin{aligned}
(\bar{F}_2)_{CSR} &= \int_{-a}^{a} \sqrt{(\bar{x}+a)-(\bar{x}-\xi_{\bar{x}})}\frac{\partial}{\partial\bar{x}} g(\bar{x}-\xi_{\bar{x}},\bar{y})\,\mathrm{d}\xi_{\bar{x}} \\
&= \frac{1}{4\pi(1-\nu)}\left[2\sqrt{2a}+\sqrt{\bar{r}_2}Z_4\right] \\
(\bar{F}_3)_{CSR} &= \int_{-a}^{a} \sqrt{(\bar{x}+a)-(\bar{x}-\xi_{\bar{x}})}\frac{\partial}{\partial\bar{y}} g(\bar{x}-\xi_{\bar{x}},\bar{y})\,\mathrm{d}\xi_{\bar{x}} \\
&= -\frac{1}{4\pi(1-\nu)}\sqrt{\bar{r}_2}Z_1 \\
(\bar{F}_4)_{CSR} &= \int_{-a}^{a} \sqrt{(\bar{x}+a)-(\bar{x}-\xi_{\bar{x}})}\frac{\partial^2}{\partial\bar{x}\partial\bar{y}} g(\bar{x}-\xi_{\bar{x}},\bar{y})\,\mathrm{d}\xi_{\bar{x}} \\
&= \frac{1}{4\pi(1-\nu)}\left[\frac{\sqrt{2a}\bar{y}}{\bar{r}_1^2}+\frac{1}{2\sqrt{\bar{r}_2}}Z_2\right] \\
(\bar{F}_5)_{CSR} &= \int_{-a}^{a} \sqrt{(\bar{x}+a)-(\bar{x}-\xi_{\bar{x}})}\frac{\partial^2}{\partial\bar{x}^2} g(\bar{x}-\xi_{\bar{x}},\bar{y})\,\mathrm{d}\xi_{\bar{x}} \\
&= \frac{1}{4\pi(1-\nu)}\left[\frac{\sqrt{2a}(\bar{x}-a)}{\bar{r}_1^2}+\frac{1}{2\sqrt{\bar{r}_2}}Z_3\right] \qquad (8.229)\\
(\bar{F}_6)_{CSR} &= \int_{-a}^{a} \sqrt{(\bar{x}+a)-(\bar{x}-\xi_{\bar{x}})}\frac{\partial^3}{\partial\bar{x}\partial\bar{y}^2} g(\bar{x}-\xi_{\bar{x}},\bar{y})\,\mathrm{d}\xi_{\bar{x}} \\
&= \frac{1}{4\pi(1-\nu)}\left[\frac{\sqrt{2a}}{\bar{r}_1^2}\left(\frac{(\bar{x}-a)^2-\bar{y}^2}{\bar{r}_1^2}+\frac{\bar{y}^2-(\bar{x}+a)(\bar{x}-a)}{2\bar{r}_2^2}\right)\right. \\
&\qquad \left. -\frac{1}{4\sqrt{\bar{r}_2}\bar{r}_2^2}\left(\bar{r}_2Z_4-2(\bar{x}+a)Z_3\right)\right] \\
(\bar{F}_7)_{CSR} &= \int_{-a}^{a} \sqrt{(\bar{x}+a)-(\bar{x}-\xi_{\bar{x}})}\frac{\partial^3}{\partial\bar{y}^3} g(\bar{x}-\xi_{\bar{x}},\bar{y})\,\mathrm{d}\xi_{\bar{x}} \\
&= \frac{1}{4\pi(1-\nu)}\left[\frac{\sqrt{2a}\bar{y}}{\bar{r}_1^2}\left(\frac{2(\bar{x}-a)}{\bar{r}_1^2}-\frac{\bar{x}}{\bar{r}_2^2}\right)+\frac{1}{4\sqrt{\bar{r}_2}\bar{r}_2^2}\left(\bar{r}_2Z_1+2(\bar{x}+a)Z_2\right)\right]
\end{aligned}$$

$$
\begin{aligned}
(\bar{F}_2)_{LSR} &= \int_{-a}^{a} (\bar{x} - \xi_{\bar{x}})\sqrt{(\bar{x}+a) - (\bar{x} - \xi_{\bar{x}})}\frac{\partial}{\partial \bar{x}} g(\bar{x} - \xi_{\bar{x}}, \bar{y})\, \mathrm{d}\xi_{\bar{x}} \\
&= -\frac{1}{4\pi(1-\nu)}\left[\frac{4a\sqrt{2a}}{3} - \sqrt{\bar{r}_2}\bar{y}Z_1\right] \\
(\bar{F}_3)_{LSR} &= \int_{-a}^{a} (\bar{x} - \xi_{\bar{x}})\sqrt{(\bar{x}+a) - (\bar{x} - \xi_{\bar{x}})}\frac{\partial}{\partial \bar{y}} g(\bar{x} - \xi_{\bar{x}}, \bar{y})\, \mathrm{d}\xi_{\bar{x}} \\
&= \frac{1}{4\pi(1-\nu)}\left[2\sqrt{2a}\bar{y} + \sqrt{\bar{r}_2}\bar{y}Z_4\right] \\
(\bar{F}_4)_{LSR} &= \int_{-a}^{a} (\bar{x} - \xi_{\bar{x}})\sqrt{(\bar{x}+a) - (\bar{x} - \xi_{\bar{x}})}\frac{\partial^2}{\partial \bar{x}\partial \bar{y}} g(\bar{x} - \xi_{\bar{x}}, \bar{y})\, \mathrm{d}\xi_{\bar{x}} \\
&= \frac{1}{4\pi(1-\nu)}\left[\frac{\sqrt{2a}(\bar{x}-a)\bar{y}}{\bar{r}_1^2} + \frac{1}{2\sqrt{\bar{r}_2}}\left(2\bar{r}_2 Z_1 + \bar{y}Z_3\right)\right] \\
(\bar{F}_5)_{LSR} &= \int_{-a}^{a} (\bar{x} - \xi_{\bar{x}})\sqrt{(\bar{x}+a) - (\bar{x} - \xi_{\bar{x}})}\frac{\partial^2}{\partial \bar{x}^2} g(\bar{x} - \xi_{\bar{x}}, \bar{y})\, \mathrm{d}\xi_{\bar{x}} \qquad (8.230) \\
&= -\frac{1}{4\pi(1-\nu)}\left[\frac{\sqrt{2a}\left(2(\bar{x}-a)^2 + 3\bar{y}^2\right)}{\bar{r}_1^2} + \frac{1}{2\sqrt{\bar{r}_2}}\left(2\bar{r}_2 Z_4 + \bar{y}Z_2\right)\right] \\
(\bar{F}_6)_{LSR} &= \int_{-a}^{a} (\bar{x} - \xi_{\bar{x}})\sqrt{(\bar{x}+a) - (\bar{x} - \xi_{\bar{x}})}\frac{\partial^3}{\partial \bar{x}\partial \bar{y}^2} g(\bar{x} - \xi_{\bar{x}}, \bar{y})\, \mathrm{d}\xi_{\bar{x}} \\
&= \frac{1}{4\pi(1-\nu)}\left[\frac{2\sqrt{2a}}{\bar{r}_1^2}\left\{\frac{(\bar{x}-a)^3}{\bar{r}_1^2} + \frac{\bar{x}\bar{y}^2}{2\bar{r}_2^2}\right\}\right. \\
&\qquad \left. + \frac{1}{4\sqrt{\bar{r}_2}\bar{r}_2^2}\left\{\left(2\bar{r}_2^2 - \bar{y}^2\right)Z_3 - (\bar{x}+a)\bar{y}Z_2\right\}\right] \\
(\bar{F}_7)_{LSR} &= \int_{-a}^{a} (\bar{x} - \xi_{\bar{x}})\sqrt{(\bar{x}+a) - (\bar{x} - \xi_{\bar{x}})}\frac{\partial^3}{\partial \bar{y}^3} g(\bar{x} - \xi_{\bar{x}}, \bar{y})\, \mathrm{d}\xi_{\bar{x}} \\
&= \frac{1}{4\pi(1-\nu)}\left[\frac{\sqrt{2a}\bar{y}}{\bar{r}_1^2}\left\{1 + \frac{2(\bar{x}-a)^2}{\bar{r}_1^2} + \frac{\bar{y}^2 - (\bar{x}+a)(\bar{x}-a)}{2\bar{r}_2^2}\right\}\right. \\
&\qquad \left. + \frac{1}{4\sqrt{\bar{r}_2}\bar{r}_2^2}\left\{2\left(\bar{r}_2^2 + (\bar{x}+a)^2\right)Z_2 + \bar{r}_2\bar{y}Z_4\right\}\right]
\end{aligned}
$$

The preceding results can be used to compute the boundary influence coefficients for crack tip elements in exactly the same way as was done for quadratically varying displacement discontinuity elements in (6.61) and (6.62). The same system of algebraic equations (6.63) can therefore be used to represent both types of elements.

8.11.1 Stress Intensity Factors

The primary reason for seeking an accurate representation of the displacement discontinuity components in the neighborhood of a crack tip is to obtain values for the stress components in this same neighborhood. If r is the distance from the crack tip into the solid, Williams (1957) showed that, for r small in relation to the length of the crack, the stresses can be written in plane polar coordinates as

$$
\begin{aligned}
\sigma_{rr} &= \frac{1}{\sqrt{2\pi r}} \cos\frac{\theta}{2}\Big[K_I(1+\sin^2\frac{\theta}{2}) + K_{II}(\frac{3}{2}\sin\theta - 2\tan\frac{\theta}{2})\Big] \\
\sigma_{\theta\theta} &= \frac{1}{\sqrt{2\pi r}} \cos\frac{\theta}{2}\Big[K_I\cos^2\frac{\theta}{2} - \frac{3}{2}K_{II}\sin\theta\Big] \\
\sigma_{r\theta} &= \frac{1}{2\sqrt{2\pi r}} \cos\frac{\theta}{2}\Big[K_I\sin\theta + K_{II}(3\cos\theta - 1)\Big]
\end{aligned}
\tag{8.231}
$$

where θ is the angle measured from the local $\bar{x}$ direction, parallel to the crack at its tip, and where K_I and K_{II} are the mode I (opening) and mode II (sliding) stress intensity factors.[8] The stress intensity factors can in turn be expressed in terms of the near-tip displacement discontinuity components as

$$
\begin{aligned}
K_I &= \frac{\mu}{4(1-\nu)} \lim_{\varepsilon\to 0} \sqrt{\frac{2\pi}{\varepsilon}}\, \Delta u_{\bar{y}}(\varepsilon) \\
K_{II} &= \frac{\mu}{4(1-\nu)} \lim_{\varepsilon\to 0} \sqrt{\frac{2\pi}{\varepsilon}}\, \Delta u_{\bar{x}}(\varepsilon)
\end{aligned}
\tag{8.232}
$$

where ε is the distance within the crack to the crack tip. In practice, the limiting process is not followed and the stress intensity factors are calculated for a small fixed value of ε.

8.11.2 Example

As an example, we consider the case of a traction-free line crack of length 2ℓ in a plane under a biaxial stress field $\sigma_{yy} = p$ and $\sigma_{xx} = \alpha p$ at infinity. The crack is inclined at an angle β from the x axis and the local $\bar{x}$ and $\bar{y}$ coordinates are parallel and perpendicular to the line of the crack. The stress intensity factors for this simple problem are (Tada et al. 2000)

[8] A mode III (tearing) stress intensity factor is also defined for antiplane shear in the third dimension.

$$K_I = p(\sqrt{\pi\ell}(\cos^2\beta + \alpha\sin^2\beta)$$

$$K_{II} = p\sqrt{\pi\ell}(1-\alpha)\sin\beta\cos\beta \tag{8.233}$$

If we take $\alpha = 1/3$, $\beta = 30°$, $p = 10$ MPa, and $\ell = 10^{-2}$ m, the stress intensity factors are $K_I = 1.477$ and $K_{II} = 0.512$, both with units of MPa-m$^{1/2}$.

For a numerical solution to this problem we first consider a relatively coarse approximation with $N = 5$; two crack tip elements and three quadratic elements as in Sect. 6.12. Each element then has length $2\ell/5 = 0.4 \times 10^{-2}$ m and the value of ε in (8.232) is $\varepsilon = 0.268 \times 10^{-3}$. For these parameters and with elastic constants $\nu = 0.1$ and $\mu = 10^4$ MPa, the near tip displacement discontinuities are found to be $\Delta u_{\bar{x}}(\varepsilon) = 0.1195 \times 10^{-5}$ and $\Delta u_{\bar{y}}(\varepsilon) = 0.3445 \times 10^{-5}$, giving stress intensity factors of $K_I = 1.467$ and $K_{II} = 0.508$, which underestimate the exact values by 0.35 and 0.41 percent, respectively. If the number of elements is doubled (i.e., if $N = 10$) the errors reduce to 0.20 percent in both cases.

8.12 Crack Tip Elements II—A Quarter Point Displacement Discontinuity Element

A special version of the quadratically varying displacement discontinuity element considered in Sect. 6.12 can be constructed to approximate the square-root-of-distance behavior of the components of displacement discontinuity in the vicinity of a crack tip. The basis for this approximation was established by Henshell and Shaw (1975) for the finite element method and has been adopted by other investigators over the years for boundary element analyses of crack problems. We will illustrate the general approach in the following discussion.

The shape functions that define the geometry for a three-noded quadratic element can be written in terms of a dimensionless coordinate η as in (7.77). If, for an element $|\ \bar{x}\ | \leq a,\ \bar{y} = 0$, the quantities $\bar{x}_1 = -a$, $\bar{x}_2 = 0$, and $\bar{x}_3 = a$ are the physical nodal points corresponding to coordinates $\eta = -1$, $\eta = 0$, and $\eta = 1$, the parameter η is equal to $\bar{x}/a$. Henshell and Shaw (1975) showed that if the element is distorted by moving its midpoint coordinate to a location one-fourth of the element length away from the $\bar{x} = -a$ end the dimensionless coordinate η becomes

$$\eta = 2\Big(\frac{\bar{x}+a}{2a}\Big)^{1/2} - 1 \tag{8.234}$$

which introduces the desired square root of distance into the analysis. Substituting this value of η into (7.77) we find that

$$\psi_1(\bar{x}) = -3\Big(\frac{\bar{x}+a}{2a}\Big)^{1/2} + \frac{(\bar{x}+a)}{a} + 1$$

$$\psi_2(\bar{x}) = 4\left(\frac{\bar{x}+a}{2a}\right)^{1/2} - \frac{2(\bar{x}+a)}{a} \tag{8.235}$$

$$\psi_3(\bar{x}) = -\left(\frac{\bar{x}+a}{2a}\right)^{1/2} + \frac{(\bar{x}+a)}{a}$$

If these shape functions were used to define a displacement discontinuity along the element it can be seen that the behavior in the vicinity of the tip would be governed by quantity $(\bar{x}+a)^{1/2}$, as desired. Gray et al. (2003) modified these expressions effectively to replace the $\bar{x}+a$ terms by $(\bar{x}+a)^{3/2}$, leading to a representation for the displacement discontinuity components comparable to (8.216) but without the $(\bar{x}+a)^2$ terms. The interested reader is referred to their paper for details.

A quarter point element does not actually employ functions (8.235) directly. Instead, a quadratic element is used with its midpoint node shifted to the point $\bar{x} = -a/2$, i.e., the element "quarter point." It can be seen that this point corresponds to the value $\eta = 0$ in (8.234). The other two nodes for the element are located at points corresponding to $\eta = -1$ and $\eta = b/a$, where $b < a$. Using (8.234), these points are found to be $\bar{x} = -a$ (the crack tip) and $\bar{x} = (b^2 + 2ab - a^2)/(2a)$. The displacement discontinuity is zero at the crack tip so only two shape functions are involved in the calculations for this special discontinuous quadratic element. We will denote these shape functions as $\alpha_2^{**}(\bar{x})$ and $\alpha_3^{**}(\bar{x})$. The superscripts ** are used to distinguish these shape functions from those used previously.

According to the preceding discussion the displacement discontinuity components for a discontinuous quarter point element are of the form

$$\Delta u_i(\bar{x}) = \mathsf{a}_i\bar{x}^2 + \mathsf{b}_i\bar{x} + \mathsf{c}_i \tag{8.236}$$

where a_i, b_i, and c_i are constants and $i = \bar{x}$ or $\bar{y}$. The constants in this expression can be found by letting $\Delta u_i(-a) = (\Delta u_i)_1 = 0$, $\Delta u_i(-a/2) = (\Delta u_i)_2$, and $\Delta u_i(\bar{x}_3) = (\Delta u_i)_3$, where $\bar{x}_3 = (b^2 + 2ab - a^2)/(2a)$, and solving the resulting system of equations for a_i, b_i, and c_i. After grouping together the terms involving $(\Delta u_i)_2$ and $(\Delta u_i)_3$ the solution can be written as

$$\begin{aligned} \Delta u_{\bar{x}}(\bar{x}) &= (\Delta u_{\bar{x}})_2\alpha_2^{**}(\bar{x}) + (\Delta u_{\bar{x}})_3\alpha_3^{**}(\bar{x}) \\ \Delta u_{\bar{y}}(\bar{x}) &= (\Delta u_{\bar{y}})_2\alpha_2^{**}(\bar{x}) + (\Delta u_{\bar{y}})_3\alpha_3^{**}(\bar{x}) \end{aligned} \tag{8.237}$$

in which

$$\alpha_2^{**}(\bar{x}) = -\frac{4}{b(2a+b)}\left[(\bar{x}+a)^2 - \frac{(a+b)^2}{2a}(\bar{x}+a)\right] \tag{8.238}$$

$$\alpha_3^{**}(\bar{x}) = \frac{4}{b(2a+b)} \frac{a^2}{(a+b)^2}\Big[(\bar{x}+a)^2 - \frac{a}{2}(\bar{x}+a)\Big]$$

Both of these shape functions vanish at the crack tip, $\bar{x} = -a$, as required. In addition, it is easy to show that $\alpha_2^{**}(-a/2) = 1, \alpha_3^{**}(-a/2) = 0$ and $\alpha_2^{**}(\bar{x}_3) = 0, \alpha_3^{**}(\bar{x}_3) = 1$ for $\bar{x}_3 = (b^2 + 2ab - a^2)/(2a)$.

Following the same approach as in Sect. 6.12 (see also Sect. 5.10) we now write

$$\alpha_k^{**}(\xi_{\bar{x}}) = \mathsf{A}_k^{**}(\bar{x} - \xi_{\bar{x}})^2 + \mathsf{B}_k^{**}(\bar{x} - \xi_{\bar{x}}) + \mathsf{C}_k^{**} \tag{8.239}$$

where $k = 2$ or 3, and where

$$\begin{aligned}
\mathsf{A}_2^{**} &= -\frac{4}{b(2a+b)}, \qquad \mathsf{B}_2^{**} = -\Big[2(\bar{x}+a) - \frac{1}{2a}(a+b)^2\Big]\mathsf{A}_2^{**} \\
\mathsf{C}_2^{**} &= (\bar{x}+a)\Big[\bar{x} - \frac{1}{2a}(b^2 + 2ab - a^2\Big]\mathsf{A}_2^{**} \\
\mathsf{A}_3^{**} &= \frac{4}{b(2a+b)} \frac{a^2}{(a+b)^2}, \qquad \mathsf{B}_3^{**} = -\frac{1}{2}\big[4(\bar{x}+a) - a\big]\mathsf{A}_3^{**} \\
\mathsf{C}_3^{**} &= \frac{1}{2}(\bar{x}+a)(2\bar{x}+a)\mathsf{A}_3^{**}
\end{aligned} \tag{8.240}$$

The required solution for the quarter point element is obtained by using (8.239) in place of (5.44) for $k = 2$ and 3. The system of algebraic equations (6.63) must also be modified by eliminating reference to node 1 for all crack tip elements.

A numerical solution to the inclined crack problem considered in the previous section for $N = 5$ gives stress intensity factors of $K_I = 1.469$ and $K_{II} = 0.509$, which underestimate the actual values by 0.53 and 0.59 percent, respectively. Interestingly, the values computed for $N = 10$ overestimate the exact values by similar amounts: $K_I = 1.486$ (+0.64 percent) and $K_{II} = 0.515$ (+0.59 percent).

8.13 Crack Growth Simulation

The displacement discontinuity method is well suited to the task of simulating the growth of a crack in the vicinity of a stress raiser such as a cavity, an elastic inclusion, or another crack. It is a simple matter to add new displacement discontinuity elements to an existing network of elements, but it must first be determined whether a crack will propagate and, if so, in which direction. We will discuss these questions below as an example of a crack growth simulation procedure. We will simulate propagation incrementally in a step-by-step fashion. It is easy to arrange a boundary element computer program to carry out this process, including the addition new elements as required.

The simplest two-dimensional crack growth algorithm uses a maximum tensile strength criterion to determine the direction of growth θ^* at a crack tip. The local crack tip hoop stress $\sigma_{\theta\theta}$ has its maximum value when the shear stress $\sigma_{r\theta}$ is zero. Referring to equations (8.231) it can be seen that the latter condition gives the following equation for angle θ^*:

$$K_I \sin\theta^* + K_{II}(3\cos\theta^* - 1) = 0 \tag{8.241}$$

The stress intensity factors K_I and K_{II} are calculated from the near-tip values of the displacement discontinuity components using (8.232). Equation (8.241) can then be solved using the substitutions $\sin\theta^* = 2\tan(\theta^*/2)(1+\tan^2(\theta^*/2))^{-1}$ and $\cos\theta^* = (1-\tan^2(\theta^*/2))(1+\tan^2(\theta^*/2))^{-1}$ to find

$$\theta^* = 2\arctan\frac{K_I \pm \sqrt{K_I^2 + 8K_{II}^2}}{4K_{II}} \tag{8.242}$$

The sign in this expression is chosen such that the corresponding value of θ^* gives the larger value of $\sqrt{2\pi r}\sigma_{\theta\theta}$ in (8.231). If $K_{II} = 0$ and $K_I \neq 0$ a simple application of L'Hopital's rule gives that $\theta^* = 0$, a result that can also be seen by inspection of (8.241).

In order for a crack to propagate in the direction θ^*, the local mode I stress intensity factor $K_I = \sqrt{2\pi r}\sigma_{\theta\theta}$ must be equal to a critical value, K_{Ic}, known as the fracture toughness.[9] That is, using the value of $\sigma_{\theta\theta}$ in (8.231), the propagation criterion is

$$\cos\frac{\theta^*}{2}\left[K_I\cos^2\frac{\theta^*}{2} - \frac{3}{2}K_{II}\sin\theta^*\right] \geq K_{Ic} \tag{8.243}$$

If this condition is met, the crack should be extended by an increment δ in the direction θ^* given by (8.242). In the absence of any direct physical guidance for determining the crack advance distance δ, it is convenient to make the assumption that it is a constant, equal to the length $(2a)$ of a crack tip element. This assumption appears to produce reasonable results provided the crack tip elements are sufficiently small in relation to the overall crack length(s).

As an illustration of the way that a stress field affects the direction of crack propagation we consider again a straight line crack of length $2\ell = 2\times 10^{-2}$m inclined at an angle of 30° from the x axis. The elastic constants of the material are $\mu = 10^4$ MPa and $\nu = 0.1$ and the stresses at infinity are $(\sigma_{yy})_\infty = p$, $(\sigma_{xx})_\infty = \alpha p$, and $(\sigma_{xy})_\infty = 0$, where $p = 100$ MPa and $\alpha \geq 0$. We will examine the direction of crack growth for various values of the parameter α, supposing for simplicity that propagation criterion (8.243) is always satisfied, without incrementing the stresses at infinity. We divide the crack into 20 elements of length $2a = 10^{-3}$ m and let the first and

[9] Values of K_{Ic} must be found experimentally and are tabulated in numerous references. For steel alloys K_{Ic} is in the range 80–200 MPa-m$^{1/2}$ while for rocks it is much lower, generally in the range 0.2–5 MPa-m$^{1/2}$.

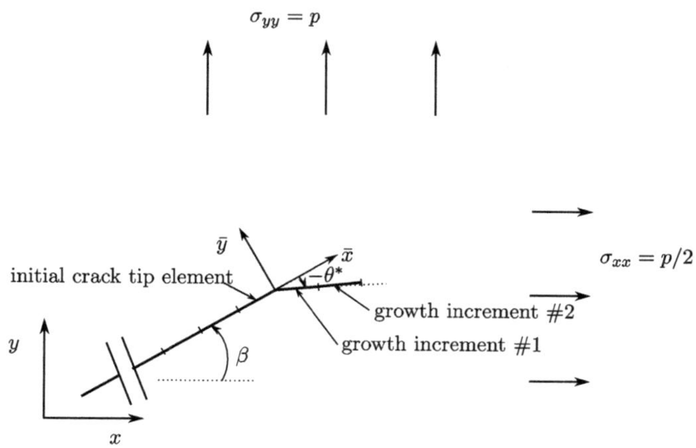

Fig. 8.25 Crack growth increments for an inclined crack in biaxial tension

Table 8.1 Crack growth increment angles θ^* (in degrees) for various values of α

α	$(\theta^*)_1$	$(\theta^*)_2$	$(\theta^*)_3$
0	−43.22	5.45	1.45
1/4	−35.34	3.48	0.61
1/2	−25.13	1.28	−0.16
3/4	−12.84	−0.21	−0.54
1	0	0	0

last elements be the square root variation crack tip elements described in Sect. 8.11. Each crack growth increment is taken as the same size (10^{-3} m) and each time a new crack tip element is added the previous tip elements are changed to ordinary quadratic elements.

Figure 8.25 shows the first two crack growth increments at the right-hand end of the crack for the case that $\alpha = 1/2$. The first increment is at an angle of $\theta^* = -25.13°$ from the local $\bar{x}$ direction (parallel to the line of the crack and pointing away from the tip[10]. The angles for the next two growth increments are $\theta^* = 1.28°$ and $\theta^* = -0.16°$, each of which is measured relative to the direction of the previous increment. The results of calculations for various values of α are given in Table 8.1. The notation $(\theta^*)_k$ in the table stands for the crack growth angle θ^* for increment k.

More elaborate problems involving situations where parts of one or more cracks are closed and subject to frictional sliding can be modeled using the iterative procedure described in Sects. 6.8 and 6.9. The incremental loading process needed to model crack closure and sliding can be combined with the crack extension algorithm discussed above to examine crack growth under a variety of loading conditions.

[10] The exact value of θ^* for the first increment is found by substituting (8.233) into (8.242) to find $\theta^* = -25.12°$.

Examples of this are given in papers by Napier (1990), Napier and Hildyard (1992), Scavia (1992, 1995), Thomas and Pollard (1993), Bobet and Einstein (1998), and others.

8.14 A Galerkin Formulation of the Displacement Discontinuity Method

An important extension of the displacement discontinuity method allows the system of algebraic equations to be written in terms of the displacement discontinuities at the endpoints of the elements, even though the stresses are strongly singular at these points. The key to this *Galerkin formulation* of the method is that the boundary conditions are satisfied in an average sense along the elements, thereby avoiding the need for collocation at points removed from the element ends. The averaging process, in turn, employs a second integration that removes the stress singularities at the element nodal points, at least insofar as setting up the system of algebraic equations is concerned. This approach is described in, for example, Sutradhar et al. (2008), in which a more general symmetric Galerkin method is also considered.

To illustrate the basic methodology a simplified treatment of the Galerkin formulation of the displacement discontinuity method is given here for the specialized problem of a pressurized line crack of length 2ℓ in an infinite, isotropic elastic solid. The crack is situated along the line $-\ell \leq \xi_x \leq \ell, \xi_y = 0$ and is subjected to a uniform internal pressure p, i.e., $\sigma_{yy} = -p$. By symmetry, the shear component of displacement discontinuity $\Delta u_x(\xi_x)$ along the crack is zero. We will divide the crack into N elements of length $2a = 2\ell/N$ and assume that the displacement discontinuity $\Delta u_y(\xi_x)$ varies linearly over each element, with $\Delta u_y(\xi_x)$ equal to zero at the "tip" ends of the first and last elements.

The double layer elastic potential (4.300) reduces to the following expression for our simplified problem:

$$t_y(\boldsymbol{x}) = \int_{-\ell}^{\ell} \Delta u_y(\boldsymbol{\xi}) H_{yy}(\boldsymbol{x}, \boldsymbol{\xi}) \mathrm{d}\xi_x \tag{8.244}$$

in which $\boldsymbol{x}$ is a field point outside the crack, the traction $t_y(\boldsymbol{x})$ at this point is calculated on a line with normal $\boldsymbol{n}(\boldsymbol{x})$, and $H_{yy}(\boldsymbol{x}, \boldsymbol{\xi})$ is given by (4.297). For a line parallel to the crack we have $n_x(\boldsymbol{x}) = n_x(\boldsymbol{\xi}) = 0$, $n_y(\boldsymbol{x}) = n_y(\boldsymbol{\xi}) = 1$, $\partial r/\partial \boldsymbol{n}(\boldsymbol{\xi}) = -\partial r/\partial \boldsymbol{n}(\boldsymbol{x}) = -\partial r/\partial y$, and the displacement discontinuity $\Delta u_y(\boldsymbol{\xi})$ is defined as $\Delta u_y(\xi_x, 0) = u_y(\xi_x, 0_+) - u_y(\xi_x, 0_-)$. Using these relations in (4.297), (8.244) can be put in the form

$$t_y(\boldsymbol{x}) = \frac{2\mu}{4\pi(1-\nu)} \int_{-\ell}^{\ell} \Delta u_y(\boldsymbol{\xi}) \frac{1}{r^2}\Big[1 + 4\Big(\frac{\partial r}{\partial y}\Big)^2 - 8\Big(\frac{\partial r}{\partial y}\Big)^4\Big] \mathrm{d}\xi_x \tag{8.245}$$

From (4.389) we know that $t_y(\boldsymbol{x})$ is continuous as a hypersingular elastic potential so we can take the limit of (8.245) for $y = 0$ and write (see Sect. 2.1.5)

$$t_y(x) = \frac{2\mu}{4\pi(1-\nu)} \fint_{-\ell}^{\ell} \Delta u_y(\xi_x) \frac{1}{(x-\xi_x)^2} \mathrm{d}\xi_x \tag{8.246}$$

A typical element J of the crack has nodal points j and $j+1$ and the displacement discontinuity along the crack is given as

$$\Delta \overset{J}{u}_y(\xi_x) = \psi_1(\xi_x)\Delta \overset{j}{u}_y + \psi_2(\xi_x)\Delta \overset{j+1}{u}_y \tag{8.247}$$

in which the shape functions $\psi_1(\xi_x)$ and $\psi_2(\xi_x)$ are

$$\begin{aligned} \psi_1(\xi_x) &= \frac{1}{2a}(a + x_J - x + x - \xi_x) \\ \psi_2(\xi_x) &= \frac{1}{2a}(a - x_J + x - x + \xi_x) \end{aligned} \tag{8.248}$$

where x_J is the midpoint coordinate of element J. It can be seen from these expressions that when $\xi_x = x_J - a$ (the left-hand end of element J) $\psi_1(\xi_x) = 1$ and $\psi_2(\xi_x) = 0$; when $\xi_x = x_J + a$ (the right-hand end) $\psi_1(\xi_x) = 0$ and $\psi_2(\xi_x) = 1$.

The traction for element J can therefore be written as

$$\begin{aligned} \overset{J}{t}_y(x) &= \frac{2\mu}{4\pi(1-\nu)} \Delta \overset{j}{u}_y \fint_{x_J-a}^{x_J+a} \psi_1(\xi_x) \frac{1}{(x-\xi_x)^2} \mathrm{d}\xi_x + \frac{2\mu}{4\pi(1-\nu)} \Delta \overset{j+1}{u}_y \fint_{x_J-a}^{x_J+a} \psi_2(\xi_x) \frac{1}{(x-\xi_x)^2} \mathrm{d}\xi_x \\ &= -\frac{\mu}{4\pi(1-\nu)a} \Delta \overset{j}{u}_y \Big[(a + x_J - x) \fint_{x_J-a}^{x_J+a} \frac{1}{(x-\xi_x)^2} \mathrm{d}(x-\xi_x) + \fint_{x_J-a}^{x_J+a} \frac{(x-\xi_x)}{(x-\xi_x)^2} \mathrm{d}(x-\xi_x) \Big] \\ &\quad - \frac{\mu}{4\pi(1-\nu)a} \Delta \overset{j+1}{u}_y \Big[(a - x_J + x) \fint_{x_J-a}^{x_J+a} \frac{1}{(x-\xi_x)^2} \mathrm{d}(x-\xi_x) - \fint_{x_J-a}^{x_J+a} \frac{(x-\xi_x)}{(x-\xi_x)^2} \mathrm{d}(x-\xi_x) \Big] \\ &= -\frac{\mu}{2\pi(1-\nu)} \Delta \overset{j}{u}_y \Big[\frac{1}{x - x_J + a} + \frac{1}{2a} \ln\Big|\frac{x - x_J - a}{x - x_J + a}\Big| \Big] \\ &\quad + \frac{\mu}{2\pi(1-\nu)} \Delta \overset{j+1}{u}_y \Big[\frac{1}{x - x_J - a} + \frac{1}{2a} \ln\Big|\frac{x - x_J - a}{x - x_J + a}\Big| \Big] \end{aligned} \tag{8.249}$$

It is easy to see that the traction is singular at the endpoints of the element, $x_J - a$ and $x_J + a$; collocation at these points is therefore not possible.

By superposition, the traction due to the contributions of all N elements is

$$t_y(x) = \sum_{J=1}^{N} \overset{J}{t}_y(x) \tag{8.250}$$

This result can also be written as a summation over the $N + 1$ nodal points as follows:

$$t_y(x) = \sum_{j=1}^{N+1} \overset{j}{t}_y(x) \tag{8.251}$$

In this expression, $\overset{j}{t}_y$ is found by adding the contributions from the two elements sharing nodal point j, elements $J - 1$ and J. The displacement discontinuities at the first and last nodal points of the crack are zero, i.e., $(\Delta u_y)_1 = 0$ and $(\Delta u_y)_{N+1} = 0$, so the limits of the summation in (8.251) could as well be taken as $j = 2$ and $j = N$. Using (8.249) and a companion expression for element $J - 1$ and setting $x_J + a = x_{j+1}$, $x_J - a = x_j$, etc., it is easy to show that $\overset{j}{t}_y$ is given as

$$\overset{j}{t}_y(x) = \frac{-\mu}{4\pi(1-\nu)a} \Delta \overset{j}{u}_y \Big[\ln|x - x_{j-1}| - 2\ln|x - x_j| + \ln|x - x_{j+1}| \Big] \tag{8.252}$$

which is singular at the three nodal points $j - 1$, j, and $j + 1$.

In the Galerkin formulation of the problem, the ith equation in system of algebraic equations is obtained by multiplying both sides of (8.251) by a weight function $\overset{i}{\psi}(x)$ associated with nodal point i and then integrating the resulting expressions along the crack. The weight function $\overset{i}{\psi}(x)$ for the Galerkin formulation is defined in terms of shape functions for the two elements connected at nodal point i, as follows:

$$\begin{aligned} \overset{i}{\psi}(x) &= 1 + \frac{1}{2a}(x - x_i); \ x_{i-1} \le x \le x_i \\ \overset{i}{\psi}(x) &= 1 - \frac{1}{2a}(x - x_i); \ x_i \le x \le x_{i+1} \\ \overset{i}{\psi}(x) &= 0; \ -\ell \le x < x_{i-1} \text{ and } x_{i+1} < x \le \ell \end{aligned} \tag{8.253}$$

Thus, $\overset{i}{\psi}(x_{i-1}) = \overset{i}{\psi}(x_{i+1}) = 0$ and $\overset{i}{\psi}(x_i) = 1$, as depicted in Fig. 8.26. Because of the last condition for $\overset{i}{\psi}(x)$ in (8.253) the integrals along the crack need only extend from x_{i-1} to x_{i+1}. We then write

Fig. 8.26 Weight function $\psi(x)$ for Galerkin formulation

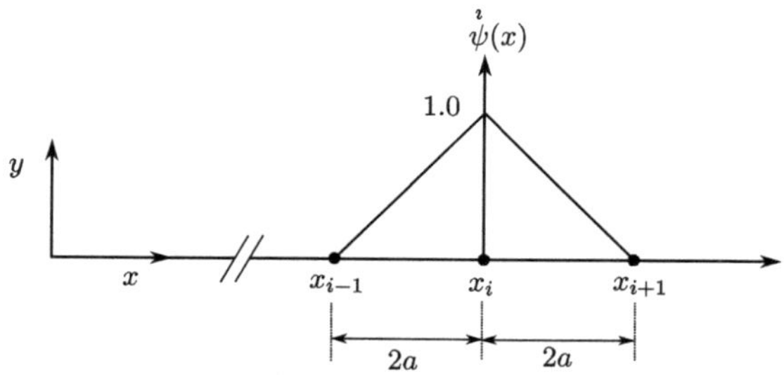

$$\int_{x_{i-1}}^{x_{i+1}} \overset{i}{\psi}(x) t_y(x)\,\mathrm{d}x = \sum_{j=1}^{N-1} \int_{x_{i-1}}^{x_{i+1}} \overset{i}{\psi}(x) \overset{j}{t}_y(x)\,\mathrm{d}x \tag{8.254}$$

The traction $t_y(x)$ is a constant, $-p$, for the problem under consideration, and hence the left-hand side of (8.254) is equal to $-2ap$ for $i = 1$ to $N - 1$. Evaluation of the integrals on the right-hand side of the expression is straightforward, and, after some simple algebra, we obtain the following system of $N - 1$ algebraic equations in $N - 1$ unknowns (the nodal point displacement discontinuities):

$$\sum_{j=2}^{N} \overset{ij}{A}_{yy} \Delta \overset{j}{u}_y = -2ap; \qquad i = 2, ..., N \tag{8.255}$$

where

$$\begin{aligned}\overset{ij}{A}_{yy} = \frac{-\mu}{16\pi(1-\nu)a^2} \Big[& 4(x_i - x_j)^2 \ln|x_i - x_j| + (x_{i-1} - x_{j-1})^2 \ln|x_{i-1} - x_{j-1}| \\ & + (x_{i+1} - x_{j+1})^2 \ln|x_{i+1} - x_{j+1}| + (x_{i-1} - x_{j+1})^2 \ln|x_{i-1} - x_{j+1}| \\ & + (x_{i+1} - x_{j-1})^2 \ln|x_{i+1} - x_{j-1}| - 2(x_{i-1} - x_j)^2 \ln|x_{i-1} - x_j| \\ & - 2(x_i - x_{j+1})^2 \ln|x_i - x_{j+1}| - 2(x_i - x_{j-1})^2 \ln|x_i - x_{j-1}| \\ & - 2(x_{i+1} - x_j)^2 \ln|x_{i+1} - x_j| \Big]\end{aligned} \tag{8.256}$$

The coefficient matrix $\overset{ij}{A}_{yy}$—which is symmetric—is well defined for all values of i and j, but the computations within a computer program must be arranged so as to exclude terms of the form $z^2 \ln z$ when $z = 0$ (these terms are zero). By inspection, it can be seen that nine separate occurrences of this will arise for each combination of i and j.

As a specific example we consider the case $\ell = 1, \nu = 0.1, p/\mu = 10^{-3}$, and $N = 20$. The computed value of the displacement discontinuity at the midpoint of the crack is $\Delta u_y(0) = 1.781 \times 10^{-3}$, compared with 1.800×10^{-3} for the analytical solution (e.g., Sneddon 1951), a difference of about 1 percent. A similar calculation using 5 discontinuous quadratic elements gives a value of 1.808×10^{-3}, which is slightly more accurate than the result for the Galerkin formulation with linear elements, even though fewer degrees of freedom are involved—15 for the quadratic elements versus 19 (i.e., $N - 1$) for the Galerkin formulation. It therefore appears that, for equivalent numbers of degrees of freedom, the Galerkin approach and the internal element collocation procedure yield comparable crack opening displacement values.

It is possible to include quarter-point crack tip elements in the Galerkin formulation presented above but, for the simple specialized crack problem considered herein, the additional effort is not warranted. A Galerkin formulation for general two-dimensional crack problems using three-noded quadratic elements and modified quarter-point crack tip elements is given by Gray et al. (2003) and the interested reader is referred to that reference for details.

Finally, it must be realized that the ability to express the boundary solution in terms of the displacement discontinuities at the element endpoints is not a panacea: results calculated inside the domain (using (8.246), for example) still exhibit singularities at the nodal points, and these locations must be avoided when carrying out general computations. In this respect, there is not an appreciable difference between the Galerkin formulation of the method and the collocation approach.

8.15 Exercises/Projects

1. Show that the stresses due to a concentrated force in an isotropic, linearly elastic half-plane (Melan's problem), reduce to those for Flamant's problem, (8.26) and (8.28), when $\ell = 0$.
2. Referring to Sect. 4.15, derive the solution to the problem of a line of centers of dilatation $\mathcal{F}$ at the point $x = 0$, $y = -\ell$ in an isotropic, linearly elastic half-plane $y \le 0$.
   Answer:

$$u_x = -(1-2\nu)\frac{\mathcal{F}}{\mu}\left[\overset{A}{g}_{,x} + (3-4\nu)\overset{I}{g}_{,x} + 2(1-2\nu)y\overset{I}{g}_{,xy}\right]$$

$$u_y = (1-2\nu)\frac{\mathcal{F}}{\mu}\left[-\overset{A}{g}_{,y} + (3-4\nu)\overset{I}{g}_{,y} + 2(1-2\nu)\overset{I}{g}_{,xx}\right]$$

$$\sigma_{xx} = -2(1-2\nu)\mathcal{F}\left[\overset{A}{g}_{,xx} + 3\overset{I}{g}_{,xx} + 2y\overset{I}{g}_{,xxy}\right]$$

$$\sigma_{yy} = 2(1-2\nu)\mathcal{F}\left[\overset{A}{g}_{,xx} - \overset{I}{g}_{,xx} + 2y\overset{I}{g}_{,xxy}\right]$$

$$\sigma_{xy} = -2(1-2\nu)\mathcal{F}\left[\overset{A}{g}_{,xy} + \overset{I}{g}_{xy} - 2y\overset{I}{g}_{,xxx}\right]$$

3. The force pair F_y at $x = c$, $y = 0$ and $-F_y$ at $x = -c$, $y = 0$ is statically equivalent to a couple $C = 2cF_y$. Derive the solution for the case that the couple is concentrated on the surface of an isotropic, linearly elastic half-plane $y \le 0$.

Answer: $\sigma_{xx} = 4Cxy(y^2 - x^2)r^{-6}/\pi$, $\sigma_{yy} = -8Cxy^3r^{-6}/\pi$, $\sigma_{xy} = 2Cy^2(y^2 - 3x^2)r^{-6}/\pi$

4. Prepare a computer program based on the stress discontinuity method to solve problems for a piecewise homogeneous elastic solid, as in Sect. 8.4.
5. Explain how the results of Sect. 8.5 can be modified for the case that the concentrated force is located in the upper half-plane, $y \geq 0$.
   Answer: Replace all occurrences of ℓ by $-\ell$, interchange ν, μ and ν^*, μ^*, and designate the displacements and stresses in (8.108) as u_i^* and σ_{ij}^*, and those in (8.135) and (8.136) as u_i and σ_{ij}. The modified expressions for u_i^* and σ_{ij}^* now apply for the upper half-plane, $y \geq 0$, while those for u_i and σ_{ij} apply for the lower half-plane, $y \leq 0$.
6. For the bonded half-planes in Sect. 8.5, suppose that the stresses at infinity are $\sigma_{xx}^{(\infty)}$, $\sigma_{xx}^{*(\infty)}$, $\sigma_{yy}^{(\infty)} = \sigma_{yy}^{*(\infty)}$, and $\sigma_{xy}^{(\infty)} = \sigma_{xy}^{*(\infty)} = 0$. Use expressions comparable to (6.38) for global coordinates to derive a relationship between the stresses $\sigma_{xx}^{(\infty)}$, $\sigma_{xx}^{*(\infty)}$, and $\sigma_{yy}^{(\infty)}$ for plane strain conditions to assure continuity of the displacements and tractions across $y = 0$.
   Answer: $\sigma_{xx}^{(\infty)} = \frac{\mu}{1-\nu}\left[\frac{1-\nu^*}{\mu^*}\sigma_{xx}^{*(\infty)} + \left(\frac{\nu}{\mu} - \frac{\nu^*}{\mu^*}\right)\sigma_{yy}^{(\infty)}\right]$
7. Consider the following problem for a straight line crack of length 2ℓ in an infinite, isotropic elastic plane:

$$\sigma_{xy} = 0$$

$$\sigma_{yy} = \begin{cases} -p & |\, x \,| < \ell,\ y = 0_+ \\ 0 & |\, x \,| < \ell,\ y = 0_- \end{cases}$$

Use a combination of the stress discontinuity and displacement discontinuity methods to devise a numerical procedure for solving this problem.
Answer: Divide both crack surfaces into N elements of length $2a$. Let the the unknowns be the stress discontinuities $\overset{i}{P}_x$ and $\overset{i}{P}_y$ on $y = 0_+$ and the displacement discontinuities $\overset{i}{D}_x$ and $\overset{i}{D}_y$ on $y = 0_-$. Show that the stresses at the midpoint of element i can be expressed in terms of the stress and displacement discontinuities at element j as follows:

$$\overset{i}{\sigma}_{xy} = -\frac{1}{2\pi}\Delta\theta \overset{ij}{I} \overset{j}{P}_x + \frac{(1-2\nu)}{8\pi(1-\nu)}\left[\ln(\overset{i}{x} - \overset{j}{x} - a)^2 - \ln(\overset{i}{x} - \overset{j}{x} + a)^2\right]\overset{j}{P}_y$$
$$- \frac{a\mu}{\pi(1-\nu)}\frac{1}{(\overset{i}{x} - \overset{j}{x})^2 - a^2}\overset{j}{D}_x$$

$$\overset{i}{\sigma}_{yy} = -\frac{(1-2\nu)}{8\pi(1-\nu)}\left[\ln(\overset{i}{x} - \overset{j}{x} - a)^2 - \ln(\overset{i}{x} - \overset{j}{x} + a)^2\right]\overset{j}{P}_x - \frac{1}{2\pi}\Delta\theta \overset{ij}{I} \overset{j}{P}_y$$
$$- \frac{a\mu}{\pi(1-\nu)}\frac{1}{(\overset{i}{x} - \overset{j}{x})^2 - a^2}\overset{j}{D}_y$$

where $\overset{ij}{I}$ is the $N \times N$ identity matrix and where $\Delta\theta = +\pi$ on $y = 0_+$ and $\Delta\theta = -\pi$ on $y = 0_-$. These expressions can be used to write a system of $4N$ algebraic equations in $4N$ unknowns, which solves the problem.

8. Use (7.77) and (7.79) to show that (8.234) holds for a quarter point element $-a \leq \bar{x} \leq a$, $\bar{y} = 0$ with nodes at $(-a, 0)$, $(-a/2, 0)$, and $(a, 0)$.
9. Prepare a Galerkin formulation of the displacement discontinuity method for the pressurized crack problem of Sect. 8.14 using quadratic variations of the displacement discontinuity. What are the shape functions comparable to (8.253)? Answer: $\psi(x) = 1 - \dfrac{1}{4a^2}(x - x_i)^2$, $x_{i-1} \leq x \leq x_{i+1}$ and $\psi(x) = 0$, $-\ell \leq x \leq x_{i-1}$ and $x_{i+1} \leq x \leq \ell$.

References

Bobet, A., and H.H. Einstein. 1998. Fracture coalescence in rock-type materials under uniaxial and biaxial compression. *International Journal of Rock Mechanics and Mining Sciences & Geomechanics Abstracts* 35: 263–88.

Brady, B.H.G., and J.W. Bray. 1978. The boundary element method for determining stresses and displacements around long openings in a triaxial stress field. *International Journal of Rock Mechanics and Mining Sciences & Geomechanics Abstracts* 15: 21–28.

Crouch, S.L., and S.G. Mogilevskaya. 2006. Loosening of elastic inclusions. *The International Journal of Solids and Structures* 43: 1638–68.

Crouch, S.L., and A.M. Starfield. 1983. *Boundary element methods in solid mechanics*. George Allen & Unwin.

Eshelby, J.D. 1957. The determination of the elastic field of an ellipsoidal inclusion, and related problems. *Proceedings of the Royal Society A* 241: 376–96.

Frasier, J.T., and L. Rongved. 1957. Force in the plane of two joined semi-infinite plates. *The Journal of Applied Mechanics* 24 (4): 582–4.

Gray, L.J., and G.H. Paulino. 1998. Crack tip interpolation, revisited. *SIAM Journal on Applied Mathematics* 58 (2): 428–55.

Gray, L.J., A.-V. Phan, G.H. Paulino, and T. Kaplan. 2003. Improved quarter-point crack tip element. *Engineering Fracture Mechanics* 70: 269–83.

Henshell, R.D., and K.G. Shaw. 1975. Crack tip finite elements are unnecessary. *The International Journal for Numerical Methods in Engineering* 9: 495–507.

Hong, H., and J. Chen. 1988. Derivation of integral equations of elasticity. *Journal of Engineering Mechanics ASCE* 114: 1028–44.

Jaeger, J.C. 1962. *Elasticity, fracture and flow*, 2nd ed. Methuen.

Linkov, A., and S. Mogilevskaya. 1998. Complex hypersingular BEM in plane elasticity problems. In *Singular integrals in boundary element methods*, ed. V. Sladek and J. Sladek, 239–364. Computational Mechanics Publications.

Melan, E. 1932. Point force at internal point in a semi-infinite plate. *Zeitschrift für Angewandte Mathematik und Mechanik* 12: 343–6.

Mi, Y., and M.H. Alibadi. 1992. Dual boundary element method for three-dimensional fracture mechanics analysis. *Engineering Analysis with Boundary Elements* 10: 161–71.

Mura, T. 1987. *Micromechanics of defects in solids*. Kluwer Academic Publishers.

Muskhelishvili, N. I. 1963. *Some basic problems of the mathematical theory of elasticity*, 2nd (English) edn. P. Noordhoff

Napier, J.A.L. 1990. Modeling of fracturing near deep level gold mining excavations using a displacement discontinuity approach. In *Mechanics of jointed and faulted rock*, ed. H.P. Rossmanith, 709–15. Balkema.

Napier, J.A.L., and M.W. Hildyard. 1992. Simulation of fracture growth around openings in highly stressed, brittle rock. *Journal of the Southern African Institute of Mining and Metallurgy* 92: 159–68.

Petit Bois, G. 1961. *Tables of indefinite integrals*. Dover.

Portella, A., M.H. Aliabadi, and D.P. Rooke. 1992. The dual boundary element method: Effective implementation for crack problems. *The International Journal for Numerical Methods in Engineering* 33: 1269–87.

Rizzo, F.J., and D.J. Shippy. 1968. A formulation and solution procedure for the general non-homogeneous elastic inclusion problem. *The International Journal of Solids and Structures* 4: 1161–79.

Rongved, L., and J.T. Frasier. 1958. Displacement discontinuity in the elastic half-space. *The Journal of Applied Mechanics* 25: 125–8.

Scavia, C. 1992. A numerical technique for the analysis of cracks subjected to normal compressive stresses. *The International Journal for Numerical Methods in Engineering* 33: 929–42.

Scavia, C. 1995. A method for the study of crack propagation in rock structures. *Géotechnique* 45: 447–63.

Sneddon, I.N. 1951. *Fourier transforms*. McGraw-Hill.

Sokolnikoff, I.S., and R.M. Redheffer. 1966. *Mathematics of physics and modern engineering*, 2nd ed. McGraw-Hill.

Sutradhar, A., G. H. Paulino, and L. J. Gray. 2008. *Symmetric Galerkin boundary element method*. SpringerLink

Tada, H., P. C. Paris, and G. R. Irwin. 2000. *The stress analysis of cracks handbook*, 3rd edn. ASME Press

Timoshenko, S. P., and J. N. Goodier. 1970. *Theory of elasticity*, 3rd. edn. McGraw-Hill

Thomas, A.L., and D.D. Pollard. 1993. The geometry of en echelon fractures in rock: implications from laboratory and numerical experiments. *Journal of Structural Geology* 15: 323–34.

Toya, M. 1974. A crack along the interface of a circular inclusion embedded in an infinite solid. *Journal of the Mechanics and Physics of Solids* 22: 325–48.

Williams, M.L. 1957. Stress singularities resulting from various boundary conditions in angular corners of plates in extension. *The Journal of Applied Mechanics* 24: 109–14.

Appendix A
Integration Formulas

In this appendix we present indefinite integrals used to develop higher order versions of the boundary element methods described in the main body of this book.

A.1 Integrals for Plane Isotropic Elasticity

First, we consider integrals of the function

$$g(u, v) = -C \ln \sqrt{u^2 + v^2} \tag{A.1}$$

that are needed to evaluate shape function-kernel products for the direct boundary integral and displacement discontinuity methods. Three sets of integration formulas are listed, corresponding to integrals with respect to u of function $g(u, v)$ and its derivatives, as well as integrals with respect to u of the products of these same functions with u and u^2. The latter integrals are required for quadratic variations of boundary parameters and are provided here for reference purposes.

The following indefinite integrals all contain the term $\arctan u/v$. When used throughout the book to evaluate definite integrals this term is replaced by $\pi/2 - \arctan v/u$ in order to obtain expressions of the form $\arctan y/(x-a) - \arctan y/(x+a)$, which can be written as $\theta_1 - \theta_2$. The earliest occurrence of this is in Eqs. (1.9) and (1.10).

S. L. Crouch and S. G. Mogilevskaya, *A First Course in Boundary Element Methods*, Mathematical Engineering, https://doi.org/10.1007/978-3-031-63341-6

(1) Integrals of g, $g_{,u}$, $g_{,v}$, etc.:

$$
\begin{aligned}
\int g\,\mathrm{d}u &= -C\left[u\ln\sqrt{u^2+v^2} + v\arctan\frac{u}{v} - u\right]\\
\int g_{,u}\,\mathrm{d}u &= -C\ln\sqrt{u^2+v^2}\\
\int g_{,v}\,\mathrm{d}u &= -C\arctan\frac{u}{v}\\
\int g_{,uv}\,\mathrm{d}u &= -C\frac{v}{u^2+v^2}\\
\int g_{,uu}\,\mathrm{d}u &= -C\frac{u}{u^2+v^2}\\
\int g_{,uvv}\,\mathrm{d}u &= -C\frac{(u^2-v^2)}{(u^2+v^2)^2}\\
\int g_{,vvv}\,\mathrm{d}u &= -C\frac{2uv}{(u^2+v^2)^2}
\end{aligned}
\tag{A.2}
$$

(2) Integrals of ug, $ug_{,u}$, $ug_{,v}$, etc.:

$$
\begin{aligned}
\int ug\,\mathrm{d}u &= -\frac{1}{2}C(u^2+v^2)\left(\ln\sqrt{u^2+v^2} - \frac{1}{2}\right)\\
\int ug_{,u}\,\mathrm{d}u &= C\left(-u + v\arctan\frac{u}{v}\right)\\
\int ug_{,v}\,\mathrm{d}u &= -Cv\ln\sqrt{u^2+v^2}\\
\int ug_{,uv}\,\mathrm{d}u &= -C\left[\frac{uv}{u^2+v^2} - \arctan\frac{u}{v}\right]\\
\int ug_{,uu}\,\mathrm{d}u &= -C\left[\frac{u^2}{u^2+v^2} - \ln\sqrt{u^2+v^2}\right]\\
\int ug_{,uvv}\,\mathrm{d}u &= -C\frac{2u^3}{\left(u^2+v^2\right)^2}\\
\int ug_{,vvv}\,\mathrm{d}u &= -C\frac{v(3u^2+v^2)}{\left(u^2+v^2\right)^2}
\end{aligned}
\tag{A.3}
$$

(3) Integrals of $u^2 g$, $u^2 g_{,u}$, $u^2 g_{,v}$, etc.:

$$\begin{aligned}
\int u^2 g \, \mathrm{d}u &= -\frac{1}{3} C \left[u^3 \ln \sqrt{u^2+v^2} - v^3 \arctan \frac{u}{v} + uv^2 - \frac{u^3}{3} \right] \\
\int u^2 g_{,u} \, \mathrm{d}u &= -C \left[\frac{(u^2+v^2)}{2} - v^2 \ln \sqrt{u^2+v^2} \right] \\
\int u^2 g_{,v} \, \mathrm{d}u &= -C \left[uv - v^2 \arctan \frac{u}{v} \right] \\
\int u^2 g_{,uv} \, \mathrm{d}u &= -C \left[\frac{u^2 v}{u^2+v^2} - 2v \ln \sqrt{u^2+v^2} \right] \\
\int u^2 g_{,uu} \, \mathrm{d}u &= -C \left[\frac{u^3}{u^2+v^2} - 2u + 2v \arctan \frac{u}{v} \right] \\
\int u^2 g_{,uvv} \, \mathrm{d}u &= -C \left[\frac{2u^4}{(u^2+v^2)^2} + \frac{u^2}{u^2+v^2} - 2 \ln \sqrt{u^2+v^2} \right] \\
\int u^2 g_{,vvv} \, \mathrm{d}u &= -2C \left[\frac{u^3 v}{(u^2+v^2)^2} + \frac{uv}{u^2+v^2} - \arctan \frac{u}{v} \right]
\end{aligned} \tag{A.4}$$

A.2 Integrals for Plane Anisotropic Elasticity

$$\begin{aligned}
\int \arctan \frac{Mu+N}{Pu+Q} du &= \left(u + \frac{MN+PQ}{M^2+P^2}\right) \arctan \frac{Mu+N}{Pu+Q} \\
&\quad - \left(\frac{MQ-NP}{M^2+P^2} \right) \ln \sqrt{(Mu+N)^2 + (Pu+Q)^2}
\end{aligned} \tag{A.5}$$

The following integrals are for the case that $4AC - B^2 > 0$.

$$\begin{aligned}
\int \ln \sqrt{Au^2+Bu+C} \, \mathrm{d}u &= -u + \left(u + \frac{B}{2A} \right) \ln \sqrt{Au^2+Bu+C} \\
&\quad + \frac{\sqrt{4AC-B^2}}{2A} \arctan \frac{2Au+B}{\sqrt{4AC-B^2}}
\end{aligned}$$

$$\int \frac{1}{Au^2+Bu+C} \mathrm{d}u = \frac{2}{\sqrt{4AC-B^2}} \arctan \frac{2Au+B}{\sqrt{4AC-B^2}}$$

$$\int \frac{u}{Au^2+Bu+C} \mathrm{d}u = \frac{1}{A} \ln \sqrt{Au^2+Bu+C} - \frac{B}{A\sqrt{4AC-B^2}} \arctan \frac{2Au+B}{\sqrt{4AC-B^2}}$$

$$\int \frac{u^2}{Au^2+Bu+C}\,du = \frac{1}{A}u - \frac{B}{A^2}\ln\sqrt{Au^2+Bu+C} + \frac{B^2-2AC}{A^2\sqrt{4AC-B^2}}$$

$$\arctan\frac{2Au+B}{\sqrt{4AC-B^2}}$$

$$\int \frac{u^3}{Au^2+Bu+C}\,du = \frac{1}{2A}u^2 - \frac{B}{A^2}u + \frac{B^2-AC}{A^2}\int\frac{u}{Au^2+Bu+C}\,du + \frac{BC}{A^2}\int\frac{1}{Au^2+Bu+C}du$$

(A.6)

$$\int \frac{1}{(Au^2+Bu+C)^2}\,du = \frac{2Au+B}{4AC-B^2}\frac{1}{Au^2+Bu+C} + \frac{2A}{4AC-B^2}\int\frac{1}{Au^2+Bu+C}\,du$$

$$\int \frac{u}{(Au^2+Bu+C)^2}\,du = -\frac{1}{2A}\frac{1}{Au^2+Bu+C} - \frac{B}{2A}\int\frac{1}{(Au^2+Bu+C)^2}\,du$$

$$\int \frac{u^2}{(Au^2+Bu+C)^2}\,du = -\frac{1}{A}\frac{u}{Au^2+Bu+C} + \frac{C}{A}\int\frac{1}{(Au^2+Bu+C)^2}\,du$$

$$\int \frac{u^3}{(Au^2+Bu+C)^2}\,du = \frac{(3AC-B^2)Bu+(2AC-B^2)C}{A^2(4AC-B^2)}\frac{1}{Au^2+Bu+C} + \frac{1}{A^2}\ln\sqrt{Au^2+Bu+C} - \frac{B(6AC-B^2)}{2A^2(4AC-B^2)}\int\frac{1}{Au^2+Bu+C}\,du$$

$$\int \frac{u^4}{(Au^2+Bu+C)^2}\,du = \frac{1}{A^2}u + \frac{2}{A^3}(AC-B^2)\frac{u}{Au^2+Bu+C} + \frac{B^3}{2A^4}\frac{1}{Au^2+Bu+C} - \frac{2B}{A^3}\ln\sqrt{Au^2+Bu+C} - \frac{(B^4-6AB^2C+3A^2C^2)}{2A^4}\int\frac{1}{(Au^2+Bu+C)^2}du$$

A.3 Integrals for Crack Tip Elements

Implementation of the crack tip elements described in Sect. 8.11 involves integrals of the form $\int\sqrt{B-u}(\bullet)du$ and $\int u\sqrt{B-u}(\bullet)du$, where $(\bullet)$ stands for a derivative, through third order, of function $g(u,v) = -C\ln\sqrt{u^2+v^2}$. The results can be expressed in terms of the following parameters and functions:

$$k^2 = (B^2 + v^2)^{1/2}; \qquad \alpha = \arctan \frac{v}{B}; \qquad \sin\alpha = \frac{v}{k^2}; \qquad \cos\alpha = \frac{B}{k^2} \tag{A.7}$$

$$\Omega_1(u, v) = \ln\left[\frac{(B-u) - 2k(B-u)^{1/2}\cos\alpha/2 + k^2}{(B-u) + 2k(B-u)^{1/2}\cos\alpha/2 + k^2}\right]^{1/2}$$

$$\Omega_2(u, v) = \arctan \frac{2k(B-u)^{1/2}\sin\alpha/2}{k^2 - (B-u)}$$

$$\begin{aligned} Z_1(u, v) &= \sin\frac{\alpha}{2}\Omega_1(u, v) + \cos\frac{\alpha}{2}\Omega_2(u, v) \\ Z_2(u, v) &= \sin\frac{\alpha}{2}\Omega_1(u, v) - \cos\frac{\alpha}{2}\Omega_2(u, v) \\ Z_3(u, v) &= \cos\frac{\alpha}{2}\Omega_1(u, v) + \sin\frac{\alpha}{2}\Omega_2(u, v) \\ Z_4(u, v) &= \cos\frac{\alpha}{2}\Omega_1(u, v) - \sin\frac{\alpha}{2}\Omega_2(u, v) \end{aligned} \tag{A.8}$$

(1) Integrals of $\sqrt{B-u}g_{,u}$, $\sqrt{B-u}g_{,v}$, etc.:

$$\begin{aligned} \int \sqrt{B-u}g_{,u}\,\mathrm{d}u &= -2C\sqrt{B-u} - CkZ_4 \\ \int \sqrt{B-u}g_{,v}\,\mathrm{d}u &= CkZ_1 \\ \int \sqrt{B-u}g_{,uv}\,\mathrm{d}u &= -C\sqrt{B-u}\frac{v}{u^2+v^2} - \frac{C}{2k}Z_2 \\ \int \sqrt{B-u}g_{,uu}\,\mathrm{d}u &= -C\sqrt{B-u}\frac{u}{u^2+v^2} - \frac{C}{2k}Z_3 \\ \int \sqrt{B-u}g_{,uvv}\,\mathrm{d}u &= -C\frac{\sqrt{B-u}}{u^2+v^2}\left(\frac{u^2-v^2}{u^2+v^2} - \frac{(Bu-v^2)}{2k^4}\right) \\ &\quad + \frac{C}{4k^5}\left(k^2Z_4 - 2BZ_3\right) \end{aligned} \tag{A.9}$$

$$\int \sqrt{B-u}\,g_{,vvv}\,\mathrm{d}u = -C\frac{\sqrt{B-u}}{u^2+v^2}\left(\frac{2uv}{u^2+v^2} - \frac{v(u+B)}{2k^4}\right) - \frac{C}{4k^5}\left(k^2 Z_1 + 2BZ_2\right)$$

(2) Integrals of $u\sqrt{B-u}g_{,u}$, $u\sqrt{B-u}g_{,v}$, etc.:

$$\int u\sqrt{B-u}\,g_{,u}\,\mathrm{d}u = \frac{2C}{3}(B-u)^{3/2} - CkvZ_1$$

$$\int u\sqrt{B-u}\,g_{,v}\,\mathrm{d}u = -2Cv\sqrt{B-u} - CkvZ_4$$

$$\int u\sqrt{B-u}\,g_{,uv}\,\mathrm{d}u = -C\sqrt{B-u}\frac{uv}{u^2+v^2} - \frac{C}{2k}\left(2k^2 Z_1 + vZ_3\right)$$

$$\int u\sqrt{B-u}\,g_{,uu}\,\mathrm{d}u = C\sqrt{B-u}\frac{(2u^2+3v^2)}{u^2+v^2} + \frac{C}{2k}\left(2k^2 Z_4 + vZ_2\right) \quad \text{(A.10)}$$

$$\int u\sqrt{B-u}\,g_{,uvv}\,\mathrm{d}u = -\frac{2C\sqrt{B-u}}{u^2+v^2}\left(\frac{u^3}{u^2+v^2} + \frac{v^2(u+B)}{4k^4}\right) - \frac{C}{4k^5}\left((2k^4-v^2)Z_3 - BvZ_2\right)$$

$$\int u\sqrt{B-u}\,g_{,vvv}\,\mathrm{d}u = -C\sqrt{B-u}\frac{v}{u^2+v^2}\left(\frac{3u^2+v^2}{u^2+v^2} - \frac{(Bu-v^2)}{2k^4}\right) - \frac{C}{4k^5}\left(2(k^4+B^2)Z_2 + k^2vZ_4\right)$$

Appendix B
Kernels for Somigliana's Formulas

The values of the tensor fields $U_{ij}(\boldsymbol{x}, \boldsymbol{\xi})$ and $T_{ij}(\boldsymbol{x}, \boldsymbol{\xi})$ in (4.328) and $D_{kij}(\boldsymbol{x}, \boldsymbol{\xi})$ and $S_{kij}(\boldsymbol{x}, \boldsymbol{\xi})$ in (4.332) are summarized in this appendix for plane problems in both isotropic and orthotropic elasticity. All quantities are functions of the distance of the sample point $\boldsymbol{x}$ from the load point $\boldsymbol{\xi}$; the abscissa of this distance is $x - \xi_x$ and the ordinate is $y - \xi_y$. In the expressions for $T_{ij}(\boldsymbol{x}, \boldsymbol{\xi})$ and $S_{kij}(\boldsymbol{x}, \boldsymbol{\xi})$ the angle β specifies the tangent to the boundary S at point $\boldsymbol{x}$, as illustrated, for example, in Fig. 7.1.

B.1 Infinite (or Finite) Isotropic Elastic Region

For isotropic elasticity, the function $g(x, y)$ is defined as in (4.252), i.e.,

$$g(x, y) = -\frac{1}{4\pi(1-\nu)} \ln\left[(x-\xi_x)^2 + (y-\xi_y)^2\right]^{1/2} \tag{B.1}$$

The values of the tensor fields $U_{ij}(\boldsymbol{x}, \boldsymbol{\xi})$, etc., are given in terms of this function and its derivatives with respect to x and y as follows:

$$\begin{aligned}
U_{xx} &= \frac{1}{2\mu}\left[(3-4\nu)g - (x-\xi_x)\frac{\partial g}{\partial x}\right] \\
U_{yx} &= \frac{1}{2\mu}\left[-(y-\xi_y)\frac{\partial g}{\partial x}\right] \\
U_{xy} &= \frac{1}{2\mu}\left[-(x-\xi_x)\frac{\partial g}{\partial y}\right] = U_{yx} \\
U_{yy} &= \frac{1}{2\mu}\left[(3-4\nu)g - (y-\xi_y)\frac{\partial g}{\partial y}\right]
\end{aligned} \tag{B.2}$$

S. L. Crouch and S. G. Mogilevskaya, *A First Course in Boundary Element Methods*, Mathematical Engineering, https://doi.org/10.1007/978-3-031-63341-6

$$
\begin{aligned}
T_{xx} &= -2(1-\nu)\sin\beta\frac{\partial g}{\partial x} + (1-2\nu)\cos\beta\frac{\partial g}{\partial y} + (x-\xi_x)\left[\sin\beta\frac{\partial^2 g}{\partial x^2} - \cos\beta\frac{\partial^2 g}{\partial x\partial y}\right]\\
T_{yx} &= -2\nu\sin\beta\frac{\partial g}{\partial y} + (1-2\nu)\cos\beta\frac{\partial g}{\partial x} + (y-\xi_y)\left[\sin\beta\frac{\partial^2 g}{\partial x^2} - \cos\beta\frac{\partial^2 g}{\partial x\partial y}\right]\\
T_{xy} &= 2\nu\cos\beta\frac{\partial g}{\partial x} - (1-2\nu)\sin\beta\frac{\partial g}{\partial y} + (x-\xi_x)\left[\cos\beta\frac{\partial^2 g}{\partial x^2} + \sin\beta\frac{\partial^2 g}{\partial x\partial y}\right]\\
T_{yy} &= 2(1-\nu)\cos\beta\frac{\partial g}{\partial y} - (1-2\nu)\sin\beta\frac{\partial g}{\partial x} + (y-\xi_y)\left[\cos\beta\frac{\partial^2 g}{\partial x^2} + \sin\beta\frac{\partial^2 g}{\partial x\partial y}\right]
\end{aligned}
\tag{B.3}
$$

$$
\begin{aligned}
D_{xxx} &= -2(1-\nu)\frac{\partial g}{\partial x} + (x-\xi_x)\frac{\partial^2 g}{\partial x^2}\\
D_{yxx} &= -2\nu\frac{\partial g}{\partial y} + (y-\xi_y)\frac{\partial^2 g}{\partial x^2}\\
D_{xyy} &= -2\nu\frac{\partial g}{\partial x} - (x-\xi_x)\frac{\partial^2 g}{\partial x^2}\\
D_{yyy} &= -2(1-\nu)\frac{\partial g}{\partial y} - (y-\xi_y)\frac{\partial^2 g}{\partial x^2}\\
D_{xxy} = D_{xyx} &= -(1-2\nu)\frac{\partial g}{\partial y} + (x-\xi_x)\frac{\partial^2 g}{\partial x\partial y}\\
D_{yxy} = D_{yyx} &= -(1-2\nu)\frac{\partial g}{\partial x} + (y-\xi_y)\frac{\partial^2 g}{\partial x\partial y}
\end{aligned}
\tag{B.4}
$$

$$
\begin{aligned}
S_{xxx} &= 2\mu\left[\sin\beta\frac{\partial^2 g}{\partial x^2} - (x-\xi_x)\left(\sin\beta\frac{\partial^3 g}{\partial x^3} - \cos\beta\frac{\partial^3 g}{\partial x^2\partial y}\right)\right]\\
S_{yxx} &= 2\mu\left[\cos\beta\frac{\partial^2 g}{\partial x^2} + 2\sin\beta\frac{\partial^2 g}{\partial x\partial y} - (y-\xi_y)\left(\sin\beta\frac{\partial^3 g}{\partial x^3} - \cos\beta\frac{\partial^3 g}{\partial x^2\partial y}\right)\right]\\
S_{xyy} &= 2\mu\left[\sin\beta\frac{\partial^2 g}{\partial x^2} - 2\cos\beta\frac{\partial^2 g}{\partial x\partial y} + (x-\xi_x)\left(\sin\beta\frac{\partial^3 g}{\partial x^3} - \cos\beta\frac{\partial^3 g}{\partial x^2\partial y}\right)\right]\\
S_{yyy} &= 2\mu\left[\cos\beta\frac{\partial^2 g}{\partial x^2} + (y-\xi_y)\left(\sin\beta\frac{\partial^3 g}{\partial x^3} - \cos\beta\frac{\partial^3 g}{\partial x^2\partial y}\right)\right]\\
S_{xxy} = S_{xyx} &= 2\mu\left[-\cos\beta\frac{\partial^2 g}{\partial x^2} - (x-\xi_x)\left(\sin\beta\frac{\partial^3 g}{\partial x^2\partial y} + \cos\beta\frac{\partial^3 g}{\partial x^3}\right)\right]\\
S_{yxy} = S_{yyx} &= 2\mu\left[-\sin\beta\frac{\partial^2 g}{\partial x^2} - (y-\xi_y)\left(\sin\beta\frac{\partial^3 g}{\partial x^2\partial y} + \cos\beta\frac{\partial^3 g}{\partial x^3}\right)\right]
\end{aligned}
\tag{B.5}
$$

B.2 Infinite (or Finite) Anisotropic Elastic Region

For orthotropic elasticity, the tensor fields $U_{ij}(\boldsymbol{x}, \boldsymbol{\xi})$, etc., can be developed from the results given in Sect. 4.11. In order to achieve some similarity with the corresponding expressions for the isotropic case we express these results in terms of a function $h(x, y_i)$, defined as

$$h(x, y_i) = \frac{1}{2\pi(q_1 - q_2)} \ln\left[(x - \xi_x)^2 + (y - \xi_y)^2/\gamma_i^2\right]^{1/2} \tag{B.6}$$

The tensor fields of interest are then given as follows:

$$\begin{aligned}
U_{xx} &= \frac{1}{c_{66}}\left[\frac{\gamma_1}{q_1}h(x, y_1) - \frac{\gamma_2}{q_2}h(x, y_2)\right]\\
U_{yx} &= -\frac{1}{c_{66}}\left[\int \frac{\partial}{\partial x}h(x, y_1)\,\mathrm{d}y_1 - \int \frac{\partial}{\partial x}h(x, y_2)\,\mathrm{d}y_2\right]\\
U_{xy} &= U_{yx}\\
U_{yy} &= -\frac{1}{c_{66}}\left[\frac{q_1}{\gamma_1}h(x, y_1) - \frac{q_2}{\gamma_2}h(x, y_2)\right]
\end{aligned} \tag{B.7}$$

$$\begin{aligned}
T_{xx} = &-\frac{(1+q_1)}{q_1}\left[\frac{1}{\gamma_1}\frac{\partial}{\partial x}h(x, y_1)\sin\beta - \frac{\partial}{\partial y_1}h(x, y_1)\cos\beta\right]\\
&+ \frac{(1+q_2)}{q_2}\left[\frac{1}{\gamma_2}\frac{\partial}{\partial x}h(x, y_2)\sin\beta - \frac{\partial}{\partial y_2}h(x, y_2)\cos\beta\right]\\
T_{yx} = &-\frac{(1+q_1)}{\gamma_1}\left[\frac{1}{\gamma_1}\frac{\partial}{\partial y_1}h(x, y_1)\sin\beta + \frac{\partial}{\partial x}h(x, y_1)\cos\beta\right]\\
&+ \frac{(1+q_2)}{\gamma_2}\left[\frac{1}{\gamma_2}\frac{\partial}{\partial y_2}h(x, y_2)\sin\beta + \frac{\partial}{\partial x}h(x, y_2)\cos\beta\right]\\
T_{xy} = &-\frac{\gamma_1(1+q_1)}{q_1}\left[\frac{1}{\gamma_1}\frac{\partial}{\partial y_1}h(x, y_1)\sin\beta + \frac{\partial}{\partial x}h(x, y_1)\cos\beta\right]\\
&+ \frac{\gamma_2(1+q_2)}{q_2}\left[\frac{1}{\gamma_2}\frac{\partial}{\partial y_2}h(x, y_2)\sin\beta + \frac{\partial}{\partial x}h(x, y_2)\cos\beta\right]\\
T_{yy} = &\ (1+q_1)\left[\frac{1}{\gamma_1}\frac{\partial}{\partial x}h(x, y_1)\sin\beta - \frac{\partial}{\partial y_1}h(x, y_1)\cos\beta\right]\\
&- (1+q_2)\left[\frac{1}{\gamma_2}\frac{\partial}{\partial x}h(x, y_2)\sin\beta - \frac{\partial}{\partial y_2}h(x, y_2)\cos\beta\right]
\end{aligned} \tag{B.8}$$

$$
\begin{aligned}
D_{xxx} &= -\frac{(1+q_1)}{\gamma_1 q_1}\frac{\partial}{\partial x}h(x, y_1) + \frac{(1+q_2)}{\gamma_2 q_2}\frac{\partial}{\partial x}h(x, y_2) \\
D_{yxx} &= -\frac{(1+q_1)}{\gamma_1^2}\frac{\partial}{\partial y_1}h(x, y_1) + \frac{(1+q_2)}{\gamma_2^2}\frac{\partial}{\partial y_2}h(x, y_2) \\
D_{xyy} &= \frac{\gamma_1(1+q_1)}{q_1}\frac{\partial}{\partial x}h(x, y_1) - \frac{\gamma_2(1+q_2)}{q_2}\frac{\partial}{\partial x}h(x, y_2) \\
D_{yyy} &= (1+q_1)\frac{\partial}{\partial y_1}h(x, y_1) - (1+q_2)\frac{\partial}{\partial y_2}h(x, y_2) \\
D_{xxy} = D_{xyx} &= -\frac{(1+q_1)}{q_1}\frac{\partial}{\partial y_1}h(x, y_1) + \frac{(1+q_2)}{q_2}\frac{\partial}{\partial y_2}h(x, y_2) \\
D_{yxy} = D_{yyx} &= \frac{(1+q_1)}{\gamma_1}\frac{\partial}{\partial x}h(x, y_1) - \frac{(1+q_2)}{\gamma_2}\frac{\partial}{\partial x}h(x, y_2)
\end{aligned}
\tag{B.9}
$$

$$
\begin{aligned}
S_{xxx} = {} & c_{66}\frac{(1+q_1)^2}{\gamma_1^2 q_1}\left[\frac{1}{\gamma_1}\frac{\partial^2}{\partial x^2}h(x, y_1)\sin\beta - \frac{\partial^2}{\partial x \partial y_1}h(x, y_1)\cos\beta\right] \\
& - c_{66}\frac{(1+q_2)^2}{\gamma_2^2 q_2}\left[\frac{1}{\gamma_2}\frac{\partial^2}{\partial x^2}h(x, y_2)\sin\beta - \frac{\partial^2}{\partial x \partial y_2}h(x, y_2)\cos\beta\right] \\
S_{yxx} = {} & c_{66}\frac{(1+q_1)^2}{\gamma_1 q_1}\left[\frac{1}{\gamma_1}\frac{\partial^2}{\partial x \partial y_1}h(x, y_1)\sin\beta + \frac{\partial^2}{\partial x^2}h(x, y_1)\cos\beta\right] \\
& - c_{66}\frac{(1+q_2)^2}{\gamma_2 q_2}\left[\frac{1}{\gamma_2}\frac{\partial^2}{\partial x \partial y_2}h(x, y_2)\sin\beta + \frac{\partial^2}{\partial x^2}h(x, y_2)\cos\beta\right] \\
S_{xyy} = {} & -c_{66}\frac{(1+q_1)^2}{q_1}\left[\frac{1}{\gamma_1}\frac{\partial^2}{\partial x^2}h(x, y_1)\sin\beta - \frac{\partial^2}{\partial x \partial y_1}h(x, y_1)\cos\beta\right] \\
& + c_{66}\frac{(1+q_2)^2}{q_2}\left[\frac{1}{\gamma_2}\frac{\partial^2}{\partial x^2}h(x, y_2)\sin\beta - \frac{\partial^2}{\partial x \partial y_2}h(x, y_2)\cos\beta\right] \\
S_{yyy} = {} & c_{66}\frac{\gamma_1(1+q_1)^2}{q_1}\left[\frac{1}{\gamma_1}\frac{\partial^2}{\partial x \partial y_1}h(x, y_1)\sin\beta + \frac{\partial^2}{\partial x^2}h(x, y_1)\cos\beta\right]
\end{aligned}
\tag{B.10}
$$

$$- c_{66}\frac{\gamma_2(1+q_2)^2}{q_2}\left[\frac{1}{\gamma_2}\frac{\partial^2}{\partial x\partial y_2}h(x, y_2)\sin\beta + \frac{\partial^2}{\partial x^2}h(x, y_2)\cos\beta\right]$$

$$S_{xxy} = S_{xyx} = c_{66}\frac{(1+q_1)^2}{\gamma_1 q_1}\left[\frac{1}{\gamma_1}\frac{\partial^2}{\partial x\partial y_1}h(x, y_1)\sin\beta + \frac{\partial^2}{\partial x^2}h(x, y_1)\cos\beta\right]$$
$$- c_{66}\frac{(1+q_2)^2}{\gamma_2 q_2}\left[\frac{1}{\gamma_2}\frac{\partial^2}{\partial x\partial y_2}h(x, y_2)\sin\beta + \frac{\partial^2}{\partial x^2}h(x, y_2)\cos\beta\right]$$

$$S_{yxy} = S_{yyx} = -c_{66}\frac{(1+q_1)^2}{q_1}\left[\frac{1}{\gamma_1}\frac{\partial^2}{\partial x^2}h(x, y_1)\sin\beta - \frac{\partial^2}{\partial x\partial y_1}h(x, y_1)\cos\beta\right]$$
$$+ c_{66}\frac{(1+q_2)^2}{q_2}\left[\frac{1}{\gamma_2}\frac{\partial^2}{\partial x^2}h(x, y_2)\sin\beta - \frac{\partial^2}{\partial x\partial y_2}h(x, y_2)\cos\beta\right]$$

MIX
Papier aus verantwortungsvollen Quellen
Paper from responsible sources
FSC® C105338

If you have any concerns about our products,
you can contact us on
ProductSafety@springernature.com

In case Publisher is established outside the EU,
the EU authorized representative is:
Springer Nature Customer Service Center GmbH
Europaplatz 3, 69115 Heidelberg, Germany

Printed by Libri Plureos GmbH
in Hamburg, Germany